General Genetics

THE JONES AND BARTLETT SERIES IN BIOLOGY

Genetics
John R. S. Fincham
University of Edinburgh

Genetics of Populations
Philip W. Hedrick
University of Kansas

Genetic Principles: Human and Social Consequences
  (nonmajors text)
Gordon Edlin
University of California, Davis

Molecular Biology: A Comprehensive Introduction to
  Prokaryotes and Eukaryotes
David Freifelder
University of California, San Diego

Essentials of Molecular Biology
David Freifelder
University of California, San Diego

Molecular Evolution: An Annotated Reader
Eric Terzaghi, Adam S. Wilkins, and David Penny
All of Massey University, New Zealand

Viral Assembly
Sherwood Casjens
University of Utah College of Medicine

The Molecular Biology of Bacterial Growth (a symposium
  volume)
M. Schaechter, Tufts University Medical School;
  F. Neidhardt, University of Michigan; J. Ingraham,
  University of California, Davis; N. O. Kjeldgaard,
  University of Aarhus, Denmark, editors

Population Biology
Philip W. Hedrick
University of Kansas

General Genetics

LEON A. SNYDER
University of Minnesota, St. Paul

DAVID FREIFELDER
University of California, San Diego

DANIEL L. HARTL
Washington University School of Medicine

Jones and Bartlett Publishers, Inc.
BOSTON   PORTOLA VALLEY

Editorial offices: Jones and Bartlett Publishers, Inc., 30 Granada Court, Portola Valley, CA 94025.

Sales and customer service offices: Jones and Bartlett Publishers, Inc., 20 Park Plaza, Boston, MA 02116.

Library of Congress Cataloging in Publication Data

Snyder, Leon A.
  General genetics.

  Bibliography: p.
  Includes index.
  1. Genetics.   I. Freifelder, David Michael,
1935–   II. Hartl, Daniel L.   III. Title.
QH430.S69   1985   575.1   84-21292
ISBN 0-86720-050-2

ISBN: 0-86720-050-2

Production	Bookman Productions
Designer	Hal Lockwood
Editor	Kirk Sargent
Illustrator	Donna Salmon, assisted by Alden Erickson, Kelly Solis-Navarro, Evanell Towne, and Jack Vitkus
Composition	Typothetae
Printing and binding	Halliday Lithograph

Printed in the United States of America

Printing number (last digit):   10 9 8 7 6 5 4 3

CONTENTS

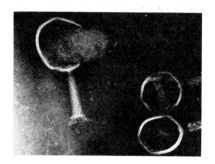

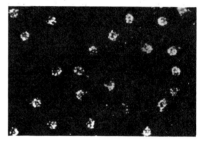

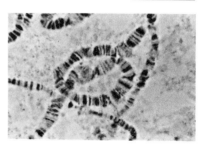

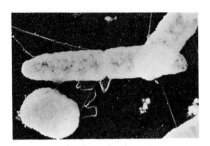

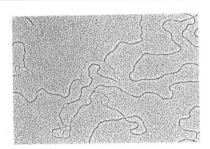

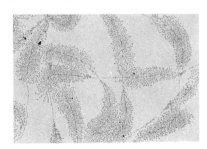

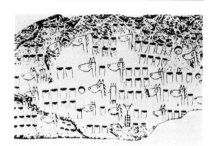

Publishers Note: A separate *Student Study Guide* (ISBN 0-87620-054-5) is available to accompany *General Genetics* (ISBN 0-86720-050-2). The *Student Study Guide* contains the following: • Complete worked out solutions to each problem in the text • Complete chapter summaries • Drill questions with answers.

PREFACE

General Genetics is intended for a one-semester or one-quarter introduction to the principles of genetics. Studies of the undergraduate biology curriculum have identified genetics as one of the most difficult subjects, for two reasons: certain aspects of genetics are abstract and include unfamiliar concepts from probability and statistics, and genetics is quite broad, ranging from molecular biology to population genetics. Furthermore, many instructors have been concerned that available textbooks are too long for all but a full-year course, a perception that is in accord with our own experience. The need for a comprehensive, but not encyclopedic, book has led us to create *General Genetics*. The project has brought together three teachers of complementary expertise—in classical genetics, molecular genetics, and population genetics. Each topic has been written by an author in command of it, and each chapter has been reviewed and criticized by the other two and by teachers throughout the United States, in order to achieve suitable coverage and uniformity of level and style. We have all labored to make this a useful book, one written with a sympathetic concern for the typical undergraduate, who is easily overwhelmed by unnecessary detail. We hope that our efforts have been successful.

The organization of *General Genetics* is the following. First, in Chapters 1–3 we present the most basic phenomena, methods of analysis, and rules of inheritance that make up what is often referred to as classical genetics, the genetics of diploid organisms that began with Mendel's experiments and their interpretations and was extended in the early years of the 20th century by Thomas Hunt Morgan and his *Drosophila* school. This introduction to the formal aspects of the science provides students with the background for subsequent consideration of the molecular basis of the phenomena. The elements of Mendelism and the basis for the chromosome theory of inheritance are introduced in the first two chapters. The third chapter is a discussion of linkage, crossing over and chromosome mapping, and complementation. These first chapters also introduce the basic concepts of probability and statistics that underlie much of classical genetics.

In the next chapters, 4–6, we describe the structure of DNA, chromatin, and chromosomes, the organization of genes in chromosomes, and various mechanisms by which DNA and chromatin are replicated. In these chapters the student will learn about repetitive sequences, Cot analysis, and transposable elements in eukaryotes. These chapters lead logically to a presentation of general cytogenetics in Chapter 6, in which are discussed variations in chromosome number and structure, such as polyploidy, inversions, and translocations. Chapter 6 also includes a discussion of the human chromosome set, important chromosome abnormalities in humans, chromosomal dosage compensation, positions effects, and the molecular basis of the relation between chromosome abnormalities and certain types of cancer.

Chapter 7 departs from the genetics of diploid organisms and presents an analysis of the genetics of bacteria and viruses, with the main emphasis on the systems of genetic exchange in *E. coli* and the genetic properties of selected *E. coli* bacteriophages. Phage biology is presented in terms of five phenomena: examples of the cellular takeover phenomenon, temporal regulation in the phage life cycle, special features of the DNA of some phages, genetic exchange, and lysogeny. The emphasis is primarily on the classical phages T4, T7, and λ.

The next set of chapters (8–12) is concerned with molecular genetics, including gene expression (DNA → RNA → proteins, and many of the subtleties in the flow of genetic information), recombination mechanisms, mutation, and regulation of gene expression in both prokaryotes and eukaryotes. Regulation of gene expression can be a complicated topic to teach because so much detail is known, especially in bacterial systems. In order to be concise and to focus on basic principles, the discussion of regulation centers on the following subjects: general principles of regulation; the λ*lac* and *trp* operons of *E. coli*, as they exemplify basic phenomena that are easily interpreted; a brief description of translational regulation, autoregulation, and antitermination; and those features of regulation in eukaryotes that are fairly well understood. The chapter on recombination mechanisms (Chapter 8) presents the molecular models for transformation, site-specific exchange, transposition, and recombination. Gene conversion and related phenomena are also first presented there. This chapter is discretionary for the instructor, because no information contained in it is required to understand subsequent chapters. The chapter on mutation (Chapter 10) is concerned with three topics: the random nature of mutation, genetic phenomena resulting from mutations, and the molecular basis of mutagenesis. The molecular part of the book is completed by a presentation of genetic engineering (Chapter 12) and some of its more exciting uses, such as gene therapy and the engineering of plants.

Chapter 13 presents somatic cell genetics. It includes discussions of cell fusion as a technique and its importance in gene mapping, the genetics of blood groups, the genetics of the major histocompatibility complex and other antigens implicated in the rejection of transplanted tissue, and the remarkable mechanisms by which antibody diversity is produced.

The final three chapters (14–16) are concerned with the application of genetic principles to entire populations of organisms. These chapters present population genetics and the inheritance of quantitative characters. The use of mathematics has been kept to a minimum, and no mathematical background other than elementary algebra is required. Population genetics and quantitative genetics are active, evolving fields of research, but this is not adequately reflected by the treatment of these subjects in many textbooks. In *General Genetics* these topics are not only up to date but should provide the student with a feeling for the important phenomena and an understanding of how the phenomena can be studied.

Chapter 14 focuses on polymorphisms and their detection, on the influence of mating systems and population structure on genotype frequencies, and on the measurement and phenotypic effects of inbreeding. Chapter 15 presents the processes of mutation, migration, natural selection, and random genetic drift as they affect allele frequencies, with an emphasis on observational and experimental examples of the phenomena. Chapter 16 concerns multifactorial inheritance and its analysis by means of correlations between relatives, different types of heritability and their uses in animal and plant breeding, and the applications (and misapplications) of the principles of quantitative genetics to the analysis of human behavior.

Each chapter in *General Genetics* is supplied with numerous problems that range in difficulty. Most of the problems are straightforward, but some are challenging. Answers to all problems are given in the back of the book, though without explanation in order to conserve space. Ken Jones (California State University, Northridge) checked all answers to problems of the text. Many colleagues and friends have graciously provided problems that they found to be instructive through their own teaching experience.

An extensive glossary is also provided at the end of the text, as well as a short reference list, as well as a list of additional readings and references. We have emphasized general articles, such as those in *Scientific American*, that provide additional information without being so specialized and advanced as to overwhelm the beginner.

A *Student Guide and Solutions Manual* is also available as a separate volume. This manual contains detailed solutions to each problem in *General Genetics*, a summary of each chapter for purpose of emphasis and review, a listing of all key terms that appear in boldface in the text, simpler and more extensive explanations of certain phenomena presented in the text, a set of drill questions (simple questions that only test whether the student has read and understood the very basic material), additional questions of the same level as those in the text, and a few more challenging questions. Complete answers are given to all questions in the manual. Ed Simon (Purdue University) provided many of the problems in the *Student Guide* from his extensive collection of excellent problems collected over more than a decade of teaching genetics at the undergraduate level.

We are indebted to many reviewers, who read and criticized portions of the book ranging from single chapters to nearly the entire book. These reviewers represent research scientists who are knowledgeable in the minutiae of individual chapters, teachers of large and small classes in genetics, and undergraduate and graduate students. The reviewers were the following: Adelaide Carpenter (Univ. of California, San Diego), Rowland Davis (Univ. of California, Irvine), Frank Enfield (Univ. of Minnesota), Seymour Fogel (Univ. of California, Berkeley), David Fox (Univ. of Tennessee), Rachel Freifelder and Jon Greene (both students at Univ. of California, San Diego), Jeff Hall (Brandeis Univ.), Phil Hedrick (Univ. of Kansas), James Higgins (Michigan State Univ.), Ben Hochman (Univ. of Tennessee), Karen Hughes

(Univ. of Tennessee), Richard Imberski (Univ. of Maryland), Ken Jones (California State Univ., Northridge), James McGhee (Univ. of Calgary), Ross MacIntyre (Cornell Univ.), John McReynolds (Univ. of Florida), Joseph O'Tousa (Purdue Univ.), John Parkinson (Univ. of Utah), Charles Radding (Yale Univ.), R. H. Richardson (Univ. of Texas, Austin), T. M. Rizke (Univ. of Michigan), Richard Russell (Univ. of Pennsylvania), Patricia St. Lawrence (Univ. of California, Berkeley), Frank Stahl (Univ. of Oregon), Dale Steffenson (Univ. of Illinois), Suresh Subramani (Univ. of California, San Diego), Irwin Tessman (Purdue Univ.), and Glenys Thomson (Univ. of California, Berkeley).

We wish to acknowledge the support and encouragement of Jones and Bartlett Publishers, and the production staff that produced this book. Much of the credit for a beautiful book should go to them. In particular, we would like to thank Kirk Sargent, whose expertise in editing made our writing clearer; Donna Salmon, our artist, well known for her work in Stryer's *Biochemistry;* and Hal Lockwood, book designer and production director. We also thank the many people who contributed photographs, drawings, and micrographs from their own research and publications.

January 1985

Leon Snyder
David Freifelder
Daniel Hartl

General Genetics

C H A P T E R 1

The Elements of Heredity and Variation

Two important concepts unify modern biology. The first is the idea that all living things are highly ordered systems with properties determined to a large extent by this organization. The second is the persistence of ordered living systems over long periods of time through heredity and evolution—the continuity of life, or descent with modification. **Heredity** is the process by which all living things produce offspring unquestionably like themselves. This capacity of complex biological systems for self-reproduction involves the transmission from parent to offspring of information that specifies a particular pattern of growth and organization. **Genetics** is the science of heredity; it is concerned with the physical and chemical properties of the hereditary material, how this material is transmitted from one generation to the next, and how the information it contains is expressed in the development of an individual.

Organisms resemble one another because they have received hereditary information from some common ancestor. Since species continue to exist through many generations, the hereditary material must be inherently stable. However, within most species, immense hereditary diversity is present. A familiar example is the variation that exists among humans, even those as closely related as sisters or brothers. Hereditary variability is an essential factor in evolution, a term that means cumulative change in the genetic characteristics of populations of organisms through time. Evolution is the most important generalization in biology; it is the process that enables us to understand the origin of the enormous number of species of living things and the relation between them. Evolution provides the logical basis for understanding how the complex organisms existing today were derived through continuous lines of descent, with modification, from the first primitive organisms.

The development of modern ideas concerning the processes involved in evolution began in 1858, with the proposal by Charles Darwin that new

Facing page: The ancient practice of artificial pollination. An Assyrian relief, dating from the ninth century B.C., showing masked priests pollinating date palms. (Courtesy of Hans Stubbe.)

species are formed through the action of environmental forces on hereditary variation present in populations. This mechanism of evolutionary change, which Darwin called **natural selection,** is an idea based on three observations:

1. Almost all species produce more offspring than can possibly survive and reproduce.

2. A great deal of variability affecting survival and reproduction is present in most populations.

3. At least some of this variability is inherited.

From these observations Darwin reasoned that individuals differing even slightly in any way that makes them better adapted to their environment will have a higher rate of survival and reproduction. Furthermore, to the extent that the advantageous variations are hereditary, they will tend to increase in frequency in succeeding generations. The role of natural selection in evolutionary change was clear to Darwin, but he was unable to explain either the source of the inherited variations or the process of hereditary transmission.

In 1866 Gregor Mendel published the results of experiments in which he had investigated inheritance in garden peas. From these findings he deduced both the existence of discrete hereditary elements and the rules determining their behavior in reproduction. The principles of inheritance that Mendel recognized were to become the foundation for genetics and a key to development of modern biology, but their significance was not appreciated until 1900. The results of independent experiments published during that year by three botanists clearly confirmed Mendel's results and led to the recognition of their importance. In the early decades of this century, the processes involved in the transmission of inherited traits became the first well-understood aspect of the new science of genetics. The interactions of genetics with other scientific disciplines have resulted in a constant flow of new ideas and approaches to understanding heredity and evolutionary change. Applications during the late 1940s of the concepts and techniques of physics and chemistry to the investigation of genetic materials and genetic mechanisms initiated rapid and exciting advances in our understanding of the molecular basis of genetic phenomena. We now know in detail the chemical and physical properties of deoxyribonucleic acid (DNA), the molecule that carries hereditary information, and understand many aspects of the conversion of that information into the biological properties of cells and organisms. This knowledge of the physical nature of heredity has led to another important biological generalization:

> All living things, from those as simple as viruses to those as complex as humans, are based on a common system of information storage and expression.

Our consideration of genetics will begin with the development of the fundamental concept of the science—namely, that the hereditary determinant deduced by Mendel and now called a **gene** is a unit of inheritance transmitted from parent to progeny during reproduction. The ideas and methods of analysis that were initiated by Mendel and later evolved in the investigation of phenomena related to the transmission of hereditary differences are basic to our understanding of what genes are, how they are inherited, and how they are expressed.

1.1 Mendel and His Experiments

During the period in which he developed his theory of the basis of heredity, Gregor Mendel was an Augustinian monk at a monastery in what is now Brno, Czechoslovakia, and a teacher of mathematics in the local secular high school. Although he entered the University of Vienna in 1851 to study science and served for a time there as a teaching assistant in physics, the study of natural phenomena seems to have been a dominant interest in his life. At the time, a major concern in biology was how evolution occurred. In the paper in which he presented his theory of inheritance based on hybridization experiments with peas, Mendel noted that the formulation of the general principles of inheritance would have great importance in understanding evolution. The principal difference between Mendel's approach and that of other scientists who were interested in inheritance is that he thought in quantitative terms. He proceeded by clearly stating simple questions to be answered by experiments, carefully doing the experiments, and then looking for statistical regularities that might identify general rules. However, the intellectual framework and mathematical models permitting people to think about statistical regularities had not yet been established. It is possible that Mendel's uses of numerical relations, as simple as they were, may have been the primary reason for the failure of nineteenth-century biologists to appreciate his results.

Mendel selected peas for his experiments for two reasons: (1) he had access to varieties that differed by easily distinguished alternative characteristics, and (2) his earlier studies of flower structure (Figure 1-1) indicated that peas usually reproduce by self-pollination (in which pollen produced in a flower is transferred to the stigma of the same flower). To produce hybrids by

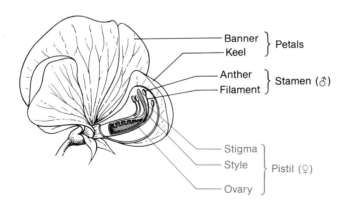

Figure 1-1 A pea flower from which a section has been removed to show the reproductive structures. Each flower has a single ovary, containing up to ten ovules, which develops into the seed pod. The pistil is shown in red.

cross-pollination one needed only to open the keel petal enclosing the reproductive structures, remove the anthers before they had shed pollen, and then dust the stigma with pollen from a flower on a second plant.

Mendel was a capable experimentalist at a time when biology was largely a descriptive science. He recognized the need for being certain that the characteristics he studied were constant in inheritance and at the beginning of his experimentation he established **true-breeding** lines in which the plants produced only progeny like themselves when allowed to self-pollinate normally. These lines—which bred true for flower color, pod shape, or one of the other well-defined characters that Mendel had selected for investigation (Figure 1-2)—provided the parents for hybridization and were also the controls for

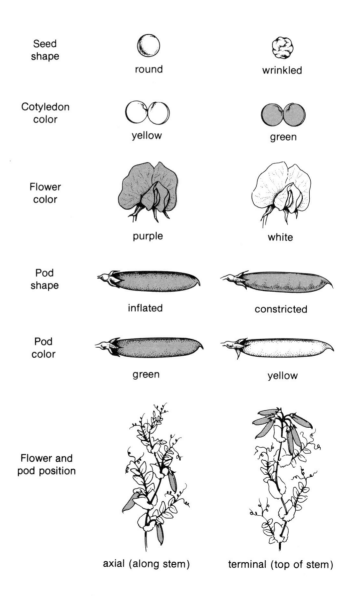

Figure 1-2 Six of the seven character differences in peas studied by Mendel (the seventh difference was long versus short stems). In each case the characteristic on the left is seen in a hybrid.

the hybridization experiments. A **hybrid** is the offspring of a cross between inherently unlike individuals.

We need to consider only a few representative experiments to illustrate the methods Mendel used and his interpretation of the results. One pair of characters that he studied was round versus wrinkled seeds. When pollen from plants of a line with wrinkled seeds was used to cross-pollinate plants from a round-seeded line, all of the hybrid (abbreviated as F_1, for **first filial generation**) seeds produced were round. Hybrid seeds obtained from the **reciprocal cross,** in which plants from the round-seeded line were used as the pollen parents and those from the line with wrinkled seeds as female parents, also were round. In the F_2 **(second filial generation),** obtained by self-pollination of the F_1 hybrid plants, seeds like both of the original parental types appeared. Mendel counted 5474 F_2 seeds that were round and 1850 that were wrinkled and noted that this ratio was approximately $3:1$. The results of this experiment can be summarized in the following way:

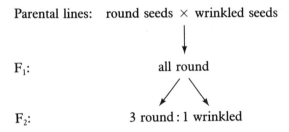

Parental lines: round seeds × wrinkled seeds

F_1:   all round

F_2:   3 round : 1 wrinkled

Remarkably similar results were obtained when Mendel made crosses between plants differing in six other pairs of alternative characteristics. The results of some of these experiments are summarized in Table 1-1. The principal observations were:

1. Only one of the pair of characters distinguishing the parental lines appeared in the F_1 hybrids.

Table 1-1   Results of several of Mendel's experiments

Parental characteristics	F_1	Numbers of F_2 progeny	F_2 ratio
round × wrinkled (seeds)	round	5474 round, 1850 wrinkled	2.96:1
yellow × green (cotyledons)	yellow	6022 yellow, 2001 green	3.01:1
purple × white (flowers)	purple	705 purple, 224 white	3.15:1
inflated × constricted (pods)	inflated	882 inflated, 299 constricted	2.95:1
long × short (stems)	long	787 long, 277 short	2.84:1

2. In the F_2, both of the characters from the original parental lines were present.

3. The character that appeared in the F_1 was always present in the F_2 about three times as frequently as the alternative character.

These observations represented a general numerical (statistical) regularity of the kind Mendel sought. In the succeeding sections we will see how he followed up this basic observation, performing critical experiments that led to his concept of discrete genetic units and the principles governing their inheritance.

Particulate Hereditary Determinants

The prevailing concept of heredity in Mendel's time was that the process consisted of a blending of the traits of the parents in a hybrid, as though the hereditary material consisted of fluids that became permanently blended when combined in the hybrids. An important observation made by Mendel was that one of the parental characteristics was absent in F_1 hybrids and reappeared in unchanged form in the F_2. From this observation, which he saw was inconsistent with the notion of blending, Mendel inferred that the characters from the parental lines were transmitted as two different elements of a particulate nature that retained their purity in the hybrids. The element associated with the character seen in the hybrids (round seeds, in the example used earlier) he called **dominant,** and the other element, associated with the character not seen in the hybrids but seen in their progeny (wrinkled seeds), he called **recessive.**

Mendel's conclusions were reinforced by the results of experiments in which F_3 progeny produced by the self-pollination of individual F_2 plants were observed. In the experiment with round versus wrinkled seeds, for example, F_2 plants grown from wrinkled seeds produced only wrinkled F_3 seeds. When 565 F_2 plants were grown from round seeds, 193 of them produced round seeds but the other 372 plants produced both round and wrinkled seeds in a proportion very close to $3:1$. In quantitative terms,

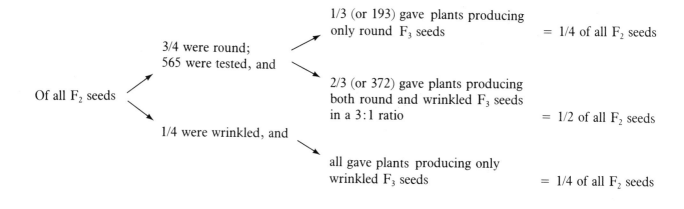

Of all F_2 seeds

3/4 were round; 565 were tested, and

1/3 (or 193) gave plants producing only round F_3 seeds   = 1/4 of all F_2 seeds

2/3 (or 372) gave plants producing both round and wrinkled F_3 seeds in a $3:1$ ratio   = 1/2 of all F_2 seeds

1/4 were wrinkled, and

all gave plants producing only wrinkled F_3 seeds   = 1/4 of all F_2 seeds

The same underlying 1:2:1 ratio was observed each time progeny from individual F_2 plants were obtained to determine the composition of a 3:1 ratio.

The Principle of Segregation

From the regularities observed in his experimental results and the deduction that the hereditary determinants are distinct particulate entities, Mendel formulated the following simple explanation of the 1:2:1 ratio in an F_2:

1. A pea plant has two hereditary determinants for each observed characteristic.

2. Each reproductive cell (**gamete**) of a plant has only one of the two determinants; this one may be either member of the pair present in the particular plant. The two determinants of the pair occur with equal frequencies in the reproductive cells.

3. The union of male and female reproductive cells in the formation of new **zygotes** (fertilized eggs) is a random process.

Application of this explanation to the 1:2:1 ratios observed by Mendel in the F_2 from crosses between plants differing in any pair of alternative characters is shown diagrammatically in Figure 1-3, using the symbols A and a to represent the dominant and recessive determinants, respectively.

The key element in this explanation is the separation, or **segregation,** in unaltered form, of the two hereditary determinants in a hybrid plant during

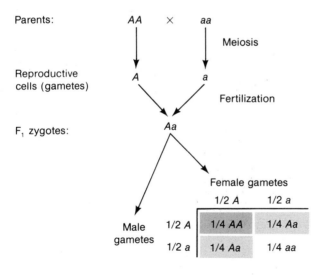

Figure 1-3 A diagrammatic explanation of the 1:2:1 ratio observed by Mendel in the F_2 progeny from a cross between a homozygous dominant and a homozygous recessive pea plant.

the processes leading to the formation of the reproductive cells. This **principle of segregation** is also sometimes called Mendel's first law. Note that the assumption that the hereditary elements are present in pairs applies as well to the parents and to the progeny of the hybrids. Mendel used the concept of segregation to predict the outcome of experiments in which other types of crosses were performed. Before we consider examples of these crosses, we will introduce some additional terminology that is commonly used to simplify the description of genetic phenomena.

Some Genetic Terminology

Mendel's particulate hereditary elements are called **genes,** a term that was introduced soon after his results and interpretations were first confirmed at the beginning of this century. The various forms of a given gene are called **alleles.** Organisms in which the members of a pair of alleles are different, as in the Aa hybrids in Figure 1-3, are said to be **heterozygous,** and those in which the two alleles are alike are said to be **homozygous.** An individual may be homozygous for the dominant (AA) or for the recessive (aa) allele, and in either case will be true-breeding for the characteristic determined by the particular allelic form of the gene. The genetic constitution of an organism is called its **genotype,** of which AA, Aa, and aa are examples. Since a gamete carries only one allele of a given gene, its genotype will be either A or a, in the case of an Aa organism. The genotype of an individual is usually designated incompletely in that only those pairs of alleles of immediate interest are specified. The *observable* properties of an organism make up its **phenotype,** which is determined by interaction of the genotype with the environment. Dominance, as Mendel observed, results in the expression of the same phenotypic character—for example, round seeds—by both the homozygous dominant (AA) and heterozygous (Aa) genotypes. These few terms are the rudiments of a useful genetic vocabulary, which we will supplement as additional phenomena and concepts are considered.

The Principle of Independent Assortment

In the experiments from which the important principle of segregation was formulated, Mendel was concerned with the inheritance of pairs of alleles affecting the expression of single characteristics. To determine whether the same pattern of inheritance applied to each pair of alleles when more than one pair was present in hybrids, he next crossed parental plants that differed in two or three pairs of alleles affecting the expression of different characters. For example, plants from a true-breeding line having round seeds with yellow cotyledons (embryo leaves) were crossed with plants from a line having wrinkled seeds and green cotyledons. The F_1 seeds from this cross, hybrid for both characteristics, or **dihybrid,** were round and yellow. This phenotype was expected from the results of the individual **monohybrid** crosses (Table 1-1), in which the genes resulting in round seed shape and yellow cotyledon color were dominant over their respective alleles. The F_2 seeds obtained when F_1 plants were grown and allowed to self-pollinate had the four possible combinations of phenotypic characteristics with the following frequencies:

round, yellow	315
wrinkled, yellow	101
round, green	108
wrinkled, green	32
	556

Mendel saw that when the pairs of alternative phenotypes were considered separately, the ratios 423 round to 133 wrinkled seeds and 416 yellow to 140 green seeds were in each case very close to the same ratio (3:1) that he had observed in the F_2 populations from the monohybrid crosses. He also noticed that the four phenotypes in the F_2 from the dihybrid cross occurred approximately in the proportions 9/16 round and yellow, 3/16 wrinkled and yellow, 3/16 round and green, and 1/16 wrinkled and green. Similar 9:3:3:1 ratios of F_2 phenotypes were found in the progeny of other dihybrid crosses.

From the recognition that a 9:3:3:1 ratio is the expected result if two independently occurring 3:1 ratios are combined, Mendel formulated the **principle of independent assortment,** sometimes called his second law. This principle states that *segregation of the members of a pair of alleles is independent of the segregation of other pairs during the processes leading to formation of the reproductive cells.* To illustrate with the dihybrid F_1 we have considered, we can represent the dominant and recessive alleles of the pair affecting seed shape as W and w, respectively, and the allelic pair affecting cotyledon color as G and g. Then, the genotype of the F_1 is

$$\frac{W}{w} \frac{G}{g}$$

The result of independent assortment is that W is as likely to be included in a gamete with G as it is with g, and w is equally likely to be included with G or g. The expectation from this behavior of the two pairs of alleles is that gametes produced by such a dihybrid are of the kinds and relative proportions

$$1/4 \ WG, \ 1/4 \ Wg, \ 1/4 \ wG, \ \text{and} \ 1/4 \ wg$$

We will see in subsequent chapters that there are important exceptions to Mendel's principle of independent assortment.

Mendel's hypothesis for the transmission of inherited characters—encompassing the concepts of segregation, independent assortment, and the random union of male and female gametes in the formation of progeny zygotes—explained with beautiful simplicity the 9:3:3:1 ratio of F_2 phenotypes in the dihybrid cross. This can be seen in Figure 1-4, in which the format used to show which combinations of F_1 female and male gametes produce which F_2 genotypes is called a **Punnett square.** The hypothesis

The relation between numbers of segregating gene pairs and numbers of phenotypic and genotypic classes in F_2, assuming dominance

Gene pairs	Phenotypes	Genotypes
1	2	3
2	4	9
3	8	27
⋮	⋮	⋮
n	2^n	3^n

Parents: round, yellow × wrinkled, green   Phenotypes
          $WWGG$ $wwgg$ Genotypes

Gametes: WG wg

F₁ progeny: round, yellow
                     $WwGg$

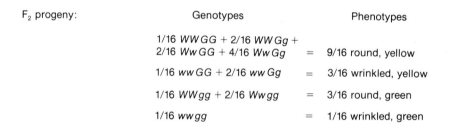

Figure 1-4 Diagram showing the basis for the 9:3:3:1 ratio of F₂ phenotypes resulting from a cross between parents differing in two independently assorting characters.

Female gametes

	1/4 WG	1/4 Wg	1/4 wG	1/4 wg
1/4 WG	$WWGG$	$WWGg$	$WwGG$	$WwGg$
1/4 Wg	$WWGg$	$WWgg$	$WwGg$	$Wwgg$
1/4 wG	$WwGG$	$WwGg$	$wwGG$	$wwGg$
1/4 wg	$WwGg$	$Wwgg$	$wwGg$	$wwgg$

Male gametes

F₂ progeny: Genotypes   Phenotypes

1/16 $WWGG$ + 2/16 $WWGg$ +
2/16 $WwGG$ + 4/16 $WwGg$ = 9/16 round, yellow

1/16 $wwGG$ + 2/16 $wwGg$ = 3/16 wrinkled, yellow

1/16 $WWgg$ + 2/16 $Wwgg$ = 3/16 round, green

1/16 $wwgg$ = 1/16 wrinkled, green

accounted equally well for the more complex ratio of phenotypes that occurred in the F₂ progeny from a trihybrid, in which three pairs of genes were segregating. Mendel tested the predicted genotypes from the crosses by growing plants from the F₂ seeds and obtaining progenies of F₃ seeds by self-pollination. The basis for the procedure was the expectation that a plant grown from a round seed with the genotype WW, for example, would produce only round seeds, whereas a plant grown from a round seed with the genotype Ww would produce both round and wrinkled seeds when self-pollinated.

Note from Figure 1-4 that round green F₂ seeds would be expected to have the genotype $WWgg$, or, twice as frequently, $Wwgg$. To test this prediction, Mendel grew 102 plants from such seeds and found that 35 of them produced only round green seeds (indicating that they had the genotype $WWgg$), whereas the other 67 produced both round and wrinkled green seeds (indicating that they must have been $Wwgg$), a good agreement with the expected frequencies of the two genotypes. Similar agreement with the predicted relative frequencies of the different genotypes was found when plants were grown from round green and also from wrinkled yellow F_2 seeds. As expected, plants grown from wrinkled green seeds, with the predicted homozygous recessive genotype *ww gg*, produced only wrinkled green seeds.

Testcrosses

A second way in which Mendel tested his hypothesis was by crossing plants of the F_1 dihybrid *Ww Gg* with plants that were homozygous recessive for both genes, *ww gg*. As shown in Figure 1-5, the prediction from this hypothesis is that the dihybrid plants should produce four types of gametes—*WG*, *Wg*, *wG*, and *wg*—in equal frequencies, whereas the *ww gg* plants should produce only *wg* gametes. Thus, the progeny phenotypes are expected to consist of the round yellow, round green, wrinkled yellow, and wrinkled green phenotypes in a 1:1:1:1 ratio, a direct reflection of the kinds of gametes produced by the dihybrid because no dominant alleles are contributed by the *ww gg* parent to obscure the results. The progeny Mendel obtained were 55 round yellow, 51 round green, 49 wrinkled yellow, and 53 wrinkled green, in good agreement with the predicted 1:1:1:1 ratio. The results were the same when the cross was repeated with the dihybrid as the female parent and the homozygous recessive as the male parent. This observation confirmed the important assumption in Mendel's hypothesis that in the gametes of the hybrids each possible genotype was present in both female and male gametes in approximately equal proportions.

A cross between a heterozygote, such as the *Ww Gg* dihybrid, and an individual homozygous for the recessive alleles of the genes concerned is called a **testcross.** Because both the genotypes and their relative frequencies in the gametes produced by the heterozygote are immediately apparent from examining the phenotypes of the progeny, testcrosses are exceedingly useful in the analysis of genetic processes. The testcross used as an example in Figure 1-5 is also a **backcross,** a cross between a hybrid and an individual with the same genotype as one or the other of its parents. Backcrosses have particular value in some of the analyses and methods used by geneticists and by plant and animal breeders.

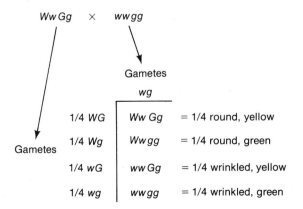

Figure 1-5 Genotypes and phenotypes resulting from a testcross of a dihybrid.

1.2 Mendelian Inheritance and Probability

The approach used to illustrate the kinds of predictions Mendel made to test his scheme of inheritance tends to be intuitive but it is actually based on simple probability. Mendel was a capable mathematician and realized that the proportions of the different types of progeny obtained from a cross represent an average result of numerous repeated events. Furthermore, the combinations of dominant and recessive alleles in the zygotes produced by the random union of gametes represent events that are subject to chance variation. Thus, some knowledge of the rules of probability for predicting the outcome of chance events is basic to understanding the transmission of hereditary characteristics.

A Definition of Probability

A useful way to think about probability is in terms of the proportion of times an event is expected to occur in repeated trials. The proportion of occurrences of an event expected in numerous trials is regarded as the same as the probability of its occurrence in a single trial; this is a concept we have already used in considering the interpretation of Mendel's results. An assessment of the probability of an event may be based on experience or on an understanding of the mechanism allowing the event to be one possible outcome. When n outcomes of an experiment are possible and equally likely, and in m of these the event of interest occurs, the probability of the event is m/n. For example, there are 52 possible, equally likely outcomes of drawing a card from a standard deck. Since 13 of the cards are spades, the probability of drawing a spade is 13/52, or 1/4. Similarly, from the self-pollination of an Aa pea plant, or from the equivalent cross $Aa \times Aa$, four equally likely outcomes—AA, Aa, aA, and aa—are possible. The probability of a heterozygous genotype, represented by two of the four possible outcomes, is 2/4, or 1/2.

The Addition Rule

The addition rule states that the probability P of the occurrence of either of two events, A or B, is the sum of their individual probabilities minus the probability of their joint occurrence, A and B. In the form of a simple expression,

$$P(\text{A or B}) = P(\text{A}) + P(\text{B}) - P(\text{A and B})$$

The rule is applicable to the probability that a card drawn from a deck will be either an ace or a spade. A deck contains 4 aces and 13 spades, but one of the aces is also a spade; thus, P(ace or spade) = 4/52 (ace) + 13/52 (spade) − 1/52 (ace of spades) = 16/52 = 4/13.

In Mendelian genetics we are usually concerned with events that are *mutually exclusive*. Events are mutually exclusive if the occurrence of one event precludes the occurrence of others of the same set of events in the same trial, so the probability of their joint occurrence $P(A \text{ and } B)$ is zero. For example, the dominant and recessive phenotypes for a particular trait are mutually exclusive, as are kings and queens in a deck of cards. From the addition rule, *the probability of the occurrence of one or another of a set of mutually exclusive events is the sum of the probabilities of the separate events*. Thus, the probability of a dominant phenotype in the progeny from the cross $Aa \times Aa$ is $1/4 + 2/4$, or $3/4$, which is the sum of the probabilities of a dominant homozygote and a heterozygote. Similarly, the probability of drawing a face card—a king, a queen, or a jack—is $4/52 + 4/52 + 4/52 = 3/13$.

Conditional Probability

It is sometimes necessary to consider the probability of an event B, given that event A has occurred—that is, the probability of B conditional on the occurrence of A, or $P(B, \text{ given } A)$. For example, assume that a card is drawn and not returned to the deck, and then a second card is drawn. If the first card is an ace, the probability that the second card is an ace is $3/51$, because removing the first ace leaves three aces in a deck of 51 cards. By the same reasoning, the conditional probability of drawing a king, if the first two cards drawn are aces, is $4/50$.

The Multiplication Rule

The multiplication rule states that the probability of the joint occurrence of events A and B is the probability of the occurrence of A times the conditional probability of B given that A has occurred, or

$$P(A \text{ and } B) = P(A) \cdot P(B, \text{ given } A)$$

Using another example of cards drawn from a deck and not replaced, the probability of drawing an ace followed by a second ace is $(4/52) \cdot (3/51)$, or $1/221$. The rule can be extended to any number of events. For example, there are four aces in a deck, so the probability of drawing four aces in successive trials is $(4/52) \cdot (3/51) \cdot (2/50) \cdot (1/49) = 1/270,725$.

In genetics it is more common that the events of interest are independent—that is, the occurrence of the first event has no effect on the probability of occurrence of the second. When this condition applies, the multiplication rule states that *the probability of two or more independent events occurring together is the product of their individual probabilities*. Figure 1-6 shows the application of this rule to determining the probabilities, or expected relative frequencies, of the nine different genotypes among the F_2 progeny produced by self-pollination of a $Ww\,Gg$ dihybrid. The rule provides a simple way to determine the probability of a specific genotype among the progeny

F_1: $Ww\,Gg$

F_2 genotypes:

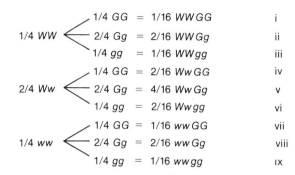

F_2 phenotypes:

Figure 1-6 An example of the use of the addition and multiplication rules to determine the probabilities of the nine genotypes and four phenotypes in the F_2 produced by self-pollination of a dihybrid F_1. The roman numerals are arbitrary labels identifying the F_2 genotypes.

from a cross involving numerous pairs of alleles undergoing independent assortment. For example, the probability of the genotype $Aa\,Bb\,Cc\,Dd$ among the progeny from the cross $Aa\,Bb\,Cc\,Dd \times Aa\,Bb\,Cc\,Dd$ is $(1/2)\cdot(1/2)\cdot(1/2)\cdot(1/2) = (1/2)^4$, or 1/16.

The reasoning illustrated in Figure 1-6 by which the probabilities of the four different phenotypes among the F_2 progeny of a $Ww\,Gg$ dihybrid can be determined is an example of the use of both the addition and multiplication rules. Recall that the probability of a dominant phenotype for a character is the sum of the probabilities of a homozygote and a heterozygote, or $1/4 + 2/4 = 3/4$, since these are mutually exclusive events. Then, for example, the probability of a dominant phenotype for one character occurring with a dominant phenotype for a second, independent character is the product of their separate probabilities, or $(3/4)(3/4) = 9/16$. The probabilities of other phenotypic combinations are calculated similarly, as shown in the lower part of Figure 1-6.

1.3 Segregation in Pedigrees

Determination of the genetic basis of a character from the kinds of crosses we have considered requires the production of large numbers of offspring from the mating of selected parents. The analysis of segregation by this method is

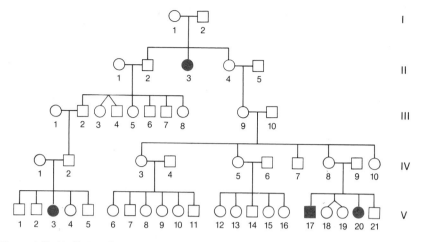

Figure 1-7 Pedigree of a human family showing the inheritance of albinism. Females and males are denoted by circles and squares, respectively. Red symbols indicate individuals with the albino phenotype.

not possible in humans and is not usually economically feasible for some traits in large domestic animals. However, the mode of inheritance of a trait can sometimes be deduced by examining the segregation of alleles in several generations of related individuals represented phenotypically in a diagram of ancestral history called a **pedigree.** An important application of probability in genetics is its use in pedigree analysis.

In the construction of a pedigree such as the one for the human family shown in Figure 1-7, females and males are commonly represented by circles and squares, respectively, and a diamond is used if the sex of an individual is unknown. Individuals having the phenotype of interest are indicated by shaded symbols (in this case, red). The symbols of parents are joined by a horizontal line, which is connected vertically to a second horizontal line that extends above the symbols for their offspring. The offspring of two parents are called **siblings,** or **sibs,** regardless of sex, and are represented from left to right in order of their birth. Successive generations in a pedigree are designated by Roman numerals and the individuals in a generation by Arabic numbers. Twins are indicated by diagonal lines, which converge at the horizontal line above the sibship in the case of nonidentical twins (see III-3,4 in Figure 1-7) or at the end of a short vertical line from the sibship line for identical twins (see V-18,19).

It is important to recognize that the history contained in a given pedigree may not always permit an interpretation of the genetic basis of a trait. However, analysis of the same trait in a different pedigree may permit a conclusive genetic interpretation. We can illustrate this point with examples of two small family pedigrees in which the alleles of genes determining simply inherited conditions are segregating (Figure 1-8). From the information in the first of these pedigrees (panel (a)), it is not possible to determine the manner in which the particular trait is inherited. The ambiguity becomes apparent if we repre-

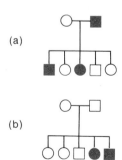

Figure 1-8 Two hypothetical small pedigrees for simply inherited traits. Females and males are denoted by circles and squares, respectively. Red symbols indicate individuals having the phenotype of interest. See text for discussion of each panel.

sent the segregating dominant and recessive alleles as A and a, respectively. An interpretation of the trait as the result of homozygosity for the recessive allele is seen to be plausible if we assume that the female parent is heterozygous (Aa) and the affected male parent is homozygous recessive (aa). However, a second and equally possible interpretation is that the condition is inherited as a dominant, with the phenotypically normal female parent being homozygous for the recessive allele and the affected male parent being heterozygous.

The pattern of segregation in the second pedigree (panel (b)) enables one to make a more definite interpretation. The key observation is that affected individuals are found among the offspring of two phenotypically normal parents; when such a situation exists, one can be reasonably certain that (1) both parents are heterozygous and (2) the trait of interest is determined by homozygosity for the recessive allele. A phenotypically normal individual produced by two such heterozygous parents must, of necessity, have either the homozygous dominant or heterozygous genotype; also, the probabilities of these alternative constitutions are 1/3 and 2/3, respectively. In making these inferences we are assuming that a particular genotype always results in the same phenotype, and it is necessary to recognize that the assumption is not always valid.

Pedigree analysis in humans is frequently concerned with traits, such as albinism (Figure 1-7), that are quite rare in the population. The recessive inheritance of a rare condition is indicated by several features of its pattern of occurrence among related individuals. Specifically, the trait usually does not appear in every generation; it may reappear in a family after being absent in several previous generations. Children with the trait usually will have phenotypically normal parents, who must therefore be heterozygous. (Such a parent is often termed a **carrier.**) In the infrequent case in which both parents are affected, all of their offspring will be affected.

If a rare trait is inherited as a dominant, it would be expected that most affected individuals will be heterozygous. Most matings of these individuals will be with homozygous recessive partners and result in about equal numbers of affected and normal offspring. Thus, the trait will appear in every generation, though chance occurrence of only phenotypically normal offspring in a family will eliminate it within that particular line of descent. In the case of dominant inheritance, of course, the trait will never be seen among the offspring of normal parents.

In the next section, we will see an example of a pedigree analysis that stands out in the development of genetics as a science.

1.4 Functional Aspects of Dominance

The first suggestion of a relationship between genes and specific cellular products was made by Archibald Garrod not long after the rediscovery of Mendel's principles. Garrod was an English physician interested in the rare disease **alkaptonuria** and several other human disorders. He proposed that these disorders result from inherited defects in body chemistry and called them **inborn errors of metabolism.** From the examination of pedigrees of human families in which alkaptonuria occurred, he correctly deduced that alkaptonuria is determined by a recessive gene. The disease, which is actually quite harmless, is characterized by (1) the accumulation and excretion of homogentisic acid, an innocuous substance that causes the urine of an affected individual to turn black upon exposure to air, and (2) a pigmentation of the cartilage and other connective tissues. Normal individuals do not excrete homogentisic acid, though it is a product regularly formed in metabolism in the degradation of the amino acids phenylalanine and tyrosine. Garrod's suggestion, which proved to be correct, was that the defect in alkaptonuria is the absence of an enzyme required for the breakdown of homogentisic acid. That the active enzyme (homogentisic acid oxidase) is absent in homozygous recessive (aa) alkaptonurics and is present in liver and kidney tissues of both homozygous dominant (AA) and heterozygous (Aa) individuals provides the basis for interpreting the dominant-recessive relation between these alleles.

From the observation that AA and Aa individuals have the same phenotype and show none of the symptoms of alkaptonuria, it is reasonable to infer that a single A allele results in sufficient enzyme activity to break down all the homogentisic acid produced. An enzyme is a catalytic protein whose function is to increase the rate of a chemical reaction in which specific covalent bonds (strong chemical bonds in which electrons are shared) are formed or broken. An essential point relevant to understanding dominance relations is that an enzyme is a catalyst and hence is not changed during a reaction; therefore, it can function repeatedly. A single enzyme molecule may be able to catalyze a particular reaction hundreds (occasionally, millions) of times per second. Thus, it is not surprising to find that heterozygotes and homozygous dominant individuals may have the same phenotype with respect to a character determined by the action of a particular enzyme, even though the total number of active molecules of the enzyme produced per cell in the heterozygote is expected to be only one-half that in the homozygote because of the inactive enzyme molecules produced from the a allele. This relation between the amount of an enzyme produced and the number of dominant alleles present has been proved for the genes related to several enzyme-deficiency diseases in humans. However, technical problems in assaying homogentisic oxidase in human tissue have precluded an accurate determination of concentrations of the enzyme in Aa heterozygotes and AA homozygotes.

The Absence of Dominance of Some Alleles

In contrast to the genetic situations studied by Mendel, in which pairs of characters exist as clearly defined alternatives, instances of heterozygotes being phenotypically intermediate between the two homozygotes were found in some of the earliest tests of his principles. These exceptions to the dominance of one member of a pair of alleles over the other are quite common in most organisms. For example, in crosses between snapdragon plants from a

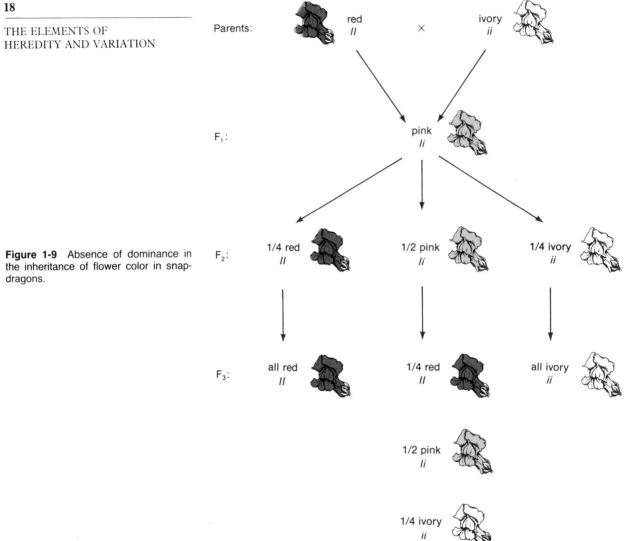

Figure 1-9 Absence of dominance in the inheritance of flower color in snapdragons.

red-flowered variety and a variety with ivory-colored flowers, the F_1 plants produce only pink flowers intermediate in color between those of the parental varieties. In one experiment, the F_2 obtained by self-pollination of the F_1 hybrids consisted of 22 plants with red flowers, 52 with pink flowers, and 23 with ivory flowers. The numbers agree with the Mendelian monohybrid ratio of 1 dominant homozygote : 2 heterozygotes : 1 recessive homozygote expected in the absence of dominance (Figure 1-9). In agreement with the predictions from this interpretation, the red-flowered F_2 plants produced only red-flowered progeny, the ivory-flowered plants also were true-breeding, and the pink-flowered plants again produced progeny of all three phenotypes in the proportions 1/4 red, 1/2 pink, and 1/4 ivory.

In this example of flower-color inheritance the absence of dominance is explained if one supposes that, in *Ii* heterozygotes, concentrations of the enzyme required for synthesis of the red pigment are reduced. At the reduced enzyme concentration (probably one-half that present in *II* homozygotes), pigment synthesis is also reduced during development of the flower, because the pigment is produced in fixed and limiting amounts per *I* allele. Note that the situation is different in alkaptonuria, where the amount of enzyme produced by heterozygotes is sufficient to maintain a low enough concentration of homogentisic acid that an alkaptonuric phenotype does not occur. This fact does not imply that the homozygous dominant individual and the heterozygous carrier of the alkaptonuria allele make the same amount of the enzyme. Rather, if the activity of the enzyme in a heterozygote is great enough, the dominant phenotype will prevail; when it is insufficient in relation to the time in which the reaction must occur, an intermediate phenotype may be seen.

Even in cases in which dominance seems unmistakable, it is not unusual to find an effect of a recessive allele evident in heterozygotes. Recall that when Mendel crossed a pea plant having wrinkled seeds with a plant from a round-seeded variety, the F_1 seeds were round—indicating complete dominance of the allele (*W*) from the parent with round seeds. Microscopic examination later revealed differences in the number and form of the starch grains in seeds of the three genotypes. Homozygous *WW* peas contain many large and well-rounded starch grains, with the result that the seeds retain water and shrink uniformly as they ripen, and do not become wrinkled. The grains in wrinkled (*ww*) peas are irregular in shape and much less numerous, so the ripening seeds lose water more rapidly and shrink unevenly. In heterozygous peas, the grains are intermediate in shape and number, but the starch content is high enough to result in uniform shrinking of the seeds and no wrinkling. Thus, the basic physiological phenomenon that determines the shape of pea seeds is the enzyme-mediated synthesis of starch, with the heterozygotes having a capacity for starch synthesis and a starch content intermediate between the two homozygotes.

For characters such as flower pigment in snapdragons and the starch content of pea cotyledons, the expression in heterozygotes appears to be almost exactly intermediate between the homozygotes. This is the expected condition if there is *no dominance* of one allele over the other (Figure 1-10).

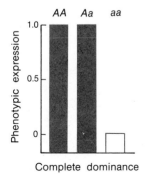 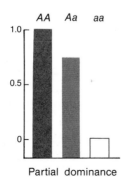 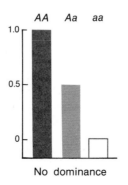

Figure 1-10 Levels of phenotypic expression in heterozygotes with varying degrees of dominance of one allele over the other.

The expression of other phenotypic characters in heterozygotes, though also intermediate between the respective homozygotes, may be more similar to one than to the other. Phenotypic expression of this type is the result of *partial dominance* of the effects of one of the alleles.

Codominance

Another exception to the simple dominance described by Mendel occurs when the two different alleles in a heterozygote are both fully expressed, resulting in a phenotype that is qualitatively different from those of the homozygotes. The effects of the genes that determine human blood groups, of which the ABO group is the best known, provide the clearest examples of the phenomenon of **codominance.**

The blood group genes determine the synthesis of enzymes that produce certain complex molecules located on the surface of red blood cells (see Chapter 13). The phenotype associated with these molecules becomes evident when red blood cells from one individual are either put into a second individual (for example, in a transfusion) or mixed with cell-free blood serum from a second individual. In certain combinations of cells and serum the red cells form visible clumps and precipitate. To understand the phenomenon, one must recognize that the surface of a red cell from a donor may differ from that of the red cells of the recipient and hence the immune system of the recipient may view the donor cells as a foreign substance (that is, a substance not normally present in the recipient individual). A specific characteristic of many complex molecules is that they are able to elicit an immune response; a substance with this capability is called an **antigen.** When an antigen enters the circulatory system of a vertebrate, the animal will respond by producing an **antibody,** a specific protein that is capable of binding to the antigen and inactivating it. An antigen is defined by two attributes: (1) the capacity to stimulate the formation of corresponding antibodies, and (2) the ability to react specifically with these antibodies. When human red blood cells are injected into a rabbit, the surface antigens of the human cells, which are different from those of the rabbit, stimulate antibody synthesis. Consequently, a preparation of blood serum (**antiserum**) taken from the rabbit a few days later will contain antibodies against the numerous molecules on the surface of that particular type of human red blood cell. If some of the rabbit antiserum is then mixed with a sample of cells of the type injected earlier into the rabbit, a precipitate will form. Precipitation occurs because the antibody molecule has several sites that bind to molecules of the specific antigen (in this case, red blood cells), resulting in formation of an antibody-antigen network and clumping (**agglutination**) of the cells. If the red cells are not the type originally administered to the rabbit, agglutination will not occur (Figure 1-11).

The antigen-antibody relations active in the ABO blood group were discovered in 1900. When red blood cells from one person were mixed with serum from another, agglutination was observed to occur in some cases and not in others. Four classes of people could be distinguished on the basis of

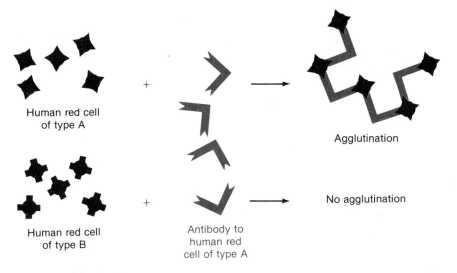

Figure 1-11 Reaction of antibody to type A red blood cells with type A, but not type B, cells.

whether agglutination occurred, and the classes were explained on the basis
of the presence or absence of two antigens and the corresponding antibodies.
The antigens, arbitrarily designated A and B, are associated with the red cells,
and the cells of a particular individual may possess one, or both, or neither
antigen. The antibodies against each of these two antigens are regularly
present in the serum of an individual, but only if the corresponding antigen
is not present on the individual's own red cells. How the formation of these
antibodies is stimulated in an individual who has not been exposed to the
antigen is not clear, but they are probably produced in response to A- and
B-like substances located on the cell wall of one type of intestinal bacteria or
brought into the body in some foods. The key point is that if either the A or
B antigen is present on the red cells, the corresponding antibody will be
absent. This observation was important in establishing the idea that normal
organisms are self-tolerant—they have a mechanism (which is not understood) that prevents the formation of antibodies against their own antigens.
The antigen and antibody constitutions of the four phenotypes in the ABO
blood group are summarized in Table 1-2.

Table 1-2   Relationships between cellular antigens, serum antibodies, and
genotypes of the four phenotypes in the ABO blood group

Phenotype	Red-cell antigens	Serum antibodies	Genotype
A	A	Anti-B	$I^A I^A$ or $I^A I^O$
B	B	Anti-A	$I^B I^B$ or $I^B I^O$
AB	A and B	—	$I^A I^B$
O	—	Anti-A, anti-B	$I^O I^O$

walnut: *R–P–*

rose: *R–pp*

pea: *rrP–*

single: *rrpp*

Figure 1-12 Comb types characteristic of different breeds of chickens and the corresponding genotypes. A dash indicates that the phenotype is unaffected by the nature of the second allele.

In previous examples of inherited characters, we were concerned with genes represented by only two alternative alleles. The genetic basis for the ABO phenotypes illustrates that a gene can have more than two allelic forms, a phenomenon that was recognized early in the development of genetics. Although a particular gene may have numerous different alleles, only two of these alleles can be present in the genotype of any one individual in most higher organisms. The four phenotypes of the ABO group are determined by three alleles: I^A, I^B, and I^O. The I^A and I^B forms of the gene function in the production of similar enzymes that are required for the synthesis of the A and B antigens, respectively. These genes and the genes responsible for the synthesis of other antigens are dominant—that is, the antigen is formed whether the particular gene is homozygous or heterozygous. Thus, both A and B antigens are produced by heterozygous $I^A I^B$ cells, and these alleles are described as codominant. The I^O allele is recessive; it does not determine the synthesis of a specific red cell antigen and its presence in a heterozygote cannot be detected. The genotype of a person with type A blood can be either $I^A I^A$ or $I^A I^O$, and the genotype of a person with type B blood either $I^B I^B$ or $I^B I^O$ (Table 1-2).

1.5 Effects of Genes on the Expression of Other Genes

Some apparent exceptions to inheritance based on Mendel's rules of segregation and independent assortment arise from unusual phenotypic ratios that can be shown to be the result of more than a single pair of alleles affecting the expression of a given character. In the early years of genetics many different modifications of the $9:3:3:1$ dihybrid ratio of F_2 phenotypes were observed and found to result from a masking effect of one gene on the expression of another, or from the occurrence of novel phenotypes produced by the combined effects of two genes. The general term for such interactions in the expression of genes that are not allelic with one another is **epistasis.** Recall that *dominance*, on the other hand, always refers to modification of the expression of one member of a pair of alleles by the other. We can illustrate some of the phenomena of epistatic interaction by considering two early examples of modified dihybrid ratios.

The combs of chickens have four alternative shapes—rose, pea, single, and walnut (Figure 1-12). When the crosses rose × single and pea × single were made, using true-breeding parents of each type, the following results were obtained:

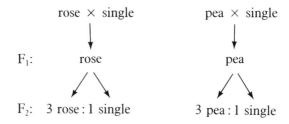

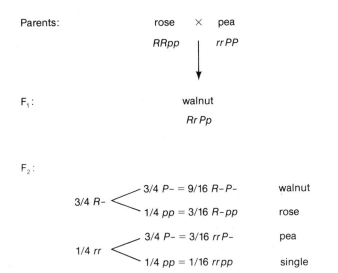

Parents: rose × pea

$RRpp$ $rrPP$

F_1: walnut

$RrPp$

F_2:

3/4 R- 3/4 P- = 9/16 R-P- walnut

1/4 pp = 3/16 R-pp rose

1/4 rr 3/4 P- = 3/16 rrP- pea

1/4 pp = 1/16 $rrpp$ single

Figure 1-13 The inheritance of comb type in chickens.

The results of the cross rose × pea were not as straightforward: the F_1 hybrids were walnut combed, and all four phenotypes appeared among the F_2 progeny. That the F_2 phenotypes occurred in the approximate proportions 9/16 walnut, 3/16 rose, 3/16 pea, and 1/16 single provided the important key to the interpretation of the experiments—namely, that walnut comb is the product of the combined effects of the alleles of the dominant genes for rose and pea combs, as summarized in Figure 1-13. Note that in the grouping of F_2 genotypes into phenotypic classes in the figure, a dash is used to indicate the presence of either a dominant or a recessive allele, since the homozygous dominant and heterozygous genotypes are phenotypically indistinguishable. The two-gene interpretation of comb shape could be tested by a procedure analogous to the one Mendel used to determine the genotypes of F_2 peas having the identical phenotype. For example, one prediction from the explanation shown in Figure 1-13 is that there will be twice as many genotypically $Rrpp$ chickens as $RRpp$ chickens among the F_2 progeny with rose combs. Testcrossing a rose-combed chicken with a single-combed ($rrpp$) chicken would produce equal numbers of rose-combed and single-combed offspring in the first case, but only rose-combed progeny in the second.

A second example of epistasis is the determination of flower color in sweet peas. When a variety that breeds true for colored flowers is crossed with either of two different white-flowered varieties, the F_1 plants have colored flowers in both cases. Self-pollination of the F_1 hybrids from each cross results in an F_2 in which three-fourths of the plants have colored flowers and one-fourth are white flowered. The simplest explanation for these results is that the two white-flowered varieties were homozygous for a recessive allele of the same gene, in which case they would be expected to produce a white-flowered F_1 when crossed. However, when the two varieties were crossed, the F_1 plants produced colored flowers. In the F_2 obtained in one experiment by self-pollination of the hybrids, 382 plants with colored flowers and 269 with white flowers were found—a ratio close to 9:7. The colored flowers in the F_1 hybrids and the occurrence of a 9:7 ratio in the F_2 can both be explained by the effects of two independently segregating pairs of genes, with the presence of at least one dominant allele of both genes required for the production of flower color (Figure 1-14). This interpretation is testable by examining the self-pollinated progeny of individual F_2 plants with colored flowers. (To demonstrate the predictions for yourself, determine the actual genotypes of the F_2 plants grouped as $C–P–$ in Figure 1-14, the expected relative frequencies of these genotypes, and the relative proportions of colored and white-flowered progeny expected from the self-pollination of plants with each genotype.)

The pigments responsible for the colors of sweet pea flowers belong to a class of compounds called anthocyanins. An important consideration that will be examined in more detail in later chapters is that these compounds and other complex molecules formed by living organisms are the products of biosynthetic pathways having numerous enzyme-catalyzed steps controlled by different genes.

The general relationship of genes C and P to synthesis of the anthocyanins in sweet peas can be illustrated in the following way:

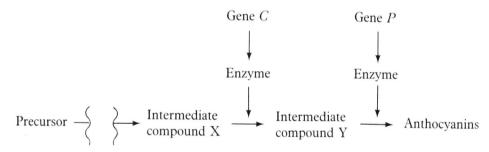

The particular steps in the pathway mediated by the enzymes determined by genes C and P are not known, but this level of detail is not important for an understanding of the overall relations. In the synthetic pathway represented in the diagram, early steps, which are controlled by enzymes, result in the production of an intermediate compound X. This compound represents the substrate molecule acted on by the enzyme determined by gene C, which converts it into intermediate compound Y. In the absence of gene C (that is, in cc homozygotes), little or no Y would be produced and the pathway would be blocked at that point. In turn, compound Y is the substrate for the gene-P enzyme, which converts it into anthocyanin. Therefore, the absence of gene P (in pp homozygotes) would block the final step in anthocyanin synthesis. The genes are related to the extent that pigment production requires both enzymes, and these are determined by dominant alleles of the genes.

Additional cases of epistatic interactions, resulting in other types of modifications of Mendelian dihybrid ratios, are considered in the problems at the end of this chapter. The term epistasis was originally used to describe interactions in the expression of two different genes, as in the examples of

Parents: white variety 1   ×   white variety 2

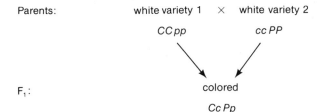

F₁:

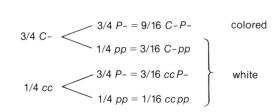

F₂:

Figure 1-14   A cross showing epistasis in the determination of flower color in sweet peas. Color formation requires at least one dominant allele of each of two genes.

comb shape in chickens and flower color in sweet peas. Soon it was recognized that these cases are just examples of a more general phenomenon—that the phenotype of an individual is sometimes determined by the interaction of many genes with each other. That is, *the characters we observe in organisms are the products of numerous interrelated processes.* Each of these processes may require many individual steps that are usually influenced by the activity of specific genes. Certain genes may be identified because they have striking effects on a phenotypic characteristic, but this characteristic is nevertheless determined by the combined action of many genes.

The phenotypic expression of *some* genes, including those that determine the antigens on the surface of human blood cells, is the same in all individuals having the same genotype. However, the phenotypes determined by *most* genes are more variable. This variation may result from the effects of other genes or from sensitivity of the biological processes, which produce the particular phenotype, to environmental conditions. Such genes are said to have **variable expressivity,** meaning that they vary in expression in different individuals and sometimes in different parts of the same individual. The different degrees of expression often form a continuous series from full expression to no expression of phenotypic characteristics. A second form of variation in the expression of a gene is **variable penetrance.** Examples of this phenomenon include cases of identical human twins in which a genetically determined abnormal character occurs in one twin but not in the other. Penetrance is defined as the proportion of individuals whose phenotype matches their genotype for a given character. A genotype that is always expressed has a penetrance of 100 percent.

Problems

1. In tomato plants, tall growth habit (D) is dominant to dwarf (d), hairy stems and leaves (H) is dominant to hairless (h), and yellow flower color (W) is dominant to white (w). A true-breeding tall, hairy, white-flowered variety is crossed with a dwarf, hairless, yellow-flowered variety.
   (a) What types of gametes will be produced by the resulting F_1 plants, and in what relative proportions, if it is assumed that the three pairs of alleles segregate independently?
   (b) What proportion of the F_2 progeny will be genotypically like the tall, hairy, white-flowered parental variety? Phenotypically like that parent? What proportion will be genotypically like the dwarf, hairless, yellow-flowered parental variety? Phenotypically like that parent?

2. Assume that the trihybrid cross $AA\,BB\,rr \times aa\,bb\,RR$ is made in a plant species in which A and B are dominant alleles of their respective genes and there is no dominance in the case of gene R. In the F_2 from this cross:
   (a) How many phenotypic classes are expected?
   (b) What is the probability of the parental $aa\,bb\,RR$ phenotype?
   (c) What proportion would be expected to be homozygous for alleles of all three genes?

3. Achondroplasia is a type of dwarfism in humans in which the bones of the arms and legs do not elongate normally. The condition is rare, affecting about one person in 10,000, and is inherited as a simple monogenic trait (that is, determined by the alleles of one gene). Two achondroplastics are married and have a dwarf child and a normal child.
   (a) Is achondroplasia determined by a recessive or a dominant allele? Why?
   (b) What is the probability that a third child produced by this couple will be normal?

4. The pedigree in the figure below represents a human family in which the daughter indicated by the solid circle is deaf. This form of deafness is determined by a recessive allele. What is the probability that the phenotypically normal son, indicated by the open square, is heterozygous for the gene?

5. Assume that a rare degenerative disease in humans is determined by the dominant allele A, and that this disorder is never manifested until after age 45. A young man has learned that his father has developed the disease.
   (a) What is the probability that the young man will later develop the disorder?
   (b) What is the probability that a recently born son of the younger man has received the dominant allele A?

6. Phenylketonuria (also called PKU) is a human metabolic disorder in which affected individuals, homozygous for the recessive allele p, lack a liver enzyme required for the initial step in the breakdown of excess phenylalanine. A woman whose paternal grandfather had phenylketonuria is married to a phenotypically normal man with two phenotypically normal parents and a phenyketonuric brother. The trait is rare and we can assume that the paternal grandmother and both maternal grandparents of the woman were genotypically PP. What is the probability that a child produced by this couple will have the disorder?

7. The most common form of albinism in humans is determined by homozygosity for the recessive allele c. Individuals homozygous for this allele do not produce the enzyme tyrosinase, which catalyzes the first steps in the synthesis of melanin pigment from the precursor tyrosine. A normally pigmented man and woman, each of whom had an albino parent, are married. If this couple has two children, what is the probability that both will be albino? What is the probability that at least one of the children will be albino?

8. Red kernel color in wheat results from the presence of at least one dominant allele at each of two independently segregating genes; that is, $R-B-$. Kernels on doubly homozygous recessive plants are white, and the genotypes $R-bb$ and $rr\,B-$ result in brown kernel color. If plants of a variety that is true-breeding for red kernels are crossed with plants true-breeding for white kernels,
   (a) What is the expected phenotype of the F_1 plants?
   (b) What are the expected phenotypic classes and their relative proportions in the F_2 progeny?

9. White Leghorn chickens are known to be homozygous for a dominant allele C of a gene responsible for colored feathers, and also for a dominant allele I that prevents the expression of C. The White Wyandotte breed is homozygous recessive for both genes. What proportion of the F_2 progeny from White Leghorn $\times$ White Wyandotte F_1 hybrids can be expected to have colored feathers?

10. One of the crosses Mendel made to test his idea of independent assortment was between pea plants true-breeding for round yellow seeds with gray-brown coats and plants true-breeding for wrinkled green seeds with white coats. He represented the genotypes of these plants as $AA\,BB\,CC$ and $aa\,bb\,cc$, respectively. In the F_2 from this cross,

   (a) What proportion of the peas is expected to have white coats—that is, the phenotype determined by the c allele?

   (b) What proportion is expected to have the same genotype as the F_1? The same phenotype as the F_1?

   (c) What is the probability that a pea selected at random will have the genotype $aa\,bb\,cc$? The genotype $aa\,Bb\,cc$?

11. The production of color in the outer layer of the endosperm of a corn (maize) kernel requires the presence of at least one dominant allele of each of the three independently segregating genes; A_1, C_1, and R. The dominant allele Pr of a fourth independently segregating gene determines the synthesis of purple pigment, and the recessive allele, pr, the synthesis of red pigment. In this problem we will be concerned only with the development of color in the kernels and not with its intensity or mottling. (In endosperm tissue the genes are present in three doses: the number of A_1 alleles and C_1 alleles determine color intensity, and the presence of only a single R allele may result in color mottling.) A red-kerneled strain with the genotype $AA\,CC\,RR\,pr\,pr$ is crossed with a colorless strain having the genotype $aa\,cc\,rr\,Pr\,Pr$.

   (a) What will be the phenotype of the F_1 seeds?

   (b) What proportion of the F_2 seeds, obtained by self-pollination of the F_1 plants, is expected to be red? What proportion is expected to be colorless?

   (c) What proportion of the testcross seeds, obtained by crossing the F_1 hybrids to homozygous recessive plants, is expected to be red? What proportion is expected to be colorless?

12. Colored coats in mice require the presence of the dominant allele C; cc homozygotes are albino. In laboratory strains, the genotype $B-C-$ results in black and the genotype $bb\,C-$ in brown coat color. Two alleles of a third gene determine whether the coat color is solid $(S-)$ or spotted (ss). The three pairs of alleles segregate independently. What is the expected phenotypic ratio in F_2 from crossing homozygous solid black $(CC\,BB\,SS)$ with homozygous recessive albino $(cc\,bb\,ss)$ mice?

13. Black wool in sheep, indicated by the solid symbol in the pedigree in the figure below, is determined by homozygosity for the recessive allele b, and white wool (open symbols) by the dominant allele B. Unless there is evidence to the contrary, assume that individuals 1 and 5 in the second generation of the pedigree do not carry the recessive allele. What is the probability that an offspring from a mating between the two third-generation individuals (III-1 and III-2) will be a black sheep?

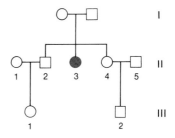

14. The coat colors of Labrador Retriever dogs are black (determined by the genotype $B-E-$), brown $(bb\,E-)$, or yellow $(--ee)$. If dihybrid black dogs are mated—that is, $Bb\,Ee \times Bb\,Ee$, what phenotypes can be expected in the progeny, and in what proportions?

15. The pedigree shown in the figure below shows the inheritance of coat color in a group of Labrador Retrievers. Dogs with black coats (genotypically $B-E-$) are represented by solid symbols, and those with brown coats (genotypically $bb\,E-$) by shaded symbols. Homozygous ee individuals are always yellow (open symbols).

   (a) What are the genotypes of the initial parents at the top of the pedigree?

   (b) What is the most probable genotype of individual III-1?

   (c) If a single pup is produced from the mating of III-4 × III-7, what is the probability that the pup will be brown?

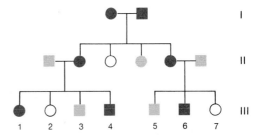

16. Andalusian fowls are black, splashed white, or blue (resulting from an uneven sprinkling of black pigment through the feathers). Black and splashed white are true-breeding, and blue is a hybrid that segregates in the ratio 1 black: 2 blue: 1 splashed white. If a pair of blue Andalusians is mated and the hen lays three eggs, what is the probability that the chicks hatched from these eggs will be one black, one blue, and one splashed white?

CHAPTER 2

Genes and Chromosomes

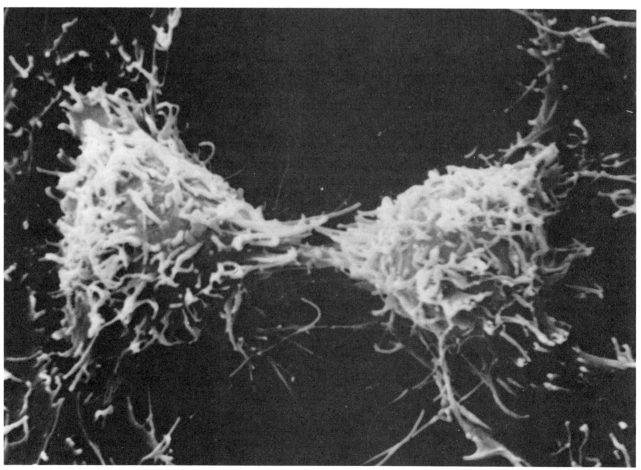

The dominant influences in biology during the last third of the ninteenth century were the elaboration of the theory of natural selection and concern with the difficult problem of how a complex organism develops from a fertilized egg. For the understanding of heredity—which is essential in the consideration of both evolution and development—this period was important because it produced major advances in knowledge of the cell as a living system and established the basis for understanding hereditary continuity between generations. Mendel's idea that the units of heredity are stable and particulate resulted from observing the phenomenon of segregation and was confirmed by his ability to predict the outcome of a cross. However, understanding the physical basis of the behavior of these hereditary elements required knowledge of the organization of cells and the details of cell division, which were elucidated during the thirty years after publication of Mendel's work. These advances were responsible for the immediate appreciation of Mendel's theory of inheritance when it was rediscovered in 1900 and were the foundation for the rapid development of genetic concepts that followed.

A second development that occurred toward the end of the nineteenth century and contributed to the acceptance of Mendel's scheme of inheritance was the increased use of mathematical reasoning as a tool in biology. We have seen in Chapter 1 that an appreciation of some of the basic assumptions made by Mendel requires knowledge of the rules of probability. Concepts of probability also underlie the statistical methods used in the analysis of genetic data, which will be introduced in the last part of this chapter and in later chapters.

2.1 The Stability of Chromosome Complements

The **cell theory** states that cells are the fundamental units of all living systems and that the formation of cells by means of the division of other cells is the

Facing page: A cell in the act of division, as visualized by scanning electron microscopy. (From G. Shih and R. Kessel. 1982. *Living Images.* Jones and Bartlett.)

basis for the continuity of life. Although these generalizations resulted from observations of the cells of higher organisms, they apply also to the simpler forms of life, such as protozoa and bacteria. During the lengthy period in which the cell theory was being formulated, an understanding of the importance of the nucleus and its contents developed after the recognition in the 1870s that the nuclei of two gametes fuse in the process of fertilization and that nuclei reproduce by an orderly division occurring before division of the cell itself. The **chromosomes** of cell nuclei, stained by basic dyes and visible by light microscopy, were identified and described.

Three important regularities characterize the chromosome complements (the complete set of chromosomes) of higher plants and animals:

1. The nucleus of each somatic cell (a cell of the body, in contrast with a germ cell or gamete) contains a fixed number of chromosomes typical of the particular species. However, the numbers vary tremendously among species and have little relationship to the complexity of the organism (Table 2-1). The largest and most common variations in chromosome number are found in higher plants; the species in many genera have numbers that are multiples of a basic number characteristic of the genus. This phenomenon, called **polyploidy,** is related to evolution of the species and will be explained more fully in Chapter 6.

2. The chromosomes in the nuclei of somatic cells usually occur in pairs of morphologically alike members. Thus, the 46 chromosomes of humans consist of 23 pairs and the 14 chromosomes of peas consist of 7 pairs. Furthermore, one chromosome of each pair comes from the maternal parent and the other from the paternal parent of the or-

Table 2-1 Somatic chromosome numbers of some plant and animal species

Plants		Animals	
Organism	Chromosome number	Organism	Chromosome number
Yeast (*Saccharomyces cerevisiae*)	34	Fruit fly (*Drosophila*)	8
Field horsetail	216	House fly	12
Bracken fern	116	Scorpion	4
Giant sequoia	22	Geometrid moth	224
Redwood	44	Common toad	22
Speltoid wheat	14	Leopard frog	26
Macaroni wheat	28	Chicken	78
Bread wheat	42	Dog	78
Garden pea	14	Cat	38
Corn (*Zea mays*)	20	Mouse	40
Tobacco	48	Rat	42
Lily	12	Gibbon	44
Cabbage	18	Gorilla	48
Radish	18	Human	46

ganism. Cells with nuclei of this sort, containing two similar sets of chromosomes, are said to be **diploid.**

3. The germ cells or gametes that unite in fertilization to produce the diploid state of somatic cells have nuclei with only one set of chromosomes, consisting of one member of each pair—these nuclei are termed **haploid.**

The presence of a constant diploid chromosome number in cells of complex organisms that develop from single cells, and the formation of gametes with the haploid chromosome number indicate that there are two processes of nuclear division through which the number of chromosomes is maintained. The basic morphological features of the two division processes—**mitosis** and **meiosis**—have been understood since the early years of this century. An understanding of these processes of nuclear division is fundamental to the understanding of many genetic phenomena.

2.2 Mitosis

Mitosis is a precise process of nuclear division that ensures that each of two daughter cells receives a complement of chromosomes identical with the complement of the parent cell. The essential details of the process are the same in all organisms—the same, for example, in the division of a unicellular green alga as in the repeated divisions through which the approximately 10^{13} cells that make up a human adult are derived from a fertilized egg. Moreover, the basic process is remarkably simple: each chromosome, present as a doubled structure at the beginning of nuclear divison, divides into identical halves that are separated from each other, and one of them goes into each of the two daughter nuclei that are formed.

Chromosome replication, resulting in the two halves that subsequently separate, takes place when the chromosomes are uncondensed during **interphase,** the interval between mitotic divisions. In almost all higher organisms, the synthesis of DNA associated with chromosome replication occurs during a discrete period in interphase known as the S (synthetic) period. Before and after S, there are periods, called G_1 (first gap) and G_2 (second gap), respectively, in which DNA synthesis does not occur. The **cell cycle,** or the life cycle of a cell, is commonly described in terms of these three interphase periods followed by mitosis, M—that is, $G_1 \rightarrow S \rightarrow G_2 \rightarrow M$—as shown in Figure 2-1. This is a somewhat arbitrary representation in which the final event in cell reproduction, the division of the cytoplasm into two approximately equal parts containing the daughter nuclei, is included in the M period.

The length of time required for a complete life cycle varies with cell type. In higher organisms the majority require 18–24 hr, but with optimal conditions of nutrition and temperature some cells divide every 10–12 hr. The relative duration of the different periods in the cycle also varies considerably with cell type. Mitosis is usually the shortest period, requiring between 1/2

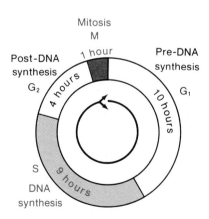

Figure 2-1 The cell cycle of a representative mammalian cell growing in tissue with a generation time of 24 hours.

and 2 hr. The duration of the G_1 period is the principal difference between cells that divide rapidly and those that divide slowly. Cells in which the division cycle has been arrested, as by the onset of differentiation, tend to remain in the G_1 period, the cycle stopping just before S.

The essential features of mitosis are illustrated in Figure 2-2. The four stages—**prophase, metaphase, anaphase,** and **telophase**—into which mitosis is conventionally divided have the following characteristics:

1. Prophase. During interphase, the chromosomes have the form of extended filaments and cannot be seen as discrete bodies with a light microscope. Except for the presence of one or more conspicuous dark bodies (**nucleoli**), the nucleus has a diffuse, granular appearance.

The beginning of prophase is marked by the condensation of chromosomes to form visibly distinct, thin threads within the nucleus. Occasionally,

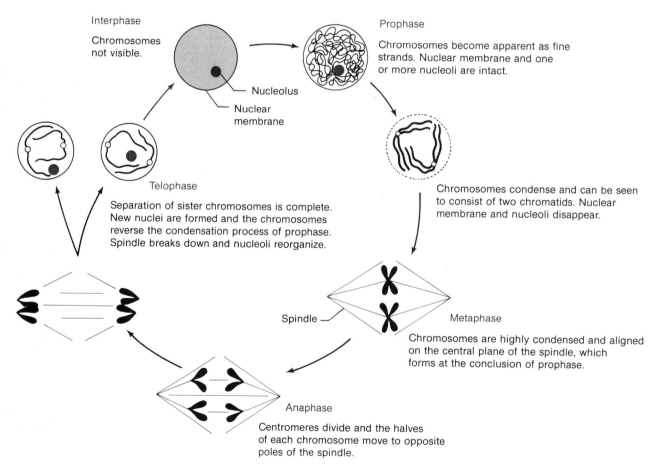

Interphase
Chromosomes not visible.

Nucleolus

Nuclear membrane

Prophase
Chromosomes become apparent as fine strands. Nuclear membrane and one or more nucleoli are intact.

Chromosomes condense and can be seen to consist of two chromatids. Nuclear membrane and nucleoli disappear.

Telophase
Separation of sister chromosomes is complete. New nuclei are formed and the chromosomes reverse the condensation process of prophase. Spindle breaks down and nucleoli reorganize.

Spindle

Metaphase
Chromosomes are highly condensed and aligned on the central plane of the spindle, which forms at the conclusion of prophase.

Anaphase
Centromeres divide and the halves of each chromosome move to opposite poles of the spindle.

Figure 2-2 Schematic diagram of mitosis in an organism with two chromosomes. Only the nucleus of the cell is shown. Interphase is the state of a resting cell and is not usually considered part of mitosis. The red arrow indicates the first step in the cycle. Recall from Figure 2-1 that the interphase stage is much longer than the mitotic phase.

the chromosomes are seen rather early to be longitudinally double, consisting of two closely associated subunits called **chromatids.** Each pair of chromatids is the product of replication of one chromosome in the S period of interphase. The chromatids in a pair are held together at a specific region of the chromosome called the **centromere.** As prophase progresses, the chromosomes become shorter and thicker, as a result of intricate coiling within each chromatid strand, which will be discussed in Chapter 5.

In later prophase, three phenomena occur: (1) The nucleoli, which are associated with specific regions of particular chromosomes, gradually decrease in size, their staining density diminishes, and ultimately they disappear; (2) the membrane enclosing the nucleus disintegrates; and (3) the **mitotic spindle** forms. The spindle is a bipolar structure and consists of fiberlike bundles, composed primarily of aggregates of hollow cylinders of protein (microtubules) that extend between the poles. Each chromosome becomes attached to one or more fibers at its centromere.

2. Metaphase. After the chromosomes are attached to spindle fibers they migrate, until all centromeres lie on a plane equidistant from the spindle poles. The period during which the chromosomes are located at the central plane of the spindle is called metaphase (the interval just before this stage, from spindle attachment to alignment of the centromeres, is called prometaphase). At metaphase the chromosomes have reached their maximum contraction and are easiest to count or examine for differences in structure. The members of different pairs often differ in length, position of the constriction that marks the location of the centromere, and the presence of other constrictions.

3. Anaphase. The centromeres divide at the beginning of anaphase, and the two sister chromatids that were attached begin to move toward opposite poles of the spindle. Centromere division does not require a spindle: it occurs even when formation of a spindle is prevented experimentally by the drug colchicine, which binds to the microtubules, though in this case no polar movement of the chromatids occurs. With the separation of sister chromatids each of the half-chromosomes becomes an independent chromosome. The subsequent movement of these chromosomes to the poles gives the impression, probably correct, that they are pulled by the spindle fibers attached to their centromeres. If the centromere is near the center of a chromosome, the chromosome is pulled into the shape of a V. At the completion of anaphase, the chromosomes lie in two groups near the poles of the spindle. Each group contains the same number of chromosomes that was present in the original interphase nucleus.

4. Telophase. Telophase is the period during which nuclei re-form. As at other stages in the cycle, with the exception of the onset of anaphase, a sharp transition to this final stage in mitosis does not exist. During telophase, a nuclear envelope forms around each compact group of chromosomes, the spindle disappears, nucleoli reorganize, and the chromosomes undergo a reversal of the condensation process that occurred in prophase. Gradually, the two daughter nuclei assume a typical interphase appearance, and the cell itself divides.

The details of mitosis and the basis for its precision, just described, should not obscure the simplicity of the process. It is difficult to conceive of a simpler and more certain mechanism for the precise division of a complement of long, fragile chromosomes into two complements identical to the original one and to each other.

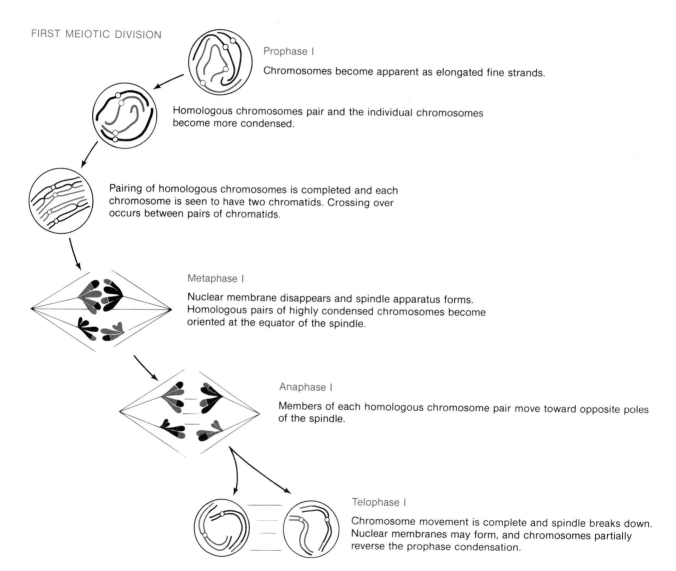

FIRST MEIOTIC DIVISION

Prophase I

Chromosomes become apparent as elongated fine strands.

Homologous chromosomes pair and the individual chromosomes become more condensed.

Pairing of homologous chromosomes is completed and each chromosome is seen to have two chromatids. Crossing over occurs between pairs of chromatids.

Metaphase I

Nuclear membrane disappears and spindle apparatus forms. Homologous pairs of highly condensed chromosomes become oriented at the equator of the spindle.

Anaphase I

Members of each homologous chromosome pair move toward opposite poles of the spindle.

Telophase I

Chromosome movement is complete and spindle breaks down. Nuclear membranes may form, and chromosomes partially reverse the prophase condensation.

Figure 2-3 Schematic diagram illustrating the major features of meiosis in an organism with four chromosomes. For clarity, realistic changes in the dimensions of the chromosomes during condensation are not shown.

2.3 Meiosis

Meiosis is a mode of cell division in which cells form that have half the number of chromosomes present in the premeiotic cell; typically cells with the haploid chromosome number are formed by division of a cell with the diploid number. Meiosis consists of two successive nuclear divisions, the essential details of which are shown in Figure 2-3. During the first division, the two members of each pair of chromosomes, which are said to be **homologous** to each other, become closely associated along their length. The homologous chromosomes, each consisting of two sister chromatids that remain joined at the centromere, are then separated from one another into two nuclei, each containing a haploid set of duplex chromosomes. Without chromosome replication, a second division (resembling a mitotic division) occurs in which the chromatids of each chromosome are separated into different daughter nuclei.

SECOND MEIOTIC DIVISION

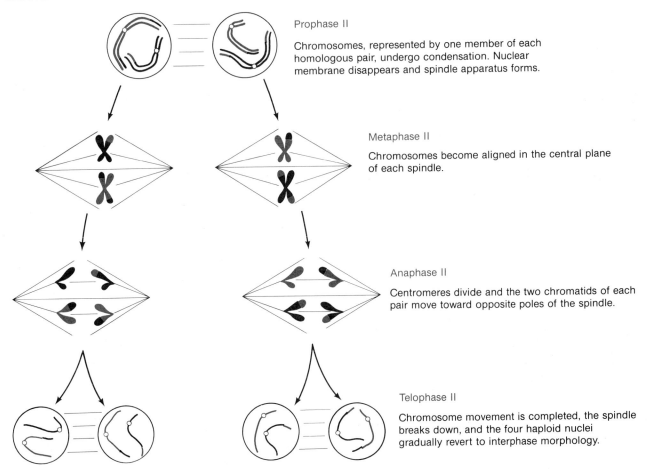

Prophase II

Chromosomes, represented by one member of each homologous pair, undergo condensation. Nuclear membrane disappears and spindle apparatus forms.

Metaphase II

Chromosomes become aligned in the central plane of each spindle.

Anaphase II

Centromeres divide and the two chromatids of each pair move toward opposite poles of the spindle.

Telophase II

Chromosome movement is completed, the spindle breaks down, and the four haploid nuclei gradually revert to interphase morphology.

Figure 2-3 (continued)

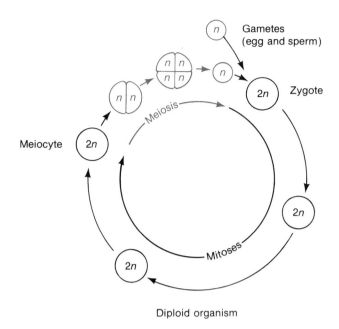

Figure 2-4 The life cycle of a higher animal; n is the haploid chromosome number. In males the four products of meiosis develop into functional sperm, whereas in females only one of the four products develops into an egg.

Consequently, the products of the two divisions are four daughter nuclei, each containing one complete haploid set of chromosomes that consists of a single chromatid from each pair of homologous chromosomes.

Although these details of meiosis are similar in all sexually reproducing organisms, only one of the four products develops into a functional cell in females of both animals and plants; meiosis occurs in specific cells called **meiocytes,** a general term for the primary oocytes and spermatocytes in the gamete-forming tissues of animals (Figure 2-4) and for the primary megasporocytes and microsporocytes in the reproductive (spore-forming) tissues of plants (Figure 2-5). In plants the haploid spores produced by meiosis typically undergo one or more mitotic divisions to produce a gametophyte in which gametes are produced by mitotic division of a haploid nucleus (Figure 2-5). Thus, in animals the products of meiosis are gametes, whereas in plants the products of these nuclear divisions (which occur in the sporophyte generation) are **spores.** These spores develop into gametophytes (the gametophyte generation), in which the gametes are produced by mitosis. In Protista and lower plants haploid spores are produced almost immediately after a diploid zygote is formed by gametic fusion—that is, the sporophyte in these organisms consists only of the zygote.

Meiosis is a more complex and considerably longer process than mitosis and usually requires days or even weeks. The essence of meiosis, which we will now examine in more detail, is that *it consists of two divisions of the nucleus but only one duplication of the chromosomes.* The nuclear divisions—called the

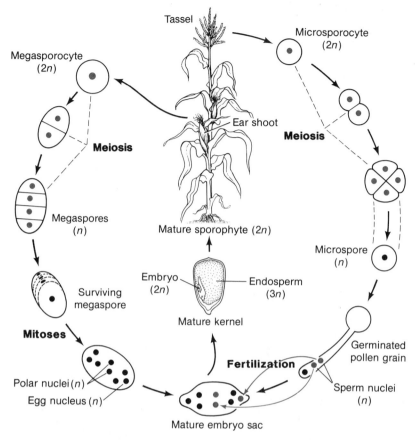

Figure 2-5 The life cycle of corn *Zea mays*. As is typical of higher plants, the diploid spore-producing (sporophyte) generation is conspicuous, whereas the gamete-producing (gametophyte) generation is microscopic. Nuclei participating in meiosis and in fertilization are shown in red.

first meiotic division and the **second meiotic division**—can be separated into a sequence of stages similar to those used to describe mitosis, as shown in Figure 2-2. The distinctive events of this important process occur during the first division of the nucleus; these events are described in the following section.

The First Meiotic Division

The four main stages of the first meiotic division are called **prophase I, metaphase I, anaphase I,** and **telophase I.** These stages are generally more complex than their counterparts in mitosis.

1. Prophase I. This is a long stage, lasting several days in most higher organisms and commonly divided into five substages—**leptotene, zygotene, pachytene, diplotene,** and **diakinesis.**

In the **leptotene** substage the chromosomes become visible as long, threadlike structures, appearing in stained preparations to be single even though they were replicated in the premeiotic interphase. With the electron microscope the pairs of sister chromatids can be distinguished and they are closely associated with a structure called an **axial element** that is formed between them. During this initial phase in condensation of the chromosomes numerous dense granules occur at irregular intervals along their length. These localized contractions, called **chromomeres,** have a characteristic number, size, and position in a given chromosome (Figure 2-6(a)).

The **zygotene** substage is marked by the pairing, or **synapsis,** of homologous chromosomes. This pairing begins at one or more points along the length of the chromosomes and results in a precise chromomere-by-chromomere association (Figure 2-6(b)). The ends of the chromosomes often can be seen to be associated with a specific region of the nuclear envelope as the pairing begins, but the forces that produce the pairing are not known.

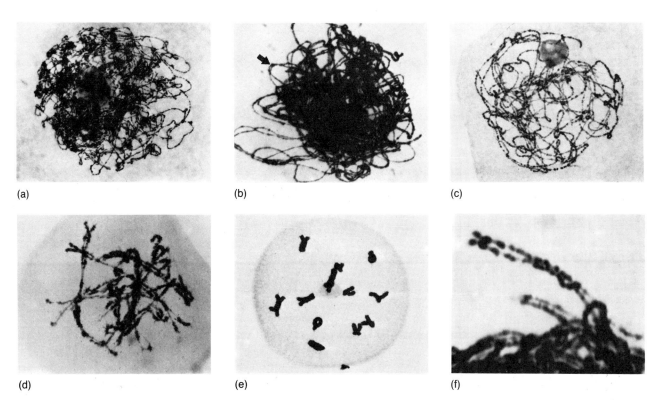

(a)

(b)

(c)

(d)

(e)

(f)

Figure 2-6 Substages of prophase of the first meiotic division in microsporocytes of lily (*Lilium longiflorum*). (a) Leptotene, in which condensation of the chromosomes is initiated and beadlike chromomeres are visible along the length of a chromosome. (b) Zygotene, in which chromosome synapsis occurs. The arrow in the zygotene stage indicates chromosomes beginning to pair. (c) Pachytene, during which crossing over between homologous chromosomes occurs. (d) Diplotene, characterized by separation of the paired homologues, which remain held together at one or more points along their length; (e) Diakinesis, during which the chromosomes reach their maximum contraction. (f) Zygotene, at higher magnification (in another cell), showing synapsed homologues and matching of chromomeres during pairing. (Courtesy of Marta Walters (a, b, c, e, f) and Herbert Stern (d).

Between the synapsed chromosomes a well-defined structure—the **synaptonemal complex**—can be seen by electron microscopy; it is composed of the two axial elements formed in the leptotene substage and a central region (Figure 2-7). Each pair of synapsed homologous chromosomes is referred to as a **bivalent.**

During the **pachytene** substage (Figure 2-6(c)) condensation of the chromosomes continues, and the chromosome complement is represented by the haploid number of bivalents. Each bivalent consists of a **tetrad** of four chromatids, but the two sister chromatids of each chromosome usually cannot be distinguished. A genetically important event called **crossing over** occurs during pachytene, but it does not become evident until the transition to diplotene, the next substage.

At the onset of **diplotene** the synapsed chromosomes begin to separate (Figure 2-6(d)). However, they remain held together at regions along their length called **chiasmata** (singular, **chiasma**). A chiasma, literally meaning a crosspiece or cross, is the result of breakage and rejoining between nonsister chromatids—that is, *a physical exchange between chromatids of homologous chromosomes* (Figure 2-8). In normal meiosis each bivalent will have at least one chiasma, and bivalents of long chromosomes often have three or more. The total number of chiasmata in the 23 bivalents present in diplotene nuclei of humans is usually between 50 and 60.

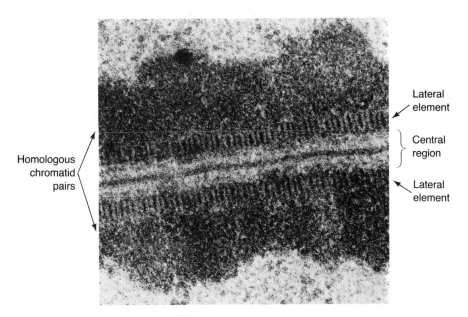

Figure 2-7 Electron micrograph of a portion of a synaptonemal complex associated with synapsed chromosomes in the first meiotic prophase in the fungus *Neotiella rutilans.* The roles of the components of the complex are unknown; what is significant is that a complicated structure is used in chromatid pairing. (From *Proc. Nat. Acad. Sci. USA,* 1971. 68: 851; courtesy of D. von Wettstein.)

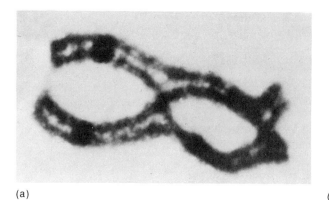

(a)

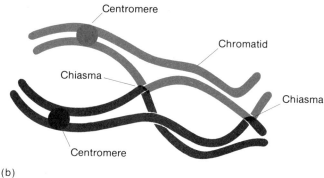

(b)

Figure 2-8 A pair of homologous chromosomes of the salamander *Oedipina poelzi* at late diplotene in a spermatocyte, showing chiasmata where chromatids of the two chromosomes appear to exchange pairing partners. (a) Light micrograph. (b) Interpretive drawing. (From F. W. Stahl. 1964. *The Mechanics of Inheritance*. Prentice-Hall, Inc.; courtesy of James Kezer.)

During the final substage of prophase I—**diakinesis**—the chromosomes attain their maximum condensation (Figure 2-6(e)). The homologous chromosomes in a bivalent remain connected by one or two terminal chiasmata, which persist until the first meiotic anaphase. Near the end of diakinesis the formation of a spindle is initiated and the nuclear envelope breaks down.

2. Metaphase I. There is a general similarity between this stage and a mitotic metaphase. The bivalents become positioned with the centromeres of the two homologous chromosomes on opposite sides of the plane through the middle of the spindle (Figure 2-9(a)). The co-orientation of the undivided centromeres of each bivalent relative to the two poles of the spindle occurs *at random* and determines the member of each pair of chromosomes that will subsequently move to a particular pole.

3. Anaphase I. During this stage the homologous chromosomes, each composed of two chromatids joined at an undivided centromere, separate from one another and move to opposite poles of the spindle (Figure 2-9(b)).

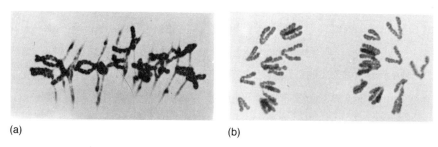

(a)

(b)

Figure 2-9 Metaphase I (a) and anaphase I (b) in microsporophytes of the lily *Lilium longiflorum*. (Courtesy of Herbert Stern.)

4. Telophase I. At the completion of anaphase I a haploid set of chromosomes consisting of one homologue from each bivalent is located near each pole of the spindle. During telophase the spindle breaks down and, depending on the species, either a nuclear envelope briefly forms around each group of chromosomes or the chromosomes enter the second meiotic division after only a limited uncoiling. In many species the cytoplasm divides during the interval between the two meiotic divisions. In the oocytes of most animals (that is, in the female) the first meiotic division occurs near the cell surface, with one pole of the spindle oriented toward the exterior and the other toward the interior of the oocyte. During telophase the formation of a new cell membrane separates the group of chromosomes nearest the surface and a small amount of cytoplasm into a tiny sphere called the first **polar body.** This process is repeated after the second meiotic division of the nucleus in the developing oocyte and a second polar body is formed. The polar bodies disintegrate at a later time.

The Second Meiotic Division

In some species the chromosomes pass directly from telophase I to **prophase II** without loss of condensation; in others, there is an interkinesis stage between the two meiotic divisions. *Chromosome replication never occurs between the two divisions;* the chromosomes present at the beginning of the second division are identical to those present at the end of the first division. After a short prophase (prophase II) and the formation of second-division spindles, the centromeres of the chromosomes in each nucleus become aligned on the central plane of the spindle at **metaphase II** (Figure 2-10(a)). During **anaphase II** the centromeres (replicated in the *first* division) separate and the chromatids of each chromosome move to opposite poles of the spindle (Figure 2-10(b)). **Telophase II** (Figure 2-10(c)) is marked by a transition to the interphase condition of the chromosomes in the four haploid nuclei accom-

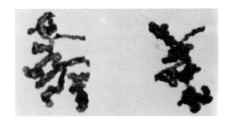

(a)

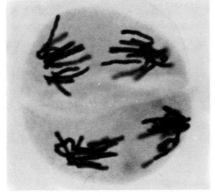

(b)

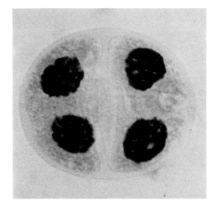

(c)

Figure 2-10 Metaphase II (a), anaphase II (b), and telophase II (c) in microsporocytes of the lily *Lilium longiflorum.* Cell walls have begun to form in telophase, whch will lead to the formation of four pollen grains. (Courtesy of Herbert Stern.)

panied by division of the cytoplasm. Thus, the second meiotic division superficially resembles a mitotic division. There is an important difference, however: the chromatids of a chromosome are usually not identical sisters along their entire length because of the occurrence of crossing over associated with the formation of chiasmata during prophase of the first division.

2.4 Chromosomes and Heredity

The first correct interpretation of meiosis, made in 1903 from studies of grasshoppers, was possible because the somatic chromosomes of this organism occur in pairs, all of which are morphologically different. It was proposed that the segregation of a pair of genes would be explained simply if they were carried by the maternal and paternal members of a pair of chromosomes that synapse and are then separated in the first meiotic division. Accounting for the independent assortment of two pairs of genes associated with different pairs of chromosomes required the additional assumption that each bivalent becomes oriented at random in the first division spindle.

This hypothesis, based on observations of meiosis in grasshoppers, presented a convincing but indirect case for the relation between genes and chromosomes, a relation later established directly in experiments of several kinds. The first clear proof that genes are parts of chromosomes was obtained in experiments concerned with the pattern of transmission of the **sex chromosomes,** the chromosomes responsible for the determination of the separate sexes in some plants and in almost all higher animals.

The Chromosomal Determination of Sex

The sex chromosomes represent an important exception to the rule that all chromosomes of diploid organisms are present in pairs of morphologically similar homologues. It was observed in early examinations of chromosomes that one of the chromosomes in males of some insect species does not have a homologue. This unpaired chromosome, called the **X chromosome,** because its nature and function were unknown, was found to divide in only one of the meiotic divisions and to be present in half of the sperm cells produced. The biological significance of these observations became clear when females of the same species were shown to have two X chromosomes. In other species in which the females have two X chromosomes, it was observed that a morphologically different chromosome, referred to as the **Y chromosome,** is present with the X in males and pairs with it during meiosis. These differences in the chromosomal constitution of males and females were recognized to represent a method for the determination of sex at the time of fertilization. That is, whereas every egg will contain an X chromosome, only half of the sperm will have an X and the other half either a Y or no sex chromosome. Fertilization of an egg by a sperm carrying an X results in an XX zygote, which develops into a female, and fertilization by a sperm having no X

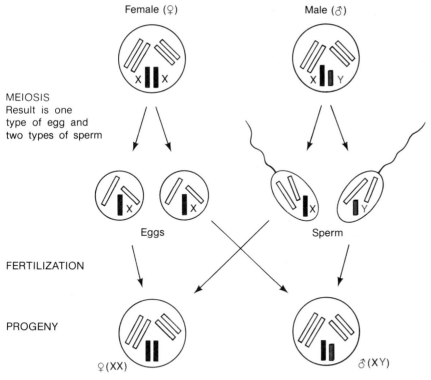

Female (♀)   Male (♂)

MEIOSIS
Result is one
type of egg and
two types of sperm

Eggs   Sperm

FERTILIZATION

PROGENY

♀(XX)   ♂(XY)

Sex of the progeny is determined by the
sex chromosome contained in the sperm

Figure 2-11 The chromosomal basis of sex determination found in mammals, many insects, and other animals.

produces an XY or an X0 zygote, which develops into a male (Figure 2-11). The result is a criss-cross pattern of inheritance of the X chromosome, in which a male receives his only X chromosome from his mother and transmits it only to his daughters. In organisms with this type of chromosomal sex determination—which is now known to occur in mammals including humans, many insects and other animals, and some flowering plants—the female is called the **homogametic** sex and the male the **heterogametic** sex because they produce one and two types of gametes, respectively.

Assuming that the union of gametes in fertilization occurs at random, as Mendel had supposed, the inheritance of the sex chromosomes (one X and one Y) neatly explains the 1:1 sex ratio usually observed. On the other hand, although the determination of sex is a simply inherited trait, the embryonic development of the sexual phenotype is actually inherited in a much more complex way: many genes participate and the zygotes of most organisms have the potential for the expression of either male or female phenotypes. Those organisms in which the sexes are separate almost certainly evolved from ancient ancestral forms that were **hermaphroditic**—that is, able to produce both male and female gametes in the same individual. The demonstration of the relation between sex chromosomes and sex determination was the first step toward an understanding that the genes that trigger differentiation of the two sexes have come to be restricted to one pair of chromosomes, the X and Y. However, other genes controlling various processes involved in sexual development occur in the other chromosome pairs. These nonsex chromosomes are collectively called **autosomes,** and they are present equally in the two sexes. Many genes with functions unrelated to sex are also located in the X chromosomes. In most organisms, including humans, the Y chromosome carries few genes other than those associated with sex determination.

X-linked Inheritance

Crosses of the kind considered in Chapter 1 yield similar progeny when the genotypes of the male and female parents are reversed—that is, in reciprocal crosses. One of the earliest observations of an exception to this behavior was made by Thomas Hunt Morgan in 1910, in an early study of the fruit fly, *Drosophila melanogaster* (Figure 2-12).

The normal eye color of the fly is brick red. In a laboratory population that had been maintained for many generations, Morgan found a single male with white eyes. In a mating of this male with red-eyed females, all the F_1 progeny had red eyes, as would be expected if the allele for white is recessive. In the F_2 produced by the mating of F_1 males and females there were 2459 red-eyed females, 1011 red-eyed males, and 782 white-eyed males. These numbers represent a rather poor fit to a 3:1 ratio of red- versus white-eyed phenotypes, but the surprising result was that all of the white-eyed flies were males. However, when a white-eyed male was crossed with his red-eyed female offspring, the progeny consisted of both red- and white-eyed males and red- and white-eyed females in approximately equal numbers, clearly indicating that the white-eyed phenotype is not limited to males. Another unexpected result was obtained when white-eyed females were mated with red-eyed males—the *reciprocal* of the cross with the original white-eyed male. In this cross the F_1 females had red eyes and the males had white eyes.

Morgan summarized the results of these crosses by stating that this trait (not all eye-color variants) is **sex linked** in inheritance, meaning simply that it is associated with the inheritance of sex. It later became known that *Drosophila* females have two X chromosomes and males have an unequal XY pair, from which the inheritance of the white-eye character could be explained by assuming that the alleles for red (w^+) and white (w) are located on the X chromosome and not present at all on the Y. The resulting chromosomal interpretation of the reciprocal crosses, which can be diagrammed as in Figure 2-13, readily accounts for the drastically different phenotypic ratios in the F_1 and F_2 progenies from the two initial crosses. Support for Morgan's interpretation soon came from the discovery of several other genes that follow this same X-linked pattern of inheritance. As has been shown, reciprocal crosses that result in different phenotypic ratios are diagnostic of sex linkage. The term **hemizygous** is used to describe the condition in which only one member

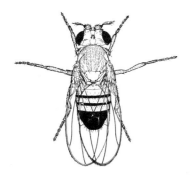

Male (♂)

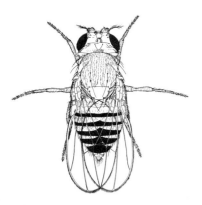

Female (♀)

Figure 2-12 Drawings of a male and female fruit fly *Drosophila melanogaster*. (Original drawings by Edith Wallace, artist for Thomas Hunt Morgan.)

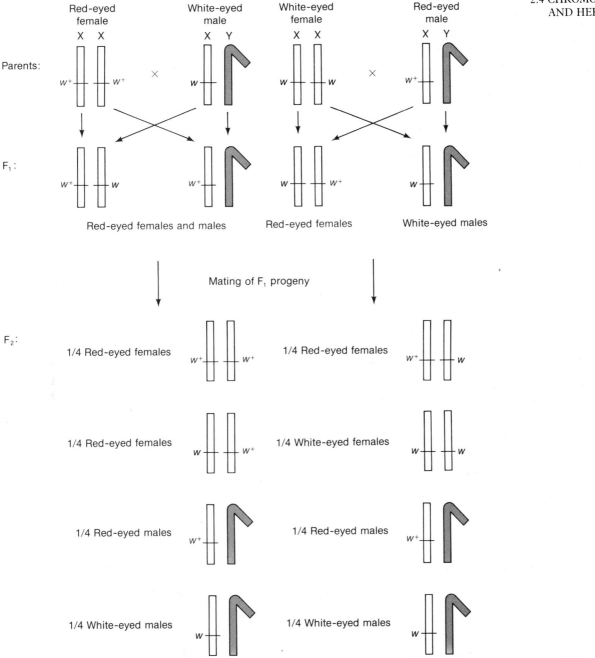

Figure 2-13 A chromosomal interpretation of the results obtained in F_1 and F_2 progenies when a *Drosophila* female with red eyes is crossed with a white-eyed male (Cross 1) and when the reciprocal cross of white-eyed female with red-eyed male (Cross 2) is made. In the X chromosome the w^+ genotype is shown in red; w is black. The Y chromosome (gray) does not carry either allele of the gene.

of a pair of alleles is present, as is the case for genes located on the X chromosome present in a *Drosophila* male.

One of the best known human traits with an X-linked pattern of inheritance is **hemophilia A,** a severe disorder of blood clotting determined by a recessive allele. Affected individuals are unable to synthesize a blood protein that is required for normal clotting. A famous pedigree of this disease starts with Queen Victoria of England. One of her sons was hemophilic, and two of her daughters were heterozygous carriers of the gene who produced three hemophilic sons and four heterozygous daughters. Through two of these carrier granddaughters the gene was introduced into the royal families of Russia and Spain. The present royal family of England, having descended from a normal son of Victoria, is free of the disease. These observations are consistent with the predicted genetic pattern of a trait controlled by a sex-linked gene. A heterozygous female, as we have seen from the *Drosophila* example, is expected to transmit the recessive allele to half of her daughters and half of her sons. Since a male has only one X chromosome, he will show the defective trait if the recessive allele is present in that chromosome, and will transmit the allele to all of his daughters but none of his sons. A male who does not show the defective trait does not carry the recessive allele and cannot transmit it to any of his children. Other examples of X-linked inheritance in humans will be examined in the problems at the end of this chapter.

A different pattern of sex-linked inheritance occurs in organisms in which the male is homogametic and the female is heterogametic. This pattern was found in the moth *Abraxas* and in chickens several years before Morgan's experiments with *Drosophila*. When females from a strain of the moth with light-colored wings were crossed with dark-winged males, only dark-winged progeny were produced—indicating that the allele for light wing color is recessive. In the progeny from the reciprocal cross, all females were light and all males were dark. Similar observations were made in experiments concerned with the inheritance of a feather coloration pattern in chickens. Some breeds have feathers with alternating transverse bands of light and dark color, resulting in a phenotype referred to as barred. The feathers are uniformly colored in the nonbarred phenotypes of other breeds. Reciprocal crosses between true-breeding barred and nonbarred types gave the following results:

Later microscopic examination showed that in moths, butterflies, and birds the females have two morphologically dissimilar sex chromosomes (equivalent to the X and Y) and the males have two morphologically identical sex chromosomes. This type of sex determination also occurs in reptiles, and in some groups of fish and amphibia.

The parallelism between the inheritance of a particular allele and the distribution of the X chromosome carrying it implied that genes must be parts of chromosomes. Other experiments with *Drosophila*, also instructive, provided the definitive proof. Before examining this work, we will consider the symbols used to designate the alleles of a gene in *Drosophila* genetics. A superscript + (as in a^+) is used to identify the **wildtype** allele of any gene. Wildtype means that the allele is the one found most commonly in natural populations of the organism. Recessive deviations from this standard type are indicated by a small letter, and dominant deviations with a capital letter, so the dominance relations of the allelic forms of the gene are immediately clear. For example, y^+ is the symbol for the wildtype counterpart of recessive allele *y* determining yellow body color; whereas Cy^+ is the symbol for the wildtype counterpart of another gene for which the dominant allele, *Cy* (Curly), results in wings that are curled. The letters used for gene symbols come from a usually descriptive term for the trait determined by the altered form of the gene. A common practice is to represent the wildtype allele of a gene only with a + sign (omitting the letter), as in the heterozygous genotype $y/+$, when it is clear from the context which gene is being designated. These conventions are also used with many other species, though most plant geneticists elect to represent recessive alleles by small letters and dominant alleles by initial capital letters without designating the wildtype. In this book the notations used will be those in general use for whichever organism is being considered.

One of Morgan's students discovered rare exceptions to the expected pattern of inheritance in crosses with several X-linked genes, including the one for white eyes. For example, when white-eyed females were mated with red-eyed males, most of the progeny consisted of the expected red-eyed females and white-eyed males. However, about one in every 2000 F_1 flies was an exception, either a white-eyed female or a red-eyed male. It was proposed that these rare exceptional offspring—each phenotypically like the parent of its own sex—could be accounted for by occasional failure of the two X chromosomes in the mother to separate from each other during meiosis—a phenomenon called **nondisjunction.** The predictable consequences of this unusual meiotic behavior are the formation of some eggs with two X chromosomes and others with none. Four classes of zygotes are expected from the fertilization of these abnormal eggs (Figure 2-14). Individuals with no X chromosome were not detected because embryos lacking this chromosome are not viable; most progeny with three X chromosomes also die during early development. Examination by microscopy indicated that the exceptional white-eyed females had two X chromosomes and a Y chromosome, and exceptional red-eyed males had a single X but no Y. These X0 males were sterile and could not be used for further genetic analysis.

Additional evidence was obtained that whenever eye color was not inherited in the expected way, the chromosome makeup of the resulting individual also differed from what was expected. In crosses between the exceptional

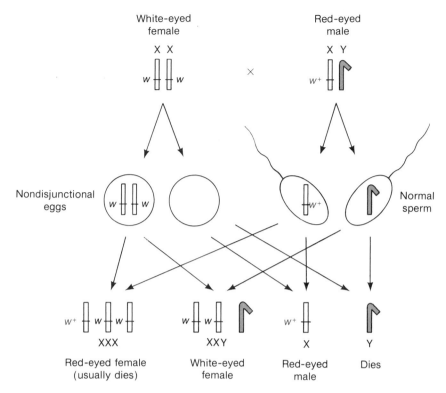

Figure 2-14 The results of meiotic nondisjunction of the X chromosomes in a *Drosophila* female.

white-eyed females and normal red-eyed males, about 4 percent of the viable offspring were again exceptional white-eyed females and red-eyed males; microscopic examination showed these exceptions to be XXY and XY, respectively, and they were fully fertile. The regular occurrence of such secondary exceptions among the progeny of primary exceptional females could be accounted for by a meiotic pairing and segregation of the XXY chromosomes resulting in the production of four kinds of eggs: XX, Y, XY, and X. Then, the two types of exceptional offspring would be the expected products from fertilization of the XX and Y eggs that result from nondisjunction of the maternal X chromosomes in XX ↔ Y meiotic segregations (Figure 2-15). Flies with either three X chromosomes or two Y chromosomes and no X, which would also be expected, were not observed and were therefore assumed to be inviable. From the observation that secondary exceptions constitute about 4 percent of the progeny from such crosses but represent only half of the offspring expected from fertilization of nondisjunctional eggs, it could be concluded that the XX ↔ Y meiotic segregations producing these eggs occur with a frequency of about 8 percent in XXY females.

The results of the experiments just described were of great importance in the development of genetics. The precise correspondence between the genetic and chromosomal constitution of the exceptional offspring obtained—with a

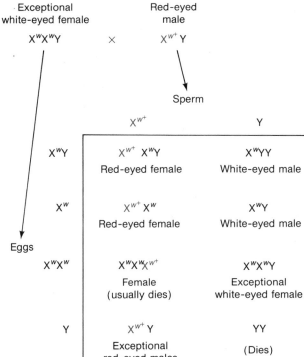

Figure 2-15 The origin of secondary exceptional progeny as a result of nondisjunction of the X chromosomes in an exceptional XXY female, as depicted in Figure 2-14. The symbol X^w indicates an X chromosome carrying the *w* allele.

frequency of 1 in 2000 among the progeny of white-eyed females crossed with red-eyed males, and with a substantially higher frequency in crosses between the primary exceptional XXY white-eyed females and red-eyed males—provided overwhelming evidence that genes are associated with chromosomes. The exceptional progeny from the experiments were also the basis for a better understanding of sex determination in *Drosophila*.

Sex Determination in Drosophila

The sexual phenotypes resulting from nondisjunction in *Drosophila melanogaster* established that the Y chromosome has no effect on sex determination in this organism. That is, females usually have two X chromosomes and males have usually one X and one Y, but a female may in addition have a Y and a male may lack a Y and the organisms will still be female and male, respectively. However, the Y is essential for male fertility, since X0 males are unable to produce normal sperm.

In *D. melanogaster* a complete haploid set of autosomes, represented by A, consists of two large chromosomes with median centromeres and a very

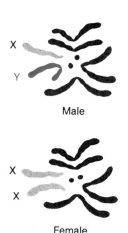

X

Y

Male

X

X

Female

Figure 2-16 The diploid chromosome complements of a *Drosophila melanogaster* male and female. The centromere of the Y chromosome is not centrally located, which gives the Y a characteristic J shape; the X chromosome has a nearly terminal centromere. Chromosomes II and III are not easily distinguishable. Chromosome IV, which is very small, appears as a dot.

small, dotlike chromosome (Figure 2-16). Studies of the sexual phenotypes of individuals with various combinations of X chromosomes and sets of autosomes provided the evidence that the primary determinant of sex in *Drosophila* is the X:A ratio. Individuals with half as many X chromosomes as sets of autosomes (1X:2A) are males and those with equal numbers of X chromosomes and complete sets of autosomes (2X:2A, 3X:3A) are females. The genotypes 1X3A (0.33:1.0) and 3X2A (1.5:1.0) result in normal male and normal female somatic sexual phenotypes, respectively, though these individuals are weak, have underdeveloped reproductive organs, and are sterile.

When two X chromosomes are present with three sets of autosomes (0.67:1.0), as happens in some offspring of 3X3A females mated with XYAA males, the individuals are **intersexual.** They display a wide variation in overall sexual phenotype and have a mixture of male and female structures in both the external genitalia and internal sex organs. The basis for understanding the phenotypes of these intersexes is provided by evidence that the X:A ratio determines sex in *Drosophila* at an early stage in embryo development and in a manner that is irreversible for an individual cell and its descendants. If some cells in an early embryo undergo determination as male and others as female in response to the 0.67:1.0 ratio, the structures and patches of tissue derived from these cells would result in the mixed sexual phenotypes observed in the adults.

In some organisms with homogametic females and heterogametic males the Y chromosome is male-determining. An early observation of the flowering plant *Melandrium album* showed that flowers produced by both 2X2A and 2X4A plants are pistillate (female), and those of XY2A and XY4A plants are invariably staminate (male). When another X chromosome is present, in XXY2A or XXY4A plants, most of the flowers are staminate, though an occasional one is bisexual (having both stamens and a functional pistil). This correspondence between the expression of sex and chromosome constitution indicates that in this plant the X is female-determining and, unlike *Drosophila*, the Y is male-determining. A balance between the effects of the two sex chromosomes is approached in XXXXY4A plants, which produce mostly bisexual flowers, though a few are staminate. Many other variations in the primary sex-determining mechanisms of both plants and animals are known.

Sex Determination in Mice and Humans

In mice the Y is male-determining, as it is in humans. The evidence for this conclusion in mice came from experiments in which the allele *Ta* (Tabby) of a gene on the X chromosome was used to detect unusual sex-chromosome constitutions among the progeny of crosses. The *Ta* allele produces a dark coat color in both *Ta/Ta* females and *Ta*/Y males, whereas wildtype +/+ females and +/Y males have light-colored coats. Heterozygous *Ta*/+ females have a distinctive phenotype with irregular patches of light and dark hair. When many matings of *Ta*/+ females with +/Y males are carried out, a few of the female offspring are phenotypically like *Ta/Ta* females. If sex deter-

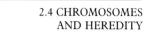

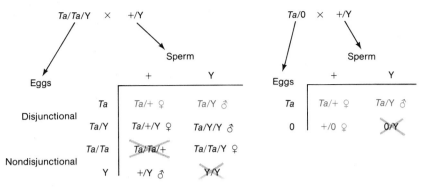

Figure 2-17 Progeny expected from crossing exceptional *Ta* females with light-colored $+/Y$ male mice, if the females are *Ta/Ta/Y* (left panel) or *Ta/0* (right panel). The progeny observed are indicated in red, and the crossed-out genotypes are expected to be lethal. From these results, the exceptional Tabby females clearly have the genotype *Ta/0*.

mination in mice were similar to that in *Drosophila*, these exceptional Tabby females would be genetically *Ta/Ta/Y*, the result of unusual maternal nondisjunction. However, were X0 individuals to develop as females, then their genotype would be *Ta/0* owing to the loss of a paternal sex chromosome. The progeny expected from mating the exceptional females with $+/Y$ males provided the basis for distinguishing between these alternatives, as can be seen in Figure 2-17. Three kinds of offspring are produced: phenotypically heterozygous (*Ta/+*) females, wildtype (+) females, and Tabby (*Ta/Y*) males. From the recovery of wildtype females and absence of homozygous Tabby females it was concluded that the exceptional Tabby mothers probably are *Ta/0*. Confirmation of the interpretation that the Y chromosome is male-determining in mice came from the microscopic demonstration that exceptional Tabby females have a single X chromosome (X/0) and the finding of sterile XXY males.

The first evidence that the Y chromosome in humans is male-determining came from analyses of the chromosome complements associated with two abnormal sexual phenotypes. Males with a condition known as **Klinefelter syndrome** have 47 chromosomes rather than the normal 46 and are XXY. A similar phenotype characterized by underdeveloped testes and sterility has been found to result from XXYY, XXXY, XXXYY, and XXXXY sex-chromosome constitutions. Females with the aberrant condition known as **Turner syndrome** have only 45 chromosomes and are X0. These sexually underdeveloped individuals are sterile.

Humans with an XXX sex-chromosome complement are fertile females, and individuals with an X and two Y chromosomes are fertile males. An unexpected and interesting finding is that only a small percentage of the children of XXX or XYY parents have abnormal sex-chromosome constitutions. For example, the process of meiosis in XXX females would be expected to result in a substantial fraction of eggs with two X chromosomes, and a similarly large fraction of sperm from XYY males would be expected to carry two Y chromosomes or an X and a Y. It is apparent from the offspring that this does not occur in the case of the human sex chromosomes, but whether the extra chromosomes are eliminated prior to meiosis or by preferential segregation of only a single X or Y into functional gametes during meiosis is not known.

The role of the Y chromosome in sex determination in humans, mice, and other mammals is to induce a rudimentary embryonic gonad with both male and female potentials to develop into a testis instead of an ovary. A major factor in this induction appears to be related to a substance called H-Y antigen, determined by a gene on the Y chromosome. It is not yet known whether the Y-chromosome gene determines the substance directly or in some way acts to regulate the expression of the antigen-forming gene located on another chromosome. A newly formed testis secretes both a substance that suppresses the development of the primitive oviduct and the hormone testosterone, which is required to induce development of both the genital duct system and secondary sexual characteristics of the male. In the absence of a Y chromosome, the rudimentary gonad differentiates into an ovary and the primitive oviduct into Fallopian tubes and a uterus. Ovarian development occurs normally in mice with only a single X chromosome but not in X0 humans.

2.5 Probability in Genetic Prediction and in Analysis of Genetic Data

Genetic ratios are the products of chance operating in the assortment of genes into gametes and in the combination of gametes into zygotes. The result is that exact predictions are not possible for a particular event. However, it is possible to state in advance that the event has a certain probability of occurring, as we have seen from application of the rules of probability considered in Chapter 1. A probability statement of this kind requires certain assumptions about the genetic basis of the traits concerned, and these assumptions are likewise the basis for expected ratios, and sometimes the deviation is large. In this section we consider some of the methods used in interpreting genetic data.

Use of the Binomial Distribution in Predicting Genetic Outcomes

The probability that each of three children in a family will be of the same sex is an example of a simple problem that uses both the addition and multiplication rules of probability. The probability that all three will be girls is $(1/2)(1/2)(1/2) = (1/2)^3 = 1/8$, and the probability that all three will be boys is likewise 1/8. Since these outcomes are mutually exclusive, the probability of one or the other is the sum of the two probabilities, which is 1/4. The remaining possible outcomes are that two of the children will be girls and the other a boy, or two will be boys and the other a girl. For each of these outcomes only three different orders of birth are possible—for example, GGB, GBG, and BGG—each having a probability of 1/8. Thus, the probability of two girls and a boy, disregarding birth order, is the sum of the probabilities for the three possible orders, or 3/8; likewise, the probability of two boys and a girl is also 3/8. Therefore, the distribution of probabilities for the sex ratio in families with three children is:

GGG	GGB	GBB	BBB
	GBG	BGB	
	BGG	BBG	

$(1/2)^3$ $\quad$ $3(1/2)^2(1/2)$ $\quad$ $3(1/2)(1/2)^2$ $\quad$ $(1/2)^3$

$1/8 \quad + \quad 3/8 \quad + \quad 3/8 \quad + \quad 1/8 \ = \ 1$

This sex-ratio information can be obtained more directly by expanding the binomial expression $(p + q)^n$, in which p is the probability of the birth of a girl ($= 1/2$), q is the probability of the birth of a boy ($= 1/2$), and n is the number of children; in our example,

$$(p + q)^3 = p^3 + 3p^2q + 3pq^2 + q^3$$

Similarly, the binomial distribution of probabilities for the sex ratios in families of five children is

$$(p + q)^5 = p^5 + 5p^4q + 10p^3q^2 + 10p^2q^3 + 5pq^4 + q^5$$

Each term tells us the probability of a particular combination. For example, the third term tells the probability of three girls and two boys in a family having five children—namely,

$$10(1/2)^3(1/2)^2 = 10/32 = 5/16$$

There are $n + 1$ terms in a binomial expansion. The exponents of p decrease from n in the first term to 0 in the last term, and the exponents of q increase by one from 0 in the first term to n in the last term. The coefficients generated by successive values of n can be arranged in a regular triangle known as **Pascal's triangle** (Figure 2-18). Note that the horizontal rows of the

n	Coefficients
0	1
1	1 1
2	1 2 1
3	1 3 3 1
4	1 4 6 4 1
5	1 5 10 10 5 1
6	1 6 15 20 15 6 1

Figure 2-18 Pascal's triangle.

triangle are symmetrical, and that each number is the sum of the two numbers on either side of it in the row above.

In general, if the probability of event A is p and that of event B is q, and the two events are independent and mutually exclusive (see Chapter 1), the probability that A will occur four times and B three times—in a specific order—is p^4q^3, by the multiplication rule. However, suppose that we are interested in the occurrence of this combination of events: four of A and three of B, irrespective of order. In that case, we multiply the probability that the combination 4A:3B will occur in any one specific order by the number of possible orders. The number of different combinations of seven events, four of one kind and three of another, is

$$\frac{7!}{4!\,3!} = \frac{1 \times 2 \times 3 \times 4 \times 5 \times 6 \times 7}{1 \times 2 \times 3 \times 4 \times 1 \times 2 \times 3} = 35$$

(The symbol ! is for factorial, or the product of all positive integers from one through a given number.) This calculation provides the coefficient of the p^4q^3 term in the expansion of the binomial $(p + q)^7$. Therefore, the probability that event A will occur four times and event B three times is $35p^4q^3$.

The probability rule for repeated trials of events with constant probabilities states that if the probability of occurrence of event A is p, and the probability of the alternative event B is q, the probability that in n trials event A will occur s times and event B will occur t times is

$$\frac{n!}{s!\,t!}\,p^sq^t$$

in which $s + t = n$, and $p + q = 1$. To use this expression, one must remember that 0! is defined to be 1, and that any number raised to the zero power (e.g., 2^0) also equals 1. Note that each individual term in the expansion of the binomial $(p + q)^n$ is the same as the probability expression just given.

The probabilities of the events considered can be any fractions whose sum equals 1. If we are concerned with the offspring of two heterozygous parents, for example, the probability p of a child showing the trait determined by the dominant allele is 3/4, and the probability q of a child showing the trait determined by the recessive allele is 1/4. Suppose that we wanted to know in what proportion of such families having eight children, six of the children would be expected to have the trait determined by the dominant allele and two children the trait determined by the recessive allele. In this case, $n = 8$, $s = 6$, $t = 2$, and the probability of this combination of events is

$$\frac{8!}{6!\,2!}\,(p)^6(q)^2 = \frac{6! \times 7 \times 8}{6! \times 2!}\,(3/4)^6(1/4)^2 = 0.31$$

That is, in slightly less than one-third of families of eight children will the offspring show the ideal 3:1 phenotypic ratio; the other families will deviate in one direction or the other because of chance variation.

Evaluating the Fit of Observed
Results to Theoretical Expectations

An important question in dealing with genetic ratios, or with other events whose outcomes are subject to the rules of probability, is whether an observed result is in agreement with the theoretical expectation based on a particular hypothesis. That is, how far can observed results deviate from those expected before they must be considered an indication of a real difference and not just an accident of sampling? For example, suppose that 100 seedlings from a cross between a green F_1 corn plant and a yellow-green plant of one of the parental strains were grown to test for a 1:1 ratio, and that 65 of the seedlings were green and 35 yellow-green. Does this mean the expected ratio is wrong and that the genetic basis for the color difference is more complex than a single pair of alleles, or might the deviation from the 1:1 ratio merely result from chance? The answer depends on how often, in progenies of size 100, a deviation as large or larger would be expected to occur by chance if the expectation were correct.

The probability of obtaining results that differ from an expected 50:50 by 15 individuals or more *in either direction* can be determined from the expansion of the binomial $(p + q)^{100}$. This tedious chore would yield 101 terms, representing the probabilities of all possible outcomes—that is, from 100 green and 0 yellow-green to 0 green and 100 yellow-green. Summing the appropriate terms would tell us that the chance of a deviation as large or larger than 15 is about 1 in 300—a probability so small that we would most likely conclude that the 1:1 expectation is wrong. There is, of course, still the one chance in 300 that the expected ratio is correct. Two kinds of mistakes are possible in decisions about the agreement between observed and expected results: (1) a correct hypothesis may sometimes be rejected, or (2) an incorrect hypothesis may sometimes be accepted. These two types of errors are called type I and type II errors, respectively.

Testing Genetic Hypotheses

Observed and expected results are rarely identical because of the statistical nature of experimental outcomes. In this section we consider how to interpret these differences.

The agreement between observed and expected results can be evaluated in terms of the probability that chance alone is responsible for the discrepancy between them. A genetic hypothesis is rejected whenever the probability is sufficiently small that chance alone may have produced a discrepancy as large as or larger than the one actually observed. How small the probability must be to indicate unsatisfactory agreement is a matter of judgment, but there are some generally accepted guidelines. When the probability is smaller than 0.05 (1 in 20), the difference between the results observed and those expected is said to be **significant.** Such a small value is conventionally regarded as an unacceptable level of agreement, and the hypothesis is rejected. One must accept the fact that if the genetic hypothesis is correct, results that are significantly different from expectation will nevertheless occur by chance once in every 20 experiments. Rejection at the 5 percent level does not guarantee that the hypothesis is wrong. The hypothesis can be correct but the sample from which the data are obtained may be an unusual (improbable) sample. A stronger basis for rejection is a probability of less than 0.01 (1 in 100), in which case the difference between observed and expected result is said to be **highly significant.** With this level of significance, the chance of rejecting a correct hypothesis is reduced to 1 percent, but the chance of failing to reject a false hypothesis is increased.

When the probability of a discrepancy as large as or larger than the one observed is greater than 0.05, then the difference between observed and expected results is said to be **nonsignificant.** In this case the agreement between observed and expected is judged to be satisfactory, which does not necessarily mean that the hypothesis is true, but only that the hypothesis is consistent with the observed results. The reason for the ambiguity is that the same data may also be consistent with other hypotheses, and deciding among these alternatives may require additional experiments.

The Chi-square Method

Calculation of exact probabilities for assessing goodness of fit between observed values and those predicted from a specific genetic hypothesis is extremely laborious when the number of observations is large. Several statistical methods give approximate probabilities that are accurate enough for most purposes, and the calculations are less cumbersome. A particularly useful one is called the chi-square (χ^2) method, or sometimes the χ^2 test for goodness of fit.

The value of χ^2 is given by the expression

$$\chi^2 = \Sigma \frac{(\text{Observed} - \text{expected})^2}{\text{Expected}}$$

in which Σ means the summation over all the classes. For each class the difference between the observed number and expected number is squared and then divided by the expected number. To illustrate, suppose we have an F_2

Table 2-2   Calculation of χ^2 for a monohybrid ratio

Phenotype (class)	Observed number	Expected number	Deviation from expected number	$\frac{(\text{Deviation})^2}{\text{expected number}}$
Wildtype	99	108	−9	0.75
Mutant	45	36	+9	2.25
Totals	144	144	0	$\chi^2 = 3.00$

generation consisting of 99 wildtype and 45 mutant individuals, and we wish to know whether this set of observed numbers is in satisfactory agreement with the expected $3:1$ ratio resulting from the segregation of a single pair of alleles. Calculation of the value of χ^2 is carried out in Table 2-2. Note that the sum of the deviations of observed from expected numbers equals zero; when there are only two classes, as in this example, the deviations are always equal in magnitude but opposite in sign. Note also that a deviation of 7 would be a very large one if the expected number were 12, but a relatively small one if the expected number were 120. Since sample size is important, the measure of the size of the deviation from expectation—(observed − expected)^2— is expressed as a proportion of the expected number. These values are the components of χ^2, and their summation gives a value of 3.00 in the example.

It is necessary to relate the χ^2 value to the probability of obtaining, by chance, deviations as large or larger than those observed, if in reality the $3:1$ expectation is correct. The larger the difference between expected and observed numbers, the larger χ^2 is for a given sample size. The larger χ^2 is, the less likely it is that the deviations are simply a result of chance. In the interpretation of a χ^2 value, the number of classes of data must be taken into account. Every χ^2 test has associated with it a number called its **degree of freedom,** which is used in assessing the significance of the χ^2 value. For the type of χ^2 test illustrated in Table 2-2, the number of degrees of freedom is simply the number of classes of individuals minus 1. In general terms, degrees of freedom refers to the number of classes that can vary independently. Table 2-2 contains two classes of individuals, and since the total sample size is fixed at 144, there is only one degree of freedom. Once the number of individuals in either class has been determined, the number in the other class is automatically set. Similarly, when there are four classes of data, three of them can have any frequency, but the frequency of the fourth class must equal the difference between the sum of the three and the total. That is, with four classes of individuals there are three degrees of freedom—the total number of classes minus 1.

When a χ^2 value has been calculated and its degrees of freedom determined, the final step in assessing goodness of fit between expected and observed data is straightforward. One uses either a graph or a table of χ^2 values (Table 2-3) to determine the approximate probability of obtaining, by chance alone, a deviation as great as or greater than that actually observed. In the example in Table 2-2, $\chi^2 = 3.00$ with one degree of freedom. In Table 2-3, the top line corresponds to one degree of freedom, and the probability associated with a χ^2 value of 3.00 is somewhere between 0.05 (3.841) and 0.20 (1.642). This range implies that a deviation from the $3:1$ ratio at least as great as that observed would be expected to occur by chance in more than 5 percent, but fewer than 20 percent, of similar experiments. Recall that a probability greater than 5 percent is conventionally considered to represent satisfactory agreement between observed and expected results. Thus, the data in Table 2-2 provide no basis for rejecting the hypothesis. Other examples in the application of the χ^2 test are included in the problems at the end of this chapter.

Table 2-3 Selected critical values in the distribution of χ^2

Degrees of freedom	Probability (P)						
	0.95	0.80	0.50	0.20	0.05	0.01	0.005
1	0.004	0.064	0.455	1.642	3.841	6.635	7.879
2	0.103	0.446	1.386	3.219	5.991	9.210	10.597
3	0.352	1.005	2.366	4.642	7.815	11.345	12.838
4	0.711	1.649	3.357	5.989	9.488	13.277	14.860
5	1.145	2.343	4.351	7.289	11.070	15.086	16.750
6	1.635	3.070	5.348	8.558	12.592	16.812	18.548
7	2.167	3.822	6.346	9.803	14.067	18.475	20.278
8	2.733	4.594	7.344	11.030	15.507	20.090	21.955
9	3.325	5.380	8.343	12.242	16.919	21.666	23.589
10	3.940	6.179	9.342	13.442	18.307	23.209	25.188
15	7.261	10.307	14.339	19.311	24.996	30.578	32.801
20	10.851	14.578	19.337	25.038	31.410	37.566	39.997
25	14.611	18.940	24.337	30.675	37.652	44.314	46.928
30	18.493	23.364	29.336	36.250	43.773	50.892	53.672

Two precautions must be observed in carrying out χ^2 tests:

1. The test must be calculated in terms of the observed *numbers*, not their *proportions*.

2. The approximation is invalid when the expected numbers are too small. A widely accepted rule of thumb is to avoid use of the χ^2 test whenever the smallest expected number is 5 or less.

Alternative statistical procedures are available for testing goodness of fit in such cases, but they are beyond the scope of this book.

Problems

1. The following pedigree shows the incidence of two inherited traits in a human family. Individuals with a progressive form of deafness are represented by solid black symbols and those with a rare congenital cataract condition by red symbols.

(a) What is the probable mode of inheritance of each of these traits (that is, dominant or recessive, X-linked or autosomal)?

(b) What were the genotypes of the two original parents in the pedigree?

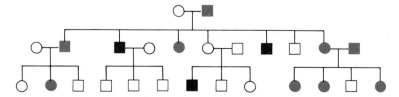

2. The pedigree shown below relates to the segregation in a human family of a sex-linked recessive allele that results in severe mental retardation (solid symbol).

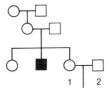

What is the probability that a child produced by individuals 1 and 2 will be retarded?

3. The black and yellow pigments in the coats of cats are determined by an X-linked pair of alleles. Males are either black or yellow, and females are black, calico (with patches of black and patches of yellow), or yellow. The alleles for black and yellow coat color can be represented by the symbols c^b and c^y, respectively.
  (a) What phenotypes would be expected among the offspring from a cross between a black female and a yellow male?
  (b) In a litter of eight kittens there are two calico females, one yellow female, two black males, and three yellow males. What are the probable phenotypes of the mother and father?
  (c) Calico males are occasionally found and are almost always sterile. What would you predict for the sex-chromosome makeup of these males?

4. Assuming that the human sex ratio at birth is $1:1$, consider two separate families, A and B, each having three children.
  (a) What is the probability that all of the children in family A will be girls and that all children in family B will be boys?
  (b) What is the probability that one or the other of the families will have only boys and the remaining family will have only girls?

5. You know that a couple has two children and that one of the children is a girl. What is the probability that the other child is also a girl?

6. The X chromosome of *D. melanogaster* has a centromere very close to one end. In some laboratory stocks two X chromosomes have become joined to one another at a single centromere, resulting in their permanent nondisjunction. Females with attached-X chromosomes ($\overline{XX}$) usually carry a Y chromosome in addition. Vermilion eye color in the fly is determined by the X-linked recessive allele v. If an $\overline{XX}Y$ female with vermilion eyes is crossed with a wildtype (v^+) male, what are the expected genotypes and phenotypes of the viable offspring?

7. The diploid chromosome number of maize (*Zea mays*) is 20; that is, the nuclei of somatic cells contain two haploid sets, each consisting of 10 chromosomes. Occasionally, an unfertilized egg develops into a haploid plant. The 10 chromosomes of a haploid nucleus are unpaired during the first meiotic division and are distributed at random to the two poles of the spindle. What is the probability that all 10 chromosomes will be distributed to the same pole?

8. The individuals indicated by solid symbols in the following pedigree of a human family have a common form of red–green color blindness determined by an X-linked recessive allele.

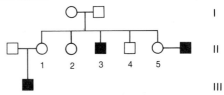

  (a) If the woman identified as II-1 has two more children, what is the probability that both of these children will have normal color vision?
  (b) What is the probability that the first child of II-5 will have this form of red–green color blindness?

9. In *Drosophila* the progeny from the mating of a single pair of flies consist of individuals having the characteristics listed below and in approximately the proportions shown:

Very narrow eyes	1 female : 1 male
Intermediate eyes	All females
Normal round eyes	All males

Only one pair of alleles is responsible for these differences. What can you conclude about dominance and whether the gene in question is X-linked or autosomal? Using convenient symbols, indicate the genotypes of the parents.

10. A recessive allele w of a gene on the X chromosome of *D. melanogaster* determines white eye color; the wildtype eye color is brick red. A second recessive allele b of a gene on one of the autosomes determines black body color; the wildtype body color is gray. What are the expected genotypes and phenotypes of the F_1 progeny produced when homozygous white-eyed, gray-bodied females are mated with red-eyed, black-bodied males?

11. Feather coloration in chickens requires the presence of the dominant allele C of an autosomal gene. The dominant allele I of another autosomal gene acts as an inhibitor preventing the expression of C; that is, only individuals with the genotype $ii\,C–$ have colored feathers. White Leghorn and White Wyandotte chickens have the genotypes

II CC and *ii cc*, respectively. The recessive allele *k* of a sex-linked gene results in slow growth of the primary wing feathers, and the corresponding dominant allele *K* results in fast feather growth. If slow-feathering White Leghorn males are mated with fast-feathering White Wyandotte females and the resulting F_1 birds are intercrossed, what are the expected phenotypes and their expected frequencies in the F_2?

12. In considering human families with six children,
    (a) What is the probability that three of the children will be boys and three will be girls?
    (b) What is the probability of this birth order: girl–boy–girl–boy–girl–boy?
    (c) In what proportion of such families will there be *at least* two boys?

13. Phenylketonuria in humans is a severe metabolic disorder resulting from a defect in the enzyme required to convert the amino acid phenylalanine to tyrosine. Unless the disease is detected early and a low phenylalanine diet is used, affected children accumulate the phenylalanine and phenylpyruvic acid (a product of its breakdown) and sustain serious damage to the central nervous system. The defect in enzyme activity is determined by a recessive allele of an autosomal gene. Heterozygous individuals can be detected because they accumulate phenylalanine in their blood serum after ingesting excessive amounts of the amino acid. If two heterozygous parents have four children,
    (a) What is the probability that none of the children will have phenylketonuria?
    (b) What is the probability that two of the children will be affected?

14. The following data were obtained when corn seedlings were grown from two samples of seeds produced by crossing a green plant with one having yellow-striped leaves.

	Sample 1	Sample 2
Green	40	140
Yellow-striped	20	160
	60	300

Test each of these observed results for agreement with a 1:1 expectation. What can you conclude about the effect of increasing sample size when the absolute deviation from expectation remains constant?

15. A cross was performed with *D. melanogaster* in which two pairs of alleles were segregating; dp^+ and *dp* determine long versus short wing, and e^+ and *e* determine gray versus ebony body color. The F_2 data below were obtained:

Long wing, gray body	95
Long wing, ebony body	36
Short wing, gray body	24
Short wing, ebony body	5

Test these data for agreement with a 9:3:3:1 ratio.

16. The probability expression for repeated trials of events with constant probabilities, $P = (n!/s!t!)p^s q^t$, can be directly extended to three or more events. If the probabilities of three events are *p*, *q*, and *r* (in which $p + q + r = 1$), the probabilities of the possible outcomes are represented by the terms of the expanded trinomial $(p + q + r)^n$. In this case, the expression for each term is $(n!/s!t!u!)p^s q^t r^u$.
    (a) When two blue Andalusian chickens are crossed, the probability of a black offspring is 1/4, of a blue one is 1/2, and of a splashed white one is 1/4. What is the probability that eight eggs produced by such a cross will yield exactly two black, four blue, and two splashed white chicks?
    (b) If two doubly heterozygous walnut-combed chickens are crossed, the probabilities of a walnut-combed, pea-combed, rose-combed, and single-combed offspring are 9/16, 3/16, 3/16, and 1/16, respectively. What is the probability that a dozen eggs from such a cross will yield exactly six walnut-, two pea-, three rose-, and one single-combed offspring?

17. Mendel studied the inheritance of phenotypic characters determined by seven pairs of alleles. It is an interesting coincidence that the pea plant also has seven pairs of chromosomes. What is the probability that no two of the traits studied by Mendel were determined by genes located in the same pair of chromosomes?

C H A P T E R 3

Gene Linkage and Chromosome Mapping

Early proponents of the concept that chromosomes represent the physical basis of heredity recognized that numerous genes must be present in a chromosome. Acceptance of the idea that chromosomes behave as the units of segregation in meiosis led to the expectation that the alleles of all genes located in a chromosome should be transmitted as a unit and therefore show complete **linkage.** This would seem to be an exception to the principle of independent assortment. Such exceptions were soon discovered in crosses with certain combinations of genes in several organisms. However, the observed linkages between genes were never entirely complete.

We will see in the discussion of linkage in this chapter that incomplete linkage between two genes in the same chromosome is usual and is the result of exchange of segments between homologous chromosomes, which may occur when the chromosomes are paired during the first meiotic prophase. The occurrence of such an exchange event between homologous chromosomes, called **crossing over,** results in **recombination** of genes in the pair of chromosomes. The probability of crossing over between any two genes serves as a measure of genetic distance between them. Thus, the frequency of recombination between different genes in a chromosome provides the basis for construction of a **genetic map** of the chromosome, which is a diagram showing the relative positions of the genes.

3.1 Linkage and Recombination of Genes in a Chromosome

A simple and direct test of independent assortment, as we have seen in Chapter 1, is to testcross an F_1 hybrid ($Aa\,Bb$) with a doubly recessive ($aa\,bb$)

Facing page: The stigma of a flower covered with numerous pollen grains (spiked spheres), as visualized by scanning electron microscopy. (From G. Shih and R. Kessel. 1982. *Living Images.* Jones and Bartlett.)

individual. If the two pairs of alleles assort independently, as expected if they are on different chromosomes,

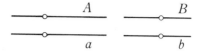

the hybrid will produce the four possible types of gametes—*AB*, *Ab*, *aB*, and *ab*—in equal proportions:

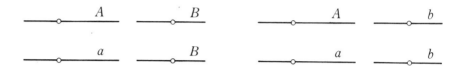

Note that the result will be the same if the *parents* of the hybrid have the genotypes *AA BB* and *aa bb* or the genotypes *AA bb* and *aa BB*. The four phenotypes of the testcross progeny, determined directly by the genotypes of the gametes produced by the hybrid, are therefore expected to occur in a 1:1:1:1 ratio. Thus, 50 percent of the testcross progeny will consist of two phenotypes with the same combinations of traits that were present in the parents of the hybrid (**parental combinations**), and 50 percent of two phenotypes with new combinations of the traits (**recombinants**). In this chapter we will examine phenomena that cause deviation from the 1:1:1:1 ratio.

Linkage

In early experiments with *Drosophila*, several X-linked genes were identified that provided ideal materials for studying the inheritance of genes in the same chromosome. One of these genes, whose w^+ and w alleles govern the alternative red versus white eye color, was discussed in Chapter 2; another such gene has the alleles m^+ and m, which determine normal versus miniature wings. When females with white eyes and normal wings were mated with red-eyed and miniature-winged males,

$$w\,m^+/w\,m^+ \times w^+\,m\,/Y$$

and the wildtype females and white-eyed, normal-winged males of the F_1 were intercrossed,

$$w\,m^+/w^+\,m \times w\,m^+/Y$$

the progeny consisted of females with normal wings and either red eyes or white eyes, and males that were

white-eyed, normal-winged $(w\,m^+/Y)$	226	
red-eyed, miniature-winged $(w^+\,m/Y)$	202	66.5 percent parental phenotypes
red-eyed, normal-winged $(w^+\,m^+/Y)$	114	
white-eyed, miniature-winged $(w\,m/Y)$	102	33.5 percent recombinant phenotypes
	644	

The results of this experiment show a considerable deviation from the $1:1:1:1$ ratio of the four male phenotypes expected with independent assortment of genes. The pattern of the deviation is what would be predicted if genes in the same chromosome tended to remain together in inheritance but were not completely linked. That is, the combinations of phenotypic traits in the parents of the original cross (parental phenotypes) were present in 66.5 percent (428/644) of the F_2 males and nonparental combinations (recombinant phenotypes) of these traits were present in 33.5 percent (216/644). This result means that the frequency of recombination between the genes, which occurred at meiosis in the F_1 females in forming w^+m^+ and wm gametes, was 33.5 percent instead of the 50 percent that would be expected if the genes assorted independently.

The notation we use for linked genes has the general form $w^+m/w\,m^+$. This notation is a simplification of a more graphic, but cumbersome, form

$$\frac{w^+ \qquad m}{w \qquad m^+}$$

where the horizontal lines represent the two homologous chromosomes on which the respective alleles of the genes are located. The linked genes in a chromosome are always indicated in the same order, for reasons that will become apparent. The convention makes it possible with *Drosophila*, or other organisms for which a similar system of gene notation is used, to indicate the wildtype allele of a gene with a $+$ sign; that is, $w\,m^+/w^+m$ can be written $w +/+ m$. This particular arrangement of the mutant and wildtype alleles of two genes in a heterozygote, as in the F_1 females from the cross we have described, is called the **trans** or **repulsion** configuration. The alternative arrangement, with the wildtype alleles of the two genes on the same chromosome and the mutant alleles on the homologous chromosome ($++/wm$), is called the **cis** or **coupling** configuration.

When females with white eyes and miniature wings were mated with red-eyed, normal-winged males,

$$w\,m/w\,m \ \times \ ++/Y$$

and the wildtype females and white-eyed, miniature-winged males of the F_1 were intercrossed,

$$w\,m/+\,+ \;\times\; w\,m/Y$$

the F_2 progeny were

red-eyed, normal-winged ($+\,+/w\,m$ ♀♀ and $+\,+/Y$ ♂♂)	395	62.4 percent parental phenotypes
white-eyed, miniature-winged ($w\,m/w\,m$ ♀♀ and $w\,m/Y$ ♂♂)	382	
white-eyed, normal-winged ($w\,+/w\,m$ ♀♀ and $w\,+/Y$ ♂♂)	223	37.6 percent recombinant phenotypes
red-eyed, miniature-winged ($+\,m/w\,m$ ♀♀ and $+\,m/Y$ ♂♂)	247	
	1247	

Note that, compared to the previous experiment, the classes of offspring representing parental combinations of the phenotypic traits and those recombinant for the traits thus were reversed. Crosses of these kinds led to the conclusion that recombination of linked genes occurs with about the same frequency when the alleles of the genes are in the *trans* configuration as when they are in the *cis* configuration, and that the recombination frequency is characteristic of the particular pair of genes.

The recessive allele y of a third X-chromosome gene in *Drosophila* results in yellow body color instead of the usual gray color determined by the y^+ allele. When white-eyed females were mated with yellow-bodied males, and the wildtype F_1 females were testcrossed with yellow-bodied, white-eyed males,

$$+\,w/+\,w \;\times\; y\,+/Y$$
$$\downarrow$$
$$+\,w/y\,+ \;\times\; y\,w/Y$$

the progeny were

gray-bodied, white-eyed (maternal gamete, $+\,w$)	4292	98.6 percent parental phenotypes
yellow-bodied, red-eyed (maternal gamete, $y\,+$)	4605	
gray-bodied, red-eyed (maternal gamete, $+\,+$)	86	1.4 percent recombinant phenotypes
yellow-bodied, white-eyed (maternal gamete, $y\,w$)	44	
	9027	

In a second experiment, yellow-bodied, white-eyed females were crossed with wildtype males, and the wildtype females and yellow-bodied, white-eyed males produced in the F_1 were intercrossed:

67

3.1 LINKAGE AND
RECOMBINATION OF
GENES IN A CHROMOSOME

$$yw/yw \quad \times \quad ++/Y$$

$$++/yw \quad \times \quad yw/Y$$

The F_2 progeny were

gray-bodied, red-eyed (maternal gamete, $++$)	1159	98.7 percent parental phenotypes
yellow-bodied, white eyed (maternal gamete, yw)	1017	
gray-bodied, white-eyed (maternal gamete, $+w$)	17	1.3 percent recombinant phenotypes
yellow-bodied, red-eyed (maternal gamete, $y+$)	12	
	2205	

The parental and recombinant classes of offspring were again reversed in the two crosses. However, the recombination frequency was virtually the same, and much lower than that obtained in crosses involving the genes determining red versus white eye color and normal versus miniature wing size.

Drosophila is unusual in that recombination does not occur in males—that is, all alleles located in a particular chromosome show complete linkage in males. To illustrate this point, we will consider two autosomal genes. The *pr* allele of one of these genes results in purple eye color, instead of the usual red, and the *vg* allele of the other gene results in development of vestigial (very small abnormal) wings. From a mating of vestigial-winged females and purple-eyed males, wildtype F_1 males and females having the genetic constitution $+vg/pr+$ were obtained. When F_1 females were testcrossed with homozygous recessive males,

$$+vg/pr+ \quad \times \quad prvg/prvg$$

about 13 percent of the males and females in the progeny were recombinant for the phenotypic traits—that is, they were either red-eyed and normal-winged or they were purple-eyed and vestigial-winged. However, in testcrosses of the heterozygous F_1 males, no recombinant offspring were produced. The absence of recombination in *Drosophila* males is a useful characteristic for geneticists but is atypical—recombination occurs in both sexes in most other animals and plants.

GENE LINKAGE AND
CHROMOSOME MAPPING

The typical Mendelian dihybrid ratio in the F_2 generation—the experimental basis for the principle of independent assortment—results from the formation by $AaBb$ F_1 individuals of equal numbers of gametes with the four possible combinations of genes (AB, Ab, aB, and ab). Linkage of the segregating genes, resulting in a decrease in the relative frequency of the two recombinant gametic types and a corresponding increase in frequency of the two parental gametic types, causes a deviation from the $1:1:1:1$ gametic ratio and hence also a deviation from the $9:3:3:1$ phenotypic ratio. As we have seen in previous examples of such deviations from the $1:1:1:1$ phenotypic ratio, the extent of the deviation depends on the extent of linkage between the genes. However, it is more difficult to interpret an F_2 ratio than to interpret a testcross ratio, since a given phenotype in the F_2 can result from the union of nonrecombinant gametes (for example, AB with ab), recombinant gametes (Ab with aB), or both kinds of gametes (AB with Ab or with aB) from the two hybrid parents.

An example of two linked genes in corn (maize) illustrates the effects of linkage on an F_2 ratio. The alleles C and c of the first of these genes determine colored versus colorless kernels (the color is in the outer layer of the endosperm), and the Wx and wx alleles of the second gene result in nonwaxy versus waxy endosperm, respectively. In one experiment a homozygous plant grown from a colored nonwaxy seed was crossed with a plant from a colorless waxy seed, and the F_1 plants were intercrossed (or, equivalently, self-pollinated):

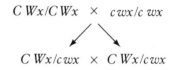

$$C\,Wx/C\,Wx \quad \times \quad c\,wx/c\,wx$$

$$C\,Wx/c\,wx \quad \times \quad C\,Wx/c\,wx$$

The F_2 seeds obtained were:

colored, nonwaxy	1774	=	64.8 percent
colored, waxy	263	=	9.6 percent
colorless, nonwaxy	279	=	10.2 percent
colorless, waxy	420	=	15.4 percent
	2736		

These data clearly represent a substantial deviation from the $9:3:3:1$ ratio of the phenotypes expected with independent assortment of the genes. Again, an excess of offspring with parental combinations of the traits is present.

From extensive testcross data the c and wx genes are known to recombine with a frequency of about 20 percent. This means that among the gametes formed by a $C\,Wx/c\,wx$ dihybrid, about 80 percent will be either $C\,Wx$ or $c\,wx$ and the remaining 20 percent will be the recombinant types $C\,wx$ or $c\,Wx$. In addition, the frequencies of the parental and recombinant phenotypes among testcross progenies indicate that the two nonrecombinant types of gametes are formed by an F_1 hybrid in approximately equal numbers, and similarly for the two recombinant types. Thus, if we know the recombination frequency between the genes, we know the relative frequencies of the gametes produced by the F_1 hybrids and can easily predict the frequencies of genotypes and phenotypes in the F_2 generation. The necessary calculations are shown in Figure 3-1 for a CWx/cwx dihybrid. The agreement between the predicted numbers of F_2 phenotypes and those observed is remarkably good:

	Colored nonwaxy	Colored waxy	Colorless nonwaxy	Colorless waxy
Predicted	1806	246	246	438
Observed	1774	263	279	420

The recombination frequency of c and wx determined from testcross data is 20 percent meaning that 80 percent of the gametes formed by a CWx/cwx dihybrid will have parental combinations and 20 percent will have recombinations of the genes:

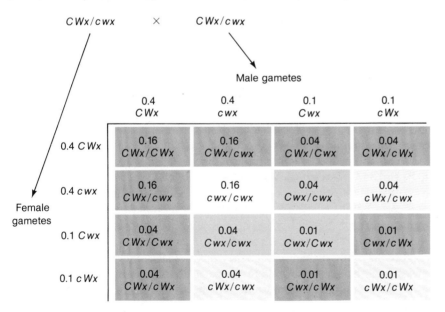

Figure 3-1 Prediction of an F_2 phenotypic ratio from the known recombination frequency between the c and Wx genes.

The predicted frequencies of phenotypes in F_2 are:

0.66 colored, nonwaxy (C-Wx-)
0.09 colored, waxy (C-$wxwx$)
0.09 colorless, nonwaxy ($ccWx$-)
0.16 colorless, waxy ($ccwxwx$)

From these data χ^2 equals 6.91 with three degrees of freedom, and P is 0.05–0.20. If the two genes were independently inherited, the corresponding numbers would be 1539, 513, 513, and 171 (that is, a 9:3:3:1 ratio).

Estimating the recombination frequencies between genes from F_2 data is more complex than doing so from testcross data, and larger numbers of offspring are required for comparable accuracy of the estimates. Methods for such analyses are valuable when it is impossible or impractical to obtain testcross data. In some cases, individuals homozygous for one of the recessive alleles may fail to survive until sexual maturity, and in self-pollinating plants such as wheat and barley the time and effort required to produce sufficient numbers of testcross seeds are usually prohibitive.

3.2 Crossing Over and Genetic Mapping

Linkage of the genes in a chromosome can be represented in the form of a **linkage map** or **chromosome map,** which portrays the linear order of the genes along the chromosome with the distances between adjacent genes proportional to the frequency of recombination between them. Morgan inferred this ordering of genes from the results of early linkage experiments with *Drosophila*, and postulated that recombination between the genes occurs by an exchange of segments between homologous chromosomes in a process called **crossing over.** This concept was based on an earlier interpretation of the origin of chiasmata, the cross-shaped configurations between homologous chromosomes seen during the first meiotic prophase (Figure 2-8). The exchange of pairing partners by two of the four chromatids of a pair of chromosomes had led to the hypothesis that a **chiasma** arises by a breaking and rejoining of these chromatids, with the result that there is an exchange of corresponding segments between them. The theory of crossing over is that there is a one-to-one correspondence between a chiasma and formation of a new association of genetic markers.

The unit of distance in a linkage map is called a **map unit;** it is defined to be equal to 1 percent recombination. Thus, two genes that recombine with a frequency of 3.5 percent are said to be located 3.5 map units apart. One map unit corresponds physically to a length of the chromosome in which, on the average, an exchange will occur once in every 50 meioses. Since a crossover results in two recombinant chromatids and two nonrecombinant chromatids (Figure 3-2), this exchange frequency means that 2 of every 200 products of meiosis will have an exchange in the particular chromosome segment.

When adjacent chromosome regions separating linked genes are sufficiently short, the recombination frequencies (and hence the map distances) between the genes are additive. This important feature of recombination and the logic used in genetic mapping is illustrated by the following example. Assume that the recombination frequency between genes a and b is 4 percent and the frequency between b and c is 7 percent. The order of the genes can be either $a\,b\,c$ or $b\,a\,c$. Which of these alternative orders is correct can be determined by a separate measurement of the recombination frequency be-

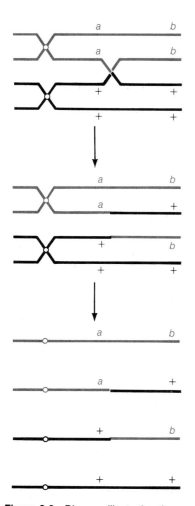

Figure 3-2 Diagram illustrating the occurrence of crossing over between two loci. The exchange between two of the four chromatids results in two recombinant and two nonrecombinant products of meiosis.

tween a and c. That is, if a and c are found to recombine with a frequency of 11 percent, we can deduce that b is located between them:

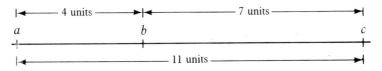

On the other hand, if a and c recombine with a frequency of 3 percent, we would conclude that the order of the genes is $b\,a\,c$:

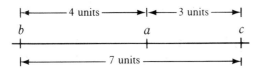

A linkage map can be expanded by this reasoning to include all of the known genes in a chromosome; these genes constitute a **linkage group.** The number of linkage groups is the same as the haploid chromosome number of the species, as expected. Abbreviated genetic maps of the four linkage groups of *Drosophila melanogaster* are shown in Figure 3-3.

Crossing Over

The orderly arrangement of genes represented by a linkage map is consistent with the conclusion that each gene occupies a well-defined site or **locus** in its chromosome, with the alleles of a gene in a heterozygote having corresponding locations in the pair of homologous chromosomes. Crossing over, which occurs by a process of exchange resulting in a new association of genes in the same chromosome, has the following features:

1. The exchange of segments between parental chromatids occurs in eukaryotes during the first meiotic prophase, after the chromosomes have replicated. The group of four chromatids (strands) of a pair of homologous chromosomes, which are closely synapsed at this stage, is called a **tetrad.**

2. The exchange process involves a breaking and rejoining of the two chromatids, resulting in the *reciprocal* exchange of equal and corresponding segments between them (see Figure 3-2). Our present understanding of the molecular details of this important process will be discussed in Chapter 8.

3. Crossing over occurs more or less at random along the length of a chromosome pair, with the result that the probability of its occurrence between two loci increases with increasing physical separation of the loci along the chromosome. This principle is the basis for determining

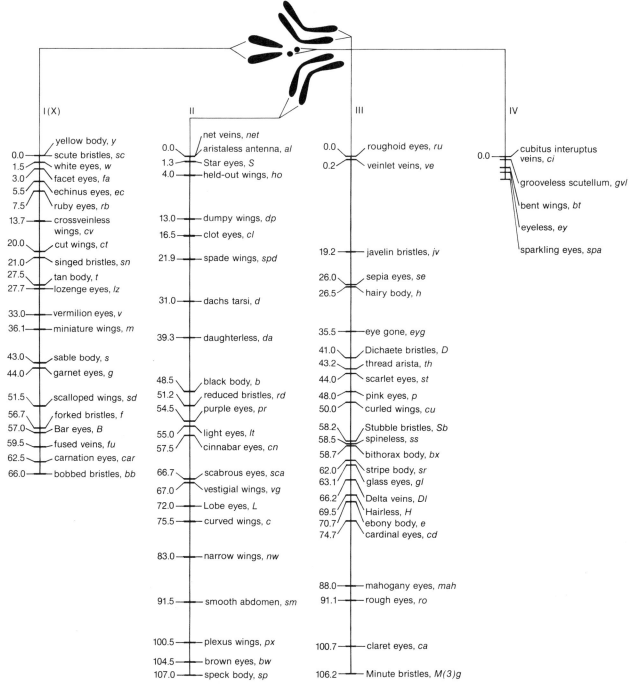

Figure 3-3 The genetic map of *Drosophila melanogaster*, showing the correspondence between each of the four linkage groups and a pair of chromosomes. The map positions of the genes in a chromosome are in map units from the gene closest to one end of the chromosome. Only a few of the many genes that have been mapped are shown. Capitalized symbols indicate a dominant allele.

the relative positions of the genes along a chromosome from linkage data.

The first proof that crossing over occurs after the chromosomes have replicated came from a study of laboratory strains of *D. melanogaster* in which the two X chromosomes in females are joined to a common centromere to form what is called an **attached-X** or **compound-X chromosome.** The normal X chromosome in this species has a centromere almost at the end of the chromosome (a subterminal centromere), and the attachment of two of these chromosomes to a single centromere results in a chromosome with two equal arms—each consisting of a virtually complete X. Females with a compound-X chromosome usually carry a Y chromosome as well. They produce two classes of offspring: females who receive the maternal compound-X chromosome and a Y from their fathers, and males with a maternal Y chromosome and a paternal X. The XXXAA and YYAA zygotes that are also expected from the mating of an attached-X female and a normal male are inviable (see Figure 2-15).

Phenotypically wildtype females heterozygous for genes in an attached-X chromosome produce some female progeny homozygous for the recessive alleles of one or more of the genes as a result of crossing over between the homologous chromosome arms. The frequency with which such homozygosity occurs increases with increased map distance of the particular gene from the centromere. From the diagrams in Figure 3-4 it can be seen that the exchange in the chromosome region between the gene and the centromere must occur after the chromosome has replicated—that is, at the *four-strand* or tetrad stage of meiosis. Crossing over before replication of the chromosome (at the *two-strand* stage) would result only in a shifting of the alleles between the chromosome arms.

The demonstration that crossing over detected by the recombination of two heterozygous markers is associated with an actual exchange of segments between homologous chromosomes required two morphological markers in the pair of chromosomes for detection of the physical exchange. Since normal homologous chromosomes are morphologically identical, it was necessary to find structurally altered chromosomes. Two X chromosomes of *Drosophila* that had undergone spontaneous structural changes making them distinguishable from each other and from a normal X were identified by Curt Stern in 1931 and used in an experiment that provided one of the classical proofs of the physical basis of crossing over (Figure 3-5). One of the X chromosomes had become physically separated into two segments by the occurrence of a translocation (a physical exchange of fragments between two nonhomologous chromosomes—see Chapter 6) between the X and chromosome 4. Two features of this variant X are significant: (1) Both segments had centromeres and stable ends and therefore were transmitted as normal chromosomes. (2) The segments do not always segregate to the same pole in the first meiotic division, but only those progeny that receive either both segments (and therefore an entire X) or none are viable. The second morphologically marked X had a small piece of a Y chromosome attached as a second arm. The mutant alleles *car* (resulting in carnation eye color, instead of red) and *B* (for bar-shaped eyes, instead of the usual round) were present in the first altered X chromosome and the wildtype alleles of these genes were in the second X. Females with the two structurally and genetically marked X chromosomes

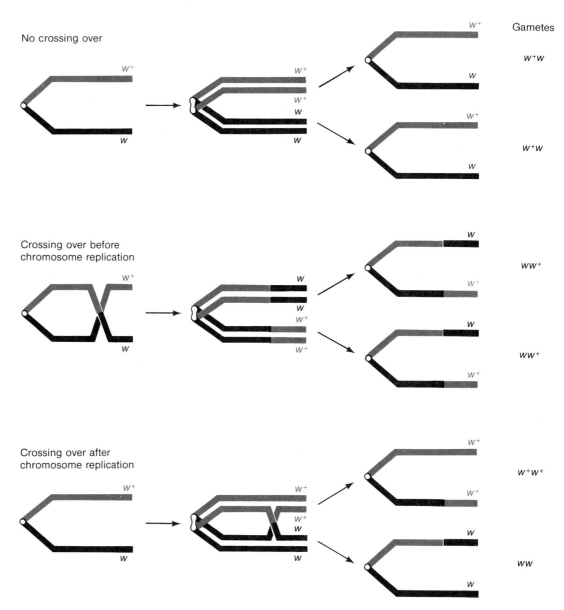

Figure 3-4 Diagram showing that crossing over must occur at the four-strand or tetrad stage in meiosis to produce a homozygous compound-X chromosome product from a compound-X chromosome that is heterozygous for a locus. Note that the exchange must occur between the centromere and the locus.

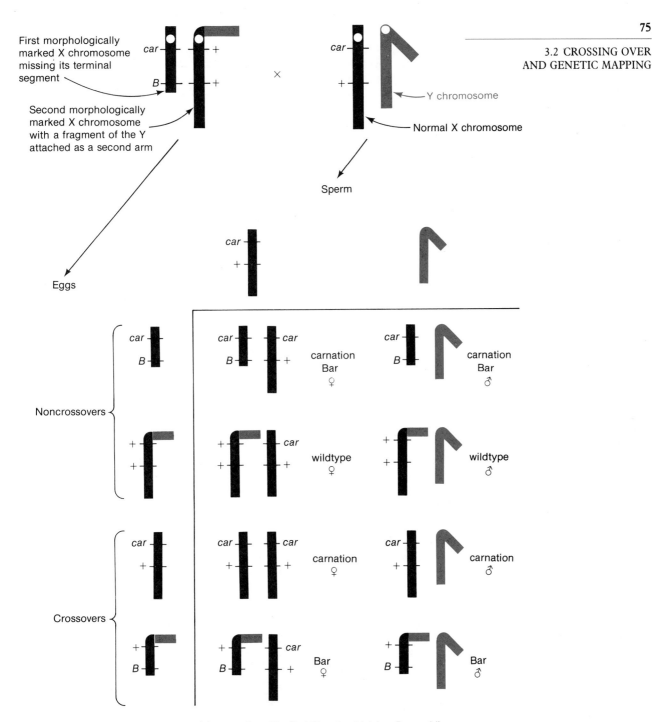

Figure 3-5 Diagrammatic summary of the experiment by Curt Stern in which two *Drosophila* X chromosomes that are morphologically distinguishable from each other and from a normal X were used to demonstrate that recombination between genetic markers is associated with the exchange of segments between homologous chromosomes. See text for details of the experiment.

were mated with males having a normal X that carried the recessive alleles of the genes. In the progeny from this cross, individuals with parental or recombinant combinations of the phenotypic traits were easily recognizable; their chromosomal makeup was determined by microscopic examination of tissue from some of their offspring. In the genetically recombinant progeny from the cross, the X chromosome had the morphology expected if recombination of the genes was accompanied by an exchange that recombined the chromosome markers. That is, the individuals with red, bar-shaped eyes had an X in two segments one of which had the attached second arm, and those with carnation-colored round eyes had an undivided X with no second arm. The nonrecombinant progeny were found to have an X chromosome morphologically identical with one in their mothers. An equally elegant demonstration of the association of genetic exchange with physical exchange between structurally marked chromosomes in corn was done, also in 1931, by Harriet Creighton and Barbara McClintock.

Multiple Crossing Over

When two genes are located far apart along a chromosome, more than one crossover can occur between them in a single meiosis and this leads to complications in interpreting recombination data. The probability of multiple crossovers usually increases with distance between the genes considered. Multiple crossing over complicates genetic mapping because map distance is based on the number of physical exchanges that occur, and some of the multiple exchanges occurring between two genes will not result in recombination of the genes and hence will not be detected. For example, if two exchanges involving the same chromatids occur between genes a and b, then their net effect will be that shown in Figure 3-6. Two of the products of this meiosis are double-crossover chromosomes, but these chromosomes are not recombinant for the alleles of the two genes and, therefore, are genetically indistinguishable from noncrossover chromosomes. The occurrence of such cancelling events means that the observed recombination value will be an *underestimate* of the true exchange frequency, and, consequently, of the map distance between the genes. Since in higher organisms double crossing over effectively never occurs in chromosome segments less than 10 map units long, the difficulty is avoided by using recombination data for closely linked genes to build up linkage maps.

The maximum frequency of recombination between any two genes is 50 percent—the same value that would be observed if the genes were on nonhomologous chromosomes and assorted independently. Fifty percent recombination will occur when the genes are so far apart in the chromosome that at least one crossover almost always occurs between them. From Figure 3-2 it is easy to see that the occurrence of a single exchange in every meiosis would result in half of the products having parental combinations and the other half having recombinant combinations of the genes. The occurrence of two exchanges between two genes has the same effect, as shown in Figure 3-7. Again, no recombination of the marker genes is detectable if the same chro-

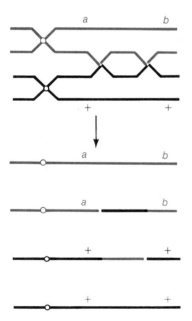

Figure 3-6 Diagram showing that the occurrence of two exchanges between the same two chromatids in the interval between two genes will not have a genetically detectable effect.

matids are involved in the two exchanges (two-strand double-exchange tetrads). When the two exchanges have one chromatid in common (three-strand double-exchange tetrads), the result is indistinguishable from that of a single exchange—two products with parental combinations and two with recombinants are produced. Note that there are two types of three-strand dou-

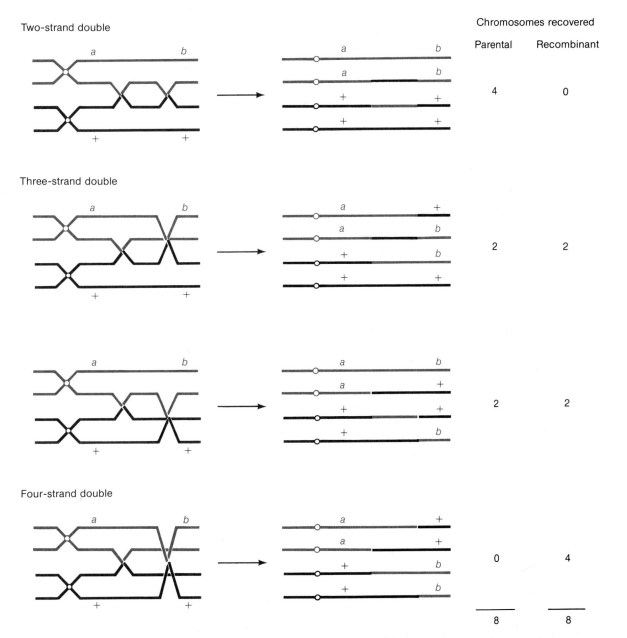

Figure 3-7 Diagram showing that when two exchanges always occur in the interval between two genes, and the chromatids participate at random in the exchanges, the result is indistinguishable from independent assortment of the genes.

bles; the types differ according to which three chromatids are involved. The third possibility is that the second exchange occurs between the chromatids that did not participate in the first exchange (four-strand double-exchange tetrads), in which case all four products will be recombined. If the chromatids involved in the two exchange events are chosen at random (which appears to be the case), the expected proportions of the three types of double exchanges will be 1/4, 1/2, and 1/4, respectively. This means that, on the average, $(1/4) \cdot 0 + (1/2) \cdot 2 + (1/4) \cdot 4 = 2$ of 4 products of meiosis in which two exchanges occur between two genes will be recombinant. This is the same proportion obtained when a single exchange always occurs between the genes. Moreover, this result (a maximum of 50 percent recombination) will be obtained for any number of exchanges.

The occurrence of double-exchange events is detectable in recombination experiments that employ **three-point crosses**—those using three pairs of segregating alleles. If a third pair of alleles, c^+ and c, is located between the two with which we have been concerned (the outermost markers), double exchanges in the region can be detected when the crossovers flank the c gene (Figure 3-8). The two crossovers, occurring in this example between the same pair of chromatids, would result in a reciprocal exchange of the c^+ and c alleles between the chromatids. A three-point cross is an efficient way to obtain recombination data, and it also provides a simple method for determining the order of the three genes, as will be seen in the next section.

3.3 Gene Mapping from Three-Point Testcrosses

The data in Table 3-1, from a testcross in corn in which alleles of three genes on chromosome 4 were segregating, illustrate the basis for the analysis of data from a three-point cross. The recessive alleles of the genes in this cross were lz (for lazy or prostrate growth habit), gl (for glossy leaf), and su (for sugary endosperm), and the trihybrid parent in the cross had the genotype $Lz\,Gl\,Su/lz\,gl\,su$. Therefore, the two classes of segregants among the progeny that represent noncrossover (parental-type) gametes are the normal individuals and those with the lazy, glossy, sugary phenotype. These classes are far

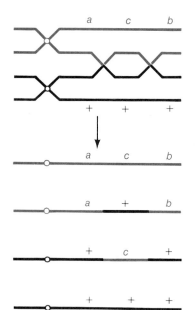

Figure 3-8 Diagram showing that two exchanges occurring between the same chromatids and spanning the middle pair of alleles in a triple heterozygote result in a reciprocal exchange of that pair of alleles between the two chromatids.

Table 3-1 Progeny from a three-point testcross in corn

Phenotype of testcross progeny	Genotype of gamete from hybrid parent			Number
normal (wildtype)	Lz	Gl	Su	286
lazy	lz	Gl	Su	33
glossy	Lz	gl	Su	59
sugary	Lz	Gl	su	4
lazy, glossy	lz	gl	Su	2
lazy, sugary	lz	Gl	su	44
glossy, sugary	Lz	gl	su	40
lazy, glossy, sugary	lz	gl	su	272

larger than any of the crossover classes, which is consistent with the fact that recombination between any two linked genes is less than 50 percent. Thus, if the combination of dominant and recessive alleles in the chromosomes of the heterozygous parent were unknown, we would deduce from their relative frequency in the progeny that the noncrossover gametes were *Lz Gl Su* and *lz gl su*.

In mapping experiments the gene sequence is usually not known—the order in which we have represented the three genes in this example is entirely arbitrary and may not be the actual order. However, there is an easy way to determine the correct order from three-point data—namely, by identifying the genotypes of the reciprocal double-crossover gametes produced by the heterozygous parent and then comparing them to the parental gametes. Since the probability of the simultaneous occurrence of two exchanges is considerably smaller than that of either single exchange, the double-crossover gametes will be the least frequent types. It can be seen from Table 3-1 that the classes composed of four individuals with the sugary phenotype and two individuals with the lazy and glossy phenotype—products of the *Lz Gl su* and *lz gl Su* gametes, respectively—are the least frequent and therefore represent the double-crossover progeny. The effect of double-exchange events, as Figure 3-8 shows, is to interchange the members of the *middle* pair of alleles between the chromosomes. This means that if the parental chromosomes are

$$Lz\,Gl\,Su \quad \text{and} \quad lz\,gl\,su$$

and the double-crossover chromosomes are

$$Lz\,Gl\,su \quad \text{and} \quad lz\,gl\,Su$$

Su and *su* are interchanged by double crossing over and must be the middle pair of alleles. Therefore, the genotype of the heterozygous parent in the cross, represented correctly with respect to both the order of the genes and the combination of alleles in the homologous chromosomes, is

Lz	*Su*	*Gl*
lz	*su*	*gl*

When double crossing over between chromatids of these parental types is diagrammed, the products can be seen to correspond to the two gametic types indicated by the data as the double crossovers:

From this diagram it can also be seen that the reciprocal products of crossing over occurring only between *lz* and *su* would be *Lz su gl* and *lz Su Gl* and the products of a single exchange between *su* and *gl* would be *Lz Su gl* and *lz su Gl*.

We can now summarize the data in a more informative way, writing the genes in correct order and identifying the numbers of the different chromosome types produced by the heterozygous parent that are represented in the progeny:

$$
\begin{array}{llll}
Lz & Su & Gl & 286 \\
lz & su & gl & 272
\end{array} \Big\} \text{ Parental types}
$$

$$
\begin{array}{llll}
Lz & su & gl & 40 \\
lz & Su & Gl & 33
\end{array} \Big\} \text{ Single crossovers between } lz \text{ and } su
$$

$$
\begin{array}{llll}
Lz & Su & gl & 59 \\
lz & su & Gl & 44
\end{array} \Big\} \text{ Single crossovers between } su \text{ and } gl
$$

$$
\begin{array}{llll}
Lz & su & Gl & 4 \\
lz & Su & gl & 2
\end{array} \Big\} \text{ Double-crossover types}
$$

$$740$$

Calculating the frequency of crossing over from such data requires correcting the data for the fact that the double-crossover chromosomes result from two exchanges, one in each of the adjacent chromosome regions defined by the three genes. Therefore, crossovers between *lz* and *su* are represented by the following chromosome types:

$$
\begin{array}{llll}
Lz & su & gl & 40 \\
lz & Su & Gl & 33 \\
Lz & su & Gl & 4 \\
lz & Su & gl & 2
\end{array}
$$

$$79$$

That is, 79/740, or 10.7 percent, of the chromosomes recovered in the progeny are crossovers between the *lz* and *su* loci, so the map distance between these genes is 10.7 units. Similarly, crossovers between *su* and *gl* are represented by the following chromosome types:

$$
\begin{array}{llll}
Lz & Su & gl & 59 \\
lz & su & Gl & 44 \\
Lz & su & Gl & 4 \\
lz & Su & gl & 2
\end{array}
$$

$$109$$

The crossover frequency between this second pair of loci is 109/740, or 14.8 percent, so the map distance between them indicated by these data is 14.8 units. The chromosome segment in which the three genes are located may be represented as

Consider next a three-point cross in *Drosophila* in which the segregating
X-linked genes are identified by the recessive alleles *cv* (crossveinless wings),
ec (echinus, or roughened, eye surfaces), and *ct* (cut wing edges). The progeny
summarized in Table 3-2 were obtained from a mating of wildtype females
heterozygous for each pair of alleles with wildtype ($+++/Y$) males. Note that
in this cross it is possible to determine the constitution of progeny produced
by the heterozygous female parent in the male progeny only, since the female
progeny receive a $+++$ gamete from their fathers.

From the two largest classes of male progeny we can infer that the
noncrossover gametes produced by the heterozygous female parents were
cv $++$ and $+$ *ec ct*. The two classes with the lowest frequency identify the
double-crossover gametes as $+++$ and *cv ec ct*. Comparison of these gametes
with the noncrossover (parental) gametes shows that the alleles interchanged
between the chromosomes by double crossing over—and therefore representing the middle gene—are *cv* and its wildtype allele. Thus, the genotype of the
female parents in the cross, represented correctly with respect to gene order
and the combination of alleles in the two X chromosomes, is

$$\frac{+ \qquad cv \qquad +}{ec \qquad + \qquad ct}$$

The data in Table 3-2 can then be summarized by the genotypes of the
gametes from the female parents as follows:

+	cv	+	2207	Parental types
ec	+	ct	2125	
+	+	ct	265	Single crossovers between *ec* and *cv*
ec	cv	+	273	
+	cv	ct	223	Single crossovers between *cv* and *ct*
ec	+	+	217	
+	+	+	5	Double-crossover types
ec	cv	ct	3	
			5318	

Table 3-2 Male progeny from the mating of *Drosophila melanogaster* females
heterozygous for three X-linked genes with wildtype males

Phenotype	Genotype of female gamete			Number
wildtype	+	+	+	5
crossveinless	cv	+	+	2207
echinus	+	ec	+	217
cut	+	+	ct	265
crossveinless, echinus	cv	ec	+	273
crossveinless, cut	cv	+	ct	223
echinus, cut	+	ec	ct	2125
crossveinless, echinus, cut	cv	ec	ct	3

The frequency of crossing over between *ec* and *cv* calculated from the data is (265 + 273 + 5 + 3)/5318, or 10.3 percent, and between *cv* and *ct* it is (223 + 217 + 5 + 3)/5318, or 8.4 percent. That is, the map distances between the pairs of loci are 10.3 and 8.4 units, respectively:

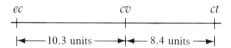

The double-crossover classes, which are not recombinant for the outside markers, are not included if the distance between *ec* and *ct* is calculated from the data:

+	+	ct	265
ec	cv	+	273
+	cv	ct	223
ec	+	+	217
			978

The observed *ec–ct* map distance of 18.4, or (978/5318) × 100, is in good agreement with the predicted value of 18.7 obtained by adding the *ec–cv* and *cv–ct* distances. Note that the difference between the observed and predicted distances is equal to twice the 0.15 percent frequency of the double crossovers, since this value was used twice—once in determining the *ec–cv* distance and again in determining the *cv–ct* distance. The agreement between the two values for the *ec–ct* distance confirms the deduced map order of the three genes, and it also affirms that the map order of three genes can be determined by calculating all three of the distances; the largest will be between the outside markers.

Interference in the Occurrence of Double Crossing Over

The detection of double crossing over makes it possible to determine whether exchanges in two different regions of a pair of chromosomes occur independently of each other. Using the information from our previous example, we know from the recombination frequencies that the probability of exchange is 0.103 between *ec* and *cv* and 0.084 between *cv* and *ct*. If crossing over occurs independently in the two regions (that is, if the occurrence of one exchange does not alter the probability of the second exchange), the probability of the simultaneous occurrence of an exchange in both regions is the product of these individual probabilities: that is, 0.103 × 0.084 = 0.0086, or 0.86 percent. This means that in a sample of 5318 the expected number of double crossovers would be 5318 × 0.0086, or 46, whereas the number actually observed was only 8. Such deficiencies in the observed number of double crossovers are common and identify a phenomenon called **interference**—the occurrence of crossing over in one region of a chromosome associated with a reduced probability of simultaneous crossing over in a second region. An explanation that has been given (but not proved) for interference is that it is mechanically difficult, either because of some feature of the structure of chromosomes or the mechanism of pairing, for two chiasmata to form close together in the paired chromosome. However, it is now believed that interference may result from a regulatory process that limits the number of exchange events that can occur in a cell and in a chromosome.

The **coefficient of coincidence** is the observed number of double recombinants divided by the expected number; its value is a simple measure of the degree of interference, the measure of interference being defined as

$$\text{Interference} = 1 - \text{coefficient of coincidence}$$

From the data in our example, the coefficient of coincidence is $8/46 = 0.174$, meaning that the number of double crossovers that occurred was only 17.4 percent of the number expected if crossing over in the two regions were independent. It has been found experimentally that interference usually increases as the distance between the two outside markers becomes smaller, until a point is reached at which double crossing over does not occur; that is, no double crossovers are found and the coincidence is 0. The distance is about 10 map units in a variety of organisms. Conversely, when the total distance between the gene loci is greater than about 45 map units, interference disappears and the coincidence becomes 1.

Correction for Multiple Crossing Over in Mapping

Map distance is a concept based on the average frequency of crossing over occurring in a chromosome interval, whereas recombination frequency measures crossovers resulting in exchange of the genes used as markers. In the absence of multiple crossing over, there would be a direct linear relationship between recombination percentage and map distance, as is represented by line (a) in Figure 3-9. Experimentally, this proportionality occurs only when map distances are small; therefore, the most accurate linkage maps are obtained by using recombination data for a sequence of short intervals. Since the proportionality between recombination frequency and map distance no longer holds when multiple crossing over occurs, accurate estimates of larger map distances are difficult to obtain directly. Mathematical expressions called **mapping functions** are sometimes useful for correcting observed recombination values for multiple crossing over.

Numerous mapping functions are possible, depending on the assumptions made concerning interference. A mapping function should convert recombination percentages into map distances that are additive. One of the first of these functions, based on the simple but seldom correct assumption of no interference, is shown by curve (b) in Figure 3-9. Experimental curves that relate recombination frequency to map distance can be obtained in genetically well-characterized organisms. The observed recombination value for a given

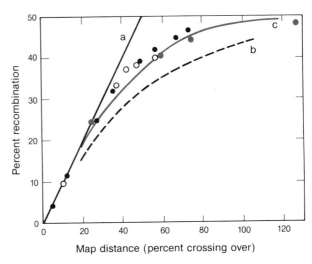

Figure 3-9 The relation between recombination and crossing over as expected (a) with complete interference, (b) with no interference, and (c) from the assumption that interference is inversely proportional to recombination percentage. The experimental data are from *Drosophila melanogaster* (solid black circles), corn (solid red circles), and *Neurospora crassa* (open black circles).

pair of genes is compared with their actual map distance obtained by adding the distances of short intervening regions. Examples of such data from *Drosophila*, corn, and the common bread mold *Neurospora crassa* are shown in the figure. Curve (c), for a mapping function based on the assumption that coincidence increases linearly with the percentage of recombination, fits these data reasonably well.

Mapping functions are particularly useful in any map region in which the number of genes available for mapping is small. In such a region map distances determined directly between two distant genes are inaccurate because of the occurrence of multiple crossovers. A mapping function allows one to use recombination data from such distant markers by correcting observed values.

3.4 Mapping by Tetrad Analysis

In some species of fungi and unicellular algae each tetrad is contained in a saclike structure and can be recovered as an intact group. The advantage of these organisms for the study of recombination is the potential for analyzing all of the products of an individual meiosis. Two other features of the organisms are especially useful for genetic analysis: (1) they are haploid, so dominance is not a complicating factor, since the genotype is expressed directly in the phenotype, and (2) they produce very large numbers of progeny, making it possible to detect rare events and to estimate their frequencies accurately.

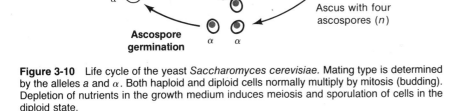

Figure 3-10 Life cycle of the yeast *Saccharomyces cerevisiae.* Mating type is determined by the alleles *a* and *α*. Both haploid and diploid cells normally multiply by mitosis (budding). Depletion of nutrients in the growth medium induces meiosis and sporulation of cells in the diploid state.

The life cycles of these organisms tend to be short. The diploid stage consists only of the zygote; it undergoes meiosis soon after it is formed, and the haploid meiotic products, called **spores,** germinate to regenerate the vegetative stage (Figure 3-10). In some species the four products of meiosis subsequently undergo a mitotic division, but this means only that each member of the tetrad yields a pair of genetically identical spores. In most organisms the meiotic products, or their derivatives, are usually not arranged in a particular order in the spore sac. However, bread molds of the genus *Neurospora*, and several other species also belonging to the class of fungi called Ascomycetes, have the very useful characteristic that these products are arranged in an order that is directly related to the planes of the meiotic divisions. We will examine the ordered system after first looking at unordered tetrads.

Analysis of Unordered Tetrads

When two pairs of alleles are segregating, three patterns of segregation are possible in the tetrads. For example, in a cross $AB \times ab$ three types of tetrads can occur:

AB AB ab ab, referred to as **parental ditype,** or PD. Only two genotypes are represented and their alleles have the same
combinations found in the parents.

Ab Ab aB aB, referred to as **nonparental ditype,** or NPD. Only two
genotypes are represented but their alleles have nonparental combinations.

AB Ab aB ab, referred to as **tetratype,** or TT. All four of the possible
genotypes are present.

Tetrad analysis is an effective way to determine whether two genes are
linked, because *in the absence of linkage the parental ditype tetrads and nonparental ditype tetrads will occur in equal frequencies* (Figure 3-11). If all genotypes represented in the three types of tetrads shown above are considered
simply as random products of meiosis, half of them will be parental (*AB* and

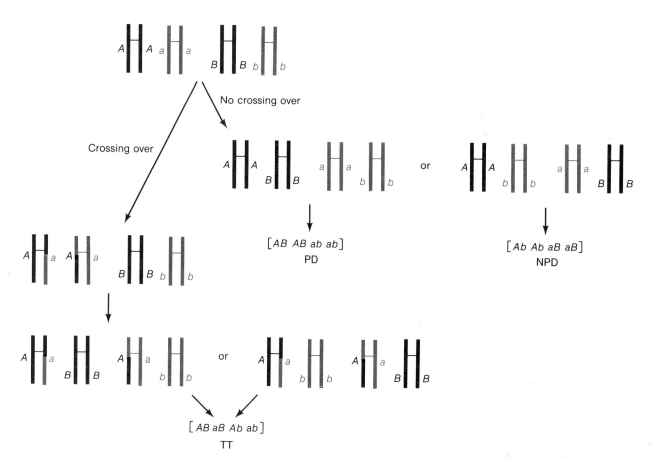

Figure 3-11 Types of unordered asci produced with unlinked
genes, in this case on different chromosomes. In the absence of
crossing over, the random arrangement of chromosome pairs at
metaphase I yields two different combinations of chromatids. One
produces PD asci; the other produces NPD asci. When crossing
over occurs between either gene and its centromere, the two arrangements both yield TT asci. If the genes are closely linked to the
centromere (so crossing over is rare), there will be few tetratypes.

ab) and the other half will be recombinant (*Ab* and *aB,*) as long as PD = NPD. This conclusion is independent of the relative frequency of tetratypes, since each tetratype includes two parental genotypes and two recombinant genotypes. The frequency of tetratype tetrads will depend on the location of the genes. For example, if the genes are on nonhomologous chromosomes and each of them is closely linked to its centromere, there will be few if any tetratypes. At the other extreme, the tetratypes will constitute two-thirds of all the tetrads produced if the two pairs of alleles are distributed at random in meiosis. This would be the case if the two genes are on nonhomologous chromosomes and not linked to their respective centromeres, or if they are on a pair of homologous chromosomes and there is no linkage between the genes or between either of them and the centromere. In all of the conditions described, the parental ditype tetrads and nonparental ditype tetrads will occur with equal frequencies. The effect of linkage, as we will see, is to reduce the frequency of nonparental ditype tetrads relative to that of parental ditype tetrads.

Let us now assume that the genes are linked and consider the events required for production of the three types of tetrads. If no more than two exchanges occur between the gene loci, parental ditype tetrads arise when either no crossing over or two-strand double crossing over occurs between *a* and *b* (Figure 3-7, top); nonparental ditype tetrads result only from the occurrence of four-strand double crossing over between the loci (Figure 3-7, bottom); and tetratype tetrads result from either a single exchange (Figure 3-2) or a three-strand double exchange (see Figures 3-7). Because of the nature and number of exchanges required to produce them, nonparental ditype tetrads are predictably infrequent relative to the other types of tetrads when the marker genes are linked. Therefore, *an observation that nonparental ditype tetrads occur with less frequency than parental ditype tetrads is a sensitive indicator of linkage.*

In calculating the recombination frequency of two genes from tetrad data it is necessary to take into consideration the fact that only half of the meiotic products in a tetratype tetrad are recombinant, whereas all products in a nonparental ditype tetrad are recombinant. The expression

$$\frac{1/2(\text{TT}) + (\text{NPD})}{\text{Total tetrads}} \times 100$$

is used to estimate the recombination frequency between two genes.

Analysis of Ordered Tetrads

In *Neurospora crassa,* a species used extensively in genetic investigations, the products of meiosis are contained in an *ordered* array of spores (Figure 3-12). The vegetative body of the organism consists of filaments (hyphae) that are segmented by partial septa through which nuclei can pass. Each segment normally contains many nuclei. Asexual reproduction occurs by means of spores called conidia or by fragments of the unspecialized hyphae. There are

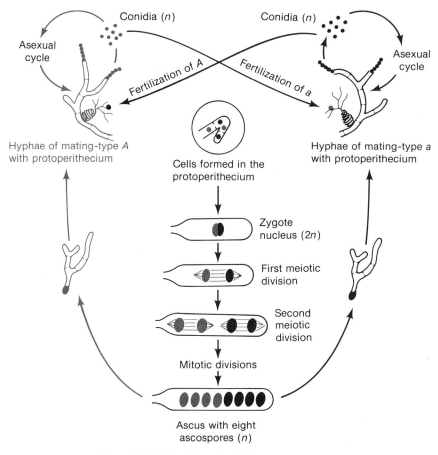

Figure 3-12 The life-cycle of *Neurospora crassa*.

two mating types, A and a (determined by a single pair of alleles), either of which will produce fruiting bodies (protoperithecia) under appropriate conditions. The sexual cycle is initiated by fertilization with conidia or hyphae of the mating type opposite to that of the protoperithecium. After a nucleus has migrated into the protoperithecium, a large number of cells containing nuclei of the two mating types are produced. The zygote nucleus formed by the fusion of one A and one a nucleus is the only diploid stage in the life cycle. These nuclei, contained in saclike structures called **asci** (singular **ascus**), almost immediately undergo meiosis. The four meiotic products, maintained in a linear order that provides an accurate record of the events that occurred in the two meiotic divisions, then divide by mitosis to form eight **ascospores.** The eight ascospores consist of four pairs, each pair representing one of the products of meiosis. The spores may be removed in sequence from an ascus and germinated in individual culture tubes to determine their genotypes.

This ordered arrangement of the meiotic products makes it possible to determine the recombination frequency between a gene and its centromere—

that is, to map the centromere. This is usually the first step in mapping genes in *Neurospora*. The mapping technique is based on two features of meiosis: (1) sister chromatids do not divide until the second meiotic division, and (2) alleles segregate either in the first meiotic division, if crossing over does not occur between the gene and its centromere (**first-division segregation**), or in the second meiotic division if such crossing over does occur (**second-division segregation**). This distinction is illustrated in Figures 3-13 and 3-14. First-division segregation occurs because each chromatid pair contains identical alleles and the chromatid pairs segregate during this division. A second-division segregation pattern results if a crossover occurs prior to metaphase I, because both alleles are present in both chromatid pairs and are not distributed to separate cells until the second division. Because of the random arrangement possible for the chromosomes on the division spindle in metaphase I and for the chromatids in metaphase II, four different arrangements of spores are possible, as shown in Figure 3-14.

The percentage of asci having second-division segregation patterns for a gene can be used to determine the number of map units between a gene and its centromere. For example, let us assume that 30 percent of a sample of asci from a cross have a second-division segregation pattern for the *A* and *a* alleles. This means that 30 percent of the cells undergoing meiosis had a crossover between the *A* gene and its centromere, and of the four meiotic products generated from each of those cells, two were recombinant. In other words, half of the spores produced by cells in which a crossover occurs between a gene and its centromere are recombinant spores; the other half are non-recombinant or parental types. Thus, if a crossover occurs between a gene and its centromere in 30 percent of the cells, half that frequency, or 15 percent of the meiotic products, will be recombinant. By convention, map distances reflect the frequency of recombinant meiotic products rather than the frequency of cells in which the crossover occurred. That is, the recombinant frequency is one-half the frequency of asci that show a second-division segregation pattern. In the example given, the map distance between the gene and its centromere is 15 units. The map distance between a gene and its centromere is given by the equation

$$\frac{1/2 \ (\text{Asci with second-division segregation patterns})}{\text{Total number of asci}} \times 100$$

Ordered tetrads can also be classified as PD, NPD, and TT, just as unordered tetrads can. As a first step in mapping, one usually maps genes with respect to the centromere and then determines whether the number of PD tetrads equals the number of NPD tetrads; if so, the genes are on separate chromosomes or are so far apart as to be unlinked, and no further analysis of the data is warranted. However, if the number of NPD tetrads is significantly less than the number of PD tetrads, the genes are not assorting independently and their degree of linkage can be determined; that is, they can be mapped with respect to one another. An example of how this is done is given in the following paragraph.

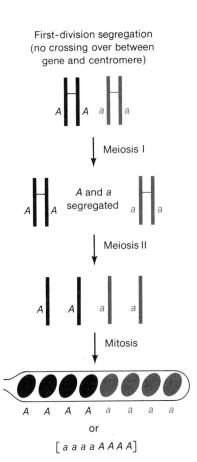

Figure 3-13 Diagram showing the first-division segregation ascus patterns produced when crossing over between the gene and centromere does not occur. The alleles separate (segregate) in meiosis I. Two spore patterns are possible, depending on the orientation of the pair of chromosomes on the first-division spindle and of the sister chromatids of each chromosome on the second-division spindle.

Second-division segregation
(single crossover between
gene and centromere)

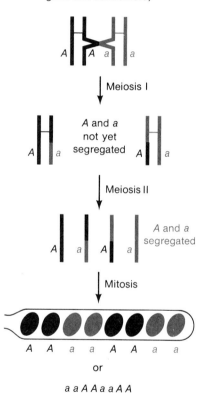

or

$a\ a\ A\ A\ a\ a\ A\ A$
$A\ A\ a\ a\ a\ a\ A\ A$
$a\ a\ A\ A\ A\ A\ a\ a$

Figure 3-14 Diagram showing the second-division segregation ascus patterns produced when crossing over between the gene and the centromere prevents segregation in meiosis I. Several spore patterns are possible, depending on the orientation of the pair of chromosomes on the first-division spindle and of the chromatids of each chromosome on the second-division spindle.

We now use some hypothetical data to map two *Neurospora* genes, *A* and *B*, with respect to each other and the centromere. The data for a cross $AB \times ab$ are given in Table 3-3. It should be noted that data from different second-division segregation patterns (for example, the genotype sequences *a A a A*, *A a a A*, and *a A A a*) have been combined and listed as *A a A a*, because on the *number* of asci formed by second-division segregation, not the particular pattern, is important. The NPD are represented by the type-4 asci. Two NPD are present, a small number, indicating that the genes are linked. Thus, we may proceed to prepare a map of the genes and the centromere. The genes and the centromere can have three possible orders: (I) centromere–*A*–*B*, (II) centromere–*B*–*A*, or (III) *A*–centromere–*B*. Examination of Figure 3-13 shows that ascus pattern 1 is the result of first-division segregation of *both* markers and hence need not be considered for the mapping, because in first-division segregation crossing over has not occurred between either gene and the centromere. Type-2 asci are useful; they show first-division segregation for alleles at the *A* loci and second-division segregation at the *B* locus and hence must result from a single crossover between *B* and the centromere with no crossing over between *A* and the centromere. The existence of type-2 asci eliminates the possibility of order II, because with that order any crossover between the centromere and *B* must also be between the centromere and A. Type-3 asci are also informative; these show second-division segregation for both *A* and *B*, indicating that a single crossover must have occurred between *both* loci and the centromere. The existence of these asci eliminates order III, since in this order a single crossover can only be in one of the intervals, centromere–*A* or centromere–*B*, but not both intervals. Order I is consistent with the existence of both type-2 and type-3 asci: a crossover between the centromere and *B* need not be between the centromere and *A*, for it can be between *A* and *B* (type 2); and a crossover can be between both loci and the centromere by being between the centromere and *A* (type 3). Thus, the order is centromere–*A*–*B*.

Calculation of the distance between the two genes requires consideration of the 60 TT and 2 NPD asci: $[(1/2)60 + 2/200] \times 100 = 16.0$ map units. Since the 60 type-2 asci and 30 type-3 asci exhibit second-division segregation for the alleles at the *B* locus, the length of the interval centromere–*B* is

Table 3-3 Asci from a cross $AB \times ab$

Spore pair	Types of asci			
	1 (PD)	2 (TT)	3 (PD)	4 (NPD)
1	AB	AB	AB	aB
2	AB	Ab	ab	aB
3	ab	aB	AB	Ab
4	ab	ab	ab	Ab
Total number	108	60	30	2

$[(1/2)(60 + 30)/200] \times 100 = 22.5$ map units. Similarly, the 30 type-3 asci exhibit second-division segregation for the alleles at the A locus, and the length of the interval centromere–A is $[(1/2)30/200] \times 100 = 7.5$ map units. Thus, the linkage map obtained from these data is centromere–7.5 map units–A–16.0 map units–B.

3.5 Mitotic Recombination

Crossing over occurs during mitosis, though with much less frequency than in meiosis. The first evidence for mitotic recombination was obtained from experiments in which *Drosophila* females heterozygous for sex-linked genes

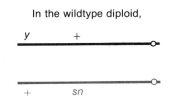

In the wildtype diploid,

crossing over in the region between

Figure 3-15 The mitotic exchange events required to produce single singed patches of cells (left) and twin patches of yellow and singed cells (right) in the diploid body of a *Drosophila* female heterozygous for the genes.

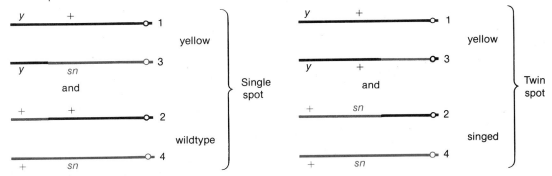

Alternatively, the segregation of chromatids 1 and 4 into one nucleus and chromatids 2 and 3 into the other nucleus would in both cases result only in wildtype products.

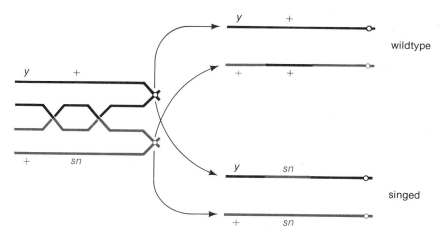

Figure 3-16 The mitotic exchange events required to produce single singed patches of cells in a heterozygous *Drosophila* female with the same genotype as in Figure 3-15.

with easily recognizable recessive phenotypes were produced. Females heterozygous for the genes *y* (yellow body color) and *sn* (singed bristles) with the recessive alleles on different X chromosomes—*y* +/+ *sn*—were phenotypically wildtype, as expected. Close examination of the flies revealed occasional spots (patches of cells) with either the yellow or singed phenotype. These spots often occurred in pairs (twin spots) consisting of adjacent patches of yellow and of singed tissue. The developmental origin of individual patches of cells with a recessive phenotype might be explained either by mutation of the corresponding dominant allele or by the loss from a cell line of the chromosome carrying the dominant allele. However, because twin spots occurred with a frequency far greater than would be expected on the basis of occurrence of the yellow spots or singed patches alone, it was reasoned that the pairs must represent reciprocal products of the same event. The interpretation was that crossing over first occurred between chromatids of homologous chromosomes and that this was followed by normal mitotic division of the centromeres and disjunction of chromatids (Figure 3-15). The relative frequencies of the different kinds of spots were consistent with this explanation. More twin spots were found than single yellow spots, as would be expected from the greater map distance (determined by meiotic recombination) between the centromere and *sn* than between *sn* and *y*. Similarly, the very low frequency of single singed spots would be expected because they would require the occurrence of double crossing over (Figure 3-16).

Mitotic Analysis in Fungi

Mitotic recombination was first demonstrated directly in the ascomycete *Aspergillus nidulans* and has been used extensively in the genetic analysis of other fungi, including some species that lack a sexual cycle. *Aspergillus* has a life cycle very similar to that of *Neurospora* but differs in not having separate mating types. The normal vegetative state of the fungus is haploid. The diploid state required for mitotic crossing over and segregation arises spontaneously but with a low frequency. Formation of diploids can be illustrated with two genetically dissimilar strains. When hyphae of the two strains come into contact, they tend to fuse. This fusion is followed by a transfer of nuclei from one hypha to the other, forming **heterokaryons,** in which the hyphae contain nuclei from both strains. During growth of the heterokaryotic hyphae, the two types of nuclei multiply independently. However, in the formation of asexual spores each initial sporogenous cell receives only a single nucleus of one type or the other. Chains of spores are produced by division of these cells, and within each chain the spores have the same genotype. On rare occasions, two of the nuclei in a heterokaryon fuse to form a diploid nucleus, and subsequent nuclear divisions result in a diploid sector of hyphae. The asexual spores formed on these hyphae will be diploid and will yield diploid individuals when germinated.

A selective isolation of the rare diploids is possible when the parental haploid strains differ in complementary nutritional requirements determined by recessive genes. For example, an ad^+pro haploid will not grow on a solid medium that lacks proline, and one with the genotype $ad\,pro^+$ will not grow in the absence of adenine. The $ad^+pro/ad\,pro^+$ diploid formed by the fusion of these haploid nuclei will have a wildtype allele for synthesis of each of the amino acids. Thus, diploid spores can be detected, and cultures from them can be obtained by spreading a mixture of spores on a medium lacking both adenine and proline, so that only the diploids will be able to grow.

The analysis of mitotic recombination in *Aspergillus* and similiar fungi is done with strains heterozygous for a number of genes used as markers. Mitotic crossing over is a very rare event—it occurs with less frequency than in meiosis by perhaps 1000 times—and double exchanges are not a factor in most analyses. In 50 percent of the subsequent segregation events, the result will be two daughter nuclei, each homozygous for one of the genes, as in the previous examples with *Drosophila*. The use of marker genes with detectable effects on morphology, nutrition, or resistance to harmful agents makes it possible to select hyphae carrying these recombinant nuclei.

To illustrate, consider a diploid in *Aspergillus* heterozygous for several genes located in one of the eight pairs of chromosomes:

ad		+	+	+	bi

+		pro	paba	y	+

The recessive alleles *ad, pro, paba,* and *bi* confer requirements for adenine, proline, *para*-aminobenzoic acid, and biotin, respectively. These are compounds that are required for growth and that the heterozygous diploid would be able to synthesize. The *y* and y^+ pair of alleles, determining yellow versus green color in the asexual spores, respectively, permits the visual detection of hyphae with recombinant nuclei homozygous for the recessive allele. Small sectors with yellow spores are easy to detect in a colony of the heterozygous diploid producing green spores.

A single mitotic exchange followed by the appropriate type of segregation results in homozygosity for all genes between the position of the exchange and the end of the chromosome arm, as can be seen in Figure 3-17. Note that the genetic marker (y) used for selection of the rare recombinants is located near the end of the chromosome arm. This location is an important consideration in analyses based on mitotic crossing over, because it means that exchanges occurring in the long chromosome segment between the centromere and the selective marker can be detected and used to determine the order and the relative positions of intervening genes.

In experiments with the diploid considered above some of the selected homozygotes for y will also be homozygous for *paba* (Figure 3-17), others for both *paba* and *pro*, but none will be homozygous for *pro* without also being homozygous for *paba*. These genotypes are consistent with the order: centromere–*pro*–*paba*–y. In addition, the frequencies of the genotypes—homozygous only for y, for both y and *paba*, and for y, *paba*, and *pro*—make it possible to estimate the relative frequency of crossing over in the three intervals between the centromere and y. In species having a sexual cycle, the gene order determined by mitotic analyis is found to correspond to the genetic map based on meiotic recombination, as expected. However, large differences in the relative lengths of the intervals in a chromosome arm are sometimes found when mitotic and meiotic maps are compared.

Occasional Reversion of Diploid Nuclei to Haploidy

Some of the phenotypically recessive sectors in heterozygous diploid colonies of *Aspergillus* and other fungi are haploid, and in some cases the proportion may be as high as 50 percent. The formation of haploids is rarely associated with recombination of genes on the same chromosome, so it is necessary in the analysis of mitotic crossing over to exclude the haploid segregants. Various ways of sorting them out from the diploids are possible. One is by measuring the diameter of the spores, since spores with haploid nuclei have only half the volume and a correspondingly smaller diameter than those with diploid nuclei.

The production of haploids is apparently initiated by the accidental loss of one member of a pair of chromosomes, which is followed in subsequent mitotic divisions by the loss of one member of other chromosome pairs until the fully haploid condition is attained. The haploids produced have all possible combinations of nonhomologous chromosomes and can be used to determine whether two genes are on the same or on different chromosomes. For example, a diploid heterozygous for genes a and b will form haploids in which the genotypes a^+b^+, a^+b, ab^+, and ab are equally likely if the genes are on nonhomologous chromosomes. On the other hand, if the genes are on the same chromosome, they will segregate as pairs in haploid formation. That is, the haploids will be a^+b^+ and ab if the diploid has the constitution a^+b^+/ab;

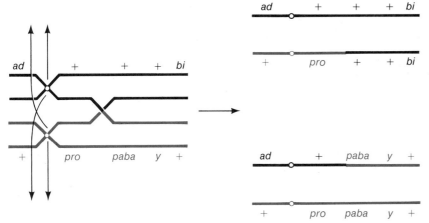

Product can be detected as a sector
producing yellow spores

Figure 3-17 The origin of a diploid sector homozygous for the recessive *y* allele from a
heterozygous diploid colony of *Aspergillus nidulans*. A selected sector is homozygous for any
genes between the position of the crossover event and the end of the chromosome arm.

or a^+b and ab^+ if the diploid is a^+b/ab^+. This efficient method of determining whether two genes are located on the same chromosome was used to
establish the eight known linkage groups in *Aspergillus*.

What we have described in considering mitotic recombination in fungi is
a sequence of events called the **parasexual cycle.** To summarize, this cycle
consists of four steps: (1) the formation of a heterokaryon, (2) the rare fusion
of two haploid nuclei within the heterokaryon to form a diploid nucleus, (3)
occasional crossing over and segregation during mitotic division of the diploid
nuclei, and (4) the occasional reversion of diploid nuclei to the haploid
condition. These events constitute a system of genetic recombination that is
quite distinct from the sexual cycle. The parasexual cycle represents a potentially important alternative to sexual reproduction in fungi that are not known
to have a sexual cycle. From the study of mitotic recombination in fungi it was
anticipated that parasexual events would be useful in the genetic analysis of
other organisms. Perhaps the most important application, which will be considered in Chapter 13, has been in the mapping of human chromosomes by
means of hybridization and mitotic segregation in cultured cells.

3.6 Recombination Within Genes

Genetic analyses and microscopic observations made prior to 1940 led to the
view of a chromosome as a linear array of particulate units—the genes—
joined in some way and resembling a string of beads. Neither the chemical nor
physical structure was known, though on occasion hypotheses were
presented. The gene was believed to be the smallest unit of genetic material
capable of alteration by mutation and the smallest unit of inheritance. Cross-

ing over had been seen in all regions of the chromosome but not within genes, so the idea of indivisibility of the gene had also developed. This classical concept of a gene began to change in the early 1940s. One important change was the realization that mutant alleles of a gene might represent alterations at different sites. The evidence that led to this idea was the finding of rare recombination events within several *Drosophila* genes. The initial observation came from investigations of the lozenge locus (lz) in the X chromosome of *D. melanogaster*. Mutant alleles of lozenge are recessive, and their phenotypic effects include disturbed arrangement of the facets of the compound eye and a reduction in the red pigment. Numerous lz alleles having recognizably different phenotypes are known. Females heterozygous for two different lz alleles, one in each homologous chromosome, have lozenge eyes. When such heterozygous females were crossed with males having one of the lz mutant alleles in the X chromosome and large numbers of progeny were examined, normal-eyed individuals were occasionally found. For example, in one cross

$$
\begin{array}{c}
\text{X} \quad \underline{\quad + \quad\quad lz^{BS}+ \quad\quad + \quad\quad} \\
\text{X} \quad \underline{\quad\quad\quad\quad\quad\quad\quad\quad\quad} \\
\quad\quad ct \quad\quad +lz^{g} \quad\quad v
\end{array}
\times
\begin{array}{c}
\quad + \quad\quad +lz^{g} \quad\quad v \quad \\
\underline{\quad\quad\quad\quad\quad\quad\quad\quad\quad} \text{X} \\
\underline{\quad\quad\quad\quad\quad\quad\quad\quad\quad} \text{Y}
\end{array}
$$

in which lz^{BS} and lz^{g} are mutant alleles of lozenge, and ct (cut wing) and v (vermilion eye color) are genetic markers that map 7.7 units to the left and 5.3 units to the right, respectively, of the lozenge locus, 134 males and females with wildtype eyes were found among more than 16,000 progeny, a frequency of less than 0.1 percent. These rare individuals might have resulted from a reverse mutation of one or the other of the mutant lozenge alleles to lz^{+}, but the observed frequency, though very low, was significantly higher than the known frequency of reverse mutation. The genetic constitution of the maternally-derived X chromosomes carried by the normal-eyed offspring was also inconsistent with such an explanation: the male offspring had cut wings and the females were $ct/+$ heterozygotes (as determined by an appropriate mating.) That is, all of the rare progeny had an X chromosome with the constitution

$$
\underline{\quad ct \quad\quad + + \quad\quad\quad + \quad\quad}
$$

which could be accounted for by crossing over between the two lozenge alleles. Convincing support for this interpretation was provided by detection of the reciprocal crossover chromosome

$$
\underline{\quad + \quad\quad lz^{BS}lz^{g} \quad\quad v \quad}
$$

in five male progeny having vermilion eyes and a lozenge phenotype distinctly different from the phenotype resulting from the presence of either an lz^{BS} or lz^{g} allele alone. This observation of intragenic crossing over indicated that genes indeed have fine structure and that the multiplicity of allelic forms of some genes might represent mutations at different sites in the gene. Later, additional observations of the same type were made with *Drosophila* and fungi, and in the mid 1950s the most detailed analysis of the fine structure of a locus ever carried out—that of the bacteriophage T4 *rIIA* and *rIIB* genes—was completed; this system will be described in Chapter 7.

3.7 Complementation

In the 1940s the study of the genetics of metabolism in fungi led to the important conclusion that a particular phenotype is often the result of the activities of a number of genes (these observations will be presented in detail in Section 9.2). A genetic investigation of a system determining, for example, the synthesis of a pigment or the breakdown of a sugar commonly requires the isolation of numerous mutations that produce a particular phenotype. Invariably it becomes important to know whether particular mutations are allelic— that is, whether they are located in the same or different genes. The genetic test used to determine both allelism among a group of mutations and the number of different genes represented is called **complementation.** It can be used only with recessive mutations.

Complementation tests are done with the mutations in the *trans* configuration

$$\frac{m^1 \qquad\qquad +}{+ \qquad\qquad m^2}$$

in which m^1 and m^2 represent two mutations. This configuration is obtained in *Drosophila* and other diploid organisms simply by crossing individuals homozygous for the respective mutations. A comparable arrangement of the genetic material in *Neurospora* or *Aspergillus* is achieved by the production of a heterokaryon containing haploid nuclei of each of the two mutant strains. How it is done with viruses and with bacteria is explained in Chapter 7. A cell or organism with two mutations in the *trans* configuration (one on each chromosome) is called a ***trans*-heterozygote.**

Consider a *trans*-heterozygote in which the two mutations are in different genes and both genes must be active for an organism to have a wildtype phenotype. The two mutations impair different functional units, but since each chromosome has one wildtype allele of each gene, the organism will have the wildtype phenotype. When a *trans*-heterozygote has the wildtype phenotype, the mutations are said to *complement* one another, because the phenotype is that associated with the normal expression of both genes. The genetic constitution of such a heterozygote and its relation to the synthesis of functional products of each gene is illustrated in Figure 3-18(a). If the two mutations are in the same gene, the state of the organism is the same as that with homozygosity for a particular recessive mutation—that is, synthesis of a normal, functional product of a gene is prevented because both members of the pair of genes are mutant. If the two mutations are in the same gene (panel

(a) *Trans*-heterozygote for two mutations in different genes.

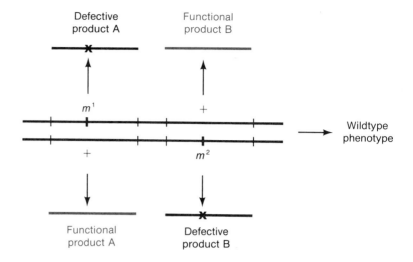

(b) *Trans*-heterozygotye for two mutations in the same gene.

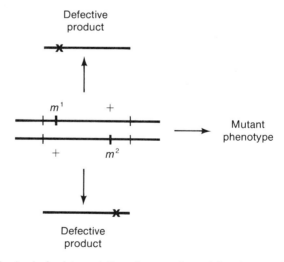

Figure 3-18 The basis for interpretation of a complementation test used to determine whether two mutations m^1 and m^2 are (a) in different genes or are (b) alleles of the same gene.

(b)), no functional product will be made and the *trans*-heterozygote has a mutant phenotype—complementation will be absent. In summary, in the absence of complicating interactions,

> a *trans*-heterozygote for two recessive mutations will have the wildtype phenotype if the mutations are in different genes and a mutant phenotype if the mutations are allelic.

Mutations that complement each other are said to belong to different **complementation groups,** whereas alleles are members of the same complementation group.

Complementation and recombination should not be confused. Complementation between two mutations is independent of the relative locations of the respective genes, and its occurrence provides no information either about linkage of the genes or about whether the mutations are located in homologous or nonhomologous chromosomes. Complementation is a property of *trans*-heterozygotes that have mutations in two different genes, whereas recombination is detected by genetic analysis of the progeny of heterozygotes.

The complementation test for the allelism of two recessive mutations is one part of the *cis-trans* **test.** The *cis*-**heterozygotes**

$$
\frac{m^1 \qquad\qquad m^2}{+ \qquad\qquad +}
$$

have normal (nonmutant) copies of the relevant gene or genes in one of their chromosomes and will usually be phenotypically wildtype. If the mutations are allelic, a nonfunctional product will be produced by the doubly-mutant member of the pair of genes and a functional product by the normal member. If the mutations are not allelic, functional products will be produced by the normal member of each pair of genes, as in the case of a *trans*-heterozygote. However, rare exceptions to this pattern occur when the product related to one of the mutations acts in some way to interfere with or inhibit the expression of the corresponding normal allele; in such a case, a heterozygote $(m/+)$ for the particular mutation will have a phenotype that is not fully normal. For mutations of this kind, complementation does not occur in *trans*-heterozygotes with other mutant alleles of the gene. Thus, for successful testing of mutation by the *trans* test (that is, testing for complementation), the mutation must behave normally in the *cis* test. Two important classes of mutations that behave abnormally are (1) those that affect the expression of adjacent genes, and (2) those that occur in regions of the genetic material that regulate the activities of adjacent genes. These will be examined in Section 9.9 (Chapter 9) and in a problem at the end of this chapter. Construction of a *cis*-heterozygote often requires a great amount of effort, and mutations that fail the *cis* test are rare; thus, the *cis* test is usually bypassed in routine complementation analysis.

To complete the discussion, an example of complementation analysis is presented. Eight different mutations are analyzed pairwise in *trans*-heterozygotes. The results are shown in Table 3-4. Each + or − entry designates the result of a single test of a pair of mutations. A + entry denotes complementation, and a − entry indicates noncomplementation. To begin with, notice any − entries that can be aligned in a single direction (that is, that align down one column or across one row). These indicate noncomplementing (allelic) mutations. For example, in column 1, mutations 1 and 5 fail to complement. Three genes can be identified by the lack of complementation:

Table 3-4 An example of complementation data

		Mutation number							
		1	2	3	4	5	6	7	8
Mutation number	1	−							
	2	+	−						
	3	+	+	−					
	4	+	−	+	−				
	5	−	+	+	+	−			
	6	+	+	−	+	+	−		
	7	+	+	−	+	+	−	−	
	8	+	−	+	−	+	+	+	−

Note: + = complementation; − = no complementation. An entry at the intersection of a horizontal row and a vertical column represents the result of one complementation test between two mutations. For example, the + entry at the intersection of row 3 and column 2 indicates that mutations 2 and 3 complement; the corresponding entry for row 2 and column 3 is not given since it is the same as the one just stated—that is, the table is symmetric about the diagonal.

the first contains mutations 1 and 5; the second contains mutations 2, 4, and 8; and the third contains mutations 3, 6, and 7. Single mutations (a row or column) that intersect at a + sign are complementing, so to confirm the three genes just identified, tests of pairs of mutations in different genes should complement one another. This is confirmed—for example, testing mutations 1 and 8 yields a + entry. Thus, the three complementation groups, often denoted A, B, and C, are the following:

Group	Mutation
A	1, 5
B	2, 4, 8
C	3, 6, 7

The simplest explanation of the data is that the phenotype being studied is the result of the activity of at least three genes. One cannot say that there are only three genes, because mutations in other genes may not have been isolated.

Problems

1. A phenotypically normal woman is known to have the X-chromosome constitution *gd* +/+ *hemA*. The recessive allele *gd* is associated with a deficiency of the enzyme glucose 6-phosphate dehydrogenase in the red blood cells, and *hemA* is the recessive allele determining classical hemophilia. If the genetic distance between the two loci is 16 map units, what is the probability that a son born to this woman will be phenotypically normal with respect to both of these traits?

2. Two loci in chromosome 7 of corn are identified by the recessive alleles *gl* (glossy), determining glossy leaves, and *ra* (ramosa), determining branching of ears. When a plant heterozygous at each of these loci was crossed with a homo-

zygous recessive plant, the progeny consisted of the following genotypes with the numbers of each indicated:

$Gl\,ra/gl\,ra$	88
$Gl\,Ra/gl\,ra$	6
$gl\,Ra/gl\,ra$	103
$gl\,ra/gl\,ra$	3

Calculate the percentage of recombination between these genes.

3. The recessive alleles b and cn of two genes in chromosome 2 of *Drosophila* determine black body color and cinnabar eye color, respectively. If the loci are 9 map units apart, what are the expected genotypes and their relative frequencies among the progeny of the cross $++/b\,cn \times ++/b\,cn$? Remember that crossing over does not occur in *Drosophila* males.

4. Dark eye color in rats requires the presence of a dominant allele at each of two genes r and p. Animals homozygous for the recessive alleles of either or both of these genes have light-colored eyes. Homozygous dark-eyed rats were crossed with doubly recessive light-eyed rats, and some of the resulting F_1 individuals were then testcrossed with rats of the homozygous light-eyed strain. The progeny from this testcross consisted of 628 dark-eyed and 889 light-eyed rats. When Rp/Rp rats were crossed with those having the genotype rP/rP and F_1 individuals from this cross were crossed with animals from an rp/rp strain, the progeny consisted of 86 dark-eyed and 771 light-eyed rats. What is the genetic map distance between the two loci?

5. In corn the alleles C and c result in colored versus colorless seeds, Wx and wx in nonwaxy versus waxy endosperm, and Sh and sh in plump versus shrunken endosperm. When plants grown from seeds heterozygous for each of these pairs of alleles were crossed with plants from colorless, waxy, shrunken seeds, the testcrossed seeds were

colorless, nonwaxy, shrunken	116
colorless, nonwaxy, plump	626
colorless, waxy, shrunken	2
colorless, waxy, plump	2708
colored, waxy, shrunken	601
colored, waxy, plump	113
colored, nonwaxy, shrunken	2538
colored, nonwaxy, plump	4
	6708

Determine the order of the three genes and construct a map showing the genetic distances between them.

6. In the linkage map of chromosome 9 of corn, three genes, Sh, Wx, and Gl, easily identified by the phenotypic effects of their recessive alleles, have the positions

```
 Sh                          wx        gl
 +————————————————————————————+————————+
 29                           59       69
```

From the cross

$$\frac{Sh \qquad wx \quad Gl}{sh \qquad Wx \quad gl} \times \frac{sh \qquad wx \quad gl}{sh \qquad wx \quad gl}$$

how many homozygous recessive individuals would be expected among 2000 progeny if the coefficient of coincidence is 0.5?

7. The recessive alleles k (kidney-shaped versus wildtype round eyes), cd (cardinal-colored versus wildtype red eyes), and e (ebony versus wildtype gray body color) identify three genes in chromosome 3 of *Drosophila*. Females with kidney-shaped, cardinal-colored eyes were mated with ebony males. When wildtype females from this cross were then mated with homozygous recessive males, the progeny obtained were:

k	cd	e	3
k	cd	$+$	876
k	$+$	e	67
k	$+$	$+$	49
$+$	cd	e	44
$+$	cd	$+$	58
$+$	$+$	e	899
$+$	$+$	$+$	4
			2000

(a) Determine the order of the three loci and construct a map showing their relative separations.

(b) How does the frequency of double crossovers obtained in this experiment compare with the frequency expected if crossing over occurs independently in the two chromosome regions?

8. In an early experiment with *Drosophila*, females heterozygous for the dominant eye mutation Star on chromosome 2 were mated with males homozygous for the chromosome-2 recessives aristaless (reduced bristlelike parts of the antennae) and dumpy (shortened wings). When the Star F_1 females were testcrossed to homozygous aristaless dumpy males, the following phenotypes were observed in the progeny:

Star	956
aristaless, dumpy	918
aristaless, Star	7
dumpy	5
aristaless	132
Star, dumpy	100

(a) Determine the recombination frequencies and the order of the genes.

(b) Which phenotypic classes are missing, and why?

9. When *Drosophila* females having vermilion (bright scarlet) eyes were mated with males having garnet (brownish) eye color, approximately equal numbers of wildtype (red-eyed) females and vermilion-eyed males were produced in the F_1. The F_2 progeny obtained by crossing these F_1 males and females consisted of:

wildtype females	382
vermilion females	368
wildtype males	45
vermilion males	331
garnet males	335
orange males	39

Explain these results, using appropriate diagrams to show the locations of the genes.

10. The following classes and frequencies of ordered tetrads were obtained from the cross $a^+ b^+ \times a\, b$ in *Neurospora:*

Spore pair				Numbers of asci
1–2	3–4	5–6	7–8	
$a^+ b^+$	$a^+ b^+$	$a\, b$	$a\, b$	1766
$a^+ b^+$	$a\, b$	$a^+ b^+$	$a\, b$	220
$a^+ b^+$	$a\, b^+$	$a^+ b$	$a\, b$	14

What is the order of the genes in relation to the centromere?

11. The following spore arrangements, in the frequencies indicated, are from ordered tetrad analysis of a cross between a *Neurospora* strain ($c\, v$), which exhibits a compact growth form and is unable to synthesize the amino acid valine, and a wildtype ($+\,+$) strain. Note that only one member of each pair of spores is shown.

Spore pair	Ascus composition				
1–2	$c\ \ v$	$c\ \ +$	$c\ \ v$	$+\ \ v$	$c\ \ v$
3–4	$c\ \ v$	$c\ \ +$	$c\ \ +$	$c\ \ +$	$+\ \ v$
5–6	$+\ \ +$	$+\ \ v$	$+\ \ v$	$c\ \ v$	$c\ \ +$
7–8	$+\ \ +$	$+\ \ v$	$+\ \ +$	$+\ \ +$	$+\ \ +$
Number:	34	36	20	1	9

What can you conclude about the linkage and the location of the genes?

12. A *Neurospora* strain ($t\, +$), unable to synthesize the vitamin thiamine, was crossed with a strain ($+\, m$), unable to synthesize the amino acid methionine, and the following classes of spore arrangements were produced in the frequencies indicated.

Spore pair	Ascus composition					
1–2	$t\ \ +$	$t\ \ +$	$t\ \ +$	$t\ \ +$	$t\ \ m$	$t\ \ m$
3–4	$t\ \ +$	$t\ \ m$	$+\ \ m$	$+\ \ +$	$t\ \ m$	$+\ \ +$
5–6	$+\ \ m$	$+\ \ +$	$t\ \ +$	$t\ \ m$	$+\ \ +$	$t\ \ +$
7–8	$+\ \ m$	$+\ \ m$	$+\ \ m$	$+\ \ m$	$+\ \ +$	$+\ \ m$
Number:	260	76	4	54	1	5

(a) Determine the map distance between the two genes and between each gene and the centromere; then, construct a linkage map showing these relations.

(b) If the spores represented by the data are treated as random products of meiosis, what is the recombination frequency between t and m?

(c) In calculating the map distance between a gene and the centromere it is necessary to divide the percentage of second-division segregations for the gene by 2, if the map units are to be comparable to those used, for example, in chromosome mapping in corn, *Drosophila*, or humans. Why is this division by 2 necessary?

13. In yeast, which has unordered tetrads, the following tetrads were obtained when the cross $a^+ b\, c \times a\, b^+ c^+$ was made to investigate possible linkages among the three genes:

Tetrad composition				Number
$a\ \ +\ \ +$	$a\ \ +\ \ +$	$+\ \ b\ \ c$	$+\ \ b\ \ c$	142
$a\ \ b\ \ +$	$a\ \ b\ \ +$	$+\ \ +\ \ c$	$+\ \ +\ \ c$	130
$a\ \ +\ \ +$	$a\ \ +\ \ c$	$+\ \ b\ \ +$	$+\ \ b\ \ c$	62
$a\ \ b\ \ +$	$a\ \ b\ \ c$	$+\ \ +\ \ +$	$+\ \ +\ \ c$	74
$a\ \ +\ \ c$	$a\ \ +\ \ c$	$+\ \ b\ \ +$	$+\ \ b\ \ +$	1
$a\ \ b\ \ c$	$a\ \ b\ \ c$	$+\ \ +\ \ +$	$+\ \ +\ \ +$	1
				$\overline{410}$

From these data determine which of the genes are linked, if any, and the appropriate linkage distances.

14. In *Neurospora* the a locus is 5 map units from the centromere in chromosome 5 and the b locus is 10 map units from the centromere in chromosome 7. From the cross $a\, + \times +\, b$, what are the expected frequencies of

(a) Parental ditype, nonparental ditype, and tetratype asci?

(b) Recombinant ascospores?

15. The order and linkage relations of three genes in chromosome 9 of corn are represented by the map

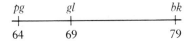

The recessive alleles of these genes result in pale green seedlings, glossy leaves, and brittle leaves, respectively. Assume that the coefficient of coincidence for this region of the chromosome is 0.6. Predict the kinds of phenotypes and the frequencies expected to occur in a progeny of 2000 individuals from the testcross

$$Pg\,gl\,Bk\,/pg\,Gl\,bk \times pg\,gl\,bk/pg\,gl\,bk$$

16. The following genetic map summarizes the data obtained in a large experiment in which the recombination frequencies between three pairs of alleles have been calculated:

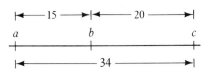

What was the frequency of double crossing over in this experiment?

17. In the nematode *Caenorhabditis elegans*, *dpy* (dumpy) and *unc* (uncoordinated) are recessive alleles at two linked genes having a recombination frequency of *P*. The heterozygote *dpy* +/+ *unc* is obtained and allowed to self-fertilize (self-fertilization is a normal mode of reproduction in this organism). In terms of *P*, what fraction of the progeny is expected to be both dumpy and uncoordinated if crossing over in spermatogenesis is independent of crossing over in oogenesis?

18. An *Aspergillus* diploid was obtained from a heterokaryon formed by the haploid strains $ad^+leu^+ribo^+fpa^+w^+$ and $ad\,leu\,ribo\,fpa\,w$. The recessive alleles of these genes determine requirements for adenine, leucine, and riboflavin; resistance to fluorophenylalanine; and white asexual spores, respectively. When asexual spores produced by the diploid strain were spread on a medium containing fluorophenylalanine, 93 colonies resistant to this compound were selected.
(a) It was determined that 38 of the fluorophenylalanine-resistant colonies were haploid; 22 with the genotype $fpa\,leu\,ribo\,ad^+w^+$, and 16 with the genotype $fpa\,leu\,ribo\,ad\,w$. What do these data indicate about the linkage relations between the genes?
(b) Of the 55 diploid colonies resistant to the selective compound, 24 required neither leucine nor riboflavin, 17 required riboflavin but not leucine, and 14

required both. Diagram the relation between the three genes and the centromere indicated by these data.

19. A complementation test is sometimes called an "allelism test." Why?

20. Why do complementation tests require the use of recessive mutations?

21. A *Drosophila* geneticist exposes flies to a mutagenic chemical and obtains nine mutations in the X chromosome that are lethal when homozygous. The mutations are tested in pairs in complementation tests, with the results shown in the table below. A + indicates complementation (that is,

	1	2	3	4	5	6	7	8	9
1	−	+	−	+	+	+	−	+	+
2		−	+	+	+	+	+	+	+
3			−	+	+	+	−	+	+
4				−	+	−	+	+	+
5					−	+	+	+	−
6						−	+	+	+
7							−	+	+
8								−	+
9									−

flies carrying both mutations survive), and a − indicates noncomplementation (flies carrying both mutations die). How many genes (complementation groups) are represented by the mutations and which mutations belong to each complementation group?

22. In *Drosophila*, individuals carrying both lz^k and lz^{46} alleles or carrying both lz^k and lz^{BS} alleles have eyes that are more abnormal than flies carrying any of the mutations alone. Among 12,666 male offspring produced by females of genotype $sn\,lz^{46}\,v/lz^k$, two were phenotypically v and extreme lz, and one was phenotypically sn but otherwise wildtype. Among 12,595 male offspring produced by females of genotype $sn^3\,lz^{BS}\,v/lz^k$, one was phenotypically v, one was phenotypically sn and extreme lz, and one was phenotypically extreme lz. The markers sn and v flank the lz markers, and sn and v confer the sn and v phenotypes, respectively. Specify the genotype of each class of exceptional males in each cross, and determine the order of the mutations lz^k, lz^{BS}, and lz^{46}.

23. Refer to Table 3-3 in the text. A ninth mutation is isolated and tested against the eight mutations already classified. The results of complementation with mutations 1–8 are the following: 1, ±; 2, −; 3, ±; 4, −; 5, ±; 6, ±; 7, ±; 8, −. The symbol ± represents weak complementation. Explain the properties of mutation 9.

CHAPTER 4

The Chemical Nature and Replication of the Genetic Material

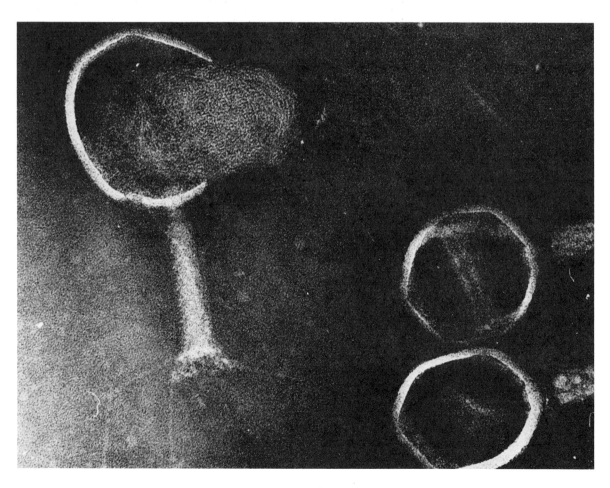

Phenomena such as segregation and linkage imply that genes are discrete factors located at definite positions on chromosomes and transmitted in predictable ways during reproduction. Moreover, the stability of these units of hereditary information persisting through many generations implies that each time a chromosome is replicated, its genes must be copied exactly. Since different allelic forms of a gene exist (many examples of which have been seen in the preceding chapters), these forms must arise by rare changes in structure (mutations) that alter the ability of the gene to function. However, the properties that can be ascribed to genes from analysis of their phenotypic expression and patterns of inheritance reveal nothing about their structure, how they might be copied to yield exact replicas, or how they might function to determine cellular characteristics. Understanding these basic features of heredity depends on identification of the chemical nature of the genetic material and the processes involved in its replication. These topics are the subjects of this chapter.

4.1 The Importance of Bacteria and Viruses in Genetics

Cells are organized in two fundamentally different ways (Figure 4-1). In bacteria and blue-green algae, which comprise the group of unicellular organisms called **prokaryotes** (Greek: *pro*, before; *karyon*, nucleus), the genetic material is located in a region that lacks clear boundaries and is called the **nucleoid.** In other single-celled organisms and in all cells of multicellular organisms, a nuclear envelope encloses the nucleus, separating it from the cytoplasm. Organisms whose cells have nuclei are called **eukaryotes** (meaning true nucleus). In eukaryotic cells other membrane systems also subdivide the cytoplasm into regions of specialized organization and function. Mitosis, and

Facing page: An electron micrograph of *E. coli* phage T4 with a broken head, releasing its single DNA molecule as a tightly tangled mass. The two particles at the right are phage lacking DNA; such particles, which are frequently observed, are probably formed by attachment of tails to empty heads. (Courtesy of Robley Williams.)

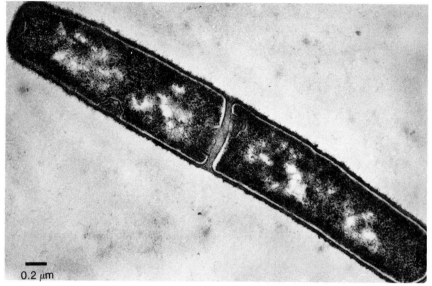

0.2 μm

(a)

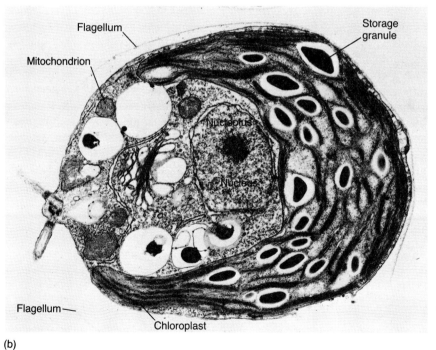

Flagellum

Storage
granule

Mitochondrion

Nucleolus

Nucleus

Flagellum—

Chloroplast

(b)

Figure 4-1 The organization of a prokaryotic cell and a eukaryotic cell. (a) Prokaryote. An electron micrograph of a dividing bacterium, showing the dispersed genetic material (light areas). (b) Eukaryote. Electron micrograph of a section through a cell of the alga *Tetraspora,* showing the membrane-bound nucleus, the nucleolus (dark central body), and other membrane systems (chloroplasts and mitochondria) that subdivide the cytoplasm into regions of specialized function. The dark sharply bounded regions are starch granules, which store carbohydrate. This complex structure should be compared to the simpler organization of the bacterium shown in (a). (Courtesy of (a) A. Benichou-Ryter and (b) Jeremy Pickett-Heaps.)

the doubling and halving of total chromosome number caused by alternation of nuclear fusion during fertilization with meiosis during gamete formation, are also attributes of eukaryotes.

Viruses are small particles, considerably smaller than eukaryotic cells, able to infect susceptible cells and multiply within them to form large numbers of progeny virus particles. Few, if any, organisms are not subject to viral infection, though most viruses are severely restricted in the organisms they can infect. Many human diseases are caused by different viruses—for example, encephalitis, influenza, measles, and the common cold. Other viruses are able to infect bacteria, and these are called **bacteriophages** (Greek: *phagos*, one that eats), which is usually shortened to **phages.** Most viruses consist of a single molecule of genetic material enclosed in a protective coat composed of one or more kinds of protein molecules; however, their size, molecular constituents, and structural complexity vary greatly (Figure 4-2). Viruses have neither the capability for the storage and conversion of energy required in metabolic processes nor the enzymes and other components required for the synthesis of proteins. Thus, a virus can multiply only within a cell and is "living" only in the sense that its genetic material directs its own multiplication; outside its host cell a virus is an inert particle.

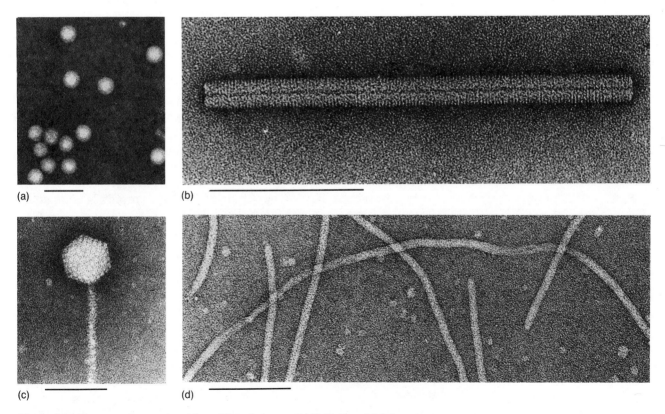

(a) (b) (c) (d)

Figure 4-2 Electron micrographs of four different viruses. (a) Poliovirus. (b) Tobacco mosaic virus. (c) *E. coli* phage λ. (d) *E. coli* phage M13. The length of the bar is 1000 Å in each case. (Courtesy of Robley Williams.)

Bacteria and viruses bring to traditional types of genetic experiments four important advantages over multicellular plants and animals: (1) They are haploid, so dominance or recessiveness of alleles is not a complication in identifying genotype. (2) A new generation is produced in minutes rather than weeks or months, which vastly increases the rate of accumulation of data. (3) They are easy to grow in enormous numbers under controlled laboratory conditions, which facilitates biochemical studies and the analysis of rare genetic events. (4) The individual members of these large populations are genetically identical; that is, the population is a **clone.** Later chapters will show the extent to which work with these organisms has contributed to our present understanding of the organization and expression of the hereditary material.

4.2 Evidence That the Genetic Material is DNA

Deoxyribonucleic acid (DNA) was discovered in 1869 as a weakly acidic phosphorus-rich compound isolated from the nuclei of human white blood cells. This compound, subsequently called **nucleic acid,** could also be isolated from the nuclei of salmon sperm and other animal cells. Chemical analysis done in 1910 identified two classes of nucleic acids—DNA and ribonucleic acid (RNA).

The demonstration that DNA is a major constituent of chromosomes resulted from development of the Feulgen staining procedure in 1924. In Feulgen staining a red compound is formed by a chemical that combines with the deoxyribose sugar in DNA. Other staining reagents identified protein as the second major constituent of chromosomes. In addition to the chromosomal location of DNA, other indirect evidence suggested a close relation between DNA and the genetic material. For example, almost all somatic cells of a given species contain a constant amount of DNA, whereas the RNA content and the amount and kinds of proteins differ greatly in different cell types. Also, nuclei resulting from meiosis in both plants and animals have only half the DNA content of nuclei in their somatic cells. Nevertheless, DNA was not considered likely to be the genetic material, mainly because crude chemical analyses had suggested that DNA lacks the chemical diversity needed by a genetic substance. In contrast, proteins form an exceedingly diverse collection of molecules, so it was widely believed that proteins comprised the genetic material. Since DNA usually occurs in association with protein, it was instead thought to be the structural framework or skeleton of chromosomes. That DNA is indeed the genetic material was shown directly in two experiments with bacterial transformation and phage infection. These key experiments in genetics are described in the following sections.

Transformation of Bacterial Cells by DNA

Bacterial pneumonia in mammals is caused by strains of *Streptococcus pneumoniae* (also called *Pneumococcus*) able to synthesize a polysaccharide (complex carbohydrate) capsule around itself; this capsule protects the bacterium from the defense mechanisms of the infected animal and enables the bacterium to cause disease. When a bacterium is grown on solid medium, the enveloping capsule gives the bacterial colony a glistening, smooth (S) appearance. Some mutant strains of *Pneumococcus* have lost an enzyme activity required to synthesize the capsular polysaccharide, and these bacteria form colonies that have a rough (R) surface when grown in the same way. R strains do not cause pneumonia, because without their capsules the cells are inactivated by the immune system of the host.

Both R and S strains of *Pneumococcus* breed true, except for rare mutations in R strains that give rise to the S phenotype and rare mutations in S strains that give rise to the R phenotype. Since simple mutations can cause conversion between the forms, the S and R phenotypes are likely to be determined by different allelic forms of a simple genetic unit.

When mice are injected with either living R pneumococci or with heat-killed S cells, they remain healthy. However, in 1928 it was discovered that mice injected with a mixture containing a small number of R bacteria and a large number of heat-killed S cells often died of pneumonia. Bacteria isolated from blood samples of the dead mice produced *pure* S cultures having a capsule typical of the heat-killed S cells (Figure 4-3). The purity is significant because if S cells had arisen by a rare reverse mutation in one of the injected R cells, one would expect that at least some R cells would be obtained from the dead animal. Evidently, dead S cells can in some way restore to the living R bacteria the ability to withstand the immunological system of the mouse, multiply, and cause pneumonia. Furthermore, this acquired ability is inherited by descendants of the changed or **transformed** bacteria.

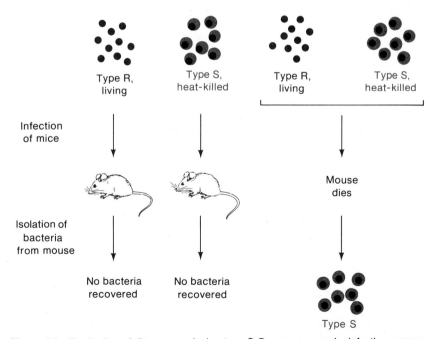

Figure 4-3 Production of disease-producing type S *Pneumococcus* by infecting a mouse with a mixture of living type R cells, which alone do not cause disease, and heat-killed type S cells.

Later experiments showed that living S cells can also be produced (at low frequency) when R cells are grown in a liquid medium containing heat-killed S cells. Extracts of S cells, from which intact cells and large pieces of cellular debris have been removed by filtration, are as effective as intact S cells in causing specific transformation of the R phenotype to the S phenotype. In detailed studies with such transforming extracts, Oswald Avery, Colin Mac-Leod, and Maclyn McCarty purified DNA from S cells and found that minute amounts of this material added to growing cultures of R cells consistently resulted in production of some transformed cells with the capsular poly-saccharide characteristic of type S bacteria. The DNA preparations contained traces of protein, but the transforming activity was not altered by treatment with enzymes that degrade proteins or by treatment with ribonuclease, an enzyme that degrades RNA. On the other hand, transforming activity was completely destroyed by treatment with deoxyribonuclease, an enzyme that degrades DNA (Figure 4-4). Such experiments implied that the substance responsible for genetic transformation was the DNA of the cell and, hence, that DNA is the genetic material.

Phage Infection Experiments

A second kind of experiment that indicated more directly that DNA is the material basis of heredity was reported by Alfred Hershey and Martha Chase in 1952. In these experiments reproduction of bacteriophage T2 in the bacterium *Escherichia coli* was studied. Infection by T2 (Figure 4-5) had earlier been shown to proceed by sequential attachment of a phage particle by the tip of its tail to the bacterial cell wall, entry of phage material into the cell, multiplication of this material to form a hundred or more progeny phage, and

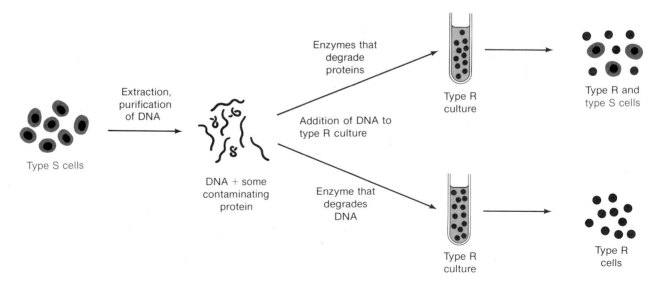

Figure 4-4 A diagram of the experiment that demonstrated that DNA is the active material in bacterial transformation.

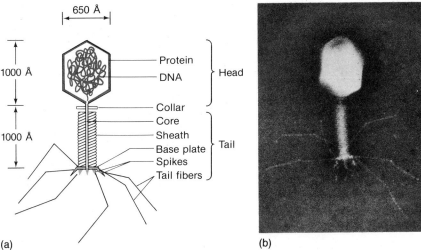

(a)   (b)

Figure 4-5 (a) Drawing of *E. coli* phage T2, showing various components. The arrangement of the DNA in the phage head is unknown. (b) An electron micrograph of phage T4, a nearly identical phage. (Courtesy of Robley Williams.)

release of progeny by disruption of the host cell (see Chapter 7). T2 particles were also known to be composed of DNA and protein in approximately equal amounts.

DNA contains phosphorus but no sulfur, and proteins generally contain some sulfur and no phosphorus. Consequently, it is possible to label DNA and proteins differentially, with radioactive isotopes of the two elements. Hershey and Chase produced particles with labeled DNA by infecting *E. coli* cells that had been grown for several generations in a medium containing ^{32}P (a radioactive isotope of phosphorus) and then collecting the phage progeny. Particles with labeled proteins were obtained in the same way, using medium containing ^{35}S (a radioactive isotope of sulfur).

In the experiments summarized in Figure 4-6, unlabeled *E. coli* cells were infected with phage labeled with *either* ^{35}S or ^{32}P in order to follow the proteins and DNA during infection. Infected cells were separated from unattached particles by centrifugation, resuspended in fresh medium, and then agitated in a kitchen blendor to shear attached phage material from the cell surfaces. This treatment was found to have no effect on the subsequent course

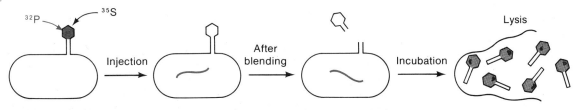

Figure 4-6 A diagrammatic summary of the Hershey-Chase ("blendor") experiment, which demonstrated that DNA and not protein is responsible for directing the reproduction of phage T2 in infected *E. coli* cells.

of the infection in that the number of progeny was the same with and without blending. When intact bacteria were separated from phage debris by a second centrifugation, most of the radioactivity from ^{32}P-labeled phage was found to be associated with the bacteria; on the other hand, if the infecting phage was labeled with ^{35}S, a very small fraction of the radioactivity was associated with the bacterial cells. Thus, it appears that a T2 phage transfers most of its DNA, but very little of its protein, to the cell it infects. A critical finding was that about 50 percent of the transferred ^{32}P-labeled DNA was inherited in the *progeny* phage particles, but less than one percent of the transferred ^{35}S-labeled protein was found in the progeny. These results strongly suggested that DNA is the genetic material in bacteriophage T2.

These two experiments—the transformation experiment and the "blendor experiment"—were widely accepted as proof that DNA is the genetic material in all organisms. In later years some exceptions were found: certain viruses, such as tobacco mosaic virus and influenza virus, were found to lack DNA and to utilize RNA as their genetic material. Despite these exceptions, the generalization that DNA is the genetic material has proved to be very powerful and a driving force for further research in molecular genetics.

Soon after the T2 experiments of Hershey and Chase were performed, increased knowledge of DNA chemistry was used in proposing a physical structure for the DNA molecule—the double helix. This model made it easy to comprehend the role of DNA molecules as stable carriers of genetic information, the manner of their precise replication, and the probable nature of some kinds of mutations. These foundations of molecular genetics are the subject of the following sections.

4.3 Chemical Composition of DNA

DNA is a polymer (a molecule containing repeating units) composed of a five-carbon sugar (2′-deoxyribose), phosphoric acid, and four nitrogen-containing bases. Two of these nitrogenous bases are **purines,** which have a double ring structure; the other two are **pyrimidines,** which contain a single ring. The purine bases are **adenine (A)** and **guanine (G),** and the pyrimidine bases are **thymine (T)** and **cytosine (C)** (Figure 4-7). In certain organisms

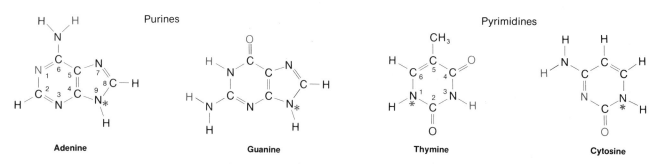

Figure 4-7 The four nitrogen-containing bases of DNA. The N linked to deoxyribose is denoted by a red asterisk. The red atoms are engaged in hydrogen bonds in the DNA base pairs.

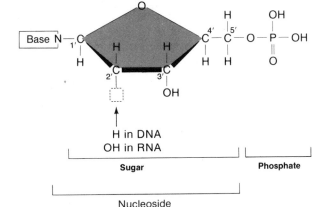

Figure 4-8 A typical nucleotide showing the three major components, the difference between DNA and RNA, and the distinction between a nucleoside and a nucleotide.

modified forms of cytosine, adenine, and guanine are sometimes present in small amounts. When a base is covalently linked to a deoxyribose, it forms a compound called a **nucleoside.** (A covalent bond is a strong chemical bond in which atoms share their outer electrons.) In nucleic acids a phosphate group is also attached to the sugar of a nucleoside, yielding a **nucleotide** (Figure 4-8); the terminology is that a nucleotide is a nucleoside phosphate. Note particularly the numbering of the carbon atoms in the sugar in Figure 4-8, beginning with the carbon atom to which the base is attached (the 1′ carbon). The numbers are conventionally represented with affixed primes. In the structure of DNA, as we shall see, the 3′ and 5′ carbon atoms of the sugar play a central role in linking one nucleotide to the next in line. In any case, DNA contains four common nucleotides—deoxyadenylic acid, deoxyguanylic acid, thymidylic acid, and deoxycytidylic acid—which differ from one another only in the base attached to the sugar. It is common to name nucleotides as monophosphates—that is, deoxyadenylic acid is called deoxyadenosine 5′-monophosphate and abbreviated dAMP.

In nucleic acids, the nucleotides are covalently joined to form a **polynucleotide chain,** in which the 5′ carbon of one sugar is linked by its phosphate group to the 3′ carbon of the next sugar (Figure 4-9). The chemical bonds by which the sugar components of adjacent nucleotides are linked through the phosphate groups are called **phosphodiester bonds.** The 5′-3′-5′-3′ orientation of these linkages continues throughout the chain, resulting in a polarity (orientation) that is now known to be important both in replication of the molecules and in their genetic function. Note that the terminal groups of each polynucleotide chain are a 5′-phosphoryl (**5′-P**) group at one end and a 3′-hydroxyl (**3′-OH**) group at the other.

The molar concentrations (denoted by [ ]) of the bases in DNA exhibit three important features:

THE CHEMICAL NATURE
AND REPLICATION OF THE
GENETIC MATERIAL

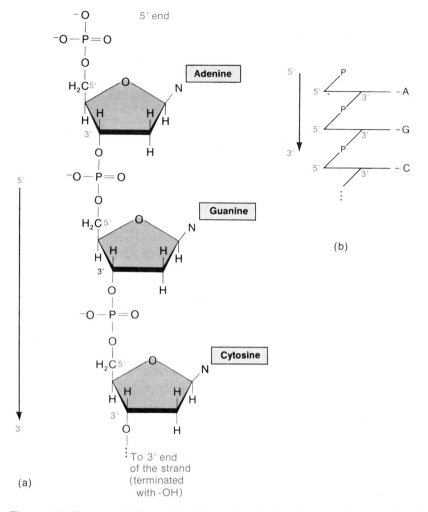

Figure 4-9 Three nucleotides at the 5′ end of a single polynucleotide strand. (a) The chemical structure of the sugar-phosphate linkages showing the 5′-to-3′ orientation of the strand. The red numbers are the carbon-atom numbers shown in Figure 4-8. (b) A common schematic way to depict a polynucleotide strand.

Molar concentration, M A measure of the number rather than the weight of molecules per unit volume.

1. The concentration of purine bases equals that of the pyrimidine bases; that is,

[total purines] = [A] + [G] = [total pyrimidines] = [T] + [C]

2. The concentrations of adenine and thymine are equal, as are the concentrations of guanine and cytosine; that is,

[A] = [T]   and   [G] = [C]

3. The **base composition,** represented by the fraction

$$\frac{([G] + [C])}{([G] + [C] + [A] + [T])}$$

sometimes called the **percentage G + C,** is constant in all cells of an organism and within a species, though it varies between species. Data on the base composition of DNA from some representative organisms are given in Table 4-1. The equality between purines and pyrimidines, or adenine and thymine, or guanine and cytosine is usually not perfect because of slight errors of measurement in the analyses; this experimental error also accounts for the apparent differences in base composition of human DNA obtained from different tissues. Higher plants and animals all have a deficiency of G + C compared to A + T in their DNA, with a base composition between 38 and 47 percent G + C. The base composition of viruses, bacteria, and lower plants shows significant species variation, ranging from about 31 percent G + C ("high AT") to 72 percent G + C ("high GC"). Generally, the base compositions of closely related organisms are similar.

4.4 Physical Structure of DNA: The Double Helix

An essentially correct three-dimensional structure of the DNA molecule was first proposed by James Watson and Francis Crick based on physical data from x-ray diffraction analysis and the chemical data described in items 1 and 2 of the preceding section; the model has subsequently been refined only in

Table 4-1 Base composition of DNA from different organisms

Organism	Base (and percent of total bases)				Base composition (percent G + C)
	Adenine	Thymine	Guanine	Cytosine	
Bacteriophage T7	26.0	26.0	24.0	24.0	48.0
Bacteria					
Clostridium perfringens	36.9	36.3	14.0	12.8	26.8
Streptococcus pneumoniae	30.3	29.5	21.6	18.7	40.3
Escherichia coli	24.7	23.6	26.0	25.7	51.7
Sarcina lutea	13.4	12.4	37.1	37.1	74.5
Fungi					
Saccharomyces cerevisiae	31.7	32.6	18.3	17.4	35.7
Neurospora crassa	23.0	23.3	27.1	27.6	54.7
Higher plants					
Wheat	27.3	27.1	22.7	22.8*	45.5
Maize	26.8	27.2	22.8	23.2*	46.0
Animals					
Drosophila melanogaster	30.7	29.4	19.6	20.2	39.8
Pig	29.4	29.7	20.5	20.5	41.0
Salmon	29.7	29.1	20.8	20.4	41.2
Human, sperm	30.7	31.2	19.3	18.8	38.1
thymus	29.8	31.8	20.2	18.2	39.0
liver	30.3	30.3	19.5	19.9	39.4

*Includes one-fourth 5-methylcytosine, a modified form of cytosine found in most plants more complex than algae and in some animals.

THE CHEMICAL NATURE
AND REPLICATION OF THE
GENETIC MATERIAL

Units of length
1 angstrom (Å) = 10^{-8} cm
(about the diameter of a hydrogen atom)
1 nanometer (nm) = 10 Å
1 micrometer (μm) = 1000 nm
1 millimeter (mm) = 1000 μm
1 centimeter (cm) = 10 mm

terms of detail. DNA consists of two polynucleotide chains twisted around one another forming a symmetric double-stranded helix. Each chain makes one complete turn every 34 Å (Figure 4-10). The helix is right-handed, meaning that each chain follows a clockwise path as it progresses. The bases are spaced 3.4 Å apart along each chain, so there are ten bases per helical turn in each strand, and, correspondingly, ten base pairs per turn of the double helix. Each base is paired to a base in the other strand by hydrogen bonds, thereby holding the strands together. (A hydrogen bond is a *noncovalent* weak bond in which a positively charged atom and a negatively charged atom are joined together by a hydrogen atom that is linked covalently to one of the atoms.) The paired bases are planar, parallel to one another, and perpendicular to the long axis of the double helix. When discussing a DNA molecule, one frequently refers to the individual strands as single strands or single-stranded DNA, and to the double helix as double-stranded DNA or a duplex molecule.

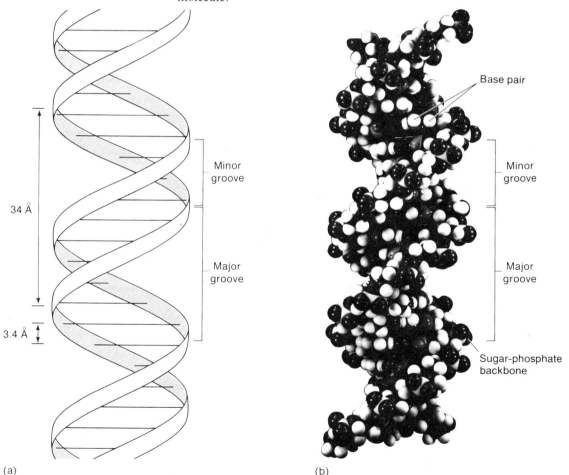

(a)

(b)

Figure 4-10   The DNA molecule. (a) A diagrammatic model of the double helix. (b) Space-filling model of the DNA double helix.

A central feature of DNA structure is the pairing between specific bases:

the purine adenine pairs with the pyrimidine thymine (forming an AT pair) and the purine guanine pairs with the pyrimidine cytosine (forming a GC pair).

Three factors are responsible for this specificity in base pairing:

1. Hydrogen bonds are stable only if the atoms form an array that is in the same plane as the covalent bonds.

2. The positions in the bases of the atoms that participate in hydrogen bonds allow only particular hydrogen bonds to form.

3. The dimensions of the purine bases are such that a purine-purine pair is too large to fit into a double helix without distorting the sugar-phosphate backbone. Pyrimidines can be situated opposite one another, but they cannot reach one another to form hydrogen bonds if they are surrounded by purine-pyrimidine pairs.

The adenine-thymine base pair and the guanine-cytosine pair are illustrated in Figure 4-11. Note that an AT pair has two hydrogen bonds and a GC pair has three hydrogen bonds.

The stability of the double-stranded helix is a consequence of three factors:

1. Hydrogen bonds between the bases hold the strands together.

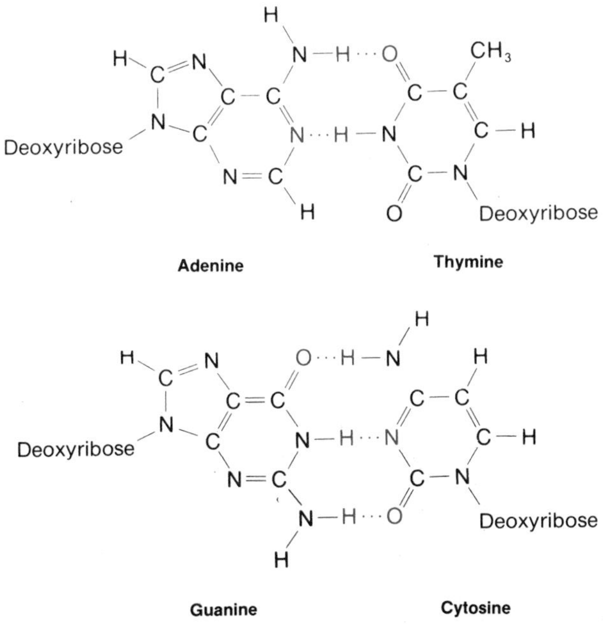

Figure 4-11   The two common base pairs of DNA. Hydrogen bonds (dotted lines) and the joined atoms are shown in red.

2. Purine and pyrimidine bases are quite insoluble (that is, they interact poorly with water) and hence tend to become arranged in a way that minimizes contact with water, namely, with the planes of the flat bases parallel and nearly in contact. Such a tendency is called a **hydrophobic interaction,** and it leads, in this case, to **base stacking,** an arrangement that orients the bases such that their hydrogen bonding is maximized.

3. Positive ions (such as Na^+, K^+, Mg^{2+}, and Ca^{2+}) present in solution neutralize the negatively charged phosphate groups. In the absence of such ions, electrical repulsion drives the sugar-phosphate chains apart and double-stranded DNA is unstable.

Parallel   ↑ ↑

Antiparallel ↑ ↓

The two polynucleotide strands of the double helix are oriented in opposite directions in the following sense. Recall that the backbone of a chain (Figure 4-10) consists of deoxyribose molecules alternating with phosphate groups that link the 3′ carbon atom of one sugar to the 5′ carbon of the next in line. The two strands of a double helix are **antiparallel** with respect to this linkage. That is, if the strands are followed from the 5′ end to the 3′ end (the 5′ → 3′ direction), the structure . . .-P-5′-sugar-3′-P. . . occurs in one direction in one strand and in the opposite direction in the other strand. Note that this means that each terminus of the double helix possesses one 5′-P group (on one strand) and one 3′-OH group (on the other strand).

The specificity of base pairing means that each base along one polynucleotide strand of the DNA determines the base in the opposite position on the other strand; one says that the sequences of bases along the two strands

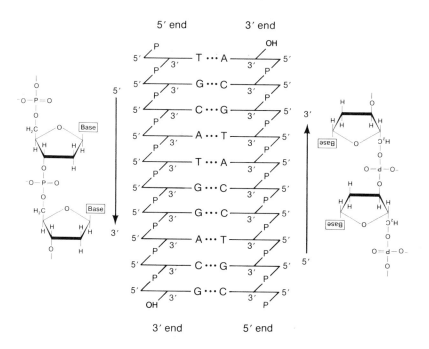

Figure 4-12 A segment of a DNA molecule showing the antiparallel orientation of the complementary strands. The arrows indicate the 5′-to-3′ direction of each strand. The phosphates (P) join the 3′ carbon atom of one deoxyribose (horizontal line) to the 5′ carbon atom of the adjacent deoxyribose. The orientations of the chemical structures are shown on either side of the sugar-phosphate chain.

are **complementary.** Nothing restricts the sequence of bases in a single strand; that is, any sequence could occur along one strand.

The principal features of the double helix are shown diagrammatically in Figure 4-12.

4.5 Requirements of the Genetic Material Possessed by DNA

Not every polymer would be useful as genetic material. However, DNA is admirably suited to this function, as it satisfies the following three essential requirements:

1. *A genetic material must carry all of the information needed to direct the specific organization and metabolic activities of the cell.* The product of a gene is a protein molecule—a polymer composed of hundreds of copies of 20 different molecular units called amino acids. The sequence of amino acids in the protein determines its chemical and physical properties. Inasmuch as a gene is expressed when its protein product is synthesized, a requirement of the genetic material must be to direct the order of addition of the amino acid units to the end of a growing protein molecule. If a polymer were used as genetic material, the complete amino acid sequence of the protein product would need to be encoded in the sequence of different monomers making up the genetic material. A polymer consisting of only a single repeated unit—for example, polyadenine—or only a single repeated sequence—for example, AGCT AGCT AGCT. . .—could carry very little genetic information. However, since chemically the four bases in a DNA molecule can be arranged in any sequence, and since the sequence can vary from one part of the molecule to another and from organism to organism, it can contain a great many unique subsequences, each of which can be a distinct gene. Thus, synthesis of a variety of different protein molecules can be directed by a long DNA chain. The problem of containing genetic information in a DNA molecule is one of how a linear sequence of four bases (or base pairs) can determine a linear sequence of hundreds of amino acids of 20 different types in a protein chain. With DNA as genetic material this information-transfer process is accomplished simply by having a system in a cell that "reads" the base sequence in order—three at a time—and translates this sequence into an amino acid sequence. The set of relations between triplets of bases and amino acids is called the **genetic code;** it will be discussed further in Chapter 9, when the synthesis of proteins is examined.

2. *The genetic material must replicate accurately, so that the information it contains is precisely inherited by daughter cells.* The basis for exact duplication of a DNA molecule is the complementarity of the AT and GC pairs in the two nucleotide chains. Unwinding and separation of the chains, with each free chain then serving as a template for the synthesis of a new complement, results

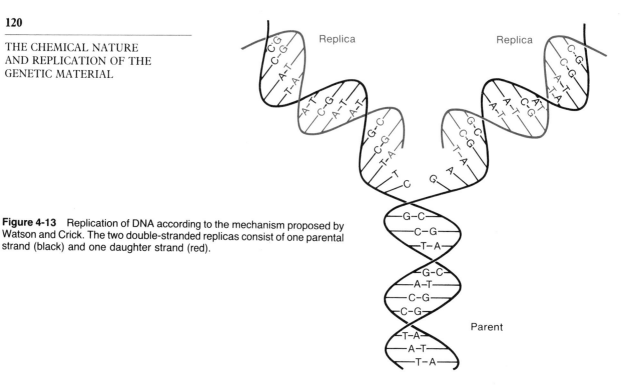

Figure 4-13 Replication of DNA according to the mechanism proposed by Watson and Crick. The two double-stranded replicas consist of one parental strand (black) and one daughter strand (red).

in the formation of two identical double helices (Figure 4-13). We will see in the next section that the structure is ideally adapted to accurate self-replication of a very long molecule, but the actual process is not as simple as the figure suggests.

3. *The genetic material must be capable of undergoing occasional mutation, so that the information it carries is altered in a heritable way.* Specificity in pairing of the purine and pyrimidine bases in the structure of DNA requires that the hydrogen atoms in the bases have fixed locations. However, transient movements of these atoms occasionally occur. For example, adenine and cytosine are usually unable to pair, but a shift in the hydrogen atom at the 6-amino position in adenine to the N-1 position would permit hydrogen bonding between these bases (Figure 4-14). If this rare pairing occurred during replication of the DNA, one of the polynucleotide strands would have cytosine instead of thymine at that point in its base sequence. The normal pairing of this cytosine with guanine in the next round of replication would result in replacement of the original AT base pair by a GC pair. The effect would be that the information in the molecule would be altered by a change in one letter of the genetic message Such reversible changes in location of the hydrogen atoms in the bases are called **tautomeric shifts.** They are generally quite rare, consistent with the stability of genetic information. The molecular basis and value of mutation will be discussed in more detail in Chapter 10.

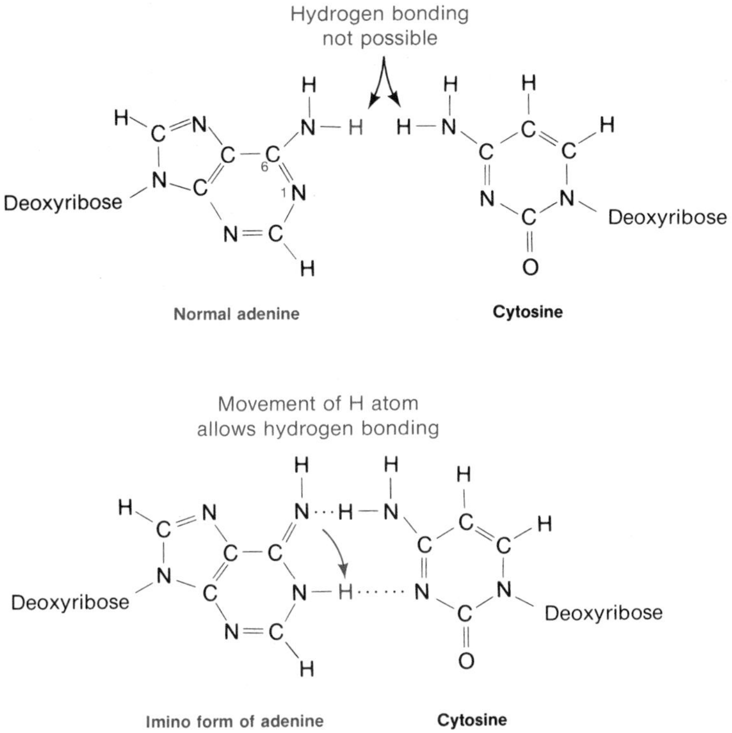

Figure 4-14 Formation of an adenine-cytosine pair when a hydrogen atom in adenine moves from the 6-amino position to the 1-nitrogen position.

4.6 The Replication of DNA

The process of replication, in which each strand of the double helix serves as a template for synthesis of a new strand (Figure 4-13), is simple in principle. It requires only that the hydrogen bonds joining the bases break to allow separation of the chains, and that appropriate free nucleotides of the four types be available to pair with the newly accessible bases in each strand. Although the hydrogen bonds that hold the strands of a parental double helix together are highly specific, they are weak and easily broken by a variety of agents. Hydrogen bonds will also form spontaneously under the proper conditions. However, within cells the conditions for spontaneous breaking and re-forming of these bonds are not met and a barrage of enzymes and other proteins is needed to assist in these processes. When adjacent free nucleotides are correctly paired with their complementary partners in the chain being copied (the **template chain**), they will be properly positioned for the enzyme-catalyzed joining of their sugar and phosphate components in the formation of a new chain along the old one (Figure 4-15). This process of replication has the property of being **semiconservative,** meaning that one chain of the original molecule is conserved in each of the two newly formed double helices.

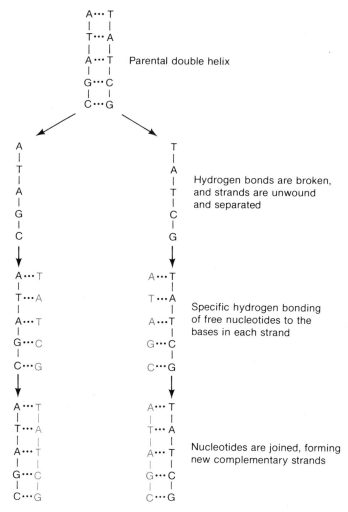

Figure 4-15 Schematic drawing of DNA replication showing how bases are selected and joined in forming a replica. The template strands are black; the new strands are red.

Evidence Supporting Semiconservative Replication

Semiconservative replication was demonstrated in 1958 by Matthew Meselson and Franklin Stahl by an experiment in which the heavy ^{15}N isotope of nitrogen was used for physical separation of parental and daughter DNA molecules. DNA isolated from the bacterium *E. coli* grown in a medium containing ^{15}N as the only available source of nitrogen is denser than DNA from bacteria grown under normal conditions with ^{14}N. ^{15}N- and ^{14}N-containing DNA can be separated by means of **equilibrium centrifugation in a density gradient.** DNA molecules have about the same density as very concentrated cesium chloride solutions; for example, 5.6 *M* CsCl has a density of 1.700 g/cm^3, and *E. coli* DNA containing ^{14}N in the purine and pyrimidine rings has a density of 1.708 g/cm³. Total replacement of the ¹⁴N with ¹⁵N increases the density of *E. coli* DNA to 1.722 g/cm³.

When 5.6 *M* CsCl is centrifuged at a speed exerting a force of about 100,000 times gravity on the solution, the Cs⁺ ions gradually sediment toward the bottom of the centrifuge tube. This movement is counteracted by diffusion (the random movement of molecules), which prevents complete sedimentation. After about 48 hours the processes of sedimentation and diffusion are in equilibrium, no net movement of ions occurs, and a linear gradient of increasing CsCl concentration—and, thus, of density—is present from the top to the bottom of the centrifuge tube. The density of the solution is about 1.65 g/cm³ at the top and 1.75 g/cm³ at the bottom of the tube. DNA added to the CsCl solution before centrifugation will also move—upward or downward—to a position in the gradient at which the density of the solution is precisely equal to its own density. At equilibrium, a mixture of ¹⁴N-containing ("light") and ¹⁵N-containing ("heavy") *E. coli* DNA will separate into two distinct zones in such a density gradient, with the denser species of DNA nearer the bottom of the centrifuge cell (Figure 4-16).

In the Meselson-Stahl experiment summarized in Figure 4-16, bacteria were grown for many generations in a ¹⁵N-containing medium. The cells were then transferred to a ¹⁴N-containing medium, and DNA was isolated from

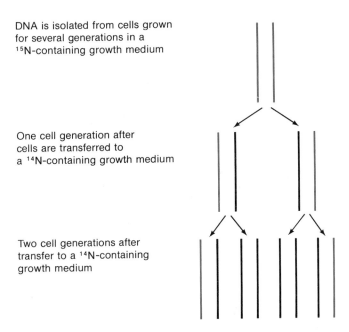

DNA is isolated from cells grown for several generations in a ¹⁵N-containing growth medium

One cell generation after cells are transferred to a ¹⁴N-containing growth medium

Two cell generations after transfer to a ¹⁴N-containing growth medium

Figure 4-16 The DNA replication experiment of Meselson and Stahl. DNA extracted from cells is mixed with a CsCl solution having a density approximately equal to that of the DNA molecules. The DNA-CsCl solution is centrifuged. During this time the Cs⁺ and Cl⁻ ions are distributed, and at equilibrium a stable density gradient is formed. The DNA molecules also move during centrifugation and come to rest at positions in the centrifuge tube at which their density equals the density of the CsCl. The experiment begins with cells grown in ¹⁵N-containing medium. The state of DNA molecules after various periods of growth in ¹⁴N-containing medium is shown.

samples of cells taken from the culture at intervals after this density transfer. After one generation of growth (one round of replication of the DNA molecules and a doubling of the number of cells), all of the DNA had a density exactly intermediate between the densities of [^{15}N]DNA and [^{14}N]DNA, indicating that the replicated molecules contained equal amounts of the two nitrogen isotopes. After a second generation of growth in the ^{14}N medium, half of the DNA had the density of DNA with ^{14}N in both strands ("light" DNA) and the other half had the intermediate ("hybrid") density. After a third generation of growth in the ^{14}N medium, three-quarters of the DNA had the density of [^{14}N]DNA and one-quarter retained the intermediate density. This distribution of ^{15}N atoms in successive replications of the molecules in the ^{14}N-containing medium is precisely the result predicted from semiconservative replication of the Watson-Crick structure (Figure 4-13).

In semiconservative replication the DNA molecules of intermediate density should contain one parental, fully heavy (^{15}N) strand and one newly synthesized, fully light (^{14}N) strand. This prediction was confirmed in a second experiment. An aqueous solution of intermediate-density DNA was held at 100°C for 30 minutes, a procedure that breaks the hydrogen bonds between paired bases (but not the covalent bonds between adjacent nucleotides in a chain) and thereby results in separation of the complementary chains. When the solution was cooled, using conditions in which complementary chains could not reassociate to form double helices, and then analyzed by density-gradient centrifugation, the DNA was found to contain two components in equal amounts. One component had a density equal to that of a single chain containing only ^{14}N, and the second component had the density of fully heavy ^{15}N single chains, as expected with semiconservative replication. Subsequent experiments with replicating DNA from numerous viruses, bacteria, and higher organisms have indicated that semiconservative replication is universal.

Replication Visualized

DNA molecules are very long and fragile, subject to both physical and enzymatic breakage during extraction from cells and viruses. Although breakage did not detract from the validity of the conclusion that DNA replication is semiconservative, the *E. coli* DNA molecules used in the experiment in Figure 4-16 had been extensively fragmented during isolation. Entire replicating molecules from a bacterium can be observed by isolating DNA and using the technique of autoradiography to visualize the DNA. **Autoradiography** is a procedure in which photographic images of the distribution of radioactive atoms within cellular structures or large molecules are produced by radiation emitted during decay of radioactive atoms incorporated into the molecules. The tritium (^{3}H) isotope of hydrogen is especially useful in autoradiography because it emits a low-energy electron capable of moving only a very short distance. Radioactive DNA can be made by growing cells in a medium containing [^{3}H]thymidine (in which some hydrogen atoms have been replaced by tritium). Since thymidine is incorporated exclusively into DNA and not other macromolecules, the DNA will be the only labeled large molecule in a cell—that is, [³H]thymidine is a specific label for DNA. In the experiment shown in Figure 4-17 cells that had been grown for various periods of time in a medium containing [³H]thymidine were disrupted, their DNA was collected on very thin filters, and these filters carrying now-radioactive DNA were placed on glass slides. The filters were then coated with a thin photographic emulsion sensitive to the electrons emitted by ³H nuclei and stored in the dark for two months to allow the decay of numerous ³H atoms. In the autoradiograms obtained after developing the emulsion, the geometry of the tritium-labeled DNA was traced by dark silver grains (Figure 4-17). These autoradiograms showed that intact *E. coli* DNA molecules are *circular* structures, each apparently containing a single DNA molecule, in which the two nucleotide chains separate during the replication process. DNA molecules at intermediate stages of replication were seen to have the general form of the Greek letter theta (θ). Electron microscopy has also shown that a multitude of phage, viral, and bacterial DNA molecules and those of chloroplasts replicate as θ molecules.

The circularity of the replicating molecules in Figure 4-17 brings out an important geometric feature of semiconservative replication. There are about 400,000 turns in an *E. coli* double helix, and since the two chains of a replicating molecule must make a full rotation to unwind each of these gyres, some kind of swivel must exist to avoid tangling the entire structure (Figure 4-18). Current evidence suggests that the axis of rotation for unwinding is

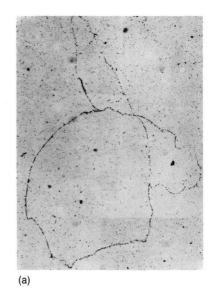

(a)

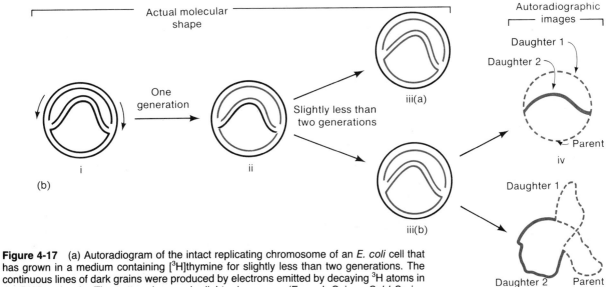

Figure 4-17 (a) Autoradiogram of the intact replicating chromosome of an *E. coli* cell that has grown in a medium containing [³H]thymine for slightly less than two generations. The continuous lines of dark grains were produced by electrons emitted by decaying ³H atoms in the DNA molecule. The pattern is seen by light microscopy. (From J. Cairns, *Cold Spring Harbor Symp. Quant. Biol.* 1963. 28: 44). (b) Interpretation of the autoradiogram of part (a). (i) The original nonradioactive molecule. (iv) How (iii(b)) would appear in an autoradiogram. The heavy line has twice the density of grains as the dashed line. (v) How (iv) would appear when oriented like the molecule in part (a).

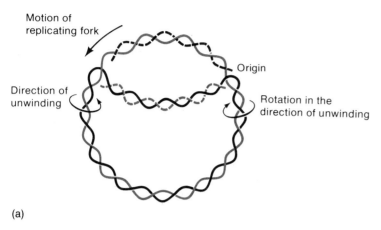

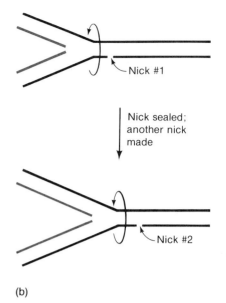

(a)

(b)

Figure 4-18 Replication of a circular DNA molecule. (a) The un-winding motion of the daughter branches of a replicating circle lacking positions at which free rotation can occur causes over-winding of the unreplicated portion. (b) Mechanism by which a single-strand break (a nick) ahead of a replication fork allows rotation.

provided by cuts made in the backbone of *one* strand of the double helix during replication; these cuts are rapidly repaired after unwinding. Enzymes capable of making such cuts and then rapidly repairing them have been isolated from both bacterial and mammalian cells; they are called **topo-isomerases** or **gyrases.** The particular enzyme in *E. coli* having this function is called **DNA gyrase.**

Origin and Direction of Movement of Replication Forks

The position along a molecule at which DNA replication begins is called a **replication origin,** and the region in which parental strands are separating and new strands are being synthesized is called a **replication fork.** The process of generating a new replication fork is **initiation.** At one extreme, initiation may occur at random positions; alternatively, the origin may be a unique site. Analysis of phage DNA replication—described in this section—has shown that for all known species of double-stranded DNA phages, a unique origin is used. Once initiated, replication may occur in a single direction, or two bidirectionally moving replication forks may be generated. Replication is almost always bidirectional, a few phage species being the only exceptions.

Determination of the position of the origin and the direction of move-ment of replication forks can be illustrated most simply with two viruses that infect *E. coli,* using a method of treating DNA that provides an easy way to distinguish the two ends of linear molecules or to define positions within circular molecules accurately. Most DNA molecules have local regions in which either AT or GC base pairs predominate. This observation is the basis for the electron microscopic technique of **partial denaturation mapping.** Exposure of DNA molecules to high temperature or high alkalinity causes hydrogen bonds to break and the single strands to separate. At a temperature or pH at which unwinding just begins, preferential unwinding of regions rich in AT pairs (which have two hydrogen bonds) occurs. Those regions in which GC pairs predominate remain intact, because these pairs are more strongly hydrogen-bonded (they are joined by three hydrogen bonds). Such a molecule is said to be partially denatured. If the temperature or pH is lowered, the denatured regions of a partially denatured molecule will again form the double-stranded structure. However, if the molecules are treated with formaldehyde, the single-stranded regions will be stabilized because the chemical combines with the amino ($-NH_2$) groups in the adenine bases and prevents reestablishment of the AT pairs. When molecules treated in this way are examined by electron microscopy, the unwound regions appear as single-stranded "bubbles" of different sizes occurring at specific positions in a particular kind of DNA molecule (Figure 4-19); these bubbles provide a series of physical markers for each molecule.

The first DNA molecule in which replication was shown to occur in both directions from a specific origin is that of the lambda (λ) phage of *E. coli*. The DNA of this phage is linear when in a phage particle but inside a host cell it forms a circular molecule. The molecular mechanism by which a linear molecule is circularized is shown in Figure 4-20. The linear form possesses single-stranded extensions 12 nucleotides long at the 5′-P end of each chain. The base sequences in the two single-strand terminal regions are complementary and can pair with each other, so the molecules are said to have **cohesive ends.** Base pairing between the terminal complementary nucleo-

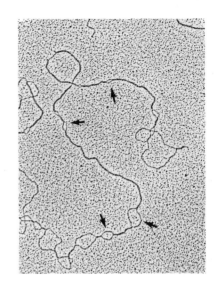

Figure 4-19 An electron micrograph of a portion of a partially denatured *E. coli* phage λ DNA molecule. The arrows indicate several unwound "bubbles." (Courtesy of Ross Inman.)

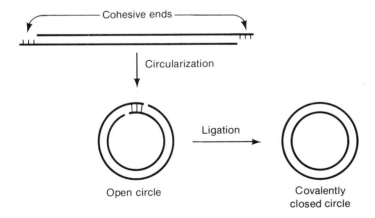

Figure 4-20 A diagram of a linear λ DNA molecule showing the cohesive ends (complementary single-stranded ends), circularization by means of base-pairing between the cohesive ends forming an open or nicked circle, and formation of a covalently closed (uninterrupted) circle by sealing of the single-strand breaks (ligation).

tides results in a hydrogen-bonded circular structure with staggered single-strand breaks. In the host cell these structures are converted to **covalently closed circles** by the enzyme **DNA ligase,** which seals single-strand breaks in DNA double helices (a function that is also essential in recombination and replication).

The λ DNA molecule is replicated in its circular form, resulting in the same kind of θ-shaped intermediates found in the replication of *E. coli* DNA. Exposure to pH 11.05 at room temperature produces a characteristic pattern of landmark denaturation bubbles. Examination of many replicating molecules, in which the positions of the replication branch points in the molecules could be related to the positions of the landmark bubbles, shows that neither branch point has a fixed position, so both branch points are replication forks that move in opposite directions around the circular molecule (Figure 4-21). Furthermore, the origin is unique because for all sizes of the replication loop (the segment of a replicating molecule consisting of two daughter branches),

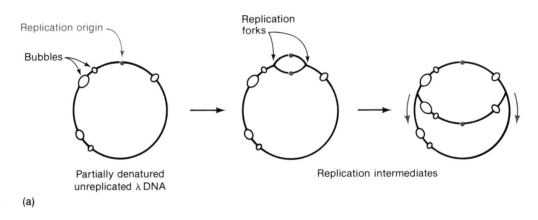

(a)

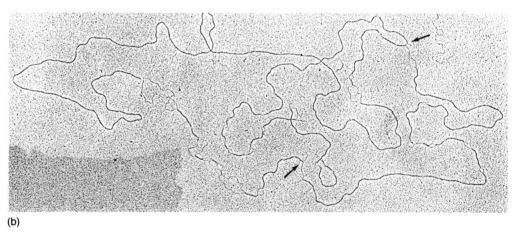

(b)

Figure 4-21 Partially denatured λ DNA molecule, replicated to varying degrees. (a) Heavy and light lines represent double- and single-stranded DNA, respectively. The dot is the replication origin. The red arrows indicate the direction of movement of the replication forks. (b) An electron micrograph of a partially denatured replicating intermediate. The arrows indicate the growing forks. (Courtesy of Manuel Valenzuela.)

the position of the midpoint between the forks is fixed with respect to the landmark denaturation bubbles; thus, the two replication forks must move at the same rate.

Denaturation mapping has been used to show that replication of the linear DNA molecule of phage T7, another *E. coli* phage, also occurs bidirectionally from a fixed origin. Phage T7 contains a double-stranded DNA molecule, in which replication always begins at a point 17 percent from the end conventionally identified as the left end. DNA synthesis in both directions from this origin generates a linear molecule with a single replication loop (Figure 4-22(a)). This structure becomes a Y-shaped intermediate when the replication fork meets the nearest end of the molecule (Figure 4-22(b)). Two progeny DNA molecules are produced when the remaining replication fork reaches the other end of the molecule.

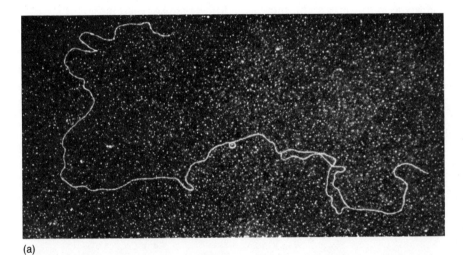

(a)

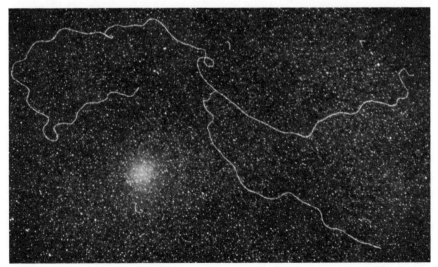

(b)

Figure 4-22 Two replicating DNA molecules of *E. coli* phage T7. (a) Early stage with single replication loop. (b) Later stage showing Y-shaped intermediate. (Courtesy of David Dressler.)

With some viruses and in mammals, *Drosophila*, and all other eukaryotes that have been examined, replication is initiated at many sites in the DNA. The structures resulting from the numerous initiations are seen as multiple loops along a DNA fiber in electron micrographs (Figure 4-23). Multiple initiation is a means of reducing the total replication time of a large molecule.

Replication of an *E. coli* DNA molecule, which contains about 4×10^6 nucleotide pairs, requires 40 minutes under favorable conditions. Thus, each of the two replication forks generated at the single origin moves at a rate of approximately 50,000 nucleotide pairs per minute; similar rates are found in the replication of all phage DNA molecules studied. Movement of a replication fork in eukaryotic DNA molecules is much slower—in *Drosophila melanogaster* about 2600 nucleotide pairs per minute at 24°C. The DNA molecule in the largest chromosome of this organism contains about 7×10^7

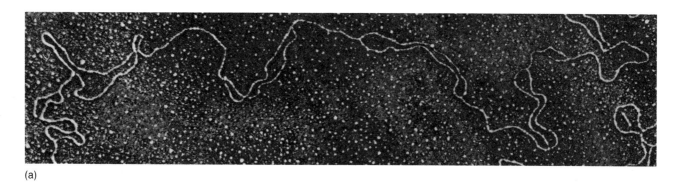

(a)

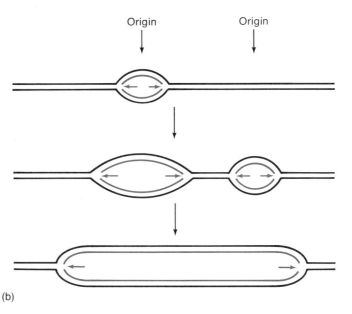

(b)

Figure 4-23 Replicating DNA of *Drosophila melanogaster.* (a) An electron micrograph of a 30,000-nucleotide-pair segment showing seven replication loops. (Courtesy of David Hogness.) (b) An interpretive drawing showing how loops merge. Two replication origins are shown. The red arrows indicate the direction of movement of the replication forks.

nucleotide pairs. In each of the 12 early cleavage divisions in *Drosophila* embryos, which undergo one of the most rapid mitotic cycles known, replication is completed in 2–3 minutes and the nuclei divide synchronously every 10 minutes. From the spacing of the replication loops in DNA molecules isolated from these nuclei, it is apparent that the sites at which replication is initiated have an average spacing of about 8000 nucleotide pairs; thus, more than 8500 replication origins must be used in the largest molecules.

The number of replication-initiation sites that are actively used is not the same in all cell types and also varies with the developmental stage of an organism. In the DNA of early cleavage nuclei of *Drosophila*, about five times as many replication origins are active as in dividing nuclei at later stages of development. The average spacing between adjacent origins in dividing adult cells grown in tissue culture is about 40,000 nucleotide pairs; thus, in these cells only a subset of the potential origins is used. The factors that determine how many or which potential origins are active in different conditions are not known.

The replication origin of a number of viral and other small DNA molecules, in which the entire molecule replicates from a single site, consists of specific nucleotide sequences that are often similar in different species. Some of these molecules also have secondary replication origins that are normally inactive. For example, if the usual replication origin of the phage T7 DNA is deleted from the molecule, a specific secondary site becomes the active origin. The composition of the numerous origins in eukaryotes has not been determined for any eukaryotic species; presumably, these sites consist of nearly identical nucleotide sequences.

Rolling Circle Replication

Some circular DNA molecules, including those of a number of bacterial and eukaryotic viruses, replicate by a process different from that in which θ-shaped intermediates occur. This replication mode is called **rolling circle replication.** In this process replication starts with a cut made by a nuclease (an enzyme that cuts sugar-phosphate bonds) at a specific site in a double-stranded circle (Figure 4-24). Two chemically distinct ends are generated—a

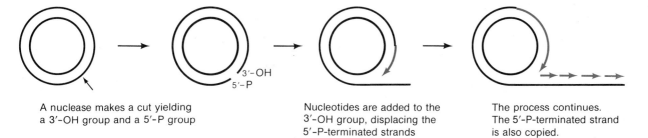

A nuclease makes a cut yielding a 3'-OH group and a 5'-P group

Nucleotides are added to the 3'-OH group, displacing the 5'-P-terminated strands

The process continues. The 5'-P-terminated strand is also copied.

Figure 4-24 Rolling circle replication. Newly synthesized DNA is red. The displaced strand is replicated in short fragments, as explained in Section 4.7.

3′ end, at which the nucleotide has a free 3′-OH group on its sugar component, and a 5′ end, at which the nucleotide has a free 5′-P group. The synthesis of new DNA occurs by addition of deoxynucleotides to the 3′ end with concomitant displacement of the 5′ end from the circle. As replication proceeds around the circle, the 5′ end is rolled out as a tail of increasing length.

In most cases, as extension occurs, a complementary chain is synthesized, which results in a double-stranded DNA tail. Since the displaced strand is covalently linked to the newly synthesized DNA in the circle, termination does not occur, and extension proceeds without interruption, forming a tail that may be many times longer than the circumference of the circle. Rolling circle replication is a common feature in late stages of replication of double-stranded DNA phages having circular intermediates, such as *E. coli* phage λ. An example of rolling circle replication will also be seen in Chapter 7, where matings between male and female *E. coli* are described.

Rolling-circle intermediates, though without a copy of the displaced strand, are present in the replication cycle of the circular single-stranded DNA molecules of several groups of small viruses, of which *E. coli* bacteriophage φX174 is representative. Replication of these viral chromosomes in infected host cells occurs in two distinct stages (Figure 4-25):

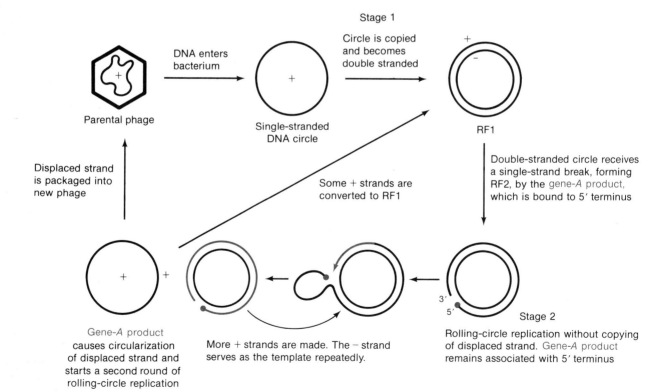

Figure 4-25 The DNA replication cycle of *E. coli* phage φX174.

1. The infecting viral strand (called the "+" strand) is converted to a circular double helix—called the **replicative form** or **RF**—by synthesis of a "−" strand complementary to the infecting strand.

2. A modified rolling-circle replication (sometimes called looped rolling circle replication), in which the displaced strand (a + strand) is not converted to double-stranded DNA, then ensues. This stage uses the product of ϕX174 gene A, which makes a single-strand break and binds to the 5′-P terminus produced. When one round of rolling circle replication is complete, the gene-A product cuts off a DNA unit, forming a progeny +-strand circle, and a second round of rolling circle replication begins. Some of these +-strand circles are converted to additional copies of the double-stranded RF by synthesis of a − strand, whereas others are incorporated into progeny viruses. The choice is determined by whether viral structural proteins (produced in the infected cells) succeed in coating the progeny circle before synthesis of the − strand is initiated.

So far, we have discussed only certain geometrical features of DNA replication. In the next section, the enzymes and protein factors used in DNA replication are described.

4.7 DNA Synthesis

Nucleic acids are synthesized in chemical reactions controlled by enzymes, as is the case with most metabolic reactions in living cells. Enzymes are catalytic proteins (Chapter 1)—that is, they have the ability to speed up reactions. In general, the molecules that participate in chemical reactions require a certain amount of starting energy (activation energy), and also must become properly oriented with respect to each other. Enzymes function by recognizing and binding to one or more of the molecules participating in the reaction; the bound molecules are called **substrate** molecules and the region of the enzyme at which binding occurs is called the **active site** of the enzyme. The binding results in proper positioning of the atoms that participate in the making or breaking of bonds. Furthermore, binding may alter the substrate molecule somewhat, and in so doing reduce the energy required for the reaction. One mechanism by which an enzyme positions the reactants is shown in Figure 4-26.

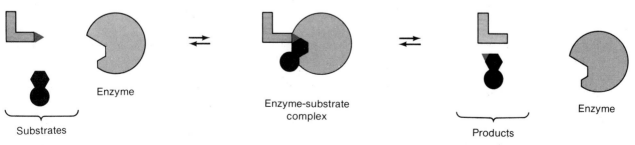

Enzyme

Substrates

Enzyme-substrate complex

Products

Enzyme

Figure 4-26 A schematic diagram showing how an enzyme can aid in the transfer of a chemical group (red triangle) from one molecule (pink) to another molecule (black).

Many enzymes are highly specific and act on only a particular substrate—for example, an enzyme that can cut lactose (milk sugar) but not sucrose (cane sugar). Other enzymes can act on different molecules having some common structural features—for example, all alcohols. Several of the enzymes required for the synthesis of DNA or RNA are of this latter type.

The enzymes that form the sugar-phosphate bond (the phosphodiester bond) between adjacent nucleotides in a nucleic acid chain are called **DNA polymerases.** A variety of DNA polymerases have been purified and DNA synthesis has been carried out in a cell-free system prepared by disrupting cells and combining purified components in a test tube under precisely defined conditions. (These are called *in vitro* experiments. The term *in vitro*—meaning "in glass"—pertains to experiments done with such reconstituted systems; they have been used in the study of various metabolic processes.) The best-understood polymerases are those isolated from *E. coli*.

Three principal requirements must be fulfilled before DNA polymerases can catalyze synthesis of DNA:

1. *The 5′-triphosphates of the four deoxyribonucleosides must be present.* These are deoxyadenosine triphosphate (dATP), deoxyguanosine triphosphate (dGTP), deoxycytidine triphosphate (dCTP), and thymidine triphosphate (dTTP) (Figure 4-27). Synthesis does not occur with the 3′-triphosphates or 5′-diphosphates or if one of the 5′-triphosphates is omitted from the mixture.

Figure 4-27 Two deoxynucleoside triphosphates used in DNA synthesis. The red portion of each molecule is removed during synthesis.

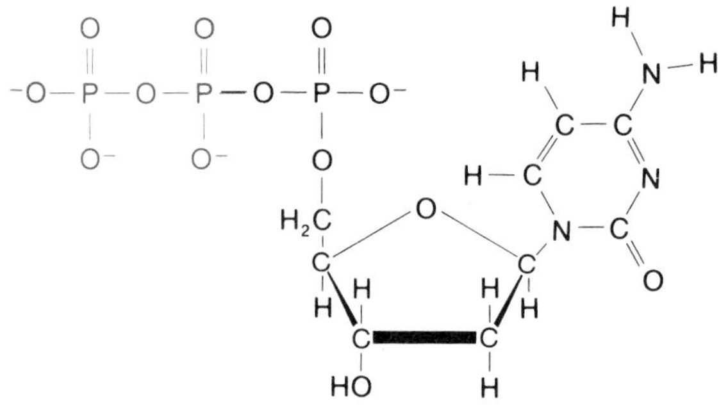

Deoxycytidine 5′-triphosphate (dCTP)

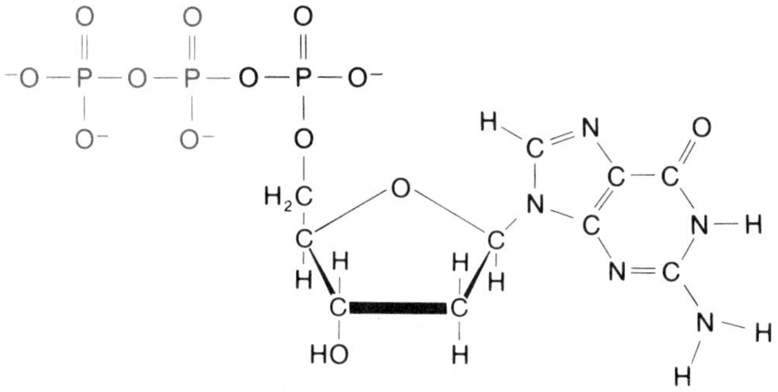

Deoxyguanosine 5′-triphosphate (dGTP)

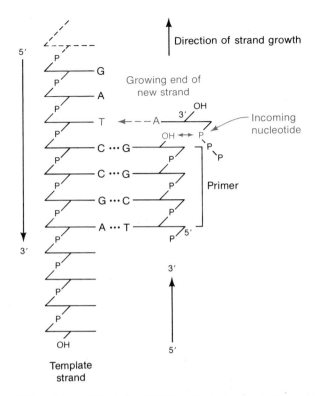

Figure 4-28 Addition of nucleotides to the 3′-OH terminus of a primer. The recognition step is shown as the formation of hydrogen bonds between the red A and the red T. The chemical reaction is between the red 3′-OH group and the red phosphate of the triphosphate.

2. *A preexisting single strand of DNA to be copied must be present.* Such a strand is called a **template.**

3. *A nucleic acid segment, which may be very short and either of the deoxyribo or ribo type, hydrogen-bonded to a parental DNA strand must be present.* Such a segment is called a **primer** (Figure 4-28). The need for a primer arises because *none of the known DNA polymerases is able to initiate chains.* Thus, to have a primer chain with a free 3′-OH group is absolutely necessary for initiation of replication. The reaction catalyzed by the DNA polymerases is the formation of a phosphodiester bond between the free 3′-OH group of the primer and the innermost phosphorus atom of the nucleoside triphosphate being incorporated at the new primer terminus (Figure 4-29). Thus, DNA synthesis occurs by the elongation of primer chains, *always in the 5′ → 3′ direction.* Recognition of the appropriate incoming nucleoside triphosphate during growth of the primer chain depends on base pairing with the opposite nucleotide in the template chain. A DNA polymerase usually will catalyze the polymerization reaction incorporating the new nucleotide at the primer terminus only when the correct base pair is present within the active site; in this reaction the two terminal phosphate groups of the nucleoside triphosphate are released as a molecule of inorganic pyrophosphate (PP_i). The polymerization reaction is

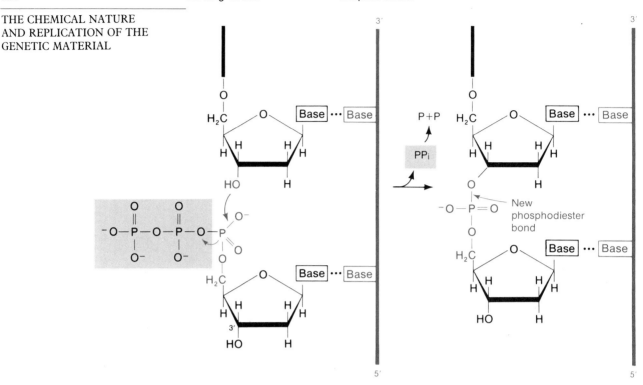

Figure 4-29 The chemistry of addition of a nucleotide to the growing end of a DNA strand. The DNA polymerase is not shown but extends over about 15 nucleotides of the template strand, though the region active in nucleotide addition is much smaller.

reversible (that is, the polymerase can also act like a nuclease and cleave a DNA strand). Reversal of the reaction does not occur because another enzyme cleaves the pyrophosphate, thereby eliminating one reactant needed for the reverse reaction. The same DNA polymerase is used to add each of the four deoxynucleotide phosphates to the 3′-OH terminus of the growing strand.

DNA (and RNA) polymerases are unusual enzymes in that they can act on a variety of templates. For example, if a test-tube mixture contains template molecules from any organism, an *E. coli* DNA polymerase will catalyze the synthesis of that DNA molecule whether it is plant or animal DNA. Similarly, a human DNA polymerase will synthesize bacterial DNA if a bacterial DNA template is provided. The conclusion that the reactions catalyzed by the enzymes under these conditions actually represent DNA replication rather than random polymerization of nucleotides is based on the absolute requirement that DNA templates and all four nucleoside 5′-triphosphates be present, and on chemical analyses demonstrating that the template used determines the composition and sequence of the polynucleotide synthesized. In fact, except for rare errors, the polynucleotide strands synthesized by a DNA polymerase from any organism always has a base sequence complementary to that of the template strand.

The first DNA polymerase purified was that of *E. coli* and is known as **DNA polymerase I,** often written PolI. Two other polymerases (II and III) from *E. coli* have also been purified. **Polymerase III** is the major replication enzyme. Polymerase I plays a secondary role in replication and is required for repairing certain types of DNA damage (for example, that caused by ultraviolet light). The function of polymerase II, which probably does not play a role in either replication or repair, has not been determined. Eukaryotes also possess three DNA polymerases (α, β, and γ). The eukaryotic DNA polymerase responsible for replication of chromosomal DNA is polymerase α.

In addition to their ability to polymerize nucleotides, most DNA polymerases possess nuclease activities that break phosphodiester bonds in the sugar-phosphate backbones of nucleic acid chains. Many other enzymes have nuclease activity, and these are of two types: (1) **exonucleases,** which can only remove a nucleotide from the end of a chain, and (2) **endonucleases,** which break bonds within the chains. DNA polymerases I and III of *E. coli* both have exonuclease activity, which acts only at the 3′ terminus (a 3′ → 5′ exonuclease activity) and provides a built-in mechanism for correcting errors in the polymerization process. This so-called **proofreading** or **editing function** ensures a high degree of fidelity in DNA replication; proofreading occurs as an excision of a nucleotide from the 3′-OH end of the growing chain if it is not correctly base-paired with the corresponding nucleotide in the template chain. A 5′ → 3′ exonuclease activity, required for the polymerization process in a way we will describe below, is also an integral property of DNA polymerase I but not polymerase III. Note the versatility of polymerase I: two or more enzymatic activities necessary in the polymerization process are catalyzed by different parts of this one large protein molecule. All DNA polymerases do not possess such complexity; those of higher eukaryotes have no intrinsic nuclease activities and in these organisms nuclease activity is catalyzed by separate enzymes.

Two properties of all known DNA polymerases—(1) synthesis only in the 5′ → 3′ direction and (2) the inability to initiate new chains—create problems in replication. These problems and their solutions are described in the following section.

Synthesis of DNA in Short Pieces

Recall that the two chains of a double helix have opposite chemical polarities, and that study of the replicating molecules by autoradiography seems to show that the new complementary chains being synthesized at a replication fork grow in the same overall direction. How could these new chains elongate in the same direction if polymerase can only catalyze chain growth in the 5′ → 3′ direction (Figure 4-28)? The solution is that replication does not proceed simultaneously at corresponding locations along the antiparallel chains of a double helix.

THE CHEMICAL NATURE
AND REPLICATION OF THE
GENETIC MATERIAL

Figure 4-30 (a) A replicating θ molecule of phage λ DNA. The arrows show the two replication forks. The segment between each pair of thick lines at the arrows is single-stranded; note that it appears thinner and lighter. (b) An interpretive drawing. (Courtesy of Manuel Valenzuela.)

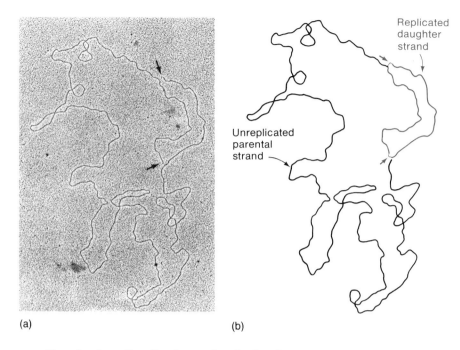

(a)

(b)

Examination of replicating molecules by electron microscopy shows that single-stranded regions are present in one arm of each Y-shaped replication fork (Figure 4-30). This and other observations indicate that replication of one chain, called the **leading strand,** occurs continuously, whereas replication of the other, or **lagging strand,** occurs discontinuously (Figure 4-31). Extension of the lagging strand in the overall $3' \rightarrow 5'$ direction is accomplished by means of $5' \rightarrow 3'$ (normal) synthesis of short segments that are rapidly joined to the growing chain by DNA ligase. The short segments are often called **Okazaki fragments** after their discoverer, or, descriptively, **precursor fragments.** They are 1000–2000 nucleotides long in replicating bacterial or viral DNA and only about one-tenth as long in eukaryotes. The principal evidence for the

Figure 4-31 Short fragments in the replication fork. For each tract of base pairs the lagging strand is synthesized later than the leading strand.

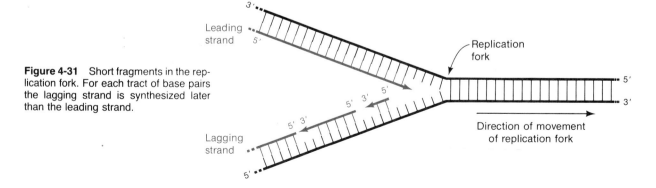

existence of short fragments as intermediates in DNA synthesis was obtained from experiments of the kind summarized in Figure 4-32. Growing bacterial cells were placed for short periods of time (a few seconds) in a medium containing [³H]thymidine. A sample was taken, and the DNA was extracted by treatment of the cells with alkali, which separates the strands of the double-stranded molecules; most of the radioactivity was then found to be present in short strands less than 2000 nucleotides long. Another sample of the cells exposed to such a short pulse of [³H]thymidine was taken and subsequently washed, and then allowed to grow for several minutes in a medium that contained nonradioactive thymidine. The DNA was extracted with alkali; in this case, most of the radioactivity was found in long (normal length) chains, indicating that the short chains had been joined together. Proof that joining is accomplished by DNA ligases came from experiments with cells containing a mutant DNA ligase that was inactive at temperatures greater than 40°C. When the experiment just described was repeated at 40°C, the short fragments persisted, because no enzyme was present to join them.

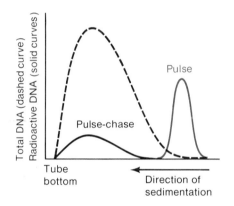

(a)

Figure 4-32 The experiment that demonstrated short fragments in the replication fork. (a) A comparison of the structure of DNA synthesized just before it was isolated with DNA that had replicated for a slightly longer time. The most recently made DNA is that which was "pulse-labeled" with [³H]thymine for a very short period of time and isolated before further replication occurred (red curve). Continued replication in nonradioactive medium leads to an increase in size ("pulse-chase," black curve). The DNA has been sedimented in alkali, which causes strand separation to occur, so that the location of the radioactivity in single polynucleotides could be determined. Total DNA is the sum of the nonradioactive and radioactive DNA. Larger molecules sediment further than smaller molecules. (b) An interpretation of the data showing the location of radioactive DNA (red) at the time of pulse-labeling and after incubation in nonradioactive medium. The radioactive molecules produced by treatment with alkali are shown.

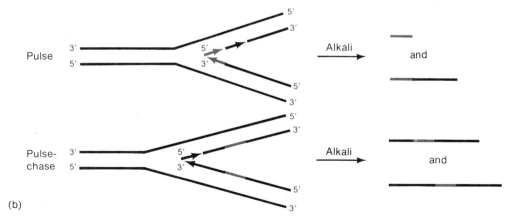

(b)

We mentioned earlier that no known DNA polymerase can initiate synthesis of a new strand, so a free 3'-OH is needed. In almost all organisms examined, initiation is accomplished by an RNA-polymerizing enzyme. RNA is usually a single-stranded nucleic acid having the same basic structure as DNA: it consists of four types of nucleotides joined together by $3' \rightarrow 5'$ phosphodiester bonds. Two chemical differences distinguish RNA from DNA (Figure 4-33). The first difference is a small one in the sugar component. RNA contains **ribose,** which is identical to the deoxyribose of DNA except for the presence of an —OH group on the 2' carbon atom. The second difference is in one of the four bases: the thymine found in DNA is replaced in RNA by the closely related pyrimidine **uracil (U).** RNA is synthesized by copying the base sequence of a DNA template strand and forming a complementary strand in which the complements are guanine and cytosine (as in DNA) and adenine and uracil. Synthesis is catalyzed by enzymes called **RNA polymerases.** These enzymes differ from DNA polymerases in that they can initiate synthesis of RNA chains—they do not need a primer.

Normally, when RNA is synthesized, it dissociates immediately from the DNA template. However, in the initiation of DNA synthesis a short piece of RNA is made that remains bound to the DNA template after the RNA polymerase dissociates from the DNA. This segment of RNA provides a primer onto which a DNA polymerase can add deoxyribonucleotides (Figure 4-34). The first evidence for a requirement for synthesis of RNA in DNA

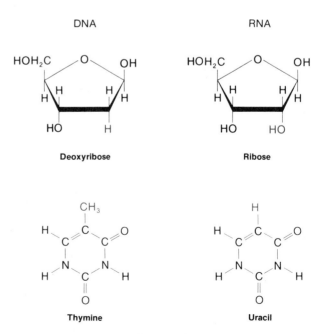

Figure 4-33 Differences between DNA and RNA. The red groups are the distinguishing features of deoxyribose and ribose and of thymine and uracil.

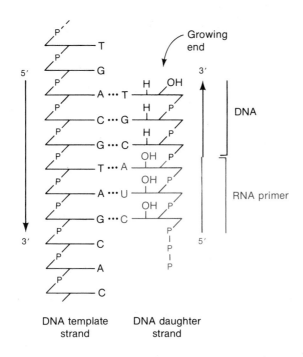

Figure 4-34 Priming of DNA synthesis with an RNA segment. The DNA template strand and newly synthesized DNA are shown in black. The RNA segment, which has been made by an RNA-polymerizing enzyme, is shown in red.

chain initation came from the study of conditions required for replication of the circular single-stranded DNA of the small *E. coli* phage M13. Replication of M13 DNA in an *in vitro* system requires all four of the ribonucleoside 5'-triphosphates (adenosine triphosphate, guanosine triphosphate, cytidine triphosphate, and uridine triphosphate), and is also strongly inhibited by the antibiotics rifampicin and actinomycin, which block the activity of many RNA polymerases. Subsequent direct evidence for the RNA priming of DNA chains has been provided by the detection of short ribonucleotide sequences covalently joined to the 5'-ends of DNA fragments synthesized during replication of the chromosomes of several viruses, *E. coli* and the bacterium *Bacillus subtilis*, and animal lymphocytes. In the synthesis of the lagging strand, precursor fragments are also initiated by synthesis of RNA; a special RNA polymerase called **primase** is used.

Fully replicated DNA molecules do not contain any DNA-RNA hybrid regions, as the RNA primers do not persist. The short RNA primers are recognized and excised as part of the replication process. Excision of RNA primers of the precursor fragments in *E. coli* is best understood (Figure 4-35). These ribonucleotide sequences are removed by the 5' → 3' exonucleolytic activity of DNA polymerase I. The polymerase then replaces the excised ribonucleotides with deoxyribonucleotides, and DNA ligase covalently joins the fragments, thus completing the replication. With some phages and in eukaryotic systems distinct RNA-removing enzymes are used. A common mechanism in eukaryotes is the use of ribonuclease H (RNase H), an enzyme

THE CHEMICAL NATURE
AND REPLICATION OF THE
GENETIC MATERIAL

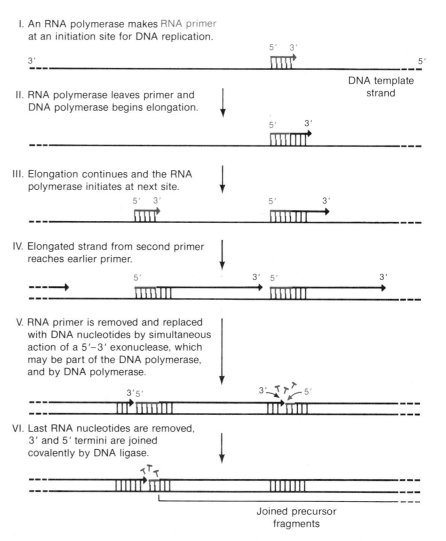

Figure 4-35 Removal of RNA nucleotides from the lagging strand in *E. coli*. In bacteria the polymerizing and 5'-to-3' exonuclease activities are possessed by the same enzyme. With some phages and in eukaryotes different enzymes are used. In circular DNA molecules the RNA primer for the leading strand is removed when replication of the leading strand is completed and the growing strand reaches the primer.

I. An RNA polymerase makes RNA primer at an initiation site for DNA replication.

DNA template strand

II. RNA polymerase leaves primer and DNA polymerase begins elongation.

III. Elongation continues and the RNA polymerase initiates at next site.

IV. Elongated strand from second primer reaches earlier primer.

V. RNA primer is removed and replaced with DNA nucleotides by simultaneous action of a 5'–3' exonuclease, which may be part of the DNA polymerase, and by DNA polymerase.

VI. Last RNA nucleotides are removed, 3' and 5' termini are joined covalently by DNA ligase.

Joined precursor fragments

that recognizes and cuts out ribonucleotides that are contained in DNA molecules. Free single-stranded RNA molecules are unaffected by RNase H.

The antiparallel orientation of the complementary strands of DNA and the requirement of DNA polymerases for a primer leads to a problem in the completion of replication of a linear DNA molecule, such as those found in certain phages and in eukaryotic chromosomes. At the termini of a linear molecule whose daughter molecules are almost completed, the leading strand is able to run to the end of the template and to complete synthesis of one daughter molecule. However, the final segment of the lagging strand, which will be part of the other daughter, cannot be completed unless it is initiated by an RNA primer. Thus, when the daughter molecule is completed, RNA nucleotides will be present at the terminus (Figure 4-36). In a bidirectionally replicating molecule the opposite end of the double-stranded molecule will also be in the same state, so each daughter molecule will have one fully-DNA end and one RNA-containing end. Several mechanisms exist for removing

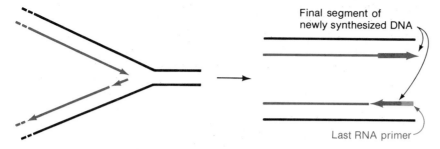

Figure 4-36 Origin of RNA in a completed daughter molecule of a linear DNA molecule. Only the right terminal regions of the molecules are shown.

these terminal RNA nucleotides. However, it is not obvious how they become replaced with DNA nucleotides once they have been removed, because no known DNA polymerase can initiate replication without a primer. The alternatives are clear: (1) the RNA nucleotides are not removed; (2) they are removed and the ends remain single-stranded; (3) they are removed and a DNA polymerase, not yet discovered, fills in the gap; or (4) something totally different occurs. Information is meager at present. For phage DNA that replicates linearly, daughter molecules have blunt double-stranded ends free of RNA, thus eliminating alternatives (1) and (2). For *E. coli* T7 a complex mechanism based on special features of T7 DNA has been proposed; if correct, this mechanism would enable the termini to be completed without additional enzymes. However, little evidence is available to support the proposal, so the idea must be considered speculative. With eukaryotic chromosomes alternatives (1) and (2) are also incorrect, and the proposal for T7 is unlikely. At present, the mechanism remains a mystery.

4.8 Other Components of the Replication System

Mutations in numerous genes block DNA replication, which indicates that the proteins specified by a number of genes are essential to the replication process. The products of more than fifteen genes are essential to the bacterial DNA replication system, and mutations in numerous other genes block such related functions as synthesis of the nucleoside 5′-triphosphates. The use of these mutations to study the complex replication system is complicated by the difficulty that DNA replication is vital to the cell; moreover, it is not possible to provide mutant cells with required replication proteins from the growth medium. The solution has been to obtain mutant cells in which a mutation results in proteins that are inactive at one temperature but active at another. Some mutants of *E. coli*, for example, replicate their DNA at 30°C but not at 42°C, though other metabolic functions of the cells are not impaired at the higher temperatures.

Most of the gene products required for the replication of *E. coli* DNA have been purified, and reconstituted *in vitro* systems have been used to study replication. In addition to the proteins of the *E. coli* DNA replication system which we have already considered—DNA polymerase I, DNA polymerase III, DNA ligase, and the enzyme (primase) required for synthesis of RNA primers in the initiation of short DNA fragments—the functions of several other proteins are known. One class of proteins binds to DNA and is responsible for unwinding the double helix. Proteins with similar functions affecting the physical form of DNA molecules have been isolated from other replication systems, including those in the nuclei of human and higher plant cells. DNA-binding proteins of one type, called **DNA-unwinding proteins** or **helicases,** catalyze the unwinding of duplexes. Molecules of a second type—**single-stranded DNA-binding proteins**—bind tightly to single-stranded DNA but only weakly to double-stranded DNA and maintain the single-stranded regions. A third class—the untwisting enzymes, such as DNA gyrase (see also Section 4.6)—catalyze the breaking and rejoining of one strand of a duplex and relieve the tangling that occurs in the replication of uninterrupted circles. However, the functions of some proteins required for replication are still unknown.

DNA-binding proteins of the types we have considered are an important group in relation to other aspects of DNA function. Some of them have essential roles in recombination between different DNA molecules and in the processes utilized in the expression of genes. In addition, other proteins that bind to and stabilize DNA are structurally necessary in the organization of chromosomes, as we will see in the next chapter.

4.9 Determination of the Sequence of Bases in DNA

A great deal of information about gene structure and gene expression can be obtained by direct determination of the sequence of bases in a DNA molecule. Several techniques are available for base sequencing; one is described in this section. No technique can determine the sequence of bases in an entire chromosome in a single experiment, so chromosomes are first cut into fragments about 100 base pairs long, a size that can be sequenced easily. (Procedures for fragmentation are described in Chapter 12.) To obtain the sequence of an entire chromosome a set of overlapping fragments is prepared, the sequence of each is determined, and all sequences are then combined. The procedures are straightforward, but time-consuming, and the longest sequence determined to date is that of *E. coli* phage λ, which contains 48,502 nucleotide pairs.

Several procedures are available for determining the base sequence of DNA. Only one will be described, the **Maxam-Gilbert sequencing method.** This procedure is based on the ability of certain chemical reagents to induce cleavage of the sugar-phosphate backbone at sites occupied by particular DNA bases. The size of the fragments produced by the cleavage is determined by gel electrophoresis, and the base sequence is then determined by the following rule:

> If a fragment containing n nucleotides is generated by a chemical treatment that causes cleavage at the site of a particular base, then that base is present in position $n + 1$ of the DNA strand, the position being counted from the 5' end.

For example, if a 23-base fragment results from the protocol that identifies guanine, then guanine is the 24th base in the original molecule.

Gel Electrophoresis

DNA molecules are electrically charged and thus can move in an electric field. For example, if the terminals of a battery are connected to the opposite ends of a horizontal tube containing a solution of DNA molecules, the molecules, which are negatively charged, will move from the negative end of the tube to the positive end. The rate of movement depends on the magnitude of the charge and on the friction encountered as the molecules move through the solvent. The mass of the molecule plays no direct role on the rate of movement but determines the rate indirectly by its effect on the surface area of the molecule, which affects the friction.

The most common type of electrophoresis used in genetics is **gel electrophoresis.** This procedure can be performed in such a way that the rate of movement depends only on the molecular weight of the molecule, as long as all molecules are linear, as will be seen below.

An experimental arrangement for gel electrophoresis of DNA is shown in Figure 4-37. A thin slab of a gel called **agarose** (an extract of agar) is prepared containing small slots (called wells) into which samples are placed. An electric field is applied and the negatively charged DNA molecules penetrate and move through the agarose. A gel is a complex network of molecules, and migrating macromolecules must squeeze through narrow tortuous pas-

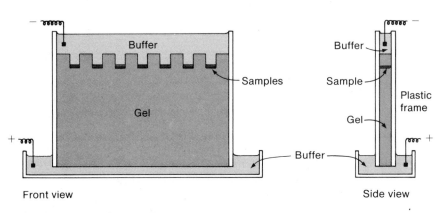

Front view   Side view

Figure 4-37 Apparatus for slab-gel electrophoresis capable of handling seven samples simultaneously. Liquid gel is allowed to harden in place, with an appropriately shaped mold placed on top of the gel during hardening in order to make "wells" for the sample (red). After electrophoresis, the sample is made visible by removing the plastic frame and immersing the gel in a solution containing a reagent that binds to or reacts with the molecules that were electrophoresed, which are located at various positions in the gel. The separated components of the sample appear as bands, which may be either visibly colored or fluorescent when illuminated with fluorescent light, depending on the particular reagent used. The region of a gel in which the components of one sample can move is called a *lane*. Thus, this gel has seven lanes.

sages. The result is that smaller molecules pass through more easily and, hence, the migration rate increases as the molecular weight M decreases. For unknown reasons, the distance moved D depends logarithmically on M, obeying the equation $D = a - b \log M$, in which a and b are empirically determined constants. Figure 4-38 shows the result of electrophoresis of a collection of double-stranded DNA molecules. Each discrete region containing DNA is called a band.

Gel electrophoresis can be carried out with both double-stranded and single-stranded DNA. With short single-stranded fragments, molecules differing in size by a single base can be separated. If a collection of molecules, each containing 100 bases, is fragmented so that, on the average, every molecule receives one strand break, electrophoresis of the fragments so generated will yield 99 bands—one for a single base, one for a fragment with two bases, and so forth. This extraordinary sensitivity to size is made use of in the base-sequencing procedure.

The Sequencing Procedure

The procedure for sequencing a DNA fragment is diagrammed in Figure 4-39. First, a radioactive ^{32}P atom is chemically added to the two 5′ ends of each fragment, and then the complementary single strands are separated and purified. Each complementary strand is sequenced independently and then the sequences are compared for confirmation. A sample containing only one of the complementary strands is divided into two portions. To portion I a reagent is added that reacts with guanine and adenine, but only to the extent that, on the average, one A or one G per single strand is affected and at random positions. This portion is next divided into two parts (Ia and Ib), each of which is subjected to a treatment that, first, removes from the sugar-phosphate chain either primarily G or primarily A and, second, cleaves the chain at the site of the base that has been removed. Because of the random location of the affected base this cleavage generates a set of fragments of varying size, and the differing number of nucleotides in each fragment is a result of the different positions of the affected G or A. Parts Ia and Ib contain primarily G- and A-identifying fragments, respectively.

Portion II is used to identify the positions of cytosine and thymine. This portion is also divided into two parts IIa and IIb, which are subjected to a procedure that again causes cleavage at the site of either base. By this procedure one part contains fragments resulting from reaction with both C and T and the other with C only, so the C-identifying fragments are present in both parts.

All four parts are then electrophoresed and the bands are located by placing autoradiographic film (a film that responds to radioactivity) on the gel. (Note that when a 5′-^{32}P-labeled molecule is cleaved, only one of the two fragments produced contains ^{32}P and only that one is detected.) The positions of A and G in the single strand are determined by the following rules (special cases of the rule stated in the introduction to Section 4.9): *if a band containing*

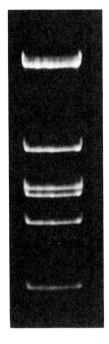

Figure 4-38 Gel electrophoresis of DNA. Molecules of seven different sizes were mixed and placed in a well. Electrophoresis was in the vertical direction. The DNA has been made visible by addition of a dye (ethidium bromide) that binds only to DNA and that fluoresces when the gel is illuminated with short-wavelength ultraviolet light.

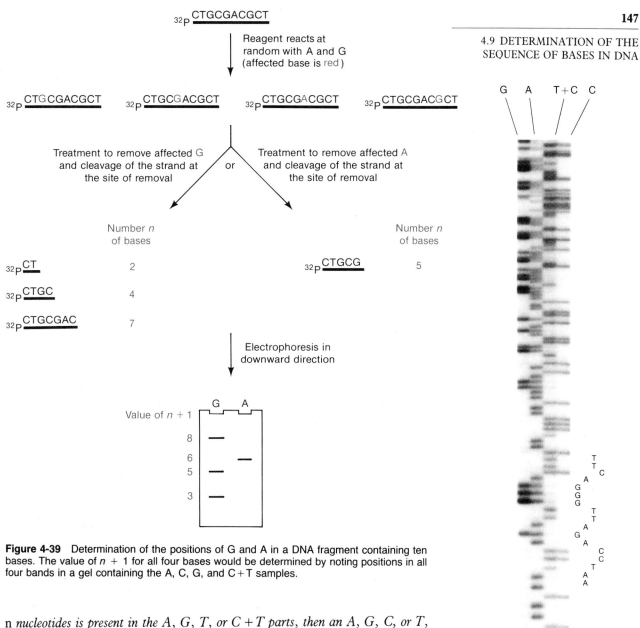

Figure 4-39 Determination of the positions of G and A in a DNA fragment containing ten bases. The value of $n + 1$ for all four bases would be determined by noting positions in all four bands in a gel containing the A, C, G, and C+T samples.

Figure 4-40 A section of a sequencing gel. The sequence is read from the bottom to the top. Each horizontal row represents a single base. Each vertical column is formed by a specific chemical treatment that cleaves the DNA strand at a site of either A, C, G, or a pyrimidine (C or T). The sequence from a part of the gel is indicated.

n *nucleotides is present in the A, G, T, or C + T parts, then an A, G, C, or T, respectively, exists at position* $n + 1$ *in the original molecule.*

All four samples Ia, Ib, IIa, and IIb are electrophoresed simultaneously so that all bands are seen in a single gel. The sequence is read directly from the gel. Figure 4-40 is an autoradiograph of a sequencing gel. The shortest fragments are those that move the fastest and farthest. Each fragment contains the original 5′-^{32}P group; the sequence can therefore be read from the bottom to the top of the gel. Where the figure is labeled with base letters, this means that the base was removed from the 3′ end of the molecule to generate the fragment.

Problems

1. Which nucleic acid base is unique to DNA? to RNA?

2. What is the difference between a nucleoside and a nucleotide?

3. When the base composition of a DNA sample from *Mycobacterium tuberculosis* was determined, 18 percent of the bases were found to be adenine.
   (a) What is the percentage of cytosine?
   (b) Calculate the base composition of this bacterial DNA.

4. The double-stranded DNA molecule of a newly discovered virus was found by electron microscopy to have a length of 34 μm.
   (a) How many nucleotide pairs are present in one of these molecules?
   (b) How many complete turns of the two polynucleotide chains are present in such a double helix?

5. Chemical analysis of the DNA extracted from a previously unknown virus indicates the molar composition: 40 percent adenine, 20 percent thymine, 15 percent guanine, and 25 percent cytosine.
   (a) What is the base composition of the DNA, based on this analysis?
   (b) How would you interpret this result, in terms of the structure of the viral DNA?

6. When a DNA solution is boiled, the two strands of each of the DNA molecules separate because the hydrogen bonds break. When the solution is cooled again, the two strands do not usually rejoin, because the probability is very low that complementary strands will collide in such a way that complementary base sequences match. In other words, boiled DNA remains single-stranded, though under certain conditions (which are not considered in this problem) the strands will reassociate. Consider a DNA molecule containing only adenine and thymine, in which the bases alternate along each strand—that is, the sequence is ATA TATAT.... If such a molecule were boiled and then cooled, what types of base-paired structures might be formed by the individual strands?

7. When a DNA solution is boiled and then cooled slowly for a long period of time (using the appropriate concentrations of salts), the complementary strands ultimately reassociate, forming proper double-stranded DNA. The process is called **renaturation.**
   (a) ^{15}N-labeled DNA from phage T4 is mixed with T4 DNA of normal density and then heated together and slowly cooled. The resulting DNA is analyzed by centrifugation in a CsCl density gradient of the type used by Meselson and Stahl. How many bands will be observed and what will their relative proportions be?
   (b) If ^{15}N-labeled T4 DNA is mixed with ^{14}N-labeled DNA from phage P2 and then boiled and cooled, two bands are seen in a CsCl density gradient. Explain this number of bands.

8. An elegant combined chemical and enzymatic technique enables one to isolate "nearest neighbors" of bases. That is, the DNA is broken at random, and all possible adjacent nucleotides are isolated. For example, if the single-stranded tetranucleotide 5'-AGTC-3' were treated in this way, the nearest neighbors would be AG, GT, and TC (they are always written with the 5' terminus at the left). Before techniques were available for determining the base sequence of DNA, nearest-neighbor frequency analysis was used to determine sequence relationships. Nearest-neighbor analysis also indicates an important feature of DNA structure—namely, its antiparallel structure—which you are asked to examine in this problem by predicting some nearest-neighbor frequencies. Assume that you have determined the frequencies of the following nearest neighbors: AG, 0.15; GT, 0.03; GA, 0.08; TT, 0.10.
   (a) What must the nearest-neighbor frequencies of CT, AC, TC, and AA be?
   (b) If DNA had a parallel, rather than an antiparallel, structure, what nearest-neighbor frequencies would you know and what would they be?

9. What is meant by the terms primer and template?

10. State whether each of the following statements is true or false:
   (a) In the synthesis of DNA the covalent bond that forms is between a 3'-OH and a 5'-P group.
   (b) In general, the DNA replicating enzyme in *E. coli* is polymerase I.
   (c) A single strand of DNA can be copied if the four nucleoside triphosphates and either polymerase I or polymerase III are provided.
   (d) If a DNA polymerase were added to a mixture of the four nucleoside triphosphates, but without any DNA, it is likely that some DNA would be synthesized but the base sequence would be random.
   (e) To initiate DNA synthesis, an RNA primer must be present that is complementary in base sequence to some region of the DNA.

11. An early structural model for DNA was the so-called tetranucleotide hypothesis: four nucleotides—one each of adenine, cytosine, guanine, and thymine—were thought to be covalently linked to form a planar unit. It was suggested that these units were linked together to yield a repeating polymer of the tetranucleotide. Explain why, if DNA had such a structure, it could never have become the genetic material of cells.

12. A criticism of the early transformation experiments was that DNA might somehow participate in the biosynthesis of the polysaccharide capsule of the virulent *Pneumococcus*. Thus, the nonvirulent mutant would lack some step required for polysaccharide synthesis and this step would be bypassed by the addition of DNA. By this argument, DNA would not be the genetic substance. What simple genetic experiments might be done to counter such an argument?

13. What is the chemical group (3'-P, 5'-P, 3'-OH, 5'-OH) at the sites indicated by the dots labeled a, b, and c in the figure below?

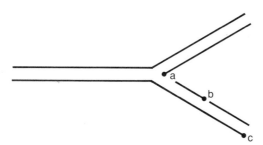

14. What is the chemical group that is at the indicated terminus of the daughter strand of the extended branch of the rolling circle in the figure below?

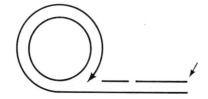

15. Assume that the following sequence of bases occurs along one chain of a DNA duplex that has opened up at a replication fork, and that synthesis of an RNA primer on this template begins by copying the base printed in red.

3'-...TCTGATATCAGTACG...-5'

(a) If the RNA primer synthesized consists of six nucleotides, what is its base sequence?

(b) In the intact RNA primer, which nucleotide will have a free OH terminus and what will be the free chemical group on the nucleotide at the other end of the primer (see Figure 4-34)?

(c) If replication of the other chain of the original DNA duplex proceeds continuously (with few or no intervening RNA primers), in which direction is the replication fork most likely moving?

16. In experiments that confirmed that DNA replication is semiconservative, *E. coli* DNA was labeled by growing cells for a number of generations in a medium containing the heavy ^{15}N isotope and then observing the densities of the molecules during subsequent generations of growth in a medium containing a great excess of ^{14}N. (See Figure 4-16.) If DNA replication were conservative (that is, if parental single strands never separated but remained associated), what pattern of molecular densities would have been observed after one and after two generations of growth in the ^{14}N-containing medium?

17. Assume that *E. coli* cells grown for many generations in a medium containing ^{14}N are transferred for exactly one generation (one replication of DNA) to a ^{15}N-containing medium. The cells are then transferred back to a medium with an excess of ^{14}N. What are the expected density classes of the molecules and the relative proportions of these classes, if the DNA is isolated and examined after one final generation of growth in the ^{14}N-containing medium? After two final generations of growth in the ^{14}N-containing medium?

18. Assume that you have a culture of *E. coli* cells grown for many generations in a ^{15}N-containing medium. The cells are washed and transferred to a ^{14}N-containing medium. After exactly two chromosome replications in this second medium, the DNA is extracted *without any breakage whatsoever* (usually not possible, but it has been done). What density classes would result and in what proportions?

19. A DNA fragment containing 17 base pairs is sequenced by the Maxam-Gilbert procedure. The figure that follows shows the data; panels 1 and 2 correspond to the two complementary strands. Note that the 5'-terminal base does not appear in either lane, as it would have to be identified by a sugar-phosphate lacking a base; such molecules do not electrophorese with the nucleotides. What are the complete sequences of the two complementary strands, including the 5'-terminal bases?

19. (continued).

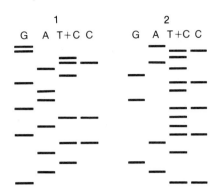

20. RNA polymerase can only synthesize RNA in a single direction. Suppose you have a double-stranded RNA molecule that is being replicated by RNA polymerase. Would you expect to find precursor fragments of newly synthesized RNA?

21. Suppose you have a DNA molecule with a gap in one strand 5000 nucleotides long and terminated with a 3'-OH group and a 5'-P group.
   (a) If a DNA polymerase is added to this molecule *in vitro* (with everything else needed to make DNA), will the DNA filling the gap be a single piece or consist of short fragments?
   (b) Would you expect fragments of any kind if the gap was filled *in vivo*?

22. In early studies of the properties of Okazaki fragments, it was shown that these fragments could hybridize to both parental strands of *E. coli* DNA. This was taken as evidence that they are synthesized on both branches of the replication fork. Even without any rigorous evidence contradicting the idea that precursor fragments are found in both daughter strands, a particular characteristic of the replication of *E. coli* DNA invalidates a conclusion based solely on the hybridization data. What is that characteristic?

23. Answer the following questions about rolling circle replication.
   (a) If the circle has unit length, what is the maximum length of the linear branch that can be formed?
   (b) What is the primer?

24. Following publication of the transformation experiments of Avery, MacLeod, and McCarty, opponents of the DNA-gene theory argued that the transformation was caused by proteins that were contaminating the DNA sample.

(a) If transformation was indeed carried out by protein rather than DNA molecules and if the DNA preparation used contained at most 0.02 percent protein, how many protein molecules (each consisting of about 300 amino acids) would have been present in 1 milliliter of a DNA solution at a concentration of 10^{-7} mg/ml?
(b) If protein was the active agent in transformation, would the number calculated in part (a) account for the fact that in a typical transformation experiment 1000 transformants result from 0.0001 g of *Pneumococcus* DNA?

25. When DNA is placed in distilled water, the two strands come apart. Explain why. *Hint:* Consider the effect of ions in solution on the interaction between charged groups.

26. The gene *dnaE* codes for the polymerase that makes DNA in *E. coli*. The only known *dnaE*⁻ mutants are temperature-sensitive, that is, replication is normal at 30°C, but at 42°C, there is no DNA synthesis.
   (a) Assume a culture of the cells is pulse-labeled with ³H-thymidine at 30°C. Then, the ³H-thymidine is removed from the growth medium and at the same time the cells are shifted to 42°C. What happens to the precursor fragments during the period at 42°C?
   (b) Would you expect to find radioactive precursor fragments in a *dnaE*⁻ mutant grown at 42°C in the presence of ³H-thymidine?

27. The enzyme S1 nuclease acts preferentially on single-stranded DNA.
   (a) The activity against double-stranded DNA molecules is weak but increases with temperature even when the temperature is much less than that producing separation of the strands. Explain.
   (b) If a double-stranded DNA molecule has a missing nucleotide (a gap), S1 nuclease can break the phosphodiester bond opposite the gap. Explain. How will the rate of breakage be affected by increasing temperature?
   (c) S1 nuclease is active against a supercoiled DNA molecule. Why? Will it make a double-strand break or a single-strand break?
   (d) If a double-stranded DNA molecule is denatured and renatured, it remains a relatively poor substrate for S1 nuclease. However, if the DNA molecules isolated from a phage A are mixed with DNA molecules from a phage B that contains a point mutation (that is, one base pair in B differs from the base pair in the corresponding position in A and the mixture is denatured and renatured, approximately half of the renatured DNA molecules are broken by S1 nuclease. Why? Is

a particular fraction of the renatured DNA subject to S1 activity? Which one?

28. Why do you think most nucleases fail to work in 1 M NaCl?

29. A double-stranded hexanucleotide, for example, 3 guanines in one strand and 3 cytosines in the other, has very low thermal stability compared to a polynucleotide containing 1000 guanines in one strand and 1000 cytosines in the other. Explain the difference.

C H A P T E R 5

The Molecular Organization of Chromosomes

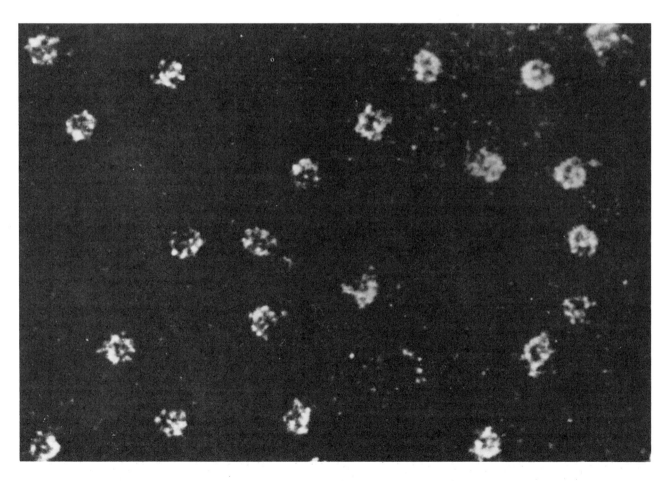

Chromosomes, which function as units of inheritance, are major components of the apparatus that ensures the stability and precise distribution of the hereditary material in cell division. A chromosome is an intricate structure in which DNA and various proteins are bound to one another in a highly organized way. We saw in Chapter 4 that free DNA molecules appear in electron micrographs as linear or circular molecules that may be folded or twisted to some extent but are generally extended. However, in a chromosome a DNA molecule is in a very different state in that it is repeatedly twisted and folded; this so-called **condensed state** of the DNA is caused by the binding of particular chromosomal proteins. Other classes of DNA-binding proteins, which contribute little to the folding, are also present in the chromosome; these proteins participate in replication, recombination, and gene expression.

To understand genetic processes requires knowledge of the organization of the genetic material. In this chapter we will see that chromosomes are diverse in size and structural properties and that their DNA differs in the composition and arrangement of nucleotide sequences. The most pronounced differences in structure and genetic organization are between the chromosomes of eukaryotes and those of prokaryotes and viruses. Some viral chromosomes are especially noteworthy in that they consist of one single-stranded (rather than double-stranded) DNA molecule and, in a small number of viruses, of RNA instead of DNA. Also, eukaryotic cells contain several chromosomes, each of which is one condensed DNA molecule, whereas prokaryotes contain a single major chromosome (plus, occasionally, several copies of a small circular DNA molecule called a plasmid). A few RNA-containing animal and plant viruses have several chromosomes.

The genetic complement of a cell or virus is referred to as a **genome,** though in eukaryotes the term is commonly used to refer to one complete (haploid) set of chromosomes.

Facing page: A dark-field electron micrograph of individual nucleosomes, at high magnification. See also Figures 5-6 and 5-7. (Courtesy of Ada Olins.)

5.1 Genome Size and Evolutionary Complexity

Measurement of the nucleic acid content of viruses and bacteria, which have a single chromosome, and of the DNA content of the haploid set of chromosomes of eukaryotes has led to the following generalization: *genome size increases roughly with evolutionary complexity*. That is, the single nucleic acid molecule of a typical virus is smaller than the DNA molecule in a bacterial chromosome; unicellular eukaryotes, such as the yeasts, contain more DNA than a typical bacterium and the DNA is organized in several chromosomes; and multicellular eukaryotes have the greatest amount of DNA. However, among the eukaryotes no correlation exists between evolutionary complexity and the number of chromosomes. We will see shortly that although the amount of DNA is certainly a reflection of the number of biochemical activities of the organism, it is not directly proportional to the number of genes.

Bacteriophage MS2 is representative of the smallest viruses. It has only four genes in a single-stranded RNA molecule consisting of a known sequence of 3569 nucleotides. The small RNA viruses, which include some additional bacteriophages and a diverse group of viruses that cause tumors in animals, have about the simplest genetic organization conceivable in a biological system that is capable of directing the synthesis and assembly of its own components. Some DNA viruses are almost as simple in organization. Simian virus 40 (SV40), which infects monkey and human cells, has a genetic complement of five genes in a circular double-stranded DNA molecule consisting of 5224 nucleotide pairs (Table 5-1). The more complex phages and animal

Table 5-1   DNA content of some representative viral, bacterial, and eukaryotic genomes

Genome	Number of nucleotide pairs (thousands)*	Form
Virus		
SV40	5.224	Circular double-stranded
ØX174	5.375	Circular single-stranded;
M13	6.408	double-stranded replicative form
λ	48.6	
Herpes simplex	152	
T2, T4, T6	165	Linear double-stranded
Smallpox	267	
Bacteria		
Mycoplasma hominis	760	Circular double-stranded
Escherichia coli	4500	
Eukaryotes		Haploid chromosome number
Saccharomyces cerevisiae (yeast)	13,000	17
Caenorhabditis elegans (nematode)	80,000	6
Drosophila melanogaster (fruit fly)	165,000	4
Homo sapiens (humans)	2,900,000	23
Zea mays (maize)	4,500,000	10
Amphiuma sp. (salamander)	76,500,000	14

*The approximate molecular length in μm can be calculated by dividing by 3000.

viruses have genomes of up to 250 genes in DNA molecules more than 50 times the size of the DNA of the simplest viruses. Bacterial genomes are substantially larger. For example, the chromosome of *E. coli* illustrated in Figure 4-17 contains about 1500 genes in a DNA molecule composed of about 4×10^6 nucleotide pairs.

Eukaryotes have a more complex genetic apparatus. Their genomes are packaged in the chromosomes of a haploid set whose number is characteristic of the particular species, as we have seen in Chapter 2. In moving up the evolutionary scale of animals or plants, the DNA content per haploid genome generally increases, though the number of chromosomes shows no pattern. One of the smallest genomes in a multicellular animal is that of the nematode worm *Caenorhabditis elegans*, with a DNA content about 20 times that of the *E. coli* genome. The *D. melanogaster* and the human haploid sets of chromosomes have about 40 times and 700 times as much DNA, respectively, as the *E. coli* genome. However, exceptions to the general correspondence between genome size and evolutionary complexity exist; for example, among the amphibia and fish, several species have genomes many times the size of mammalian genomes.

The genomes of higher eukaryotes are extraordinarily large. With an average gene size of a few thousand nucleotide pairs, enough DNA is present in the *D. melanogaster* genome for more than 60,000 genes and in the human genome for more than a million genes. These estimates of gene number seem unduly large to most biologists on the basis of other methods for arriving at such estimates. Furthermore, among the amphibia and flowering plants examples are known of more than 30-fold differences in DNA content per genome between related species that have comparable levels of biological complexity (Figure 5-1). Such observations led to the suspicion that not all base sequences of the DNA in the genomes of higher eukaryotes are used for genes. As will be seen later in this chapter, in these organisms a considerable fraction of the genome has functions other than those usually attributed to genes.

A remarkable feature of the genetic apparatus of eukaryotes is how the enormous amount of genetic material contained in the nucleus of each cell is

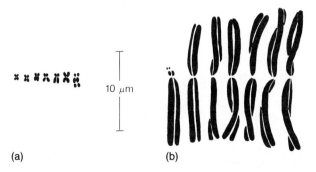

10 μm

(a)   (b)

Figure 5-1 Haploid chromosome complements of (a) *Aquilegia caerulea* × *chrysantha* and (b) *Anemone fasciculata* in the family Ranunculaceae. The DNA content of these related species differs by more than 40-fold. (Courtesy of Klaus Rothfels.)

precisely divided in each cell division. A haploid human genome, which is contained in a gamete, has a DNA content equivalent to a linear DNA molecule one meter (10^6 μm) in length. The largest of the 23 chromosomes in the genome contains DNA equivalent to a molecule 82 mm (8.2×10^4 μm) long. However, at the metaphase of a mitotic division such a chromosome is condensed into a compact structure about 10 μm long and less than 1 μm in diameter. The genomes of prokaryotes and viruses, though much smaller than eukaryotic genomes, are also very compact. For example, an *E. coli* chromosome, which contains a DNA molecule about 1300 μm long, is contained in a cell about 2 μm long and 1 μm in diameter. How this compactness is attained will be described in Section 5.4.

5.2 The Supercoiling of DNA

An important feature of the DNA of prokaryotic and eukaryotic chromosomes is that it is **supercoiled**—that is, double-stranded segments are twisted around one another. The geometry of supercoiling can be illustrated by a simple example. Consider first a linear double-stranded DNA molecule whose ends are joined in such a way that each strand forms a continuous circle. Such a DNA molecule is called a **covalent circle,** and it is said to be **relaxed** if no further twisting is present (Figure 5-2(a)). The individual polynucleotide strands of a relaxed circle form the usual right-handed (positive) helical structure with ten nucleotide pairs per turn of the helix. If before the ends of the

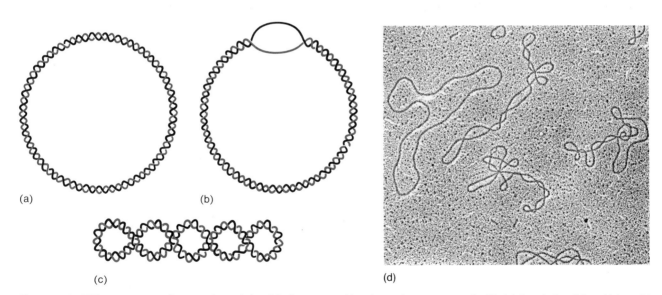

(a) (b)

(c) (d)

Figure 5-2 Different states of a covalent circle. (a) A non-supercoiled relaxed covalent circle having 36 helical turns. (b) An underwound covalent circle having only 32 helical turns. (c) The molecule in (b), but with four twists to eliminate the underwinding.

Note that no bases are unpaired in (c). In solution, (b) and (c) would be in equilibrium. (d) Electron micrograph showing nicked circular and supercoiled DNA of phage PM2. (Courtesy of K. G. Murti.)

linear molecule are joined one end is rotated one or more times through 360° with respect to the other end in a direction that produces unwinding of the double helix, then after the ends are joined the molecule will be an underwound circular helix. A DNA molecule has a strong tendency to maintain its standard helical form with ten nucleotide pairs per turn and therefore will respond to the underwinding in one of two ways: (1) by forming regions in which the bases are unpaired (Figure 5-2(b)), or (2) by twisting the circular molecule in the opposite sense from the direction of underwinding (Figure 5-2(c)). This twisting is called **supercoiling,** or **superhelicity,** and a molecule with this sense of twisting is **negatively supercoiled.** An example of a supercoiled molecule is shown in Figure 5-2(d). The two responses to underwinding are not independent, and underwinding is usually accommodated by a combination of the two processes—namely, an underwound molecule contains some unpaired bases and some supercoiling, with the supercoiling predominating. If, instead, the molecule were overwound, supercoiling in the opposite (positive) sense would result. The supercoiling of naturally occurring molecules is always of the negative type, meaning that helices are underwound. In addition, the degree of underwinding—that is, the number of twists per ten nucleotide pairs—is the same (0.05) in all known naturally supercoiled molecules, whether in a viral, prokaryotic, or eukaryotic chromosome.

The supercoiling of natural DNA molecules is not produced by the unwinding of a linear molecule before it is joined into a circle. In bacteria it is the result of the activity of DNA gyrase, one of a general class of enzymes called topoisomerases, described in Chapter 4. This enzyme, which causes negative supercoiling in DNA molecules, has been isolated from several organisms—both prokaryotes and eukaryotes. The DNA gyrase from *E. coli* has been most extensively studied and found to be required for DNA replication. The specific role of the enzyme in DNA replication is not known. However, when a circular DNA molecule replicates, an unwinding of the daughter strands causes a twisting force in the opposite direction within the unreplicated portion of the molecule—that is, that portion becomes positively supercoiled. Without the removal of this positive supercoiling, the molecule would become so tightly twisted that a complete round of replication of the circle would not occur. DNA gyrase, which introduces negative supercoiling, removes these positive twists. Supercoiling is also essential for certain stages of recombination, and there is evidence that it affects the expression of certain genes.

The DNA of eukaryotic chromosomes from which the proteins have been chemically removed is also supercoiled; the mechanism of supercoiling is more complex than in bacteria and is not clearly known.

The introduction of one single-strand break (a **nick**)—for example, by a deoxyribonuclease (DNase), an enzyme that breaks sugar-phosphate bonds in DNA strands—into a typical supercoiled DNA molecule eliminates all supercoiling because the constraint of underwinding can be removed by a free rotation of the intact strand about the sugar-phosphate bond opposite the break.

5.3 Structure of the Bacterial Chromosome

The chromosome of *E. coli* is a condensed unit—called the **nucleoid** or **folded chromosome**—containing a single circular DNA molecule. It is located in a poorly delimited region of the cell called the nuclear region. The most striking feature of the nucleoid is that the DNA is organized into a set of looped domains (Figure 5-3), a feature that is also characteristic of eukaryotic chromosomes. As isolated, the nucleoid contains, in addition to DNA, small amounts of several proteins, which are thought to be responsible in some way for the multiply-looped arrangement of the DNA. Some RNA is also present in the isolated nucleoid, but it may not be an essential component, possibly becoming associated with the structure only during the isolation procedure. The degree of condensation of the isolated nucleoid (that is, its physical dimensions) is affected by a variety of factors, and some controversy exists about the state of the nucleoid within a cell. However, several lines of evidence indicate that the multiple loops are also present intracellularly and in roughly the same number (40–50) as in isolated material.

Figure 5-3 also shows that loops of the DNA of the *E. coli* chromosome are supercoiled. Notice that some loops are not supercoiled; this is a result of the action of several DNases during isolation and indicates that the loops are

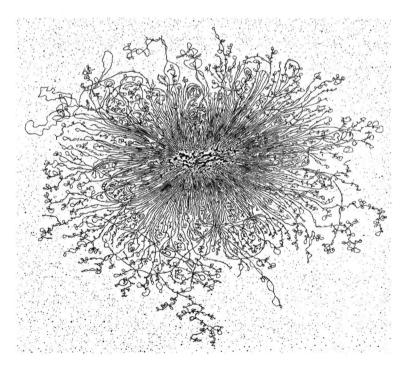

Figure 5-3 An electron micrograph of an *E. coli* chromosome showing the multiple loops emerging from a central region. (Bluegenes #1. 1983. © All rights reserved by Designergenes Posters Ltd., 106 11 St., Del Mar, CA, 92014, from which posters, postcards, and shirts are available.)

in some way independent of one another. In the previous section, we pointed out that supercoiling is generally eliminated in a DNA molecule by one single-strand break. However, such a break in the *E. coli* chromosome does not eliminate all supercoiling. If nucleoids all of whose loops are supercoiled are treated with a DNase and examined at various times after single-strand breaks are introduced, it is observed that a single-strand break removes the supercoiling of only one loop, not all loops (Figure 5-4). Thus, the loops must be isolated from one another in such a way that rotation in one loop is not transmitted to other loops. How this occurs is unknown.

It is possible experimentally to eliminate all supercoiling (by introduction of many single-strand breaks or by treatment with certain types of topo-isomerases). When this is done, the overall looped structure of the chromosome is not lost, which indicates that the folding and supercoiling of the DNA are independent phenomena.

5.4 Structure of Eukaryotic Chromosomes

Convincing evidence from work with several organisms indicates that each eukaryotic chromosome contains a single DNA molecule of enormous length. For example, the largest chromosome in the *D. melanogaster* genome has a DNA content of 6.5×10^7 nucleotide pairs, equivalent to a continuous linear duplex about 18 mm long. A rather complicated physicochemical technique called viscoelastometry has been used to measure the size of DNA molecules isolated from the nuclei of *Drosophila* cells by very gentle procedures. The technique provides a reliable estimate of the size of the largest molecule in the solution, even when these molecules represent only a small fraction of the total number of molecules. The largest molecules from the nuclei of wildtype *Drosophila* have been found to contain 6.2×10^7 nucleotide pairs, a value that agrees well with the measured DNA content of the largest chromosome. The concept that the DNA in a eukaryotic chromosome consists of a single molecule is also supported by the observed length of DNA molecules extracted

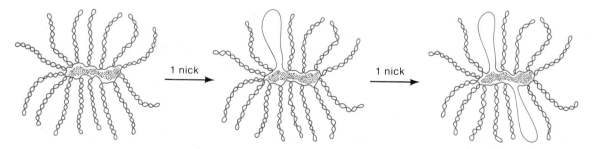

Figure 5-4 A schematic drawing of the folded supercoiled *E. coli* chromosome, showing only 15 of the 40–50 loops attached to a putative protein core (shaded area) and the opening of a loop by a nick.

from *Drosophila* cells grown in culture and labeled with the radioactive DNA precursor [³H]thymidine (Figure 5-5). The longest of these molecules are about two-thirds the length of the largest molecules indicated by viscoelastometry.

The DNA of all eukaryotic chromosomes is associated with numerous protein molecules in a stable ordered aggregate called **chromatin.** Some of the proteins present in chromatin determine chromosome structure and the changes in structure that occur during the division cycle of the cell. Other chromatin proteins appear to have important but not well-understood roles in regulating chromosome functions.

The Chemical Composition of Chromatin

The simplest form of chromatin is that present in nondividing eukaryotic cells in which chromosomes are not sufficiently condensed to be visible by light microscopy. Chromatin isolated from such cells is a complex aggregate of DNA and two classes of protein molecules. One class, the **histones,** are very basic proteins (positively charged at neutral pH) and are tightly bound to DNA. The other proteins associated with chromatin are called **nonhistone chromosomal proteins.** The precise composition of chromatin varies somewhat from one species to another and between different cell types in an organism, and also depends on the method used to isolate the chromatin. Some RNA is usually found in isolated chromatin, but it has a transient association with the protein components, and there is no evidence that it has either a structural or regulatory function.

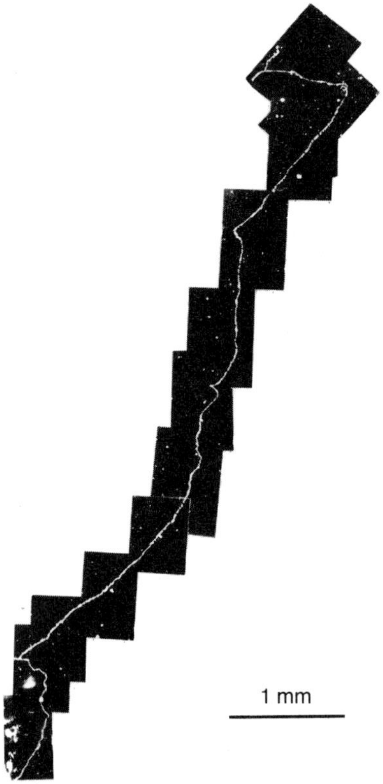

The nonhistone protein fraction of chromatin may include more than 100 different proteins with molecular weights from 10,000 to 300,000. This fraction can be represented in amounts from about 30 to 120 percent of the mass of the DNA, depending on the source of the chromatin. Many of these proteins are present in all species and tissues, which suggests that they have structural or metabolic functions common to all chromatin. Included are various enzymes that act on DNA—such as DNA and RNA polymerases, and nucleases—and several others associated with the chemical modification of chromosomal proteins. Other nonhistone proteins have properties that suggest that they may be responsible for certain facets of gene expression and its regulation. Proteins of this latter class are usually species- and tissue-specific—that is, the kinds present in one cell type may differ markedly from those found in comparable cells of a different species, or in cells of a different type in the same organism.

Histones are largely responsible for the structure of chromatin. Five major types—**H1, H2A, H2B, H3, and H4**—occur in the chromatin of almost all eukaryotes and are present in amounts about equal in mass to that of the DNA. These are small proteins with molecular weights of 11,300 to 21,000. One or more of the histones (usually H1) may be absent from the chromatin of some lower eukaryotes. The nuclei of the sperm of birds, most fish, and some invertebrates lack histones altogether; in these cells the corre-

1 mm

Figure 5-5 Autoradioagram of a DNA molecule from *D. melanogaster.* The molecule is 1.2 cm long. (From R. Kavenoff, L. C. Klotz, and B. H. Zimm. 1974. *Cold Spring Harbor Symp. Quant. Biol.,* 38: 4.)

sponding proteins are other small basic protein molecules called **protamines.** Recall that proteins are long chains consisting of numerous small molecules called amino acids, which are joined together to form the chain. Histones differ from most other proteins in that 20–30 percent of the amino acids are lysine and arginine, both of which have a positive charge. (Only a few percent of the amino acids of a typical protein are lysine and arginine.) The positive charges enable histone molecules to bind to DNA, primarily by electrostatic attraction to the negatively charged phosphate groups in the sugar-phosphate backbone of DNA. Placing chromatin in a solution with a high salt concentration (for example, 2 M NaCl) to eliminate the electrostatic attractions causes the histones to dissociate from the DNA. Not only are histones able to bind to DNA, but also certain histones bind tightly to each other; both DNA-histone and histone-histone binding are important for chromatin structure.

The histones from different organisms are remarkably similar to one another, with the exception of H1. In fact, the amino acid sequences of H3 molecules from widely different species are almost identical. For example, the H3 of cow chromatin and pea chromatin differ by only four of 135 amino acids. The H4 proteins of all organisms are also quite similar; again, cow and pea H4 differ by only two of their 102 amino acids. There are few other proteins whose amino acid sequences vary so little from one species to the next. When the variation is very small between organisms, one says that the sequence is highly **conserved.** The extraordinary conservation in histone composition over millions of years of evolutionary divergence is consistent with the important role of these proteins in the structural organization of eukaryotic chromosomes.

Nucleosomes: The Basic Structural Unit of Chromatin

If the nonhistone proteins and most salt ions are removed from chromatin and the remaining material is viewed with an electron microscope, a structure looking like a regularly beaded thread is seen (Figure 5-6). Evidence for such a repeating particulate structure also comes from observing the product of

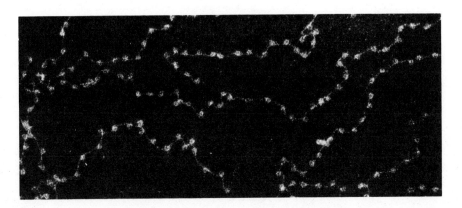

Figure 5-6 Dark-field electron micrograph of chromatin showing the beaded structure at low salt concentration. The beads have diameters of about 100 Å. (Courtesy of Ada Olins.)

digestion of both isolated chromatin and intact nuclei by certain nucleases. Brief treatment with some DNases yields a collection of small particles of quite uniform size consisting only of histones and DNA. When the histones are removed from these particles, the DNA fragments are found to be of lengths equal to about 200 nucleotide pairs or small multiples of that unit size (the precise size varies with species and tissue). Convincing evidence for the relation between the beaded structure of chromatin seen by electron microscopy and the 200-nucleotide-pair repeating unit of DNA came from a simple experiment. Particles of chromatin produced by incomplete nuclease digestion of the nuclei of rat liver cells were first fractionated according to size by sedimentation through a sucrose concentration gradient. Four size fractions of digested chromatin were obtained in this experiment, and when the different fractions were examined by electron microscopy, they were found to contain particles with either one, two, three, or four beads (Figure 5-7). Dissociation of the histones from these particles yielded DNA fragments consisting of roughly 200, 400, 600, and 800 nucleotide pairs, respectively.

The beadlike units in chromatin are called **nucleosomes.** Each unit has a definite composition—namely, one molecule of H1, two molecules each of H2A, H2B, H3, and H4, and one segment of DNA containing about 200 nucleotide pairs. Extensive digestion of these units with a nuclease removes some of the DNA and causes the loss of H1. The resulting structure, called a **core particle,** consists of an octamer of pairs of H2A, H2B, H3, and H4, around which the remaining 145-nucleotide-pair length of DNA is wound in about 1-3/4 turns (Figure 5-8). Thus, a nucleosome is composed of a core particle, additional DNA that links adjacent core particles (the DNA that is removed by nuclease digestion), and one molecule of H1; the H1 binds to the histone octamer and to the linker DNA, causing the linkers extending from both sides of the core particle to cross and draw nearer to the octamer, though some of the linker DNA does not come into contact with any histones. The size of the linker ranges from 20–100 nucleotide pairs for different species and even in different cell types in the same organism (200 − 145 = 55 nucleotide

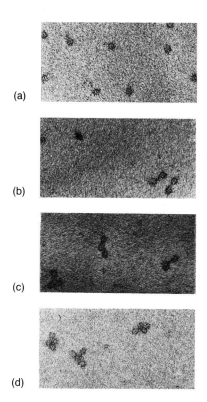

Figure 5-7 Electron micrograph of incompletely digested chromatin fractionated by centrifugation through a sucrose concentration gradient, showing (a) nucleosome monomers, (b) dimers, (c) trimers, and (d) tetramers. (Courtesy of Roger Kornberg.)

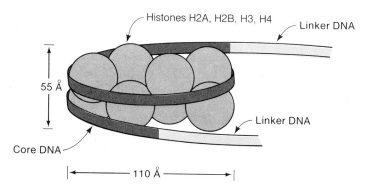

Figure 5-8 Diagram of a nucleosome core particle. The DNA molecule is wound 1-3/4 turns around a histone octamer. If H1 were present, it would bind to the octamer surface and to the linkers, causing the linkers to cross.

pairs is usually considered an average size). Little is known about the structure of the linker DNA or whether it has a special genetic function, and the cause of the variation in its length is also unknown. The term **chromatosome** is coming into use for the nucleosome containing H1 and in which the DNA makes almost two full turns around the octamer.

Arrangement of Chromatin Fibers in a Chromosome

The DNA molecule of a chromosome is folded and refolded in such a way that it is convenient to think of chromosomes as having several levels of organization, each responsible for a particular degree of shortening of the enormously long strand (Figure 5-9). Assembly of DNA and histones represents the first level—namely, a sevenfold reduction in length of the DNA and the formation of a beaded flexible fiber 110 Å (11 nm) wide, roughly five times the width of free DNA (Figure 5-6). The structure of chromatin varies with the concentration of salts, and the 110-Å fiber is present only when the salt concentration is quite low. If the salt concentration is increased slightly, the fiber becomes shortened somewhat by forming a zigzag arrangement of closely spaced beads between which the linking DNA is no longer visible in electron micrographs (panel (b)). If the salt concentration is further increased to that present in living cells, a second level of compaction occurs—namely, the organization of the 110-Å nucleosome fiber into a shorter, thicker fiber with an average diameter of 300–350 Å, called the **30-nm fiber** (panel (c)). In forming this structure the 110-Å fiber apparently coils in a somewhat irregular left-handed superhelix or solenoidal supercoil with six nucleosomes per turn (Figure 5-10). It is believed that most intracellular chromatin has the solenoidal supercoiled configuration.

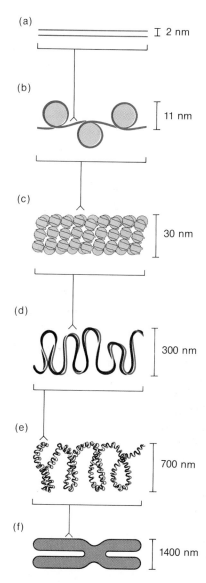

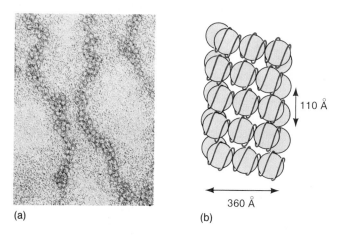

Figure 5-10 (a) Electron micrograph of the 30-nm component of mouse metaphase chromosomes. (Courtesy of Barbara Hamkalo.) (b) A proposed solenoidal model of chromatin. The DNA (red) is wound around each nucleosome. It is unlikely that the real structure is so regular. (After J. T. Finch and A. Klug. 1976. *Proc. Nat. Acad. Sci.*, 73: 1900.)

Figure 5-9 Various stages in the condensation of (a) DNA and (b–e) chromatin in forming (f) a metaphase chromosome. The dimensions indicate known sizes of intermediates, but the detailed structures are hypothetical.

THE MOLECULAR ORGANIZATION
OF CHROMOSOMES

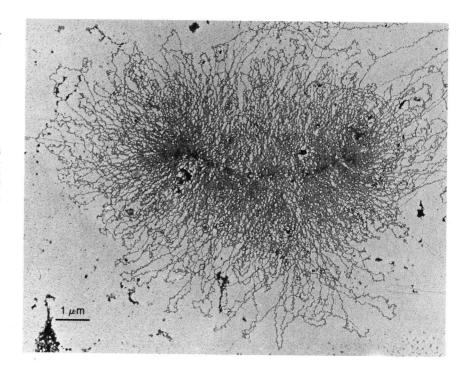

Figure 5-11 Electron micrograph of a partially disrupted anaphase chromosome of the milkweed bug *Oncopeltus fasciatus*, showing multiple loops of 30-nm chromatin at the periphery. (From V. Foe, H. Forrest, L. Wilkinson, and C. Laird. 1982. *Insect Ultrastructure*, I: 222.)

The final level of organization is that in which the 30-nm fiber condenses into a chromatid of the compact metaphase chromosome (Figure 5-9(d–f)). Little is known about this process other than it seems to precede in stages. In electron micrographs of isolated metaphase chromosomes from which histones have been removed, the partially unfolded DNA has the form of an enormous number of loops that seem to extend from a central core or **scaffold** composed of nonhistone chromosomal proteins (Figure 5-11). Electron microscopic studies of chromosome condensation during mitosis and meiosis suggest that the scaffold extends along the chromatid and that the 30-nm fiber becomes arranged into a helix of loops radiating from the scaffold. Details are not known about the additional folding that is required of the fiber in each loop to produce the fully condensed metaphase chromosome.

The compaction of DNA and protein into chromatin and ultimately into the chromosome greatly facilitates the distribution of the genetic material during nuclear division. We will see in later chapters that certain variations in chromatin structure affect gene expression.

Replication of Chromatin

During DNA replication DNA polymerase and other proteins important to the replication process move along the DNA molecule. It seems that the nucleosome structure would interfere with this movement; furthermore, as new DNA is synthesized, it must combine with histones to form chromatin. Therefore, it is reasonable to ask what happens to the nucleosomal organization of chromatin during DNA replication. Electron micrographs of replicating chromatin show that the beaded structure is not extensively disassembled ahead of a DNA replication fork and is present in the two daughter strands a short distance behind the fork. Several experiments have made clear what occurs. First, it has been shown that histones are synthesized only during the replication cycle and as they are needed—that is, the amount of histone made is just sufficient for forming new chromatin. Second, blockage of protein synthesis by the drug cycloheximide, which does not stop movement of the replication fork, causes the formation of forks in which one of the newly synthesized double-stranded branches is organized into a beaded fiber, and the other branch of the fork is free DNA (Figure 5-12). Removal of cycloheximide results in rapid synthesis of histones and the conversion of the new strand from free DNA to one with a nucleosomal organization. The results of other experiments, in which the histones are labeled with heavy atoms (^{13}C and ^{15}N), indicate that the octamer of core histones is conserved during DNA replication—that is, the histones do not dissociate and then reassemble in mixed complexes of old and newly synthesized molecules. In addition, old and new octamers segregate conservatively. Thus, as the replication machinery moves along a chromosome fiber and the DNA strands separate at the growing fork, the preexisting octamers remain associated with one of the parental DNA single strands, and octamers of newly synthesized histones are assembled along the other branch of the fork after replication of the DNA.

DNA made during a period of
inhibition by cycloheximide

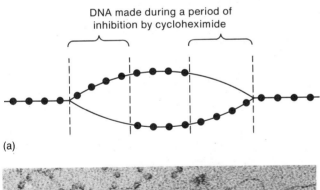

(a)

Figure 5-12 Replication of chromatin in the presence of cycloheximide, an inhibitor of protein synthesis. (a) A schematic diagram of a replication loop. (b) Electron micrograph showing replicating chromatin from cycloheximide-treated cells (left) and an interpretive drawing (right). (Courtesy of Harold Weintraub.)

(b)

(c)

5.5 Polytene Chromosomes

A typical eukaryotic chromosome contains only a single DNA molecule. However, in the nuclei of cells of the salivary glands and certain other tissues of the larvae of *Drosophila* and other two-winged (dipteran) flies there are giant chromosomes, called **polytene chromosomes,** which contain many DNA molecules (Figure 5-13). Each of these chromosomes has a volume about 1000 times greater than that of the corresponding chromosome at mitotic metaphase in ordinary somatic cells, and a constant and distinctive pattern of transverse banding (Figure 5-14). The polytene structures are formed by repeated replication of the DNA in a closely synapsed pair of homologous chromosomes unaccompanied by separation of the replicated chromatin strands or of the two chromosomes. Polytene chromosomes are atypical chromosomes and are formed in "terminal" cells; that is, the larval cells containing them do not divide further during development of the fly and are later discarded during formation of the pupa. However, they have been especially valuable in the genetics of *Drosophila*, as will become apparent in Chapter 6.

In polytene nuclei of some species, of which *D. melanogaster* is an example, large blocks of heterochromatin (a particular type of chromatin described in the following section) adjacent to the centromeres are aggregated into a single compact mass called the **chromocenter.** Since the two largest chromosomes (numbers 2 and 3) have centrally located centromeres, the chromosomes appear in the configuration shown in Figure 5-13: the paired X chromosomes (in a female), the left and right arms of chromosomes 2 and 3, and a short chromosome (4) project from the chromocenter. In a male the Y

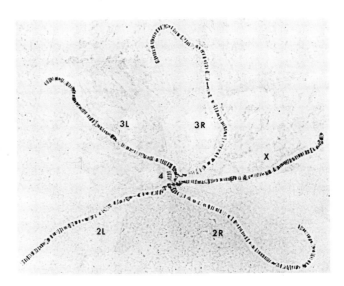

Figure 5-13 Polytene chromosomes from a larval salivary gland cell of *Drosophila melanogaster.* The centromeres of all chromosomes are united in the common chromocenter. (Courtesy of George Lefevre.)

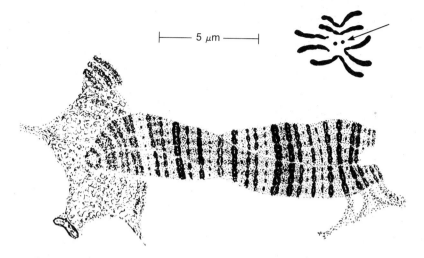

├── 5 μm ──┤

Figure 5-14 The polytene fourth chromosome of *Drosophila melanogaster* adhering to the chromocenter, shown at the left. Above at the right, drawn to the same scale, are the somatic chromosomes as they appear at mitotic prophase, with the pair of dotlike fourth chromosomes indicated by the arrow. (From C. Bridges. 1935. *J. Heredity*, 26: 60.)

chromosome, which consists almost entirely of heterochromatin, is incorporated in the chromocenter.

The darkly staining transverse bands in polytene chromosomes have about a tenfold range in width. These bands result from side-by-side alignment of tightly folded regions of the individual chromatin strands that are often visible in mitotic and meiotic prophase chromosomes (see Figure 2-6) and are called chromomeres. More DNA is present within the bands than in the interband (lightly stained) regions. More than 5000 bands have been identified in the *D. melanogaster* polytene chromosomes. This linear array of bands, which has a pattern that is constant and characteristic for each species, provides a finely detailed **cytological map** of the chromosomes. The banding pattern is such that short regions in any of the chromosomes can be identified, as can be seen in Figure 5-14.

Because of their large size and finely detailed morphology, polytene chromosomes are exceedingly useful for *in situ* nucleic acid hybridization (Section 5.6) to determine the location of specific DNA sequences (Figure 5-15). Other uses of these large chromosomes in genetic investigations will be discussed in later chapters.

5.6 The Organization of Nucleotide Sequences in Chromosomal DNA

The information content of a DNA molecule resides in its base sequence, and an understanding of many features of genetics has been derived from both the overall organization of DNA sequences and the details of the sequences of individual genes.

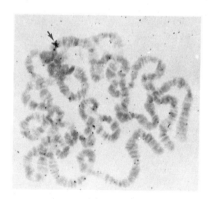

Figure 5-15 Autoradiogram of *Drosophila melanogaster* polytene chromosomes hybridized *in situ* with radioactive (^{3}H) RNA copied from the histone genes, showing localization of the RNA in particular bands. The arrow points to the region in which hybridization has occurred. (Courtesy of Mary Lou Pardue.)

Early experiments to explore the gross organization of DNA sequences revealed two striking differences between prokaryotic and eukaryotic DNA: (1) The base composition of eukaryotic DNA varies markedly from one part of the molecule to another, and (2) eukaryotic DNA contains multiply repeated sequences. In Chapter 4 it was explained that the DNA of a species has a characteristic base composition, expressed as the G + C content. Base composition is conveniently measured by centrifuging a DNA sample in concentrated cesium chloride (Figure 4-17) because the density of DNA in a CsCl chloride solution increases linearly with the G + C content. When bacterial DNA is isolated, broken randomly into fragments of about 10^4 nucleotide pairs, and then sedimented to equilibrium in a CsCl density gradient, the DNA forms a single band at a position corresponding to the mean base composition of the DNA. Formation of a single band means that the base composition of bacterial DNA does not vary significantly from one region of the DNA to another. However, several bands often appear in a CsCl density gradient when the fragmented DNA of a eukaryote is centrifuged. Most of the DNA forms a single **main band** of fairly uniform density, but other bands, called **satellite bands,** which contain G + C in amounts distinctly different from the bulk of the DNA and thus have discrete densities, are frequently found. The separation of mouse main-band and satellite-band DNA in a CsCl density gradient is shown in Figure 5-16. Approximately ten percent of the total DNA of the mouse is in the satellite fraction. In other eukaryotes, the amount of satellite DNA ranges from 1 to 40 percent of the total DNA in up to five bands, which may be either more or less dense than the main band. Even more noteworthy is the fact that satellite DNA fractions consist of sequences ranging from 2 to more than 1000 nucleotide pairs, and each sequence may be repeated *as many as a million times in a haploid genome.* Other techniques of compositional analysis show that numerous other **repetitive sequences** are present in eukaryotic DNA, though their densities are usually not sufficiently different from the bulk of the DNA for them to form distinct bands in a CsCl gradient. In the first part of this section we examine some of the most significant findings about sequence organization that are revealed by these other techniques. Later we will return to the satellite sequences.

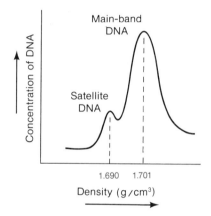

Figure 5-16 Separation of mouse main-band and satellite DNA by equilibrium centrifugation in a CsCl density gradient.

Denaturation and Renaturation of DNA

The double-stranded helical structure of DNA is maintained by forces that include hydrogen bonding between the bases of complementary pairs. When solutions of DNA are exposed to temperatures considerably higher than those normally encountered by most living cells or to excessively high pH, the hydrogen bonds break and the paired strands separate. Unwinding of the helix occurs rapidly, the time depending on the length of the molecule. When the ordered structure of DNA is disrupted and the strands are separated, the molecule is said to be **denatured.** A common way to detect denaturation is by measuring the capacity of DNA in solution to absorb ultraviolet light of 260-nm wavelength. The absorbance at 260 nm (A_{260}) of a solution of single-

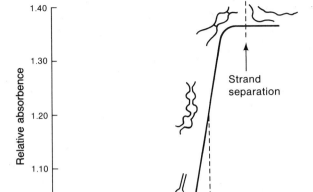

Figure 5-17 A melting curve of DNA showing T_m and possible shapes of a DNA molecule at various degrees of denaturation.

stranded molecules is 37 percent higher than the absorbence of the double-stranded molecules at the same concentration. Thus, when a DNA solution is slowly heated and the value of A_{260} is recorded at various temperatures, a curve called a **melting curve** is obtained. An example is shown in Figure 5-17. The melting transition is usually described in terms of the temperature at which the increase in the value of A_{260} is half complete. This temperature is called the **melting temperature** and denoted by T_m. The value of T_m increases with G + C content, because GC pairs, joined by three hydrogen bonds, are stronger than AT pairs, which are joined by two hydrogen bonds.

The single strands in a solution of denatured DNA can, under certain conditions, re-form double-stranded DNA. The process is called **renaturation** or **reannealing.** For renaturation to occur, two requirements must be met: (1) The salt concentration must be high enough ($> 0.25\ M$) to neutralize the negative charges of the phosphate groups, which would otherwise cause the complementary strands to repel one another; and (2) the temperature must be sufficiently high to disrupt hydrogen bonds that form at random between short sequences of bases within the same strand, but not so high that stable base pairs between the complementary strands would be disrupted. A temperature about 20°C below T_m is usually optimal. Renaturation is a fairly slow process and its rate is limited by the initial step in the process—namely, a precise collision between two complementary strands—which permits a short sequence of correct base pairs to form. This initial pairing step is followed by a rapid pairing of the remaining complementary bases and rewinding of the helix. Rewinding occurs in a matter of seconds and *its rate is independent of DNA concentration.* In contrast, correct initial base pairing of all molecules in a sample is concentration-dependent and may require several minutes to many hours when standard conditions are used.

An example utilizing a hypothetical DNA molecule will enable us to understand some of the molecular details of renaturation. Consider the following double-stranded DNA molecule containing 30,000 base pairs, in which a specific sequence of five base pairs appears only twice. (In general, such a short sequence would occur several times in a molecule of that length.)

This molecule is first heat-denatured and then the temperature is lowered to promote renaturation. In the solution of these denatured molecules a random collision between noncomplementary base sequences cannot initiate renaturation, but a collision that brings together sequences I and II′ or I′ and II can result in base pairing. However, this pairing will be transient at the elevated temperatures used for renaturation, because the paired region is short and the adjacent bases in the two strands are out of register and unable to pair, as is required to form a double-stranded molecule. However, if a collision results in the pairing of sequence I with I′ or II with II′—or any other short complementary sequences—pairing of the adjacent bases and in fact all other bases in the strands will occur in a zipperlike action. The main point is that only base pairing that brings the complementary sequences into register will cause renaturation to occur.

Another important point is that because a solution of denatured DNA usually contains a large number of identical molecules, a double-stranded molecule formed by renaturation is rarely composed of the same two strands that were paired before denaturation. This phenomenon is the basis of one of the most useful tools in molecular genetics—**molecular hybridization.** In this context, molecular hybrids are molecules formed when two single nucleic acid strands obtained by denaturing DNA from *different* sources have sufficient sequence complementarity to pair. An example of the use of DNA-DNA hybridization is the determination of the fraction of the DNA of two different species that have common base sequences. To illustrate one procedure that is used, let us assume that two bacterial species have been grown in media that cause the DNA of one species to contain only the nonradioactive isotope ^{12}C and the DNA of the second species to contain the radioisotope ^{14}C. The DNA molecules of the two species are isolated and experimentally broken into many small fragments. The nonradioactive molecules are immobilized on a nitrocellulose filter and the radioactive DNA is added. The temperature and salt concentration are raised to promote renaturation and then, after a suitable period of time, the filter is washed. The existence of common base sequences will be detected by the presence of renatured fragments of radioactive DNA on the filter. The extent of common base sequences determined in this way in the genomes of two species generally agrees with the evolutionary relatedness of the species indicated by other criteria.

Information about the size of repeated sequences and the number of copies of a particular sequence can be obtained by studies of the rate of renaturation. Since the initial step in renaturation is a collision between two complementary single strands, the rate of reassociation increases with DNA concentration (Figure 5-18(a)). That is, an increase in DNA concentration results in a corresponding increase in the number of potential pairing partners for a given strand. The study of this dependence of renaturation rate on concentration has contributed greatly to our understanding of the organization of the genetic material.

In order to see how the analysis works, we begin by comparing the renaturation rates of two solutions of DNA molecules, A and B, having no common base sequences and different molecular weights, with A smaller than B. In solutions in which the number of grams of DNA per milliliter is the same, the molar concentration of A is greater than that of B. Thus, if each solution is separately denatured and renatured, the molecules in A will renature more rapidly than those in B. If the two solutions are instead mixed, A and B will renature independently of one another (because they are non-homologous), and a curve such as that in Figure 5-18(b) will be obtained. Note that the curve consists of two steps, one for the more rapidly renaturing A molecules and the other for the B molecules. Each step accounts for half of the change in A_{260} because the initial concentration (in g/ml, which is proportional to A_{260}) of each type of molecule was the same.

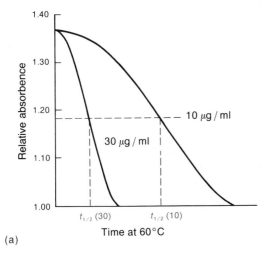

(a)

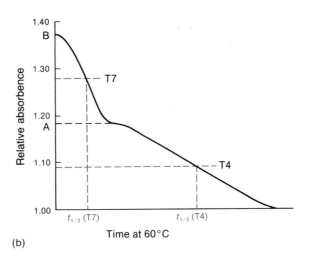

(b)

Figure 5-18 (a) Dependence of renaturation time on the concentration of phage T7 DNA. After a period at 90°C to separate the strands, the DNA was cooled to 60°C. Renaturation is complete when the relative absorbence reaches 1. (b) Renaturation of a mixture of T4 and T7 DNA, each at the same temperature. Extrapolation (black dashed line) yields the early portion of the T4 curve; the ratio of the absorbences at points A and B yields the fraction of the total DNA that is T4 DNA. The times required for half-completion of renaturation, $t_{1/2}$, are obtained by drawing the red horizontal lines, which divide each curve equally in the vertical direction, and then extending the red vertical lines to the time axis.

Now consider a molecule containing 50,000 base pairs and consisting of 100 copies of a tandemly repeated sequence of 500 base pairs. The molecules are broken into about 100 fragments of roughly equal size—that is, about 500 base pairs each. A renaturation curve for the fragments will have a single step and a renaturation rate characteristic of molecules 500 nucleotides in length.

The renaturation curve of a molecule containing no repeating base sequences will obviously consist of only a single step and the renaturation rate will correspond to the size of the molecule. If such molecules are fragmented into many components of equal size, the molar concentration of each component will be the same as that of the unbroken molecule, so the renaturation rate should be unchanged by fragmentation. The rate will actually decrease somewhat, but the important point is that *if a DNA molecule contains only a single nonrepetitious sequence of base pairs, breakage of the molecule will not yield a renaturation curve consisting of steps.* In contrast, if a molecule with an overall sequence that includes both a unique component and several copies of different repeated sequences is fragmented, *fragments containing the more numerous repeated sequences will renature more rapidly than the fragments containing portions of the unique sequence.* Thus, if the frequencies of the families of repeated sequences are different, the renaturation curve will have steps, one step for each repeating sequence. This principle is the basis of the analysis of renaturation kinetics.

Renaturation kinetics can be described in a simple mathematical form, because the reaction is one in which the rate-limiting step is the collision of two molecules. In such a case, the fraction of single strands remaining denatured at a time t after the start of renaturation is given by the expression

$$C/C_0 = 1/(1 + kC_0t)$$

in which C is the concentration of single-stranded DNA in moles of nucleotide per liter, C_0 is the initial concentration, and k is a constant. The expression C_0t is commonly called **Cot** and a plot of C/C_0 versus $\log C_0t$ is called a **Cot curve.** When renaturation is half-completed, $C/C_0 = 1/2$ and

$$C_0t_{1/2} = 1/k$$

The value of $1/k$ depends on experimental conditions, but for a particular set of conditions *the value is proportional to the number of bases in the renaturing sequences.* The longer the sequence, the greater will be the time to achieve half-complete renaturation for a particular starting concentration—because the number of molecules will be smaller. The equation just stated applies to a single molecular species. In a mixture of several such species C and C_0 refer to each species, a point to which we will return. If a molecule consists of several subsequences, one needs to know C_0 for each subsequence, and a set of values of $1/k$ will be obtained (one for each step in the renaturation curve), each value depending on the length of the subsequence. What is meant by the length of the sequence that determines the rate is best described by example. A DNA molecule containing only adenine in one strand (and thymine

in the other) has a repeating length of one. The repeating tetramer
...GACTGACT... has a repeating length of four. A nonrepeating DNA
molecule containing n nucleotide pairs has a unique length of n.

Experimentally, the number of bases per repeating unit is not determined directly. Generally, renaturation curves for a series of molecules of
known molecular weight and *with no repeating elements in their sequences* (that
is, one-step curves) serve as standards. Molecules composed of short repeating
sequences are also occasionally used. The first set of curves of this kind that
was obtained is shown in Figure 5-19. Note that two of these simple curves
represent the entire genomes of *E. coli* and phage T4. With the conditions
used to obtain this set of curves, which have become standard conditions for
Cot analysis, the sequence length N (in base pairs) yielding a particular value
of $C_0t_{1/2}$ is

$$N = (5 \times 10^5)C_0t_{1/2}$$

in which t is in seconds and C_0 is in nucleotides per liter, and 5×10^5 is a
constant dependent on the conditions of renaturation. Notice again that C_0 is
not the overall DNA concentration, but the concentration of the individual
sequence producing a particular step in a curve. How one obtains the necessary value of C_0 will become clear when we analyze a Cot curve. Such an
analysis begins by first noting the number of steps in the curve (each of which
represents a sequence or class of sequences of a particular length) and the
fraction of the material represented by each step. The observed value of $C_0t_{1/2}$
for each step must be corrected by *calculating* the value of C_0 for each sequence
class. The lengths of the sequences are then determined from these corrected

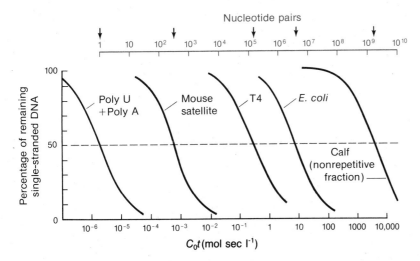

Figure 5-19 A set of Cot curves for various DNA samples. The black arrows on the red scale
indicate the number of nucleotide pairs for each sample and also point to the intersection of
each curve with the horizontal red line (point of half renaturation). This graph can be related
to Figures 5-17 and 5-18 by noting that maximum absorbence represents totally single-
stranded DNA and minimum absorbence (1.0) represents totally double-stranded DNA.

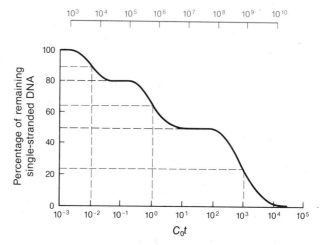

Figure 5-20 The Cot curve analyzed in the text. The scale at the top is the same as that in Figure 5-19. The black dashed lines indicate the fractional contribution of each class of molecules to the total DNA.

values by comparison to standards, and the sequence lengths and sequence fractions are compared to obtain the number of copies of each sequence. This analysis is best understood by example.

Figure 5-20 shows a Cot curve typical of those obtained in analyses of the renaturation kinetics of eukaryotic genomes. Three discrete steps are evident, with 50 percent of the DNA having a $C_0t_{1/2} = 10^3$; 30 percent, 1; and 20 percent, 10^{-2}. The scale at the top of the figure was obtained from Cot analysis of molecules having unique sequences of known lengths, as in Figure 5-19. The sequence sizes cannot be determined directly from the observed $C_0t_{1/2}$ values, because each value of C_0 used in plotting the horizontal axis in the body of the figure is the *total* DNA concentration in the renaturation mixture. Multiplying each $C_0t_{1/2}$ value by the fraction of the total DNA that it represents yields the necessary corrected $C_0t_{1/2}$ values—that is, 0.50×10^3, 0.30×1, and 0.20×10^{-2}. From the size scale at the top of Figure 5-19, the corresponding sequence sizes are about 3.0×10^8, 2.2×10^5, and 1000 base pairs, respectively.

To determine the number of copies of each sequence we make use of the fact that the number of copies of a sequence having a particular renaturation rate is inversely proportional to $t_{1/2}$ and hence to the observed (uncorrected) $C_0t_{1/2}$ values for each class. Thus, if the haploid genome contains only one copy of the longest sequence (3.0×10^8 base pairs), it would contain 10^3 copies of the sequence(s) with 2.2×10^5 base pairs and 10^5 copies of the 1000-base-pair sequence. An estimate of the total number of base pairs per genome would in this case be $3.0 \times 10^8 + 1000(2.2 \times 10^5) + 10^5(1000) = 6.2 \times 10^8$.

The different sequence components of a eukaryotic DNA molecule can be isolated by procedures that recover the double-stranded molecules formed at different times during a renaturation reaction. The method used is to allow renaturation to proceed only to a particular C_0t value. The reassociated molecules present at that point are then separated from the remaining single-stranded molecules, usually by passing the solution through a tube filled with a form of calcium phosphate crystals (hydroxylapatite) that preferentially binds double-stranded DNA.

5.7 Nucleotide Sequence Composition of Eukaryotic Genomes

The Cot analysis of the DNA of many eukaryotic organisms has shown that eukaryotes vary widely in the proportion of the genome consisting of repetitive DNA sequences and in the types of these sequences that are present. A eukaryote genome typically consists of three fractions:

1. **Unique** or **single-copy sequences.** This is usually the major component and typically represents 30–75 percent of the chromosomal DNA in most organisms. The fraction is identified by the most slowly renaturing component of a Cot curve.

2. **Highly repetitive sequences.** This component, which constitutes 5–45 percent of the genome, is the most rapidly renaturing component of a Cot curve. Many of these sequences may be detected as satellite DNA in CsCl density gradients. The sequences in this class are individually repeated more than 10^5 times.

3. **Middle-repetitive sequences.** This component represents 1–30 percent of a eukaryotic genome and includes sequences that are repeated from a few times to 10^5 times per genome.

It should be noted that the dividing line between many middle-repetitive sequences and highly repetitive sequences is arbitrary.

Unique Sequences

Most gene sequences and the adjacent nucleotide sequences required for their expression are contained in the unique-sequence fraction. With minor exceptions (for example, the repetition of one or a few genes) the genomes of viruses and prokaryotes are composed entirely of single-copy sequences; in contrast, such sequences constitute only 38 percent of the total genome in some sea urchin species, a little more than 50 percent of the human genome, and about 70 percent of the *D. melanogaster* genome.

Highly Repetitive Sequences

The highly repetitive sequences, which are often fairly short, are usually arranged in blocks of tandem repeats. Sequences of this type make up about

6 percent of the human genome and 18 percent of the *D. melanogaster* genome, but 45 percent of the DNA of another *Drosophila* species, *D. virilis*. One of the simplest possible repetitive sequences is composed of an alternating . . .ATAT. . . sequence with about 3 percent G and C interspersed and makes up 25 percent of the genomes of three species of land crabs. In the *D. virilis* genome, the major components of the highly repetitive class are three sequences of seven base pairs, which have the following compositions in one of the complementary strands:

$$5'\text{-ACAAACT-}3'$$
$$5'\text{-ATAAACT-}3'$$
$$5'\text{-ACAAATT-}3'$$

DNA fragments composed of these closely related simple sequences are resolved as distinct satellite fractions in CsCl density gradients and are easily purified. Many highly repetitive sequences are both long and complex, often containing short repeats that are irregularly spaced within the long repeating unit.

Blocks of satellite (highly repetitive) sequences in the genomes of several organisms have been located by *in situ* or **cytological hybridization,** a technique that is a simple extension of the nucleic acid renaturation methods we have described. Cells, some of which are in metaphase, are squashed on a glass cover slip (or simply grown on the glass surface) and treated with an alkaline solution that denatures the cellular DNA. The preparation is then immersed in a solution containing denatured, radioactively labeled copies of a particular repetitive sequence of DNA and subjected to conditions that favor renaturation. The labeled sequences hybridize with complementary se-

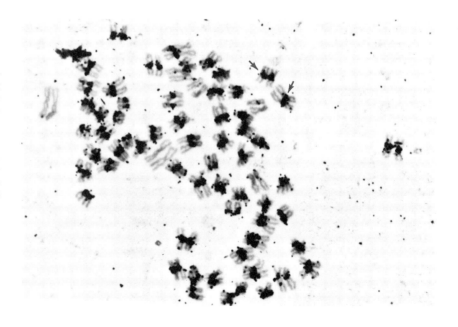

Figure 5-21 Autoradioagram of metaphase chromosomes of the kangaroo rat *Dipodomys ordii;* [³H]RNA copied from purified satellite HS-β DNA sequences have been hybridized to the chromosomes to show the localization of the satellite DNA. Hybridization occurs principally in the regions adjacent to the centromeres (arrows). Note that some chromosomes are apparently free of this satellite DNA. They contain a different satellite DNA not examined in this experiment. (Courtesy of David Prescott.)

quences in the genome of the cell. After the cells have been washed to remove excess radioactive DNA that has not hybridized, the sites at which hybridization has occurred can be determined by the technique of autoradiography (Figure 5-21). The satellite sequences located by this technique have been found to be in the regions of the chromosomes called **heterochromatin.** These are regions that condense earlier in prophase than the rest of the chromosome and are darkly stainable by many standard dyes used to make chromosomes visible (Figure 5-22); sometimes the heterochromatin remains highly condensed throughout the cell cycle. **Euchromatin,** which makes up most of the genome, is visible only during the mitotic cycle. The major heterochromatic regions are adjacent to the centromere; smaller blocks occur at the ends of the chromosome arms (the **telomeres**) and interspersed with the euchromatin. In many species an entire chromosome, such as the Y, is almost completely heterochromatic. Different highly repetitive sequences have been purified from *D. melanogaster* and *in situ* hybridization with *Drosophila* cells has shown that each chromosome has its own distinctive types and distribution of these sequences.

Few genes have been located in heterochromatic regions of chromosomes, so this substantial fraction of the chromatin was formerly considered to be genetically inert. However, detailed genetic and cytogenetic experiments have indicated that heterochromatin has well-defined, though not well-understood, functions in such processes as the pairing and segregation of homologous chromosomes during meiosis, the structural rearrangment of chromosomes, and the regulation of gene expression. The specific DNA sequences that are responsible for these phenomena have not yet been determined.

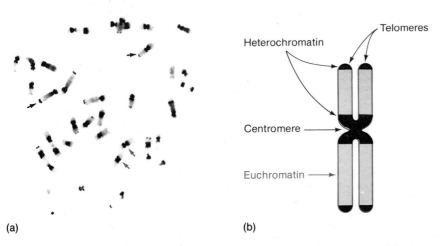

(a) (b)

Figure 5-22 (a) Metaphase chromosomes of the ground squirrel *Ammospermophilus harrissi,* stained to show the heterochromatic regions near the centromere of most chromosomes (red arrows) and the telomeres of some chromosomes (black arrows). (Courtesy of T. C. Hsu.) (b) An interpretive drawing.

Middle-repetitive sequences are a diverse and interesting group constituting about 12 percent of the *D. melanogaster* genome and 40 percent or more of the human and other eukaryotic genomes. These sequences differ greatly in the number of copies and the distribution of these within a genome. They represent many **families** of related sequences and include several groups of genes. For example, the genes for the RNA components of the ribosomes—the particles on which proteins are synthesized (Chapter 9)—and the genes for tRNA molecules, which are also integral in protein synthesis, are repeated in the genomes of all organisms. Genes for the two major ribosomal RNA molecules occur as a tandem pair that is repeated seven times in the *E. coli* genome, and several hundred times in eukaryotes. The genomes of all eukaryotes also contain multiple copies of the histone genes. Each histone gene is repeated about 10 times per genome in chickens, 20 times in mammals, about 100 times in the *D. melanogaster* genome, and as many as 600 times in some sea urchin species. In most eukaryotes the histone components of chromatin are synthesized only during the short interval of the cell cycle in which DNA is synthesized. The organisms in which histone genes are repeated many times are those with extremely rapid nuclear division during early embryo development, and correspondingly short periods of DNA and histone synthesis in which the genome must be replicated.

A substantial proportion of the middle-repetitive DNA of many eukaryotic genomes consists of families of related nucleotide sequences. The individual members of such a group of sequences are quite similar, though some differences in length and nucleotide composition are found among members within a given species. Each species appears to have a distinctive collection of these sequences, with the result that there are often striking differences in this component of the genome between even closely related species. The sequences of these families are typically distributed throughout a genome, where they are interspersed in a more or less regular manner with longer segments of nonrepetitive DNA and also with satellite sequences in the heterochromatin. An abundant family of such interspersed repeated sequences occurring in primate genomes is composed of sequences about 300 base pairs in length. The members of this family are present more than 100,000 times, and possibly 500,000 times, in the human genome. A family of 130-base-pair sequences, distantly related to the abundant family in primates, is distributed throughout the genomes of rodents. Some families of interspersed repeated sequences present in mammalian genomes are composed of segments more than 5000 base pairs long.

The dispersed middle-repetitive-DNA component of the *D. melanogaster* genome consists of at least 40 families of sequences, each with 20–60 copies that are widely scattered throughout the chromosomes. The sequences in a family are usually closely matched, as determined by the formation of stable hybrids in renaturation, but extensive homology does not appear to exist between the sequences of different families. In fact, their size ranges considerably—from 2900 to 7000 base pairs. Many of these sequences are able to move from one location to another in a chromosome and between chromosomes; they are said to be **transposable elements.** Analogous nucleotide sequences are also found in the genomes of yeast, maize, and bacteria (in which they have been extensively studied; see Chapters 7 and 8), and probably occur in all organisms. An important dimension has been added to our understanding of the genome as a structural and functional unit by detection of these mobile elements, for they can in some cases cause chromosome breakage, chromosome rearrangements, modification of the expression of genes, and stable mutations.

5.8 Transposable Elements

The existence of elements that are capable of transposition within a genome was deduced in the 1940s by Barbara McClintock, who discovered them in the analysis of unusual bursts of phenotypic variation in maize. The endosperm of a kernel of maize is derived from the fusion of three nuclei at the time of fertilization and hence is triploid (Figure 2-5). An early example of the exceptional genetic behavior observed by McClintock was the occurrence of numerous sectors of endosperm tissue that were both colored and waxy in kernels of the genotype $C\,wx/C\,wx/I\,Wx$. C and I are alleles at a locus in the short arm of chromosome 9. The C allele is one of several dominant alleles at different loci required for the expression of plant and kernel color, and I acts as a dominant inhibitor of color. The Wx and wx alleles of a second gene in the short arm of chromosome 9 determine nonwaxy versus waxy endosperm, respectively. Cytological study of the meiotic chromosomes of the maize variety in which these phenotypic changes were noticed revealed a chromosome-breakage phenomenon in the short arm of chromosome 9. The simultaneous loss of the I and Wx alleles could be explained by the loss of part of the chromosome arm. The occurrence of such breakage in some cells of the developing endosperm, permitting expression of the C and wx alleles in cell lines that received the broken chromosome and two intact chromosomes, would result in the sectors of colored and waxy tissue observed in the mature kernel (Figure 5-23). Most significantly, the breakage of chromosome 9 was found to occur at or near the site of an element called **Dissociation (Ds).**

Occasional movement (**transposition**) of Ds was detected by observing either the occurrence of chromosome breakage or the modified expression of a gene at or near the new site of the element, accompanied by the disappearance of chromosome breakage at the original site. Furthermore, transposition of Ds was found to occur only when a second element called **Activator (Ac)** is present in the genome. The Ac element has a dosage effect on Ds transposition in that the frequency of transposition of the latter element increases as the number of copies of Ac in the genome increases. In addition, the Ac element itself undergoes movement within the genome and can cause alterations in the expression of genes at or near its insertion site similar to the modifications resulting from the presence of Ds. The movement of either of these elements away from an affected gene often results in restoration of the

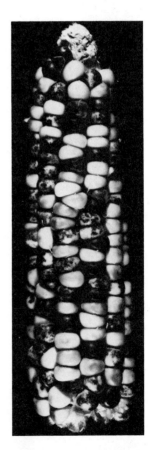

Figure 5-23 Sectors of colored tissue in the endosperm of maize kernels resulting from the presence of the transposable elements Ds and Ac. (Courtesy of Barbara McClintock and the Cold Spring Harbor Laboratory Library Archives. Photo by David Greene.)

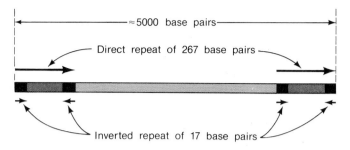

Figure 5-24 Sequence organization of a *copia* transposable element of *Drosophila melanogaster.*

gene to its original type and level of expression, but sometimes results in the production of a stable mutation.

Additional transposable elements with characteristics and genetic effects similar to those of *Ac* and *Ds* are known in maize. Much of the color variegation seen in kernels of the varieties used for decorative purposes are attributable to the presence of one or more of these elements.

Transposable nucleotide sequences are widespread in eukaryotes. In *D. melanogaster* they constitute 5–10 percent of the genome and represent 30–40 distinct families of sequences. A well-studied family of about 30 closely related, but not identical, sequences called ***copia*** is representative of these elements, having structural features similar to those of several different mobile elements found in this organism. *Copia* elements (Figure 5-24) contain about 5000 base pairs with two identical sequences of 267 base pairs that are located terminally and in the same orientation (they are called **direct repeats**). The ends of each of these terminal repeats contain two segments of 17 base pairs, whose sequences are also nearly identical. These shorter segments have opposite orientations and are called **inverted repeats.** Other transposable elements have a similar organization of direct and terminal repeats, as do many such elements in other organisms—for example, the Ty1 element of the yeast *Saccharomyces cerevisiae* and many of the transposable elements in *E. coli* described in Chapter 7. However, some elements in both prokaryotes and eukaryotes lack the long direct repeats, though a short inverted terminal repeat of 10–35 base pairs seems to be universal. The maize *Ac* elements consist of about 4500 base pairs with a terminal inverted repeat of only 11 base pairs. The *Ds* elements, which cannot promote their own transposition, are shorter than *Ac*, but closely related to it, and retain the 11-base-pair terminal repeats.

The molecular processes responsible for the movement of transposable elements are not yet understood (some information will be presented in Chapters 7 and 8). A common feature is that relocation of the element is accompanied by the duplication of 4–12 base pairs originally present at the insertion site, with the result that a copy of this short chromosomal sequence is found immediately adjacent to both ends of the inserted element (Figure 5-25). This sequence is called the **target sequence.** When a DNA sequence is obtained, observation of a pair of inverted-repeat sequences (the terminal segments) separated by a long sequence and flanked by a short direct-repeat sequence (the target sequence) is considered to be the hallmark of a transposable element, even when movement of the element has not been observed. Insertion of a transposable element is not a sequence-specific process, in that at each location the element is flanked by a distinct target sequence; however, the *number* of base pairs in the target sequence is the same at all locations and is characteristic of a particular transposable element. Experimental deletion of part of the base sequences of several different elements has shown that the short terminal inverted repeats are essential for transposition, though the reason is not known. Transposition also requires an enzyme called a **transposase.** Many transposable elements contain a transposase gene, which is located in the central region between the terminal repeats, and hence these elements are able to promote their own transposition. Elements in which the gene has been lost (or inactivated by mutation) are transposable only if a related element is present in the genome to provide this activity. Thus, the inability of the maize *Ds* element to transpose without *Ac* presumably results from the absence of a functional transposase gene in *Ds*.

Recognition in recent years that transposable DNA sequences exist in most genomes and are quite numerous has greatly altered our perception of the organization and stability of the genetic material. In later chapters some of the important genetic consequences of the insertion and excision of these elements will be discussed.

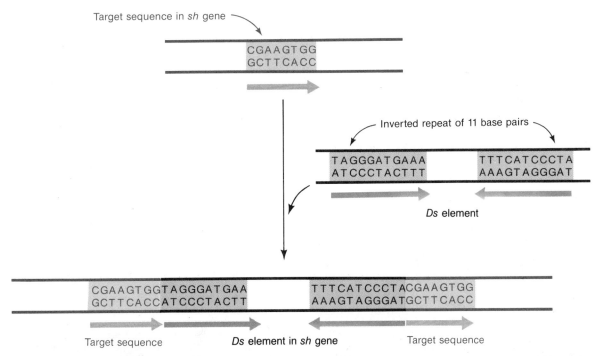

Figure 5-25 The sequence arrangement of a typical transposable element—in this case, *Ds* of maize—and the changes that occur during insertion. *Ds* is inserted into the maize *sh* gene next to a sequence eight base pairs long—the target sequence (direction arbitrarily chosen). In the insertion process the target sequence is duplicated and flanks *Ds*.

5.9 Centromere and Telomere Structure

The centromere is a specific region of the eukaryotic chromosome that becomes visible as a distinct morphological entity along the chromosome during condensation. It is responsible for chromosome movement during both mitosis and meiosis, functioning, at least in part, by providing an attachment site for one or more spindle fibers. Electron microscopic analysis has shown that in some organisms—for example, the yeast *Saccharomyces cerevisiae*—a single spindle-protein fiber is attached to centromeric chromatin. The chromatin segment of the centromeres of yeast has a unique structure in that it is exceedingly resistant to the action of various DNases and has been isolated as a protein-DNA complex containing 220–250 base pairs. The nucleosomal constitution and DNA base sequences of four different yeast chromosomes have been determined; several common features of the base sequences are shown in Figure 5-26(a). There are four regions—labeled I–IV; the sequences in regions I, II, and III are nearly identical, but that of region IV varies from one centromere to another. Region II is noteworthy in that 93 percent of the 82–89 base pairs are AT pairs. The centromeric DNA is contained in a structure (the centromeric core particle) that contains more DNA than a typical yeast nucleosome core particle (160 base pairs) and is larger. This structure is responsible for the resistance of centromeric DNA to DNase. It is not known whether histones or other proteins form the centro-

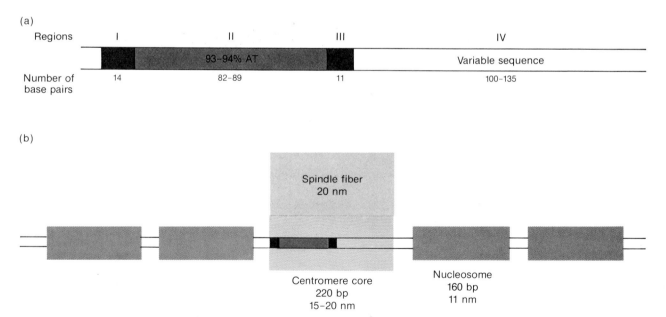

Figure 5-26 A yeast centromere. (a) Diagram of the DNA showing the major regions. The base sequences of regions I–III are nearly the same; those of region IV vary from one centromere to the next. (b) Positions of the centromere core and the nucleosomes on the DNA. The DNA is wrapped around histones in the nucleosomes; the organization and composition of the centromere core are unknown. (After K. S. Bloom, M. Fitzgerald-Hayes, and J. Carbon. 1982. *Cold Spring Harbor Symp. Quant. Biol.,* 47: 1175.)

meric particle. The spindle fiber is believed to be attached directly to this particle (Figure 5-26(b)).

Whether the base-sequence arrangement of the yeast centromeres is typical of eukaryotic centromeres remains to be determined. In higher eukaryotes the chromosomes are about 100 times larger than yeast chromosomes and several spindle fibers are usually attached to each chromosome; thus, it is possible that the centromeres are larger and more complex than those of yeast. Furthermore, the yeast genome is free of highly repetitive sequences, whereas the centromeric regions of the chromosomes of many higher eukaryotes contain large amounts of heterochromatin, consisting of repetitive satellite DNA, as described in Section 5.6.

Telomeres, the complex structures at the ends of eukaryotic chromosomes, are essential for chromosome stability, based on genetic and microscopic observation. For example, in both maize and *Drosophila*, chromosomes that lack telomeres (broken chromosomes) often fuse end to end. Furthermore, if the one end of a single metaphase chromosome is broken away, the chromatids often fuse, forming a chromosome with two centromeres (a dicentric, see Chapter 6).

In Chapter 4 the problem of replication of a linear DNA molecule was described. Telomeres are thought to be responsible for the completion of the replication of the linear DNA molecule contained in a eukaryotic chromosome. A telomere of the protozoan *Tetrahymena* has been isolated. It has an unusual structure, with between 20 and 70 repeats of the sequence 5'-CCCCAA-3', a hairpin terminus, and several gaps in which the first cytosine of the adjacent repeated sequence is absent (Figure 5-27). Various models have been proposed that purport to show how this structure enables replication of the chromosome to be completed, but little evidence is available concerning the validity of any of these models.

5.10 Genomes of Organelles

Some genes are found outside the cellular nucleus. Initial evidence of this was obtained from examples of inheritance in plants that did not follow the patterns predicted from Mendelian principles. Two phenomena distinguish such non-Mendelian inheritance:

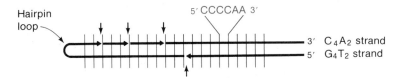

Figure 5-27 A telomere sequence of the protozoan *Tetrahymena*. The vertical red lines separate the repeated six-base-pair sequence. The arrows point to sites of missing nucleotides.

1. Genotypic contributions from the parents in a cross between a mutant and a wildtype are unequal. In extreme cases the gene determining the trait is inherited from only one parent, generally the female, in which case the phenomenon is called **maternal inheritance.**

2. When the genotype of a heterozygous plant has been transmitted from both parents, segregation of the mutant and wildtype phenotypes occurs during growth of the plant. The result is that some sectors of tissue have one parental phenotype and other sectors have the alternative parental phenotype.

A classic example of maternal inheritance occurs in the four-o'clock plant *Mirabilis jalapa*, in which some varieties produce branches with leaves that are green, white, or a mixture of green and white sectors (variegated). Controlled pollination of flowers formed on these three kinds of branches of the same plant indicate that the phenotype of the leaves on a young plant is determined only by the contribution of the female parent; the source of the pollen used is unimportant. That is, seeds formed on green branches produce only green plants, those formed on white branches produce only white plants (which die because they are incapable of carrying out photosynthesis), and those formed on variegated branches produce irregular ratios of green, white, and variegated plants.

Phenomena of this kind, in which the progeny inherit the phenotype of the female parent, can often be explained by the segregation of genes contained in the **chloroplasts** (photosynthetic particles) of green plant cells or in the **mitochondria** (particles that function in respiration) of most eukaryotic cells. Both of these cytoplasmic particles, or **organelles,** contain DNA. The representation in a zygote of the genetic information of an organelle depends on the relative contributions by the parents and sometimes on mechanisms that select against the contribution of one parent. An example of differential contribution by the parents is the maternal inheritance in higher animals of traits determined by mitochondrial genes. Inheritance in this case is to a large extent uniparental because the mitochondria in the zygote tend to be contributed exclusively by the ovum. In the example of *Mirabilis,* in which the color of a cell depends on whether it contains normal green chloroplasts or defective chloroplasts with little chlorophyll, inheritance is again maternal because the plastids are transmitted only through the female gametophyte. When a mixture of the normal and defective plastids are contained in the gametophyte, the resulting plant will be variegated. The patches of green and white tissue and the formation of branches that are all green or all white are the result of a sorting out of the normal and defective plastids in their random distribution during somatic cell divisions. In the green alga *Chlamydomonas reinhardii* the gametes are of the same size and contribute equal amounts of cytoplasm to the zygote formed by their fusion; however, the chloroplast DNA from one parent is degraded (for unknown reasons) soon after formation of the zygote, so only one parent contributes the chloroplast genes.

Most organelle genomes are single circular double-stranded DNA molecules having a unique nucleotide sequence (Figure 5-28). The mitochondrial

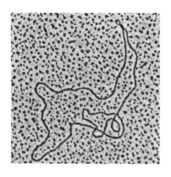

Figure 5-28 Electron micrograph of mouse mitochondrial DNA. (Courtesy of H. Kasamatsu.)

genomes of higher plants are exceptional in that they consist of two or more circular DNA molecules of different sizes. Several copies of the genome are usually present in an organelle, and there are multiple copies of the organelle per cell. In a *Chlamydomonas* cell there is only a single chloroplast with as many as 100 copies of the genome. The chloroplast genome of this unicellular alga contains about 195,000 nucleotide pairs, and those in higher plants contain 120,000–180,000 nucleotide pairs, sizes that are roughly the same as the larger phage DNA molecules (Table 5-1). Mitochondrial genomes are usually smaller but have a considerable range in size. In the mouse, cow, and human, these genomes contain about 16,500 nucleotide pairs, in *D. melanogaster* about 18,500, and in some higher plants more than 100,000. Some basic histonelike proteins are associated with the DNA of these genomes, though organelle DNA is not organized in chromatinlike fibers. It is not known how the DNA is packaged, but the genome structure is likely to resemble that in bacteria (Figure 5-3).

Certain components required for the structure and biological activities of mitochondria and chloroplasts are determined by genes unique to the organelle genomes, but many components are determined by nuclear genes and enter the organelles after being synthesized in the surrounding cytoplasm. The number of genes in mitochondria is not particularly large. For example, in the human mitochondrial genome (16,569 nucleotide pairs) there is evidence for about 37 genes, and in the yeast *S. cerevisiae* (84,000 nucleotide pairs) there is evidence for only about 40. The larger genomes of chloroplasts are thought to contain greater numbers of genes.

Problems

1. Answer the following.
   (a) Consider a long linear DNA molecule, one end of which is rotated four times with respect to the other end in the unwinding direction. The two ends are then joined. If the molecule remains in the underwound state, how many base pairs will be broken?
   (b) If the molecule is allowed to form a supercoil, how many nodes will be present?
   (c) Suppose ten protein molecules are bound to a circular DNA molecule having a nick. Each bound protein molecule breaks one base pair. The nick is then sealed with DNA ligase and the protein molecules are removed by treatment with a protease. What will be the shape of the DNA molecule after removal of the protein?
   (d) A covalently closed circular DNA molecule that is relaxed is partially denatured and 50 base pairs are broken. How does the degree of supercoiling change?

2. Endonuclease S1 makes a strand break only in single-stranded DNA but does not break double-stranded linear DNA. However, S1 can cleave supercoiled DNA, usually making a single break. Why does that occur?

3. Order the DNA molecules shown below from highest to lowest melting temperature.

   (1) AATGTCTTCGAA
       TTACAGAAGCTT

   (2) TAGCTGCATACGAG
       ATCGACGTATGCTC

   (3) AGGCCTCTCGGA
       TCCGGAGAGCCT

4. Which of the following two DNA molecules would have the lower temperature for strand separation? Why?

(1) $\overline{\text{AGTTGCGACCATGATCTG}}$
$\overline{\text{TCAACGCTGGTACTAGAC}}$

(2) $\overline{\text{ATTGGCCCCGAATATCTG}}$
$\overline{\text{TAACCGGGGCTTATAGAC}}$

5. When melting curves of DNA solutions are obtained, the temperature indicated on the x axis is generally that at which the absorbence was measured. However, another kind of melting curve can be obtained by heating the DNA to a temperature T, cooling the DNA to 25°C, and plotting the absorbence at this temperature versus T. Such a melting curve is called an irreversibility curve, since an increase in absorbence is observed only if a DNA molecule is irreversibly denatured—that is, if it remains single-stranded at 25°C. For a melting curve having the shape of that in Figure 5-17, draw an irreversibility curve for the following DNA samples: (1) Unbroken identical DNA molecules isolated from a phage sample. (2) Randomly fragmented bacterial DNA.

6. DNA from organism A, labeled with ^{14}N and randomly fragmented, is renatured with an equal concentration of DNA from organism B, labeled with ^{15}N and randomly fragmented, and then centrifuged to equilibrium in CsCl. Five percent of the total renatured DNA has a hybrid density. What fraction of the base sequences are common to the two organisms?

7. A Cot curve is shown in the figure below.
(a) What fraction of the DNA is contained in components that are unique? What fraction is redundant? What fraction might be satellite?
(b) If the unique sequences are represented once per haploid genome, how many copies per haploid genome are there of the other two classes?

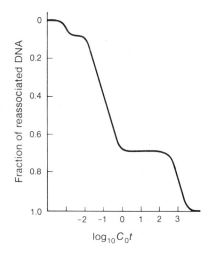

8. Define the terms chromatin, histone, nucleosome, core particle, and chromatosome.

9. A DNA-binding protein binds tightly to double-stranded DNA and very poorly to single-stranded DNA. It can bind to all base sequences with equal affinity. In 1 M NaCl binding is poor. What is the probable binding site of the DNA?

10. A sample of identical DNA molecules each containing about 3000 base pairs per molecule is mixed with histone octamers under conditions that allow formation of chromatin. The reconstituted chromatin is then treated with a nuclease and enzymatic digestion is allowed to occur. The histones are removed and the positions of the cuts in the DNA are identified by sequencing the fragments. It is found that the breaks have been made at *random* positions, and, as expected, at about 200-base-pair intervals. The experiment is then repeated with a single variation. A protein P known to bind to DNA is added to the DNA sample before addition of the histone octamers. Again, reconstituted chromatin is formed and digested with the nuclease. In this experiment it is found that the breaks are again at intervals of about 200 base pairs but they are localized at particular positions in the base sequence. At each position the site of breakage can vary only over a 2–3 base range. However, examination of the base sequences in which the breaks have occurred does not indicate that breakage occurs in a particular sequence—that is, each 2–3-base region in which cutting occurs has a different sequence. Explain the difference between the two experiments.

11. What is meant by the terms "direct repeat" and "inverted repeat?" Use the sequence ABCD as an example.

12. Since at present there is no detailed understanding of the mechanism of transposition, what is the evidence that compels one to conclude that DNA replication is an essential step?

13. The denaturation of two unusual DNA molecules, one consisting of 99 percent AT pairs (molecule I) and the other consisting of 99.9 percent AT pairs (molecule II), is being studied. Both molecules have the same value of T_m. However, physical analysis shows that the individual strands of molecule II separate completely at a temperature 12°C higher than molecule I. Suggest an explanation for this phenomenon.

14. A molecule containing 5000 nucleotide pairs consists only of the sequence ACGTAG. How would the melting temperature of this molecule be related to a molecule with the same base composition (50 percent AT pairs) but whose sequence is not repeated?

15. A bacterial species is known to contain a supercoiled DNA molecule. The supercoil is isolated from a particular bacterial strain, and examination of the supercoil by electron microscopy indicates that the molecule is twice as long as normal. However, two other possibilities are that (1) it is a dimer or (2) it consists of two supercoiled monomers linked as in a chain. A simple experimental procedure will distinguish these alternatives: the isolated DNA is treated with a slow-acting DNase and at various times the molecules are examined by electron microscopy. What observations would confirm one model or the other?

16. A Cot analysis is carried out with DNA obtained from a particular plant. An identical plant is kept in the dark for several weeks until it loses all of its chlorophyll—that is, the leaves become white. DNA isolated from this plant has the same Cot curves as that from green plants. What does this finding tell you about the effect of continued darkness on chloroplasts in this species?

17. Most DNases make single-strand breaks in DNA. However, if a DNAse is added to DNA contained in a solution containing 1 M NaCl, the DNA molecule is completely resistant to the enzyme. Suggest an explanation.

18. Mutations have not been observed that give rise to altered histones. Why should this result be expected?

19. A fraction of middle-repetitive DNA is isolated and purified. It is used as a template in a polymerization reaction with DNA polymerase and radioactive substrates, and highly radioactive DNA is prepared. The radioactive DNA is then used in an *in situ* hybridization experiment with cells containing polytene chromosomes obtained from 100 different flies of the same species. With most of the flies autoradiography indicates that the radioactive material is localized in 22 particular bands of the chromosomes. However, in one fly one of the radioactive sites is absent but there is another site in a different band; in another fly there are 23 labeled bands. What does this observation suggest about the DNA sequence being studied?

CHAPTER 6

Variation in Chromosome Number and Structure

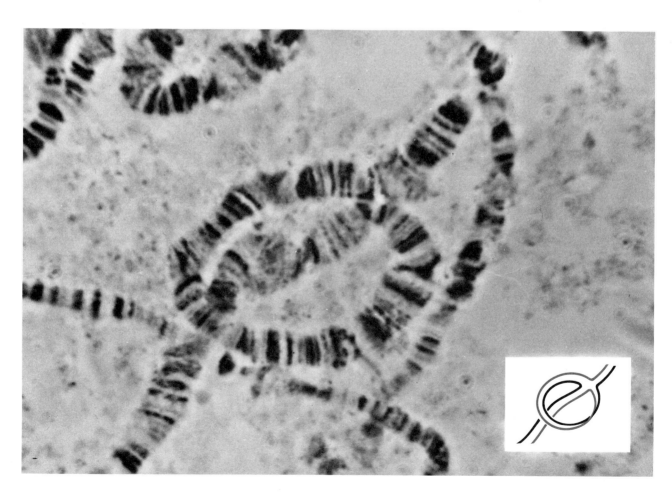

Chromosomes may be thought about in two complementary ways—as physical objects having a particular molecular composition and organization (the perspective of Chapter 5) and as genetic abstractions that cannot always be studied directly but that can be understood indirectly through their effects on inherited characteristics. Solely by genetic observations, one can infer the following: (1) Chromosomes carry genes. (2) Chromosomes occur in homologous pairs. (3) During the formation of gametes the homologous chromosomes undergo pairing and crossing over. (4) Crossing over results in recombination of genes between homologous chromosomes and permits the construction of genetic maps. (5) Genetic maps indicate that genes in chromosomes are arranged in a line. (6) Each organism has its own characteristic chromosome number corresponding to the number of linkage groups. (7) Genes that are located in different linkage groups show independent assortment. In short, because of inferences derived from genetic analysis, one would know that chromosomes exist and would understand most *genetic* characteristics even if microscopes had never been invented.

Genetic analysis has also been used to identify and study abnormalities in chromosomes. In all species, individuals are occasionally found that have extra chromosomes or that lack chromosomes. These represent abnormalities in chromosome *number*. Other rare individuals are found to have an alteration in the arrangement of genes in the genome, such as by having a chromosome with a particular segment missing, reversed in orientation, or attached to a different chromosome. These represent abnormalities in chromosome *structure*. In this chapter we will discuss the genetic effects of both numerical and structural chromosome abnormalities. It will be seen that plants are much more tolerant of such changes than animals are, and that in animals numerical alterations often produce stronger effects on phenotype than do structural alterations.

Facing page: An inversion of the distal half of the X chromosome of *Drosophila melanogaster* in heterozygous combination with a normal X. Inset: An interpretive drawing. (Courtesy of Dan Lindsley.)

6.1 Chromosomes as Physical Objects

In Chapter 5 several features of the physical structure of chromosomes were described, and the chromosome was seen to be a highly folded unit with numerous stages of packing. A key structure is the centromere, because the spindle fibers attach to the centromere during cell division and enable the sister chromatids to be separated and moved to opposite poles. Occasionally chromosomes arise having an abnormal number of centromeres, as diagrammed in Figure 6-1(a). The upper chromosome has two centromeres and is said to be **dicentric.** Such aberrant chromosomes are unstable because they are frequently lost from cells when the two centromeres proceed to opposite poles during cell division; in this case the chromosome is stretched and forms a *bridge* between the daughter cells, which may not be included in either daughter nucleus or may break, with the result that each daughter nucleus receives a broken chromosome. The lower chromosome in the figure is an **acentric** chromosome, which lacks a centromere. Acentric chromosomes are also unstable because they cannot be maneuvered properly during cell division and tend to be lost. Because acentric and dicentric chromosomes are frequently lost as a result of problems during cell division, only chromosomes that have a single centromere are regularly transmitted from parents to offspring.

Chromosomes are often described in terms of their form during anaphase movement. Three distinct shapes are seen, resembling a V, or a J, or an I. The shape is determined by the position of the centromere and the relative length of the lagging chromosome arms (Figure 6-1(b)). A V-shaped chromosome has its centromere approximately in the middle, forming arms of about equal length, and is called a **metacentric** chromosome. A J-shaped chromosome has an off-center centromere forming arms of unequal length; such chromosomes are called **submetacentric.** When the centromere is very close to one end, the chromosome appears I-shaped at anaphase because the arms are grossly unequal in length; such a chromosome is termed **acrocentric.**

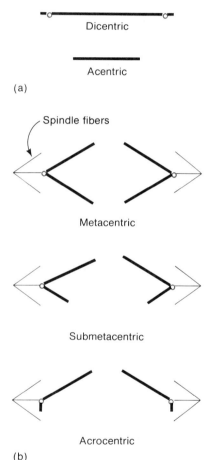

(a)

(b)

Figure 6-1 (a) A diagram of a dicentric (two centromeres) and an acentric (no centromere) chromosome. Dicentric and acentric chromosomes are frequently lost during cell division, the former because the two centromeres may bridge between the daughter cells, and the latter because the chromosome cannot attach to the spindle fibers. (b) The three possible shapes of chromosomes in anaphase. The centromeres are shown in red.

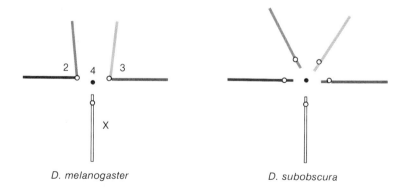

Figure 6-2 The haploid chromosome complement of two species of *Drosophila.* Shading indicates homology of chromosome arms. The large metacentric chromosomes of *Drosophila melanogaster* (chromosomes 2 and 3) correspond arm for arm with the four large acrocentric autosomes of *Drosophila subobscura.*

The distinction between a metacentric, submetacentric, and acrocentric chromosome is somewhat arbitrary, but the terms are useful because they provide a physical image of the chromosome. More important, chromosome evolution often tends to conserve the number of chromosome *arms* without conserving the number of *chromosomes*. For example, *Drosophila melanogaster* has two large metacentric autosomes, but many other *Drosophila* species have four acrocentric autosomes instead. Detailed comparison of the genetic maps of these species reveals that the acrocentric chromosomes in the other species correspond, arm for arm, with the large metacentrics in *D. melanogaster* (Figure 6-2). Also, chimpanzees and humans have 22 pairs of chromosomes that are morphologically similar, but chimpanzees have two pairs of acrocentrics not found in humans, and humans have one pair of metacentrics not found in chimpanzees. In this case, each arm of the human metacentric is homologous to one of the chimpanzee acrocentrics. Chromosome evolution often occurs through the fusion of two acrocentrics to produce a single metacentric, or the splitting of a metacentric to produce two acrocentrics. One way in which chromosomal fusion can occur will be discussed later in this chapter.

6.2 Polyploidy

Species of the genus *Chrysanthemum* illustrate an important phenomenon found frequently in flowers and other higher plants. One *Chrysanthemum* species has 18 chromosomes, whereas a closely related species has 36. However, comparison of chromosome morphology indicates that the 36-chromosome species has two complete sets of the chromosomes found in the 18-chromosome species. This phenomenon is known as **polyploidy.** The basic haploid chromosome number in the group is nine, which is the chromosome number found in gametes of the 18-chromosome species. That is, the 18-chromosome species has two copies of each of the nine chromosomes of the haploid set and so is a normal diploid. The 36-chromosome species has four copies of each of the nine basic chromosomes ($4 \times 9 = 36$) and is called a **tetraploid.** Other species of *Chrysanthemum* have 54 chromosomes (6×9), 72 chromosomes (8×9), and 90 (10×9).

During meiosis the chromosomes of all *Chrysanthemum* species pair to form bivalents (Section 2.3). The 18-chromosome species forms nine bivalents, the 36-chromosome species forms 18 bivalents, the 54-chromosome species forms 27 bivalents, and so on. Gametes receive one chromosome from each bivalent, so the number of chromosomes in the gametes of any species will be exactly half of the number of chromosomes found in its somatic cells. For example, the 90-chromosome species forms 45 bivalents, so gametes will carry 45 chromosomes. When two 45-chromosome gametes come together during fertilization, the complete set of 90 chromosomes in the species is restored. Thus, the gametes of a polyploid organism are not always haploid, as they are in a diploid; a tetraploid organism has diploid gametes.

Polyploidy is widespread in certain plant groups, and it occurs in many valuable crop plants, such as wheat, oats, cotton, potato, banana, coffee, and sugar cane. Among flowering plants, at least one-third of existing species originated as some form of polyploid. Polyploidy often leads to an increase in the size of individual cells, and polyploid plants are often larger and more vigorous than their diploid ancestors; however, there are many exceptions to these generalizations. Polyploidy is rare in vertebrate animals, but it is common in a few groups of invertebrates.

Polyploid plants occurring in nature almost always have an even number of sets of chromosomes, because organisms having an odd number have low fertility. If a diploid and a tetraploid *Chrysanthemum* were crossed, the resulting hybrid would have three sets of chromosomes (one set from the diploid parent and two sets from the tetraploid), or 27 chromosomes altogether (3 × 9). Organisms with three complete sets of chromosomes are known as **triploids.** As far as growth is concerned, the hybrid will be quite normal because the triploid condition does not interfere with mitosis; in mitosis in triploids (or any other type of polyploid), each chromosome replicates and divides just as in a diploid. However, chromosome segregation is severely upset during meiosis. In a triploid, each chromosome is represented by three homologues, which can pair in several ways. For example, two of the homologues may pair and separate normally, one going to each pole, with the remaining unpaired chromosome proceeding randomly to one pole or the other. However, two unpaired chromosomes may go either to the same pole or to different poles, so a variety of gametic types will be produced. This situation is illustrated in Figure 6-3 for a triploid with three copies of each of two chromosomes. The paired chromosomes in the middle will segregate normally, but segregation of the unpaired chromosomes in panel (a) will produce the gametes shown at the left, and segregation of the type represented in panel (b) will produce the gametes shown at the right. More commonly, different regions along one chromosome may pair with different partners, as diagrammed in Figure 6-4,

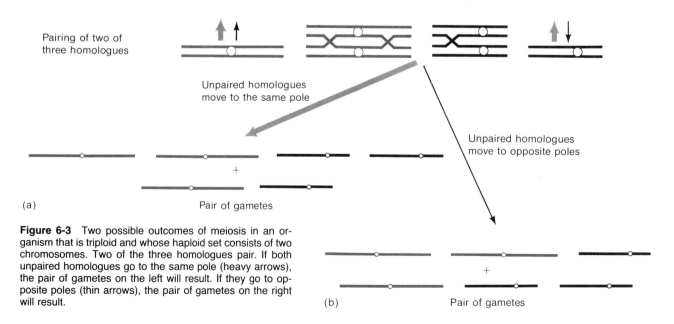

Pairing of two of three homologues

Unpaired homologues move to the same pole

Unpaired homologues move to opposite poles

(a)   +   Pair of gametes

Figure 6-3 Two possible outcomes of meiosis in an organism that is triploid and whose haploid set consists of two chromosomes. Two of the three homologues pair. If both unpaired homologues go to the same pole (heavy arrows), the pair of gametes on the left will result. If they go to opposite poles (thin arrows), the pair of gametes on the right will result.

(b)   +   Pair of gametes

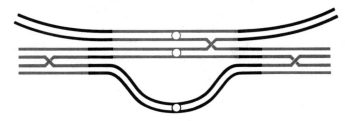

Figure 6-4 Multivalent association of homologous chromosomes in a triploid in meiosis, caused by a switch in pairing partners along the length of the chromosome. Paired regions are shown in red.

forming a **multivalent** (instead of a bivalent plus one unpaired chromosome as in Figure 6-3). A multivalent with three chromosomes will typically segregate with two chromosomes going to one pole and one chromosome going to the opposite pole. The result is that, however the chromosomes pair, irregular segregation during meiosis in a triploid organism will produce a diversity of gametic types with different chromosome complements, many of which will have an unequal number of each type of chromosome. Such inequalities often result in inviability of the gametophyte (the usual outcome in plants) or inviability of the zygote (the usual outcome in animals), and this is the cause of the reduced fertility. The same types of irregularities in chromosome segregation can occur whenever an organism has an odd number of each homologous chromosome. Unless the organism can perpetuate itself by means of asexual reproduction, it will eventually become extinct. This is why the vast majority of sexually reproducing polyploid species found in nature have an even number of chromosome sets (four, six, or eight).

In nature, triploid organisms can be formed from diploids in several ways. For example, exposure of the germ cells of some plants to heat results in a failure of chromosome separation during one of the meiotic divisions and leads to the formation of diploid gametes. Failure of chromosome reduction resulting in diploid gametes can also occur spontaneously. When fertilized by a normal gamete, a diploid gamete will produce a triploid plant. In animals, a triploid organism can occur by fertilization of a single female gamete by two male gametes. However formed, triploids can seldom perpetuate themselves sexually for many generations because of their greatly reduced fertility.

Tetraploid organisms can also be produced in several ways. The simplest mechanism is a failure of chromosome separation during mitosis, which instantly doubles the chromosome number. In a plant species that can undergo self-fertilization, such an occurrence creates a new, genetically stable species, because the chromosomes in the tetraploid can pair two by two during meiosis and therefore segregate regularly, each gamete receiving a full diploid set of chromosomes. Self-fertilization of the tetraploid restores the chromosome number, so the tetraploid condition can be perpetuated. Theoretically, chromosome pairing in tetraploids can also result in the formation of multivalents and aberrant chromosome segregation, and this is often observed in experimentally produced tetraploids. However, naturally occurring tetraploids form only bivalents, which is thought to result from the action of specific genes that control chromosome pairing. Therefore, tetraploids can be fully fertile and produce tetraploid offspring when self-fertilized or crossed with other tetraploids. However, a cross between a tetraploid and a diploid results in a triploid. An **octoploid** (eight sets of chromosomes) can be generated by failure of chromosome separation during mitosis in a tetraploid. If only bivalents form during meiosis, an octoploid organism can be perpetuated sexually by self-fertilization or through crosses with other octoploids. Furthermore, cross-fertilization between an octoploid and a tetraploid results in a **hexaploid** (six sets of chromosomes), which can also form bivalents and produce hexaploid offspring by means of self-fertilization or through crosses with other hexaploids. Such repeated episodes of polyploidization and cross-fertilization may ultimately produce an entire polyploid series of closely related organisms differing in chromosome number, as exemplified in *Chrysanthemum*.

Chrysanthemum represents just one type of polyploidy. In this case, all chromosomes in the polyploid species derive from a single diploid ancestral species. Polyploidy derived from the multiplication of a single ancestral set of chromosomes is known as **autopolyploidy.** However, in many cases of polyploidy, the polyploid species have complete sets of chromosomes from two or more different ancestral species. Such polyploids are known as **allopolyploids.** The presence of genomes from multiple ancestral species indicates that allopolyploids derive from rare hybrids that occur between distinct diploid species.

Hybridization between species occurs when pollen from one species germinates on the stigma of another species and sexually fertilizes the ovule. The pollen may be carried to the wrong flower by wind, insects, or other pollinators. Figure 6-5 illustrates hybridization between species A and B leading to the formation of an allopolyploid (in this case an *allotetraploid*), which carries a complete diploid genome from each of its two ancestral species. Hybridization and the formation of allopolyploids is an extremely im-

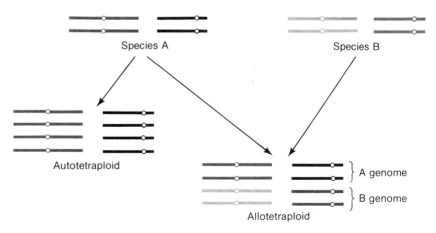

Figure 6-5 Formation of an autotetraploid by doubling of a complete diploid set of chromosomes and of an allotetraploid by union of two different diploid sets.

portant process in plant evolution and plant breeding. At least half of all naturally occurring polyploids are allopolyploids. Cultivated wheat provides an excellent example of allopolyploidy. Cultivated wheat has 42 chromosomes representing a complete diploid genome of 14 chromosomes from each of three ancestral species. The 42-chromosome allopolyploid is thought to have originated by the hybridizations outlined in Figure 6-6.

Note in Figure 6-5 that hybridization between species A and B would initially produce an organism having only one haploid set of chromosomes from each ancestor; its fertility would be expected to be low because each chromosome lacks a pairing partner during meiosis and would segregate irregularly, as with the unpaired chromosomes in the triploid in Figure 6-3. Although the hybrid could reproduce asexually, formation of a sexually fertile allotetraploid requires a doubling of the chromosomes within the hybrid in order to provide each chromosome with a pairing partner. Since the species hybrid will be formed repeatedly over the course of perhaps thousands of generations, there will be many opportunities for the rare chromosome doubling to occur—for example, as a result of failure of the chromosomes to separate during cell division. Once the doubling happens, the allotetraploid can perpetuate itself sexually by means of self-fertilization or in crosses with other allotetraploids.

Many new, useful allopolyploid species have been created artificially, essentially by the process outlined in Figure 6-5. The procedure is made practical because of the availability of drugs such as **colchicine,** which, when applied to cells of the F_1 hybrid of an interspecific cross (a cross between different species), causes disruption of the mitotic spindle and thereby leads to the required chromosome doubling. Colchicine treatment of cells of an ordinary diploid results in chromosome doubling to produce the autotetraploid.

The genetics of polyploid species is more complex than that of diploid species because polyploid individuals carry more than two alleles at any locus. With two alleles in a diploid only three genotypes are possible—(AA, Aa, and aa,) whereas in a tetraploid five possible genotypes are possible—$AAAA$,

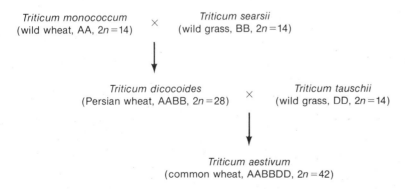

Figure 6-6 Hybridizations presumed to have occurred in the ancestry of cultivated wheat (*Triticum aestivum*), which is an allohexaploid containing complete diploid genomes (AA, BB, DD) from three ancestral species.

AAAa, AAaa, Aaaa, and *aaaa* —the middle three of which represent different types of tetraploid heterozygotes. Moreover, the diploid gametes formed by an autotetraploid occur in ratios determined not only by the genotype of the allotetraploid but by the map distance between the gene and its centromere. For example, an allotetraploid of genotype *AAaa* will produce gametes of genotype *AA, Aa,* and *aa* with a ratio ranging from $1/6 : 4/6 : 1/6$, if no crossing over occurs between the *A* gene and the centromere, to $2/9 : 5/9 : 2/9$, if recombination between gene and centromere occurs freely. Consequently, phenotypic ratios expected from tetraploid crosses are generally different from those in diploids, and many more types of matings are possible. Nevertheless, polyploid genetics is important because many cultivated plants are polyploid, and plant breeders use their knowledge of genetics to design breeding programs in order to transfer valuable genes, such as disease-resistant genes, from wild diploid species into cultivated polyploids.

6.3 Extra or Missing Chromosomes

Occasionally organisms arise in which individual chromosomes, rather than entire sets of chromosomes, are present in abnormal numbers. This situation is called **polysomy.** In contrast to polyploids, which in plants are often healthy and in some cases more vigorous than the diploid, polysomics are usually less vigorous than the diploid and are frequently abnormal in phenotype. Thus, whereas the presence of complete extra sets of chromosomes in polyploids is not necessarily harmful to the organism, the occurrence of a single extra chromosome (or a missing chromosome) may have large effects. For example, Figure 6-7 shows the seed capsule of the Jimson weed *Datura stramonium,* beneath which is a series of capsules of strains, each having an extra copy of a different chromosome. An otherwise diploid organism having an extra copy of an individual chromosome is called a **trisomic.** Note in Figure 6-7 that the seed capsule of each of the trisomics has distinctive abnormalities.

Since polysomy generally results in more severe phenotypic effects than polyploidy does, the harmful phenotypic effects in trisomics must be related to the imbalance in the number of copies of different genes. A polyploid organism has a balanced genome in the sense that the ratio in the number of copies of any pair of genes is the same as in the diploid. For example, in a tetraploid each gene is present in twice as many copies as in a diploid, so no gene or group of genes is out of balance with the others. Balanced chromosome abnormalities, which retain equality in the number of copies of each gene, are said to be **euploid.** In contrast, gene equality is upset in a trisomic because three copies of the genes located in the trisomic chromosome will be present, whereas two copies of the genes in the other chromosomes will be present. Such unbalanced chromosome complements are said to be **aneuploid.** By using these terms the principle illustrated in Figure 6-7 may be summarized by the statement that aneuploid abnormalities are usually more severe than euploid abnormalities. An example of the same principle occurs in *Drosophila,* in which triploid females are viable, fertile, and nearly normal in morphology; however, trisomics for either of the two large autosomes are invariably lethal.

In organisms in which trisomics are viable, the extra chromosome is useful in assigning unmapped genes to chromosomes. With this method, a diploid individual homozygous for an unmapped recessive mutation is crossed with a trisomic, and trisomic offspring are backcrossed with the homozygous

Diploid

1   2   3   4

5   6   7   8

9   10   11   12

Trisomics

Figure 6-7 Seed capsules of the normal diploid *Datura stramonium* (Jimson weed), which has a haploid number of 12 chromosomes, and each of the 12 possible trisomics. The phenotype of the seed capsule in trisomics differs according to the chromosome that is trisomic.

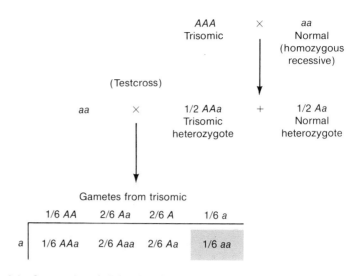

Figure 6-8 Segregation of alleles *A* and *a* in a trisomic organism of genotype *AAa* results in a phenotypic ratio of 5:1 in testcross progeny (assuming dominance of *A*) instead of the expected 1:1 ratio. Such aberrant ratios are useful in assigning unmapped recessive genes to chromosomes.

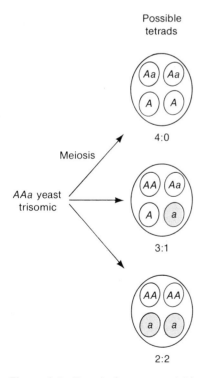

Figure 6-9 Tetrads from a yeast trisomic with the genotype *AAa*. Trisomic yeast cells produce many aberrant 4:0 and 3:1 tetrads because some of the (otherwise haploid) spores are disomic.

recessive diploid (Figure 6-8). If the unmapped gene is not located on the trisomic chromosome, then segregation will occur exactly as in a normal diploid and the ratio of phenotypes in the next generation will be the expected 1:1. However, if the unmapped gene is located on the chromosome that is trisomic, then the gametes produced by the trisomic parent (genotype *AAa*) will be as shown below:

1/6 *AA*	(disomic gamete)
2/6 *Aa*	(disomic gamete)
2/6 *A*	(normal gamete)
1/6 *a*	(normal gamete)

The disomic gametes carry two copies of the chromosome in question, and the ratio of dominant to recessive phenotypes in the next generation will be 5:1 instead of 1:1. Thus, by performing the cross in Figure 6-8 with a series of trisomics, each having an extra copy of a different chromosome, the unmapped gene can be assigned to a chromosome by noting the cross in which abnormal segregation occurs. Mapping by means of trisomics is a popular method in yeast genetics, because yeast trisomics with genotype *AAa* will produce many 3:1 and 4:0 tetrads instead of the normal 2:2 (Figure 6-9).

Just as an occasional individual may have an extra chromosome, a chromosome may also be missing. Such an individual is said to be **monosomic** for the missing chromosome. In general, a missing copy of a chromosome will result in more harmful effects than when an extra copy of the same chromosome is present, and monosomics are often lethal. Organisms in which monosomics are viable provide a particularly straightforward method of assigning unmapped genes to chromosomes. When an individual that is homozygous for an unmapped recessive is crossed with a monosomic for the same chromosome, half of the resulting progeny will be monosomic and will exhibit the recessive phenotype (see Figure 6-10). However, if the monosomic chromosome and the chromosome carrying the recessive allele are different, then all of the offspring will be phenotypically normal heterozygotes. Unfortunately, monosomics in most diploid organisms are inviable or so severely abnormal that they do not reproduce.

6.4 Human Chromosomes

The chromosome complement of a normal human male is illustrated in Figure 6-11. The chromosomes have been treated with a staining reagent, called Giemsa, which causes the chromosomes to exhibit horizontal bands that are specific for each pair of homologues. These bands permit the chromosome pairs to be identified individually. By convention, the chromosome pairs are arranged and numbered from longest to shortest and separated into seven groups designated by the letters A through G; this conventional representation of chromosomes is called a **karyotype,** and it is obtained by cutting individual chromosomes out of photographs taken during metaphase and pasting them into place. In a karyotype of a normal human female, the autosomes would not differ from those of a male and hence would be identical to those in Figure 6-11; however, there would be two X chromosomes instead of an X and a Y. Aberrations in chromosome number and morphology are made evident by a karyotype.

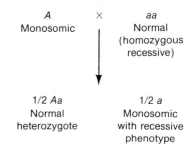

A	×	aa
Monosomic		Normal (homozygous recessive)

1/2 Aa	1/2 a
Normal heterozygote	Monosomic with recessive phenotype

Figure 6-10 Result of a cross between a homozygous recessive individual and a monosomic for the same chromosome. The recessive phenotype appears in the F_1 generation. This procedure facilitates assignment of recessive genes to chromosomes, but unfortunately in most organisms monosomics are nearly inviable or fertile.

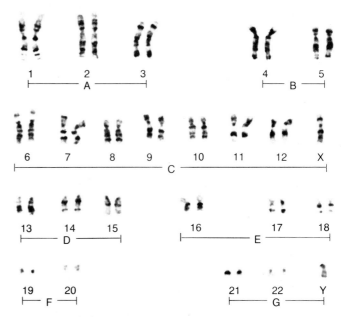

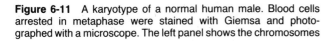

Figure 6-11 A karyotype of a normal human male. Blood cells arrested in metaphase were stained with Giemsa and photographed with a microscope. The left panel shows the chromosomes as seen in the cell by microscopy. In the right panel, the chromosomes have been cut out of the photograph and paired with their homologues. (Courtesy of Patricia Jacobs.)

Trisomy in Humans

Monosomy or trisomy of most human autosomes is usually incompatible with life, although a few exceptions are known. One of these is **Down syndrome,** which is caused by trisomy of chromosome 21 (Figure 6-12). (Modern medical terminology drops the possessive in diseases named after individuals; thus, what was once called Down's syndrome is now called Down syndrome.) Down syndrome affects about 1 in 750 live-born children. Its major symptom is mental retardation, but multiple physical abnormalities occur as well, such as major heart defects in about half the cases.

The great majority of cases of Down syndrome are caused by nondisjunction (failure in the separation of homologous chromosomes during meiosis, as explained in Chapter 2), resulting in a gamete containing two copies of chromosome 21. For unknown reasons, nondisjunction of chromosome 21 is more likely to occur during oogenesis than during spermatogenesis, so the abnormal gamete in the case of Down syndrome is usually the egg. Moreover, nondisjunction of chromosome 21 increases dramatically with the age of the mother, with the risk of Down syndrome reaching 6 percent in mothers of age 45 and over (Figure 6-13). Thus many physicians recommend that older women who are pregnant have amniocentesis performed in order to detect Down syndrome prenatally. **Amniocentesis** is a procedure in which

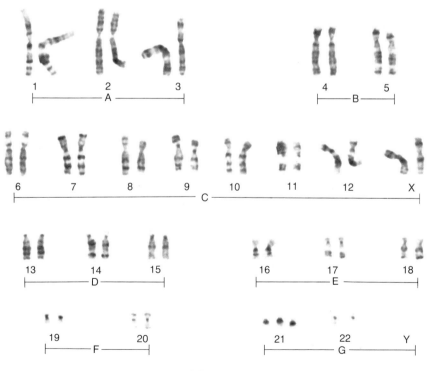

Figure 6-12 A karotype of a female with Down syndrome. Three copies of chromosome 21 are present. (Courtesy of P. Howard-Peebles.)

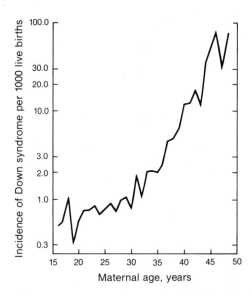

Figure 6-13 Incidence of Down syndrome per 1000 live births related to maternal age. (Based on data from E. B. Hook and A. Lindsjo. *Am. Jour. Hum. Gen.* 30 (1978): 19–27.)

cells of a developing fetus are obtained by insertion of a fine needle through the wall of the uterus and into the sac of fluid containing the fetus.

When nondisjunction is the cause of Down syndrome, the probability that a second child will be affected is approximately the same as the risk for a woman of the same age who has not had an affected child. However, in some families the probability of Down syndrome is very high—up to 20 percent of births. This high risk is caused by a chromosome abnormality in one of the parents, which will be discussed later in this chapter.

Two other human autosomal trisomies—trisomy 13 and trisomy 18—also occur in live-born children, but they are quite rare. The affected children have severe physical and mental abnormalities and usually live only a few days or weeks.

Sex-Chromosome Abnormalities and Dosage Compensation

Abnormalities in the number of sex chromosomes usually produce less severe phenotypic effects than abnormalities in the number of autosomes do, for two reasons. First, the Y chromosome in mammals carries very few genes other than those that trigger male embryonic development, so extra Y chromosomes are milder in their effects on phenotype than extra autosomes are. Second, extra X chromosomes in mammals are genetically inactivated very early in embryonic development, and consequently their effects on phenotype are reduced compared to extra autosomes.

X-chromosome inactivation is normal in female mammals. In humans, at an early stage of embryonic development, one of the two X chromosomes is inactivated in each somatic cell; different tissues undergo X inactivation at different times, but germ cells do not undergo X inactivation. The X chromosome that will be inactivated in a particular somatic cell is selected at random, but once the decision is made, in all of the descendants of the cell, that X chromosome will also be inactivated.

Inactivation is thought to proceed in both directions along the chromosome from a primary inactivation center, with the eventual result that most (but not all) genes in the chromosome become inactive. A small group of genes located near the tip of the short arm of the X chromosome escapes inactivation. A counting mechanism of unknown nature keeps one X chromosome active in each cell. Inactivation has two consequences. First, it equalizes the number of active copies of X-linked genes in females and males. Although a female has two X chromosomes and a male has only one, because of inactivation of one X chromosome in each of the somatic cells of the female, the number of *active* X chromosomes is made the same (one) in both sexes. This equalization is called **dosage compensation.** (Other organisms with XX-XY sex determination, such as *Drosophila*, also have dosage compensation, but the mechanism is different and more complex than the X chromosome inactivation found in mammals.)

Second, a normal female will be a **mosaic** for X-linked genes (Figure 6-14). That is, each somatic cell will express the genes in only one X chromo-

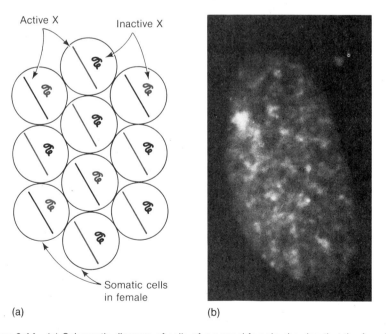

(a)

(b)

Figure 6-14 (a) Schematic diagram of cells of a normal female showing that the female is a mosaic for X-linked genes. The two X chromosomes are indicated in red and black. An active X is depicted as a straight line, and an inactive X as a tangle. Each cell has just one active X, but the particular X that remains active is a matter of chance. In humans, the inactivation includes all but a few genes in the tip of the short arm. (b) A fluorescence micrograph of a human cell showing a Barr body (bright spot at the upper left.) This cell is from a normal human female, and it has one Barr body. (Courtesy of A. J. R. de Jonge.)

some, but the X chromosome that is genetically active will differ from cell to cell. This mosaicism has been observed directly in females that are heterozygous for X-linked alleles that determine different forms of an enzyme, A and B; when cells from the heterozygous female are individually cultured in the laboratory, half of the clones are found to produce only the A form of the enzyme and the other half produce only the B form. Mosaicism can be observed directly in women who are heterozygous for an X-linked recessive mutation resulting in the absence of sweat glands; these women exhibit patches of skin in which sweat glands are present (these patches are derived from embryonic cells in which the normal X chromosome remained active and the mutant X was inactivated), and other patches of skin in which sweat glands are absent (these patches are derived from embryonic cells in which the normal X chromosome was inactivated and the mutant X remained active.) Females that are heterozygous for mutations in protein-coding genes usually produce approximately half as much protein per gram of tissue as do homozygous normal females (because the wildtype X chromosome is active in only half the cells, as is shown in Figure 6-14(a)), but this amount is usually sufficient to produce a normal phenotype.

In certain cell types the inactive X chromosome in females can be observed microscopically as a densely staining body in the nucleus of interphase cells. This is called a **Barr body** (Figure 6-14(b)). Although cells of normal females have one Barr body, cells of normal males have none.

Individuals with two or more X chromosomes have a number of Barr bodies equal to the number of X chromosomes minus one (that is, equal to the number of inactivated X chromosomes). For example, XXX individuals have two Barr bodies, XXXX individuals have three, and XXXXX individuals have four. However, extra Y chromosomes are not inactivated and do not form Barr bodies. Thus, an XYY male has no Barr bodies, and XXY or XXYY males have one Barr body. Examination of Barr bodies is a rapid and convenient test for abnormalities in the number of X chromosomes.

Many types of sex-chromosomal abnormalities have been observed; such abnormalities are usually less severe in their phenotypic effects than those of autosomes are. The four most common types are the following:

1. 47,XXX. This condition is often called the **trisomy-X syndrome,** but the term "syndrome" is misleading in that affected females do not exhibit a consistent characteristic group of symptoms. The number 47 in the chromosome designation refers to the total number of chromosomes, and XXX indicates that the female has three X chromosomes. Many 47,XXX individuals are phenotypically normal or nearly normal, though the frequency of mental retardation is somewhat greater than normal.

2. 47,XYY. This condition is often called the **double-Y syndrome,** but again syndrome is a misleading term, because there are no cardinal symptoms. Affected males tend to be tall, but are otherwise phenotypically normal. At one time it was thought that 47,XYY males developed severe personality disorders and were at a high risk of committing crimes of violence, a belief based on an elevated incidence

of 47,XYY among violent criminals. More careful study indicates that most 47,XYY males have moderately impaired mental function, and although their rate of criminality is higher than that of 46,XY (normal) males, the crimes are mainly nonviolent petty crimes such as theft. The majority of 47,XYY males are phenotypically and psychologically normal and have no criminal convictions.

3. 47,XXY. This condition is called the **Klinefelter syndrome.** Affected individuals are male. They tend to be tall, do not undergo normal sexual maturation, are sterile, and in some cases have enlargement of the breasts. Mental retardation is common among Klinefelter patients.

4. 45,X. Monosomy of the X chromosome in females is called **Turner syndrome.** Affected individuals are phenotypically female but short in stature and without sexual maturation. Mental abilities are typically within the normal range.

Chromosome Abnormalities in Spontaneous Abortion

Approximately 15 percent of all recognized pregnancies in humans terminate in spontaneous abortion, and in about half of these the fetus has a major chromosome abnormality. Table 6-1 presents a summary of chromosome

Table 6-1   Chromosome abnormalities in 100,000 recognized human pregnancies

Chromosome constitution	Number among spontaneously aborted fetuses	Number among live births
Normal	7500	84,450
Trisomy		
13	128	17
18	223	13
21	350	113
Other autosomes	3176	0
Sex chromosomes		
47,XYY	4	46
47,XXY	4	44
45,X	1350	8
47,XXX	21	44
Translocations		
Balanced	14	164
Unbalanced	225	52
Polyploid		
Triploid	1275	0
Tetraploid	450	0
Others (mosaics, etc.)	280	49
Total	15,000	85,000

complements in 100,000 recognized pregnancies, based on data from several studies. Note that no autosomal monosomies are found in spontaneous abortions. Such embryos undoubtedly occur, but abortion probably occurs so early in development that the pregnancy goes unrecognized. Autosomal trisomies and polyploids are also common in spontaneous abortions. The majority of trisomy-21 fetuses, and the vast majority of 45,X fetuses, are spontaneously aborted. Spontaneous abortion serves the biological function of eliminating many fetuses that are grossly abnormal in their development because of major chromosome abnormalities.

6.5 Deletions and Duplications

So far, abnormalities in chromosome *number* have been described. In the remainder of this chapter we discuss abnormalities in chromosome *structure*. There are several principal types of structural aberrations, each of which has characteristic genetic effects. Chromosome aberrations were initially discovered through their genetic effects, which though confusing at first, were eventually understood in terms of abnormal chromosome structure and later confirmed directly by microscopic observations.

Deletions

Chromosomes sometimes occur in which a segment is missing. Such chromosomes are said to have a **deletion.** Deletions are generally harmful to the organism, and the larger the deletion the greater the degree of harmfulness. Very large deletions are usually lethal, even when heterozygous with a normal chromosome. Small deletions are often viable when they are heterozygous with a normal chromosome; however, they are lethal when homozygous, because the deletion eliminates one or more genes that are essential to survival and must be supplied by the normal homologous chromosome.

In some cases, a deletion of a locus changes the phenotype by itself. For example, deletion of a certain locus—Notch—in the X chromosome of *Drosophila* has a dominant effect on phenotype such that the wings have an irregular margin. A deletion can also be detected genetically by making use of the fact that the deleted chromosome no longer carries the wildtype alleles of the loci that have been eliminated. For example, many Notch deletions are large enough to remove a nearby locus, white, also. When these deleted chromosomes are heterozygous with a structurally normal chromosome carrying the recessive w allele, the fly will have white eyes because no wildtype (w^+) allele is present in the deleted chromosome. This "uncovering" of a recessive allele implies that the corresponding wildtype allele is missing from the deleted chromosome. Once a deletion has been identified, its size can be assessed genetically by determining which recessive mutations in the region are uncovered by the deletion. This method is illustrated in Figure 6-15.

With the banded polytene chromosomes in *Drosophila* salivary glands, it is possible to study deletions and other chromosome aberrations physically.

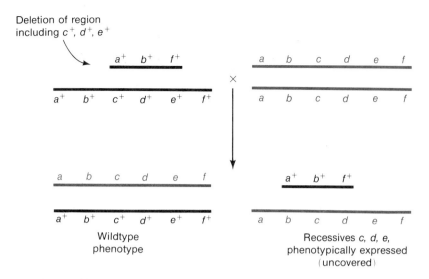

Figure 6-15 Mapping of a deletion by testcrosses. The F_1 heterozygotes carrying the deletion will express the recessive phenotype of all deleted loci. The expressed recessive alleles are said to be uncovered.

For example, all the Notch deletions cause particular bands to be missing in the salivary chromosomes. Physical mapping of deletions also allows individual genes, otherwise known only from genetic studies, to be assigned to specific bands in the salivary chromosomes.

Physical mapping of genes in a part of the *Drosophila* X chromosome is illustrated in Figure 6-16. The banded chromosome is shown near the top along with the numbering system used to refer to specific bands. Mutant chromosomes I through VI have deletions, and the deleted part is shown in color. These deletions define regions along the chromosome, some of them corresponding to specific bands. For example, the deleted region in both chromosome I and II that is present in all the other chromosomes consists of band 3A3. In crosses, only deletions I and II uncover the mutation zeste (z), so the locus of z must be in band 3A3, as indicated at the top. Similarly, the locus zw2 is uncovered by all deletions except VI; therefore, the zw2 locus must be in band 3A9. As a final example, the w mutation is uncovered only by deletions II, III, and IV; thus, the w locus must be in band 3C2. However, when a gene is assigned to a specific chromosome band by deletion mapping, the true physical location of the gene might be in an adjacent interband region rather than in the band itself.

The region of the *Drosophila* X chromosome illustrated in Figure 6-16 has been studied in exceptional detail in order to learn how many genes are present in each band of the salivary gland chromosomes. Over 100 mutations that were uncovered by deletion IV were first obtained by treatment of flies with a chemical known to cause mutations. The mutations were tested in pairs in complementation tests, as discussed in Chapter 3. A pair of mutations was brought together in the same organism by means of suitable crosses. If the

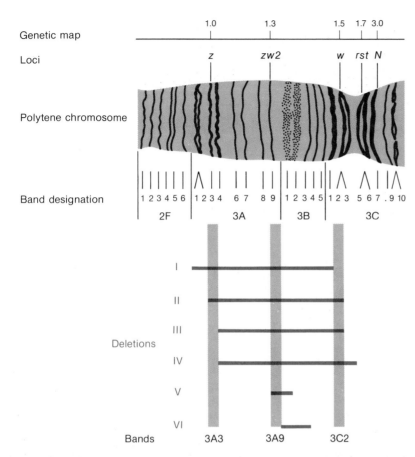

Figure 6-16 Portion of the X chromosome in salivary glands of *Drosophila melanogaster* and the extent of six deletions (I–VI). Any recessive allele that is uncovered by a deletion must be located inside the boundaries of the deletion. This principle can be used to assign genes to specific bands in the chromosome.

resulting fly had a wildtype phenotype, the mutations had complemented one another and they were assigned to separate genes (complementation groups). If instead the fly had a mutant phenotype, the mutations were assigned to the same gene. For the region shown in Figure 6-16, the mutations fell into 12 complementation groups, located between the loci of *z* and *w*. Twelve bands are also present in this region, suggesting a one-to-one correspondence between genes and bands in this region of the *Drosophila* genome. Thus, it was concluded that each band (or each interband) contains one gene. A one-to-one correspondence has also been found in several other regions of the genome. Since the salivary chromosomes of *Drosophila* contain about 5000 bands, this number is believed to represent the approximate number of genes in this organism.

Physical mapping of genes obtained by deletion studies and comparison of the physical and genetic maps has been important in proving that a genetic map accurately portrays the actual order of genes along the chromosome. Without this type of evidence, genetic maps would have remained elegant representations of genetic data but of dubious physical reality. Although genetic maps preserve the correct physical order of genes along the chromosome, the map distance between genes is not always an accurate measure of their physical separation. For example, the heterochromatic region around the centromere of each large *Drosophila* autosome accounts for approximately one-third of the total physical length of the metaphase chromosome but only about 1 percent of the total length of the genetic map of the chromosome. This discrepancy implies that crossing over within the centromeric heterchromatin during meiosis is rare, so a great physical distance corresponds to a small genetic distance. Recently, cloning techniques (Chapter 12) have allowed many regions of eukaryotic chromosomes to be mapped in molecular detail.

Duplications

Some abnormal chromosomes have a region that is present twice. These chromosomes are said to have a **duplication.** Certain duplications have phenotypic effects of their own. An example is the *Bar* duplication in *Drosophila*, which is a tandem duplication of a small group of bands in the X chromosome that produces a dominant phenotype of bar-shaped eyes. (A **tandem duplication** is one in which the duplicated segment is directly adjacent to the normal region in the chromosome.)

Tandem duplications are able to produce even more copies of the duplicated region by means of a process called **unequal crossing over,** outlined in Figure 6-17. Panel (a) illustrates the chromosomes in meiosis of an individual that is homozygous for a tandem duplication (red region). During synapsis, these chromosomes can mispair with each other, as illustrated in panel (b). A crossover within the mispaired part of the duplication (panel (c)) will thereby produce a chromatid carrying a **triplication,** and a reciprocal product (labeled "single copy" in panel (c)) that has lost the duplication. In the case of *Bar*, the triplication can be recognized because it produces an even greater reduction in eye size than the duplication.

In genetic analysis, duplications are valuable in identifying the chromosomal location of genes that are responsible for the production of particular enzymes. With such genes, the amount of enzyme present in cells is proportional to the number of copies of the corresponding gene. Thus, if a duplication includes such a gene, the amount of the enzyme will be increased by 50 percent compared with normal (because the organism with the duplication has three copies of the gene, in contrast with two copies for normal organisms). Note that this procedure finds the location of the normal allele. Unlike other methods of genetic mapping, mutant alleles are not required.

The most frequent effect of a duplication is a reduction in viability (probability of survival), and in general survival decreases with increasing size of the duplication. However, deletions are usually more harmful than duplications of comparable size. For example, in *Drosophila*, few deletions larger

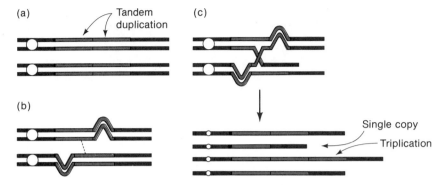

(a)

Tandem duplication

(b)

(c)

Single copy

Triplication

Figure 6-17 An increase in the number of copies of a chromosome segment resulting from unequal crossing over of tandem duplications (red). (a) Normal synapsis of chromosomes with a tandem duplication. (b) Mispairing. The right element of the the lower chromosome is paired with the left element of the upper chromosome. The dashed line indicates a potential site of crossing over. (c) Crossing over within the mispaired duplication yields the four-chromosome configuration shown. One chromosome carries a single copy of the duplicated region, and another chromosome carries a triplication.

than 3 percent of the haploid autosomal complement are compatible with survival when they are heterozygous with a structurally normal homologous chromosome; however, some duplications of up to 10 percent of the haploid autosomal complement are viable when heterozygous with a normal homologue.

6.6 Inversions

As knowledge of the genetic features of *Drosophila* became more detailed, several observations were made that were at first inconsistent with what was understood until they were later interpreted in terms of chromosomal abnormalities and confirmed by microscopy. One of these anomalous observations was an **inversion,** a segment of a chromosome in which the order of the genes is the reverse of the normal order.

Genetic Effects of Inversions

Inversions have a profound effect on the recovery of recombinants. Figure 6-18(a) shows a small part of the normal genetic map of chromosome 2 of *Drosophila*. (A series of experiments to be described shortly will relate panels (a) and (b).) The genes *pr* and *sp* are sufficiently distant that the recombination frequency between them is 50 percent. However, one anomalous chromosome carrying *sp* was found to have subnormal amounts of recombination. In a cross

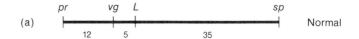

Figure 6-18 Partial genetic map of chromosome 2 of *D. melanogaster*. (a) Normal order. (b) Genetic map of same region in a chromosome that has undergone an inversion. Note that the order of *vg* and *L* are reversed in the inversion.

of $+ sp/pr +$ females with $pr\,sp/pr\,sp$ males, the following offspring were observed:

$$
\begin{array}{ccc}
+ & sp & 3946 \\
pr & + & 3921
\end{array} \Bigg\} \ 7867
$$

$$
\begin{array}{ccc}
+ & + & 6 \\
pr & sp & 6
\end{array} \Bigg\} \ 12
$$

The recombination frequency between *pr* and *sp* from these data is 12/7879 = 0.15 percent, which is far from the expected 50 percent. One possible explanation for the reduction in recombination would be the presence of a deletion; for example, the *sp*-bearing chromosome might carry a large deletion between *pr* and *sp*. However, in *Drosophila* such a large deletion would be lethal. Reduction of recombination frequency can also result from the carrying by the *sp*-bearing chromosome of a gene that suppresses crossing over; however, such a gene would suppress all recombination, and recombination is normal in the other chromosomes and in the part of chromosome 2 outside the region between *pr* and *sp*. Clearly, the *sp*-bearing chromosome is unusual, and its effects on recombination are localized to the region between *pr* and *sp*.

The events outlined in Figure 6-19 show how such a localized effect on recombination can be caused by an *inversion* located between *pr* and *sp*. When an individual is structurally heterozygous for an inversion (that is, when one homologue carries the inversion but the other homologue is normal), crossing over results in the loss of the recombinant chromatids, as shown in the figure. Synapsis between the normal and inverted homologous chromosomes results in one or the other chromosome being formed into a loop. If a single crossover occurs within the inversion loop, then the participating chromatids become physically connected. One chromatid becomes a dicentric and the reciprocal product becomes an acentric. These products can be observed most easily in the following anaphase, when the dicentric chromatid forms a chromatid bridge between two centromeres, and the acentric chromatid remains in the metaphase position. Both abnormal chromatids have deletions of some genes and duplications of others. These recombinant chromatids, if included in gametes, will produce zygotes with gross chromosomal abnormalities, and these will be lethal; consequently, progeny resulting from recombination will not be observed. In addition to the loss of recombinants resulting from the formation of dicentric and acentric chromosomes, inversions result in reduced crossing over in their vicinity because of mechanical difficulties in synapsis (stage I of Figure 6-19).

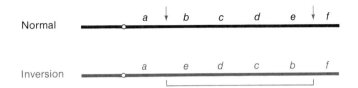

Normal

Inversion

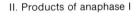

I. Paired chromosomes in prophase I, with single crossover

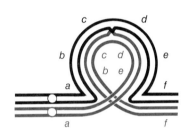

II. Products of anaphase I

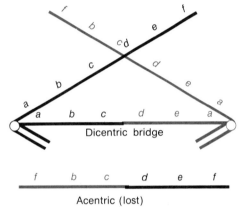

III. Products of anaphase II

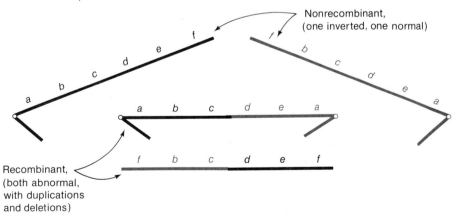

Figure 6-19 Pairing of chromosomes in an inversion heterozygote. One of the chromosomes must form a loop between the boundaries of the inversion. Crossing over within the inversion loop produces abnormal chromatids. Here the inversion is paracentric (it does not include the centromere), and the abnormal chromatids are a dicentric and an acentric (which also contain a duplication and deletion).

VARIATION IN CHROMOSOME
NUMBER AND STRUCTURE

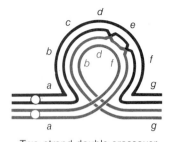

Two-strand double crossover   Products

Figure 6-20 Two-strand double crossing over within an inversion loop. Genes are exchanged between the inverted and the noninverted chromosome (right). Such events can be used to introduce genetic markers into an inversion.

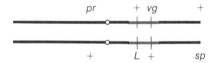

Figure 6-21 Chromosomes of a female that is homozygous for an inversion (red) and heterozygous for each of four genetic markers.

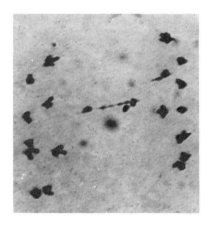

Figure 6-22 A dicentric chromosome bridge and an acentric fragment at anaphase I, resulting from a crossover in a heterozygous paracentric inversion in *Zea mays*. (Courtesy of Ronald Phillips.)

The inversion hypothesis makes a testable genetic prediction—that genetic mapping of the inverted chromosome should reveal a block of genes in reverse of the normal order. Such mapping requires (1) females that are *homozygous* for the inversion (otherwise the recombinants would be lost because of the events in Figure 6-19) and (2) that genetic markers be present in the inversion (in order to detect recombination). Genetic markers can be introduced into an inversion as a result of rare two-strand double crossing over, as illustrated in Figure 6-20.

In an experiment done with this procedure, the markers *L* (a dominant) and *vg* were then introduced into the inversion just discussed, and females of the genotype shown in Figure 6-21 were produced. These females were then crossed with males of genotype *pr vg + sp/pr vg + sp*. To simplify matters, we will ignore *sp* and focus on the other three markers, for which the results among 5876 progeny were as follows:

$$
\begin{array}{llll}
pr & vg & + & 2443 \\
+ & + & L & 2438
\end{array} \Big\} \; 4881
$$

$$
\begin{array}{llll}
pr & + & L & 310 \\
+ & vg & + & 366
\end{array} \Big\} \; 676
$$

$$
\begin{array}{llll}
pr & vg & L & 2 \\
+ & + & + & 2
\end{array} \Big\} \; 4
$$

$$
\begin{array}{llll}
pr & + & + & 160 \\
+ & vg & L & 155
\end{array} \Big\} \; 315
$$

These data can be analyzed exactly like the three-gene crosses discussed in Chapter 3. Since the rarest recombinants arise from double crossovers, the only gene order compatible with the data is *pr L vg*, so the order of *vg* and *L* have indeed been reversed in the inverted chromosome compared with the normal chromosome. The genetic map of the inverted chromosome from these data (taking *sp* into account) is shown in Figure 6-18(b). This map represents a genetic proof of an inversion; microscopic studies of the polytene chromosomes in salivary glands have confirmed the hypothesis and identified the breakpoints very precisely.

In the chromosome in Figure 6-19 the inverted region does not include the centromere. This is called a **paracentric inversion,** and the occurrence of a single crossover within the inversion loop results in a dicentric chromatid and an acentric chromatid. An example from a corn plant with a heterozygous paracentric inversion is shown in Figure 6-22.

In some cases of inversion, the centromere is included within the inverted region. This is called a **pericentric inversion,** and the consequences of a single crossover occurring within the inversion loop are outlined in Figure 6-23. With a pericentric inversion, instead of forming a dicentric and an

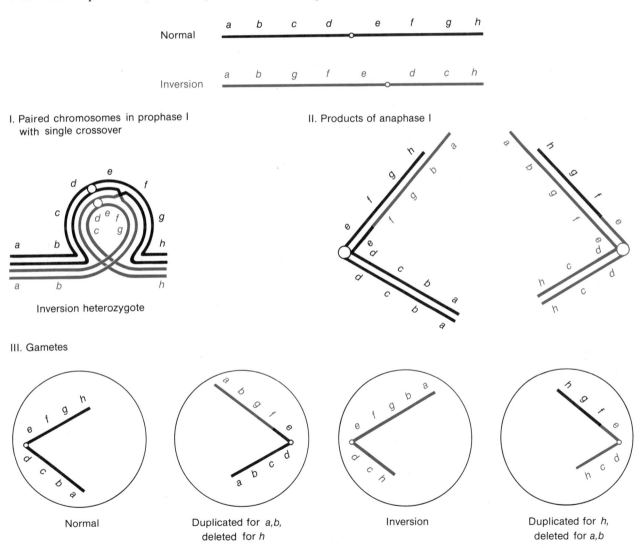

Figure 6-23 Products of a single exchange between a normal chromosome and a pericentric inversion. Crossing over within the inversion loop in heterozygotes results in chromatids each of which contains a large duplication or a deletion but has a single centromere.

acentric chromatid, the crossover chromatids both have a single centromere. The crossover chromatids are nevertheless abnormal in that each carries a duplication and a deletion, which will usually be large enough to result in lethal zygotes. Thus, a pericentric inversion also has the genetic effect of virtually eliminating the recovery of offspring resulting from crossing over within the inverted region.

6.7 Translocations

One type of chromosome aberration results in an inherited type of semi-sterility (reduced number of offspring), which is very unusual in that the reduction in fertility occurs only in heterozygotes. The homozygotes have normal fertility. This phenomenon results from the interchange of parts between nonhomologous chromosomes. An interchanged pair of nonhomologous chromosomes is called a **translocation.**

Figure 6-24(b) illustrates two pairs of homologous chromosomes in an individual that is heterozygous for a translocation (panel (a) shows the normal array). The top two chromosomes (red and black) have undergone an interchange of terminal parts. Compared to the normal individual in panel (a), the individual with a heterozygous translocation has one pair of chromosomes that is normal and one pair that is interchanged. Approximately one-half of the surviving offspring of a translocation heterozygote will inherit both parts of the translocation. Therefore, crossing two translocation heterozygotes can produce an individual that is homozygous for the translocation (panel (c)), in which both pairs of chromosomes are interchanged. The translocation in Figure 6-24 is properly called a **reciprocal translocation,** because the chromosomes have undergone a reciprocal interchange of parts.

Genetic Effects of Translocations

The genetic effects of a translocation are illustrated with an example in *Drosophila* in Table 6-2. The dominant mutations *Bl* and *D* are in chromosomes 2 and 3, respectively, and in a mating of a normal male (heterozygous *Bl/+* in the second chromosome and *D/+* in the third chromosome) with a wild-type female, four classes of offspring are produced in approximately equal frequency. The translocation male is heterozygous for a translocation between the *Bl*-bearing and the *D*-bearing chromosomes. In this case *Bl* and *D* are inherited together as a result of being on the same chromosome. No recombinant offspring are produced because in male *Drosophila* crossing over does not

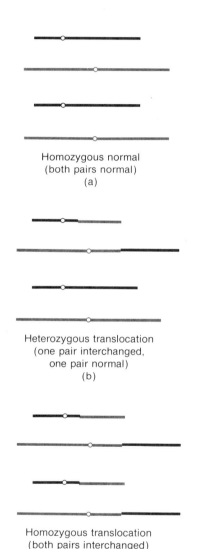

Homozygous normal
(both pairs normal)
(a)

Heterozygous translocation
(one pair interchanged,
one pair normal)
(b)

Homozygous translocation
(both pairs interchanged)
(c)

Figure 6-24 (a) Two pairs of nonhomologous chromosomes in a diploid individual. (b) Heterozygous translocation, in which two nonhomologous chromosomes (the two at the top) have interchanged terminal segments. (c) Homozygous translocation.

Table 6-2 Offspring of normal and translocation-bearing males of *Drosophila*

Parent	Genotype of offspring			
	Bl D	*Bl +*	*+ D*	*+ +*
Normal male	116	116	111	114
Translocation male	101	0	0	102

occur. Furthermore, the translocation heterozygote produces only about half as many offspring as the normal male. This semisterility occurs for reasons that will be discussed in a moment.

The semisterility resulting from a heterozygous reciprocal translocation is inherited like a simple dominant mutation, and it can be genetically mapped and assigned to a position on the linkage map. However, the genetic element that is mapped is not really a gene but rather the breakpoint of the translocation. This principle can be illustrated with the strain of maize in which semisterile individuals were discovered. These individuals had a fertility of about 50 percent of that of normal plants, which was easily recognized in the high frequency of dead pollen grains when observed with a microscope and in the many undeveloped kernels on the ears. Microscopic study of the chromosomes revealed a translocation. In this strain, the normal chromosomes 2 and 5 were recognized because each had an intensely staining heterochromatic **knob** in the long arm (Figure 6-25(a)). Individuals with the heterozygous translocation had a knobless chromosome and a two-knob chromosome, in addition to the normal homologues of chromosomes 2 and 5 (Figure 6-25(b)). The arm length in the interchanged chromosomes suggested breakpoints near the position of the arrows in panel (a).

For genetic mapping of the breakpoint in chromosome 2, plants carrying the translocation were crossed with chromosomally normal plants homozygous for the second-chromosome markers *B1* and *v4*. Semisterile translocation heterozygotes (also heterozygous for *B1* and *v4*) were testcrossed, yielding the following results among 1582 offspring (T indicates presence of the translocation, recognized as semisterility):

B1	+	*v4*	591	
+	T	+	633	1224
B1	T	+	111	
+	+	*v4*	121	232
B1	+	+	65	
+	T	*v4*	37	102
B1	T	*v4*	7	
+	+	+	17	24

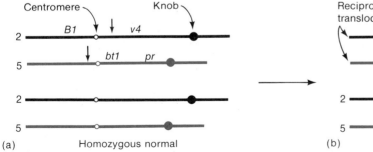 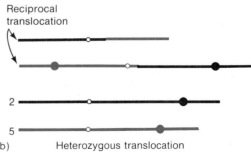

(a) Homozygous normal   (b) Heterozygous translocation

Figure 6-25 (a) Appearance of chromosomes 2 and 5 in a strain of maize in which the long arm of each chromosome contains a heterochromatic knob. Approximate locations of four genes are indicated. (b) A translocation heterozygote resulting from an inter-change of chromosome parts after breakage at the position of arrows shown in panel (a). One component of the reciprocal translocation is knobless; the other has two knobs.

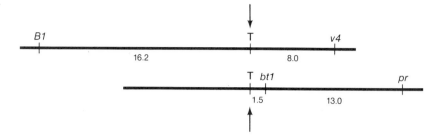

Figure 6-26 Genetic map showing position of translocation breakpoint (arrows) in chromosomes 2 and 5 in maize. (T denotes the breakpoints.)

In considering these data, note first that the semisterility is inherited as a simple Mendelian dominant; that is, there are $633 + 111 + 37 + 7 = 788$ T (semisterile) and $591 + 121 + 65 + 17 = 794 +$ (fertile) offspring, which is very close to the $1:1$ segregation expected with a dominant. Second, the semisterility maps as a single locus. The rarest class of offspring again represents the double recombinants, and therefore the gene order must be *B1* T *v4* in the chromosome. The recombination frequency between *B1* and T is $(232 + 24)/1582 = 16.2$ percent, and that between T and *v4* is $(102 + 24)/1582 = 8.0$ percent. A similar mapping experiment using the genes *bt1* and *pr* on chromosome 5 gave the gene order T *bt1 pr* with recombination frequencies of 1.5 percent and 13.0 percent for the two intervals, respectively.

Figure 6-26 shows a genetic map based on the data just presented. If T represents the breakpoint of the translocation, then in chromosome 2 it must be 16.2 map units to the right of *B1* and 8.0 map units to the left of *v4*. The breakpoint in chromosome 5 must lie 1.5 map units to the left of *bt1*. The genetic location of the translocation breakpoints is consistent with the physical data summarized in Figure 6-25.

Translocation Segregation

An individual that is heterozygous for a reciprocal translocation carries both components of the reciprocal translocation plus the normal homologue of each chromosome (Figure 6-24(b)). In such an individual, during meiosis, all four chromosomes pair to form a cross-shaped configuration (Figure 6-27(a) and Figure 6-28), in contrast with the two bivalents formed from structurally normal chromosomes (or from the chromosomes of an individual homozygous for a reciprocal translocation).

Segregation of chromosomes from the cross-shaped configuration during meiosis is irregular, because the configuration contains four centromeres instead of the normal two. At metaphase I, the configuration can align relative to the spindle in three distinct ways (Figure 6-27(b)), each of which leads to a particular type of segregation:

1. **Adjacent-1 segregation,** in which homologous centromeres go to opposite anaphase poles.

(a) Meiotic chromosome pairing in a translocation heterozygote

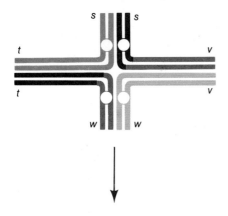

(b) First meiotic division

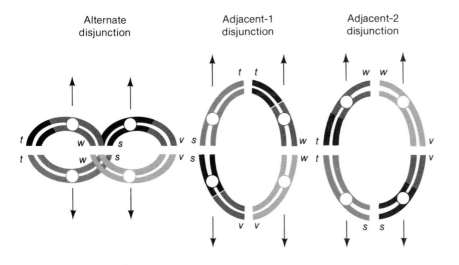

Alternate disjunction   Adjacent-1 disjunction   Adjacent-2 disjunction

(c) Gametes

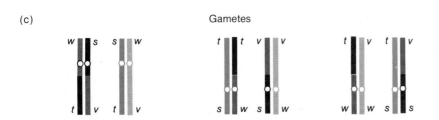

Figure 6-27 Meiosis in a translocation heterozygote. (a) The structure resulting from chromosome pairing in a translocation heterozygote. (b) Three possible orientations in metaphase I of meiosis that determine which chromosomes will go together at anaphase. (c) Chromosomes resulting from the three modes of segregation.

Figure 6-28 Chromosomes in pachytene in maize, showing the chromosome pairing in the association of four chromosomes in a plant heterozygous for a translocation (reciprocal) between chromosomes 7 and 8. (Courtesy of Gary Benzion.)

2. **Adjacent-2 segregation,** in which homologous centromeres go to the same anaphase pole.

3. **Alternate segregation,** in which both parts of the reciprocal translocation go to one pole and both normal homologues go to the other pole.

The meiotic products produced from both types of adjacent segregation are chromosomally abnormal because they have duplications of some genes and deletions of others (Figure 6-27(c)). In plants, gametophytes carrying such extensive chromosome abnormalities usually fail to function, and the abnormal chromosomes are not transmitted to the zygote. In animals, chromosomally abnormal gametes do function in fertilization, but large duplications or deletions are usually lethal in the zygote. Chromosomally abnormal gametophytes or zygotes that result from adjacent segregation inherit only part of the reciprocal translocation and are said to carry an *unbalanced* translocation. The semisterility of translocation heterozygotes results from the inviability of unbalanced plant gametophytes or animal zygotes. The occurrence of crossing over in the arms of the cross-shaped configuration in Figure 6-27(a) explains why the semisterility maps as a single genetic locus corresponding to the breakpoints of the reciprocal translocation.

With alternate segregation, the meiotic products carry either both normal chromosomes or both chromosomes that constitute the reciprocal translocation (Figure 6-27(b,c)). Either of these combinations results in a genetic complement with no duplications or deletions, and hence these combinations are transmitted to the zygotes. Zygotes that receive both parts of the reciprocal translocation are said to carry a *balanced* translocation. Because meiotic products that are chromosomally normal or that carry a balanced translocation result from reciprocal products of the same mode of segregation, these products are expected to occur with equal frequency.

Theoretically, the three types of translocation segregation should occur in the ratio ¼ : ¼ : ½ (that is, the combined frequency of both types of adjacent segregation should equal the frequency of alternate segregation). In maize and *Drosophila,* ¼ : ¼ : ½ segregation is usually observed. However, some organisms have a preponderance of alternate segregation. Certain species of *Oenothera* (evening primrose) are famous for carrying one or more heterozygous reciprocal translocations and for having evolved a mechanism of segregation that ensures that alternate segregation will almost always occur. Therefore, in *Oenothera,* unlike most other organisms, fertility is not reduced as the result of adjacent segregation of a heterozygous translocation.

Robertsonian Translocations

A **Robertsonian translocation** is a special type of nonreciprocal translocation in which the long arms of two nonhomologous acrocentric chromosomes become attached to a single centromere (Figure 6-29). The short arms of the acrocentrics become attached to form the reciprocal product, but this small

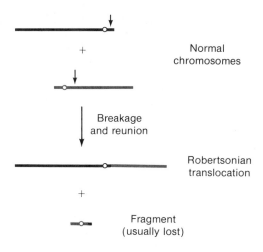

Figure 6-29 Formation of a Robertsonian translocation by chromosome breakage and reunion in two acrocentric chromosomes. The arrows indicate the breakage points.

chromosome carries nonessential genes and is usually lost within a few generations after its creation. Robertsonian translocations are important in human genetics because, when one of the acrocentrics is chromosome 21, it leads to the high-risk familial type of Down syndrome discussed in Section 6.4.

A Robertsonian translocation that joins chromosome 21 with chromosome 14 is shown in Figure 6-30. The heterozygous carrier is phenotypically

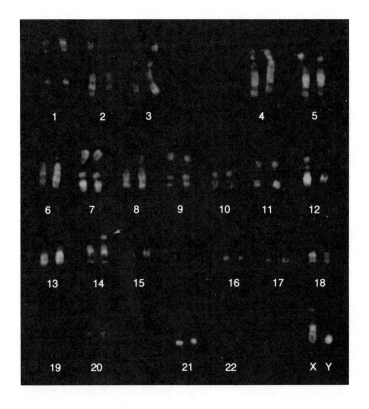

Figure 6-30 A karyotype of a child with Down syndrome, carrying a Robertsonian translocation of chromosomes 14 and 21 (arrow). Chromosomes 19 and 22 are faint in this photo; this has no significance. This figure should be compared with Figure 6-12. (Courtesy of Irene Uchida.)

normal, but a high risk of Down syndrome results from the occurrence of one type of adjacent segregation (Figure 6-31). Among the offspring that result from alternate segregation, all will be phenotypically normal, but half will again carry the Robertsonian translocation.

Earlier in this chapter (Section 6.1) an example of chromosome evolution in chimpanzees and humans was cited to emphasize that chromosome evolution often occurs through the fusion of two acrocentric chromosomes to produce a single metacentric chromosome. One mechanism of the fusion of acrocentric chromosomes is through the occurrence of a Robertsonian translocation between them.

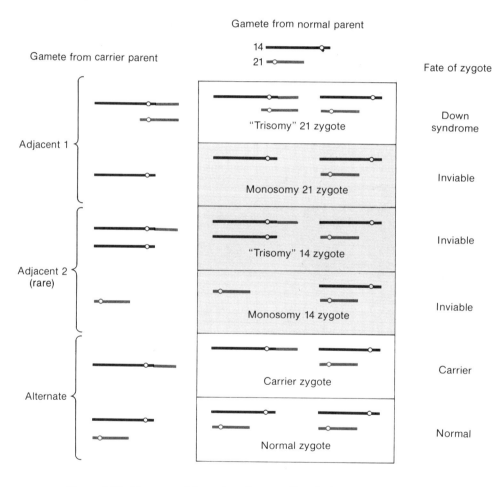

Figure 6-31 Three possible modes of segregation of a heterozygous Robertsonian translocation; these modes determine the chromosomal complements of the zygotes. In this case the zygote at the top inherits three copies of the long arm of chromosome 21 and will be affected with Down syndrome. The term "trisomy" is in quotes because the zygotes have 46 chromosomes, not 47. Among the phenotypically normal offspring from the mating, half will have normal chromosomes and half will carry the translocation.

6.8 Position Effects in Gene Expression

Genes near the breakpoints of a translocation or inversion become repositioned in the genome and are flanked by new neighboring genes. In some cases the repositioning of a gene affects the ability of the gene to function; this is called a **position effect.** These effects have been studied extensively in *Drosophila;* the most common type of effect is a mottled (mosaic) phenotype observed in heterozygotes as interspersed patches of wildtype cells, in which the wildtype gene is expressed, and mutant cells in which the wildtype gene is inactivated.

Such mosaicism often results from a chromosome aberration that moves a wildtype gene from a position in euchromatin to a new position in or near heterochromatin (Figure 6-32); such an effect on gene activity is called a **V-type** position effect. Figure 6-33 illustrates some of the patterns of wildtype (red) and white eye facets that occur in flies that are heterozygous for two X chromosomes; one X is structurally normal and carries the w allele, and the other X is a rearranged chromosome in which the w^+ gene is repositioned near heterochromatin. The patterns coincide with the known clonal lineages in the eye (that is, all of the red or white cells derive from a single ancestral cell in the embryo in which the w^+ allele was turned on or off, respectively). The determination of gene expression or nonexpression is thought to occur during early development when the precise boundary between condensed heterochromatin and euchromatin is established. Where heterochromatin is juxtaposed with euchromatin, the chromosome condensation that is characteristic of heterochromatin may spread into the adjacent euchromatin, inactivating euchromatic genes in the cell and all of its descendants. (A similar spreading of inactivation occurs in cells of female mammals when euchromatic genes are translocated to the X chromosome that becomes heterochromatic and inactive.) The spreading model is supported by the observation that, when two nearby genes both undergo a V-type position effect, the gene farther from the heterochromatin is not inactivated in a clone unless the gene closer to the heterochromatin is also inactivated. The maximum length of euchromatin that can be inactivated ranges from 1 to 50 bands in the *Drosophila* salivary chromosomes, depending on the rearrangement. However, the heterochromatin of different chromosomes varies in its ability to induce variegation of specific genes and in its response to environmental modifiers of variegation (such as temperature) and genetic modifiers (such as extra Y

w^+

(a) Normal X chromosome

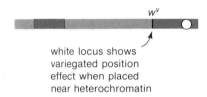

w^V

white locus shows
variegated position
effect when placed
near heterochromatin

(b) Inverted X chromosome

Figure 6-32 Occurrence of an inversion that repositions a euchromatic gene (in this case the w^+ locus of *D. melanogaster*) near heterochromatin. Such rearrangements frequently show a variegated (V-type) position effect in gene expression, observed as mottled red and white eyes in the case of *w*. The inversion is indicated by the vertical arrows.

Figure 6-33 Patterns of red and white observed in the eye of *Drosophila melanogaster* when the w^+ locus shows a V-type position effect, as illustrated in Figure 6-32. Each group of contiguous facets of the same color derives from a single cell during development.

chromosomes). Other than V-type effects, the positions of genes rarely affect gene activity. Most genes function normally irrespective of neighboring genes, but an important class of exceptions is discussed in the next section.

6.9 Chromosome Abnormalities and Cancer

Cancer refers to an unrestrained proliferation and migration of cells. In all known cases, cancer cells derive from repeated division of an individual mutant cell whose growth has become unregulated, so cancer cells constitute a clone. Many cells within such clones are found to have chromosomal abnormalities—extra chromosomes, missing chromosomes, deletions, duplications, or translocations. The chromosomal abnormalities found in cancer cells are diverse, and they may differ among cancer cells in the same individual or among individuals having the same type of cancer. The accumulation of chromosomal abnormalities is evidently one accompaniment of unregulated growth.

Amidst the large number of apparently random chromosomal abnormalities found in cancer cells, a small number of aberrations occur consistently in certain types of cancer, particularly in hematological diseases such as the leukemias. For example, chronic myelogenous leukemia is frequently associated with a deletion of part of the long arm of chromosome 22. The abnormal chromosome 22 in this disease has come to be known as the **Philadelphia chromosome,** but it is now known that the Philadelphia chromosome is one part of a reciprocal translocation in which the missing part of chromosome 22 is attached to either chromosome 8 or chromosome 9. Similarly, a deletion of part of the short arm of chromosome 11 is frequently associated with a kidney tumor called Wilms tumor, usually found in children.

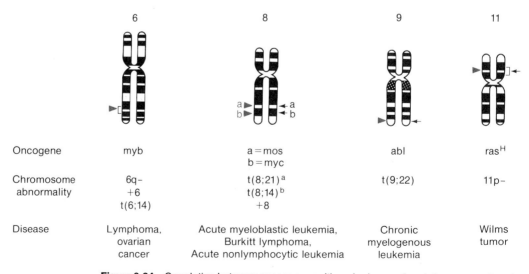

	6	8	9	11
Oncogene	myb	a = mos b = myc	abl	ras[H]
Chromosome abnormality	6q– +6 t(6;14)	t(8;21)[a] t(8;14)[b] +8	t(9;22)	11p–
Disease	Lymphoma, ovarian cancer	Acute myeloblastic leukemia, Burkitt lymphoma, Acute nonlymphocytic leukemia	Chronic myelogenous leukemia	Wilms tumor

Figure 6-34 Correlation between oncogene positions (red arrows) and chromosome breaks (black arrows) in aberrant human chromosomes frequently found in cancer cells.

Many (perhaps all) of these characteristic chromosomal abnormalities have a breakpoint near the chromosomal location of an **oncogene** (Greek: *onco*, mass or tumor). These are genes associated with cancer; approximately 20 different oncogenes are known. Their normal cellular functions are not understood, but they seem to play a role in the regulation of cell proliferation. When an appropriate type of chromosomal abnormality occurs near an oncogene, the oncogene may become expressed abnormally and promote unrestrained proliferation of the cell carrying it. However, the abnormally expressed oncogene, by itself, is not sufficient to produce cancerous growth. One or more additional mutations must also occur in the same cell, but the nature of these additional mutations is as yet poorly understood.

Characteristic chromosomal abnormalities found in certain cancers are illustrated in Figure 6-34 along with the locations of known oncogenes on the same chromosomes. The symbols are those conventionally used in human genetics, and they are as follows:

1. The long arm of a chromosome is symbolized as q, the short arm as p. For example, 11p refers to the short arm of chromosome 11.

2. A + (or −) sign *preceding* a symbol denotes an extra copy (or a missing copy) of the entire designated chromosome or arm. For example, +6 means the presence of an extra chromosome 6.

3. A + (or −) sign *following* a symbol denotes extra material (or missing material) corresponding to *part* of the designated chromosome or arm. For example, 11p− refers to a deletion of part of the short arm of chromosome 11.

4. The symbol "t" refers to a reciprocal translocation. Thus, t(9;22) refers to a reciprocal translocation between chromosome 9 and chromosome 22.

In Figure 6-34, the location of the oncogene, when known, is indicated by a triangle on the left, and the chromosomal breakpoint is indicated by an arrow on the right. In most cases for which sufficient information is available, the correspondence between the breakpoint and the oncogene location is very close. The significance of this correspondence is not understood in detail, but it is believed that the chromosomal rearrangement disturbs normal oncogene regulation and ultimately leads to the onset of cancer. This notion is supported by studies of cancer-causing viruses of a special type called **retroviruses.** These are small, *RNA*-containing viruses that on infection of animal cells are copied into DNA molecules that can be integrated into many chromosomal sites. By an unknown mechanism, retroviruses can also incorporate genes of the host cell into their own genetic material. Oncogenes were first discovered as normal host genes that had become incorporated into retroviruses and that had lost their normal control mechanisms in the process. When abnormally controlled oncogenes are introduced back into host cells by viral infection, they can contribute to the onset of cancerous growth. As Figure 6-34 demonstrates, chromosomal abnormalities can also disturb normal oncogene regulation and ultimately lead to the onset of cancer.

1. What is the shape of a chromosome that has no telomeres?

2. What kind of abnormality in chromosome structure can change a metacentric chromosome into a submetacentric chromosome? What kind can fuse two acrocentrics to make one metacentric?

3. An autopolyploid series, like *Chrysanthemum*, contains five species, and the basic haploid chromosome number is 5. What chromosome numbers would be expected among the species?

4. A spontaneously aborted human fetus was found to have the karyotype 92,XXYY. What might have happened to the chromosomes in the zygote to result in this karyotype?

5. The first allotetraploid species to be created artificially came from a cross of diploid radish (9 pairs of chromosomes) with diploid cabbage (also 9 pairs of chromosomes). The initial hybrid was semisterile, but among the offspring was a rare, fully fertile allotetraploid called the "rabbage," which unfortunately is weedlike and virtually inedible. How many chromosomes were present in the semisterile hybrid and in the fertile rabbage?

6. A plant species S coexists with two related species A and B. All species are fully fertile. In meiosis, S forms 26 bivalents, and A and B form 14 and 12 bivalents, respectively. Hybrids between S and A form 14 bivalents and 12 univalents (unpaired chromosomes), and hybrids between S and B form 12 bivalents and 14 univalents. Suggest a likely evolutionary origin of species S.

7. A human fetus, spontaneously aborted, is found to have 45 chromosomes. What is the most probable karyotype? Had the fetus lived, what genetic disorder would it have?

8. A phenotypically normal woman has a child with Down syndrome. The woman is found to have 45 chromosomes. What kind of chromosomal abnormality can account for these observations? How many chromosomes does the affected child have?

9. A strain of corn that has been maintained by self-fertilization for many generations, when crossed with a normal strain, produces an F_1 in which many meiotic cells have dicentric chromatids and acentrics, and the amount of recombination in chromosome 6 is greatly reduced. However, in the original strain no dicentrics or acentrics are found. What kind of abnormality in chromosome structure can explain these results?

10. Recessive genes *a*, *b*, *c*, *d*, *e*, and *f* are closely linked in a chromosome, but their order is unknown. Three deletions in the region are examined. One deletion uncovers *a*, *b*, and *d*; another uncovers *a*, *d*, *c*, and *e*; and the third uncovers *e* and *f*. What is the order of the genes?

11. Why are translocation heterozygotes semisterile? Why are translocation homozygotes fully fertile? If a translocation homozygote is crossed with an individual having normal chromosomes, what fraction of the F_1 will be semisterile?

12. Two species of Australian grasshoppers coexist side by side. In meiosis, each has eight bivalents. When the species are crossed, the chromosomes in the hybrid form six bivalents and one quadrivalent. What alteration in chromosome structure could account for these results?

13. Six bands in a salivary-gland chromosome of *Drosophila* are shown in the figure below, along with the extent of five deletions (Del 1–Del 5).

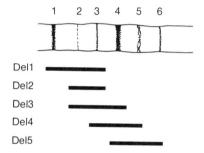

Recessive alleles *a*, *b*, *c*, *d*, *e*, and *f* are known to be in the region, but their order is unknown. When the deletions are heterozygous with each allele, the following results are obtained:

	a	*b*	*c*	*d*	*e*	*f*
Del 1	−	−	−	+	+	+
Del 2	−	+	−	+	+	+
Del 3	−	+	−	+	−	+
Del 4	+	+	−	−	−	+
Del 5	+	+	+	−	−	−

In this table, − means that the deletion is missing the corresponding wildtype allele (the deletion uncovers the recessive) and + means that the corresponding wildtype allele is still present. Use these data to infer the salivary band corresponding to each gene.

14. *Drosophila* carrying duplications (Dup) of the salivary bands shown in the following figure were examined to determine the amount of a particular enzyme.

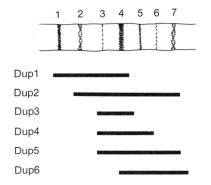

The normal diploid has 100 units of the enzyme, and in the strains with duplications the enzyme amounts are

Dup1	Dup2	Dup3	Dup4	Dup5	Dup6
143	152	160	155	148	98

Which chromosome band is most likely to carry the gene corresponding to the enzyme?

15. A wildtype strain of yeast, thought to be a normal haploid, was crossed with a different haploid strain carrying the mutation *his7*. This mutation is located in chromosome 2 and is an allele of a gene normally required for synthesis of the amino acid histidine. From this cross 15 tetrads were analyzed:

Composition		
Wildtype spores	*his7* spores	Number of tetrads
4	0	4
3	1	10
2	2	1

In contrast, when the wildtype strain was crossed with haploid strains having recessive markers on other chromosomes, segregation in the tetrads was always 2:2. What type of chromosome abnormality might account for the unusual segregation when the wildtype strain is mated with *his7*?

16. Cinnabar (*cn*) and brown (*bw*) are recessive eye-color mutations in the second chromosome of *Drosophila*. Their standard map positions are 57.5 and 104.5, respectively. A cross of females of genotype $++/cn\,bw$ ($++$ represents a wildtype chromosome isolated from a natural popu-

+	+	25200
cn	*bw*	21009
cn	+	11
+	*bw*	36

lation) with *cn bw/cn bw* males produced these progeny: What is unexpected about these results? What type of abnormality in chromosome structure is the most likely explanation?

17. Curly wings (*Cy*) is a dominant mutation in the second chromosome of *Drosophila*. A *Cy*/+ male was irradiated with x rays and crossed with +/+ females, and the *Cy*/+ sons were mated in single pairs with +/+ females. From one cross, the progeny were

Curly males	146
wildtype males	0
Curly females	0
wildtype females	163

What abnormality in chromosome structure is the most likely explanation for these results? (Remember that crossing over does not occur in male *Drosophila*.)

18. Yellow body (*y*) is a recessive mutation near the tip of the X chromosome of *Drosophila*. A wildtype male is irradiated with x rays and crossed with *y/y* females, and one y^+ son is observed. He is mated with *y/y* females, and the offspring are

yellow females	256
yellow males	0
wildtype females	0
wildtype males	231

The yellow females are chromosomally normal, and the y^+ males breed in the same manner as their father. What type of chromosome abnormality could account for these results?

19. In the genotype shown in the figure below the colored line represents an inverted segment of chromosome.

$$a^+ \quad d^+ \quad c^+ \quad b^+ \quad e^+$$

$$a \quad b \quad c \quad d \quad e$$

From this individual rare offspring were obtained that carry $a^+\,b^+\,c\,d^+\,e^+$. What events occurring during meiosis in the inversion heterozygote are required to explain these progeny? Is the gene sequence in these progeny normal or inverted?

20. In *Drosophila*, ruby eye color (*rb*) is an X-linked recessive and dumpy wings (*dp*) is a chromosome-2 recessive. A phenotypically normal male carrying a reciprocal translocation between the X chromosome and chromosome 2 was obtained among the progeny of flies exposed to x rays. The genetic and chromosomal constitution of this male is shown in Figure P-20 (Y designates the Y chromosome).

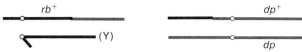

rb^+

(Y)

dp^+

dp

What phenotypes and what relative frequencies would be expected among the progeny from the mating of this male with females having ruby eyes and dumpy wings? (Remember that crossing over does not occur in male *Drosophila.*)

21. Semisterile maize plants heterozygous for a reciprocal translocation between chromosomes 1 and 2 were crossed with chromosomally normal plants homozygous for the recessive mutant virescent on chromosome 2. When semisterile F_1 plants were crossed with plants of the virescent parental type, the following phenotypes were found in the backcross progeny:

semisterile virescent	39
fertile virescent	231
semisterile nonvirescent	287
fertile nonvirescent	31

(a) What is the recombination frequency between the virescent gene and the translocation breakpoint in chromosome 2?

(b) What phenotypic ratio in the backcross progeny would have been expected if the virescent gene had not been on one of the chromosomes involved in the translocation?

22. Semisterile maize plants heterozygous for the same reciprocal translocation between chromosomes 1 and 2 referred to in Problem 21 were also crossed with chromosomally normal plants homozygous for the recessive mutants brachytic and fine stripe on chromosome 1. When semisterile F_1 plants were crossed with plants of the brachytic, fine-striped parental strain, the following phenotypes were found in a total of 682 backcross progeny:

	semisterile	fertile
wildtype	333	19
brachytic	17	6
fine stripe	1	8
brachytic and fine stripe	25	273

What is the map distance between the two genes, and between each of the genes and the translocation breakpoint in chromosome 1?

23. Four strains of *Drosophila melanogaster* are each isolated from a different country. The banding patterns of a particular region of chromosome 1 have the configurations shown below (each letter denotes a band):

(a) a b f e d c g h i j
(b) a b c d e f g h i j
(c) a b f e h g i d c j
(d) a b f e h g c d i j

Assume that (c) is the ancestral sequence. Indicate the probable order in time in which the others arose.

24. Consider an inversion heterozygote having a normal chromosome with the order $A\,B\,C$ centromere $D\,E$, and an inverted chromosome with the order $a\,d$ centromere $c\,b\,e$. Capital and lower-case letters represent homologous regions. What are the structures of the anaphase-II products if a single crossover occurs between B and C?

25. Two strains of *D. melanogaster* are being compared.

(a) In strain 1 the recombination frequency between genes A and B is 30 percent, whereas in strain 2 it is 7 percent. Neither strain carries a deletion or a duplication. A cross is carried out between the two strains; it is found that F_1 progeny give exclusively nonrecombinant viable gametes. What explanation can be given for the difference in recombination frequency between these strains?

(b) In strain 3 the recombination frequency between genes C and D is 5 percent, whereas in strain 4 it is 50 percent. F_1 progeny from a cross between these strains produce gametes with the following properties: about 50 percent are viable and contain only nonrecombinant chromosomes and about 50 percent are nonviable. How can these results be explained?

26. A particular species of *Drosophila* has a gene order on one chromosome of $A\,B\,C$ centromere $D\,E\,F\,G\,H$. A variant has the order $A\,B\,F\,E\,D$ centromere $C\,G\,H$. These two types mate and there is a crossover between nonsister chromatids within the segment E–D. What can be said about the fertility of the F_1 progeny with respect to the parents?

27. Some species of a particular plant genus are diploid and others are tetraploid and octaploid. How could you determine whether a particular tetraploid strain was the result of autopolyploidy or allopolyploidy?

28. An individual heterozygous for the inversion $A\,D\,C\,B\,E/A\,B\,C\,D\,E$ is crossed with one homozygous for the same inversion $A\,D\,C\,B\,E/A\,D\,C\,B\,E$. What percentage of the total gametes from these individuals is euploid if:

(a) The inversion is too small for synapsis to occur with the C–D region?

(b) The inversion is large enough that a synaptic loop always forms, and exactly one crossover occurs between C and the centromere?

(c) Considering only case (b), what percentage of the zygotes are euploid?

29. Why is it much easier to establish true-breeding stable lines of tetrasomic organisms than of trisomic ones?

30. The cross diploid × tetraploid is usually more fertile than diploid × triploid. Explain.

CHAPTER 7

Genetics of Bacteria and Viruses

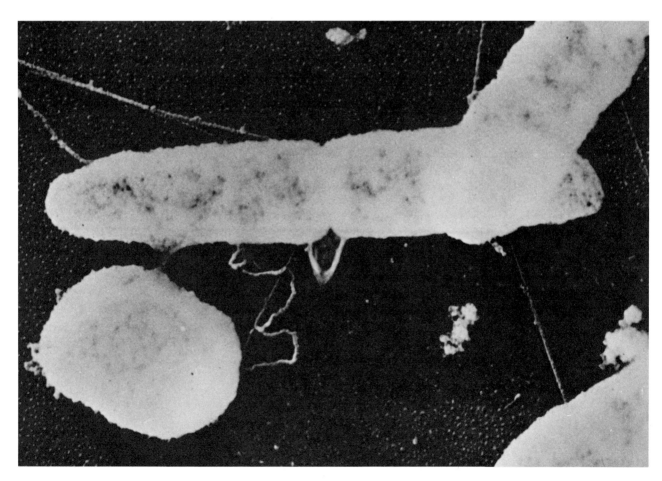

In earlier chapters we examined the genetic properties of several eukaryotic organisms. These organisms are diploid and multichromosomal, and during sexual reproduction genotypic variation among the progeny is introduced by random assortment of the chromosomes and by crossing over, processes occurring mainly during meiosis. Two important features of crossing over in eukaryotes are that it results in a reciprocal exchange of material between two homologous chromosomes and that both products of a single exchange can usually be recovered.

In this chapter we will first examine the genetics of bacteria, organisms that have only a single (major) DNA molecule, which is physically circular. Recombination mechanisms exist in many bacterial species, and these processes differ from those in eukaryotic organisms principally in that crossing over usually occurs between a chromosomal fragment and an intact chromosome, rather than between complete chromosomes. A donor–recipient relation exists—that is, a donor cell is the source of the DNA fragment, which is transferred to the recipient cell by one of several mechanisms. Exchange of genetic material occurs in the recipient. Incorporation of a portion of the transferred donor DNA into the chromosome requires at least two crossover events—and since the recipient molecule is circular, only an even number of crossovers will lead to a viable product. The usual result of these events is the recovery of only one of the crossover products. However, in some situations the transferred DNA is also circular, and a single crossover results in the incorporation of this DNA into the chromosome of the recipient.

Three major types of genetic transfer will be described in the first part of this chapter: **transformation,** which is uptake of DNA from the environment; **transduction,** which is transfer of DNA from one bacterium to another by a bacterial virus; and **conjugation,** which is transfer of DNA between two bacterial cells by direct contact.

Facing page: Electron micrograph of *E. coli* bacteria in the act of mating. The lefthand member of the pair of dividing cells is the F^+ donor and the small, nearly spherical cell at the lower left is the F^- recipient. The conjugation bridge appears as a thin fiber between the mating cells. (Courtesy of T. F. Anderson.)

Bacterial viruses (phages), particles able to multiply only within bacterial cells, possess several systems for exchanging genetic material. In the life cycle of a phage a phage particle attaches to a host bacterium and releases its nucleic acid into the cell; the nucleic acid replicates many times; finally, newly synthesized nucleic acid molecules are packaged into protein shells—forming progeny phage—and then the particles are released from the cell (Figure 7-1). Phage progeny from a bacterium infected by one phage have the parental genotype, except for infrequent mutational changes. However, if two phage particles having *different* genotypes infect a single bacterial cell, new genotypes can arise by genetic recombination. Some phages also possess systems that enable phage DNA to recombine with bacterial DNA. Phage-phage and phage-bacterium recombination will be the topic of the second part of this chapter. The best-understood bacterial and phage systems are those of *E. coli*, and we will concentrate on these.

In Chapter 5 it was pointed out that the repetitive sequences of eukaryotic DNA include transposable elements—DNA sequences capable of relocation

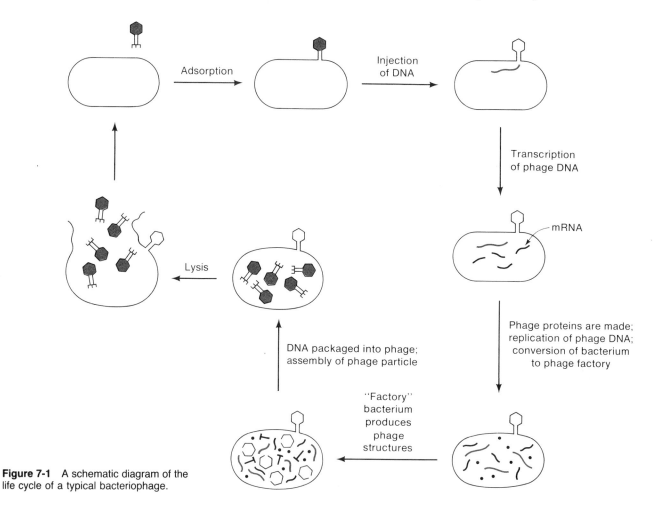

Figure 7-1 A schematic diagram of the life cycle of a typical bacteriophage.

within a genome. Transposable elements are also present in bacteria and in some phages and are responsible for a variety of genetic alterations, such as mutation, deletion, gene inversion, and fusion of circular DNA molecules. These elements have been extensively studied in the bacterium *E. coli* and are the final topic of this chapter.

7.1 Mutants of Bacteria

Bacteria can be grown both in liquid medium and on the surface of a semisolid growth medium hardened with agar. Genetic analysis of bacteria is usually carried out by growth on the latter. A single bacterium placed on a solid medium will grow and divide many times, forming a visible cluster of cells called a **colony** (Figure 7-2). The concentration of bacterial cells in a suspension (the **titer**) can be determined by spreading a known volume of the suspension on a solid medium and counting the number of colonies that form. The appearance of colonies and the ability or lack of ability to form colonies on particular media can, in some cases, also be used to identify the genotype of the cell that produced the colony.

As we have seen in earlier chapters, genetic analysis requires mutants; with bacteria three types are particularly useful:

1. **Antibiotic-resistant mutants.** These are able to grow in the presence of an antibiotic, such as streptomycin (Str) or tetracycline (Tet). For example, streptomycin-sensitive (Str-s) cells, the wildtype phenotype, fail to form colonies on medium containing streptomycin, but streptomycin-resistant (Str-r) mutants can do so.

2. **Nutritional mutants.** These cells are unable to synthesize an essential nutrient that can be produced by a wildtype strain and thus cannot grow unless the required nutrient is supplied in the medium. Such a mutant bacterium is said to be an **auxotroph** for the particular nutrient. For example, a methionine auxotroph cannot grow on a medium containing only inorganic salts and a source of energy and carbon atoms (a **minimal medium**), such as glucose minimal medium, but it can grow if the medium is supplemented with methionine. The phenotype of a methionine auxotroph is denoted Met⁻ (the wildtype is Met⁺). The Met⁻ phenotype can be caused by mutation in any one of several genes, which would usually be designated *metA*, *metB*, *metC*, and so on.

3. **Carbon-source mutants.** These cannot use particular substances as sources of energy or carbon atoms. For example, Lac⁻ mutants do not utilize the sugar lactose for growth and, therefore, cannot form colonies on minimal medium containing lactose as the only carbon source. Carbon-source mutants are particularly easy to detect with a color indicator medium, which is a highly supplemented "complete" medium (in which all cells can grow) containing a dye that imparts

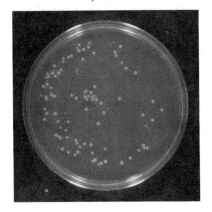

Figure 7-2 A petri dish with bacterial colonies that have formed on a solid medium. (Courtesy of Gordon Edlin.)

color to the colonies that are utilizing the main source of carbon. For example, on one type of lactose indicator medium, all bacteria can grow, but Lac⁻ colonies are white, and Lac⁺ colonies are red.

A medium on which all bacteria form colonies is called a **nonselective medium.** Mutants and wildtype may or may not be distinguishable by growth on a nonselective medium. A color indicator medium is an example of a nonselective medium on which mutants and wildtype can be distinguished. If the medium allows growth of only one type of cell (either mutant, wildtype, or a combination of alleles), it is said to be **selective.** For example, a medium containing streptomycin is selective for the Str-r phenotype and selects against the Str-s phenotype, and minimal medium containing lactose as the sole carbon source is selective for Lac⁺ cells and against Lac⁻ cells.

In bacterial genetics phenotype and genotype are designated in the following way: (1) Phenotypes are designated by three roman letters the first of which is capitalized, with a superscript + or − to denote presence or absence of the designated character, and with "s" and "r" for sensitivity and resistance, respectively. (2) Genotypes are designated by all-lowercase italicized letters. Thus, a cell unable to grow without a supplement of leucine (a leucine auxotroph) has a Leu⁻ phenotype; this would usually result from a *leu⁻* mutation. Many textbooks in bacterial genetics eliminate the − sign in genotypic symbols (for example, *leu* = *leu⁻*); we have chosen to include the − sign to avoid misreading. Individual genes in a biochemical pathway are usually given terminal capital letters (for example, *leuA* or *leuB*).

7.2 Bacterial Transformation

Bacterial transformation is a process in which genes are acquired, by a recipient cell, from free DNA molecules in the surrounding medium. Donor DNA is usually isolated from donor cells in the laboratory and then added to a suspension of recipient cells, but it can also become available in nature by spontaneous breakage of donor cells. In Chapter 4 a transformation experiment was described in which the rough-colony phenotype of *Pneumococcus* was changed to the smooth-colony phenotype by exposure of the cells to DNA from a smooth-colony strain. This experiment provided the first convincing evidence that DNA is the genetic material.

Transformation begins with uptake of a DNA fragment from the surrounding medium by a recipient cell and terminates with exchange of all or part of one strand of the donor DNA with the homologous segment of the recipient chromosome. (Transformation is, in fact, the only clearcut example of integration of single-stranded DNA.) Probably, most bacterial species are capable of the recombination step, but the ability of most bacteria to take up DNA efficiently is limited. Even in a species capable of transformation, DNA is able to penetrate only a small number of the cells in a growing population. However, incubation of cells of these species *under certain conditions* yields a population of cells—termed **competent**—most of which can take up DNA.

Competence is usually the result of synthesis of particular components of the bacterial cell wall. The conditions required to produce competence vary from species to species. Possibly the inability of some species to be transformed is only a reflection of lack of knowledge of the treatments necessary to induce competence.

Transformation is the only technique available for gene mapping in some species. The technique is different from but related to that described in Chapter 3. When DNA is isolated from a donor bacterium, it is invariably broken into small fragments. With highly competent recipient cells and excess DNA, transformation of most genes occurs at a frequency of about one cell per 10^3. If two genes a and b are so widely separated in the donor chromosome that they are always carried on two different DNA fragments, then at the same total DNA concentration the probability of simultaneous transformation (**cotransformation**) of an a^-b^- recipient to wildtype is roughly the square of that number, or one a^+b^+ transformant per 10^6 recipient cells. However, if the two genes are so near one another that they are often present on a single fragment, the frequency of cotransformation would be nearly the same as the frequency of single-gene transformation, or one wildtype transformant per 10^3 recipients. Thus, *cotransformation at a frequency substantially greater than the product of the two single-gene transformations implies physical proximity of two genes*. Studies of the ability of various pairs of genes to be cotransformed yields gene order. For example, if genes a and b can be cotransformed and genes b and c also can be, but genes a and c cannot, the gene order must be $a\,b\,c$ (Figure 7-3).

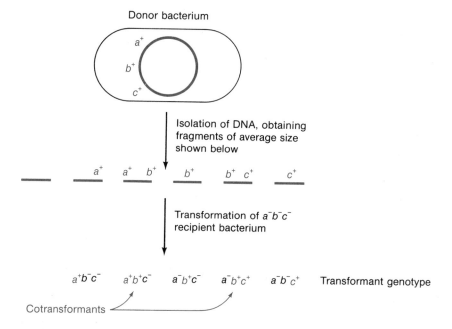

Figure 7-3. Cotransformation of linked markers. Markers *a* and *b* are near enough to each other that they will often be present on the same fragment, as are markers *b* and *c*. Markers *a* and *c* are not. The gene order must therefore be *a b c*.

It is possible that three genes may sometimes have locations such that all three will be present on a single fragment. In that case, a quantitative analysis of cotransformation frequency will also give the gene order. An analysis of this type (for transduction, rather than transformation) is presented in Section 7.3.

It is possible for an experimenter to control the size of isolated fragments; the probability of cotransformation of two genes can then be related to the average molecular weight of the transforming DNA. Thus, a physical map can be constructed from transformation data. For example, by measuring the cotransformation frequency as a function of the molecular weight of the transforming DNA, the physical separation between the genes can be determined roughly.

Transformation probably occurs in nature and may be an important evolutionary force. For example, the cells of some bacterial genera occasionally break open spontaneously, releasing their DNA molecules into the environment. At a later time these molecules may be taken up by competent cells of the same or closely related species, to yield recombinant organisms.

7.3 Transduction

Transduction is a phenomenon in which bacterial DNA is transferred from one bacterial cell to another *by a phage particle containing the DNA.* Such a particle is called a **transducing particle.** Two types of transducing phages are known—generalized and specialized. Generalized transducing phage produce some particles that contain only DNA obtained from the host bacterium, rather than phage DNA; the bacterial DNA fragment can be derived from *any* part of the bacterial chromosome. A **specialized** transducing phage produces particles containing both phage and bacterial genes linked in a single DNA molecule, and the bacterial genes are obtained from a *particular* region of the bacterial chromosome. In this section we consider *E. coli* phage P1, a well-studied generalized transducing phage. Specialized transducing particles will be discussed in Section 7.6.

DNA Transfer by Means of Transduction

During infection by P1, the phage makes a nuclease that causes fragmentation of bacterial DNA (a common event in the life of many phage species). A single fragment of bacterial DNA comparable in size to P1 DNA is occasionally packaged into a phage particle in place of P1 DNA. The positions of the nuclease cuts in the host chromosome are random, so a transducing particle may contain a fragment derived from any region of the host DNA, and a sufficiently large population of P1 phage will contain particles at least one of which carries each host gene. About one particle per 10^3 progeny is a transducing particle. On the average, any particular gene is present in roughly one transducing particle per 10^6 viable P1 phage. When a transducing particle adsorbs to a bacterium, the DNA carried within the particle head is injected into the cell and becomes available for crossing over with the homologous region of the bacterial chromosome.

Let us now examine the events that follow infection of a bacterium by a generalized transducing particle obtained, for example, by growth on wild-type *E. coli* containing a *leu*⁺ gene. If such a particle adsorbs to a bacterium whose genotype is *leu*⁻ and injects the DNA it contains into the bacterium, the cell will survive because the phage head contained only bacterial genes and no phage genes. A crossover event exchanging the *leu*⁺ allele carried by the phage for *leu*⁻ allele carried by the host will convert the genotype of the host cell from *leu*⁻ to *leu*.⁺ In such an experiment, typically about one *leu*⁻ cell in 10^6 will be converted to *leu*⁺ (Figure 7-4). Such frequencies are easily detected on selective growth medium; that is, if the infected cell is placed on a solid medium lacking leucine, it will be able to multiply and a *leu*⁺ colony will form. A colony will not form if crossing over does not result in insertion of the *leu*⁺ allele.

Transduction experiments are not performed with one bacterium and one transducing particle but with about 10^7–10^8 bacteria and a comparable number of phage of which most are virulent (contain only phage genes); only about 0.1 percent of the phage are transducing particles. Thus, when plating

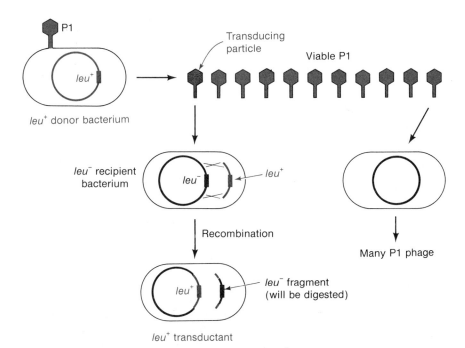

Figure 7-4 Transduction. Phage P1 infects a *leu*⁺ donor, yielding predominately viable P1 phage with an occasional one carrying bacterial DNA instead of phage DNA. If the phage population infects a bacterial culture, the viable phage will produce progeny phage and the transducing particle will yield a transductant. Notice that the recombination step requires two crossovers. For clarity, double-stranded DNA is drawn as a single line.

infected cells, it is necessary to prevent reinfection of transductants by progeny of the cells that have been infected with normal P1 phage. This is easily accomplished in several ways. A common procedure is to mix phage and bacteria, allow adsorption to occur and then to spread the infected cells on a solid medium coated with rabbit antiserum directed against P1 phage (obtained by infecting a rabbit with purified P1 phage). In this way, P1 progeny released from cells infected with normal phage will be inactivated soon after release from the cell and hence incapable of subsequent infection.

Cotransduction and Linkage

The small fragment of bacterial DNA contained in a transducing particle carries about 50 genes, so transduction provides a valuable tool for linkage analysis of short regions of the bacterial genome. Consider a population of P1 prepared from a $leu^+ gal^+ bio^+$ bacterium. This sample will contain particles able to transfer any of these alleles to another bacterium; that is, a leu^+ particle can transduce a leu^- bacterium to leu^+, or a gal^+ particle can transduce a gal^- bacterium to gal^+. Furthermore, if a $leu^- gal^-$ culture is infected with an excess of phage, both $leu^+ gal^-$ and $leu^- gal^+$ bacteria will be produced. However, if the ratio of phage particles to bacteria is much less than 1, $leu^+ gal^+$ colonies will not arise (Figure 7-5), because the *leu* and *gal* genes will be too far apart to be carried on the same DNA fragment, and few bacteria will be infected with both a leu^+ particle and a gal^+ particle.

The situation is quite different with a recipient bacterium whose genotype is $gal^- bio^-$, because the *gal* and *bio* genes are separated by only about 2.3×10^4 base pairs. A P1 particle can contain a DNA molecule with 7.7×10^4 base pairs, so both genes might be present in a single DNA fragment carried in a transducing particle, namely, a *gal-bio* particle (Figure 7-5). However, all gal^+ transducing particles will not also be bio^+, and all bio^+ particles will not also be gal^+, because the nuclease cuts that produce the bacterial DNA fragments will sometimes be made between the *gal* and *bio* genes. The probability of both markers being in a single particle and hence the probability of simultaneous transduction of both markers (**cotransduction**) depends on how close to each other the genes are. Cotransduction of the *gal-bio* pair can be detected by plating infected cells on the appropriate growth medium. If bio^+ transductants are selected (by spreading the infected cells on a glucose-containing medium lacking biotin), both $gal^+ bio^+$ and $gal^- bio^+$ colonies will be produced. If these colonies are tested for the *gal* marker by plating on galactose color-indicator plates, roughly half will be $gal^+ bio^+$ and half will be $gal^- bio^+$; similarly, if gal^+ transductants are selected (by growth on galactose-biotin medium and retested on a medium lacking biotin, half—the $gal^+ bio^+$ ones—will be able to form colonies. Thus, not only will $gal^+ bio^-$ and $gal^- bio^+$ transductants be formed, but also $gal^+ bio^+$ transductants will be produced at a frequency near that for transduction of either single allele.

Studies of cotransduction can be used to determine distances between genetic markers and can yield a crude map, whose order is unambiguous but

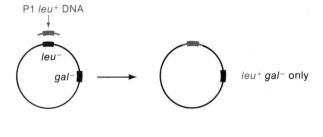

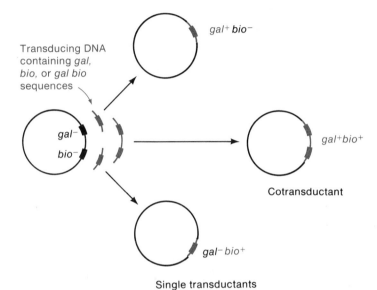

Figure 7-5 Demonstration of linkage of the *gal* and *bio* genes by cotransduction. In the upper panel a transducing particle carrying the *leu*⁺ allele can convert a *leu*⁻*gal*⁻ bacterium to the *leu*⁺*gal*⁻ genotype (but could not produce the *leu*⁺*gal*⁺ genotype. The lower panel shows the transductants that could be formed by three possible types of transducing particles—one carrying *gal*⁺, one carrying *bio*⁺, and one carrying the linked alleles *gal*⁺*bio*⁺. The third type is responsible for cotransduction.

whose distances may be inexact. This procedure is examined further in the next section.

Determining the Order of Genes by Cotransduction

The use of cotransduction to determine gene order is based on the fact that *the frequency of cotransduction increases with decreasing distance between two markers,* precisely as for cotransformation, and the order of genes can be determined by the procedures described in Section 7.2 and illustrated in Figure

7-3. When genes are very near, the method fails because of experimental uncertainty. As an example, consider three genes a, b, and c, for which the frequencies of cotransduction are: a–b, 90 percent; a–c, 33 percent; and b–c, 32 percent, each value having an error of about $\pm$ 2 percent. The values indicate that b is closer to a than to c (90 > 32), but the order of the genes cannot be deduced from the data alone, because the difference between 32 and 33 percent is not significant. That is, although the order $a\,c\,b$ can be excluded from consideration, the orders $a\,b\,c$ and $b\,a\,c$ are both consistent with the data.

A powerful method can be used to determine the order of closely linked markers, such as the a and b genes just described, with respect to a third gene, or the order of mutant sites within a single gene. This method is based on the principle that the probability of appearance of a genotype requiring four crossovers for its formation is much smaller than the probability of one formed by two crossovers (Chapter 3). In this method, the results of **reciprocal crosses**—a pair of crosses in which the + and − markers are interchanged between the parents—are compared. These crosses are shown for three hypothetical genes in Figure 7-6. In the first cross, the donor (P1) genotype is $a^+b^-c^+$ and the recipient genotype is $a^-b^+c^-$. In the second cross, + and − alleles are interchanged, so the donor genotype is $a^-b^+c^-$ and the recipient genotype is $a^+b^-c^+$. In each cross, the $+\,+\,+$ genotype is selected. The figure

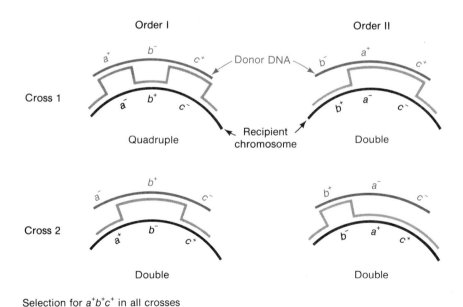

Selection for $a^+b^+c^+$ in all crosses

Expected results:

$$(a^+b^+c^+)_1 \lll (a^+b^+c^+)_2 \qquad\qquad (a^+b^+c^+)_1 \approx (a^+b^+c^+)_2$$

Figure 7-6 Use of reciprocal crosses to determine the order of markers a and b, relative to c, by transduction. The chromosome of the recipient is circular. The gray lines show the product of crossing over. The principle underlying the technique is that a double crossover has a greater probability of occurrence than a quadruple crossover.

shows that for order I, production of $+++$ in cross 1 requires a quadruple crossover, whereas in cross 2, only a double crossover is needed. Thus, for order I, the expected frequency from cross 1 is much lower than that from cross 2; in contrast, for the alternative order II the expected frequencies are roughly comparable for the two crosses, the differences depending only on the spacing between the markers selected. Let us assume that for cross 1 the frequency of $+++$ is observed to be much less than that of cross 2; then, the order must be $a\,b\,c$. Notice that we have not considered the order $a\,c\,b$, because the possibility of that order was eliminated by the two-marker cotransduction data described in the preceding paragraph. Such two-marker data are essential; without these data, results supporting order II—that is, comparable frequencies of $(a^+b^+c^+)_1$ and $(a^+b^+c^+)_2$—would not allow us to distinguish order II $(b\,a\,c)$ from $a\,c\,b$. (The reader should check this.)

The data just presented allow determination of gene order but not genetic distance. Mapping experiments can be carried out, but the analysis of such cotransduction data is not straightforward. The main problem is that even when three markers are near enough to each other that they can be carried by a single transducing particle, many particles will lack one of the outside markers and, clearly, infection by such a particle cannot lead to transduction of the marker that it lacks. Nonetheless, transduction analysis has been used to provide accurate genetic maps with closely linked markers and is, in fact, the method of choice. However, presentation of the method is beyond the scope of this book.

7.4 Conjugation

Conjugation is a process by which DNA can be transferred from a donor cell to a recipient cell by cell-to-cell contact (shown in the electron micrograph at the opening of the chapter). It has been observed in a large number of bacterial species, and is best understood in *E. coli*.

When bacteria conjugate, a clear donor–recipient relation exists: DNA is transferred to a recipient cell from a donor cell, which possesses a contiguous set of genes—called the **transfer genes**—that give the cell its donor properties. These genes may be present either in a nonchromosomal circular DNA molecule called a **plasmid** or as a block of genes in the chromosome. In the latter case the plasmid is said to have been **integrated** into the chromosome.

Conjugation begins with physical contact between a donor cell and a recipient cell. Then, a passageway of unknown structure forms between the cells, and DNA moves from the donor to the recipient. In the final stage, which requires recombination if the donor contains an integrated plasmid, a segment of the transferred DNA becomes part of the genetic complement of the recipient. If the donor contains a free plasmid, only the plasmid will be transferred and it will take up residence in the recipient in the free-plasmid form.

We will begin with a description of the genetic properties of plasmids and their transfer, and then describe plasmid-mediated chromosomal transfer.

Plasmids

Plasmids are circular DNA molecules, capable of replicating independently of the chromosome and ranging in size from 0.05–10 percent of the size of the bacterial chromosome (Figure 7-7). They have been observed in many bacterial species (and are probably present in all species). They are usually nonessential for growth of the cells and may or may not be present in a particular cell, though by virtue of the genes they carry they sometimes allow growth in special circumstances. When studying a plasmid in the laboratory, a culture of cells derived from a single plasmid-containing cell is used; since plasmids replicate and are inherited, all cells of the culture will therefore contain the plasmid of interest. (Actually, on occasion a daughter cell will fail to receive a plasmid during cell division; a small percentage of the cells in a culture—about 0.01 percent—will usually be plasmid-free.) Plasmids can carry a diversity of genes, many of which may not be present in the bacterial chromosome, and can also acquire chromosomal genes by several mechanisms. The existence of a plasmid in a bacterium is usually made evident by the phenotype conferred on the cell by the genes that the plasmid possesses. For example, a plasmid carrying the *tet-r* allele will make the bacterium resistant to tetracycline.

Plasmids rely on the DNA-replication enzymes of the host cell for their reproduction, but *initiation* of replication is controlled by plasmid genes. As a result, the number of copies of a particular plasmid in a cell varies from one plasmid to the next, depending on its particular mode of regulation of ini-

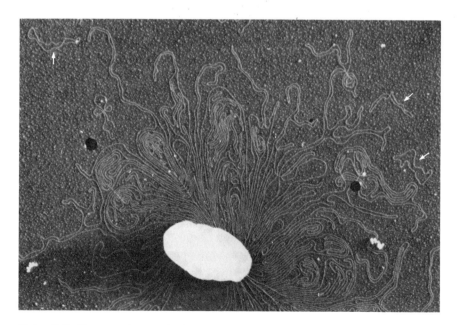

Figure 7-7 Electron micrograph of a ruptured *E. coli* cell showing released chromosomal DNA and several plasmid molecules (small circular molecules, indicated by arrows). (Courtesy of David Dressler and Huntington Potter.)

tiation. Up to 50 copies of some plasmids (**high-copy-number plasmids**) may be found in a single cell; others are present to the extent of one or two per cell. Incubation of bacteria containing some high-copy-number plasmids in media containing inhibitors of protein synthesis prevents replication of the bacterial chromosome but not replication of the plasmid, and the number of plasmid molecules can reach several thousand.

Several types of plasmids are found in *E. coli*. The best known are the R plasmids, the Col plasmids, and the F plasmid. The **R or drug-resistance plasmids** carry genes for resistance to one or more antibiotics and thereby impart an antibiotic-resistance phenotype to a cell containing them. The **Col or colicinogenic plasmids** are characterized by the ability to synthesize **colicins**—proteins capable of killing closely related bacterial strains that lack the Col plasmid. Both R and Col plasmids can be taken up by cells competent for transformation and can become permanently established in bacteria without crossing over. This ability has made plasmids important in genetic engineering (Chapter 12).

From a genetic point of view the **F plasmid** (F for *f*ertility), also called the **sex plasmid** because it confers maleness to cells, is of greatest interest, for this plasmid contains transfer genes and mediates conjugation in *E. coli*. (R plasmids also possess transfer genes, but their conjugative ability has not been studied in detail.) Cells that contain F are donors and are designated as F^+; those lacking F are recipients and are designated as F^-. F replicates once per cell cycle and segregates to both daughter cells during cell division. Consequently, the progeny of F^+ cells are also F^+. F is a low-copy-number plasmid; a typical F^+ cell contains one or two copies of F.

An F plasmid can be transferred by conjugation from an F^+ cell to an F^- cell. Transfer is always accompanied by replication of the plasmid. Pairing of an F^+ and an F^- cell initiates rolling-circle replication of F (Chapter 4), which results in transfer of a single-stranded linear branch of the rolling circle to the recipient cell. During transfer, DNA synthesis occurs in both donor and recipient (Figure 7-8). Synthesis in the donor replaces the transferred

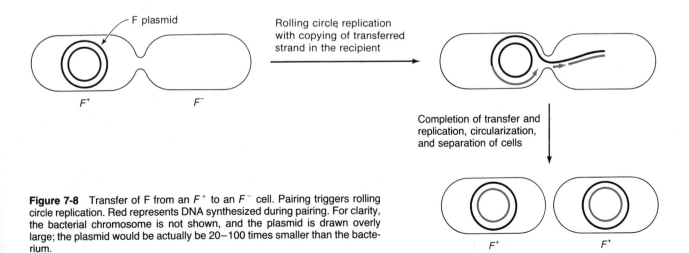

Figure 7-8 Transfer of F from an F^+ to an F^- cell. Pairing triggers rolling circle replication. Red represents DNA synthesized during pairing. For clarity, the bacterial chromosome is not shown, and the plasmid is drawn overly large; the plasmid would be actually be 20–100 times smaller than the bacterium.

single strand, and synthesis in the recipient converts the transferred single strand into double-stranded DNA. By some as-yet-unknown mechanism, the transferred linear F strand becomes recircularized in the recipient cell. Note that since one replica remains in the donor and the other is formed in the recipient, after transfer *both cells contain F and can function as donors*. Transfer of F requires only a few minutes—about five percent of the cell cycle. Thus, if a small number of donor cells is mixed with an excess of recipient cells, over a period of a few hours F will spread throughout the population and ultimately all cells will be F^+.

Hfr Cells

The F plasmid occasionally becomes integrated into the *E. coli* chromosome by an exchange between a transposable element in F and an identical transposable element in the chromosome (Figure 7-9). The bacterial chromosome remains circular, though enlarged by the F DNA. Integration of F is an infrequent event but a clone of cells containing integrated F can be purified. The cells in such a clone are called **Hfr cells.** Hfr stands for *h*igh *f*requency of *r*ecombination, which refers to the high frequency with which donor alleles are acquired by the recipient, as will be seen shortly. Integrated F mediates transfer of DNA from an Hfr cell; thus a replica of the bacterial chromosome, as well as part of the plasmid, is transferred to an F^- cell. The Hfr $\times$ F^- conjugation process is illustrated in Figure 7-10. The stages of transfer are much like those by which F is transferred to F^- cells—namely, pairing of donor and recipient, rolling-circle replication in the donor, and conversion of the transferred single-stranded DNA to double-stranded DNA by replication in the recipient—but the transferred DNA does not circularize and is not capable of further replication in the recipient. Furthermore, the replication and associated transfer of the Hfr DNA (called transfer replication), which is controlled by F, is initiated in the Hfr chromosome at the same point in F at which replication and transfer begin within an F plasmid. Thus, a portion of F is the first DNA transferred, chromosomal genes are transferred next, and the remaining part of F is the last DNA to enter the recipient.

Several differences between F transfer and Hfr transfer are notable.

1. It takes 100 minutes for an entire bacterial chromosome to be transferred, in contrast with about two minutes for transfer of F. The difference in time is a result of the relative sizes of F and the chromosome. (A common "preparatory" period of a few minutes precedes transfer in both.)

2. During transfer of Hfr DNA to a recipient cell, the mating pair usually breaks apart before the entire chromosome is transferred, owing to Brownian motion (movement resulting from bombardment by solvent molecules). On the average, several hundred genes are transferred before the cells separate.

3. In a mating between Hfr and F^- cells the F^- recipient remains F^-, because cell separation usually occurs before the final segment of F

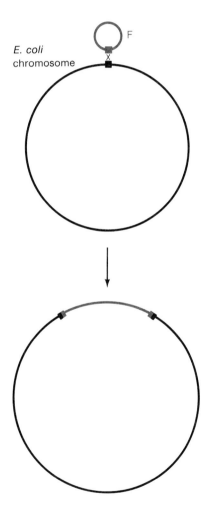

E. coli chromosome

F

Figure 7-9 Integration of F by a reciprocal exchange between a base sequence in F and a homologous sequence in the bacterial chromosome.

is transferred. The recipient does not gain the ability to engage in subsequent transfer unless it has received all of the F DNA from its donor—that is, unless chromosome transfer has been fully completed.

4. In Hfr transfer, although the transferred DNA fragment does not circularize and cannot replicate, one or more of its regions is frequently exchanged with the chromosome of the recipient, thereby generating recombinants in the F^- cell. For example, in a mating between an Hfr leu^+ culture and an F^-leu^- culture, F^-leu^+ cells arise. Note that the genotype of the donor is unchanged.

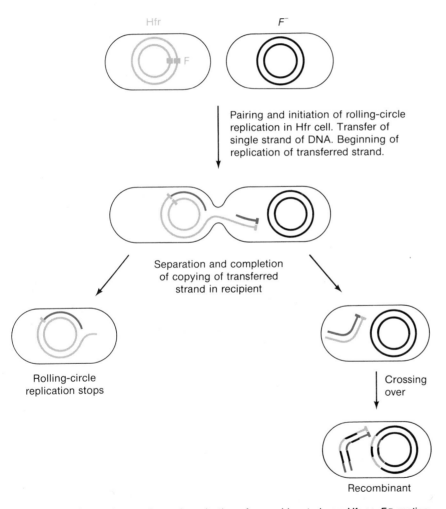

Hfr

F^-

Pairing and initiation of rolling-circle replication in Hfr cell. Transfer of single strand of DNA. Beginning of replication of transferred strand.

Separation and completion of copying of transferred strand in recipient

Rolling-circle replication stops

Crossing over

Recombinant

Figure 7-10 Stages in transfer and production of recombinants in an Hfr × F^- mating. Pairing initiates rolling-circle replication (in the F sequence) in the Hfr cell, resulting in transfer of a single strand of DNA. The single strand is converted to double-stranded DNA in the female. The mating cells break apart before the entire chromosome is transferred. Crossing over occurs between the Hfr fragment and the F^- chromosome and leads to F^- recombinants. Note that only a portion of F is transferred.

Genetic analysis requires that recombinant recipients be identified. Since recombinant recipients may have the genotype of the Hfr cell, a method is needed to distinguish donor and recipient cells. The most common procedure is to use a recipient having a recognizable genetic marker that is not present in the Hfr cell and that is located at such a place in the DNA that the donor and recipient cells will usually have broken apart before transfer of the corresponding allele in the donor has occurred. Some selective agent can then be used that allows growth only of a cell possessing the allele initially in the recipient. Markers conferring antibiotic resistance are especially useful. For instance, after a period of mating between Hfr leu^+ $str\text{-}s$ and F^- leu^- $str\text{-}r$ cells sufficient for DNA transfer, the Hfr Str-s cells can be selectively killed by plating the mating mixture on a medium containing streptomycin. Parental recipients and recombinant recipients can then be further distinguished by using a selective medium that lacks leucine. The F^- leu^- parent cannot grow in that medium but recombinant F^- leu^+ cells, which possess a leu^+ gene, can do so. Thus, only recombinant recipients—that is, cells having the genotype leu^+ $str\text{-}r$—will form colonies on a selective medium containing streptomycin and lacking leucine. When a mating is done in this way, the transferred marker that is selected by the growth conditions (leu^+ in this case) is called a **selected marker,** and the one used to prevent growth of the donor ($str\text{-}s$ in this case) is called the **counterselected marker.**

Time-of-Entry Mapping

Certain features of the transfer process in Hfr cells and information concerning the arrangement of bacterial genes in Hfr and F^- cells have been obtained by mechanically interrupting transfer during mating. The time at which a particular gene is transferred can be determined by purposely breaking the mating cells apart at various times and noting the earliest time at which breakage no longer prevents recombinants from appearing. (Violent agitation of the suspension of mating cells in a kitchen blendor is a standard method.) This procedure is called the **interrupted-mating technique.** When this is done with Hfr × F^- matings, it is observed that the number of recombinants of any particular genotype increases with the time during which the cells are in contact. The reason for the increase is slow transfer of the Hfr DNA. This phenomenon is illustrated in Table 7-1.

Greater insight into the transfer process can be obtained by observing the results of a mating with several genetic markers. Consider the mating

$$\text{Hfr } a^+b^+c^+d^+e^+ str\text{-}s \times F^- a^-b^-c^-d^-e^- str\text{-}r.$$

Again, at various times after mixing the cells, samples are taken, agitated violently, and then plated on media containing streptomycin and one of four different combinations of the five substances A through E—for example, B, C, D, and E, but no A, or A, C, D, and E, but no B. Colonies that form on the medium lacking A are $a^+ str\text{-}r$, those growing without B are $b^+ str\text{-}r$, and

Table 7-1 Data showing the production of Leu$^+$ Str-r recombinants in a cross between Hfr *leu$^+$ str-s* and *F$^-$ leu$^-$ str-r* cells when mating is interrupted at various times

Minutes after mating	Number of Leu$^+$ Str-r recombinants per 100 Hfr cells
0	0
3	0
6	6
9	15
12	24
15	33
18	42
21	43
24	43
27	43

Note: Extrapolation of the data to a value of zero recombinants indicates a time of entry of 4 min.

so forth. All of these data can be plotted on a single graph to give a set of curves, as shown in Figure 7-11(a). Four features of this set of curves are notable:

1. The number of recombinants in each curve increases with time of mating.

2. There is a time before which no recombinants are detected.

3. The number of recombinants of each type reaches a maximum, the value of which decreases with successive times of entry.

4. Each curve has a linear region that can be extrapolated back to the time axis, defining a time of entry for each locus a^+, b^+, . . . , e^+.

The explanation for the time-of-entry phenomenon is the following. Transfer begins at a particular point in the Hfr chromosome, which we now know is the replication origin of F. Genes are transferred in linear order to the recipient. The time of entry of a gene is the time at which that gene first enters a recipient in the population. All donor cells do not start transferring DNA at the same time, so the number of recombinants increases with time; separation of a mating pair prevents further transfer and limits the number of recombinants seen at a particular time. At a time much later than the time of entry, the transferred DNA undergoes genetic recombination in the recipient to form a recombinant cell.

The times of entry of the genes used in the mating just described can be placed on a map, as shown in Figure 7-11(b). Mating with a second recipient whose genotype is b^- e^- f^- g^- h^- *str-r* can then be used to locate the three genes f, g, and h. Data for the second recipient will yield a map such as that in Figure 7-11(c). Since genes b and e are common to both maps, the two maps can be combined to form a more complete map, as shown in Figure 7-11(d).

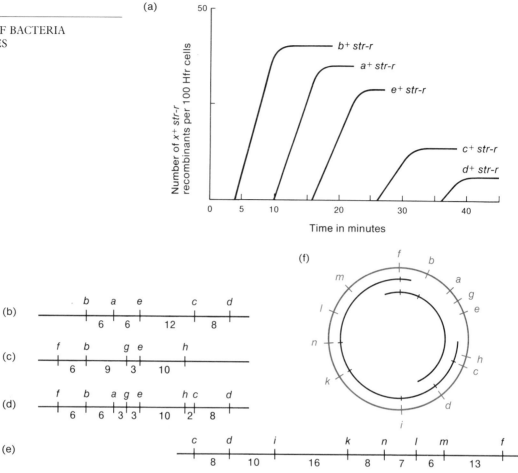

Figure 7-11 Time-of-entry mapping. (a) Time-of-entry curves for one Hfr strain. (b) The linear map derived from the data in (a). (c) A linear map obtained with the same Hfr but with a different F^- strain. (d) A composite map formed from the maps in (b) and (c). (e) A linear map from another Hfr strain. (f) The circular map (red) obtained by combining the two (black) maps of (d) and (e).

Studies with other Hfr cell lines (panel (e)) are also informative. It is usually found that different Hfr strains (for example, those of panels (c) and (e)) yield different sets of curves, distinguishable by their origins and directions of transfer, indicating that F can integrate at numerous sites in the chromosome and in two different orientations. Combining the maps obtained for different Hfr strains yields a composite map that is *circular*, as illustrated in Figure 7-11(f). The circularity of the map is a result of the circularity of the *E. coli* chromosome in F^- cells and the multiple points of integration of the F plasmid; if F could integrate at only one site, the map would be linear. Notice, however, that the circularity does not indicate whether the Hfr chromosome is circular; several physical and genetic experiments, which will not be described, indicate that it is. As we will see shortly, the linkage map of the recipient is also circular.

Several lines of evidence indicate that transfer proceeds at a fairly constant rate, and that the relative times of entry are the relative spacings between the genes on the *E. coli* chromosome. However, the time-of-entry map is not a genetic map but a physical map. A detailed genetic map can be obtained by measurement of recombination frequencies between closely linked genes (Chapter 3), but the method is not as accurate as transductional mapping (Section 7.3).

A great many mapping experiments have been carried out and the data have been combined to provide an accurate map of more than 700 genes throughout the *E. coli* chromosome. This linkage map is also circular.

F' Plasmids

F is sometimes, though rarely, excised from Hfr DNA by an exchange between the homologous sequences used in the integration event. However, in some excision events, a homologous exchange does not occur, and the process is not a precise reversal of integration; instead, a crossover occurs between two nonhomologous regions, one at one of the two boundaries between the integrated F and the chromosome, and the other in the adjacent chromosomal DNA (Figure 7-12). When such aberrant excision occurs, a plasmid containing chromosomal DNA—an **F' plasmid**—is formed. F' plasmids are isolated by the interrupted-mating technique, by selecting for early transfer of markers

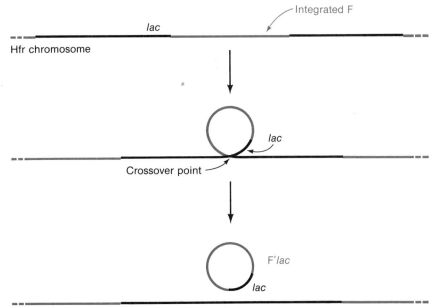

Figure 7-12 Formation of an F'*lac* plasmid by aberrant excision of F from an Hfr chromosome. Crossing over occurs between nonhomologous regions.

that would normally enter *last* from the intact Hfr strain. Thus, if an Hfr × F⁻ mating is interrupted about ten minutes after pair formation—a time insufficient for transfer of a late chromosomal marker but quite adequate for transfer of an F′ element—and a recombinant colony containing a late marker is selected, the recombinant will invariably contain an F′ plasmid. Repeated isolation of F′ plasmids from a particular Hfr strain gives a collection of plasmids containing fragments of a single region of the chromosome but of different sizes. By using Hfr strains having different transfer origins, F′ plasmids having chromosomal segments from all regions of the chromosome have been isolated. These elements are extremely useful since they render any recipient diploid for the region of the chromosome carried by the plasmid. One can then perform dominance tests and study the effects of increasing the number of copies of a gene on gene expression (gene-dosage effects). Since only a part of the genome is duplicated, cells containing an F′ plasmid are called **partial diploids.**

Plasmids also enable complementation analysis to be carried out in bacteria. For example, suppose four *prx⁻* mutants—*prx1–prx4*—each producing the Prx⁻ phenotype, had been isolated. The minimum number of *prx* genes can be determined by constructing six partial diploids: (1) F′*prx1*/*prx2*, (2) F′*prx1*/*prx3*, (3) F′*prx1*/*prx4*, (4) F′*prx2*/*prx3*, (5) F′*prx2*/*prx4*, and (6) F′*prx3*/*prx4*. Assume the phenotypes are the following: (1), +; (2), −; (3), +; (4), +; (5), −; and (6), +. These data would indicate that the four mutants fall into two complementation groups—(1, 3) and (2, 4); thus, there would be at least two *prx* genes. A real example of the use of partial diploids in this way will be seen in Chapter 11 in a discussion of the *E. coli lac* genes.

Plasmid DNA is also easily isolated, so particular plasmids provide a source of DNA of particular genes.

An F′ plasmid is described by stating the chromosomal genes it is known to possess—for example, F′*lac pro* contains the genes for lactose utilization and proline synthesis. The genes contained in an F′ plasmid are usually also present in the bacterial chromosome. To distinguish the location of the alleles, the plasmid and chromosomal alleles are separated by a diagonal line. For example, a cell containing a *lac* mutation and a streptomycin-sensitivity allele—both in its chromosome—and also containing a plasmid with a functional *lac* gene, would have the genotypic designation F′*lac⁺*/*lac⁻ str-s*.

7.5 Bacteriophage Genetics

The life cycles of phages fit into two distinct categories—the lytic and the lysogenic cycles. In the **lytic cycle** phage nucleic acid enters a cell and replicates repeatedly, the bacterium is killed, and hundreds of phage progeny result (Figure 7-1). All phage species can carry out a lytic cycle; a phage capable *only* of lytic growth is called **virulent.** In the **lysogenic cycle,** which has been observed only with phages containing double-stranded DNA, no progeny particles are produced, the bacterium survives, and a phage DNA molecule is transmitted to each daughter cell. In many phages faithful transmission of this kind is accomplished by integration of the phage chromosome into the bacterial chromosome. A phage capable of such a life cycle is called **temperate.**

Several phage particles can infect a single bacterium, and if the number of particles is not more than about ten, the DNA of each can replicate. In a multiply infected cell in which a lytic cycle is occurring, genetic exchanges occur between phage DNA molecules. If the infecting particles carry different mutations as genetic markers, recombinant phage result. Several modes of recombination—general and site-specific—occur in the lytic cycle. Only the former will be considered in this chapter. Recombination also occurs in the lysogenic cycle; however, the end result is not the production of recombinant phage chromosomes but the joining of phage and bacterial DNA molecules, as just mentioned. This process, which for most temperate phages is site-specific, is described in Section 7.6.

Plaque Formation and Phage Mutants

Phage are most easily detected by the **plaque assay,** a technique based on the fact that in a lytic cycle an infected cell breaks open (**lyses**) and releases phage particles to the growth medium. The plaque assay is performed in the following way (Figure 7-13(a)).

A large number of bacteria (about 10^8) are placed on a solid medium. After a period of growth a continuous turbid layer of bacteria, called a lawn, results. If a phage is present at the time the bacteria are placed on the medium, it will adsorb to a bacterium, and shortly afterward, the infected bacterium will lyse and release many phage; each of these progeny will adsorb to nearby bacteria, and these bacteria will in turn release phage that can infect other bacteria in the vicinity. These multiple cycles of infection continue, and after several hours each phage will have destroyed all of the bacteria in a localized area, giving rise to a clear transparent region in the otherwise turbid layer of

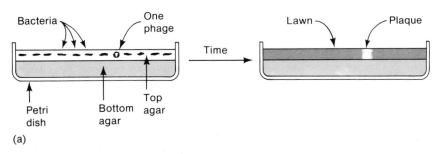 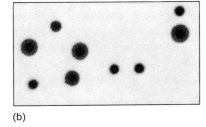

(a)   (b)

Figure 7-13 (a) Plaque formation. Bacteria grow and form a translucent lawn. Because the bacteria have lysed, there are no bacteria in the vicinity of the site of the initial phage; this empty area, which remains transparent, is called a plaque. (b) Plaques of *E. coli* phage T4. Two types of plaques are present. The smaller plaques are made by wildtype phage; the larger plaques are made by an *rII* mutant phage. Note the halo around the large plaques—it is a result of large amount of the lysis enzyme diffusing outward and lysing uninfected cells. The *rII* mutant will not form a plaque on a bacterium lysogenic for phage λ, as is explained in Section 7.6.

confluent bacterial growth—a **plaque.** Phage can only multiply in a growing bacterium, so exhaustion of nutrients in the growth medium limits phage multiplication and the size of the plaque. Since a plaque is a result of an initial infection by one phage particle, the number of individual phage put on the medium can be counted.

The genotypes of phage mutants can be determined from examining plaques. In some cases the appearance of the plaque is sufficient. For example, the plaques of most virulent phages are completely clear, whereas the plaques of certain mutants in which lysis is either inefficient or delayed may be turbid because of unlysed bacteria present in the plaque. Another type of mutation decreases the number of phage produced per cell and thereby decreases the diameter of the plaque. Plaque-morphology mutants form plaques that are visibly different from wildtype. This sort of difference can be seen in Figure 7-13(b). Host-range mutants, which can adsorb to and form a plaque on some bacterial strains to which wildtype phage cannot adsorb, are also useful in determining the genotypes of phage mutants. The most valuable type of mutant is the **conditional lethal mutant,** of which there are two kinds. One is unable to form plaques on certain bacteria (**nonpermissive** bacteria) but can form plaques on **permissive** bacteria (carrying so-called suppressor genes, see Chapter 10); the other is the **temperature-sensitive mutant,** which forms plaques at temperatures below 34°C and fails to yield plaques above about 40°C. The plaque-morphology mutant shown in Figure 7-13(b) is also a conditional mutant; the *rII* mutant, which forms the large plaques, cannot multiply on certain strains on which the wildtype grows normally. This mutant will be encountered again shortly.

Genetic Recombination in a Lytic Cycle

If two phage particles with different genotypes infect a single bacterium, genetically recombinant progeny phage form. Figure 7-14 shows the progeny of a mixed infection with *E. coli* phage T4 mutants, using the *r48* (large plaque) and *tu42* (plaque with light turbid halo) mutations as markers. The cross is

$$tu42\,r48 \text{ (turbid, large)} \times tu^{+}r^{+} \text{ (clear, small)}$$

Four plaque types can be seen—the two parental phages and the $tu42\,r^{+}$ (turbid, small) and $tu^{+}r48$ (clear, large) recombinants. When many bacteria are infected, equal numbers of complementary recombinant types are usually found. The recombination frequency is defined, as always, as

(Number of recombinant genotypes/total number of genotypes) × 100

In many experiments only one recombinant type forms a plaque (for example, ++ may make a plaque but −− may not), in which case the observed

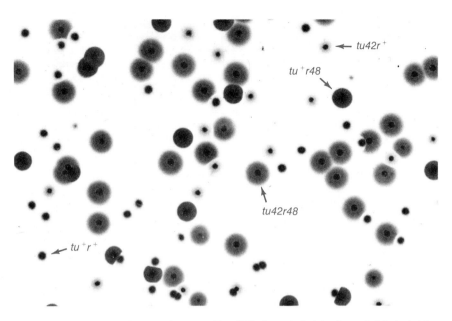

Figure 7-14 Progeny of a cross between *E. coli* T4 phage tu^+r^+ (a plaque is labeled at the lower left) and *tu42 r48* (center). Two types of recombinant plaques are found; representative plaques are labeled at the upper right. (Courtesy of A. H. Doermann.)

number of recombinants must be doubled when calculating the recombination frequency. By carrying out recombination experiments with many pairs of markers, a genetic map can be obtained, using the techniques presented in Chapter 3.

Analysis of a phage cross is an intricate problem of population genetics and is beyond the scope of this book. One problem is that among the phage progeny of an individual infected cell reciprocal recombinant types usually are produced in quite unequal numbers and in fact one of the recombinant genotypes may be absent. The reason for this is that only about half of the phage DNA molecules synthesized in the host cell are packaged into phage particles, and these are selected randomly. However, when the progeny produced by a large number of infected cells are examined, equal numbers of the reciprocal recombinants are found. A second problem is that in a phage life cycle the DNA genome undergoes repeated replication in an infected bacterium, allowing a crossover chromosome to be replicated or to participate in a second exchange event. Furthermore, recombination frequencies are affected by the ratio of the infecting parental genotypes and the ratio of phage to bacteria, necessitating the use of standard conditions (equal numbers of parental phage genotypes and a constant ratio of phage to bacteria) for obtaining reproducible data. The result of these (and other) features of phage biology is that the map unit determined from a phage cross is not comparable to the map units in eukaryotes.

Studies of genetic recombination in phage systems have been extremely important in understanding basic recombination mechanisms (see Chapter 8). Furthermore, genetic mapping has provided valuable information about the organization of phage genomes. Three features—gene clustering, terminal redundancy, and cyclic permutation—are the subjects of this section.

Several phages have been intensively studied during the past thirty years; many of their genes have been located and their gene products have been identified. A striking feature of the arrangement of the genes in each phage is that genes are generally clustered according to function. An example can be seen in the gene map of *E. coli* phage λ (Figure 7-15). Note that the right half of the map consists entirely of genes whose products—head and tail proteins—are required for assembly of the phage structure; the head genes and the tail genes also form subclusters. The left half of the map also shows several gene clusters—for example, for DNA replication, recombination, and

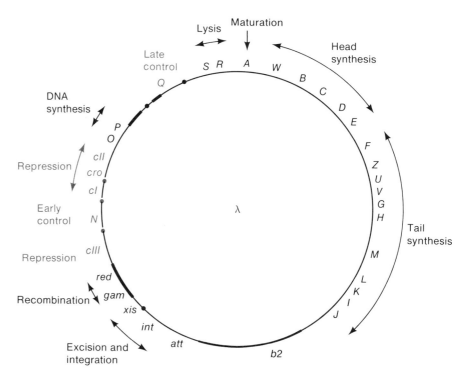

Figure 7-15 Genetic map of *E. coli* phage λ. The map is often drawn as a circle to relate it to the intracellular DNA, which is circular. Regulatory genes and functions are printed in red. All genes are not shown. Major regulatory sites are indicated by solid circles. These sites are frequently adjacent to the gene whose product acts on them. Repression refers to functions necessary for the lysogenic cycle or involved in the choice between the lytic and lysogenic cycles. Early and late control refers to genes whose activities are necessary for expression of genes early and late in the life cycle.

lysis. The genes are also clustered according to the time at which their products are synthesized. For example, the *N* gene acts early, genes *O* and *P* are active later, and genes *Q*, *S*, and *R*, and the head–tail cluster are active last. Clustering is also evident in *E. coli* phage T7.

In 1961 the genetic map of *E. coli* phage T4 was shown to be circular, whereas the T4 DNA molecule present in the phage head is linear. Additional genetic studies suggested that each of the linear T4 DNA molecules contains two copies of some genes and that the size of the duplicated segment is greater in phage mutants containing a genetic deletion, though the size of the DNA molecule is the same as that in a wildtype phage. These and several other more complex observations were explained as a consequence of **terminal redundancy**—that is, presence of the same gene at both ends of the DNA molecule. Such redundancy is generated when molecules of uniform length are cut from a DNA molecule of multiunit length, which has formed through replication and recombination (Figure 7-16). Other experiments suggested that the block of genes that is duplicated differs from one DNA molecule to the next among all progeny of the same phage. Thus, the terminally redundant segments of T4 DNA molecules do not have a unique base sequence, and, in fact, a population of T4 DNA molecules comprises a cyclically permuted set, the meaning of which is shown in the lower part of Figure 7-16.

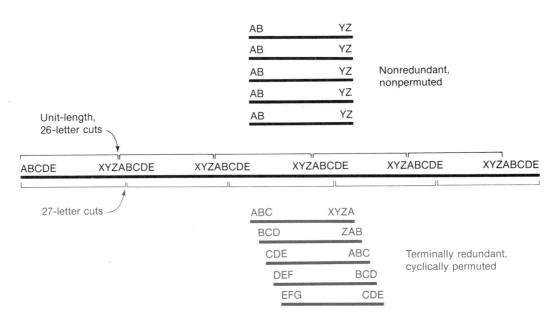

Figure 7-16 Two modes of cutting phage DNA from a molecule containing many repeated phage genomes all of the same length. The capital letters refer to genes. In the upper part of the figure, phage DNA molecules are formed by cuts made at specific sequences (between *Z* and *A*) exactly 26 letters apart; the result is a nonpermuted set of identical terminally nonredundant molecules. In the lower part of the figure, cuts are made 27 letters apart in nonspecific sites; the result is a cyclically permuted set of molecules, each having (in this case) a one-letter terminal redundancy. The DNA is shown as a single solid line. Such long DNA molecules (center of figure), consisting of repeated genomes, have been isolated from bacteria infected with a variety of phage species. They are called **concatemers** and are formed in different ways with different phages—for example, by rolling-circle replication, and by recombination.

Both terminal redundancy and cyclic permutation were later confirmed by physical analysis of T4 DNA. Several lines of evidence indicate that terminal redundancy and cyclic permutation are a result of the mode of packaging of T4 phage DNA—namely, that a constant length of T4 DNA is always packaged and that this length is cut from giant molecules in which the genome is tandemly repeated. Terminal redundancy and cyclic permutation have been observed in several other phages, though most phage species do not have either feature.

Fine Structure of the T4 rII Locus

In Chapter 3 experiments with *Drosophila* were described in which recombination within a gene was first discovered. These experiments gave the first indication that intragenic recombination could occur and that genes have fine structure. Other genes were studied in the years that followed this experiment but none could equal the fine structure mapping of the T4 *rII* locus by Seymour Benzer in which about 2400 independent mutations were mapped at 304 sites.

Figure 7-13 showed two types of T4 plaques, the wildtype (r^+) and an *rII* mutant. Plaques produced by T4 are small, because of a phenomenon (not found in many phages) called lysis inhibition. When a cell infected with wildtype T4 is reinfected, which occurs during development of a plaque, new bacterial cell wall material is synthesized, and this causes a substantial delay in lysis. Only a few cycles of infection are possible until the nutrients in the medium are exhausted, so the plaque is small. The *rII* mutants lack the ability to induce lysis inhibition, so more cycles of infection occur and the plaques are larger. Because of the appearance of their plaques, *rII* mutants are easily selected from populations of wildtype phage. Three types of mutations produce large plaques—*rI*, *rII*, and *rIII*—but these are easily distinguished because *rII* mutants fail to form plaques on a bacterial strain K12(λ) lysogenic for phage λ (Section 7.6). Crosses can be performed between *rII* mutants, and the number of recombinants can be measured simply by plating the progeny both on wildtype *E. coli*, on which both *rII* and wildtype phage grow, and on K12(λ), on which only the r^+ recombinants grow. The inhibition of growth of an *rII* mutant on K12(λ) is so severe that recombination frequencies as low as 0.00001 percent could be detected.

To initiate the study of the *rII* system Benzer performed complementation tests. In phage systems these tests are carried out by infecting a population of cells with a mixture of two phage mutants, using a ratio of phage to bacteria such that most cells are infected by both mutants (see the discussion of the Poisson distribution following this section). If the mutations are in different genes, each genome will contribute a functional product of one of the genes to the infected cell and phage progeny will form; if the mutations are in the same gene, no functional product of the mutant gene will be present and phage will not develop. Complementation tests of Benzer's collection of mutants showed that all mutants fell into two complementation groups, *rIIA* and *rIIB*, and the number of mutations used was so large that he could safely conclude that the *rII* locus consists of only two genes.

The next step was to map the mutations, a formidable task with 2400 mutations. Clearly, the direct method of carrying out two- or three-factor crosses could not be done with all mutants, because the number of required crosses would be $(1/2)(2400)(2399) = 720,000$. However, the amount of work could be reduced substantially if the *rIIA* and *rIIB* genes could be subdivided into several regions and each mutation could first be located roughly within that region. This is possible by the deletion mapping technique described for *Drosophila* in Figure 6-15. This procedure is based on the principle that if a phage with a deletion is crossed with another phage having a mutation located in the region deleted from the first phage, wildtype recombinants cannot form (because no wildtype allele will be present). Most *rII* mutants occasionally give rise to wildtype phage by a reverse mutation (Chapter 10). Benzer had in his collection several *rII* mutants that never reverted and these were presumed to be deletions. Crossing these mutants against a small collection of mapped mutations that spanned the *rII* locus indicated that these were indeed deletions. The deletions were then mapped by the following technique. Benzer reasoned that if two phage with overlapping deletions were crossed, r^+ recombinants could not form because certain regions of the gene would be missing in both phage). However, if the deletions did not overlap, wildtype recombinants would be found. Thus, by crossing all pairs of deletion mutants, the deletions could be ordered; the principle is shown in Figure 7-17. With seven ordered deletions in hand, Benzer was able to localize each of the 2400 mutations into one of seven major regions. He then used a set of smaller

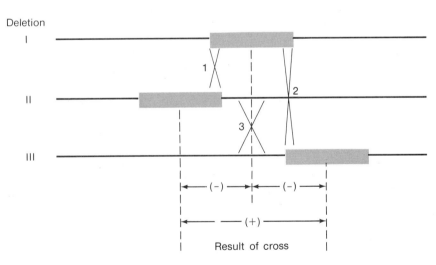

Figure 7-17 Mapping of deletions by the overlap method. The red boxes represent three deletions. In cross 1 (I × II) and cross 2 (I × III), no wildtype recombinants are produced (−), because the deletions overlap. Wildtype recombinants are produced (+) in cross 3 (II × III), because the deletions do not overlap. Indication of overlap by the lack of appearance of wildtype recombinants yields an unambiguous order of the deletions.

deletions, roughly four corresponding to each major deletion, to localize each mutation more precisely. Thus, the approximate location of each mutation could be determined by 11 crosses (seven with major deletions and four with small deletions), so with about 25,000 crosses all mutations could be localized. To reduce the labor, a special technique (called a spot test) was developed by which 10–20 crosses could be carried out on a single petri plate. Each set of mutations that had been localized in a single region was then used in individual two-factor crosses for precise mapping, using the techniques described in Chapter 3. The result of Benzer's efforts, a map with 304 sites, was a tour de force. A major conclusion of this work was that recombination can occur in all parts of a gene and that some sites are more mutable than others (see also Chapter 10). In later years a physical study of the size of each of Benzer's deletions (using the electron microscopic heteroduplexing procedure described in Section 7.6) indicated that the *rIIA* and *rIIB* genes consist of about 1800 and 850 nucleotide pairs, respectively. These numbers enabled map distances to be correlated roughly with physical distances. These data suggested that genetic exchange could probably occur between mutations in adjacent nucleotide pairs, a fact that was proved later by direct sequencing of the *E. coli trpA* protein.

The Poisson Distribution

When a collection of phage and bacteria are mixed, adsorption occurs at random. If 1000 phage are mixed with 1000 bacteria, conditions may be chosen such that all phage adsorb, but it will not be the case that each bacterium is infected by one phage—some will be infected by more and some by none. The distribution of phage over the bacterial population is governed by a statistical equation called the **Poisson distribution,** which is frequently applicable in the genetic analysis of populations.

If a large number N of balls are placed randomly into a large number r of boxes, the fraction $P(n)$ of the boxes containing $n = 0, 1, 2, \ldots N$ balls is given by

$$P(n) = (1/n!)a^n e^{-a}$$

in which e is the base of natural logarithms, ! is the factorial symbol ($4! = 4 \cdot 3 \cdot 2 \cdot 1 = 24$), and a is the average number of balls per box, N/r. Thus, if there are three times as many balls as boxes, $a = 3$ and $P(0) = 0.0497, P(1) = 0.15, P(2) = 0.22$, and so forth. An important consequence of this equation is that

$$P(0) = e^{-a}$$

This expression for the so-called "Poisson zero," which is often a measurable value, enables one to determine a and hence N.

The Poisson distribution is important in phage genetics, because some phenomena depend on the actual number of phage adsorbed per cell. For example, if equal numbers of phage and bacteria are mixed, $a = 1$ and the Poisson zero is $e^{-1} = 0.37$, which means that 37 percent of the bacteria will remain *un*infected. Similarly, if 6×10^8 phage are mixed with 2×10^8 bacteria, $a = 3$ and $P(0) = 0.0497$, or about 5 percent of the cells will be uninfected. If one were performing a recombination experiment, the value of $P(1) = 0.15$ would also be important, because a bacterium infected with one phage will not yield recombinants. Thus, the total number of bacteria not yielding recombinants would be $0.0497 + 0.15 = 0.1997$, or 20 percent. One can easily calculate from the Poisson distribution the ratio of phage to bacteria that will ensure that a particular fraction of the cells will become infected by more than one phage. Notice also that a measure of the number of uninfected cells (that is, those that survive phage infection) is a measure of the ratio of adsorbed phage to bacteria (the **multiplicity of infection**).

7.6 Lysogeny and *E. coli* Phage λ

A temperate phage has two alternate life cycles—a lysogenic cycle and a lytic cycle. The lytic cycle is that depicted in Figure 7-1. In the lysogenic cycle phage are not produced and the host survives, and a replica of the infecting phage DNA becomes **inserted** (or **integrated**) into the bacterial chromosome (Figure 7-18). The inserted DNA is called a **prophage,** and the surviving bacterium is called a **lysogen.** Many bacterial generations later, if environmental conditions are right, the prophage can be activated and excised from the chromosome, and a lytic cycle can begin. When activation occurs, the host cell is killed and progeny phage are released, exactly as in a normal lytic cycle. Two features of the lysogenic cycle will be considered here—how the DNA is integrated and how it is excised, and the discussion will be confined to the best-understood temperate phage, *E. coli* phage λ. To indicate that a bacterium contains a prophage the prophage symbol is placed in parentheses following the bacterial name—for example, *E. coli* DF4(λ) is a bacterial strain named DF4 that contains phage λ as a prophage.

Time-of-entry experiments and transduction analysis of lysogenic cells have indicated the following: (1) The *E. coli gal* and *bio* genes and the pro-

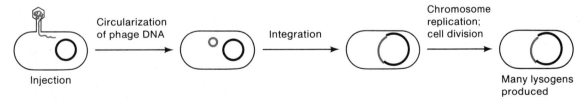

Injection   Circularization of phage DNA   Integration   Chromosome replication; cell division   Many lysogens produced

Figure 7-18 The general mode of lysogenization by insertion of phage DNA into a bacterial chromosome. Some genes—those needed to establish lysogeny—are expressed shortly after infection and are then turned off. The inserted red DNA is the prophage.

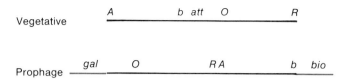

Figure 7-19 The map order of the genes as determined by phage recombination (vegetative map) and in the prophage (prophage order). The genes have been selected arbitrarily to provide reference markers.

phage are linked, with the gene order *gal* λ *bio;* the *gal* gene is linked to the prophage genes at the left end of the prophage map. (2) The *gal* and *bio* genes can be cotransduced by P1 phage grown on a nonlysogen but not by P1 grown on a lysogen. (3) The *gal* and *bio* genes are genetically and physically farther apart in a lysogen than in a nonlysogen. (4) Mapping of prophage genes with respect to the *E. coli gal* and *bio* genes by P1 transduction yields a prophage genetic map that is a permutation of the genetic map of the phage obtained from standard phage crosses. That is, the genetic map from phage crosses is *A b att O R,* and the prophage map is *gal O R A b bio* (Figure 7-19). Observations 1, 2, and 3 led to the hypothesis that the λ prophage is linearly inserted between the *gal* and *bio* loci. This notion was proved by a physical experiment in which λ DNA was inserted in a circular F′ plasmid: it was observed that plasmid DNA remained circular but its size increased by the amount of one λ molecule. Observation 4 suggested that integration is the result of a breaking-and-rejoining event that occurs between a particular site in λ DNA and a bacterial site between the *gal* and *bio* loci. The correctness of this suggestion was ascertained by studies of the structure of λ DNA and an analysis of the nature of the sites of exchange.

Mechanism of Prophage Integration

The DNA of λ is a linear molecule having **complementary single-stranded termini (cohesive ends)** 12 bases long. These termini can base-pair, forming a circular molecule, as shown in Figure 4-20. (This process, which is easily carried out in the laboratory with purified DNA, actually occurs shortly after the DNA is injected into the cell; the points of discontinuity in the single strands are then sealed to yield a covalently closed double-stranded circle, which soon becomes supercoiled.) Circularization, which occurs early in both the lytic and lysogenic cycles, is a key event in prophage integration.

The sites of breakage in the bacterial and phage DNA are called the **bacterial and phage attachment sites,** respectively. Each attachment site consists of three segments. One, the **core** or **O region,** is common to both attachment sites and is the base sequence in which the breakage and rejoining actually occurs. The phage attachment site is denoted *POP′* (P for phage) and the bacterial attachment site is denoted *BOB′* (B for bacteria). Comparison of the genetic maps of the phage and the prophage indicate that *POP′* is located near the middle of the linear form of the phage DNA molecule. A phage

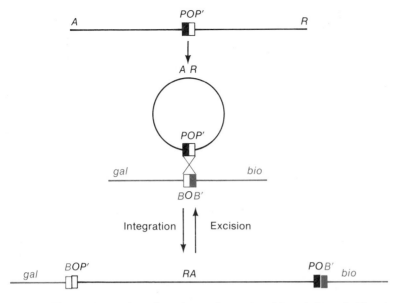

Figure 7-20 The geometry of prophage integration and excision of phage λ. The phage attachment site is *POP'*. The bacterial attachment site is *BOB'*. The prophage is flanked by two attachment sites denoted *BOP'* and *POB'*.

protein, **integrase** (or Int), catalyzes a site-specific recombination event—that is, it recognizes the phage DNA and bacterial DNA attachment sites and causes the physical exchange that results in integration of the λ DNA molecule into the bacterial DNA. The geometry of the exchange is shown in Figure 7-20. Note that the permutation of the phage and prophage maps is simply a consequence of the circularization of the phage DNA and the central location of *POP'*.

Integration can be thought of as an enzymatic reaction between attachment sites, namely,

$$BOB' + POP' \xrightarrow{\text{Integrase}} BOP' + POB'$$

Bacterium Phage   Prophage

A variety of genetic experiments indicate that the segments *B*, *B'*, *P*, and *P'* are all different, and this conclusion has been confirmed by direct determination of the base sequence of *BOB'* and *POP'* (the *O* regions are all the same). Thus, the **prophage attachment sites**—*BOP'* and *POB'*—must also be different. This conclusion suggests why integrase does not excise the prophage shortly after integration occurs by a site-specific exchange between the two prophage attachment sites—that is, integrase can recombine *BOB'* and *POP'* but (by itself) not *BOP'* and *POB'*.

Site-specific recombination, though not common, is not confined to the lysogenic cycles of phages. For example, it also occurs in the immune system in the production of specific antibodies (see Chapter 13).

Prophage Excision

A lysogenic cell can replicate nearly indefinitely without release of phage. However, prophage excision sometimes occurs followed by a lytic cycle with production of the usual number of phage progeny. This phenomenon is called **prophage induction,** and it is initiated by damage to the bacterial DNA, which can lead to cell death. The damage is sometimes spontaneously formed but is more often caused by some environmental agent, such as chemicals or radiation. The ability to be induced is advantageous for the phage, because the phage DNA can escape from a dying bacterium. The biochemical mechanism of induction is complex and will not be discussed; in contrast, the excision reaction is straightforward.

Excision is also a site-specific recombination event. It requires Int and an additional phage protein called **excisionase** (or Xis). Genetic evidence and studies of physical binding of purified excisionase, integrase, and λ DNA indicate that excisionase binds to integrase and thereby enables the latter to recognize the prophage attachment sites BOP' and POB'; once bound to these sites, integrase can make cuts in the core sequence and re-form the BOB' and POP' sites. Thus, the Xis-Int complex reverses the integration reaction, causing excision of the prophage (Figure 7-20). Note that the reactions between all attachment sites can now be written

$$BOB' + POP' \underset{\text{Int,Xis}}{\overset{\text{Int}}{\rightleftharpoons}} BOP' + POB'$$

The Xis-Int complex fails to catalyze the rightward reaction, so that when excess excisionase is present, excision is an irreversible reaction. Note that the products of excision are an intact *E. coli* chromosome and a circular λ molecule—the arrangement present in an infected cell immediately after infection.

Phenotypic Properties of Lysogens

When a cell is lysogenized, a block of phage genes becomes part of the bacterial chromosome, so it might be expected that the phenotype of the bacterium would change. Most phage genes in a prophage are kept in an inactive state by the product of one expressed phage gene called a **repressor** gene. This gene makes a protein that is synthesized initially by the infecting phage and then continually by the prophage; in fact, the repressor gene is frequently the only prophage gene that is expressed. If a lysogen is infected with a phage of the same type as the prophage—for example, λ infecting a λ lysogen—the repressor present within the cell (made from prophage genes) will prevent expression of not only the prophage genes but also the (identical) genes of the infecting phage. This resistance to infection by a phage identical to the prophage, which is called **immunity,** is the usual criterion for determining whether a bacterium contains a particular phage. For example, λ will not form a plaque on bacteria containing a λ prophage.

With some phage species genes other than the repressor gene are also expressed; the phenotypic change produced by such genes is termed **lysogenic conversion.** One example is the resistance to phage T4 *rII* mutants conferred on *E. coli* by the λ *rex* gene. The product of this gene, which is expressed continually in a λ lysogen, blocks an early step in the T4 life cycle; the product of the T4 *rII* gene is needed to overcome this blockage (in addition to its role in lysis inhibition). Another example of lysogenic conversion is found in *Corynebacterium diphtheriae*, which causes the human disease diphtheria only when lysogenized by the phage β. The phage makes the toxin responsible for the disease.

Some temperate phage species can integrate at any position in the host chromosome rather than at a defined site. These phages produce mutations in the host since they frequently insert themselves within a host gene and interrupt its continuity; an example of such a phage is *E. coli* phage Mu (for *mu*tator).

Specialized Transducing Particles

When a bacterium lysogenic for phage λ is subjected to DNA damage that leads to induction, the prophage is usually excised from the chromosome precisely. However, in about one cell per 10^6–10^7 cells an excision error is made (Figure 7-21) and a chance breakage in two nonhomologous sequences occurs—one break within the prophage and the other in the bacterial DNA. The free ends of the excised DNA are then ligated to generate a DNA circle capable of replication. The sites of breakage may not always be situated such

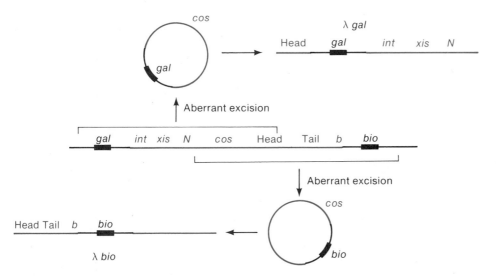

Figure 7-21 Aberrant excision leading to production of λ *gal* (upper panel) and λ *bio* (lower panel) phages.

that a length of DNA that can fit in a λ phage head is produced—the DNA may be too large or too small—but sometimes a molecule forms whose replicas can be packaged. In λ the prophage is located between the *E. coli gal* and *bio* genes and because the aberrant cut in the host DNA can be either to the right or the left of the prophage, particles can arise that carry either the *bio* genes (cut at the right) or the *gal* genes (cut at the left). These particles are called λ *bio* and λ *gal* transducing particles. Note that a transducing particle formed by aberrant excision from a lysogen can carry only particular bacterial genes, in contrast with the generalized transducing particles formed in cells infected with phage P1. Thus, these particles are called **specialized transducing particles.** Because of the aberrant excision, specialized transducing particles lack some phage genes. The λ *gal* transducing particles lack the genes for the head and tail proteins and hence are unable to infect a bacterium and produce progeny phage. The λ *bio* particles lack the genes for integration and recombination, which are nonessential, and hence can reinfect a host cell and produce progeny. Note that the λ *gal* transducing particle can be produced by the cell in which the aberrant cut occurred, because in that cell all phage genes are present. Further multipliction of a λ *gal* transducing particle is possible only in a mixed infection with a normal phage (a "helper" phage). That is, if a cell is infected with both normal λ and λ *gal*, the normal λ will provide the head and tail proteins and both normal λ and λ *gal* progeny will be produced.

The size of the bacterial DNA substituted for phage DNA in a specialized transducing particle can vary, depending on the location of the aberrant exchange. The location of the exchange can be determined by a cross between

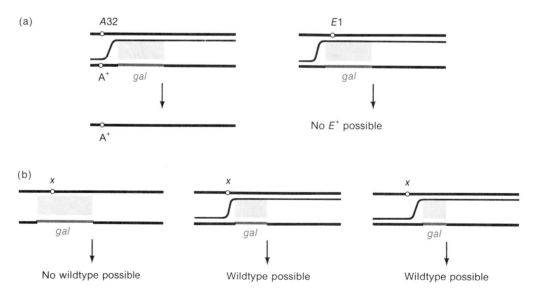

Figure 7-22   (a) Locating the left terminus of a *gal* substitution in a λ *gal* transducing particle. A cross is carried out between λ *gal* and normal λ marked either with the *A32* or *E1* mutations, whose locations are known. In the cross at the left only the relevant recombinant is shown. (b) Locating a mutation *x* by crosses with three different λ *gal* particles, selecting for *x*⁺ phage. The shaded red area indicates the nonhomologous regions in which crossing over does not occur. Note that the right end of the *gal* substitution is always the *att* site.

a particular λ *gal* phage and several normal λ phage, each carrying a different mutation with a known location. If wildtype λ cannot arise in a cross with a particular mutant, the mutation must be in the region of the substitution in that phage (Figure 7-22(a)). Similarly, mutations can be mapped by performing crosses with λ *gal* particles having substitutions of known extent (Figure 7-22(b)). The precise physical location of the substitution can also be determined by **heteroduplex analysis.** DNA of a normal phage and a transducing particle are isolated, mixed, denatured and renatured (Chapter 5), and examined by electron microscopy. Hybrid molecules can then be seen in which homologous single strands have joined to form double-stranded DNA but nonhomologous single-stranded regions remain in unrenatured bubbles (Figure 7-23). Thus, the endpoints of the substitution can be localized as the branch points of the bubble.

Transducing particles can also be used to prepare partial diploid strains in which two copies of the *gal* gene carry different markers (Figure 7-24). These strains can be used in dominance studies and for complementation analysis, a technique that will be described in Chapter 11.

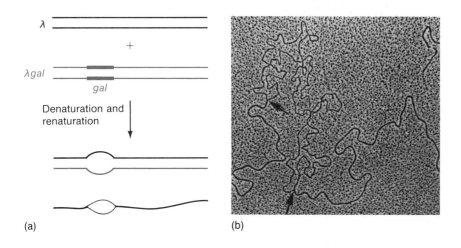

Figure 7-23 (a) A diagram showing the heteroduplexing technique for determining the size and location of a substitution. (b) An electron micrograph of λ and λ *gal* DNA. The arrows show the termini of the denaturation bubble. Single-stranded and double-stranded DNA have nearly the same width with the technique used to produce this micrograph. (Courtesy of Norman Davidson.)

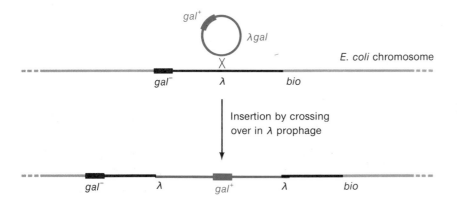

Figure 7-24 Production of an *E. coli* chromosome containing two copies of the *gal* gene. *E. coli* DNA extends in both directions (dashed lines) beyond the segment shown here.

7.7 Transposable Elements

In Chapter 5 we described DNA sequences that are present in a genome in multiple copies called **transposable elements;** these have the property of occasional movement or **transposition** to new locations in the genome. These elements, which are widespread in eukaryotes and bacteria, have the common characteristic of a short sequence of nucleotide pairs that is present in opposite orientation at the two ends (inverted repeats). In some of the elements, as we have seen from eukaryotic examples (Figure 5-24) the inverted repeats are parts of longer terminal repeats present in the same orientation.

Many transposable elements are known in bacteria, where they are easily recognized and have been extensively studied. In contrast with eukaryotes, the number of copies of a particular transposable element is rarely greater than one or two per genome. The bacterial elements first discovered—called **insertion sequences** or **IS elements**—are small and do not contain any known host genes. Other elements, which contain recognizable bacterial genes that are unrelated to the transposition of the element, are called **transposons.** The latter elements, which typically contain several thousand nucleotide pairs, are usually designated by the abbreviation Tn followed by a number (for example, Tn5). When it is necessary to refer to genes carried by such an element, standard genotypic designations for the genes are used. For example, Tn1(*amp-r*) carries the genetic locus for ampicillin resistance. Such genes provide a marker that makes it easy to detect the transposition of one of the composite elements, as is shown in Figure 7-25. An F'*lac* plasmid is trans-

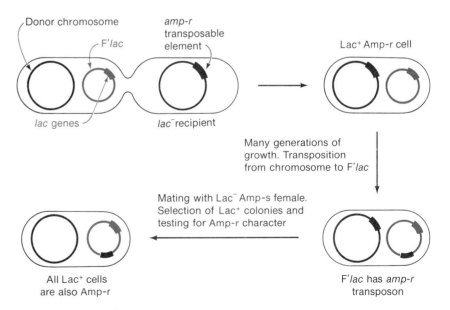

Figure 7-25 An experiment demonstrating transposition from the chromosome to a plasmid. The character used to select recombinants is not indicated. A similar experiment can be done by growing a phage on a cell containing an *amp-r* transposable element; some progeny phage particles gain the ability to transduce the *amp-r* marker.

ferred by conjugation to a bacterium containing a transposable element with the *amp-r* (ampicillin-resistance) gene. The bacterium is allowed to grow and in the course of multiplication, transposition of the *amp* gene to the F′ plasmid occasionally occurs in a progeny cell, yielding an F′ plasmid carrying both the *lac* and *amp-r* genes. In a subsequent mating to an Amp-s Lac⁻ bacterium, the *lac*⁺ and *amp-r* markers are transferred together.

When transposition occurs, the transposable element can be inserted at any one of a large number of positions. The existence of multiple insertion sites can be seen when a wildtype *E. coli* culture is infected with a phage carrying a transposable element having an antibiotic-resistance marker. Transposition events can be detected by the production of antibiotic-resistant bacteria that carry a variety of mutations in other genes (the mutations result from insertion of the element within a gene). For example, an infected *lac*⁺*leu*⁺*amp-s* culture can yield both *lac*⁻*amp-r* and *leu*⁻*amp-r* mutants. If many hundreds of Amp-r bacterial colonies are examined and tested for a variety of nutritional requirements and the ability to utilize different sugars as a carbon source, colonies can usually be found bearing a mutation in almost any gene that is examined. This observation indicates that insertion sites for transposition are scattered throughout the *E. coli* chromosome. The use of transposable elements in generating bacterial mutants will be described in Chapter 10.

Although many sites of insertion in *E. coli* are available, the locations are not randomly distributed for all transposable elements; certain positions seem to be excluded and others are used repeatedly. For example, in one experiment 22 different insertions were isolated in a *his* gene of the bacterium *Salmonella typhimurium;* direct base sequencing of the gene showed insertions at 13 different positions with one position represented nine times. One transposable element, Mu, seems to insert itself at any position and all insertion sites occur with nearly the same probability.

The end result of the transposition process is the insertion of a transposable element between two base pairs in a recipient DNA molecule. Two genetic observations reveal important features of the process. (1) Following transposition in bacteria (but not in all organisms) the element is still present at the original position—that is, *the original element has been duplicated*—which indicates that in these organisms and with the particular transposable element studied *transposition is a replicative process.* (2) In bacteria and yeast transposition occurs in cells in which the major system for homologous recombination has been eliminated by mutation. Base-sequence analysis of many transposable elements and their insertion sites extends the latter observation to eukaryotic cells, for which no such mutants are known; that is, no homology between the sequence in the recipient and any sequence in the transposable element is detectable. Evidence for replication also comes from the base-sequence analysis, inasmuch as insertion of a transposable element is always accompanied by duplication of the target sequence, as shown in Figure 5-25. The mechanism of transposition has not yet been elucidated but these observations have led to a reasonable model, which is described in Chapter 8.

Three transposable elements—IS2, IS3, and γδ—are contained in the F plasmid. These sequences are also scattered throughout the *E. coli* chromo-

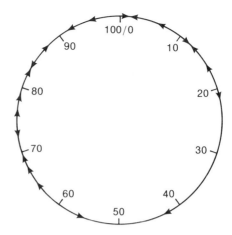

Figure 7-26 *E. coli* genetic map showing twenty different origins (tips of arrow heads) for Hfr transfer. The arrows show the direction of transfer. Note the three sites at which two Hfr strains with opposite directions of transfer have formed. The numbers represent times of entry in minutes with 0 at a position arbitrarily chosen in the standard *E. coli* map.

some, having different locations in different bacterial strains. They are responsible for the formation of many Hfr cell lines from F^+ cells. As mentioned earlier, an Hfr cell arises when F integrates into the *E. coli* chromosome by a reciprocal exchange. Two homologous sequences are needed—one in the F plasmid and the other in the chromosome. Frequently, this sequence is IS3. However, because IS3 may be located at many sites in *E. coli* and because the other two transposable elements can also be utilized, insertion of F can occur at numerous positions, generating a variety of Hfr types. More than 20 frequently occurring insertion sites and many more minor sites are known. Furthermore, since the terminal base sequences of IS3 are inverted repeats, IS3 can be transposed to a position in the chromosome with its central sequence oriented in either direction. Thus, F can integrate in both clockwise and counterclockwise orientations at different insertion sites, generating Hfr strains that transfer the chromosome in either the clockwise or counterclockwise direction, with respect to the *E. coli* genetic map (Figure 7-26).

Problems

1. An Hfr strain with genotype $met^- his^+ leu^+ trp^+$ is mated with a $met^+ his^- leu^- trp^-$ recipient. The *met* marker is known to be transferred very late. After a short mating the cells are broken apart and plated on four different growth media. The nutrients in the growth medium and the number of colonies observed on each are the following: His, Trp, 250; His, Leu, 50; Leu, Trp, 500; His only, 10. What is the order of injection of the genes? What is the purpose of the met^- mutation in this experiment? Why is the number of colonies so small for the His-only medium?

2. Suppose that phage P1 was grown on bacteria with the genotype $pur^+ pro^- his^-$. A bacterium having the genotype $pur^- pro^+ his^+$ was transduced with the P1; pur^+ transductants were selected and tested for the unselected markers pro and his. The number of pur^+ colonies having each of four genotypes is the following: $pro^+ his^+$, 102; $pro^- his^+$, 25; $pro^+ his^-$, 160; $pro^- his^-$, 1. What is the gene order?

3. A particular negative marker x^- cannot be selected for in a positive way; that is, there is no selective growth medium on which a bacterium containing the marker will grow and on which an x^+ strain will not grow. Assume that the marker has been mapped and that you wish to transfer the mutant allele to a particular x^+ cell. Suggest a way by which P1 transduction could be used to facilitate transfer and detection of the marker.

4. An Hfr donor of genotype $a^+ b^+ c^+ d^+ str\text{-}s$ is mated with an F^- recipient having genotype $a^- b^- c^- d^- str\text{-}r$. Genes a, b, c, and d are spaced equally. A time-of-entry experiment is carried out and the data shown in the table below are obtained. What are the times of entry for each gene? Suggest a possible reason for the low plateau region for $d^+ str\text{-}r$ recombinants.

Time of mating, in min.	Number of recombinants of indicated genotype per 100 Hfr			
	$a^+ str\text{-}r$	$b^+ str\text{-}r$	$c^+ str\text{-}r$	$d^+ str\text{-}r$
0	0.01	0.006	0.008	0.0001
10	5	0.1	0.01	0.0004
15	50	3	0.1	0.001
20	100	35	2	0.001
25	105	80	20	0.1
30	110	82	43	0.2
40	105	80	40	0.3
50	105	80	40	0.4
60	105	81	42	0.4
70	103	80	41	0.4

5. An Hfr donor whose genotype is $a^+ b^+ c^+ str\text{-}s$ mates with a recipient whose genotype is $a^- b^- c^- str\text{-}r$; the order of transfer is $a\,b\,c$. None of these genes are transferred early, the distance between a and b is the same as the distance between b and c, and none of the markers are near str. Recombinants are selected as usual by plating on a medium lacking particular nutrients and containing streptomycin. Which of the following are true (several answers are)? Explain.

(a) $a^+ str\text{-}r$ colonies $>$ $c^+ str\text{-}r$ colonies.
(b) $b^+ str\text{-}r$ colonies $<$ $c^+ str\text{-}r$ colonies.
(c) $a^+ b^+ str\text{-}r$ colonies $<$ $b^+ str\text{-}r$ colonies.
(d) $a^+ b^+ str\text{-}r$ colonies $=$ $b^+ str\text{-}r$ colonies.
(e) Most $a^+ c^+ str\text{-}r$ colonies will also be b^+.
(f) Most $b^- c^+ str\text{-}r$ colonies will also be a^-.
(g) $a^+ b^+ c^- str\text{-}r$ colonies $<$ $a^+ b^- c^- str\text{-}r$ colonies.

6. As shown in Figure 7-12, the plateau recombination frequency for various markers decreases continually with their time of entry. However, markers that enter *very early*—for example, one minute after mating begins—have very low plateau values rather than a maximum value. Explain this observation.

7. After a brief mating between an Hfr whose genotype is $pro^+ pur^+ lac^+$ and a recipient whose genotype is $F^- pro^- pur^- lac^- str\text{-}r$, many $lac^+ pur^+ pro^- str\text{-}r$ recombinants are found. A few $pro^+ lac^- pur^- str\text{-}r$ recombinants also arise, and all of these are Hfr donors. Explain this result and state the location of F in the Hfr. (The three genes pro, lac, and pur are very near one another.)

8. The number of phage in a suspension is being determined by plating. First 0.1 ml is diluted into 10 ml of buffer (tube A). Two successive dilutions of 0.1 ml into 10 ml are made (tubes B and C), followed by two successive dilutions of 1 ml into 9 ml (tubes D and E). Then, 0.1 ml from various dilution tubes is plated with sensitive bacteria, and the next day, plaques are counted. The number of plaques obtained from the 0.1 ml aliquot from tubes C, D, and E are 2010, 352, and 18, respectively. What is the best estimate of the number of phage per ml in the sample? Explain. Hint: Think about how many phage particles are responsible for forming each plaque on plates C, D, and E.

9. Phage T2 (a relative of T4) normally forms small clear plaques on a lawn of *E. coli* strain B. A mutant of *E. coli* called B/2 is unable to adsorb T2 phage particles, so no plaques are formed. T2*h* is a host-range mutant phage capable of adsorbing to *E. coli* B and to B/2, and it forms normal-looking plaques. If *E. coli* B and the mutant B/2 are mixed in equal proportions and used to generate a lawn, what will be the appearance of plaques made by T2 and T2*h*?

10. A temperate phage has the gene order $a\,b\,c\,d\,e\,f\,g\,h$ and a prophage gene order $g\,h\,a\,b\,c\,d\,e\,f$. What information does this give you about the phage?

11. A λ phage genetic map (as obtained in standard crosses) containing only a few of the known genes is shown in the figure below. Which of these genes would show the highest frequency of cotransduction with the *gal* gene if P1 phage grown on a *gal*⁺ λ lysogen were used to transduce a *gal*⁻ λ lysogen?

A B	J	cl	P	Q	R

12. Genetic recombination between λ phages can occur by a variety of systems. Two of these—the bacterial Rec system and the λ Red system—can be eliminated by mutation. However, recombinants still arise in a mixed infection with genetically marked phages, but only between certain markers. For example, recombination occurs between the following pairs of genes: $A \times R$, $A \times O$, $b \times O$, and $b \times R$. However, no recombinants can be detected in crosses between A and b, and between O and R. (For the position of these genes, see Figure 7-16 in the text.) Explain these results. Hint: Think about prophage integration.

13. Pairs of specialized transducing λ particles that carry genetic markers in the phage genes can yield genetically recombinant phage when crossed. Interestingly, recombinants are produced even when, owing to mutations, the general recombination system (Red) of the phage and the recombination system (Rec) of the bacterium are both absent, as long as active integrase is synthesized. This problem explores how these recombinants arise.
    (a) What attachment sites are present in λ *bio*- and λ *gal*-transducing particles?
    (b) What attachment sites can be generated by integrase-mediated recombination between a *bio* and a *gal* transducing particle? What genotypes, with respect to the *gal* and *bio* genes, are associated with each attachment site?
    (c) Could an A^+R^+ recombinant be produced in a cross between λ A^-R^+bio and λ A^+R^-bio, if both the Rec and Red systems are inactive? Explain. The A and R genes are at opposite ends of the λ DNA molecule. Hint: Think about the location of the *bio* segment in the transducing particles.

14. Why are λ specialized-transducing particles generated only by inducing a lysogen to produce phage rather than by lytic infection?

15. An Hfr *str-s* that transfers the genes in alphabetical order, $a\,b...y\,z$, is mated with $F^-\,z^-\,str\text{-}r$ cells. The mating mixture is agitated violently 15 minutes after mixing to break apart conjugating cells and then plated on a medium lacking Z and containing streptomycin. The *z* gene is far from the *str* locus. The yield of $z^+\,str\text{-}r$ colonies is about one per 10^7 Hfr cells. What are the two possible genotypes of such a colony?

16. An Amp-r bacterium is infected with λ and a suspension of phage progeny is obtained. This phage suspension is used to infect another culture of bacteria, which is lysogenic for λ, Amp-s, and lacks the bacterial general recombination system. Because the λ repressor is present in the lysogen, infecting λ molecules cannot replicate and are gradually diluted out of the culture by growth and continued division of the cells. Rare Amp-r cells are found among the lysogens. Explain how they might have arisen.

17. On continued growth of a culture of a λ lysogen rare cells become nonlysogenic by spontaneous loss of the prophage. Such loss is a result of prophage excision without subsequent development of the phage. The lysogen is said to have been **cured.** A λ variant has been found that carries the gene for tetracycline resistance. A Leu⁺Tet-s cell is infected with this phage and a Tet-r lysogen is isolated. At a later time a bacterial mutant is isolated; it is Leu⁻Tet-r and no longer contains a λ prophage. Explain how the genotype of the cured cell may have arisen.

18. A cross is done between an F′*lac*⁺*str-s tet-s* culture and an F^- *lac*⁻ *str-r tet-r* culture. F′*lac*⁺*str-r tet-r* colonies are isolated after growth on a medium containing streptomycin. As expected, cultures prepared from these recombinants are able to transfer the *lac*⁺ marker in subsequent crosses. At a much later time a culture prepared from the recombinant is used to transfer F′*lac*⁺ to a *lac*⁻ *str-r tet-s* recipient. The mated cells are plated on a color indicator medium containing steptomycin, and *lac*⁺*str-r* recombinants are selected. Although the Tet-r phenotype has not been selected, *all* recombinant colonies are found to carry the *tet-r* marker. Explain this phenomenon.

19. One milliliter of a bacterial culture containing 5×10^8 cells is infected with 2×10^9 phage. After a period of time thought to be sufficient for nearly 100 percent adsorption, phage antibody is added; the antibody inactivates unadsorbed phage but does not affect any phage that have adsorbed to bacteria. Thus, if the mixture is plated with an indicator bacterium (to produce a bacterial lawn), plaques will result only from infected cells. When 200 cells are mixed with indicator bacteria, 100 plaques form—considerably fewer than expected.
    (a) How many were expected?
    (b) What fraction of the phage failed to adsorb?

20. Consider a phage with the strange property that productive infection occurs only if at least two phage adsorb to each bacterium. If 3×10^8 phage are mixed with 10^8 bacteria, and if all phage adsorb, how many bacteria will be productively infected?

21. P2 and P4 are bacteriophages of *E. coli*. They have the following properties: (1) When one P2 phage infects a bacterium, the bacterium usually bursts, giving about 100 P2 progeny. (2) When a P4 phage infects a bacterium, the bacterium survives because P4 is a defective phage. (3) When P2 phage and P4 phage coinfect the same bacterium, lysis of the bacterium gives 100 P4 progeny and no P2 progeny (because P4 inhibits the growth of P2). If 3×10^8 P2 and 2×10^8 P4 are added to 10^8 bacteria, then:
    (a) How many bacteria will not be infected at all?
    (b) How many bacteria will survive?
    (c) How many bacteria will produce P2 progeny?
    (d) How many bacteria will produce P4 progeny?

22. When Hfr males conjugate with F^- cells lysogenic for λ, zygotes normally survive. However, when Hfr males lysogenic for λ conjugate with F^- nonlysogens, zygotes produced from matings that have lasted for almost two hours lyse, owing to the zygotic induction of λ.
    (a) How can you explain zygotic induction?
    (b) How can you determine the locus of the integrated λ prophage?

23. If an *E. coli* culture is heavily irradiated with ultraviolet light and then infected with T4 phage, the burst size of T4 is nearly normal. If it is instead infected with λ phage, very few phage are produced. Explain this difference.

24. Could you have a plasmid with no genes whatsoever?

25. How do the sizes of plasmids compare with the size of phage DNA molecules?

26. On EMB-lactose agar Lac$^+$ and Lac$^-$ colonies are purple and white, respectively. If several thousand $F'lac^+/lac^-$ cells are plated, a few sectored colonies appear. This may have a purple half and a white half or a wedge of white in a predominately purple colony. Sectored colonies that are predominately white are not found. Explain the cause of sectoring.

C H A P T E R 8

Mechanisms of Genetic Exchange

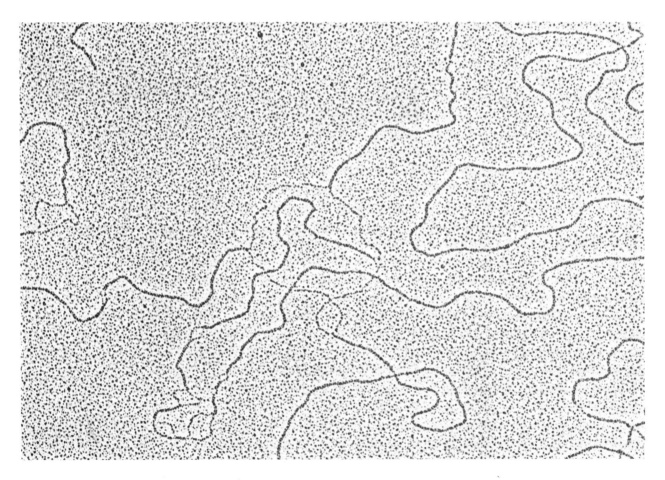

Genetic exchange, the process in which two genetic units—chromosomes, DNA molecules, or segments of DNA molecules—interact and DNA is either exchanged or rearranged, has already been the subject of several examples in Chapters 3 and 7. Meiotic recombination in eukaryotes was one such example; other examples, these in bacterial and phage systems, were transformation, transduction, Hfr $\times$ F^- conjugation, transposition, phage-phage recombination, and prophage integration and excision.

The phenomenon of recombination of genes in the same chromosome has been known since the early years of genetics, but only recently has any degree of understanding of the molecular mechanisms for that phenomenon been achieved. The elucidation of the structure of DNA in 1953 made clear how a recombinant chromosome is related to parental chromosomes, and several simple models for recombination were presented in the following two years. These models were oversimplified, mostly because neither the physical properties of DNA nor the enzymes acting on DNA were known. Almost ten years passed before this information became available. In the 1960s several models were proposed that have been significant in the development of our understanding of genetic exchange, though each has also presented difficulties. In the course of many years of research it has become clear that different types of genetic exchange utilize distinctly different mechanisms, though most mechanisms share common features regardless of what type of genetic exchange they achieve; these features include breakage of DNA molecules, participation of enzymes in a multistep process, and synthesis of a small amount of DNA. In this chapter we will examine some of the better-understood mechanisms, paying particular attention to their common features. In so doing, we will address the questions of how participating DNA molecules pair, how homologous base sequences are matched, how exchange occurs, and how the participating molecules are separated.

Facing page: Electron micrograph of a portion of two circular phage λ DNA molecules joined by a Holliday junction in the open-box configuration. (Courtesy of Ross Inman.)

8.1 Homologous and Nonhomologous Exchange

A requisite in most exchange events between two DNA base sequences is that some feature of the sequence be recognized. In prophage integration and excision (Chapter 7) and in the formation of immunoglobulins (Chapter 13) *unique* sequences of base pairs are recognized, and recombination is said to be **site-specific.** A protein that recognizes a particular base sequence must also be present—for example, the *E. coli* phage λ integrase. In meiotic crossing over, transformation, transduction, and conjugation, exchanges can probably occur anywhere within a DNA molecule, and recombination is then said to be **general.** In general recombination the two interacting base sequences must be the same or nearly the same—that is, homologous. General recombination may therefore also be called **homologous recombination.** An essential feature of homologous recombination (at least, in prokaryotes) is the necessity for a protein, which does not recognize specific base sequences, to catalyze an early step in the pairing of DNA molecules. Such a protein was first isolated in *E. coli* and called the **RecA protein.** Its activities are archetypical of similar proteins (see also Section 8.4). In transposition, the transposable element and the target sequence do not share a common sequence; hence, transposition is **nonhomologous recombination.** Furthermore, transposition does not require a RecA-type protein. The distinction between homologous and nonhomologous exchange is illustrated in Figure 8-1.

8.2 Exchange of Material during Processes of Recombination

An important feature of genetic exchange is that material is either exchanged between participating chromosomes or transferred nonreciprocally from one chromosome to another. Such exchange was first noticed in the 1930s in entire chromosomes; Stern, using *Drosophila*, and Creighton and McClintock, using maize (Chapter 3), studied recombination between two pairs of alleles located in chromosomes that were physically marked in such a way that the ends of the chromosomes were physically distinguishable. They observed that recombination between the genes was associated with physical exchange between the chromosomes. Breakage and reunion between sister chromatids was observed too, during mitotic recombination in other systems (Figure 8-2).

Exchange of material between individual DNA molecules was first demonstrated in 1962 in experiments with *E. coli* phage λ. These experiments utilized the density-labeling technique (Chapter 4) to determine the contribution of parental phage DNA molecules to recombinant progeny and showed that in forming recombinants parental molecules break and rejoin (this was one of several different theories). The original experiments were quite complex, so a variant that is more easily understood will be described. In the latter experiment, two types of λ phage were prepared: A^+R^-, whose DNA contained a heavy isotope in both strands, and A^-R^+, whose DNA contained a light isotope in both strands. In addition, the DNA of both phage

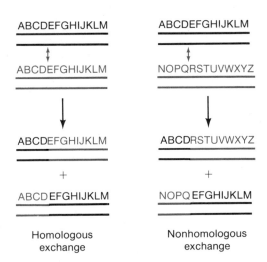

Figure 8-1 The distinction between homologous and nonhomologous recombination. Red arrows indicate points of exchange of DNA.

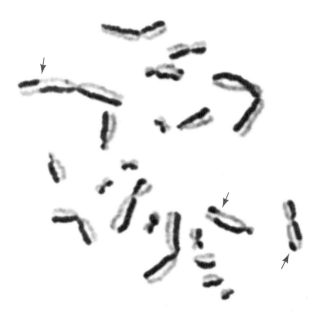

Figure 8-2 Sister-chromatid exchange in somatic cells. Chinese hamster ovary cells were allowed to grow for several generations in a medium containing 5-bromodeoxyuridine (BUDR), which can substitute for thymine (Chapter 10). Because of semiconservative DNA replication, some cells contain sister chromatids with different amounts of BUDR—namely, some chromatids have BUDR in both polynucleotide strands of the DNA, whereas others are hybrid, with thymine in one strand and BUDR in the other. Cells labeled in this way were treated with reagents that stain BUDR-containing DNA poorly. Thus, the hybrid chromatids are darker than those containing BUDR in both polynucleotide strands. Physical exchange (reciprocal breakage and rejoining) of material occasionally occurs; it can be detected in metaphase cells as pairs of chromatids whose staining intensity switches reciprocally (arrow). (Courtesy of Sheldon Wolff and Judy Bodycote.)

carried mutations to eliminate all recombination except that determined by the bacterial recombination genes; this enables one to examine the mechanism of the bacterial recombination system specifically. Mutations in the phage and in the bacteria also prevented the initiation of normal DNA synthesis (neither of these mutations were present in the original experiments) in order that density changes caused by physical exchange of parental DNA were not obscured by shifts caused by replication. Bacteria were infected with both of the phage, and the infection was allowed to proceed until lysis occurred. Progeny phage, which contained only parental material, were centrifuged to equilibrium in a CsCl solution and the density distribution of the genotypes of all phage particles was determined. Figure 8-3(a) shows that recombinant phage (A^+R^+ and A^-R^-) were found to range in density from fully heavy to fully light; that is, each recombinant particle contained material from both parents, indicating that physical exchange of material had occurred in forming the recombinants. When a central marker (in the c gene) was included (panel (b)), the density distribution showed that the c^+ marker from the A^+ parent appeared in A^+R^+ recombinants only if breakage and reunion occurred to the right of the marker. Other experiments were later carried out using phage in which only the λ *int* gene was active, again in the absence of DNA replication. Recall from Chapter 7 that Int-promoted recombination occurs only between λ attachment sites, which are not located exactly in the center of the phage DNA molecule. Phage recombinants were observed to separate into only two distinct densities, which result from breakage of each DNA molecule at its attachment site followed by rejoining the fragments (Figure 8-3(c)). Experi-

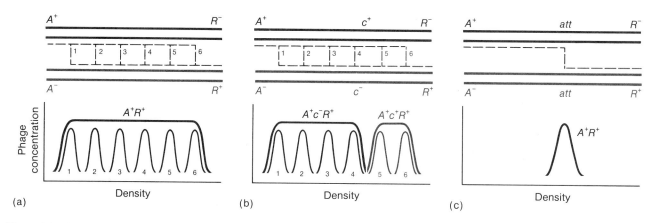

Figure 8-3 An experiment demonstrating breakage of DNA molecules and reunion to form recombinants. In the upper part of each panel, heavy lines represent high-density (black) and low-density (red) *E. coli* phage λ carrying the indicated alleles. After crossing the phage, using conditions that prevent DNA replication, as described in the text, progeny phage are centrifuged to equilibrium in CsCl. The thin lines in the distribution curve in the lower part of each panel show the expected density distribution for each of the numbered exchanges shown in the upper part of the panel. The heavy curves are the expected and observed distributions for all recombinant progeny taken together. (a) The uniform distribution obtained with two markers. (b) When three markers are used, the density distribution shows that A^+R^+ phage are also c^+ only if breakage and reunion occurs to the right of the c marker, as expected. (c) A cross in which only the λ *int* gene is active. A single narrow band is produced, because breakage and reunion occur only at *att*.

ments of a slightly different type also demonstrated breakage and reunion in bacterial transformation.

These and other experiments demonstrated that, in the course of recombination between genes, DNA molecules are both broken and rejoined. The observations have led to a proposal of a variety of mechanisms for genetic exchange, which will be discussed shortly. In each model (except for transposition, which is considered at the end of the chapter) it will be seen that an essential feature is the presence of a **heteroduplex region** in which a section of a single polynucleotide strand of one DNA molecule is base-paired to a complementary section of a polynucleotide strand of the other molecule (Figure 8-4).

8.3 Mechanism for Site-Specific Exchange

The best-understood mechanism of genetic exchange is the insertion of *E. coli* phage λ into the attachment site of the bacterium, a site-specific exchange in which long nonhomologous regions pair and exchange occurs between short homologous sequences (Chapter 7). Two proteins, the λ integrase and an accessory bacterial protein, are required for the process, and exchange occurs in the *O* region, a 15-base-pair sequence common to the phage (*POP'*) and bacterial (*BOB'*) attachment sites (Figure 7-19). The recognition step utilizes the regions *B*, *B'*, *P*, and *P'*, which are nonhomologous base sequences. Integrase and the host protein, which are sequence-specific DNA-binding proteins, bind to the *B* and *P* regions, bringing the two *O* regions together (Figure 8-5(a)). Integrase also possesses several enzymatic activities illustrated in panels (b–d). In panel (b) each *O* region has been opened by integrase and complementary strands from the two regions pair to form a four-stranded segment. Integrase acts as a site-specific nuclease, makes a pair of single-strand breaks *between particular adjacent bases* in homologous regions, and then acts as a topoisomerase (a twisting or untwisting enzyme, Section 4.6) to rotate the double-stranded region at the right and seal the breaks (panel (c)). At this stage homologous strands from the two *O* sequences have been joined. The process of breaking, rotating, and joining is repeated on the left side of the base-paired region (panel (d)), and the two newly formed *O* sequences with overlapping joints separate (panel (e)).

In prophage insertion a circular phage DNA molecule is inserted into a circular bacterial chromosome. Thus, in panel (a) the black ends of the upper molecule should be joined to form a circle, as should the ends of the lower molecule; for clarity, this detail is not included in the figure. Likewise, the black termini of panel (e) should remain joined to one another, as should the red termini; the two DNA segments in panel (e) are different parts of a single circular molecule.

Note that prophage integration is an example of reciprocal exchange (Chapter 3); that is, both molecules participate totally and equally, and neither material nor genetic information is lost or altered.

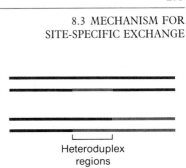

Figure 8-4 Two types of heteroduplex regions. Black and red regions refer to parental DNA molecules from which the molecules shown were derived.

Heteroduplex
regions

MECHANISMS OF
GENETIC EXCHANGE

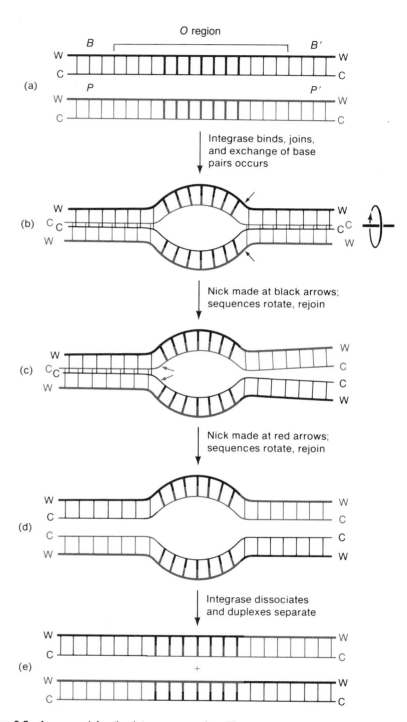

Figure 8-5 A proposal for the integrase reaction. The segments shown represent the 15-base-pair *O* region and small portions of the flanking *B*, *B'*, *P*, and *P'* sequences. The heavy bars represent the base pairs that are broken. The letters W and C are only for distinguishing the complementary strands.

8.4 The Pairing of DNA Molecules

All genetic exchange processes require pairing of DNA molecules. In prophage integration pairing occurs between specific base sequences and these sequences are recognized by site-specific proteins. In homologous exchange the base sequences that pair are complementary, and one or a small number of proteins that lack sequence specificity mediate pairing. Evidence for the general features of homologous pairing has come from studies of recombination in bacteria.

In 1965 a mutant *E. coli F*⁻ strain was isolated that was unable to serve as a recipient in an Hfr cross. The gene defined by the mutation was termed *recA*, and in 1976 the RecA protein was isolated. Proteins with similar properties have been detected in other organisms, but more is known about the *E. coli* enzyme than about the others, so our discussion will be restricted to that protein.

The RecA protein has two principal biochemical activities: (1) it binds to single-stranded DNA, and (2) it is an enzyme that cleaves proteins. Its DNA-binding activity is the feature that is relevant to recombination (the other property is regulatory.) When acting as a DNA-binding protein, the RecA protein mediates nonspecific pairing of DNA molecules and homology-dependent **strand invasion.** Figure 8-6 shows several RecA-mediated DNA-DNA interactions that have been carried out with purified RecA protein and DNA molecules at room temperature. The structures shown are very stable and are held together by AT and GC base pairs between complementary base sequences. Study of the three interactions shown in the figure (and others that

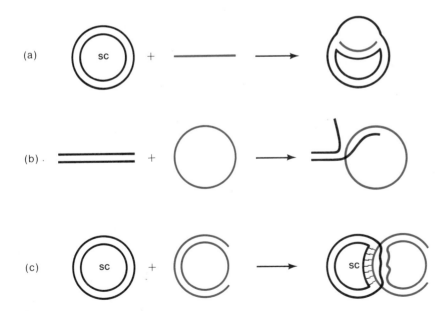

Figure 8-6   Three interactions mediated by the RecA protein; sc indicates that the circle is supercoiled.

are more complex) and of pairs of DNA molecules that will not interact has shown that stable pairing is dependent on two things:

1. One molecule must be single stranded or have a single-stranded region.

2. Either one of the molecules must have a free end.

The first requirement can be (and often probably is) provided by supercoiling of the participant lacking a free end. The second requirement eliminates the possibility that a single-stranded circle could invade a double-stranded circle.

The RecA-mediated interactions shown in Figure 8-6 are the end result of a series of steps. Each structure contains a region of complementary base pairing, but these experiments indicate nothing about the role of either the RecA protein or complementary sequences in the *initial* interaction between the two DNA molecules. The early steps have been investigated in other experiments, which suggest that the following sequence of events occurs (Figure 8-7):

1. Pairing begins by the formation of a weakly bound aggregate containing single-stranded DNA, double-stranded DNA, and RecA protein. Complementary sequences are *not* required for this interaction and, on the average, the single-stranded segment will not even be near a homologous region of the double-stranded molecule.

2. The RecA protein, by its DNA-binding activity, causes some unwinding of the double helix, and the two DNA segments can then drift back and forth with respect to one another. The molecules may separate before a stable interaction occurs, but if complementary sequences come into register, the interaction becomes more stable and a short paired region, possibly fewer than ten base pairs long, forms.

To understand how the short base-paired segment in Figure 8-7 can become an extensive base-paired segment, as in Figure 8-6, the phenomenon of **branch migration** must be examined. At elevated temperatures base pairs break, owing to disruption of hydrogen bonds (Chapter 5). At room temperature small regions of DNA are also occasionally disrupted by collisions with solvent molecules, though the base pairs quickly re-form. This phenomenon is called **breathing,** and branch migration is one consequence of this breakage and re-formation of base pairs. Branch migration can be understood by examining the DNA molecule shown in Figure 8-8. The lower strand of this molecule is continuous and two complementary single strands, I and II, having a combined length that exceeds the length of the lower strand, are both base-paired to that strand. The heavy black region and the red segment of the upper strands have the same base sequence and are both complementary to the dashed region of the lower strand. When breathing temporarily opens the right terminus of segment I, the gap can be closed either by segment I or by formation of additional base pairs between segment II and its complementary sequences, as shown in panel (b) of the figure. Breathing continues, resulting in restoration of the state of the molecule in panel (a) or in further change, as in panel (c). Breathing never stops, so the state of the molecule shifts con-

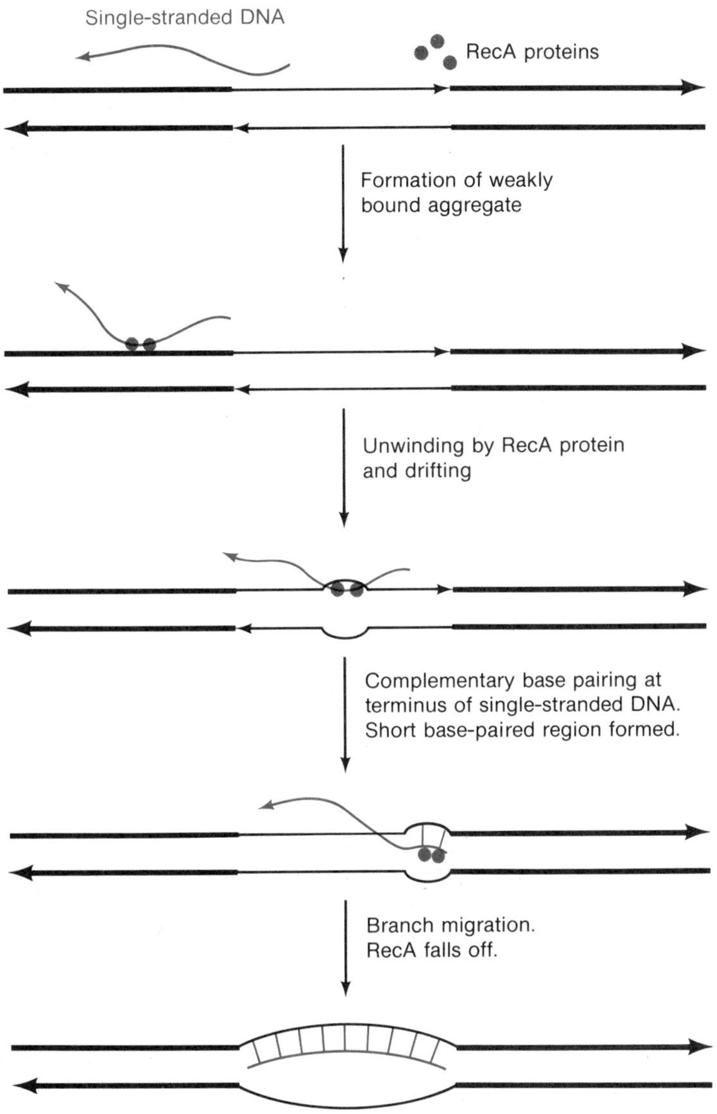

Single-stranded DNA

RecA proteins

Formation of weakly
bound aggregate

Unwinding by RecA protein
and drifting

Complementary base pairing at
terminus of single-stranded DNA.
Short base-paired region formed.

Branch migration.
RecA falls off.

Figure 8-7 A hypothetical but likely mechanism for single-strand invasion catalyzed by the RecA protein. The red strand is homologous with the regions represented by the thin lines.

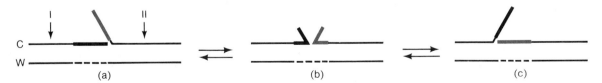

C

W

(a)   (b)   (c)

Figure 8-8 Branch migration. The heavy black and red lines are both complementary to the dashed segment of the lower strand. In panel (a) the heavy black region is hydrogen-bonded to the dashed segment. Breathing causes the free end of the black region to become occasionally unbonded, allowing the red region to pair with the dashed segment (panel (b)). By repeated breathing more of the red region can become base-paired (panel (c)) or the molecule can revert to its state in panel (a). Many intermediate states of the type shown in panel (b) are possible.

tinually, with the position of the unpaired single-stranded appendages moving both right and left at random.

The small base-paired region resulting from RecA-mediated drifting in Figure 8-7 can undergo a process similar to branch migration (called **strand exchange**) along the double-stranded molecule. The initial movement differs from branch migration in that when driven by RecA, energy is required, it can pass over regions of nonhomology, and movement is unidirectional. Once a free end is encountered, unwinding can occur from the free end and a longer double-helical region having many base pairs can be formed between one strand of each participating molecule. This phenomenon is called **assimilation.**

An essential feature of the pairing and assimilation process is the formation of a heteroduplex region through base pairing. The heteroduplex region will not always contain strictly complementary sequences. For example, if a region containing a mutation is paired with a wildtype sequence, one base from each strand will be unable to base-pair—for example, an adenine might be opposite a guanine instead of a thymine. Such a pair of bases is said to be a **mismatch.** It might be thought that a mismatched base pair will reduce the stability of a paired region, but if the heteroduplex region contains more than about ten base pairs, the region will be stable. Typically, paired regions in DNA molecules undergoing exchange contain hundreds to thousands of base pairs. The fate of the mismatched base pair will be examined in the next section.

8.5 Mismatched Base Pairs and Their Resolution

A mismatched base pair in a heteroduplex does not persist; mismatches are eliminated either by replication or **mismatch repair.** Replication removes a mismatch in a straightforward way, for in one replication cycle the mismatch is gone; for example, replication of an AG mismatch will yield an AT pair in one daughter duplex and a CG pair in the other. However, if replication does not resolve the mismatch, the mismatch repair system comes into play.

Mismatch repair was first demonstrated in an experiment summarized in Figure 8-9. DNA was isolated from two λ phage mutants. The molecules were denatured, and a technique was used that enables the complementary strands to be separately purified. Strands containing one mutant allele were mixed with complementary strands containing another mutant allele, and double-stranded DNA was then prepared by standard renaturation techniques (Chapter 5). These renatured molecules contained a mismatched pair of bases at each mutant site. The renatured DNA was mixed with bacteria in which the recombination system had been eliminated by mutation (they were recA⁻), using conditions like those in transformation that enable the DNA to enter a cell. On incubation of the infected cells in growth medium, progeny phage developed. Infected cells were examined for the production of wildtype phage. Replication alone would yield both singly mutant phages, in roughly equal numbers; however, if mismatch repair had occurred before replication, some wildtype phage would be produced. Three types of cells yielding wild-type phage were observed: those containing wildtype and the a mutant; those with wildtype and the b mutant; and those containing only wildtype phage. How these presumably arose is shown in the figure.

Mismatch repair is accomplished by excision of a segment of one strand of a heteroduplex region. A variation of the experiment of Figure 8-9 was performed to estimate the number of base pairs removed in the process. A third marker, between the pairs of markers already used in the experiment and closely linked to one of them, was included in the experiment. Phage progeny from individual infected cells were examined to see whether repair at the third marker often accompanied repair at its neighbor—that is, whether the a^+b^+ progeny of a heteroduplex with the genotype $a^+b^-c^-/a^-b^+c^+$, in which b and c are closely linked, are also often $a^+b^+c^+$ (correction of one heteroduplex strand from b^- to b^+). It was found that the frequency of co-correction depended on the separation of the two markers and their relative positions. Examination of a series of markers indicated that when mismatch repair is initiated at a particular mismatched site, repair tends to proceed along the strand in the 5′-to-3′ direction for about 3000 base pairs. *In vitro* experiments with partially purified enzymes have indicated that a single-strand break is made on the 5′ side of a mismatch and a large tract of bases is removed from the strand being repaired. The resulting gap is then filled in

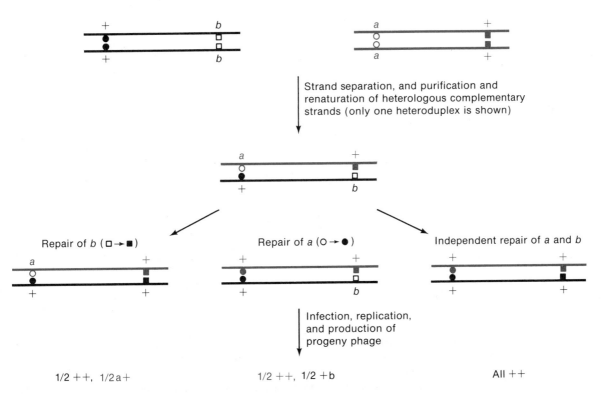

Figure 8-9 The mismatch repair experiment.

by a DNA polymerase; the intact strand is used as a template. An interesting feature of these experiments was the fact that a preferred direction of correction was observed for several of the markers; that is, for some markers, wildtype was corrected to mutant, whereas for others the correction was from mutant to wildtype. No satisfactory explanation was ever provided for this phenomenon; presumably, base sequences surrounding a mismatch determine which strand is cut.

Strand invasion, assimilation, and mismatch repair constitute almost the entire mechanism of recombination in transformation of *Pneumococcus* (Chapter 4). The currently accepted model for transformation is illustrated in Figure 8-10. A donor DNA molecule is first adsorbed to the surface of a competent recipient cell. As the DNA enters the cell, one of the two polynucleotide strands is enzymatically digested by a nuclease (panel (a)); thus, single-stranded DNA is immediately available for pairing with the recipient

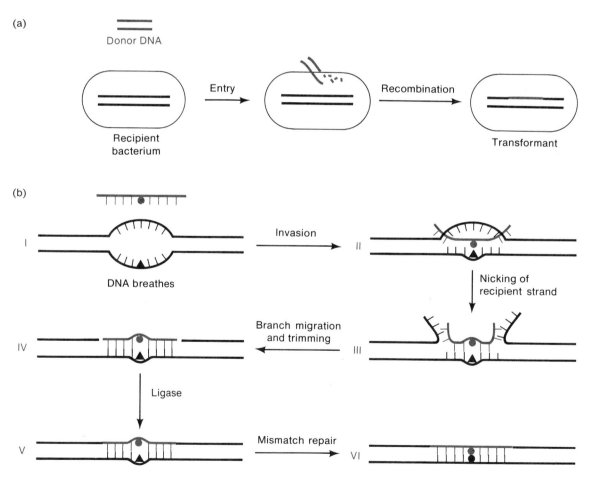

Figure 8-10 Transformation in *Pneumococcus*. (a) Conversion of double-stranded donor DNA to a single strand during penetration of the cells. (b) The accepted mechanism for integration of single-stranded donor DNA.

chromosome. The single-stranded DNA is thought to be coated with a protein that protects it from further nuclease attack. Presumably utilizing a RecA-type protein, the incoming single-stranded fragment invades the DNA of the recipient cell and is assimilated, forming a heteroduplex region (panel (b)). By an unknown mechanism the displaced single strand is cut. Branch migration then occurs, which gradually increases the fraction of the invading strand that is base-paired to the recipient strand. Trimming enzymes (exonucleases) remove the free ends, which may be on either the donor or the recipient DNA, and ultimately a DNA ligase seals the single-strand breaks. The result is an extended heteroduplex region containing a mismatched base pair. The outcome of the process—that is, whether the donor marker is or is not recovered in the progeny cells—depends on whether mismatch repair occurs and, if so, whether the donor or recipient base is removed. If mismatch repair does not occur, the cell will divide, yielding one daughter cell with the recipient genotype and one with the donor genotype. If it does occur, *both* single strands in the heteroduplex region will either carry the donor genotype or the recipient genotype. As in the experiment depicted in Figure 8-9, the probability and direction of mismatch repair varies with the mutation. In fact, certain markers are corrected from wildtype to mutant so frequently that if the donor DNA carries the wildtype allele (the usual arrangement), transformation of the wildtype allele is observed only infrequently. These markers are called low-frequency markers.

8.6 Exchange Between Homologous Double-Stranded Molecules

In the preceding sections, exchange between single- and double-stranded DNA (transformation) and between specific, mostly nonhomologous, sites in double-stranded molecules (prophage integration) has been discussed. The most prevalent type of exchange is between homologous double-stranded molecules—for example, in bacterial conjugation, transduction, phage crosses, and mitotic and meiotic crossing over—but the mechanisms are incompletely understood in these systems. In this section several genetic phenomena are described that have been important in developing theories of homologous exchange. In Section 8.7 this information will be used to develop two models of exchange.

When examining crossing over between distant markers (for example, in different genes), the mechanism of exchange seems straightforward and models such as those shown in Figure 8-11 are sufficient to explain the data. Two DNA molecules pair, and one break is made in each of the four individual polynucleotide strands. Fragments separate and rejoin, gaps are filled in by DNA polymerases, excess DNA is trimmed away by an exonuclease, and joints are sealed with DNA ligase. A major feature of intergenic recombination—namely, the additivity of map distances—is simply explained by assuming a fixed probability of strand breakage between any particular pair of adjacent bases and a probability approaching 1 for gap filling, trimming, and

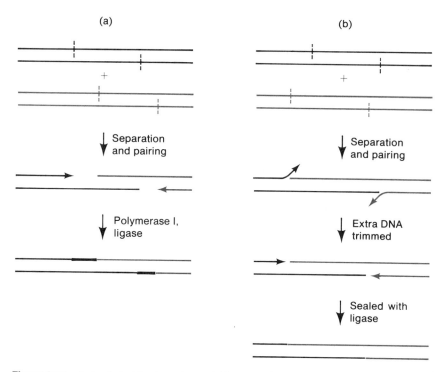

Figure 8-11 A simple but inadequate model for recombination. Trimming, polymerization, and ligation will be seen to be significant in other models.

sealing. However, when exchange between close genetic markers is studied, in *intragenic* recombination, for example, one sees that distances are no longer additive (Section 3.2), and unexpected events are observed. In looking at close genetic markers, one focuses attention on the region in which the actual genetic exchange occurs, and such a close look makes it clear that the mechanism of crossing over is considerably more complex than that shown in Figure 8-11. In the following unit, some of these observations are examined.

Evidence for a Heteroduplex Region in Meiotic Recombination

One profitable approach to understanding the molecular mechanism of meiotic exchange has been to study the genetics of the *Ascomycetes*, fungi in which the products of each meiotic event are easily available to the experimenter because they are packaged in an ascus (Section 3.2). As described in Chapter 2, upon completion of premeiotic DNA replication a diploid cell contains four double-stranded DNA molecules. In fungi having asci with four spores—for example, the yeast *Saccharomyces cerevisiae*—each of the meiotic products is present in a single spore and all four products are contained in one ascus. In *Neurospora crassa* and *Sordaria brevicollis*—two popular organisms for studying meiotic recombination—the second meiotic division is followed by a mitotic division, so eight spores are contained in a single ascus. The spores are arranged in an order determined by the successive meiotic and mitotic divisions, and the genotype of each spore is the genotype of one of the eight individual polynucleotide strands present at the end of meiosis (Section 3.2). Study of *Ascobolus immersus*, a fungus with an eight-spore unordered ascus, has also contributed significantly to our understanding. For our purposes, analysis of ordered asci will be easier to understand, and in this section theoretical examples will be confined to such systems.

The usual experimental approach is spore analysis: spores are individually dissected from an ascus and then transferred to a nonselective growth medium on which each forms a colony, and the genotype of each colony is determined by further plating tests. Occasionally, markers are used that confer visible phenotypes (color) on the colonies, eliminating the need for further testing.

Figure 8-12 reviews the composition of an ordered eight-spore ascus derived from a diploid heterozygous for a gene A. If crossing over between the

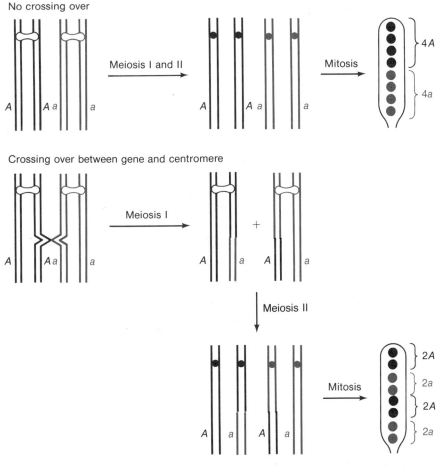

Figure 8-12 Composition of an eight-spore ordered ascus with one marker segregating, with and without crossing over between the centromere and the marked locus.

centromere and the marker does not occur, meiosis results in the formation of a tetrad of four DNA molecules—two *A* and two *a*—which after the mitotic step yields an ascus containing four *A* spores and four *a* spores. This is the classic Mendelian **4:4 segregation.** If crossing over occurs between the gene and the centromere, as shown in the lower panel of the figure, segregation is still 4:4, though the order of the spores in the ascus differs (Chapter 3).

Figure 8-13 shows the composition of an ascus when two genes *A* and *B* are considered and the zygote is heterozygous with chromatid composition *AB* and *ab*. If an ascus forms without crossing over, four *AB* spores and four *ab* spores result—again, 4:4 segregation. The lower part of the figure shows the result of crossing over in the regions between the genes. Three features of the genotypes of the spores should be noted. (1) The reciprocal recombinant genotypes—*Ab* and *aB*—are both present in the same ascus. (2) Both parents are present in equal numbers, indicating that crossing over occurs after DNA replication, as stated in Chapter 3. (3) Four copies of each of the alleles, *A*,

Figure 8-13 Composition of an eight-spore ordered ascus with two markers segregating, with and without crossing over between the markers.

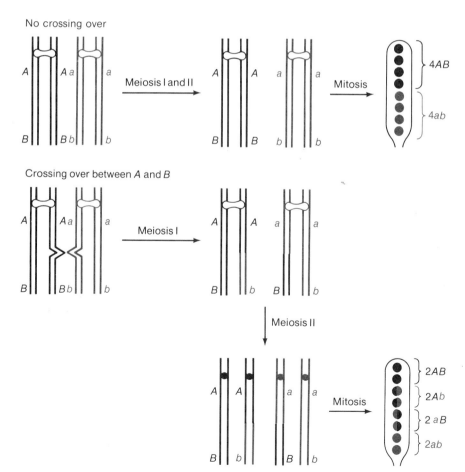

a, B, and *b,* are present, though the genotypes of the spores have been recombined; that is, each gene has segregated in the Mendelian 4:4 ratio.

"Aberrant" asci are also observed, and those formed when three closely linked markers are segregating provide insight into the exchange process. Figure 8-14 illustrates the cross *Abd* × *aBD,* in which recombination is observed to have occurred in the regions between *a* and *d.* We consider the specific case of a crossover between *b* and *d* (though the same considerations apply to a crossover between *a* and *b*). Most of the asci in which such recombination is evident have the composition shown at the left-hand side of the figure. The ratio of dominant and recessive alleles in the asci is 4:4, as expected, and each pair of spores is identical. However, some asci are also found that have the composition shown at the lower right. Again, the allelic ratios are 4:4, but two pairs of spores do *not* have identical members— namely, *AbD–ABD* and *aBd–abd.* These asci are called **aberrant 4:4** asci.

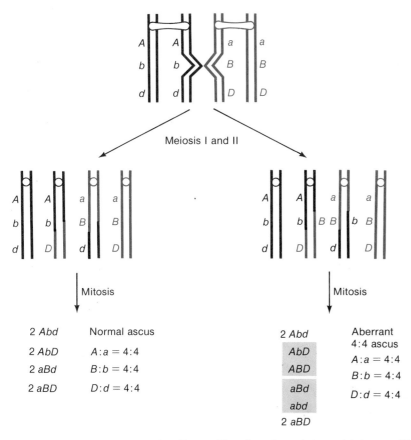

Figure 8-14 Postmeiotic segregation. Composition of one type of aberrant 4:4 ascus. The two pairs of spores shaded in pink have nonidentical members. The heteroduplex regions shown in the postmeiotic chromosomes account for the aberrant pairs; however, no mechanism for forming these regions is implied.

Recall that in an eight-spore ordered ascus adjacent spores contain the daughter DNA molecules formed by replication of each meiotic product during the final postmeiotic mitosis. *Thus, if two markers of an adjacent spore pair are different, the individual polynucleotide strands of the molecule formed after meiosis II must also have been different.* Since the *a* and *d* alleles are the same in adjacent spore pairs and only the *b* alleles are different, a simple explanation for the difference would be the following: the molecule formed after meiosis II contained a heteroduplex region, as shown in the middle part of Figure 8-14, and during postmeiotic replication the heteroduplex regions separated, giving rise to daughter molecules that differ in the alleles that were present in each heteroduplex region. Indeed, the existence of aberrant 4:4 asci is accepted as definitive evidence for the existence of heteroduplex regions, and the aberrant 4:4 asci are said to result from **postmeiotic segregation.** Furthermore, because the flanking markers are in a recombinant array, it is assumed that a heteroduplex region is formed by crossing over. Note that a heteroduplex region could also be present in the molecules that gave rise to the normal 4:4 ascus seen in the left part of the figure, but the region would not have included the *B* and *b* alleles, and hence would not be evident.

Conversion Asci and the Role of Mismatch Repair

Asci with other than normal and aberrant 4:4 segregation are observed. Figure 8-15 illustrates the phenomenon. Again, recombination has occurred

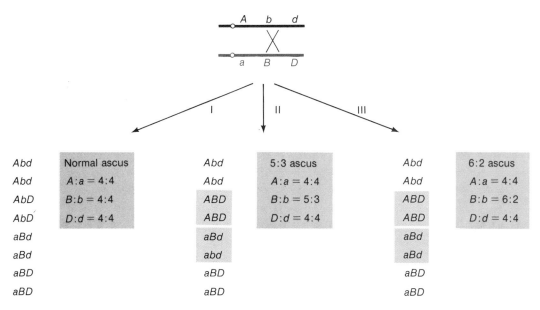

Figure 8-15 Two types of aberrant asci (II and III) in which gene conversion (*b* to *B*) has occurred. The significant spore pairs are shaded in pink. The cross and the composition of the normal asci are the same as that in Figure 8-14.

in the regions between *A* and *d*. Most of the asci are normal 4:4 asci, but some, in which a crossover has occurred between the *b* and *d* markers, are of a different type:

1. Ascus II has two important features: (i) one pair of *b* markers is missing (the pair *ABD* replaces *AbD*), and (ii) like the aberrant 4:4 asci, one spore pair has dissimilar members (*aBd* and *abd*). In this type of ascus, the *B:b* ratio is 5:3.

2. In ascus III, one of the spore pairs (the second from the top) lacks the expected *b* allele. The ratios of the outside markers are *A:a* = *D:d* = 4:4, but the ratio of *B:b* is 6:2.

In early descriptions of these unusual asci, it was said that the *b* allele had been **converted** to *B;* thus, the phenomenon of 6:2 and 5:3 segregation of an allele is termed **gene conversion.** Asci are also found in which a *B*-to-*b* conversion has occurred (note that *B* to *b* and *b* to *B* are both called conversion).

The preceding discussion indicated that conversion asci appear among asci selected for recombination of flanking markers. However, if *all* asci from a cross are examined nonselectively, without regard to the composition of the flanking markers, conversion asci are also found. A key observation in developing a theory of crossing over is that *in roughly 50 percent of the conversion asci, the flanking markers are in a recombinant array.* This observation, which has been made in several fungal species, in *Drosophila,* and for many different markers, suggests that the processes of gene conversion, as well as postmeiotic segregation, and crossing over are associated. We will see shortly that the association is a result of sharing common steps—namely, pairing, exchange of single polynucleotide strands, and formation of a heteroduplex region.

To examine some of these common steps, alternative events that follow pairing of two chromatids and exchange of DNA in the four-chromatid stage of pachytene will be considered, again with reference to the cross illustrated in Figure 8-15. Having invoked the formation of heteroduplex regions to explain postmeiotic segregation, we can use them also to explain conversion. The difference between the two phenomena is in the steps following formation of a heteroduplex region.

The heteroduplex region resulting from a crossover event is assumed to contain the markers *b* and *B*, which differ in one base pair, so a mismatched base pair is formed. Figure 8-16 illustrates three possible responses to such a mismatch:

1. In event I there is no mismatch repair. Thus, replication of each pair of mismatched strands yields one DNA molecule with the wildtype allele and one with the mutant allele. Since two molecules carry mismatches, mitosis produces two pairs of spores with nonidentical members. The ratio *B:b* is 4:4 and the ascus is of the aberrant 4:4 type.

2. In event II the mismatched base pair in the heteroduplex region is repaired from *b* to *B* in molecule 2 (that is, *b* is excised and the strand with *B* is used as a template), with the result that one of the pair of spores with different members is not present. The ratio *B:b* is 5:3.

3. In event III both mismatches are repaired in the same direction, *b* to *B*. All spore pairs have identical members and a 6:2 ascus results.

The scheme shown in Figure 8-16 has not yet been demonstrated by *in vitro* experiments but is supported by a wealth of genetic data.

Although the events in Figure 8-16 show how the association of aberrant 4:4 and conversion asci with recombination of flanking markers may be explained, recall that in half of the aberrant asci, the flanking markers are *not* in a recombinant array. How can one account for these nonrecombinant asci? The generally accepted explanation, which is presented in the following section, is that heteroduplex regions are formed during one stage of DNA pairing and that the two cuts in the individual polynucleotide strands needed to separate the paired molecules can occur in two arrangements—one yielding recombinant asci and the other yielding nonrecombinant asci.

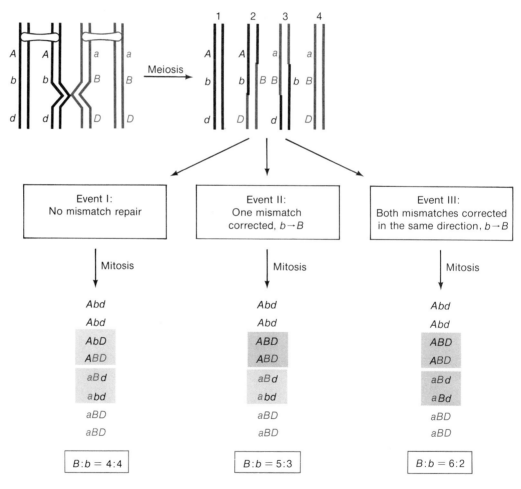

Figure 8-16 The proposed role of mismatch repair in the formation of three types of aberrant asci. Spore pairs with nonidentical members are shaded in pink; pairs that result from mismatch repair are shaded in gray.

In this section two hypotheses for the mechanism of crossing over in fungi are presented. The first is the seminal model (1964) of Robin Holliday, in which formation of a heteroduplex region followed by mismatch repair was first suggested. An important consequence of the Holliday model was that it predicted a structure, now shown to exist, that is a feature of all current thinking. As additional information was obtained, Holliday's model was modified several times by other geneticists. The second model presented in this section is one of the most important modifications (1975). Although details of this model will probably have to be adjusted as more data are obtained, the general features are thought to be a good description of the events that occur in crossing over in fungi.

The Holliday Model

The Holliday model and its later modifications attempted to account both for the existence of normal asci, gene conversion, and postmeiotic segregation in the fungus *Ustilago* and for the frequent association of the abnormal asci with crossing over. Holliday's specific model in shown in Figure 8-17.

The stages of the model are the following:

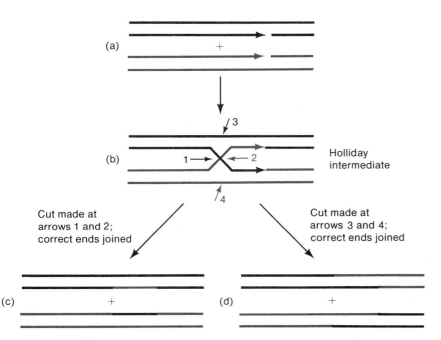

Figure 8-17 Formation of the Holliday junction (b) from two DNA molecules that received a single-strand break (a) and the structure of two sets of molecules in which exchange has occurred ((c) and (d)).

1. **Single-strand breakage.** Two DNA molecules come together and two single-strand breaks are made in homologous positions in each molecule by an unspecified mechanism.

2. **Unwinding.** At each break the double helix unwinds somewhat, releasing single polynucleotide strands (not shown in the figure).

3. **Invasion and assimilation.** The released strands are exchanged between the two DNA molecules by pairing with the unbroken complementary strand in the other molecule, forming a unit consisting of two DNA molecules connected by crossed single strands, as shown in the figure.

4. **Separation of the molecules.** A pair of cuts, symmetrically made, separates the two DNA molecules. The cuts can be made at either of two different locations. If they are made at corresponding positions in the crossed strands ("East-West" cuts), a pair of nonrecombinant molecules, shown in panel (c) of the figure, results. If, on the other hand, the cuts are made at homologous positions in the uninterrupted parent strands ("North-South" cuts), the recombinant molecules shown in part (d) will result. This cutting process, which separates the two participating DNA molecules, is called **resolution.**

The Holliday model provides a molecular basis for the association between aberrant segregation and crossing over. Gene conversion and postmeiotic segregation are assumed to result from repair of mismatches, or lack thereof, on one or both chromosomes in a symmetric structure containing heteroduplex DNA, as indicated in Figure 8-16. The model also explains how 5:3 and 6:2 segregation can occur without recombination of flanking markers (Figure 8-18). As we have just seen, a Holliday junction can be resolved by

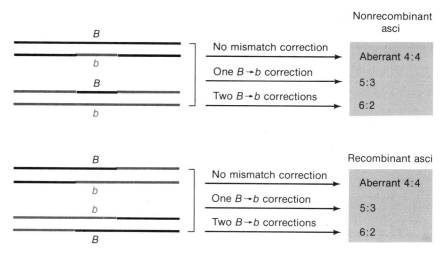

Figure 8-18 Formation of recombinant and nonrecombinant aberrant 4:4, 5:3, and 6:2 asci from two types of crossover products.

pairs of cuts that either do not produce recombination of flanking markers (East-West, cuts, type (c) in the figure) or do (North-South cuts, type (d)). If both types of cuts are equally likely, half of the 5:3 and 6:2 asci will be recombinant for flanking markers, as observed.

Existence of a crossed-strand structure such as is shown in Figure 8-17(b) is an important prediction of the Holliday model. Crossed-strand regions have been observed in DNA molecules isolated from several organisms and have come to be called **Holliday junctions.** They are most easily recognized when the junction is in an "open-box" configuration (Figure 8-19). Panel (a) shows how a 180° rotation of one component of a Holliday structure with respect to the other would yield molecules with an open-box array. A molecule of this type is shown in panel (b).

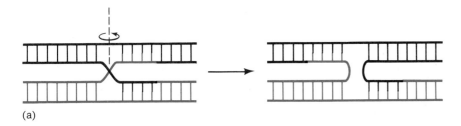

(a)

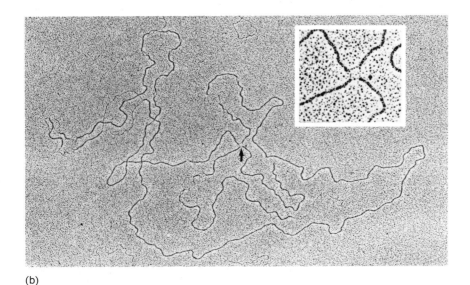

(b)

Figure 8-19 (a) Rotation of the upper half of a Holliday structure about the axis shown (consider the bottom half to be held in place) to generate an open-box structure. (b) Electron micrograph of a crossover intermediate between a phage λ DNA molecule and a linear DNA fragment. The Holliday junction is indicated by an arrow and enlarged in the inset. The molecules have been partially denatured in order to identify corresponding regions of the two molecules. (Courtesy of Manuel Valenzuela.)

After the Holliday model was first proposed, it was modified to include branch migration (not known to exist in 1964), which could allow the Holliday junction to move and the heteroduplex regions to be elongated. This modification was necessary to account for data from four-factor crosses in which two central markers are often both converted with high efficiency (**co-conversion**). In yeast, co-conversion occurs in regions that are up to 1000 base pairs long, indicating that heteroduplex regions must be at least that large.

A notable feature of the Holliday model is its symmetry. That is, *each* product of crossing over shown in Figure 8-17 possesses a heteroduplex region. This prediction has been tested in several fungal systems. The frequent occurrence of aberrant 4:4 asci in *Sordaria brevicolla* indicates that heteroduplex regions can indeed form at the same site on both chromatids. Similarly, studies at the *b2* locus in *Ascobolus immersus* showed that aberrant 4:4, 5:3, and 6:2 segregation can occur at a single site on different copies of the chromosome. However, all data have not been supportive, as can be shown for the *W17* locus of *Ascobolus* in Figure 8-20. In this cross many recombinant asci with 5:3 segregation of central markers were examined. The Holliday model predicts that to form such asci the marker must be contained in a heteroduplex region and that such a region will be present in both

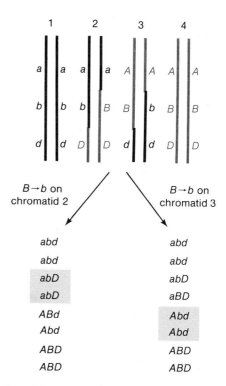

Figure 8-20 Production of two types of 5:3 asci from a pair of chromatids that have generated two heteroduplex regions. At the *W17* locus of *Ascobolus,* both types are not found.

chromatids 2 and 3. Asci with *B*-to-*b* corrections should be of two types—those in which the correction is on chromatid 2 and those in which it is on chromatid 3. However, it was observed that only one type of 5:3 ascus was found at this locus, from which it was concluded that at this locus heteroduplex regions are not present on both chromatids. A related phenomenon has been obtained with yeast—namely, for some sites 5:3 asci are frequent but aberrant 4:4 asci (the hallmark of heteroduplex regions) are very rare. A 5:3 ascus contains one nonidentical pair of spores, so at least one heteroduplex region must have been present. The Holliday model interprets the conversion pair as necessarily arising from repair of a mismatch in a heteroduplex region. However, the lack of aberrant 4:4 asci, which indicates that the formation of two heteroduplex regions is rare in yeast, suggests that the conversion pair has arisen without formation of a heteroduplex region. The conclusion from these observations is inescapable: if there is a single mechanism of crossing over, it must be capable of producing either a symmetric pair of heteroduplex regions or a single heteroduplex region. These observations were the impetus for developing the model described next, a model in which the initial interaction is asymmetric and subsequent steps determine whether the asymmetry will persist or a symmetric array of heteroduplex regions will be formed.

The Asymmetric Strand-Transfer Model

The **asymmetric strand-transfer model** of Matthew Meselson and Charles Radding retains several features of the Holliday model—namely, a joining of homologous single strands, existence of heteroduplex regions, the production of a Holliday junction, the resolution of the junction by East-West versus North-South breakage, and mismatch repair. The principal difference is that the initial pairing is asymmetric, in accord with the yeast and *Ascobolus* data. That is, in the initial interaction one molecule donates a single polynucleotide strand to a recipient molecule and a heteroduplex region is formed only in the invaded molecule. By replicative extension of the invading strand and by branch migration the heteroduplex DNA is enlarged and by a novel rearrangement (isomerization) a heteroduplex region is formed in both molecules, allowing for the reciprocal events common in other fungi. A second difference is that some conversion asci are formed without mismatch repair.

The asymmetric strand-transfer model is shown in Figure 8-21. The individual steps are the following:

1. **Nicking.** A single-strand break (nick) with a 3′-OH group and a 5′-P group is formed (in an unspecified way) on one polynucleotide strand of one of the participating double-stranded molecules. The breakage is assumed to occur in only one of the molecules, in contrast with the Holliday model, in which each molecule receives a break.

2. **Displacement**. A DNA polymerase, acting at a free 3′-OH group, adds nucleotides at the nick and displaces the original polynucleotide strand. (Displacement synthesis is a property of *E. coli* DNA polymerase I and many other polymerases and can be easily carried out in

the laboratory.) The displacement generates a single-stranded appendage that is used for complementary pairing in the next stage.

3. **Invasion.** The single-stranded appendage invades the double-stranded molecule, presumably aided by a RecA-type protein, as described in Figures 8-6 and 8-7.

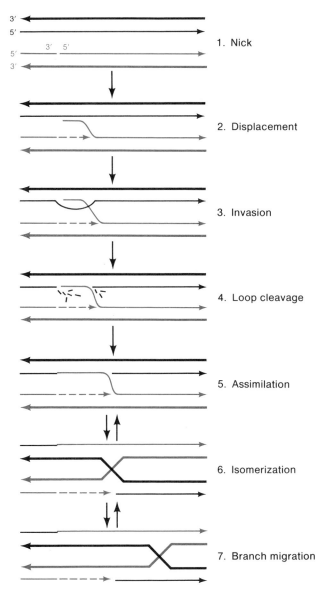

Figure 8-21 The asymmetric strand-transfer model. The dashed lines represent newly synthesized DNA. (Based on M. Meselson and C. Radding. *Proc. Nat. Acad. Sci., USA,* 72 (1975): 358).

4. **Loop cleavage.** In response to invasion the displaced strand is enzymatically removed, presumably by an initial break in the displaced polynucleotide strand of the invaded molecule; this is followed by either endonuclease or exonuclease action (or both).

5. **Assimilation.** The 3'-OH group of the gap formed in step 4 can be extended by a DNA polymerase (the reaction is the same as that in step 2) and joined to the invading strand by a DNA ligase. Furthermore, the gap can be enlarged to allow more of the invading strand to base-pair with the continuous strand of the invaded molecule. At this point the asymmetric intermediate becomes complete. Pairing is stabilized, but *only one heteroduplex region exists* (the one in the upper molecule, because replication has filled the gap that removing and transferring a strand in the lower molecule has caused. In the Holliday model the gap would be filled by a polynucleotide strand transferred downward from the upper molecule and two heteroduplex regions would be present.

6. **Isomerization.** A rearrangement of the strands of the participating DNA molecules occurs and a Holliday junction forms. (This rearrangement is difficult to visualize without three-dimensional molecular models; however, it is straightforward and is accomplished without steric problems or strain.) Note that the crossover strands in the Holliday structure become new backbone strands and former backbone strands become new crossed strands of the Holliday junction, putting the arms flanking the strand crossover in a recombinant configuration. At this point still only one heteroduplex region exists.

7. **Branch migration.** The Holliday junction migrates (for example, toward the right, as shown in the figure) and thereby generates a symmetric array with two heteroduplex regions. Note that because of the movement of the base-paired regions in the assimilation and branch-migration steps, when resolution occurs, the position of the cuts will typically not be at the site of either strand invasion or the initial break, in contrast with the Holliday model.

8. **Resolution.** Separation of the particular molecules occurs as in the Holliday model, again with a choice of North-South versus East-West pairs of cuts.

Examination of the products shows how the asymmetric strand-transfer model explains many of the genetically observable features of meiotic recombination in fungi that have been discussed. Figure 8-22(a) shows the alternative products of resolution of the asymmetric intermediate formed by isomerization without branch migration. With North-South cleavage, the flanking markers are nonrecombinant, and with East-West cleavage, they are recombinant. In both cases, the pair of molecules contains only one heteroduplex region. The recombinant array on the left is sufficient to explain the asymmetric 5:3 segregation in the *W17* locus of *Ascobolus* described in Figure

8-20: the single pair of nonidentical spores in Figure 8-20 comes from the upper molecule with the heteroduplex region (Figure 8-22(a)), and the converted pair of spores in Figure 8-20 would arise not as a result of repair of a mismatch, but rather of the DNA replication that occurs in steps 2–5 of Figure 8-21. Figure 8-22(b) shows the alternate pair of products of the symmetric intermediate formed by isomerization followed by branch migration. Two heteroduplex regions are present, and aberrant 4:4, 5:3, and 6:2 asci

(a) Asymmetric intermediate via isomerization without branch migration

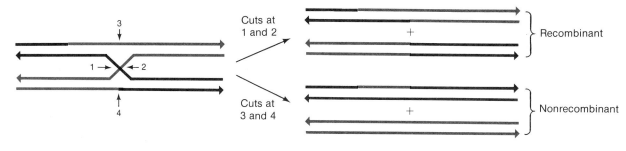

(b) Symmetric intermediate via isomerization and branch migration

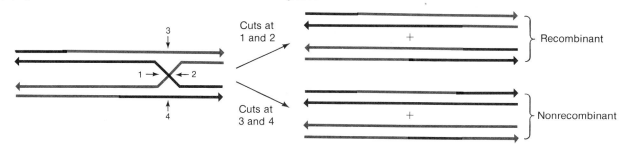

(c) Symmetric intermediate via branch migration without isomerization

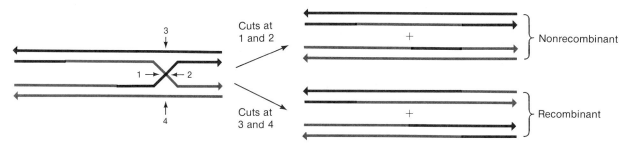

Figure 8-22 Alternative products of resolution of an (a) asymmetric intermediate formed by isomerization without branch migration, (b) a symmetric intermediate formed by branch migration of the intermediate in (a), and (c) a symmetric intermediate formed by branch migration without isomerization, according to the asymmetric strand-transfer model.

can arise by the mechanism of Figure 8-16. The asymmetric strand-transfer model does not exclude the possibility of forming the symmetric intermediate via branch migration without isomerization; in that case (panel (c)) recombinant and nonrecombinant products are formed instead by North-South and East-West cleavages, respectively.

How can one explain the difference between, on the one hand, the *Sordaria* asci in which a symmetric arrangement would have to be present and, on the other hand, the yeast data and the *Ascobolus W17* data, which requires an asymmetric intermediate? One possibility, suggested in the asymmetric strand-transfer model, is that the relative times of resolution and of branch migration determine the outcome. That is, one assumes that in yeast and at the *W17* locus, the enzymes that promote asymmetric exchange bind more tightly to their substrates and detach only late in the period of time allowed for exchange. Detachment of the enzymes is followed by rapid isomerization or branch migration, and resolution soon ensues. At other sites in *Ascobolus* and in *Sordaria* detachment is presumed to occur earlier, leaving time for the propagation of symmetric exchange before resolution occurs. Within a given species local base sequences, which could affect the activities of the enzymes and the efficiency of isomerization, branch migration, and resolution, could explain locus differences.

The asymmetric strand-transfer model describes many features of crossing over in fungi, though certain genetic observations require special modifications of the theory. As additional data accumulate in the future, modifications may need to be made to increase the understanding of meiotic crossing over. It is possible that the asymmetric strand-transfer model might prove to be incorrect and that a new explanation for crossing over would be required. In fact, recently a double-strand-break-and-repair model was proposed as an explanation of phenomena not answered simply by the asymmetric strand-transfer model (J. Szostak *et al.* 1983. *Cell*, 33: 25).

8.8 Transposition

Transposition is an exchange process in which sequence homology apparently plays no role in pairing. As mentioned in Chapters 5 and 7, two important features of transposition are that (1) the transposable element is usually replicated and (2) insertion at a new site is accompanied by a duplication of a short target sequence. For a particular transposable element, the target sequence is different for each insertion site, but the length of the duplicated sequence is almost always constant. Because of duplication of the target sequence, it is clear that some type of DNA replication occurs during transposition. The following basic mechanism, whose details are unknown and which is shown only in outline (Figure 8-23), is thought to be responsible for the duplication of the target sequence.

An enzyme, possibly encoded in the transposable element, is thought to select a target sequence (in an unknown way) in the recipient DNA and make two single-strand breaks, one in each polynucleotide strand, at opposite ends

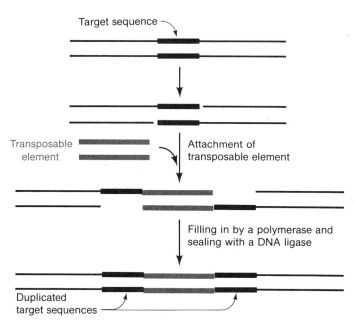

Figure 8-23 A schematic diagram indicating how a target sequence might be duplicated. The drawing is not made to scale; the transposable element would actually be much longer than the target sequence.

of the target sequence. The DNA of the transposable element is then attached to the target sequence at the free ends generated by these breaks: one strand of the transposable element is joined to one strand of the target sequence at one end of the sequence and the other strand of the transposable element is joined to the other strand at the opposite end of the target sequence. Such joining leaves two gaps, one across from each strand of the target sequence. The gap is then filled by a DNA polymerase, after which the breaks are sealed; thus, the two strands of the recipient DNA molecule are made continuous and two copies of the target sequence flank the transposable element. Note that in this model no homology is utilized in joining the terminal sequence of the transposable element to the recipient DNA.

An important phenomenon that has influenced the development of molecular models of transposition is that when transposition occurs between two circular molecules (two plasmids or a plasmid and a bacterial chromosome), a single circle consisting of both circles always arises. This process is called **replicon fusion,** and the general term for the joined molecules is a **cointegrate.** This process is shown in Figure 8-24(a). Note that *two copies of the transposable element exist in a cointegrate,* whereas only one was present before transposition occurred. Therefore, formation of a cointegrate must include a step in which the transposable element is replicated. With some transposable elements the cointegrate is a stable structure. However, with other elements, two individual circles form, so the cointegrate is considered to be an intermediate in transposition. This process of forming two circles from a cointegrate, also called resolution, is shown in panel (a) for the bacterial transposable element Tn3. This element encodes several enzymes and has a site called the **internal resolution site** (panel (b)). One enzyme, transposase, is responsible for the breakage-and-joining event and forms the cointegrate from the participants. Another enzyme produces a site-specific exchange in the resolution site and re-forms the two units, both of which contain the transposable element.

Many detailed models for transposition have been proposed—they differ primarily in how replication is utilized to create duplication. Three features common to most models are the following:

1. Transposition is initiated by two single-strand breaks at either end of a target sequence and two single-strand breaks at the termini of the transposable element.

2. Single strands are joined together without complementary base pairing.

3. A replication fork is created when single strands join.

One currently popular model for transposition is shown in Figure 8-25. Its important feature is the generation of two replication forks that advance and provide a replica of the target sequence and the transposable element. A

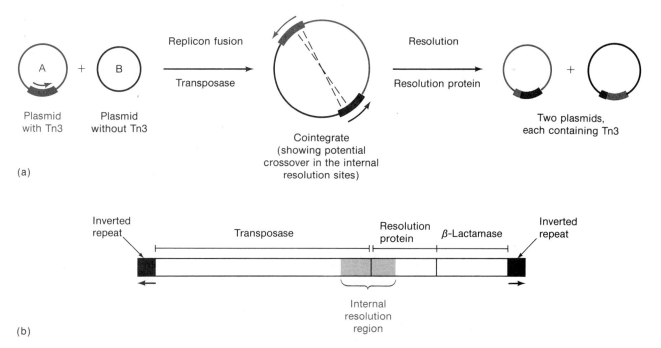

Figure 8-24 A scheme for transposition of the element Tn3. (a) Formation and resolution of a cointegrate. (b) The physical map of Tn3, showing its three genes. The internal resolution region is the site of action of the resolution protein.

site-specific exchange in the internal resolution site combines the units that
have been separated by the replication, yielding a new copy of the transposable element flanked by two copies of the target sequence and one copy of

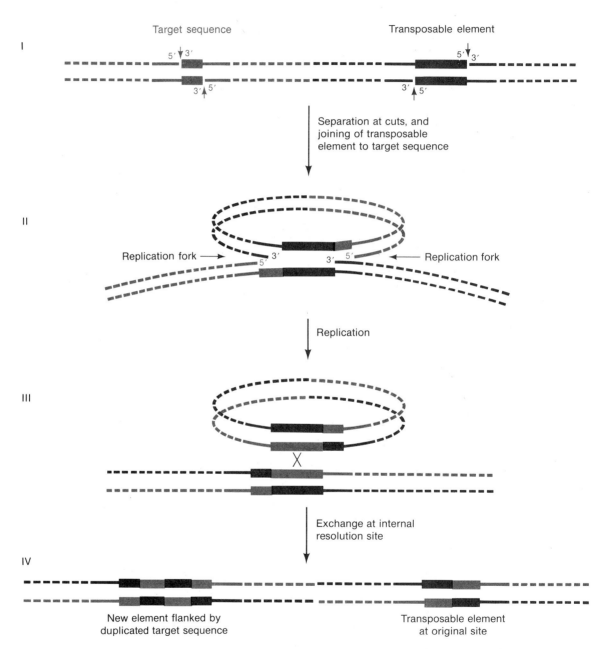

Figure 8-25 A model for transposition proposed by J. Shapiro. (I) Single-strand breaks are made at sites in the DNA molecule indicated by arrows. (II) Strands separate and nonhomologous strands are joined, linking the transposable element to the target sequence. Two replication forks result. (III) DNA synthesis (double lines) forms this structure. A site-specific strand exchange occurs between the internal resolution site (arrows). (IV) Transposition is complete. (Based on S. Cohen and J. Shapiro. *Scient. Amer.,* February 1980: 40–49.)

the transposable element at the original site. This model has recently received substantial support, though modifications will probably be necessary to account for particular features of transposition of some elements. For instance, in eukaryotes it is common that a transposable element leaves the original site and relocates at a new site.

Problems

1. What are the differences between transduction, transposition, transformation, bacterial conjugation, site-specific recombination, and meiotic recombination, with respect to requirements for homology and for a RecA-type protein?

2. Which of the following are examples of reciprocal exchanges in genetic recombination? Prophage integration, prophage excision, formation of recombinants in conjugation, transduction by phage P1, formation of an F' plasmid from an Hfr bacterium, formation of an Hfr bacterium from an F^+ cell, bacterial transformation, and meiotic crossing over.

3. In transposition what base sequences are duplicated?

4. Two homologous DNA structures are shown in the figure below. In each molecule a + strand is complementary to its own − strand and to the − strand of the other molecule. The molecules are oriented so that the *b* ends are homologous in base sequence. The upper molecule is shorter than the lower one and lacks the *a* end. The molecules are subjected to renaturing conditions during which they hybridize and branch-migrate. Which of the configurations—1 to 6—can arise?

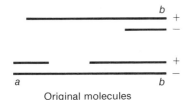

Original molecules

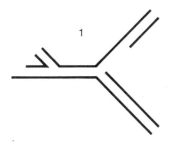

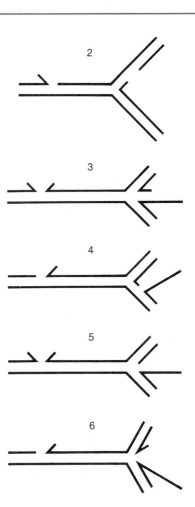

5. What observation in meiotic crossing over is one of the best arguments for the existence of heteroduplex regions in recombination intermediates?

6. In the simplest breakage-reunion models the initial exchange must occur between the markers that have been recombined. In the asymmetric strand-transfer model this is not necessary. What feature of the model eliminates this requirement?

7. In the asymmetric strand-transfer model gene conversion can arise in two ways. What are they?

8. Irradiation with ultraviolet light leads to DNA damage, which is repaired. An early stage in the repair process is the formation of a gap in the polynucleotide strand at the site of the damage. Such irradiation is also known to affect recombination frequency. If the repair system is inactive, owing to mutation, irradiation has very little effect on recombination frequency. What do you think the effect is—an increase or decrease of recombination frequency? Explain your answer.

9. Haploid yeast are of two mating types, a and α, and these can fuse to form a diploid. Their genomes are identical except for the genes of the mating-type locus. A population of diploids can be grown on certain media and sporulation will occur. A type-a yeast with the genotype Def (gene order def) is mated with a type-α dEF yeast and a diploid culture is obtained from a single zygote. The diploid culture is sporulated and asci are collected. Sixty-two asci contain 2 Def spores and 2 deF spores (these can be called type-I asci for convenience). Four asci contain 1 spore each of genotypes DeF, Def, dEF, and dEf (type-II asci). Two asci contain 1 spore each of genotypes DeF, DeF, deF, and dEf (type III). Explain how each class (I, II, and III) arose.

10. A fungal mutant is found in which aberrant 4:4 asci are observed but no 3:5, 5:3, 6:2, or 2:6 asci are ever found. What system is probably lacking in this mutant?

11. In *Pneumococcus* transformation is typically carried out by plating recipient cells, which have taken up DNA, directly on a selective growth medium that allows growth *only* of recombinants. Genetic markers in *Pneumococcus* fall into two classes—low-efficiency and high-efficiency markers. The latter, when present in the donor molecule, appear in the recombinants much more often than the low-efficiency markers do. It is known that there is no difference in the degradation of DNA molecules containing the two types of markers and that both are taken up equally well by competent cells. Suggest an explanation for the two classes.

12. Two double-stranded circular DNA molecules recombine according to the Holliday model. What is the form of the unit containing the Holliday junction?

13. Three mutations—a, b, c—of unknown map order lie within a sex-linked gene in *Drosophila*. A cross is performed using two flanking markers p and q. Females of genotype $p\,(a\,b\,+)\,q/+\,(+\,+\,c)\,+$ yield some male progeny that are wildtype with respect to genes a, b, and c; nearly all of these males are also wildtype for the flanking markers. What can be said about the map order of genes a, b, and c if gene conversion does not occur, and how will the presence of gene conversion change the conclusion?

CHAPTER 9

Gene Expression

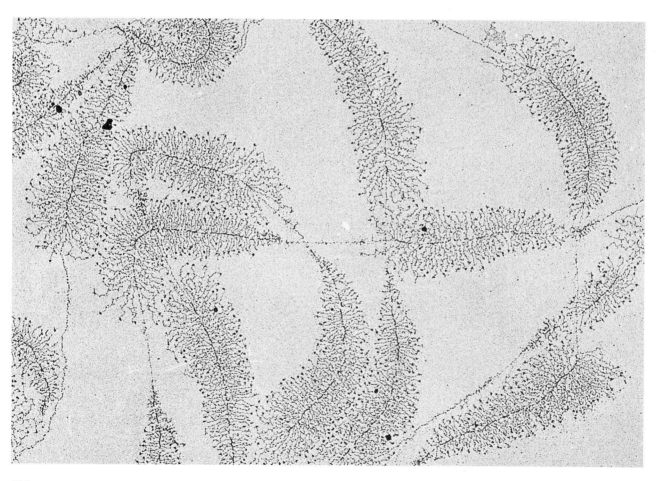

Earlier chapters have been concerned with genetic analysis—genes as units of genetic information, their relation to chromosomes, and the chemical structure and replication of the genetic material. In this chapter we shift our perspective to considering how the information contained in genes is converted to molecules that determine the metabolism, structure, and form of cells and viruses. The process by which this information is made available to a cell or virus is called **gene expression.** It is accomplished through a sequence of events in which the information contained in the base sequence of DNA is first copied into an RNA molecule and then used to determine the amino acid sequence of a protein molecule. RNA molecules are synthesized by using the base sequence of a part of a single strand of DNA as a **template** in a polymerization reaction (like that used in replicating DNA) that is catalyzed by enzymes called **RNA polymerases.** The process by which the segment corresponding to a particular gene is selected and an RNA molecule is made is called **transcription.** Protein molecules are then synthesized by using the base sequence of an RNA molecule to direct the sequential joining of amino acids in a particular order, so the amino acid sequence is a direct reflection of the base sequence. The production of an amino acid sequence from an RNA base sequence is called **translation,** and the protein made is called a **gene product.**

9.1 Proteins and Amino Acids

Proteins are the molecules responsible for catalyzing most intracellular chemical reactions (enzymes), for regulating gene expression (regulatory proteins), and for determining many features of the structures of cells, tissues, and viruses (structural proteins). A protein is a chain of covalently joined amino

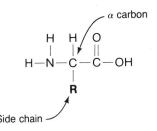

Figure 9-1 The basic structure of an amino acid.

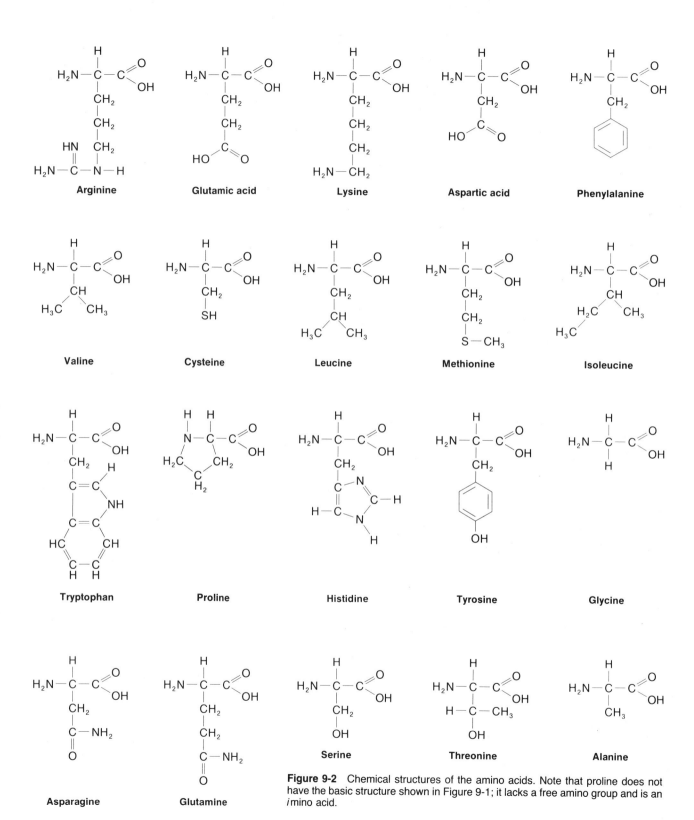

Figure 9-2 Chemical structures of the amino acids. Note that proline does not have the basic structure shown in Figure 9-1; it lacks a free amino group and is an *i*mino acid.

Arginine

Glutamic acid

Lysine

Aspartic acid

Phenylalanine

Valine

Cysteine

Leucine

Methionine

Isoleucine

Tryptophan

Proline

Histidine

Tyrosine

Glycine

Asparagine

Glutamine

Serine

Threonine

Alanine

acids. Most naturally occurring protein molecules contain 20 different amino acids. Since the number of amino acids in a protein molecule usually ranges from about 100 to nearly 1000, the number of different protein molecules that are possible is enormous.

Each amino acid contains a carbon atom (the α carbon) to which is attached one carboxyl group (—COOH), one amino group (—NH$_2$), and a side chain commonly called an **R group** (Figure 9-1). The R groups are generally chains or rings of carbon atoms bearing various chemical groups. The simplest side chains are those of glycine (a hydrogen atom) and of alanine (a methyl group or —CH$_3$). The chemical structures of all of the 20 amino acids are shown in Figure 9-2.

Protein molecules are formed when the carboxyl group of one amino acid joins with the amino group of a second amino acid; the resulting chemical bond is called a **peptide bond** (Figure 9-3(a)). Thus, a protein molecule is a **polypeptide chain** in which α-carbon atoms alternate with peptide units to form a linear chain having an ordered array of side chains (Figure 9-3(b)). The linear chain is called the **backbone** of the molecule.

The two ends of every protein molecule are distinct. One end has a free —NH$_2$ group and is called the **amino terminus;** the other end has a free —COOH group and is the **carboxyl terminus.** Proteins are synthesized by

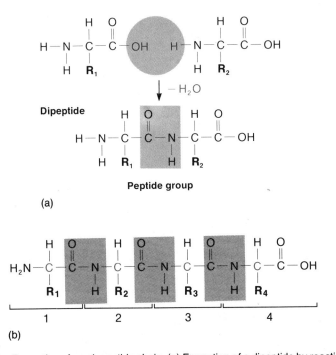

Figure 9-3 Properties of a polypeptide chain. (a) Formation of a dipeptide by reaction of the carboxyl group of one amino acid (left) with the amino group of a second amino acid (right). Water (shaded circle) is eliminated to form a peptide group (shaded rectangle). (b) A tetrapeptide showing the alternation of α-carbon atoms (solid red) and peptide groups (shaded). The four amino acids are numbered below.

adding individual amino acids to the carboxyl terminus of the growing chain. Conventionally, the amino acids of a polypeptide chain are numbered starting with the NH$_2$-terminal amino acid.

Most polypeptide chains are highly folded and a variety of three-dimensional shapes have been observed. The manner of folding is determined primarily by the sequence of amino acids—in particular, by noncovalent interactions between the side chains—so each polypeptide chain tends to fold into a unique three-dimensional shape. The rules of folding are complex and, except for the simplest proteins, shape cannot be predicted from the amino acid sequence. On the average, the molecules fold so that amino acids with charged side chains tend to be on the surface of the protein (in contact with water) and those with uncharged side chains tend to be internal. Specific folded configurations also result from hydrogen bonding between peptide groups. Two fundamental polypeptide structures are the α helix and the β structure, which are illustrated in Figure 9-4(a,b). A covalent bond (a **disulfide bond**) also may form between the sulfur atoms of some pairs of cysteines. Figure 9-5 is an idealized drawing of the three-dimensional structure of an enzyme, showing several types of folding.

Many protein molecules consist of more than one polypeptide chain. When this is the case, the protein is said to contain **subunits.** The subunits may be identical or different. For example, hemoglobin, the oxygen carrier of blood, consists of four subunits, two each of two different types. The RNA polymerase of *E. coli* has five subunits of four types, and DNA polymerase III of *E. coli* (Chapter 4) contains at least ten different subunits.

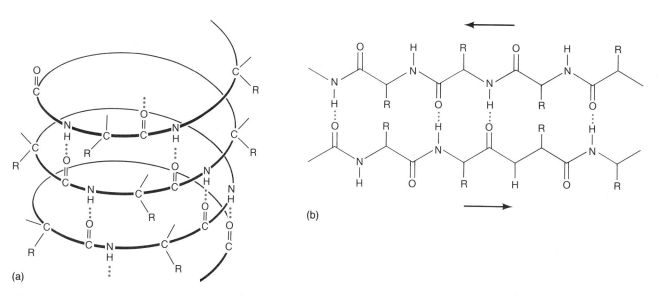

(a)

(b)

Figure 9-4 (a) An α helix drawn in three dimensions, showing how the hydrogen bonds (red dots) stabilize the structure. Hydrogen atoms other than those participating in hydrogen-bonding are omitted for the sake of clarity. (b) A β structure. Two segments of a polypeptide backbone are hydrogen-bonded (red dots) in an anti-parallel array (arrows), forming a rigid linear structure. The segments brought together may be located in nearby regions in the polypeptide chain, or in widely separated regions of the chain, or even in different chains. See Figure 9-5 for a typical relation between these segments.

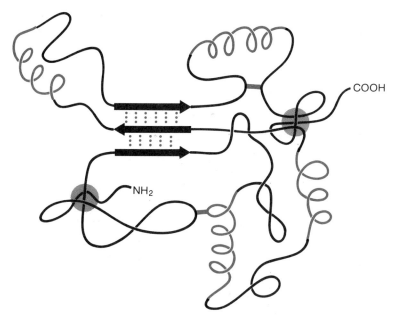

COOH

NH$_2$

Figure 9-5 A schematic diagram of the path of the backbone of a polypeptide, showing possible ways in which the polypeptide may be folded. Heavy black arrows represent β structure; the red dots joining the arrows represent hydrogen bonds. Note that the interacting regions are not always located in nearby sequences within the polypeptide chain. Helical regions are drawn in red. Two heavy red bars represent disulfide bonds. The two shaded areas are hydrophobic clusters.

9.2 Relations Between Genes and Polypeptides

A gene contains the information for synthesis of one polypeptide chain. The development of this important principle of molecular genetics began with a study of alkaptonuria and other congenital anomalies in humans (Chapter 1), which showed that each condition (1) results from an inability to synthesize a particular chemical compound specific to the condition and (2) is inherited according to Mendel's laws. These observations led to the recognition that genes and enzymes are related and suggested that each of these diseases results from a specific metabolic block, brought about by a mutation in a gene and the resulting lack of an enzyme. These suggestions were validated by George Beadle and Edward Tatum for the common bread mold *Neurospora crassa*, an organism in which both genetic and biochemical analyses could be done with ease.

Wildtype *Neurospora* is a haploid organism with simple nutritional requirements. The minimal medium required for normal growth contains only inorganic salts, a sugar, and the vitamin biotin. Beadle and Tatum recognized that *Neurospora* must be able to synthesize all metabolic components other than biotin and assumed that the synthesis of these substances must be under genetic control. If so, a mutation in a gene responsible for synthesizing an essential metabolite would be expected to render a strain unable to grow unless the strain was provided with the metabolite.

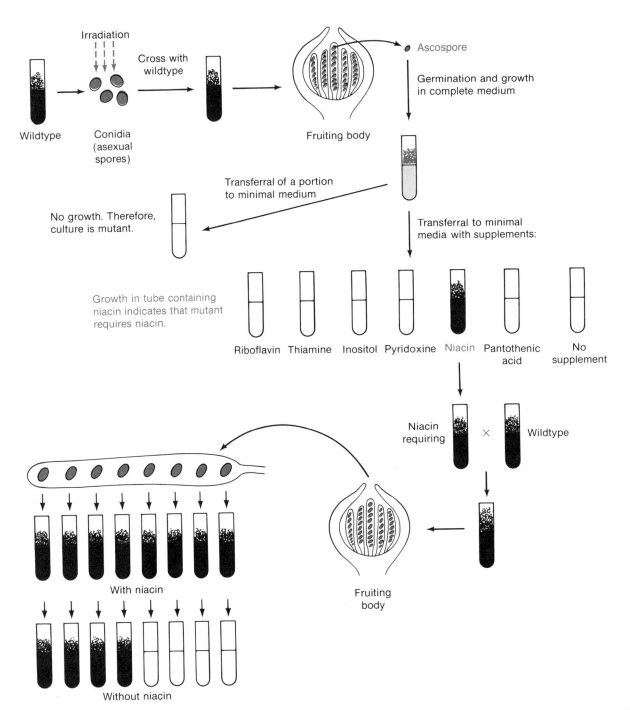

Figure 9-6 The Beadle-Tatum experiment. Conidia are irradiated to induce mutations and then crossed with a wildtype strain to produce haploid spores (ascospores). These spores are grown in complete medium. A sample is transferred to a minimal medium to identify cultures with a growth defect. The particular defect is then identified by growth in several different tubes of minimal medium, each tube containing one supplement. In the example shown, niacin added to the minimal medium allows growth; thus, the mutant is niacin-requiring. The mutant is then crossed with a wildtype strain. A 4 wildtype:4 mutant segregation ratio in the cross indicates that the phenotype results from a single mutation.

These ideas were tested in the following way. Asexual spores (conidia) of wildtype *Neurospora* were irradiated with either x rays or ultraviolet light (Chapter 10) to produce mutant strains with various nutritional requirements. In the initial step for identifying mutants, summarized in Figure 9-6, the irradiated spores were used in crosses with a wildtype strain. Ascospores produced by the sexual cycle in the resulting hybrids were individually germinated in a complete medium—one enriched with a variety of amino acids, vitamins, and other substances expected to be essential metabolites, synthesis of which could be blocked by a mutation. Conidia from each of these cultures were next transferred to a minimal medium, in which absence of growth would indicate that a mutation was present in the particular culture. Samples from a portion of each mutant culture were then transferred to a series of media to verify the ability of the mutants to grow in complete but not minimal medium, and to determine whether the mutation resulted in a requirement for a vitamin, an amino acid, or some other substance. Mutant cultures were also crossed with a wildtype strain to select those carrying only a single mutation. Single mutations were identified as those resulting in a 1-mutant:1-wildtype segregation among ascospores produced by such crosses.

Beadle and Tatum obtained numerous mutations that resulted in a requirement of a single metabolite for growth. Biochemical analyses showed that in several of these mutant strains a single mutation had caused loss of a specific enzymatic activity. From the genetic and biochemical experiments taken together the concept that one gene is responsible for the production of one enzyme evolved as a central doctrine in molecular genetics. The later discovery that numerous enzymes and other proteins, such as the hemoglobins, are composed of two or more different polypeptide chains necessitated modifying the concept to the statement that *a gene is responsible for the synthesis of one polypeptide chain*. This concept is usually called the **one gene–one polypeptide hypothesis.**

The one gene–one polypeptide concept was especially significant in view of the fact that it was known that a typical metabolite is synthesized by a **metabolic pathway** or sequence of steps, each catalyzed by a distinct enzyme. Thus, it was expected that in such a pathway each step would be controlled

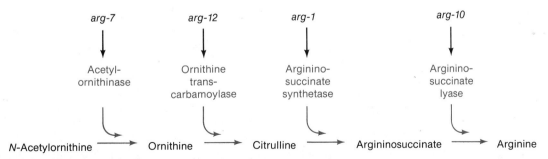

Figure 9-7 The metabolic pathway for synthesis of arginine. The names of the genes in *Neurospora* are italic black. The enzyme products of these genes and the reaction arrows are shown in red. A mutation in any gene prevents synthesis of components further along the pathway. For simplicity, several steps leading to synthesis of *N*-acetylornithine have not been shown; these are catalyzed by the genes *arg-14*, *arg-6*, and *arg-5*.

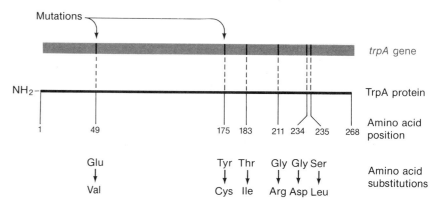

Figure 9-8 Correlation of the positions of mutations in the genetic map of the *E. coli trpA* gene with positions of amino acid substitutions in the TrpA protein.

by a particular gene. This prediction was confirmed by further genetic analysis of strains of *Neurospora* (and other organisms), which showed that a growth requirement for a particular substance may result from mutations in different genes and that each mutation caused loss of the activity of one enzyme in the pathway. A simple and well-characterized pathway is that for synthesis of the amino acid arginine (Figure 9-7). If the starting point is considered to be N-acetylornithine, the reaction proceeds through four steps. In *Neurospora* four genes *arg-7*, *arg-12*, *arg-1*, and *arg-10* are responsible for synthesizing each of the four enzymes, as indicated in the figure. For simple metabolic pathways the sequence can be worked out by identifying substances that enable a mutant to grow. For example, mutations in any of the four *arg* genes will prevent *Neurospora* from growing in a minimal medium. However, an *arg-7* mutant can grow if ornithine or citrulline, but not N-acetylornithine, is added to the growth medium. Similarly, growth of an *arg-12* mutant is possible if citrulline, but not ornithine, is provided. An *arg-1* mutant will grow only if arginine itself is provided.

The Beadle-Tatum experiment established that genes control the presence of enzymatic activity, but did not show that genes control the actual amino acid sequences of proteins. The first hint that this is the case came from a comparison of the structure of normal hemoglobin—the protein that carries oxygen in blood—and the hemoglobin found in patients with sickle-cell anemia, a human disease that causes premature death. Sickle-cell anemia is inherited according to Mendel's laws, and pedigree analysis suggested that it is caused by a recessive mutant allele. Chemical analysis showed that hemoglobin isolated from sickle-cell patients differs from normal hemoglobin in that a valine is present in the sickle-cell hemoglobin at a position occupied by glutamic acid in normal hemoglobin (Chapter 10).

More substantial evidence that the *sequence* of genetic information in a gene determines the *sequence* of amino acids in a polypeptide was provided by studies of the tryptophan synthetase gene *trpA* in *E. coli*, a gene in which many mutations had been obtained and accurately mapped. The effects of numerous mutations on the amino acid sequence of the enzyme were determined by directly analyzing the amino acid sequences of the wildtype and mutant enzymes. Each mutation was found to lead to a single *amino acid substitution* in the enzyme, and, more important, *the order of the mutations in the genetic map was the same as the order of the affected amino acids in the polypeptide chain* (Figure 9-8). This attribute of genes and polypeptides is called **colinearity,** which means that the sequence of base pairs in DNA determines the sequence of amino acids in the polypeptide in a colinear point-to-point manner. Colinearity is universally found in prokaryotes. However, we will see later (for example, Figure 9-13) that in eukaryotes noninformational DNA sequences interrupt the continuity of most genes; in these genes, the order, but not the spacing, between the mutations correlates with amino acid substitution.

In the remainder of this chapter we will see how a gene determines amino acid sequence.

9.3 Transcription

The first step in gene expression is the synthesis of an RNA molecule copied from the segment of DNA that constitutes the gene. In this section we describe the basic features of the production of RNA.

Basic Features of RNA Synthesis

The essential chemical characteristics of the enzymatic synthesis of RNA are like those of DNA synthesis (Chapter 4):

1. The precursors in the synthesis of RNA are the four ribonucleoside 5′-triphosphates (rNTP): adenosine triphosphate (ATP), guanosine triphosphate (GTP), cytosine triphosphate (CTP), and uridine triphosphate (UTP). Recall from Chapter 4 that they differ from the DNA precursors only in that the sugar is ribose rather than deoxyribose and that the base uracil, U, replaces thymine, T (Figure 4-33).

2. In the formation of RNA a 3′-OH group of one nucleotide reacts with the 5′-triphosphate of a second nucleotide; the two terminal phosphate groups are removed as inorganic pyrophosphate (PP_i), and a sugar-phosphate bond results (Figure 9-9(a)). This is the same chemical reaction that occurs in the synthesis of DNA, though the enzyme is different.

3. The sequence of bases in an RNA molecule is determined by the base sequence of the DNA template. Each base added to the growing end of the RNA chain is chosen for its ability to base-pair with the DNA template strand; thus, the bases C, T, G, and A in a DNA strand cause G, A, C, and U, respectively, to be added to the growing end of an RNA molecule.

H in DNA
OH in RNA

CH_3 in thymine
H in uracil

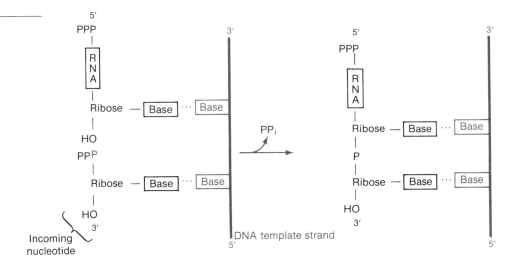

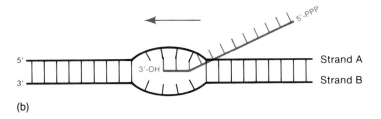

(b)

Figure 9-9 RNA synthesis. (a) The polymerization step in RNA synthesis. The incoming nucleotide forms hydrogen bonds (three dots) with a DNA base. Reaction occurs between the OH group in the upper nucleotide and the red P in the triphosphate group, leading to removal of the black phosphates (PP$_i$). (b) Geometry of RNA synthesis. RNA is copied only from strand A of a segment of a DNA molecule. It is not copied from strand B in that region of the DNA. However, elsewhere—in a different gene, for example—strand B might be copied; in that case, strand A would not be copied in that region of the DNA. The RNA molecule is antiparallel to the DNA strand being copied.

4. Nucleotides are added only to the 3′-OH end of the growing chain; as a result, the 5′ end of a growing RNA molecule bears a triphosphate group. This direction of chain growth—the 5′ → 3′ direction—is the same as that in DNA synthesis. Furthermore, the RNA strand and the DNA template strand are antiparallel to one another, like the two strands of a DNA molecule.

A significant difference between DNA polymerases and RNA polymerases is that *RNA polymerases are able to initiate chain growth without a primer*.

An important feature of RNA synthesis is that *the DNA molecule being copied is double-stranded, yet in any particular region of the DNA only one strand serves as a template*. The implications of this statement are shown in Figure 9-9(b).

The synthesis of RNA consists of four discrete stages: **promoter recognition, chain initiation, chain elongation,** and **chain termination.**

1. RNA polymerase binds to DNA within a specific base sequence (20–200 bases long), called a **promoter.** Promoter sequences have been isolated by mixing RNA polymerase with DNA and then treating the DNA with enzymes that digest DNA but leave intact the region bound to the RNA polymerase. After enzymatic digestion the RNA polymerase can be removed and the base sequence of the DNA determined. Examination of a large number of promoter sequences for different genes and from different organisms has shown that they have many features in common. For example, the sequence TATAATG (or a nearly identical sequence)—often called a **Pribnow box**—is found as part of all prokaryotic promoters. A sequence of this type is called a **consensus sequence,** because it designates a pattern of bases from which actual sequences observed in many different systems differ by usually no more than one or two bases. In eukaryotes the corresponding consensus sequence is TATAAATA (the **TATA box**). These recognition sequences instruct RNA polymerase where to start synthesis by providing a binding site. The strength of the binding of RNA polymerase to different promoters varies greatly; this variation is a fundamental mechanism for regulating gene expression.

2. After the initial binding step RNA polymerase also becomes bound to a **polymerization start site,** which is very near the initial binding site. The first nucleoside triphosphate is placed at this site and synthesis begins.

3. RNA polymerase then moves along the DNA, adding nucleotides to the growing RNA chain.

4. RNA polymerase reaches a chain-termination sequence and both the newly synthesized RNA and the polymerase are released. Two kinds of termination events are known: those that are self-terminating (dependent on the DNA base sequence only) and those that require the presence of a termination protein. Both types of events occur at specific but distinct base sequences. Self-termination usually occurs at base sequences in the template strand that consist of a series of adjacent adenines preceded by a sequence that is a **palindrome**—that is, one that reads the same forward and backwards, such as ABCDD′C′B′A′, in which a ′ represents a complementary base. The palindrome is usually interrupted by a few bases (for example, ABCDXYZD′C′B′A′), so the RNA molecule can in theory fold back on itself to form a stem-and-loop configuration (Figure 9-10). Whether stems and loops exist in cells is not known with certainty, and how the termination protein works is poorly understood.

Initiation of a second round of transcription need not await completion of the first, for the promoter becomes available once RNA polymerase has

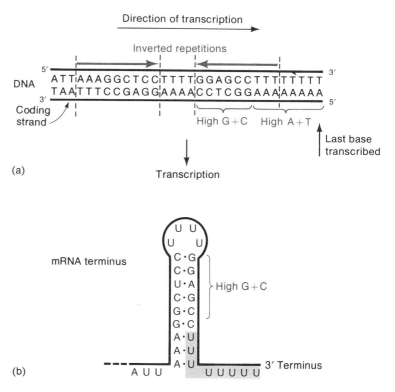

Figure 9-10 Base sequence of (a) the transcription-termination region for the set of tryptophan-synthesizing genes in *E. coli*, and of (b) the 3′ terminus of the corresponding mRNA molecule. An inverted sequence and its forward counterpart, which are characteristic of termination sites, are indicated by reversed red arrows. The mRNA is folded to form a stem-and-loop structure. The relevant regions are labeled in red; the sequence of U's found at the termini of most prokaryotic mRNA molecules is shaded in red.

polymerized 50–60 nucleotides. For a rapidly transcribed gene such reinitiation occurs repeatedly and a gene can be cloaked with numerous RNA molecules in various degrees of completion. Such an array is shown in the electron micrograph at the opening of this chapter. This micrograph shows a region of the DNA of the newt *Triturus* in which there are tandemly repeated copies of the ribosomal RNA genes. Each gene is associated with growing RNA molecules. The shortest ones are at the promoter end of the gene; the longest are near the gene terminus.

Several types of mutations affect initiation and termination of transcription. In fact, the existence of promoters was first demonstrated by the isolation of particular *E. coli* Lac⁻ mutations, denoted p^-, that eliminate activity of the *lac* gene but *only when the mutations are adjacent to the genes in the same DNA molecule*. This feature can be seen by examining a cell having two copies of the *lacZ* gene—for example, a cell containing an F′*lacZ* plasmid. The presence of the *lacZ* gene enables the cell to synthesize the enzyme β-galactosidase. Table 9-1 shows that a wildtype *lacZ* gene is inactive only if it and a p^- mutation are

Table 9-1 Effect of promoter mutations on transcription of the *lacZ* gene

Genotype	DNA molecule contains both a p^- mutation and the $lacZ^+$ gene	Transcription of $lacZ^+$ gene
1. $p^+ lacZ^+$	No p^-	Yes
2. $p^- lacZ^+$	Yes	No
3. $p^+ lacZ^+/p^+ lacZ^-$	No p^-	Yes
4. $p^- lacZ^+/p^+ lacZ^-$	Yes	No
5. $p^+ lacZ^+/p^- lacZ^-$	No	Yes

present on the same DNA molecule (either a chromosome or an F′ plasmid); this can be seen by comparing entries 4 and 5. Chemical analysis shows that in a cell with the genotype shown in entry 4 in the table, the *lacZ* gene is not transcribed, though transcription does occur if the genotype is that of entry 5. The p^- mutations are called **promoter mutations.**

In prokaryotes, mutations have been isolated that create a new termination sequence ahead of the normal one. When such a mutation is present, an RNA molecule is made that is shorter than the wildtype RNA. Some of these mutations, which always map at a single site, eliminate expression of several genes adjacent to the gene in which the mutation is mapped. Mutations of this type are of a class called **polar mutations** (see also Section 9.7). Such mutations provide evidence that a prokaryotic RNA molecule may contain more than one gene; the mutation activates a premature termination site for RNA synthesis, which prevents transcription of genes past the mutation (in the direction of RNA synthesis).

RNA Polymerases

The best-understood RNA polymerase is that of the bacterium *E. coli*. This enzyme, which consists of five protein subunits, is one of the largest enzymes known and can be easily seen by electron microscopy (Figure 9-11). The five subunits are organized in two functional groups. Four of the subunits comprise the **core enzyme,** which is responsible for the chemical activity of joining nucleoside triphosphates. The fifth subunit, the **sigma subunit,** is required for binding the complete enzyme to a promoter. The core enzyme binds weakly to DNA, but at any base sequence, rather than specifically at promoters. Once polymerization begins, the sigma subunit dissociates from the core enzyme. When transcription is completed, the core enzyme binds a different sigma subunit and is then ready to bind to a promoter again.

Eukaryotic cells have three distinct RNA polymerases, denoted I, II, and III. All are found in the eukaryotic nucleus, and in terms of chemistry and specificity, they resemble the *E. coli* enzyme. RNA polymerase I catalyzes synthesis of ribosomal RNA (rRNA), an essential component of the ribosome, a particle necessary for protein synthesis (see Section 9.7). RNA polymerase

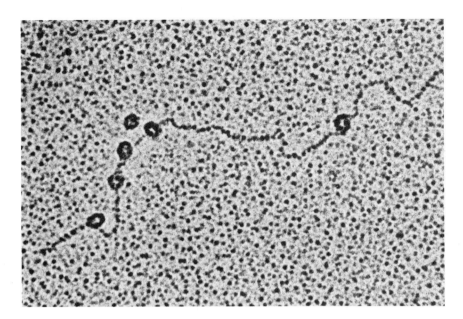

Figure 9-11 *E. coli* RNA polymerase molecules bound to DNA. (Courtesy of Robley Williams.)

II is, in the informational sense, the most important one, as it is the enzyme responsible for synthesis of all RNA molecules that contain information for amino acid sequences (that is, mRNA molecules). RNA polymerase III synthesizes 5S rRNA (another component of the ribosome) and tRNA—both of which are needed for protein synthesis—and other small RNA molecules whose functions are unknown. Minor RNA polymerases, which have not yet been studied in detail, are found in mitochondria and chloroplasts.

Messenger RNA

Amino acids do not bind to DNA. Thus, intermediate steps are needed for arranging the amino acids in a polypeptide chain in the order determined by the DNA base sequence. This process begins with transcription of the base sequence of one of the DNA strands (called the **coding strand** or **sense strand**) into the base sequence of an RNA molecule. In prokaryotes this RNA molecule, which is called **messenger RNA** or **mRNA,** is used directly in polypeptide synthesis. In eukaryotes the RNA molecule is usually modified before it becomes mRNA. The amino acid sequence is then obtained from mRNA by the protein-synthesizing machinery of the cell.

In prokaryotes mRNA molecules commonly contain information for the amino acid sequences of several different polypeptide chains; in this case, such a molecule is called **polycistronic mRNA.** (Cistron is a term commonly used at one time to mean a base sequence encoding a single polypeptide chain.) The genes contained in a polycistronic mRNA molecule often encode the different proteins of a metabolic pathway. For example, in *E. coli* the ten enzymes needed to synthesize histidine are encoded in one mRNA molecule. The use of polycistronic mRNA is an economical way for a cell to regulate synthesis of related proteins in a coordinated way. For example, in prokaryotes the usual way to regulate synthesis of a particular protein is to control the synthesis of the mRNA molecule that encodes it (Chapter 11). With a polycistronic mRNA molecule the synthesis of several related proteins can be regulated by a single signal, so that appropriate quantities of each protein are made at the same time; this is termed **coordinate regulation.** The mRNA of eukaryotic cells is almost never polycistronic (the reasons will be explained shortly); as a consequence, coordinate regulation is more complicated in eukaryotes than in prokaryotes.

Not all base sequences in an mRNA molecule are translated into the amino acid sequences of polypeptides. For example, translation of an mRNA molecule rarely starts exactly at one end of the RNA molecule and proceeds to the other end; instead, initiation of polypeptide synthesis may begin hundreds of nucleotides from the 5'-P terminus of the RNA. The section of untranslated RNA before the region encoding the first polypeptide chain is called a **leader,** which in some cases contains regulatory sequences that influence the rate of protein synthesis. Untranslated sequences are found at both the 5'-P and the 3'-OH termini. Polycistronic mRNA molecules usually contain **spacer sequences** tens of bases long, which separate the **coding sequences;** each coding sequence corresponds to a polypeptide chain. A typical coding sequence in an mRNA molecule is 300–2500 bases long (depending on the number of amino acids in the protein), but because of the accessory sequences, mRNA molecules range in size from about 400 to 20,000 nucleotides (for polycistronic molecules), the most common range being 1000–9000.

The coding sequence of each gene is obtained by transcription of only one DNA strand. It is quite rare for the two complementary base sequences in a particular gene to be transcribed, though a few exceptions are known in which two adjacent genes are transcribed from different strands, with a slight overlapping of the template segments. However, except for some small viruses and some transposable elements, *all mRNA molecules are not synthesized from the same DNA strand*, so in an extended segment of a DNA molecule, mRNA molecules would be seen growing in either of two directions (Figure 9-12), depending on which DNA strand functions as a template.

In prokaryotes most mRNA molecules are degraded within a few minutes after synthesis. In eukaryotes a typical lifetime is several hours,

Figure 9-12 A typical arrangement of promoters (black arrowheads) and termination sites (black bars) in a segment of a DNA molecule. Promoters are present in both DNA strands. Termination sites are usually located so that transcribed regions do not overlap.

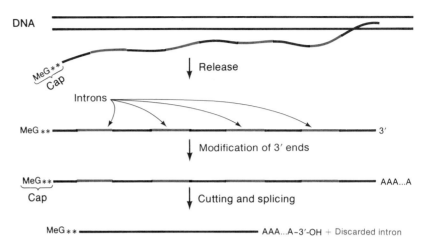

Figure 9-13 A schematic drawing showing production of eukaryotic mRNA. The primary transcript is capped before it is released. Then, its 3'-OH end is modified, and finally the introns are excised. MeG denotes 7-methylguanosine, and the two asterisks indicate the two nucleotides whose riboses are methylated.

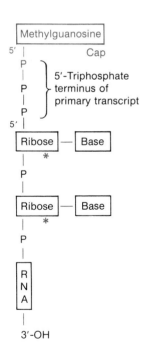

Figure 9-14 The structure of the cap at the 5' termini of eukaryotic mRNA molecules. The red asterisks indicate sites of occasional addition of a methyl group.

though some last several days. In both kinds of cells, the degradation enables cells to dispense with molecules that are no longer needed.

The mechanisms of transcription in prokaryotes and eukaryotes are, except for initiation (to be discussed shortly), nearly identical, but there are major differences in the relation between the transcript and the mRNA used for polypeptide synthesis. In prokaryotes the immediate product of transcription (called the **primary transcript**) is mRNA; in contrast, *in eukaryotes the primary transcript must be converted to mRNA.* This conversion, which is called **RNA processing,** consists of two types of events—modification of the termini and excision of untranslated sequences embedded *within* coding sequences. These events are illustrated diagramatically in Figure 9-13.

Modification of the 5' terminus forms what is called a **cap,** as is shown in Figure 9-14. The cap consists of a methylated guanosine derivative in an uncommon 5'-5' linkage to the 5'-terminal nucleotide of the primary transcript; occasionally, the sugars of one or both of the 5'-terminal nucleotides are also methylated. A cap is found on most, but not all, eukaryotic mRNA molecules. Capping occurs shortly after initiation of synthesis of the primary transcript.

The 3' terminus of a eukaryotic mRNA molecule usually consists of a polyadenosine sequence—poly(A)—up to 200 nucleotides long. The **poly(A) tail** is not added to the 3' terminus of the primary transcript, however. Instead, RNA polymerase polymerizes past the nucleotide to which poly(A) will ultimately be added; later, the RNA is cut at the **poly(A)-addition site,** which makes available a 3'-OH group to which adenines are added by a specific enzyme. The consensus sequence—AAUAAA—located 10 to 25 bases ahead of the poly(A)-addition site is important in some way in enabling this site to be recognized. The biological significance of these terminal modifications has not yet been unambiguously established, and in fact some mRNA molecules lack one modification or the other; it is thought that the

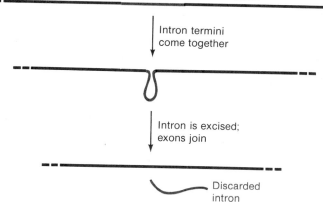

Figure 9-15 A schematic diagram showing removal of one intron from a primary transcript.

modifications may increase the resistance of eukaryotic mRNA molecules to enzymatic attack, though other roles, such as facilitating transport through the nuclear membrane, are possible.

A second important feature peculiar to the primary transcript in eukaryotes, also shown in Figure 9-13, is the presence of long segments of RNA, called **intervening sequences** or **introns,** which are excised from the primary transcript. Accompanying the excision is a rejoining of the fragments to form the mRNA molecule, as is illustrated schematically in Figure 9-15. In processing a typical primary transcript the amount of discarded RNA ranges from about 50 percent to nearly 90 percent of the primary transcript. The remaining segments (**exons**) are joined together to form the finished mRNA molecules. The excision of the introns and the joining of the exons to form the final mRNA molecule is called **RNA splicing.** The 5' segment (the cap) of a primary transcript is never discarded.

The number of introns per mRNA molecule varies considerably from one gene to the next (Table 9-2). Furthermore, within a particular mRNA

Table 9-2 Some translated eukaryotic genes in which introns have been demonstrated

Gene	Number of introns
α-Globin	2
Immunoglobulin L chain	2
Immunoglobulin H chain	4
Yeast mitochondria cytochrome *b*	6
Ovomucoid	6
Ovalbumin	7
Ovotransferrin	16
Conalbumin	17
α-Collagen	52

molecule the introns are widely distributed and have many different sizes (Figure 9-16).

The existence and the positions of introns in a particular primary transcript are readily demonstrated by annealing the transcribed DNA with the fully processed mRNA molecule. The DNA-RNA heteroduplex can then be examined by electron microscopy. An example of adenovirus mRNA and the corresponding DNA is shown in Figure 9-17. The DNA copies of the introns appear as single-stranded loops in the heteroduplex, because no corresponding RNA sequence is available for hybridization.

The mechanism of cutting and splicing is not well understood. A remarkable feature of the splicing reaction is that cuts are made at unique positions in transcripts that contain thousands of base pairs. Furthermore, each intron

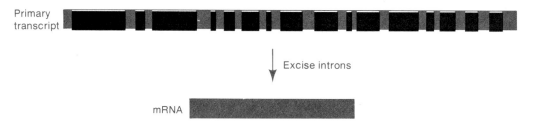

Figure 9-16 A diagram of the primary transcript and the processed mRNA of the conalbumin gene. The seventeen introns, which are excised from the primary transcript, are shown in black.

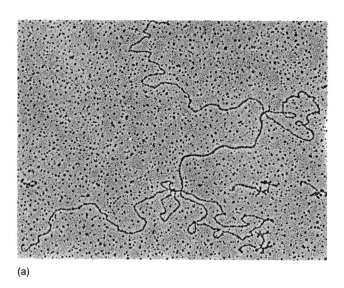

(a)

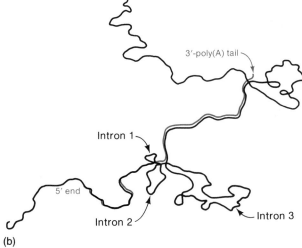

(b)

Figure 9-17 (a) An electron micrograph of a DNA-RNA hybrid obtained by annealing a single-stranded segment of adenovirus DNA with one of its mRNA molecules. The loops are single-stranded DNA. (Courtesy of Tom Broker and Louise Chow.) (b) An interpretive drawing. RNA and DNA strands are shown in red and black, respectively. Four regions do not anneal—three single-stranded DNA segments corresponding to the introns, and the poly(A) tail of the mRNA molecule.

is terminated with the sequences GU and AG, which are embedded in the consensus sequences AG*GU*AAGU and U(or C)*AG*G, respectively. Preliminary evidence suggests that a small RNA molecule (165 nucleotides long) containing sequences complementary to the consensus sequences may bring the intron termini together. This RNA is contained in an RNA-protein particle (called a **small ribonucleoprotein particle,** or **snurp**), whose proteins may mediate the enzymatic splicing reaction.

The general biological significance of splicing is unclear, for it would seem less wasteful if mRNA molecules were synthesized directly without splicing. With some viruses splicing increases the information content of DNA. The DNA molecules of these viruses are so small that each can encode only a few genes. However, a single primary transcript of each can be spliced in several ways, thereby producing different mRNA molecules. In this way different proteins can be encoded in a particular DNA segment. In Chapter 11 a few examples of regulatory mechanisms that utilize differential splicing will be given.

Many hypotheses have been presented for the origin of interrupted genes. One prominent theory is that each exon represents the coding sequence of an ancestral small protein and that in the course of evolution rearrangements occurred that brought the sequences together from various parts of the genome, forming new proteins. If no mechanism for splicing RNA had existed, production of the coding sequence of a new, functional protein would have required a precise recombination of sequences; most rearrangements would not have yielded a functional protein and in time they would have been lost. An alternative is that splicing was possible and that the sequences had instead been brought initially into a single transcription unit. With such a rearrangement the ancestral proteins would still have been synthesized and the cell would have grown and reproduced normally. If splicing had occurred occasionally, different splicing patterns would have been tried in the course of time, and any splicing event that produced an improved protein and thereby a superior organism would have been advantageous and therefore maintained by natural selection. Examination of the structure of many eu-

Figure 9-18 A diagram of a hypothetical polypeptide chain folded into two domains, one black and one red. In multifunctional proteins each domain may contain a binding site. In proteins having a single function a binding site frequently occurs at the contact point between the domains (arrow).

karyotic proteins suggests that such a scenario probably occurred repeatedly. Many proteins are known whose polypeptide chains fold to form **domains.** That is, the three-dimensional structure of these proteins is such that the molecules appear to consist of several independently folded regions, each separated by a short polypeptide segment (Figure 9-18). Furthermore, the binding sites of proteins with multiple binding sites and of enzymes having several catalytic activities are often located in separate domains. The significant point is that in many cases the amino acid sequence of each domain is encoded in one exon, suggesting that each domain is derived from an ancestral protein, as proposed.

9.4 Translation

The synthesis of every protein molecule in a cell is directed by an mRNA intermediate, which is copied from DNA, except for RNA viruses and RNA plasmids. Numerous steps follow mRNA production and these can conveniently be grouped into two categories: information-transfer processes by which the RNA base sequence determines an amino·acid sequence, and chemical processes by which the amino acids are linked together. The complete series of events is called **translation.**

The translation system consists of four major components:

1. **Ribosomes.** These are particles on which the mechanics of protein synthesis is carried out; they contain the enzymes needed to form a peptide bond between amino acids, a site for binding one mRNA molecule, and sites for bringing in and aligning the amino acids in preparation for assembly into the finished polypeptide chain.

2. **Transfer RNA,** or **tRNA.** Amino acids do not bind to mRNA, but in the synthesis of a particular polypeptide, they must be ordered by the base sequence in the mRNA molecule. This ordering is accomplished by a set of adaptor molecules, the tRNA molecules. A tRNA molecule "reads" the base sequence of mRNA. The language read by the tRNA molecules is called the **genetic code**—a set of relations between sequences of three adjacent bases on an mRNA molecule and particular amino acids.

3. **Aminoacyl tRNA synthetases.** This set of enzymes catalyzes the attachment of an amino acid to its corresponding tRNA molecule.

4. **Initiation, elongation,** and **release factors.** These molecules are proteins needed at particular stages of polypeptide synthesis. They will not be discussed in this book.

In prokaryotes all of these components are present throughout the cell; in eukaryotes, they are located primarily in the cytoplasm.

In outline, the mechanism of protein synthesis can be depicted as in Figure 9-19. An mRNA molecule binds to the surface of a protein-synthesizing particle, the ribosome. Appropriate tRNA-amino acid complexes, which have been made elsewhere by the aminoacyl tRNA synthetases,

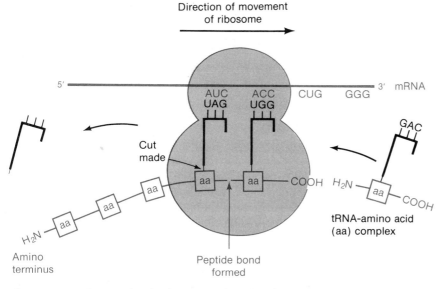

Figure 9-19 A diagram showing how a protein molecule is synthesized.

bind sequentially, one by one, to the mRNA molecule that is attached to the ribosome. Peptide bonds are made between successively aligned amino acids, each time joining the amino group of the incoming amino acid to the carboxyl group of the amino acid at the growing end. Finally, the chemical bond between the tRNA and its attached amino acid is broken and the completed protein is removed. More about protein synthesis will be presented in Section 9.7.

An important feature of translation is that it proceeds in a particular direction, obeying the following rules:

1. RNA is translated from the 5′ end of the molecule toward the 3′ end—but not from the 5′ terminus itself nor all the way to the 3′ end.
2. Polypeptides are synthesized from the amino terminus toward the carboxyl terminus, by adding amino acids one by one to the carboxyl end. For example, a protein having the sequence NH_2-Met-Pro- . . . -Gly-Ser-COOH, would have started with methionine, and serine would be the last amino acid added to the chain.

These rules are illustrated schematically in Figure 9-20.

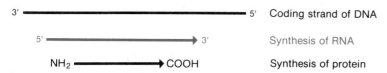

Figure 9-20 Direction of synthesis of RNA with respect to the coding strand of DNA, and of synthesis of protein with respect to mRNA.

It is conventional when writing nucleotide sequences to place the 5′ terminus at the left, and with amino acid sequences, to place the NH_2 terminus at the left. Thus, polynucleotides are generally written to show both synthesis and translation from left to right, and polypeptides are also written to show synthesis from left to right. This convention is used in all of the following sections concerning the genetic code.

9.5 The Genetic Code

Only four bases in DNA serve to specify 20 amino acids in proteins, so some combination of the bases is needed for each amino acid. Actually, more than 20 distinct combinations are needed for polypeptide synthesis, because signals are required for starting and stopping the synthesis of particular polypeptide chains. An RNA base sequence corresponding to a particular amino acid is called a **codon,** and the signal sequences are called **start codons** and **stop codons.** The genetic code is the set of all codons. Before the genetic code was elucidated, it was reasoned that if all codons were assumed to have the same number of bases, then each codon would have to contain at least three bases. Codons consisting of pairs of bases would be insufficient because four bases can form only $4^2 = 16$ pairs; triplets of bases would suffice because these can form $4^3 = 64$ triplets. In fact, the genetic code is a **triplet code** and all 64 possible codons carry information of some sort. Several different codons designate the same amino acid. Furthermore, in translating mRNA molecules the codons do not overlap but are used sequentially (Figure 9-21).

5′-AGUCAGUCAGUCAGUCAGUCAGUC-3′

⟶
Direction of reading

Figure 9-21 Bases in an RNA molecule are read sequentially in the 5′–3′ direction, in groups of three.

Genetic Evidence for a Triplet Code

We have explained the argument that each codon must contain at least three letters, but not ruled out codons with more than three letters. Biochemical experiments, described later in the chapter, clearly identified the sequences of three bases specifying particular amino acids. However, prior to this work, a strictly genetic experiment was performed that indicated that each codon consists of either three, or a multiple of three, bases. In this experiment, which is described in this section, the effect of inserting or deleting particular numbers of bases in a coding sequence was examined.

If *E. coli* phage T4 is grown in a medium containing the substance proflavin, a molecule that interleaves between the DNA base pairs, errors will occasionally be made during DNA replication, and daughter molecules will be formed that either have an additional nucleotide pair or lack a nucleotide pair. Since bases are read sequentially, a base-pair addition or deletion represents a mutation, because it upsets the phase of the units by which the code is read for a specific protein (the **reading frame**): every amino acid downstream from the added base will be different (Figure 9-22). Indeed, mutant phage, called **frameshift mutants,** are found among the progeny (see also Chapter 10). They arise at a frequency of about one mutation per gene per 10^5–10^6 phage. If a proflavin-induced mutant is grown again in a medium containing proflavin, some phage appear in which the wildtype phenotype has

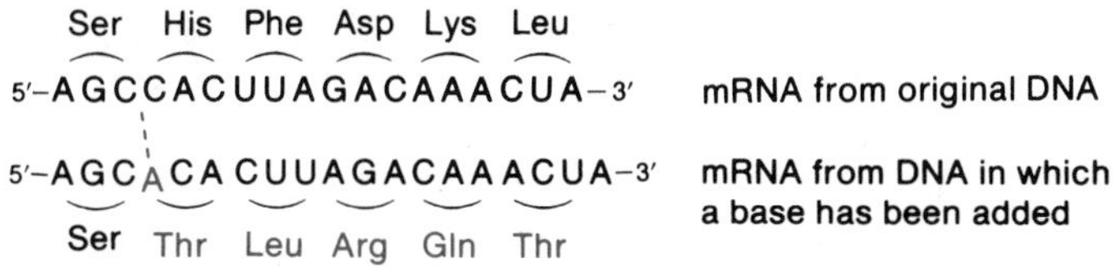

Ser His Phe Asp Lys Leu

5′–AGCCACUUAGACAAACUA–3′   mRNA from original DNA

5′–AGCACACUUAGACAAACUA–3′   mRNA from DNA in which
a base has been added

Ser Thr Leu Arg Gln Thr

Figure 9-22   The change in the amino acid sequence of a protein caused by addition of an extra base.

been restored. An experiment was undertaken to determine what mutational events would result in such restoration. A possible mechanism for this reversion might be removal of the additional pair of bases from the DNA of a base-addition mutant, but such an event would seem unlikely. Mutations are produced randomly (Chapter 10), so removal of the *particular* added base pair that gave rise to the mutation should occur with much less frequency (by about a thousandfold) than the addition that produced the original mutation. However, the observed frequencies of mutation and reversion were comparable. A mutation resulting in removal of a base pair sufficiently close to the added base pair might also restore the reading frame and in certain cases might result in production of a biologically functional protein, even though the entire amino acid sequences of the original wildtype protein and the one formed by a mutation plus a reverse mutation were not identical. This proved to be so. Study of a collection of proflavin-induced mutants in the gene called *rIIB* of *E. coli* phage T4 (Section 7.5) showed that such combinations of mutations can have this effect.

The T4 *rIIB* protein contains a region in which numerous amino acid substitutions can be tolerated without total loss of function. However, even in this region proflavin induces mutations that inactivate the protein. Normally, if two phage strains, each carrying a different mutation in the same gene, are crossed by mixed infection of a bacterium, some wildtype phage progeny arise by genetic recombination between the sites of the two mutations. However, when two randomly selected proflavin-induced *rII* mutants were crossed, wildtype phage progeny (phage able to form plaques on a λ lysogen) did not always arise. The recombination results indicated that the mutants could be placed in two distinct classes, termed (+) and (−). Crosses between two mutants in different classes (that is, one (+) and one (−) mutant) yielded recombinants having the wildtype phenotype, but no such recombinants were formed by crossing mutants in the same class ((+) × (+) or (−) × (−)). These results were interpreted in the following way. The (+) mutants were considered to have one additional pair of bases and the (−) mutants to lack one pair, and a double mutant of the type (+)(+) or (−)(−) to have two additional bases or to lack two bases, respectively. (A (+) could denote lack of a base pair and a (−) could denote an extra base pair; the assignment is arbitrary.) Each double mutation would shift the reading frame by two bases. Such a shift does not yield a wildtype phenotype, presumably because a functional protein is not made. In a (+)(−) double mutant arising by recombination, although the shifted reading frame *following* the (+) locus would be incorrect, the correct reading frame would be restored at the (−)

locus. Between the two sites of mutation, the amino acid sequence would not be wildtype, but as long as both mutations were in a region that tolerated amino acid changes, the $(+)(-)$ and $(-)(+)$ phage recombinants would be functional. That is, *if the reading frame is restored before a region that will not tolerate change is reached, a functional protein can be produced.* By this reasoning reversal of a $(+)$ mutation must often result from the production of a $(-)$ mutation at a second nearby site. Clearly, double mutants of the types $(+)(+)$ or $(-)(-)$ would never have the wildtype phenotype, since the genetic code cannot be a two-letter code; if it were a two-letter code, the reading frame would be restored in this combination. What remained was to determine whether the code is a triplet code by construction of triply mutant recombinants. It was found that the triple mutants $(+)(+)(+)$ and $(-)(-)(-)$ have the wildtype phenotype, whereas the mixed triples, $(+)(+)(-)$ and $(+)(-)(-)$, remain mutant. How the combination of three mutants of the same type can yield a wildtype phage is shown in Figure 9-23.

The conclusion that the genetic code consists of triplets rests on the assumption that $(+)$ and $(-)$ mutations are either additions and deletions (or deletions and additions), respectively, of *single* base pairs; if all mutations represented changes in *two* (or *n*) base pairs, the code would be a six- or $3n$-letter code. Several reasonable genetic arguments were given against a six-letter code, and chemical experiments at a later time verified that codons do, in fact, contain only three bases.

The genetic experiments gave additional information about the nature of the code. Since a three-letter code can produce 64 codons and only 20 are needed for the amino acids, then either 44 codons correspond to no amino acid (they are nonsense codons) or several codons correspond to the same amino acid. If most codons were nonsense codons, combination of a $(+)$ and a $(-)$

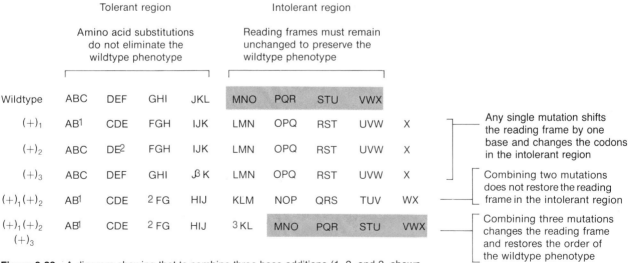

Figure 9-23 A diagram showing that to combine three base additions (1, 2, and 3, shown in red), but not two, restores the reading frame in the intolerant region (shaded red) of the T4 *rII* protein, if the code is a triplet code. An extra amino acid is present in the tolerant region of $(+)_1(+)_2(+)_3$ recombinants, but this does not lead to a mutant phenotype.

mutation would, with fairly high probability, produce a prematurely terminated, and hence, mutant protein. Instead, most (+) and (−) mutations combine to yield a recombinant phage with the wildtype phenotype, therefore indicating that several codons must correspond to the same amino acid.

Elucidation of the Base Sequences of the Codons

Polypeptide synthesis can be carried out in *E. coli* cell extracts obtained by breaking cells open. Various components can be isolated and a functioning protein-synthesizing system can be reconstituted by mixing ribosomes, tRNA and mRNA molecules, and various protein factors. Such a biochemical system is called a cell-free or an *in vitro* system (Latin, in glass, meaning in a test tube). If radioactive amino acids are added to the mixture, radioactive polypeptides are made. Synthesis continues for only a few minutes, because mRNA is gradually degraded by various nucleases in the mixture. The elucidation of the genetic code began with the observation that when the degradation of mRNA was allowed to go to completion and the synthetic polynucleotide polyuridylic acid (poly(U)) was added to the mixture as an mRNA molecule, the polypeptide polyphenylalanine (Phe-Phe-Phe-...) was synthesized. From this simple result and knowledge that the code is a triplet code, it was concluded that UUU is a codon for the amino acid phenylalanine. Variations on this basic experiment made it possible to identify other codons. When a long sequence of guanines was added at the terminus of the poly(U), the polyphenylalanine was terminated by a sequence of glycines, indicating that GGG is a glycine codon (Figure 9-24). Some leucine or tryptophan was also present in the glycine-terminated polyphenylalanine. Incorporation of these amino acids was directed by the codons UUG and UGG at the transition point between U and G. When a single guanine was added to the terminus of a poly(U) chain, the polyphenylalanine was terminated by leucine. Thus, UUG is a leucine codon, and UGG is a codon for tryptophan. Similar experiments were carried out with poly(A), which yielded polylysine, and poly(C), which produced polyproline.

Other experiments made use of a series of synthetic RNA polymers consisting of two alternating bases. For example, the polymer ACACAC is read in all reading frames as ACA CAC ACA CAC ACA CAC, so the protein made will consist of two alternating amino acids. The polypeptide Thr-His-Thr-His-Thr-His was found. From other experiments it was known that a threonine codon exists that contains two A's and one C and that a histidine

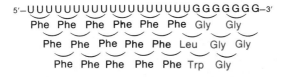

Figure 9-24 Polypeptide synthesis using UUUU . . . UUGGGGGGGG as an mRNA in three different reading frames, showing the origin of incorporation of glycine, leucine, and tryptophan.

codon exists with two C's and one A. Thus, the codon assignments must be ACA for threonine and CAC for histidine. Note that the formation of an alternating polyamino acid from an alternating polynucleotide proves that codons must contain an odd number of bases, eliminating the possibility that a codon might contain six bases, which was discussed in the preceding section. If the codon contained six bases, the alternating polynucleotide would yield either polythreonine or polyhistidine, but not both.

Polymers in which a triplet is repeated were also tested: for example, the polymer UAC UAC UAC UAC. As expected, three polyamino acids—polytyrosine, polythreonine, and polyleucine—were made, the particular one depending on the first base used to establish the reading frame. Production of all three polyamino acids indicates that in the *in vitro* synthetic system used for codon analysis, translation can begin at any site in the synthetic RNA molecule.

Stop signals are necessary for terminating synthesis of each protein. Evidence that such signals exist came from several genetic experiments, each of which demonstrated a mutational change from a codon corresponding to an amino acid to a codon that caused polypeptide chain termination. The base sequence of one of the stop codons was identified by using the polymer poly(GUAA) as mRNA. This polymer can be read in any of four ways:

GUA AGU AAG UAA GUA AGU AGG . . .

G UAA GUA AGU AAG UAA GUA AGU . . .

GU AAG UAA GUA AGU AAG UAA GUA . . .

AGU AAG UAA GUA AGU AAG UAA . . .

Only two peptides are made by this polymer—the tripeptide NH$_2$-Val-Ser-Lys-COOH and the dipeptide NH$_2$-Ser-Lys-COOH. From the known codon assignments—namely, Val = GUA, Ser = AGU, and Lys = AAG—these polypeptides can be lined up in four reading frames:

1   GUA AGU AAG UAA GUA AGU AAG . . .
    Val Ser Lys   ?

2   G UAA GUA AGU AAG UAA GUA AGU . . .
     ?   Val Ser Lys   ?

3   GU AAG UAA GUA AGU AAG UAA GUA . . .
     Lys   ?

4   AGU AAG UAA GUA AGU AAG UAA . . .
    Ser Lys   ?

Note that synthesis stops in each case just before a UAA codon. Thus, UAA must be a stop codon.

Similar experiments with other polynucleotides indicated that there are three stop codons:

UAA   UAG   UGA

It was possible to confirm all codon assignments resulting from the experiments described so far to elucidate the remaining codons through studies in which trinucleotides directed the association of amino acids with ribosomes. In these studies each of the 64 possible trinucleotides was used individually in place of polymers. Whereas a trinucleotide alone cannot direct synthesis of a polypeptide, it can cause binding of a radioactive amino acid to a ribosome (via an amino-acid-specific tRNA intermediate, as is described in a later section). It was found that, except for the stop codons, each triplet enabled a particular amino acid to bind.

Although in the *in vitro* system just described, polypeptide synthesis can start at any base in the polymer used to direct synthesis, in living cells polypeptide synthesis requires a start codon—that is, a codon that establishes the reading frame and initiates synthesis. The identity of the start codon has been determined by biochemical tests and by direct sequencing of the bases of natural mRNA molecules encoding proteins of known amino acid sequence.

The methionine codon **AUG** is the most commonly used start codon.

In rare instances, which will not be discussed, GUG is a start codon. An important question is how a particular AUG sequence is designated as a start codon, whereas others located elsewhere in a coding sequence act only as internal codons for methionine. Special base sequences in mRNA, which will be described shortly, serve this function in prokaryotes.

A Summary of the Code

The *in vitro* translation experiments, which used components isolated from the bacterium *E. coli*, have been repeated with components obtained from many species of bacteria, yeast, plants, and animals. The same codon assignments can be made for all organisms that have been examined. Thus, the genetic code is considered to be universal. The complete code is shown in Table 9-3. Note that four codons—the three stop codons and the start codon—are signals. The remaining 60 codons, as well as AUG, all correspond to amino acids (that is, 61 in total), and in many cases several codons direct the insertion of the same amino acid into a protein chain; that is, the code is highly **redundant.** (The term "degeneracy" is also used to describe this feature of the code.) The redundancy is not random; with the exception of serine, leucine, and arginine, all codons corresponding to the same amino acid are in the same box of Table 9-3. That is, *synonymous codons usually differ in only the third base.* For example, GGU, GGC, GGA, and GGG all code for glycine. Note also that all amino acids except tryptophan and methionine are

Table 9-3   The "universal" genetic code

First position (5' end)	Second position				Third position (3' end)
	U	C	A	G	
U	Phe	Ser	Tyr	Cys	U
	Phe	Ser	Tyr	Cys	C
	Leu	Ser	Stop	Stop	A
	Leu	Ser	Stop	Trp	G
C	Leu	Pro	His	Arg	U
	Leu	Pro	His	Arg	C
	Leu	Pro	Gln	Arg	A
	Leu	Pro	Gln	Arg	G
A	lle	Thr	Asn	Ser	U
	lle	Thr	Asn	Ser	C
	lle	Thr	Lys	Arg	A
	Met	Thr	Lys	Arg	G
G	Val	Ala	Asp	Gly	U
	Val	Ala	Asp	Gly	C
	Val	Ala	Glu	Gly	A
	Val	Ala	Glu	Gly	G

Note: The boxed codons are used for initiation.

specified by more than one codon. The biochemical basis for redundancy will be discussed shortly when the phenomenon of wobble is described.

In mitochondria several codons differ from those of the universal code. These differences are shown in Table 9-4. Note that the differences are not the same for mitochondria of all species. Deviations from the universal code have also been observed in chloroplasts. The significance of these deviations has been widely discussed by workers interested in evolution.

The codon assignments shown in Table 9-3 are completely consistent

Table 9-4   Differences between the universal code (Table 9-3) and two mitochondrial codes*

Codon	Universal code	Mitochondrial code	
		Mammalian	Yeast
UGA	Stop	*Trp*	*Trp*
AUA	Ile	*Met*	*Met*
CUA	Leu	Leu	*Thr*
AGA, AGG	Arg	*Stop*	Arg

*Differences are denoted by italic type.

with all chemical observations and with the amino acid sequences of wildtype and mutant proteins. In every case in which a mutant protein differs by a single amino acid from the wildtype form, the amino acid substitution can be accounted for by a single base change between the codons corresponding to the two different amino acids. For example, substitution of proline by serine, which is a common mutational change, can be accounted for by the single base changes CCC→UCC, CCU→UCU, CCA→UCA, and CCG→UCG (see also Chapter 10).

Transfer RNA and the Aminoacyl Synthetases

The decoding operation by which the base sequence within an mRNA molecule becomes translated to an amino acid sequence of a protein is accomplished by the tRNA molecules and a set of enzymes, the **aminoacyl tRNA synthetases.**

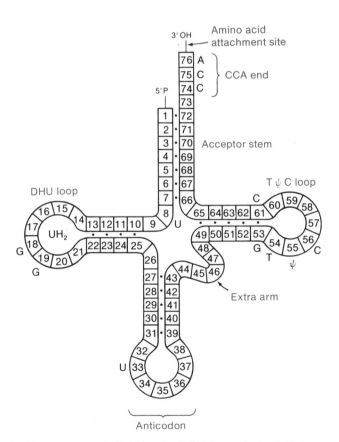

Figure 9-25 The currently accepted "standard" tRNA cloverleaf with its bases numbered. A few bases present in almost all tRNA molecules are indicated. The names of regions found in all tRNA molecules are in red. DHU refers to a base dihydrouracil found in one loop; the Greek letter ψ is a symbol for the unusual base pseudouridine.

The tRNA molecules are small, single-stranded nucleic acids ranging in size from 73 to 93 nucleotides. Like all RNA molecules, they have a 3'-OH terminus, but the opposite end terminates with a 5'-monophosphate rather than a 5'-triphosphate, because tRNA molecules are cut from a large primary transcript. Internal complementary base sequences form short double-stranded regions, causing the molecule to fold into a structure in which open loops are connected to one another by double-stranded stems (Figure 9-25). In two dimensions a tRNA molecule is drawn as a planar cloverleaf. Its three-dimensional structure is more complex, as is shown in Figure 9-26. Panel (a) shows the skeletal model of a yeast tRNA molecule that carries phenylalanine, and panel (b) shows an interpretive drawing.

Three regions of each tRNA molecule are used in the decoding operation. One of these regions is the **anticodon,** a sequence of three bases that can form base pairs with a codon sequence in the mRNA. No normal tRNA molecule has an anticodon complementary to any of the stop codons UAG, UAA, or UGA, which is why these codons are stop signals. A second site is the **amino acid attachment site,** the 3' terminus of the tRNA molecule; the amino acid corresponding to the particular mRNA codon that base-pairs with the tRNA anticodon is covalently linked to this terminus. These bound amino acids are joined together during polypeptide synthesis. A specific aminoacyl tRNA synthetase matches the amino acid with the anticodon; to do so, the

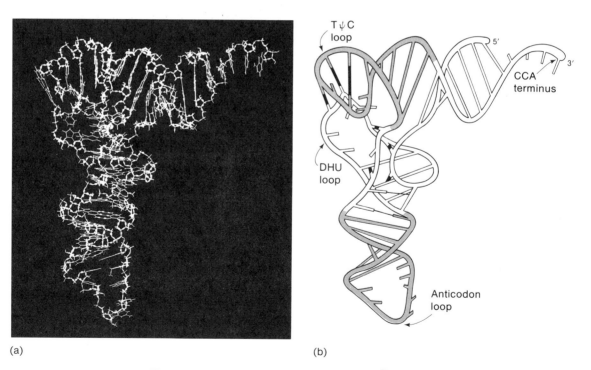

(a)

(b)

Figure 9-26 (a) A skeletal model of yeast tRNA^{Phe}. (b) A schematic diagram of the three-dimensional structure of yeast tRNA^{Phe}. (Courtesy of Sung-Hou Kim.)

enzyme must be able to distinguish one tRNA molecule from another. The necessary distinction is provided by an ill-defined region encompassing many parts of the tRNA molecule and called the **recognition region.**

The different tRNA molecules and synthetases are designated by stating the name of the amino acid that can be linked to a particular tRNA molecule by a specific synthetase; for example, leucyl-tRNA synthetase attaches leucine to tRNA^{Leu}. When an amino acid has become attached to a tRNA molecule, the tRNA is said to be **acylated** or **charged.** An acylated tRNA molecule is designated in several ways. For example, if the amino acid is glycine, the acylated tRNA would be written glycyl-tRNA or Gly-tRNA. The term **uncharged tRNA** refers to a tRNA molecule lacking an amino acid, and **mischarged tRNA** to one acylated with an incorrect amino acid.

At least one, and usually only one, aminoacyl synthetase exists for each amino acid. For a few of the amino acids specified by more than one codon, more than one synthetase exists.

Accurate protein synthesis, the placement of the "correct" amino acid at the appropriate position in a polypeptide chain, requires (1) attachment of the correct amino acid to a tRNA molecule by the synthetase and (2) fidelity in codon-anticodon binding. An important experiment showed that codon-anticodon binding is based only on base-pair recognition and that the identity of the amino acid attached to the tRNA molecule does not influence this recognition. In this experiment tRNA^{Cys} was charged with radioactive cysteine. The cysteine was then chemically converted to radioactive alanine, yielding alanyl-tRNA^{Cys}. This mischarged tRNA molecule was then used in an *in vitro* system containing hemoglobin mRNA and capable of synthesizing hemoglobin, the complete amino acid sequence of which was known. Hemoglobin containing radioactive alanine was made, but the radioactive alanine was present at the sites normally occupied by cysteine rather than at normal alanine positions. This experiment confirmed the hypothesis that tRNA is an adaptor molecule, that the amino acid recognition region and the anticodon are distinct regions, and that the anticodon determines the position of insertion of whatever amino acid is linked to tRNA.

Redundancy and the Wobble Hypothesis

Several features of the genetic code and of the decoding system suggest that something is missing in the explanation of codon-anticodon binding. First, the code is highly redundant. Second, the identity of the third base of a codon appears to be unimportant; that is, XYU, XYA, XYG, and XYC, where XY denotes any sequence of first and second bases, usually correspond to the same amino acid. (Codons, like all RNA sequences, are by convention written with the 5' end at the left; thus, the first base in a codon is at the 5' end and the third base is at the 3' end.) Third, the number of distinct tRNA molecules that have been isolated from a single organism is less than the number of codons; since all codons are used, the anticodons of some tRNA molecules

‹

Table 9-5 Allowed pairings according to the wobble hypothesis

Third position codon base	First position anticodon base
A	U,I
G	C,U
U	G,I
C	G,I

must be able to pair with more than one codon. Experiments with several purified tRNA molecules showed this to be the case.

These observations have been explained by the **wobble hypothesis,** which also provides insight into the pattern of redundancy of the code. Wobble refers to a less stringent requirement for base pairing at the third position of the codon than at the first two positions. That is, the first two bases must form pairs of the usual type (A with U or G with C), but the third base pair can be of a different type (for example, G with U). This observation was derived from the discovery that the anticodon of yeast tRNA^{Ala} contains the base **inosine, I,** in the position that pairs with the third base of the codon (the first position of the anticodon). Later analyses of other tRNA molecules showed that inosine was common in this position, though not always present. Inosine can form hydrogen bonds with A, U, and C. In the wobble hypothesis all pairs of bases that can form hydrogen bonds are considered to be possible in the third position of the codon, except purine-purine base pairs, which would cause excessive distortion in the region of the pairing. These possible base pairs are listed in Table 9-5. The wobble hypothesis, later confirmed by direct sequencing of many tRNA molecules, explains the pattern of redundancy in the code in that certain anticodons (for example, those containing U, I, and G in the first position of the anticodon) can pair with several codons.

The Arrangement of Codons in a Typical Prokaryotic mRNA Molecule

Most prokaryotic mRNA molecules are polycistronic; that is, they contain sequences specifying the synthesis of several proteins. Thus, a polycistronic mRNA molecule must possess a series of start and stop codons for use during translation. If an mRNA molecule encodes three proteins, the minimal requirement would be the sequence

AUG (start),protein 1,stop–AUG,protein 2,stop–AUG,protein 3,stop

in which the stop codons might be UAA, UAG, or UGA. Actually, such an mRNA molecule is probably never so simple in that the leader sequence preceding the first start signal may be several hundred bases long and spacer sequences that are 5–20 bases long are usually present between one stop codon and the next start codon.

9.6 Overlapping Genes

When discussing coding and signal recognition earlier in the chapter, an implicit assumption has been that the mRNA molecule is scanned for start signals to establish the reading frame and that reading then proceeds in a single direction within the reading frame. The idea that several reading frames might exist in a single segment was not considered for many years, primarily as a result of the belief in the absolute generality of the one gene-one polypeptide concept (Section 9.2); a mutation in a gene that overlapped another gene would often produce a mutation in the second gene. The notion of overlapping reading frames was rejected on the grounds that severe constraints would be placed on the amino acid sequences of two proteins translated from the same portion of mRNA. However, because the code is highly redundant, the constraints are actually not so rigid.

If multiple reading frames were used, a single DNA segment would be utilized with maximal efficiency. However, a disadvantage is that evolution could be slowed, because single-base-change mutations would be deleterious more often than if there were a unique reading frame. Nonetheless, some organisms—namely, small viruses and the smallest phages—have evolved having overlapping reading frames.

The *E. coli* phage φX174 contains a single strand of DNA consisting of 5386 nucleotides of known base sequence. If a single reading frame were used, at most 1795 amino acids could be encoded in the sequence and with an average protein size of about 400 amino acids, only 4–5 proteins could be made. However, φX174 makes eleven proteins containing a total of more than 2300 amino acids. This paradox was resolved when it was shown that translation occurs in several reading frames from three mRNA molecules (Figure 9-27). For example, the sequence for protein B is contained totally in the

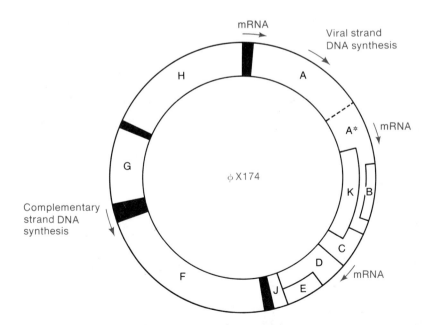

Figure 9-27 The map of *E. coli* phage φX174 showing the start and stop points for mRNA synthesis and the boundaries of the individual protein products. The solid regions are spacers.

sequence for protein A but translated in a different reading frame. Similarly, the protein-E sequence is totally within the sequence for protein D. Protein K is initiated near the end of gene *A*, includes the base sequence of B, and terminates in gene *C;* synthesis is not in phase with either gene *A* or gene *C*. Of note is protein A′ (also called A*), which is formed by reinitiation within gene *A* and in the same reading frame, so that it terminates at the stop codon of gene *A*. Thus, the amino acid sequence of A′ is identical to a segment of protein A. In total, five different proteins obtain some or all of their primary structure from shared base sequences in ϕX174. This phenomenon, known as **overlapping genes,** has been observed in the related phage G4 and in the small animal virus SV40. It has not yet been seen in any eukaryotic organism.

It should be realized that the single structural feature responsible for gene overlap is the location of each AUG initiation sequence, because these sequences establish the reading frame.

9.7 Polypeptide Synthesis

In the preceding sections how the information in an mRNA molecule is converted to an amino acid sequence having specific start and stop points has been discussed. In this section the chemical problem of attachment of the amino acids to one another is examined.

Polypeptide synthesis can be divided into three stages—(1) **initiation,** (2) **elongation,** and (3) **termination.** The main features of the initiation step are the binding of mRNA to the ribosome, selection of the initiation codon, and the binding of acylated tRNA bearing the first amino acid. In the elongation stage there are two processes: joining together two amino acids by peptide bond formation, and moving the mRNA and the ribosome with respect to one another so that the codons can be translated successively. In the termination stage the completed protein is dissociated from the synthetic machinery and the ribosomes are released to begin another cycle of synthesis.

We begin our discussion with a description of the structure of the ribosome.

Ribosomes

A ribosome is a multicomponent particle that contains several enzymes needed for protein synthesis and that brings together a single mRNA molecule and charged tRNA molecules in the proper position and orientation so that the base sequence of the mRNA molecule is translated into an amino acid sequence (Figure 9-28(a)). The properties of the *E. coli* ribosome are best understood, and it serves as a useful model for discussion of all ribosomes.

All ribosomes contain two subunits (Figure 9-28(b)). For historical reasons, the intact ribosome and the subunits have been given numbers that describe how fast they sediment when centrifuged. For *E. coli* (and for all prokaryotes) the intact particle is called a **70S ribosome** (S is a measure of the

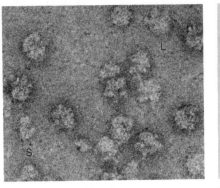
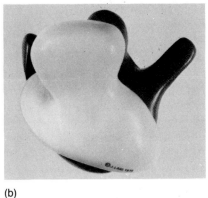

(a)   (b)

Figure 9-28 Ribosomes. (a) An electron micrograph of 70S ribosomes from *E. coli*. Some of the ribosomes are oriented as in the model shown in panel (b). A few ribosomal subunits that also lie in the field are identified by the letters S and L, which stand for small and large. (b) A three-dimensional model of the *E. coli* 70S ribosome. The 30S particle is white and the 50S particle is red. (Courtesy of James Lake.)

sedimentation rate) and the subunits, which are unequal in size and composition, are termed **30S and 50S.** A 70S ribosome consists of one 30S subunit and one 50S subunit.

Both the 30S and the 50S particles can be dissociated into RNA (called **rRNA** for *r*ibosomal RNA) and protein molecules under appropriate conditions (Figure 9-29). Each 30S subunit contains one 16S rRNA molecule and 21 different proteins; a 50S subunit contains two RNA molecules—one 5S rRNA molecule and one 23S rRNA molecule—and 32 different proteins. In

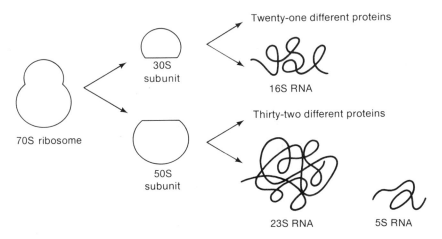

Figure 9-29 Dissociation of a prokaryotic ribosome. The configuration in which the two circles overlap will be used throughout this chapter, for the sake of simplicity; the correct configuration is that shown in Figure 9-28(b).

each particle usually only one copy of each protein molecule is present, though a few are duplicated or modified. Like tRNA molecules, rRNA molecules are cut from large primary transcripts. In many organisms some tRNA molecules are cut also from the large transcripts containing rRNA.

The basic features of eukaryotic ribosomes are similar to those of bacterial ribosomes, but all eukaryotic ribosomes are somewhat larger than those of prokaryotes. They contain a greater number of proteins (about 80) and an additional RNA molecule (four in all). The biological significance of the differences between prokaryotic and eukaryotic ribosomes is unknown. A typical eukaryotic ribosome—an **80S ribosome**—consists of two subunits, **40S** and **60S**. These sizes may vary by as much as ± 10 percent from one organism to the next, in contrast with bacterial ribosomes, which have sizes that are nearly the same for all bacterial species examined. The best-studied eukaryotic ribosome is that of the rat liver. The 40S and 60S subunits can also be dissociated. A 40S subunit, which is analogous to the 30S subunit of prokaryotes, consists of one 18S rRNA molecule and about 30 proteins, and the 60S subunit contains three rRNA molecules—one 5S, one 5.8S, and one 28S rRNA molecule—and about 50 proteins. The 5.8S, 18S, and 28S rRNA molecules of eukaryotes correspond functionally to the 5S, 16S, and 23S molecules of bacterial ribosomes. The bacterial counterpart to the eukaryotic 5S rRNA is very likely present as part of the 23S rRNA sequence.

Stages of Polypeptide Synthesis in Prokaryotes

Polypeptide synthesis in prokaryotes and eukaryotes follows the same overall mechanism, though there are differences in detail, the most important being the mechanism of initiation. The prokaryotic system is best understood and will serve as a model for our discussion.

An important feature of initiation of polypeptide synthesis is the use of a specific initiating tRNA molecule. In prokaryotes this tRNA molecule is acylated with the modified amino acid N-formylmethionine (fMet); the tRNA is often designated tRNA^{fMet} (Figure 9-30). Both tRNA^{fMet} and tRNA^{Met} recognize the codon AUG, but only tRNA^{fMet} is used for initiation. The tRNA^{fMet} molecule is first acylated with *methionine* and an enzyme (found only in prokaryotes) adds a formyl group to the amino group of the methionine. In eukaryotes the initiating tRNA molecule is charged with methionine also, but formylation does not occur. The use of these initiator tRNA molecules means that *while being synthesized*, all prokaryotic proteins have N-formylmethionine at the amino terminus and all eukaryotic proteins have methionine at the amino terminus. However, these amino acids are frequently altered or removed later (this is called **processing**), and all amino acids have been observed at the amino termini of completed protein molecules isolated from cells.

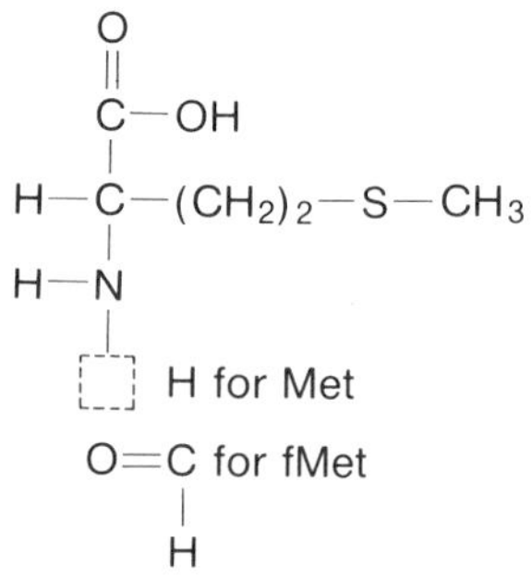

Polypeptide synthesis in bacteria begins by the association of one 30S subunit (not the entire 70S ribosome), an mRNA molecule, fMet-tRNA, three proteins known as **initiation factors,** and guanosine 5'-triphosphate

Figure 9-30 Comparison of the chemical structures of methionine and N-formylmethionine.

(GTP). These molecules constitute the **30S preinitiation complex** (Figure 9-31). Since polypeptide synthesis begins at an AUG start codon and AUG codons are found within coding sequences (that is, methionine occurs within a polypeptide chain), some signal must be present in the base sequence of the mRNA molecule to identify a particular AUG codon as a start signal. The means of selecting the correct AUG sequence differs in prokaryotes and eukaryotes. In prokaryotic mRNA molecules a particular base sequence (called the **ribosome binding site** or sometimes the **Shine-Dalgarno sequence**) near the AUG codon used for initiation forms base pairs with a complementary sequence in the 16S rRNA molecule of the ribosome. In eukaryotic mRNA molecules the 5' terminus binds to the ribosome, after which the mRNA molecule slides along the ribosome until the AUG codon *nearest the 5' terminus* is in contact with the ribosome; the consequence of this mechanism for initiation in eukaryotes will be explained at the end of this section.

Following formation of the 30S preinitiation complex, a 50S subunit joins with this complex to form a **70S initiation complex** (Figure 9-31).

The 50S subunit contains two tRNA binding sites. These sites are called the **A (aminoacyl) site** and the **P (peptidyl) site.** When joined with the 30S preinitiation complex, the position of the 50S subunit in the 70S initiation complex is such that the fMet-tRNA^{fMet}, which was previously bound to the 30S preinitiation complex, occupies the P site of the 50S subunit. Placement of fMet-tRNA^{fMet} in the P site fixes the position of the fMet-tRNA anticodon so that it pairs with the AUG initiator codon in the mRNA. Thus, *the reading frame is unambiguously defined upon completion of the 70S initiation complex.*

Once the P site is filled, the A site of the 70S initiation complex becomes available to any tRNA molecule whose anticodon can pair with the codon

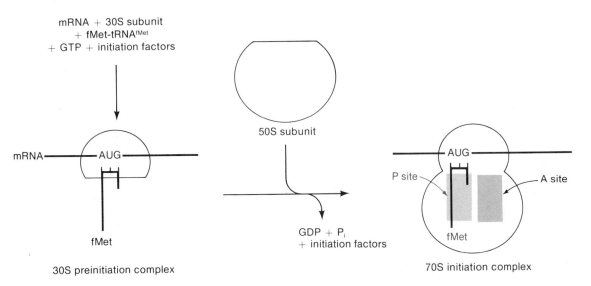

Figure 9-31 Early steps in protein synthesis: formation of the 30S preinitiation complex and of the 70S initiation complex.

adjacent to the initiation codon. After occupation of the A site a peptide bond between N-formylmethionine and the adjacent amino acid is formed by an enzyme complex called **peptidyl transferase.** As the bond is formed, the N-formyl methionine is cleaved from the fMet-tRNA in the P site.

After the peptide bond forms, an uncharged tRNA molecule occupies the P site and a dipeptidyl-tRNA occupies the A site. At this point three movements occur: (1) the tRNA^{fMet} in the P site, now no longer linked to an amino acid, leaves this site, (2) the peptidyl-tRNA moves from the A site to the P site, and (3) the mRNA moves a distance of three bases in order to position the next codon at the A site (Figure 9-32). This step requires the presence of an elongation protein **EF-G** and GTP, and it is likely that mRNA movement is a consequence of the tRNA motion. After mRNA movement has occurred, the A site is again available to accept a charged tRNA molecule having a correct anticodon.

When a chain termination codon (either UAA, UAG, or UGA) is reached, no acylated tRNA exists that can fill the A site, so chain elongation stops. However, the polypeptide chain is still attached to the tRNA occupying the P site. Release of the protein is accomplished by proteins called **release factors.** In the presence of release factors peptidyl transferase separates the polypeptide from the tRNA; then, the polypeptide chain, which has been held on the ribosome solely by the interaction with the tRNA in the P site, is

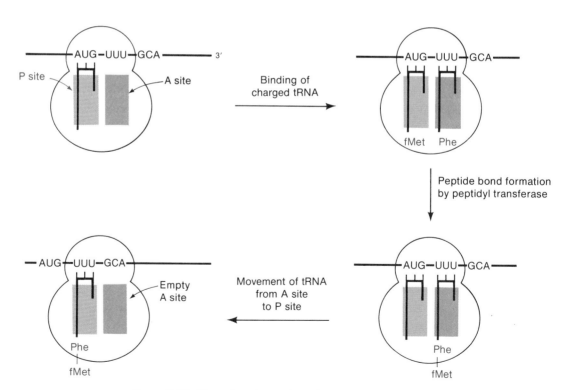

Figure 9-32 Elongation phase of protein synthesis: the binding of charged tRNA, peptide bond formation, and movement of mRNA with respect to the ribosome.

released from the ribosome. The 70S ribosome then dissociates into its 30S and 50S subunits, completing the cycle.

If the mRNA molecule is polycistronic and the AUG codon initiating the second polypeptide is not too far from the stop codon of the first, the 70S ribosome will not always dissociate but will re-form an initiation complex with the second AUG codon. The probability of such an event decreases with increasing separation of the stop codon and the next AUG codon. In some genetic systems the separation is sufficiently great that more protein molecules are always translated from the first gene than from subsequent genes, and this is a mechanism for maintaining particular ratios of gene products (Chapter 11). Mutations sometimes arise that convert a sense (amino-acid specifying) codon to a stop codon—for example, the mRNA molecule has a UAG codon at the site of a UAC codon. Such mutations, which will be discussed further in Chapter 10, cause premature termination of a polypeptide. If the mutation is sufficiently far upstream from the normal stop codon of the gene containing the mutation, the distance to the next AUG codon may be so great that separation of the translating 70S ribosome and the mRNA molecule is almost inevitable. In this case, genes downstream from the mutation will rarely be translated. Such mutations are the most common type of polar mutation (see also Section 9.3). In a *trans*-heterozygote these polar mutations are usually noncomplementing with mutations in the downstream genes. For example, if gene *B* is downstream from gene *A* in a polycistronic mRNA molecule and gene *A* contains a polar mutation, little or no gene-*B* product will be translated from the polycistronic mRNA molecule. Complementation in a *trans*-heterozygote in which the second chromosome is mutant in gene *B* (and has an active *A* allele) cannot occur, because no gene-*B* product will be made from either chromosome.

In eukaryotes reinitiation of polypeptide synthesis following an encounter of a ribosome with a stop codon does *not* occur. Also, as pointed out earlier in this section, polypeptide synthesis in eukaryotes is initiated when a ribosome binds to the 5′ terminus of an mRNA molecule and slides along to the first AUG codon. There is no mechanism for initiating polypeptide synthesis at any AUG other than the first one encountered; *eukaryotic mRNA is always monocistronic* (Figure 9-33). However, a primary transcript can contain coding

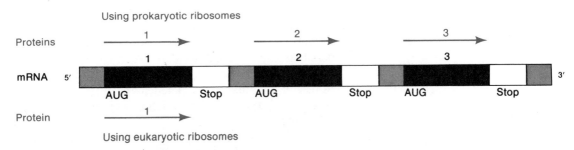

Figure 9-33 Different products are translated from a tricistronic mRNA molecule by the ribosomes of prokaryotes and eukaryotes. The prokaryotic ribosome translates all of the genes, but the eu- karyotic ribosome translates only the gene nearest the 5′ terminus of the mRNA. Translated sequences are shown in black, stop codons are white, and the leader and spacers are shaded.

sequences for more than one polypeptide chain and in fact this is a frequent arrangement in animal viruses. In these cases, differential splicing generates several different mRNA molecules from one transcript (Chapter 11). For example, if the AUG nearest the 5′ terminus is excised, the second AUG will become available. Another mechanism for producing several proteins from a single transcript is protein processing. In this case, a single giant polypeptide chain, called a **polyprotein,** is made and then cleaved into several component polypeptide chains, each constituting a distinct protein (Chapter 11).

9.8 Complex Translation Units

In most prokaryotes and eukaryotes the unit of translation is almost never simply one ribosome traversing an mRNA molecule, but is a more complex structure, of which there are several forms. Two examples are given in this section.

After about 25 amino acids have been joined together in a polypeptide chain, an AUG initiation codon will be completely free of the ribosome and a second initiation complex can form. The overall configuration is that of two 70S ribosomes moving along the mRNA at the same speed. When the second ribosome has moved along a distance similar to that traversed by the first, a third ribosome can attach to the initiation site. The process of movement and reinitiation continues until the mRNA is covered with ribosomes at a density of about one ribosome per 80 nucleotides. This large translation unit is called a **polyribosome** or a **polysome,** and this is the usual form of the translation unit. An electron micrograph of a polysome is shown in Figure 9-34.

The use of polysomes is particularly advantageous to a cell in that the overall rate of protein synthesis is increased compared to the rate that would occur without polysomes.

An mRNA molecule being synthesized has a free 5′ terminus. Since translation occurs in the 5′ → 3′ direction (Figure 9-20), the mRNA is synthesized in a direction appropriate for immediate translation. That is, the ribosome-binding site (in prokaryotes) and the 5′ terminus (in eukaryotes) are transcribed first, followed in order by the initiating AUG codon, the region encoding the amino acid sequence, and finally the stop codon. Thus, in prokaryotes, in which no nuclear membrane separates the DNA and the ribosome, the 70S initiation complex can re-form before the mRNA is released from the DNA. This process is called **coupled transcription-translation.** Figure 9-35 shows an electron micrograph of a DNA molecule with a number of attached mRNA molecules, each associated with ribosomes (panel (a)), and an interpretation (panel (b)). Transcription of DNA begins in the upper left part of the micrograph. The lengths of the polysomes increase with distance from the transcription initiation site, because the mRNA is farther from that site and hence longer.

Coupled transcription and translation does not occur in eukaryotes, because the mRNA is synthesized and processed in the nucleus and later transported through the nuclear membrane to the cytoplasm where the ribosomes

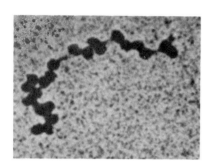

Figure 9-34 Electron micrograph of *E. coli* polysomes. (Courtesy of Barbara Hamkalo.)

are located. In prokaryotes, the coupling of transcription and translation is the rule, though a few phage mRNA molecules are first processed after release from the DNA and then translated.

9.9 Intragenic and Negative Complementation

In Chapter 3 complementation between two genes was described. Two genes *A* and *B* complement if a *trans*-heterozygote (*A b/a B*), in which the small letters represent mutant alleles, has a wildtype phenotype. This type of complementation is ***inter*genic.** For some genes complementation between individual mutations within a single gene also occurs; it is called ***intra*genic complementation** and is quite a different phenomenon from intergenic complementation.

The active forms of certain enzymes are aggregates of two or more polypeptide chains (Section 9.2). For some enzymes the subunits are products of one gene (the subunits are *identical*), and for others the products of different genes. When an enzyme is composed of identical subunits, intragenic complementation may occur. The basis for the phenomenon is shown in Figure

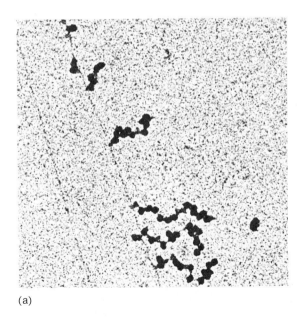

(a)

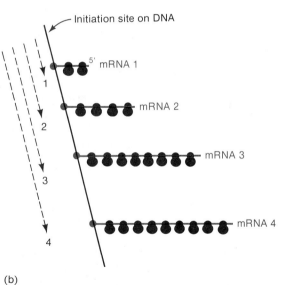

(b)

Figure 9-35 Visualization of transcription and translation. (a) Transcription of a section of the DNA of *E. coli* and translation of the nascent mRNA. Only part of the chromosome is being transcribed. The dark spots are ribosomes, which coat the mRNA. (From O. L. Miller, B. A. Hamkalo, and C. A. Thomas. 1977. *Science,* 169, 392. Copyright 1977 by the American Association for the Advancement of Science.) (b) An interpretation of the electron micrograph of panel (a). The mRNA is red and is coated with black ribosomes. The large red spots are the RNA polymerase molecules; they are actually too small to be seen in the photo. The dashed arrows show the distances of each RNA polymerase from the transcription-initiation site. Arrows 1, 2, and 3 have the same length as mRNA 1, 2, and 3; mRNA 4 is shorter than arrow 4, presumably because its 5′ end has been partially digested by an RNase.

9-36. Individuals homozygous for a mutation in the gene produce only polypeptides with the same defect, and these assemble into dimers (or higher multimers) that are not enzymatically active. For most combinations of mutant alleles of a particular gene a *trans*-heterozygote will produce only an inactive enzyme. However, for certain pairs of mutant alleles defective polypeptide chains are able to form an active protein. This occurs when the defects are located in such a way that the mutant polypeptides aggregate and form active complexes. The figure shows one example in which each mutation prevents subunit interaction but compensates for one another and allows the subunits to join and an active site to form; other types of molecular changes are also possible. The phenotypes produced by such *trans*-heterozygotes may approach the wildtype phenotype, but the total activity of the intracellular enzymes rarely exceeds half of the normal amount. The interpretation of intragenic complementation in terms of interaction between mutationally different polypeptide chains that in some way changes the three-dimensional structure of the aggregate is supported by two kinds of evidence. Active forms of the enzymes have been isolated from *trans*-heterozygotes and found to

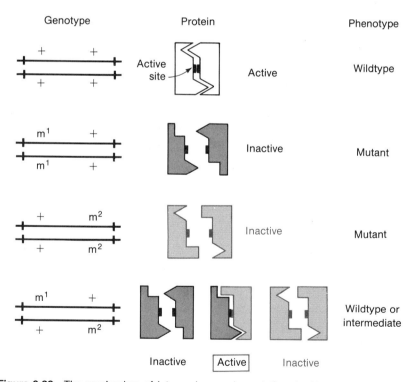

Figure 9-36 The mechanism of intragenic complementation. In this example the active enzyme consists of two identical subunits. The active site is formed by the proximity of two solid boxes. In the mutants either dimerization does not occur or dimers without active sites form. In the *trans*-heterozygote the two types of mutant polypeptides interact at random; only hybrid dimers (one red and one gray subunit) have a functional active site.

consist of two differently defective polypeptides, and inactive polypeptides isolated from two strains homozygous for different mutant alleles have been shown to assemble *in vitro* into an active enzyme.

Proteins consisting of many identical subunits occasionally exhibit **negative complementation** in *trans*-heterozygotes. That is, if gene A encodes the subunit, some mutations (a) produce the mutant phenotype in a *trans*-heterozygote (that is, the mutation is dominant or partially dominant). This phenomenon can occur if a single defective subunit can bind to wildtype subunits and either interfere with normal subunit assembly or become a subunit of the multisubunit protein and thereby eliminate the activity of the protein. Negative complementation is not a feature of all multisubunit protein systems nor of all mutant alleles, for in some cases either the defective subunit fails to interact with the wildtype subunit or the protein can tolerate a single defective subunit. Negative complementation rarely produces a strictly mutant phenotype because of the random nature of subunit aggregation—that is, statistically, some functional proteins will always form.

Problems

1. What are the substrates, the name of the catalyzing enzyme, and the template for the synthesis of RNA?

2. What groups are located at the termini of a primary transcript, a prokaryotic mRNA molecule, and a eukaryotic RNA molecule?

3. At one time the possibility was considered that the genetic code might be an overlapping code of the following type: the codons in base sequence CATCATCAT would be CAT ATC TCA CAT. How was this hypothesis affected by the observation that mutant proteins usually differ from a wild-type protein by a single amino acid?

4. A portion of the coding strand of a DNA molecule has the sequence TTTTACGGGAATTAGAGTCGCAGGATC. What is the amino acid sequence of the polypeptide encoded in this region? Designate the amino end. Assume that the entire strand is transcribed and that a start codon *is* needed for initiation of polypeptide synthesis.

5. *E. coli* DNA has a molecular weight of 2.7×10^9. If all of this nucleotide sequence encoded proteins, how many proteins of average size would *E. coli* make? The molecular weights of an average amino acid, an average nucleotide pair, and an average protein are 110, 660, and 50,000, respectively.

6. The synthetic polymer poly(A) is used as an mRNA molecule in an *in vitro* protein-synthesizing system that does not need a special start codon. Polylysine is synthesized. A single guanine nucleotide is added to one end of the poly(A). Then, polylysine with a glutamic acid at the amino terminus is made. At what end of the poly(A) has the G been added—the 3' end or the 5' end?

7. What polypeptide products are made when the alternating polymer GUGU is used in an *in vitro* protein-synthesizing system that does not need a start codon?

8. How would poly(AUAG) be used to show that UAG is a stop codon?

9. Two genes A and B are known from mapping experiments to be very near each other. A genetic deletion mutation is isolated that eliminates the activity of both genes A and B. In an attempt to study the nature of the deletion, all cellular proteins are isolated. Neither the A nor the B protein can be found, but a new protein is isolated that is larger than either of these proteins. The sequences of 30 amino acids at the amino and carboxyl termini of this are determined and found to be those at the amino terminus of the A protein and the carboxyl terminus of the B protein, respectively.

   (a) Are the two genes normally transcribed as part of a polycistronic mRNA molecule or as separate transcription units?

   (b) What can be said about the number of bases deleted? (A precise number cannot be stated, but a property of the number can.)

10. Two genes A and B are encoded in the same mRNA molecule. An added base pair in A eliminates the activity of gene B (no B protein is made). A second mutation in A, which is shown *not* to be either a base-pair addition or a base-pair deletion, restores activity of the B protein. What change was probably introduced by the second mutation?

11. The central region of a polypeptide containing a single lysine has the amino acid sequence Phe-Leu-Tyr-Ala-Lys-Gly-Glu. A mutation is found that causes the polypeptide to terminate with Phe-Leu-Tyr-Ala. Which of the lysine codons was used in synthesis of the protein?

12. The genes encoding the six enzymes required for synthesis of a substance Q in *E. coli* are transcribed as a polycistronic mRNA molecule. The genes have the order *kyu1*, *kyu2*, *kyu3*, *kyu4*, *kyu5*, and *kyu6*, which also represents the order in which the enzymes act. The product of the *kyu6*-controlled reaction is Q. A variant is isolated that has undergone a perfect reversal of the order of genes *kyu3* and *kyu4* (a genetic inversion), so the gene order is *kyu1 kyu2 kyu4 kyu3 kyu5 kyu6*. Will this variant have a Q^+ (normal) or Q^- (mutant) phenotype?

13. Explain what base changes will cause the following phenotype in yeast (a eukaryote). The amino terminus of a wild-type protein determined by a particular gene has the amino acid sequence Met-Leu-His-Tyr-Met-Gly-Asp-Tyr-Pro. A mutant (#1) is found that lacks activity of the gene, but chemical analysis indicates that the mutants contains a new protein—one having the sequence Met-Gly-Asp-Tyr-Pro at the amino terminus and the wildtype sequence at the carboxyl terminus. A second mutant (#2) is found that has also lost activity, but there is no sign of a mutant protein. A tripeptide not present in wildtype yeast is isolated in mutant #2, but you realize that it is not really necessary to determine its sequence. What changes in the base sequence have led to these mutants? What is the amino acid sequence of the tripeptide?

14. According to the wobble hypothesis the anticodon for the cysteine codons UGU and UGC could be either GCA or ICA. Consider the fact that UGA is a stop codon and deduce the cysteine anticodon.

15. Protein synthesis occurs with fairly high fidelity. That is, it is very rare that a protein is made with incorrect amino acids. What features of the process of polypeptide synthesis are responsible for this fidelity?

16. A chemical compound is found that, when added to growing cells, immediately inhibits the dissociation of a 70S prokaryotic ribosome when polypeptide synthesis is complete. It is found that in the course of time the compound also prevents initiation of protein synthesis. Why?

17. The DNA fragment containing a particular gene is isolated from a eukaryotic organism (using the methods described in Chapter 12). This DNA fragment is mixed with the corresponding mRNA isolated from the organism, denatured, renatured, and observed by electron microscopy. Heteroduplexes of the type shown in the figure below are observed. How many introns does this gene have?

18. In a segment of DNA containing overlapping genes would you expect the two proteins to terminate at the same site or have the same number of amino acids? Explain.

C H A P T E R 10

Mutation and Mutagenesis

In previous chapters numerous examples were presented in which the information contained in the genetic material had been altered in a specific and heritable way by mutation. Mutations constitute the raw material for evolution; they are the basis for the variability in a population on which natural (or artificial) selection acts to preserve those combinations of genes best adapted to a particular environment. A genetic locus that never mutated would be very difficult or impossible to detect; thus, mutations and the mutant phenotypes they cause are indispensable for genetic investigations. For example, recombination analysis depends on the use of mutant alleles as genetic markers (Chapters 3 and 8); mutations that block specific steps in a biochemical pathway make it possible to identify those steps and elucidate the pathway (Chapter 9); mutations in regulatory elements are essential for working out the mechanisms that control the expression of particular genes (Chapter 11); and the behavior of genes in populations is studied by observing the frequency of appearance of various mutant phenotypes (Chapter 14).

Mutations are abrupt, heritable changes in single genes or small regions of a chromosome. Although the term mutation is now generally used in this sense, historically it was defined more broadly to include those alterations in chromosome number and chromosome structure that were discussed in Chapter 6. In this chapter it will be seen that a mutation of an individual gene may be any of several different kinds of changes in the sequence of nucleotides in the DNA and that these changes originate in a variety of ways. Most mutations are in genes with protein products, and examination of these mutations forms the major portion of this chapter. However, it will be seen that changes in untranslated gene products, such as tRNA molecules, and in nontranscribed regions, such as regulatory sites, can also affect the phenotype. Mechanisms that repair damage to DNA that leads to potentially lethal or mutational effects will also be described.

Facing page: Scanning electron micrograph of a mutant *Drosophila melanogaster* in which legs rather than antennae emerge from the head. (Courtesy of F. R. Turner.)

10.1 General Properties of Mutations

Mutations can occur in any cell and can exhibit a wide range of phenotypic effects: minor alterations that are detectable only by biochemical methods and changes in essential processes that are sufficiently drastic to cause the death of the cell or the organism. When genetic phenomena are studied, mutations producing conspicuous and clearly defined effects are usually used, but these are the exception. The effect of a mutation is determined by the type of cell containing the allele, the stage in the life cycle or in the development of the organism in which the process affected by the mutation occurs, and, in diploid organisms, by the dominance or recessiveness of the mutant allele. A recessive mutation transmitted to the progeny of an organism in which it arises will usually not be detected until two heterozygotes happen to mate in a subsequent generation. Dominance is not a factor in the expression of mutations in bacteria and haploid eukaryotes.

Mutations can be classified in a variety of ways. In multicellular organisms an important distinction is based on the type of cell in which the mutation occurs. **Germinal mutations** are those arising in cells that ultimately form gametes; **somatic mutations** occur in all other cells. A somatic mutation yields an organism that is genotypically, and for many dominant mutations phenotypically, a mixture of normal and mutant tissue. Since reproductive cells are not affected, a mutant allele will not be transmitted to the progeny and may not be detected or be recoverable for genetic analysis. For many microorganisms the distinction between somatic and reproductive cells is genetically irrelevant, because frequently an intact organism can be grown from an asexual spore or from a small fragment of the organism. In higher plants somatic mutations often can be propagated by vegetative means such as grafting and the rooting of stem cuttings and have been the source of valuable new varieties such as the 'Delicious' apple and the 'Washington' navel orange.

Some of the most useful mutations for genetic analysis are the **conditional lethal mutations,** which impose changes that are lethal in one set of environmental condition (**restrictive** or **nonpermissive conditions**) but not in another (**permissive conditions**). Many mutations used to elucidate biochemical pathways (Section 9.2) are conditional lethals, with the mutant organisms able to survive only if provided with a compound that they are unable to synthesize. **Temperature-sensitive** mutations are a useful type of conditional lethal: these cause the organism to grow at one temperature but die at another (usually higher) temperature. Temperature-sensitive mutations are frequently used to eliminate particular biochemical reactions when required by an experimenter. With certain temperature-sensitive mutations in *Drosophila* the effect of the restrictive temperature is limited to a specific period during development; these can be used to block development at particular stages and are one of the major tools of developmental genetics.

Mutations are classified by numerous other criteria, such as the kinds of alterations that have occurred in the DNA, the kinds of phenotypic effects produced, and whether the mutational events were **spontaneous** in origin or were **induced** by exposure to a known mutagen (a mutation-causing agent). Such classifications are often useful in discussing various aspects of the mutational process. The properties of spontaneous and induced mutations will be described in Sections 10.3 and 10.4.

10.2 The Biochemical Basis of Mutation

The chemical and physical properties and the consequent biological activity of a protein are determined by its amino acid sequence, which in turn is determined by the specific sequence of nucleotides that makes up the gene encoding the protein (Chapter 9). A change in the amino acid sequence of a protein can (but will not always) alter the biological properties of a protein. The effect that substitution of a single amino acid can have on the structure of a protein is illustrated by the molecular basis of sickle-cell anemia, an inherited defect in hemoglobin structure in humans. Hemoglobin, the oxygen carrier present in red blood cells, consists of four polypeptide subunits—two identical α-globin molecules and two identical β-globin molecules, combined with the iron-containing pigment heme. Different forms of hemoglobin are present at different stages of development (Section 11.8); the major molecule of adult hemoglobin is hemoglobin A. Many variants of hemoglobin A have been identified, some associated with only minor phenotypic effects and others with severe effects. Sickle-cell hemoglobin (hemoglobin S) is one of the latter types.

Individuals homozygous for the hemoglobin S mutant allele develop a serious type of anemia. When deprived of oxygen (as in venous blood), hemoglobin S molecules precipitate into aggregates that deform the cells into an elongate, sicklelike shape. The sickle cells break down more rapidly than do normal red blood cells, producing the anemia; also, because of their shape, the sickle cells block the fine capillary networks in the circulatory system. Secondary effects of sickling are malfunctions of the lungs, heart, and kidneys. When a mutant gene imposes a primary effect (in this case, the change in hemoglobin structure) leading to a series of secondary effects, it is said to be **pleiotropic.** Individuals heterozygous for the hemoglobin S mutant allele carry the sickle-cell trait but are not seriously affected; their red blood cells undergo sickling only under conditions of severe oxygen deprivation.

The defect in the mutant hemoglobin S is in the β chain. The sixth amino acid from the amino (NH_2) end of the chain in normal hemoglobin A is glutamic acid, whereas the amino acid at that position in the β chain of hemoglobin S is valine (Figure 10-1). Substitution of lysine for the same

HbA   NH$_2$-Val-His-Leu-Thr-Pro-$\overline{\text{Glu}}$-Glu-Lys....

HbS   NH$_2$-Val-His-Leu-Thr-Pro-Val-Glu-Lys ....

HbC   NH$_2$-Val-His-Leu-Thr-Pro-$\overset{+}{\text{Lys}}$-Glu-Lys....

Figure 10-1   The amino acid sequences at the amino end of the β polypeptide chains of normal hemoglobin A (HbA), sickle cell hemoglobin (HbS), and a second abnormal hemoglobin (HbC) that causes mild anemia. The complete chain contains 146 amino acids. The amino acids that differentiate the chains are shown in red; the charges of these amino acids are also indicated.

glutamic acid in another variant called hemoglobin C results in only a mild anemia in homozygous individuals.

How an amino acid substitution can change the structure of a protein can be appreciated by considering a hypothetical polypeptide chain whose folding is determined entirely by an attraction between *one* positively charged amino acid and *one* negatively charged amino acid. Substitution of an uncharged amino acid for either of the charged ones would clearly destroy the three-dimensional structure, as would substitution of an amino acid with a different charge. Another hypothetical polypeptide might be stabilized by a disulfide bond between two cysteines, so substitution of any amino acid for one of the cysteines would also be disruptive. In a polypeptide chain whose folding is determined by interaction between many amino acids, the effects of amino acid substitutions are not so simple, and the magnitude of the effect of a substitution usually depends both on the particular substitution and its location in the polypeptide chain. For example, the alteration in hemoglobin, the replacement of a negatively charged glutamic acid by valine, which has no charge, has a more severe effect than replacement by lysine, which is positively charged, in the less affected hemoglobin C (Figure 10-1).

An amino acid substitution does not always create a mutant phenotype. For instance, substitution of one amino acid for a second one with the same charge (for example, lysine for histidine) may in some cases have no effect on either protein structure or phenotype. Whether there is an effect of substituting a like amino acid for another depends on the precise role of the particular amino acid in the structure and function of the protein. For example, any change in the active site of an enzyme will usually decrease enzymatic activity substantially. The following types of amino acid substitutions usually cause phenotypic changes (except as described below): charged to uncharged and the reverse, change of sign of a charge, small side chain to bulky side chain, hydrogen-bonding to non-hydrogen-bonding, and any change to or from proline (which changes the shape of the polypeptide backbone) or to or from cysteine (whose sulfhydryl group may be engaged in a disulfide bond). However, most proteins contain regions that are fairly tolerant of amino acid substitutions, such as those near the amino and carboxyl termini, regions connecting functional domains (Section 9.3) and particular regions such as those in the T4 *rIIB* protein used in the study of the genetic code (Section 9.5).

Every amino acid change results from a change in the base sequence of the DNA. However, the converse is not true: a base change need not result in an amino acid change because of the redundancy in the genetic code. A mutation having no phenotypic effect is called a **silent mutation.**

Mutations in which a single nucleotide pair in the DNA is changed are called **point mutations.** A point mutation may result from the substitution of one nucleotide pair for another (a **base-substitution mutation**) or from the insertion or deletion of a pair of nucleotides. When a base substitution leads to an amino acid substitution, as in hemoglobin S and hemoglobin C, the mutation is said to be a **missense mutation.** If the nucleotide substitution produces a stop codon—for example, UAU (tyrosine) to UAG (stop)

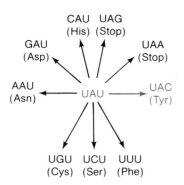

Figure 10-2 The nine codons that can result from a single base change of the tyrosine codon UAU. Tyrosine codons are shown in red. Since two changes yield stop codons, only six amino acid substitutions are possible.

—resulting in the termination of translation and production of a fragment of the protein, the mutation is called a **nonsense** or **chain termination mutation.** A base substitution in a single codon can lead to as many as nine changes (three for each base in the codon), though the number is usually less because of the redundancy of the code (Figure 10-2).

Base-substitution mutations are also classified by the relative purine-pyrimidine orientation of the wildtype and mutant base pair. If one purine is substituted for the other purine and the complementary pyrimidine for the other pyrimidine, the mutation is called a **transition** (for example, the changes $\underline{A}T \rightarrow \underline{G}C$, $T\underline{A} \rightarrow C\underline{G}$, $\underline{G}C \rightarrow \underline{A}T$, and $C\underline{G} \rightarrow T\underline{A}$, in which the purines are underlined). If the substitution is purine for pyrimidine and pyrimidine for purine, the mutation is a **transversion** (for example, $\underline{A}T \rightleftarrows T\underline{A}$, $\underline{A}T \rightleftarrows C\underline{G}$, $\underline{G}C \rightleftarrows C\underline{G}$, $\underline{G}C \rightleftarrows T\underline{A}$, again with the purines underlined).

A mutation in which a nucleotide pair is added or deleted is called a **frameshift mutation.** The effect of such a mutation is to shift the reading frame of all triplets of bases (codons) in the mRNA beyond the point of the mutation. The consequences of a frameshift can be illustrated by the insertion of an adenine in the simple *mRNA* sequence shown below:

Leu Leu Leu Leu
. . . CUG CUG CUG CUG . . .

↓

. . . CUG CAU GCU GCU G . . .
Leu His Ala Ala

An addition or deletion of any number of nucleotides that is not a multiple of three will produce a frameshift (Section 9.5), and unless it occurs very near the carboxyl terminus of a protein will result in synthesis of a nonfunctional protein.

Changes in nucleotide sequence arise spontaneously in several different ways, as will be described in the next section.

10.3 Spontaneous Mutations

Mutations are random events and there is no way of knowing when or in which cell a mutation will occur. However, every gene mutates at a characteristic rate, making it possible to assign probabilities to particular mutational events. Thus, there is a definite probability that a given gene will mutate in a particular cell, and likewise a definite probability that a mutant allele of the gene will occur in a population of a particular size. Different kinds of alterations in the DNA lead to mutations, and since these changes differ substantially in complexity, they occur with quite different probabilities.

Mutations are also random in the sense that their occurrence is not related to any adaptive advantage they may confer on the organism in its environment; in the next section the basis for this conclusion will be presented.

Most phenotypically recognizable mutations in *Drosophila* and other higher organisms, including humans, affect the adaptation of the organism in ways that range from neutral to lethal. However, the idea that mutations are spontaneous random events unrelated to adaptation was not accepted by many bacteriologists until the late 1940s. Prior to that time it was believed that mutations occur in bacterial populations *in response to* particular selective conditions. The basis for this belief was the observation that when antibiotic-sensitive bacteria are spread on a solid growth medium containing the antibiotic, some colonies form that consist of cells having an inherited resistance to the drug. The initial interpretation of this observation (and similar ones) was that these adaptive variations were *induced* by the selective agent itself.

The experiments that resulted in recognizing the spontaneous and nonadaptive nature of mutation in all organisms and that marked the birth

Table 10-1 The number of T1 phage-resistant *E. coli* mutants in small individual cultures and in samples from a large bulk culture.

Small individual cultures		Samples from large culture	
Culture	T1-r colonies per plate	Sample	T1-r colonies per plate
1	1	1	14
2	0	2	15
3	3	3	13
4	0	4	21
5	0	5	15
6	5	6	14
7	0	7	26
8	5	8	16
9	0	9	20
10	6	10	13
11	107		
12	0	Mean	16.7
13	0		
14	0	Variance	15
15	1		
16	0	Variance/Mean	0.9
17	0		
18	64		
19	0		
20	35		
Mean	11.4		
Variance	694		
Variance/Mean	60.8		

Source: Data from S. E. Luria and M. Delbrück. *Genetics* 28 (1943): 491.

of bacterial genetics were reported in 1943 by Salvador Luria and Max Delbrück, who investigated the origin of mutations in *E. coli* that confer resistance to phage T1 (T1-r mutations). In these experiments the number of T1-r mutant cells arising in different cultures of T1-s (T1-sensitive) cells was compared with the number found in repeated samples of the same size taken from a single culture. Their procedure was a statistical test called a **fluctuation test.** The data shown in Table 10-1 were obtained from an experiment in which twenty 0.2-ml cultures and one 10-ml culture, each containing a T1-s bacterial strain at an initial concentration of 10^3 cells/ml, were grown to a concentration of 2.8×10^9 cells/ml (21 generations). Each of the small cultures and ten 0.2-ml samples from the large culture were spread on individual plates containing solid medium covered uniformly with about 10^{10} T1 phage (enough to kill all T1-s cells), and the number of colonies that formed (which would be T1-r) were counted. Each of the plates received the same number of bacterial cells (5.6×10^8), but the number of T1-r colonies formed depended on whether the cells had been grown in the small individual cultures or in the large bulk culture. No T1-r cells were detected in 11 of the 20 small cultures and the numbers in the other nine of these cultures ranged from 1 to 107; in contrast, the ten samples from the large culture each had about the same number. The alternatives expected in such an experiment were the following: (1) if T1-r bacteria arise in response to the phage, there should be about equal numbers in all populations of the same size; (2) if T1-r cells arise by spontaneous mutation at different times in the growth of the cultures in the absence of the phage, the numbers in different cultures should vary greatly. The results of the experiment shown in Table 10-1 and others like it were consistent with the second alternative, which is depicted in Figure 10-3. Panel

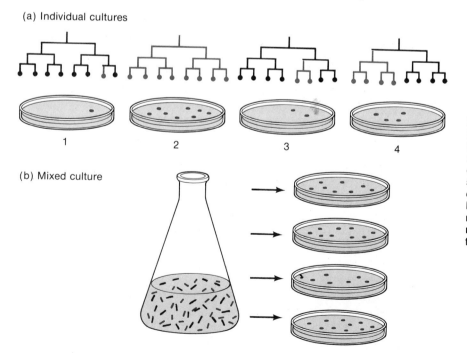

(a) Individual cultures

1   2   3   4

(b) Mixed culture

Figure 10-3 The fluctuation test. (a) Pedigrees showing the source of the mutants (red colonies) found in each of four different samples. In each pedigree, the occurrence of a mutation is indicated by a shift of the path to red. (b) Typical results of sampling from a single mixed culture. Black and red rods indicate wildtype and mutant bacteria, respectively; they are not drawn to scale, being roughly 10^8 times too large compared to the colonies.

(a) shows four separate bacterial cultures and the pedigrees by which the cells could be derived from a single ancestor. A mutation to phage resistance is shown to have occurred during a different generation in each lineage. The two extreme cases are cultures 1 and 2. Culture 1 depicts a mutation occurring just prior to sampling, resulting in only a single mutant colony. Culture 2 depicts a mutation occurring soon after addition of bacteria to the medium, yielding an entirely mutant population. Cultures 3 and 4 are intermediate cases. Thus, separate cultures will contain significantly different numbers of mutant cells, and the statistical variance among the cultures will be large. (The variance is a measure of the spread, or variation, in a set of measurement, as will be discussed in greater detail in Section 16.2.) Panel (b) shows the result of growing the four lineages in panel (a) in a single large culture. With this arrangement, the mutants from each lineage become uniformly dispersed in the medium, so sample-to-sample variation is greatly reduced.

Direct evidence for the spontaneous origin of T1-r mutations in *E. coli* cells not exposed to the phage was obtained in the early 1950s by a procedure called **replica plating** (Figure 10-4). In this procedure a suspension of bacterial cells is spread on a solid medium. After colonies have formed, a piece of

(a) The transfer process

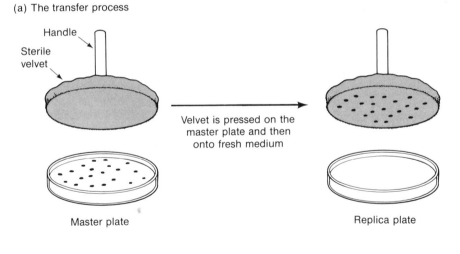

Master plate   Replica plate

Figure 10-4 Replica plating. (a) The transfer process. A velvet-covered disk is pressed onto the surface of a master plate in order to transfer cells from colonies on that plate to a second medium. (b) Detection of mutants. Cells are transferred onto two plates containing either a nonselective medium (on which all form colonies) or a selective medium (for example, one spread with T1 phage). Colonies form on the nonselective plate in the same pattern as on the master plate. Only mutant cells (for example, T1-r) can grow on the selective plate; the colonies that form correspond to certain positions on the master plate. Colonies containing mutant cells are shown in red.

(b) After incubation

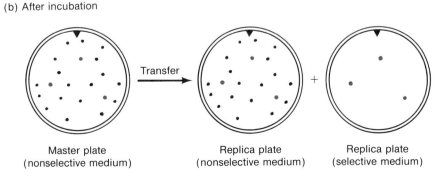

Master plate
(nonselective medium)

Replica plate
(nonselective medium)

Replica plate
(selective medium)

sterile velvet mounted on a solid support is pressed onto the surface of the plate. Some bacteria from each colony stick to the fibers, as shown in panel (a). Then, the velvet is pressed onto the surface of fresh medium, transferring the cells, which give rise to new colonies having positions identical to those on the first plate. Panel (b) shows how the method was used to demonstrate the spontaneous origin of T1-r mutants. Master plates containing 10^7 colonies growing on nonselective medium (lacking T1) were replica-plated onto a series of plates that had been spread with T1 phage. After incubation for a time sufficient for colony formation, a few colonies of phage-resistant bacteria appeared in the same positions on each of the replica plates. This meant that the T1-r cells that formed the colonies must have been transferred from corresponding colonies on the master plate. Since the colonies on the master plate had never been exposed to the phage, the mutations to resistance occurred by chance in cells not exposed to the phage.

The kind of experiment just described illustrates that *selective techniques merely select mutants that preexisted in a population*. This fact is the basis for understanding how natural populations of rodents, insects, and disease-causing bacteria become insensitive to chemical substances used to control them. A familiar example is the high level of resistance to insecticides such as DDT that now exists in many housefly populations, the result of selection for a combination of behavioral, anatomical, and enzymatic mechanisms that enable the insect to avoid or resist the chemical. In the same way, many organisms responsible for human diseases have progressively become resistant to various antibiotics. Similar problems are encountered in plant breeding. For example, the introduction of a new variety of a crop plant resistant to a particular strain of disease-causing fungus results in only temporary protection against the disease. The resistance will inevitably break down because of the occurrence of spontaneous mutations in the fungus that enable it to attack the new plant genotype. Such mutations have a clear selective advantage, and the mutant alleles rapidly become widespread in the fungal population.

Mutation Rates

Spontaneous mutations are usually rare events, and the methods used to estimate the frequency with which they arise require large populations and often special techniques. The **mutation rate** is the probability that a gene undergoes mutation in a single generation or in forming a single gamete. Measurement of mutation rates is important in population genetics, in studies of evolution, and in analyzing the effect of environmental mutagens.

In diploid organisms the easiest mutations to detect are those that cause visible dominant effects. However, among mutant genes that produce recognizable effects recessives greatly outnumber dominants. Since mutations are rare events, yielding heterozygous individuals in which recessive alleles will not be expressed, measuring the rate at which such alleles arise requires the use of special methods that are available in some organisms and for some types of mutants. Two methods that have been used for *Drosophila* are described in this section.

An important innovation in mutation research was the development in the 1920s by Hermann Muller of the **ClB method** for measuring the rate at which mutations with recessive lethal effects occur in the X chromosome of *D. melanogaster* during sexual reproduction. *ClB* refers to a special X chromosome with a large inversion (*C*) that suppresses the recovery of crossover chromosomes in the progeny from a female heterozygous for the chromosome (as described in Section 6.6); a recessive lethal (*l*); and the dominant marker Bar (*B*), which reduces the normal round eye to a narrow bar. The presence of a recessive lethal in the X means that males with the chromosome and females homozygous for it cannot survive. The technique is designed to detect mutations arising in a normal X chromosome, so detection of the mutants requires suppression of crossing over with the *ClB* chromosome.

In the *ClB* procedure females heterozygous for the *ClB* chromosome are mated with males carrying a normal X chromosome (Figure 10-5). From the progeny produced, females with the Bar phenotype are selected and then individually mated with normal males. (The presence of the Bar phenotype

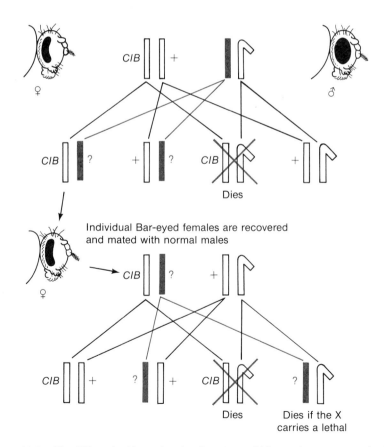

Figure 10-5 The *ClB* method for estimating the rate at which spontaneous recessive lethal mutations arise on the *Drosophila* X chromosome.

indicates that the females are heterozygous for the *ClB* chromosome and a normal X from the male parent.) A ratio of two females to one male is expected among the progeny from such a cross, as among the progeny from the initial cross. Since the recovery of crossover X chromosomes in the progeny is eliminated by the heterozygosity of the F_1 female for the inversion (C), the male offspring will be $+/Y$—because *ClB* males die. The $+/Y$ males among the progeny produced by an individual F_1 female will all carry an X chromosome derived from the X that was present in a single sperm produced by the initial normal male. Occasionally, F_2 progeny include no males, which means that the X chromosome contributed by the original male carried a new recessive lethal. The mutation in such a case would occur in a germ cell before or during gametogenesis in the male, and at a locus other than the one represented by the recessive lethal in the *ClB* chromosome—since the F_1 *ClB* female was viable. The method provides a quantitative estimate of the rate at which mutation to a recessive lethal allele occurs *at any of a large number of loci* on the X chromosome. About 0.15 percent of the X chromosomes are found to acquire such a mutation during spermatogenesis—a rate of 1.5×10^{-3} recessive lethals per X chromosome per gamete. Other versions of the *ClB* chromosome that have been developed are currently used in most quantitative studies of mutation rates in *Drosophila*.

A second method that also takes advantage of the hemizygous condition of the X chromosome in *Drosophila* males is often used to detect and measure the rate at which individual X-linked genes mutate to recessive alleles having a visible effect on the phenotype (Figure 10-6). This procedure is based on the fact that in a mating in which the female has a compound-X chromosome, male offspring will receive an X chromosome only from their father (as explained in Section 2.4). If the father is wildtype and the phenotype determined by an X-linked recessive mutant allele is expressed in one of his sons, the mutation will have occurred in a germ cell of the father. Occasionally, in

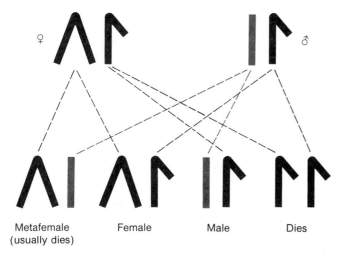

Metafemale (usually dies) Female Male Dies

Figure 10-6 The use of compound-X chromosomes to detect recessive mutations of X-linked genes occurring in germ cells of a wildtype male.

such experiments two or more sons with identical mutant phenotypes are observed. Such clusters of mutations result from a single mutational event in a germ cell of the father several cell generations before meiosis. This is known to be so because the probability that a given gene will mutate is very low, and the probability that the same mutation will occur more than once in the same parent or in the same experiment is exceedingly low.

The average spontaneous mutation rate for *D. melanogaster* genes having visibly detectable effects on the phenotype is about 1×10^{-5} per gene per gamete, and there is about a fiftyfold range in the frequency with which different genes mutate to cause conspicuous visible effects. The mutation rates for individual loci in mice and humans appear to be similar to those in *Drosophila*. For most genes in bacteria the rates per gene per generation are much lower.

Estimation of mutation rates in bacteria is complicated by the fact that a mutation can occur at any time during the growth of a culture, and division of the mutant bacteria usually will result in an increase in their number at a rate the same as that of the increase of the population as a whole. The result, shown by the occurrence of phage-resistant mutants in different cultures (Table 10-1, column 2) is that a culture may contain either many mutants, some mutants, or none. With a mutation rate per generation of μ the probability P_0 of obtaining no mutants in a culture of N cells is $e^{-\mu N}$, in which e is the base of natural logarithms. Thus, $-\mu N = \ln P_0$, or

$$\mu = -(1/N)\ln P_0$$

In the fluctuation test data in Table 10-1, 11 of the 20 small cultures contained no phage-resistant mutants, and the average number N of cells per culture was 5.6×10^8. Thus, the mutation rate could be estimated as

$$\mu = (1/5.6 \times 10^8)\ln(11/20) = 1.1 \times 10^{-9} \text{ per cell per replication}$$

The fluctuation test is one important method for estimating mutation rates in bacteria. Another method will be described in Section 15.1.

The Origin of Spontaneous Mutations

Mutation requires modification of DNA, and several mechanisms for such modification are known. The three mechanisms that are most important for spontaneous mutagenesis are (1) errors occurring during replication, (2) spontaneous alteration of a base, and (3) events related to the insertion and excision of transposable elements.

Errors in nucleotide incorporation at the end of a growing strand during DNA replication occur with sufficiently high frequency that the information content of a daughter DNA molecule would differ significantly from that of the parent were it not for two mechanisms for correcting such errors—**editing** (also called proofreading) and **mismatch repair** (Section 8.5). The editing function is a $3' \rightarrow 5'$ exonuclease activity that removes a nucleotide from the

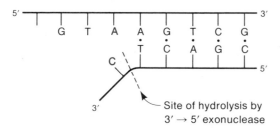

Site of hydrolysis by
3′ → 5′ exonuclease

Figure 10-7 The 3′-to-5′ exonuclease activity of the editing function; the growing strand is cleaved to release a nucleotide containing the base C (red) that does not pair with the base A (red) in the template strand.

3′-OH terminus of a growing strand in a replication fork if the nucleotide is not correctly hydrogen-bonded to the template (Figure 10-7). The editing function is performed by the DNA polymerase itself. After removal of the unpaired terminal nucleotide, polymerization continues. Occasional failure of the editing function—that is, continued polymerization without removal of the unpaired nucleotide—can lead to mutation. Evidence for the importance of the editing function in maintaining a low mutation frequency comes from studies of *E. coli* mutants in which the exonuclease activity of the DNA polymerase is impaired; the frequency of spontaneous mutations in these strains is about a thousandfold greater than the mutation frequency of wild-type *E. coli*.

Errors in incorporation that are occasionally missed by the editing function are usually corrected by the mismatch repair system described in Section 8.5. Mismatch repair is a postreplicative process. The key events are (1) the recognition of incorrectly base-paired nucleotides *within a strand* and (2) excision of a contiguous nucleotide sequence that includes one nucleotide of the unmatched pair (Figure 10-8(a)). The resulting gap is filled in by a DNA

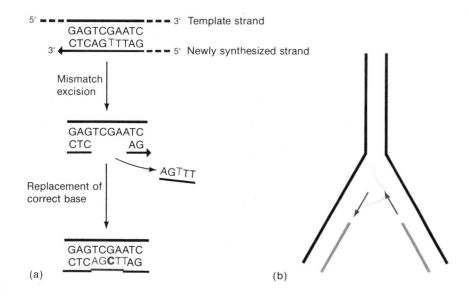

(a)   (b)

Figure 10-8 Mismatch repair. (a) Excision of a short segment of a newly synthesized strand, and repair synthesis. (b) Methylated bases in the template strand direct the excision mechanism to the newly synthesized strand containing the incorrect nucleotide. The regions in which methylation is complete are black; the regions in which methylation may not be complete are pink. Unmethylated DNA is red.

polymerase (typically not the same polymerase used in chain elongation), which in this way is given a second chance to form only correct base pairs, and sealed by a DNA ligase. If the mismatch repair system is to correct errors rather than create them, it must be able to distinguish the correct base in the template strand from the incorrectly incorporated base in the newly synthesized strand. If the system were not able to do this, the correct base might be replaced by the complement to the incorrect base, thereby producing a mutation. The distinction is made by recognition of the pattern of methylation of certain bases. DNA contains rare methylated forms of cytosine and adenine, which occur within two specific six-base sequences. Methylation of the appropriate bases in a newly synthesized strand does not occur immediately after the nucleotide is incorporated into the growing strand, but lags somewhat behind the replication fork (Figure 10-8(b)). The methylated forms of cytosine and adenine are shown in Figure 10-9. The under-methylation of this strand compared to the full methylation of the parental strands enables the excision system to distinguish the parental and daughter strands and hence the incorrect base. That is, excision occurs only in the newly synthesized (under-methylated) strand, and the parental strand, with its correct base, is always selected as the template for mismatch repair. Genetic evidence for the role of methylation in maintaining replication fidelity comes from studies with *E. coli* mutants known to be defective in the methylation of cytosine and adenine. In these mutants the mismatch repair system should not be able to distinguish between the parent strand with the correct nucleotide and the daughter strand with the incorrect nucleotide; thus, half the time the correct nucleotide should be excised from the template strand and the daughter strand containing the error would be used as a template for repair synthesis. Indeed, these mutants show a significantly increased mutation frequency for all genes. Strains lacking the excision system have also been isolated and these too have a high mutation frequency.

The mismatch repair system also counteracts the mutagenic effect of tautomeric bases (Section 4.5). Such molecules, in their rare forms, can occasionally be misincorporated into DNA. At the time of incorporation, the base will be correctly hydrogen-bonded to the template strand, so it is not recognized by the editing function as incorrect. However, when the base later assumes its normal structure, a mismatched base pair will be present. The mismatch repair system can correct such errors, but if the elapsed time is so great that the region of the daughter strand containing the incorrect base has become methylated, the mismatch repair system will be unable to distinguish parental and daughter strands and mutation can occur.

Another source of spontaneous mutation is an alteration of 5-methylcytosine (MeC), a methylated form of cytosine that pairs with guanine and is present to the extent of about five percent of the cytosines in the DNA of many organisms and viruses. Both cytosine and 5-methylcytosine are subject to occasional loss of an amino group. For cytosine, this loss yields uracil (Figure 10-10). Since uracil pairs with adenine instead of guanine, replication of a molecule containing a GU base pair would ultimately lead to substitution of an AT pair for the original GC pair (by the process GU → AU → AT in

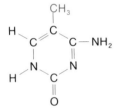

5-Methylcytosine

6-Methyladenine

Figure 10-9 Two methylated bases. The methyl groups are shown in red. A hydrogen atom would be present at the comparable position in the unmethylated base.

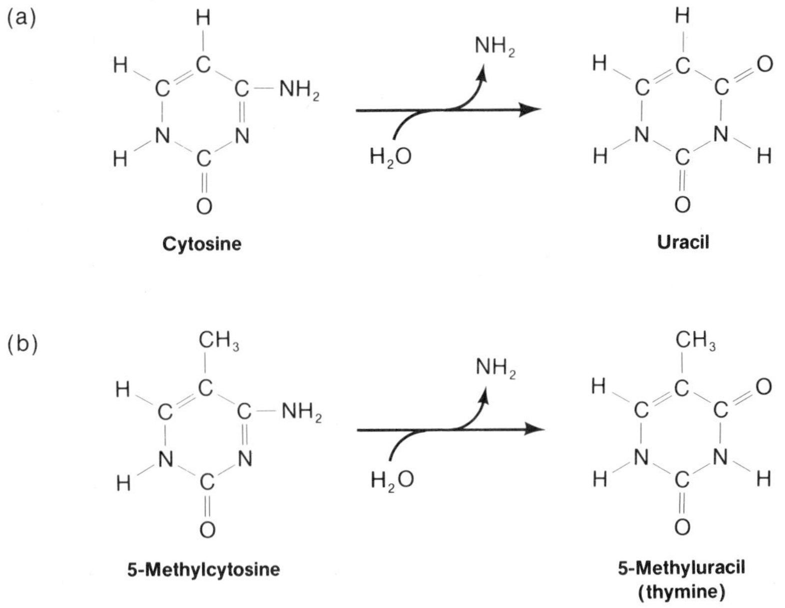

Figure 10-10 Spontaneous loss of the amino group of (a) cytosine to yield uracil and (b) 5-methylcytosine to yield thymine.

successive rounds of replication). However, cells possess an enzyme (*uracil glycosylase*) that specifically removes uracil from DNA, so the cytosine-to-uracil conversion rarely leads to mutation. Loss of the amino group of 5-methylcytosine yields 5-methyluracil, which is the same as thymine (Figure 10-10). Since thymine is a normal DNA base, no thymine glycosylase exists, so the GMeC pair becomes a GT pair. A GT pair is subject to correction by mismatch repair. However, since MeC is a methylated base and therefore is present in a methylated strand, the mismatch repair system does not recognize the thymine as incorrect. The direction of correction will be random, sometimes yielding the correct GC pair and sometimes an incorrect AT pair. Thus, MeC sites, which exist at only a few locations in a gene, constitute highly mutable sites, and the mutations are always GMeC → AT transitions.

Sites within a gene, at which mutations occur with much higher frequency than at other sites, were first observed by Benzer in the fine-structure mapping experiments of the T4 *rII* locus described in Section 7.5, in which several thousand independently isolated *rII* mutations were mapped. He called such a site a **hot spot.** In later years determination of the base sequence of several bacterial genes and of mutant alleles has shown that most hot spots are GMeC base pairs.

Some hot spots yield large deletions rather than point mutations. The base sequences of these regions indicate that the deletion is often bounded by a sequence that is repeated in the genome and may be a component of a transposable element. Two possible mechanisms for the production of a deletion in molecules having repeated nucleotide sequences are illustrated in

Figure 10-11. These mechanisms are recombinational excision and a particular type of replication error.

Transposable elements, which are widely distributed in the genomes of both prokaryotes and eukaryotes (Sections 5.8 and 7.7) and which move to new sites in a seemingly nonspecific manner, are important sources of spontaneous mutation. The mechanism is primarily simple insertion within a gene

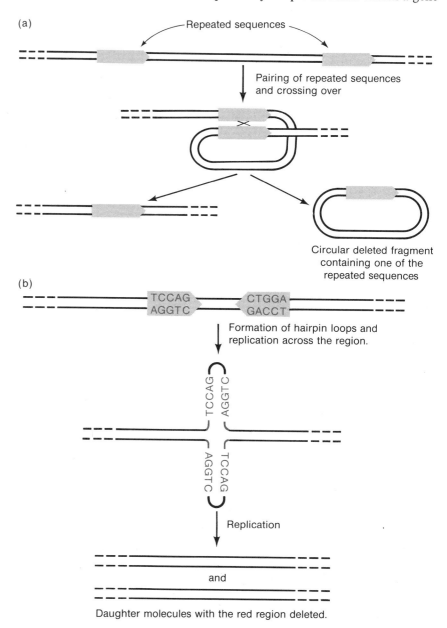

Figure 10-11   Two mechanisms for the production of deletions in molecules with repeated nucleotide sequences. (a) Crossing over between paired direct repeats. (b) Replication across an aberrant form of a molecule with an inverted repeat.

and consequent interruption of the coding sequence. In bacteria spontaneous mutations arise only occasionally by this route, but in eukaryotes transposition appears to be a more important mechanism of mutation. For example, in analyzing the base sequences of 13 spontaneous mutations at a complex locus (white) of *D. melanogaster*, it was found that in seven of them extra DNA sequences were inserted in or near the locus.

Some transposable elements contain transcription-termination sequences. When these insert in a gene between the promoter and the coding sequence, transcription of the coding region is prevented. In bacteria this event gives rise to polar mutations (Sections 9.3 and 9.7). Transposable elements also mediate the formation of deletions and inversions. For example, when such an element is present in a known location in a bacterial chromosome, the frequency of nearby deletions is 100–1000 times greater than the value observed if the element were not present. In bacteria, the hallmark of a deletion mediated by a transposable element is the presence of the transposable element at the site of the deletion and a single copy of a target sequence adjacent to the element. The mechanism of deletion formation by transposable elements is not known.

Transposable elements are also useful in induced mutagenesis; this will be described at the end of Section 10.4.

10.4 Induced Mutations

The first convincing evidence that external agents increase the mutation rate was presented in 1927 by Hermann Muller, who showed that x rays are mutagenic in *Drosophila*. Since then, a large number of physical agents and chemical reagents with diverse properties have been found that increase mutation rate. The use of these mutagens, several of which will be discussed in this section, provides a means for greatly increasing the number of mutants that can be isolated.

Base-Analogue Mutagens

A base analogue is a compound sufficiently similar to one of the four DNA bases that it can be built into a DNA molecule during normal replication. Such a substance must be able to pair with a base in the template strand or the editing function will remove it. However, if the base has two modes of hydrogen-bonding, it will be mutagenic. The basis for this mutagenesis can be illustrated with 5-bromouracil (BU) a commonly used base analogue that is efficiently incorporated into the DNA of bacteria and viruses.

The base 5-bromouracil is an analogue of thymine, the substituted bromine atom having about the same size as the methyl group of thymine (Figure 10-12(a)). If cells to be mutated are grown in a medium containing 5-bromouracil, during replication of the DNA a thymine will sometimes be replaced by a 5-bromouracil, resulting in the formation of an ABU pair at a normal AT site (Figure 10-12(b)). The subsequent mutagenic activity of the incorporated 5-bromouracil stems from a tautomeric shift from the usual keto

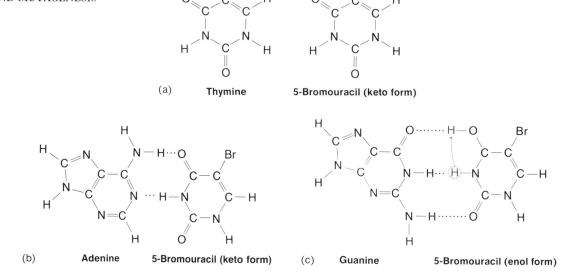

(a) **Thymine**   **5-Bromouracil (keto form)**

(b) **Adenine** **5-Bromouracil (keto form)** (c) **Guanine** **5-Bromouracil (enol form)**

Figure 10-12 Mutagenesis by 5-bromouracil. (a) Structures of thymine and 5-bromouracil. (b) A base pair between adenine and the keto form of 5-bromouracil. (c) A base pair between guanine and the rare enol form of 5-bromouracil. The red H in the dashed circle shows the position of the H when 5-bromouracil is in the keto form.

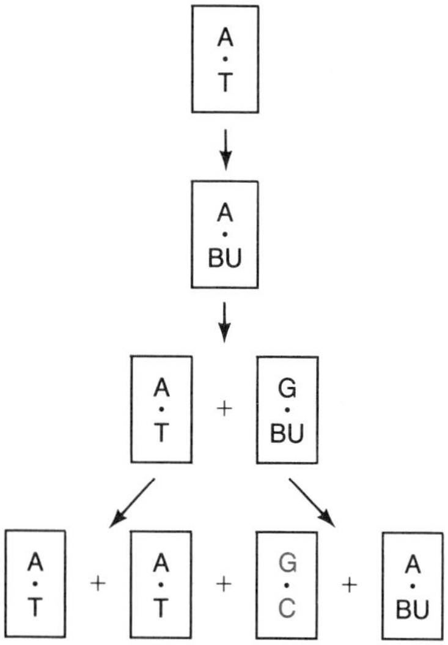

Figure 10-13 A mechanism for mutagenesis by 5-bromouracil (BU). The production of an AT → GC transition by the incorporation during DNA replication of 5-bromouracil in its usual keto form into an ABU pair. In the mutagenic round of replication the BU, in its rare enol form, pairs with G. In the next round of replication the G pairs with a C, completing the transition.

form (also the form of thymine) to the rare enol form (Section 4.5); the equilibrium between the two forms is influenced by the bromine atom and the enol form is present in 5-bromouracil more frequently than it is in thymine. In the enol form 5-bromouracil pairs with guanine (Figure 10-12(c), so in the next round of replication one daughter molecule will have a GC pair at the altered site, yielding an AT → GC transition (Figure 10-13).

Recent experiments suggest that 5-bromouracil is also mutagenic in another way. The concentrations of the nucleoside triphosphates in most cells is regulated by the concentration of thymidine triphosphate (TTP). This regulation results in appropriate relative amounts of the four triphosphates for DNA synthesis. One part of this complex regulatory process is the inhibition of synthesis of deoxycytidine triphosphate (dCTP) by excess TTP. The 5-bromouracil nucleoside triphosphate also inhibits production of dCTP. When 5-bromouracil is added to the growth medium, TTP continues to be synthesized by cells at the normal rate while the synthesis of dCTP is significantly reduced. The ratio of TTP to dCTP then becomes quite high and the frequency of misincorporation of T opposite G increases. The editing function and mismatch repair systems are capable of removing incorrectly incorporated thymine, but in the presence of 5-bromouracil the rate of misincorporation can exceed the rate of correction. An incorrectly incorporated thymine that persists will pair with adenine in the next round of DNA replication, yielding a GC → AT transition in one of the daughter molecules. Thus, 5-bromouracil induces transitions in both directions—AT → GC by the tautomerization route, and GC → AT by the misincorporation route. Thus, mutations that are induced by 5-bromouracil can also be reversed by it.

Chemical Agents that Modify DNA

Many mutagens are chemicals that react with DNA, changing the hydrogen-bonding properties of the bases. These mutagens are active on both replicating and nonreplicating DNA, in contrast with the base analogues that are mutagenic only when DNA replicates. Several of these chemical mutagens, of which nitrous acid and hydroxylamine are well-understood examples, are highly specific in the changes they produce. Others, for example, the alkylating agents, react with DNA in a variety of ways and produce a broad spectrum of effects.

Nitrous acid (HNO₂) acts as a mutagen by converting amino (NH₂) groups of the bases adenine, cytosine, and guanine to keto (=O) groups (*deamination*), which alters the hydrogen-bonding specificity of each base. Deamination of adenine yields hypoxanthine (H), which pairs with cytosine rather than thymine, resulting in an AT → GC transition (Figure 10-14).

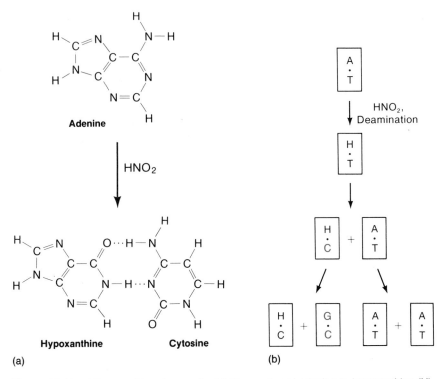

(a)

(b)

Figure 10-14   Nitrous acid mutagenesis. (a) Conversion of adenine to hypoxanthine (H), which pairs with cytosine. The cytosine → uracil conversion is shown in Figure 10-10. (b) Production of an AT → GC transition. In the mutagenic round of replication H pairs with C. In the next round of replication the C pairs with G, completing the transition.

Similarly, deamination of cytosine yields uracil (Figure 10-10), which pairs with adenine instead of guanine, producing a GC → AT transition. Guanine is converted to xanthine (X), a change that should not be directly mutagenic since both guanine and xanthine pair with cytosine. However, GC → AT transitions have been observed with certain single-stranded DNA phages at sites of a guanine, which suggests that xanthine probably has an undiscovered tautomeric form able to pair with T.

Hydroxylamine (NH₂OH) reacts specifically with cytosine, converting it to a modified base that pairs only with adenine; thus, only GC → AT transitions are produced. The chemistry of cytosine alteration by hydroxylamine is beyond the scope of this book.

The **alkylating agents** ethylmethane sulfonate (EMS) and nitrogen mustard, the structures of which are shown in Figure 10-15, are potent mutagens that have been used extensively in genetic research. Chemical mutagens such as nitrous acid and hydroxylamine (and many others), which are exceedingly useful in prokaryotic systems are not particularly useful as mutagens in higher eukaryotes (because the chemical conditions necessary for reaction are not easily obtained), whereas the alkylating agents are highly effective. Many sites in DNA are alkylated by these agents; of prime importance for the induction of mutations is the addition of an alkyl group to the hydrogen-bonding oxygen of guanine and thymine. These alkylations impair the normal hydrogen-bonding of the bases and cause mispairing of G with T, leading to the transitions AT → GC and GC → AT. EMS also reacts with adenine and cytosine. In both phage T4 and *D. melanogaster*, bases that have been altered by EMS such that their normal hydrogen-bonding is impaired mispair about half the time, so this reagent is a potent mutagen. Another phenomenon resulting from alkylation of guanine is **depurination,** or loss of the alkylated base from the DNA molecule by breakage of the bond joining the purine nitrogen and deoxyribose. Depurination is not usually mutagenic, since the gap left by loss of the purine is repaired by a multistep enzyme system that excises the free deoxyribose and replaces it with the correct nucleotide. A second important mechanism for mutagenesis by alkylating agents in many organisms results from another system of enzymes that repairs alkylation events. The altered nucleotide is first removed; then, during the resynthesis stage errors in incorporation are frequently made, often yielding transversions. Systems of this kind for the repair of DNA damage are called

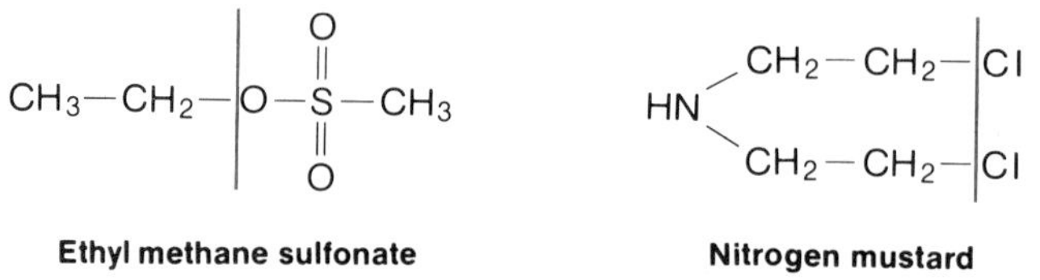

Ethyl methane sulfonate **Nitrogen mustard**

Figure 10-15   The chemical structures of two highly mutagenic alkylating agents with the alkyl group shown to the left of the red line in each case.

error-prone, and were initially discovered in *E. coli* in studies of the mutagenic effect of ultraviolet light. One such system will be described shortly in the discussion of ultraviolet irradiation. Evidence that an error-prone system leads to mutagenesis by alkylating agents in the yeast *Saccharomyces cerevisiae* comes from studies of strains in which the error-prone repair system is inactive (owing to a mutation in one of the genes responsible for activity of the system). In these strains the frequency of mutations induced by many alkylating agents is greatly reduced. The reduction is so great that it is clear that in yeast mutagenesis by alkylating agents is almost entirely a result of error-prone repair.

Many alkylating agents are known to be dangerous **carcinogens,** substances that induce tumors in animals (see also Section 10.5). Other alkylating agents are effective **carcinostatic** agents, inhibiting the growth of tumors, and some of these with low toxicity for normal cells have been useful in tumor therapy. The carcinostatic agents all have two or more reactive groups, a property illustrated by nitrogen mustard (Figure 10-15). Such molecules can alkylate two bases in complementary strands and thereby act as bridges that covalently crosslink the two strands of a DNA molecule and disrupt its replication.

Misalignment Mutagenesis

The **acridine** molecules, of which proflavin is an example, are planar three-ringed molecules, whose dimensions are roughly the same as those of a purine-pyrimidine pair (Figure 10-16). In aqueous solution, these substances bind to DNA by inserting themselves between adjacent base pairs, a process called **intercalation.** The effect of the intercalation of an acridine molecule is to cause the adjacent base pairs to move apart by a distance equal to the thickness of one pair (Figure 10-17). When DNA containing intercalated acridines is replicated, misalignment in regions of repeated sequences is induced by the intercalated acridine molecule. It has been suggested that the template strand and the newly synthesized strands are temporarily separated in local regions and that the double helix becomes misaligned as it is re-formed (Figure 10-18). In the simplest case a base is either added or deleted; the particular change depends on which strand contains an unpaired base.

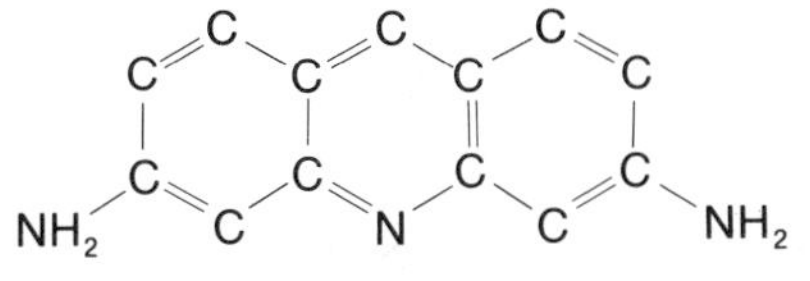

Proflavine

Figure 10-16 The structure of proflavine, an acridine derivative. Other mutagenic acridines have small substituents on the NH$_2$ group and on the C of the central ring. Hydrogen atoms are not shown.

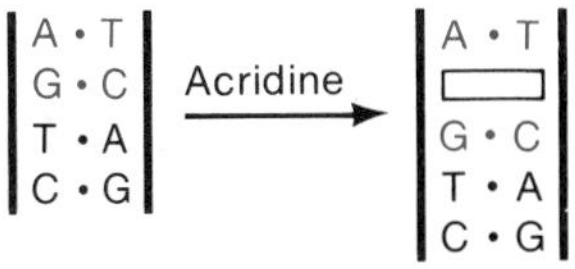

Figure 10-17 Separation of two base pairs (red) caused by intercalation of an acridine molecule (open box).

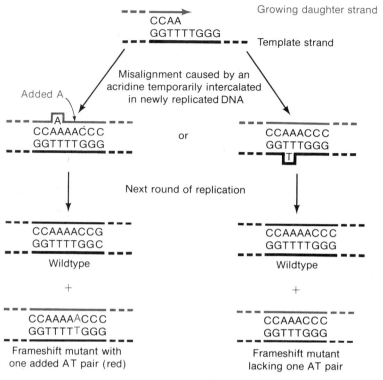

Figure 10-18 Proposed mechanism for misalignment mutagenesis generated by intercalation of an acridine molecule in the replication fork. Acridine is present only in the first round of replication; the second round serves to produce a true-breeding mutant. The left and right paths generate a base addition and deletion, respectively. The growing daughter strand is shown in red.

Ultraviolet Irradiation

Ultraviolet light (UV), which encompasses a range of wavelengths shorter than those in the blue region of the visible spectrum, produces both lethal and mutagenic effects in viruses, bacteria, and eukaryotic cells. The mutagenicity and lethality are chemical effects resulting from absorption of the energy of the light by the bases of DNA. Ultraviolet light was first shown to be mutagenic in studies of *Drosophila,* but most of what is known about its genetic effects and mechanism of action comes from studies in *E. coli.* The major products formed in DNA after UV irradiation are covalently joined pyrimidines (**pyrimidine dimers**), primarily of thymine, that are adjacent in the same polynucleotide strand (Figure 10-19(a)). This chemical linkage brings the bases closer together, causing a distortion of the helix (panel (b)), which blocks DNA replication. Most of the damage in DNA is repaired. Four different mechanisms are known in *E. coli* for repair of DNA containing pyrimidine dimers: photoreactivation, excision repair, recombination repair, and SOS repair. The first three will be described briefly in the following paragraphs; SOS repair, which is the only repair process that is mutagenic, will be presented in greater detail.

Photoreactivation is a light-dependent repair process in which an enzyme breaks the bonds that join the pyrimidines in a dimer and re-forms the original bases. The enzyme binds to the dimers, whether light is available or not, and then utilizes the energy of light in the blue part of the visible spectrum to cleave the bonds. When ultraviolet light is used to induce mutations, cells are usually protected from blue light (for example, fluorescent light and sunlight) to avoid photoreactivation and maximize the mutagenesis.

Excision repair is a multistep enzymatic process by which pyrimidine dimers are removed from the DNA and replaced by resynthesis (repair synthesis). Two distinct mechanisms have been observed. In *E. coli* a multisubunit protein consisting of the products of three genes—*uvrA, uvrB,* and *uvrC*—recognizes the distortion in DNA produced by a dimer and makes two cuts (incisions) in the sugar-phosphate backbone, each yielding a 3'-OH group (Figure 10-20). One cut is 8 nucleotides on the 5' side of the dimer and another is 4–5 nucleotides on the 3' side (for convenience the cuts are drawn nearer to the dimer in the figure). On the 5' side DNA polymerase I uses the 3'-OH group as a primer, synthesizes a new strand, and displaces the DNA segment that carries the dimer. When the second cut (the excisional cut) is reached, the displaced strand falls away, and the newly synthesized segment is covalently joined to the original strand by DNA ligase. In two systems (the bacterium *Micrococcus luteus* and *E. coli* phage T4) the incision step proceeds in several stages. The first is an enzymatic breakage of the glycosylic (base-sugar) bond in one nucleotide at the 5' end of the dimer; the enzyme

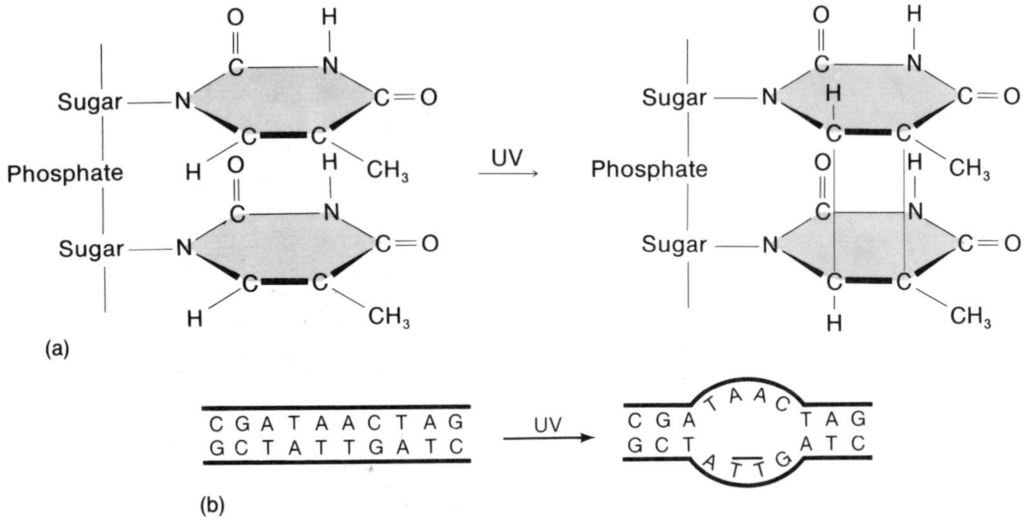

(a)

(b)

Figure 10-19 (a) The structure of a thymine dimer. Following ultraviolet (UV) irradiation, adjacent thymines in a DNA strand are joined by formation of the bonds shown in red. Although not drawn to scale, these bonds are considerably shorter than the spacing between the planes of adjacent thymines, so that the double-stranded structure becomes distorted. The shape of each thymine ring also changes. (b) The distortion of the DNA helix caused by two thymines moving closer together when joined in a dimer. The dimer is indicated as two thymines joined by a bar.

carrying out this step is a glycosylase specific for pyrimidine dimers, an enzyme of the type that removes uracil from DNA (Section 10.3). Incision is completed by action of the same enzyme at the deoxyribose lacking the base. Pol I acts at the newly formed 3'-OH group, displaces the dimer-containing strand, and fills the gap. An excisional cut is then made in the displaced strand, which falls away, and ligation occurs. The fidelity of base selection in repair synthesis is as high as that in normal DNA synthesis; thus, mutations are not often introduced by excision repair.

Mutations in any one of the *E. coli uvrA, uvrB,* or *uvrC* genes block the ability of cells to excise pyrimidine dimers and make the cells exceedingly sensitive to killing by ultraviolet light. In humans, the inherited autosomal recessive disease **xeroderma pigmentosum** is the result of a similar defect in the system that repairs ultraviolet-damaged DNA. Individuals with this disease have an extreme sensitivity to sunlight, resulting in excessive skin pigmentation and the development of numerous skin lesions that frequently become cancerous. Removal of pyrimidine dimers does not happen in the DNA of cells cultured from patients with the disease, and the cells are killed by much smaller doses of ultraviolet light than are cells from normal individuals. Both complementation analysis (using the cell-fusion technique described in Section 13.1) and *in vitro* studies of repair indicate that in mammals a larger number of enzymes are needed for excision repair than in microbial systems.

Recombination repair occurs after DNA containing unexcised pyrimidine dimers replicates and thus is often called **postreplicational repair.** A

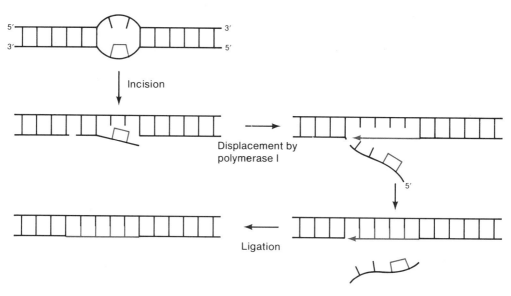

Figure 10-20 The mechanism for excision repair of a pyrimidine dimer (red) in *E. coli.* Incision occurs on both sides of the dimer, generating 3'-OH groups. One serves as a primer for DNA polymerase I, which engages in repair replication (red strand), displacing the segment containing the dimer. When the second cut is reached, the displaced strand falls away, and ligation occurs. For clarity the incisions are drawn quite near the dimer.

thymine dimer is able to form hydrogen bonds with the two adenines in the complementary strand, because the chemical change in forming a dimer does not alter the groups that participate in hydrogen-bonding. However, when an adenine nucleotide is added to a growing strand at the site of a dimer, the DNA polymerase reacts to the distorted region as if a mispaired nucleotide had been incorporated; that is, the editing function removes the nucleotide and the replication fork does not advance. Thus, the main effect of a dimer on DNA replication is to stall the replication fork. However, after a brief time, synthesis is reinitiated beyond the dimer (by a totally unknown mechanism) and chain growth continues, producing a gapped strand:

In the next round of replication of such a DNA molecule, the gapped strand is fragmented when the replication fork enters the gap, and the strand containing the dimer serves as the template for synthesis of another gapped daughter strand. Recombination repair of ultraviolet-damaged DNA has been demonstrated in several bacterial species and is best understood in *E. coli*, though the molecular details are not fully understood. The essential idea, shown in Figure 10-21, is that a gap formed in a growing strand is repaired by insertion of a segment excised from the intact homologus strand in the other branch of the replication fork. The resulting gap in the donor strand is then filled by repair synthesis. The products of this exchange and resynthesis are two intact single strands, each of which can serve in the next round of replication as a template for the synthesis of an undamaged DNA molecule. In *E. coli* recombination

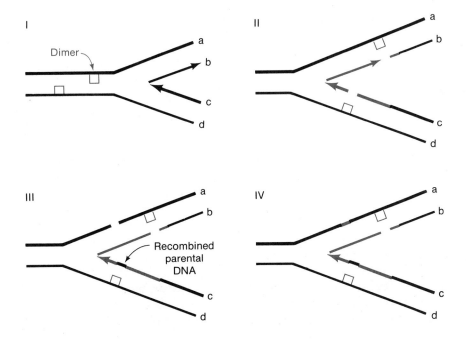

Figure 10-21 Recombinational repair. I. A molecule with two pyrimidine dimers (red boxes) in strands a and d is being replicated. II. By reinitiating synthesis beyond the dimers, gaps are formed in strands b and c. III. A segment of parental strand is excised and inserted in strand c. IV. The gap in strand a is next filled in by repair synthesis. Such a DNA molecule would probably engage in a second exchange in which a segment of c would fill the gap in b. DNA synthesized after irradiation is shown in red. Heavy and thin lines are used for purposes of identification only.

repair accounts for most of the avoidance of the lethal effects of ultraviolet radiation; evidence for this fact is that strains lacking the recombination repair system are more sensitive to ultraviolet light than strains deficient in any of the other repair systems.

SOS repair is a system that allows DNA replication to occur across pyrimidine dimers or other damaged segments, but at the cost of fidelity of replication. Even though intact DNA strands are formed, the strands are often defective. Thus, SOS repair is said to be an *error-prone repair system*. In SOS repair the editing system of DNA polymerase is relaxed in order to allow polymerization to proceed across a dimer, despite the distortion of the helix. As described earlier, the response of the editing system to incorporation of an adenine nucleotide across from a thymine in a dimer is to excise the nucleotide. This process—polymerization, then removal—continues for a while (though initiation and gap formation, described above, also occur). During this period, the SOS system, which is normally inactive, is activated in a poorly understood way by damage to the DNA; once activated, the SOS system prevents the editing system from responding to the distortion, and the growing fork then advances. The distortion produced by the thymine dimer increases the probability that nucleotides other than two adenines will be incorporated in the growing chain at the site complementary to the dimer. With the editing system turned off, misincorporated nucleotides are not removed; thus, fidelity in replication depends only on the mismatch repair system, so the mutation frequency increases. A second but less frequent event is the production of mutations downstream from the dimer owing to the editing system remaining in the off state for a while following polymerization across the dimer; this phenomenon, which has been termed untargeted mutagenesis, is illustrated in Figure 10-22.

In *E. coli* several genes participate in the SOS pathway. Of these the best understood are the *recA* and *lexA* genes, the *lexA* gene encoding a regulator of the synthesis of the *recA* product. The RecA protein has three main biochemical activities. One of these is its recombination activity (Chapter 8); another is a protease activity; and the third is a single-stranded-DNA binding activity. At least two of these activities are important for SOS repair. Although the regulation of gene activity is discussed primarily in Chapter 11, it is a useful introduction at this point to examine both the consequences of having the SOS system active at all times and the mechanism by which its activity is regulated. Clearly, high fidelity in replication demands that all error-prone replicative processes be inactive most of the time. Inactivation of the system in undamaged cells is accomplished indirectly by the *lexA*-gene product. This protein maintains synthesis of the *recA*-gene product at a low level and prevents synthesis of other enzymes of the SOS pathway by regulating transcription of the genes encoding the proteins. In an unirradiated cell, or in any cell in which DNA synthesis is occurring, the RecA protein lacks protease activity. However, when DNA is damaged or replication is prevented (for example, by a thymine dimer), the small amount of RecA protein that is present is converted, by a poorly understood mechanism, to an active protease. The proteolytic activity of the RecA protein functions only on selected proteins.

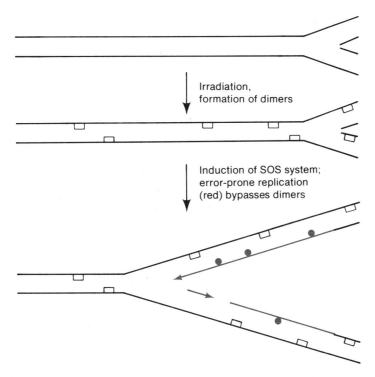

Figure 10-22 SOS repair. A DNA molecule in an early stage of replication is irradiated with ultraviolet light, and thymine dimers are formed. The SOS system is induced and all subsequent replication (shown in red) has a higher-than-usual number of misincorporated bases (red dots). Some of these bases can be removed by the mismatch repair system.

One of the proteins cleaved is the LexA protein, which becomes unable to prevent synthesis of the RecA protein or the SOS proteins, and the SOS system is turned on. When repair is complete, the RecA protein is converted back to a proteolytically inactive form; the LexA protein, which was continually synthesized and cleaved, is no longer cleaved, and synthesis of the SOS proteins is again inhibited. This activation of the SOS system only when the system is useful is the source of the name SOS, which implies an emergency.

The mutagenic effect of ultraviolet light is almost exclusively a result of error-prone SOS repair. Evidence for this conclusion comes from the almost complete absence of mutagenic effects of ultraviolet irradiation in *E. coli lexA* and *recA* mutants.

Ionizing Radiations

The high-energy radiation known as x rays has wavelengths more than a thousandfold shorter than the visible spectrum. In their biological effects they are representative of **ionizing radiation,** which includes α and β particles and γ rays, all released by radioactive elements. All forms of ionizing radiation produce both mutagenic and lethal effects in all known cells and viruses. As x rays pass through matter, they release considerable amounts of energy and thereby break covalent bonds, forming charged fragments of molecules with unpaired electrons (ions and free radicals) and numerous energetic electrons. As x rays pass through water or aqueous solutions, in the presence of molecular O_2 they produce hydrogen peroxide; when organic compounds are also present, a variety of organic peroxides and other highly reactive substances are formed.

The intensity of a beam of ionizing radiation is described quantitatively in several ways. The **roentgen (R)** is a measure of radiation exposure and is defined as the amount of radiation causing 2×10^9 ion pairs to form in 1 cm³ of dry air at 0°C at a pressure of 1 atm. Another unit is used to measure absorbed radiation; this is the **rad,** which is defined as the amount of radiation that results in the absorption of 100 ergs of energy per gram of matter. One rad is approximately 1.07 R, so the units are often used interchangeably. Different radiations have different biological effectiveness because of their varying ability to penetrate matter and the rate at which the energy of the radiation is released. Thus, to be able to compare the different forms of radiation the **rem** (*r*oentgen-*e*quivalent in *m*an) is used; this is the quantity of radiation that has the same biological effectiveness in man as one rad of high-energy γ rays. If a particular dose in rads of one type of radiation produces more damage than an equivalent dose of γ rays, the dose of the first type of radiation will have a higher value in rem.

The frequency of mutations induced by x rays is proportional to the radiation dose. This relationship is shown in Figure 10-23 as a linear increase in the frequency of X-chromosome recessive lethals in *Drosophila* with increasing x-ray dose; an exposure of 1000 R increases this frequency from the spontaneous value of 0.15 percent to about 3 percent; 2000 R increases the frequency to about 6 percent; and 4000 R increases it to about 12 percent. In such experiments the rate of increase in lethals begins to drop at doses above about 5000 R; the primary reason is that the *ClB* method (and related methods) measures the frequency of chromosomes bearing lethal mutations, rather than the frequency of the lethal mutations themselves. At higher dosages of radiation the probability that two or more lethal mutations will occur in the same chromosome increases, but these multiple mutations are counted as only a single event.

In *Drosophila* there is no threshold exposure below which mutations are not induced; even very low doses have a finite probability of inducing mutations. Moreover, in *Drosophila* the amount of mutational damage is independent of the dose rate; that is, if a population of flies is given a particular exposure, the same number of mutations will be observed whether the radiation is administered at low intensity over a long period of time (this is termed a low dose rate or chronic exposure) or at high intensity for a short period (high dose rate or acute exposure). With mice the results are quite different; for example, an early study indicated that the mutagenic effect in progeny was four times greater when adults were treated with a dose rate of 90 rad/min rather than with 90 rad/week. It is believed that dose-rate effects reflect the

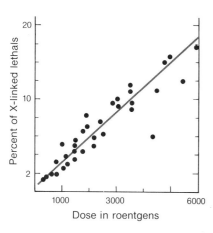

Figure 10-23 The relation between the percentage of X-linked recessive lethals in *D. melanogaster* and x-ray dose, obtained from several experiments. The spontaneous frequency of these mutations is 0.15 percent.

existence of repair processes in precursors to sperm cells and in mature eggs. If the dose rate is sufficiently low, damage can be repaired before sperm are formed and before ovulation occurs, whereas high dose rates allow insufficient time for repair; hence, mutant sperm and mutant eggs are used in forming progeny.

The mutagenic and lethal effects of ionizing radiation result primarily from damage to DNA. Three types of damage in DNA are produced by ionizing radiation—single-strand breakage (in the sugar-phosphate backbone), double-strand breakage, and alterations in nucleotide bases. Single-strand breaks are formed primarily by attack by free radicals and ions; they are rapidly repaired in all normal cells and probably have little or no genetic consequences, other than mild stimulation of crossing over. Double-strand breakage, which is less frequent, is usually lethal, though it is repairable to some extent. Base damage of many kinds occurs, particularly, ring cleavage and modification of chemical groups; some kinds are repairable, but most of the changes are either lethal or mutagenic. Both transitions and transversions are observed. It is likely that the mutagenic changes result from both subsequent mispairing of modified bases and the action of the *E. coli* SOS system or related repair systems in other organisms. The chemistry of base alteration is very complex and has yet to be understood fully. Most base damage occurs only if molecular oxygen is present, presumably because of attack by peroxides. Most cells contain peroxidases (enzymes that break down peroxides to harmless compounds), and these enzymes are presumably responsible for protection against radiation damage. Evidence for this point comes from experiments in which *Drosophila* are exposed to small quantities of cyanide during x irradiation. Cyanide is not a mutagen but causes a significant increase in the mutation rate, presumably by its known ability to inhibit the activity of peroxidases *in vitro*.

Another effect of ionizing radiation is the production of chromosome breaks, which are usually lethal. Systems exist in some organisms for reannealing of the breaks. Such reannealing often leads to translocations, inversions, duplications, and deletions. Many deletions are so small that they are not detectable by microscopy, even in the polytene chromosomes of *Drosophila*, and the resulting mutations are distinguished from base-substitution mutations only by their inability to revert to wildtype.

Ionizing radiation is widely used in tumor therapy. The basis for the treatment is the increased frequency of chromosomal breakage (and the consequent lethality) in cells undergoing mitosis compared to cells in interphase. Since tumors contain many mitotic cells and normal tissues do not, more tumor cells are destroyed than normal cells. Since all tumors are not in mitosis at the same time, irradiation is done at intervals of several days to allow interphase tumor cells to enter mitosis. Presumably, over a period of time more tumor cells will be destroyed.

Table 10-2 gives representative values of doses of ionizing radiation received by reproductive cells of humans in the United States in the course of one year. Note that with the exception of diagnostic x rays, which yield important compensating benefits, the percentage of total radiation exposure of man-made origin is much less than background. There are dangers inherent

Table 10-2 Annual exposure of humans in the United States in 1980 to various forms of ionizing radiation

Source	Dose, millirems
Natural radiation	
Cosmic rays	28
Natural radioisotopes in the body	28
Natural radioisotopes in the soil	26
	82
Man-made radiation	
Diagnostic x rays	20
Radiopharmaceuticals	2–4
Consumer products (x rays from TV, radioisotopes in clock dials) and building materials	4–5
Fallout from weapons tests	4–5
Nuclear power plants	< 1
	30–35
Total	112–117

Source: From National Research Council, Committee on the Biological Effects of Ionizing Radiations. *The Effects on Populations of Exposure to Low Levels of Ionizing Radiation.* National Academy Press. 1980.

in any exposure to ionizing radiation because of its mutagenic effects, but geneticists knowledgeable in the area of mutation have become far less concerned with the potential hazards of ionizing radiation than with those of the many mutagenic and carcinogenic chemicals introduced into the environment from a variety of sources.

Mutagenesis Induced by Transposable Elements

Transposable elements provide a powerful tool for the production of mutations in both prokaryotes and eukaryotes. The mutations result from the ability of these elements to be inserted into DNA molecules and to alter the expression of nearby genes. Sometimes they become inserted directly within a gene and interrupt the coding sequence. In prokaryotes, the presence of transposable elements can be detected not only through the mutations they cause, but also through the drug-resistance genes that many of them carry.

A transposable element can be introduced into a bacterium by infection with a defective phage that carries the element but is unable to replicate, kill the host, or lysogenize. For example, if a defective λ phage carrying a transposable element with a tetracycline-resistance (*tet-r*) gene infects a culture of Tet-s *E. coli*, the cells survive. Since the phage cannot lysogenize or replicate, most of the cells remain Tet-s. However, transposition of the element will occur in a small number of cells, enabling them to form colonies on a medium containing tetracycline. If insertion has occurred *within* a gene, the cells in a Tet-r colony will be mutant. Standard genetic tests, such as replica plating onto a solid medium lacking a particular nutrient, can be used to isolate the mutant.

An important use of transposable elements is in the transfer of genes between organisms. An example is a transposable element in *D. melanogaster* called the **P element.** When a solution of the DNA of the P element is directly injected into the gonads of *Drosophila* embryos, some of the DNA enters the germ-line cells. The genes in P that promote transposition are expressed, with the result that the P element transposes into the chromosomes of these cells. Adults developing from these embryos therefore transmit the P element to their progeny by way of the chromosomes in their gametes. The particular value of this technique is that genetic engineering methods to be described in Chapter 12 can be used to attach any DNA sequences to the DNA of the P element. Subsequently, when a hybrid P element transposes into the chromosomes of embryonic germ cells, the additional attached DNA sequence is carried along and also becomes incorporated into the chromosomes. This method has been used to transfer genes between different strains of *D. melanogaster* and even to transfer genes between species of *Drosophila* that are reproductively isolated (that is, unable to produce fertile hybrid offspring).

10.5 Reverse Mutations and Suppressor Mutations

Most of the mutations we have considered are changes from a wildtype or normal gene to a form that results in a mutant phenotype, an event called a **forward mutation.** Mutations are frequently reversible, and an event that restores the wildtype phenotype is called a **reversion, reverse mutation,** or sometimes a **back mutation.** Reversions can occur in two different ways: (1) by an exact reversal of the alteration in base sequence that occurred in the original forward mutation, restoring the wildtype sequence, or (2) by the occurrence of a second mutation, at some other site in the genome, which in one of several ways compensates for the effect of the original mutation. Reversion by the first mechanism is infrequent. The second mechanism is much more common; a mutation of this kind is called a **suppressor mutation.** A suppressor mutation can occur at a different site in the same gene as the mutation it suppresses (***intra*genic suppression**), or in a different gene in either the same chromosome or a different chromosome (***inter*genic suppression**). Most suppressor mutations do not fully restore the wildtype phenotype; the reasons will be apparent in the following discussion of the two kinds of suppression.

Intragenic Suppression

Reversion of frameshift mutations (Section 9.5) is an example of intragenic suppression; the mutational effect of the addition (or deletion) of a nucleotide pair in changing the reading frame of the mRNA is rectified by a compensating deletion (or addition) of a second nucleotide pair at a nearby site in the gene. A second type of intragenic suppression occurs when loss of activity of a protein, caused by one amino acid change, is restored by a change in a second amino acid. An example of this type of suppression in the protein product of the *trpA* gene in *E. coli* is shown in Figure 10-24. The A polypeptide, composed of 268 amino acids, is one of two polypeptides that make up the enzyme tryptophan synthetase in *E. coli*. The mutation shown, one of many that inactivate the enzyme, is a change of amino acid 210 from glycine in the wildtype protein to glutamic acid. This glycine is not in the active site; rather, the inactivation is caused by a change in the folding of the protein, which indirectly alters the active site. The activity of the mutant protein is partially restored by a second mutation, in which amino acid 174 changes from tyrosine to cysteine. A protein in which only the tyrosine has been substituted by cysteine is also inactive, again because of a change in the shape

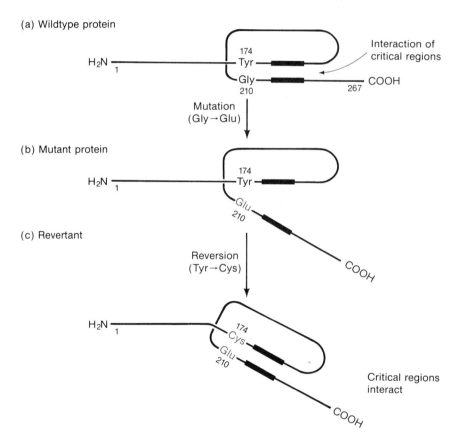

Figure 10-24 A model for the effect of mutation and intragenic suppression on the folding and activity of the A protein of tryptophan synthetase in *E. coli.* (a) The wildtype protein in which two critical regions (heavy lines) of the polypeptide chain interact. (b) Disruption of proper folding of the polypeptide chain by a substitution (red) of amino acid 210 prevents the critical regions from interacting. (c) Suppression of the effect of the original forward mutation by a subsequent change in amino acid 174 (red). The structure of the region containing this amino acid is altered, bringing the critical regions together again.

of the protein. This change at the second site, found to occur repeatedly in reversion of the original mutation, restores activity because the two regions containing amino acids 174 and 210 interact to produce a protein with the correct folding. This type of reversion can usually be taken to mean that the two regions of the protein interact, and studies of the amino acid sequences of mutants and revertants are often informative in elucidating the structure of proteins. The amino acid substitutions found in such mutant-revertant pairs often are changes between amino acids whose side chains have quite different chemical and physical properties—in this case, glycine (small, uncharged) to glutamic acid (bulky, charged), and tyrosine (bulky, weakly charged) to cysteine (small, active SH group). This is a common occurrence and indicates that the particular amino acids do not interact directly but that the intrastrand interaction is between extended regions containing many amino acids, and the changes merely form regions whose overall configuration allows an interaction to occur.

Intergenic Suppression

Intergenic suppression refers to a mutational change in a second gene that eliminates or suppresses the mutant phenotype. The most common type of intergenic suppression is one in which the product of a suppressor mutation acts to suppress the effects of mutations occurring in *many* other genes. The best-understood type of suppressor mutation occurs in tRNA genes; their effect is to change the specificity of the mRNA codon recognition by the tRNA molecules. Mutations of this type were first detected in certain strains of *E. coli*, which were able to suppress particular phage-T4 mutants that failed to form plaques on standard bacterial strains but were able to form plaques on a strain possessing a suppressor. These strains were also able to suppress mutations in numerous genes of the bacterial genome. The suppressed mutations were in each case chain termination or nonsense mutations—those in which a stop codon (UAA, UAG, or UGA) had been produced within the coding sequence of a gene, with the result that polypeptide synthesis was prematurely terminated and only an amino-terminal fragment of the polypeptide was synthesized (Section 9.5).

Suppression of nonsense mutations results from a mutation in a tRNA gene, which causes an alteration in the anticodon and permits the mutant or **suppressor tRNA** to recognize a nonsense codon. The molecular mechanism for this suppression can be illustrated by examining a chain termination mutation formed by alteration of the tyrosine codon UAC to the stop codon UAG (Figure 10-25(a)). Such a mutation can be suppressed by a mutant tryptophan tRNA molecule. In *E. coli*, tRNA^{Trp} has the anticodon 3'-ACC-5', which pairs with the codon 5'-UGG-3'. A suppressor mutation that had been isolated in the tRNA^{Trp} gene produced an altered tRNA with the anticodon AUC; this tRNA molecule could still be charged with tryptophan but responded to the stop codon UAG rather than to the normal tryptophan codon UGG. Thus, in a cell containing this suppressor tRNA, the mutant protein was completed and suppression occurred as long as the mutant protein would tolerate a substitution of tryptophan for tyrosine (panel (b)). Many suppressor tRNA molecules of this type have been observed. Inasmuch as a single base change is sufficient to alter the complementarity of an anticodon and a codon, there are (at most) eight tRNA molecules with complementary anticodons that, with a single base change, will also suppress a UAG codon. Thus, the following amino acids (whose codons are also indicated) can be put at the site of a UAG chain termination codon: lysine (AAG), glutamine (CAG), glutamic acid (GAG), serine (UCG), tryptophan (UGG), leucine (UUG), and tyrosine (UAC and UAU). Each suppressor will not be active against all muta-

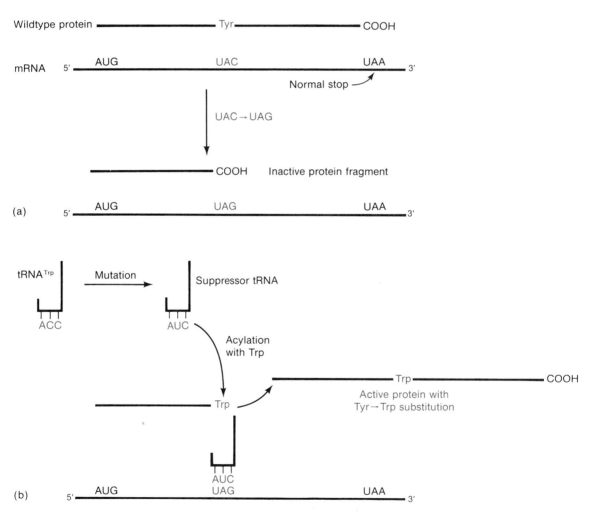

Figure 10-25 The mechanism of suppression by a suppressor tRNA molecule. (a) A UAC → UAG chain termination mutation leads to an inactive, prematurely terminated protein. (b) A mutation in the tRNA^{Trp} gene produces an altered tRNA molecule, which has a codon complementary to a UAG stop codon but which can still be acylated with tryptophan. This tRNA molecule allows the protein to be completed but with a tryptophan at the site of the original tyrosine. Suppression will occur if the substitution restores activity to the protein.

tions, because the resulting substituted amino acid may not yield a functional protein.

Four points should be noted about this type of suppression:

1. The original mutant gene will still contain the mutant base sequence—that is, the stop codon rather than the UAC tyrosine codon.

2. The suppressor tRNA suppresses not only the tyrosine → stop mutation but any chain termination mutation with a UAG at the mutant site, as long as tryptophan is an acceptable amino acid at the site.

3. A cell can survive the presence of a suppressor only if the original cell also contains two or more copies of the tRNA gene. Clearly if only one tRNA^{Trp} gene were contained in the genome and it was mutated, then the tryptophan codon UGG would no longer be read as a sense codon and chains would terminate wherever a UGG codon occurred; that is, a cell harboring such a mutant tRNA molecule would terminate virtually every protein made by the cell. However, *multiple copies of most tRNA genes exist*, so if one copy is mutated to yield a suppressor tRNA, a normal copy nearly always remains.

4. All UAG stop codons can be translated when a UAG suppressor is present. Translation of UAG by insertion of an amino acid would prevent termination of the synthesis of many proteins (those with a UAG termination codon), were it not for two facts: (i) the anticodon of the suppressor tRNA usually binds weakly to the UAG codon, so not every UAG is read as sense all of the time; and (ii) many proteins are terminated by tandem pairs of different stop codons—for example, UAG-UAA—so most polypeptide chains will still be terminated.

Suppressors also exist for chain termination mutants of the UAA and UGA type. These too are mutant tRNA molecules whose anticodons are altered by a single base change.

Suppression of missense mutations also occurs. For example, a protein that loses its activity through a mutational change from valine (an uncharged amino acid) to aspartic acid (a negatively charged amino acid) can occasionally be restored to a functional state by a missense suppressor that substitutes alanine (uncharged) for aspartic acid. Such a substitution can occur in four ways: (1) a mutation altering the anticodon enables a tRNA molecule to recognize a different codon (as in nonsense suppression); (2) a mutation changing a base adjacent to the anticodon enables a tRNA molecule to recognize two different codons (the original one and a new one); (3) a mutation outside of the anticodon loop allows a tRNA molecule to be recognized by an aminoacyl synthetase that acylates the tRNA with a different amino acid; and (4) a mutant aminoacyl synthetase charges an incorrect tRNA molecule.

In conventional notation, suppressors are given the genetic symbol *sup* followed by a number (or occasionally a letter) that distinguishes one suppressor from another. A cell lacking a suppressor is designated *sup0*.

In view of the increased number of chemicals used and present as environmental contaminants, tests for the mutagenicity of these substances have become important. Furthermore, most carcinogens are also mutagens, so that mutagenicity provides an initial screening for these hazardous agents. One simple method for screening large numbers of substances for mutagenicity is a reversion test using nutritional mutants of bacteria. In the simplest type of reversion test a compound that is a potential mutagen is added to solid growth media, known numbers of a mutant bacterium are plated, and the number of revertant colonies that arise is counted. A significant increase in the reversion frequency above that obtained in the absence of the compound tested would identify the substance as a mutagen. However, simple tests of this type fail to demonstrate the mutagenicity of a number of potent carcinogens. The explanation for this failure is that some substances are not directly mutagenic (or carcinogenic), but are converted to active compounds by enzymatic reactions that occur in the liver of animals and have no counterpart in bacteria. The normal function of these enzymes is to protect the organism from various noxious substances that occur naturally by chemically converting them to nontoxic substances. However, when the enzymes encounter certain man-made and natural compounds, they convert these substances, which may not be themselves directly harmful, to mutagens or carcinogens. The enzymes are contained in a component of liver cells called the **microsomal fraction.** Addition of the microsomal fraction of the rat liver to the growth medium as an activation system has been used to extend the sensitivity and usefulness of the reversion test system. The use of the microsomal fraction is the basis of the **Ames test** for carcinogens.

In the Ames test histidine-requiring (His$^-$) mutants of the bacterium *Salmonella typhimurium*, containing either a base substitution or a frameshift mutation, are used to test for reversion to His$^+$. In addition, the bacterial strains have been made more sensitive to mutagenesis by the incorporation of several mutant alleles that inactivate the excision-repair system and make the cells more permeable to foreign molecules. Since some mutagens act only on replicating DNA, the solid medium used contains enough histidine to support a few rounds of replication but not enough to permit formation of a colony. The procedure is the following. Rat-liver microsomal fraction is spread on the surface of the medium and bacteria are plated. Then, a paper disc saturated either with distilled water (as a control) or a solution of the compound being tested is placed in the center of the plate. The test compound diffuses outward from the disc, forming a concentration gradient. If the substance is a mutagen or is converted to a mutagen, colonies form. With a highly effective mutagen colonies will be present all over the surface of the medium, including far from the disc where the concentration is low; in contrast, with a weak mutagen colonies will form only very near the disc, where the concentration is high. The procedure is highly sensitive and permits the detection of very weak mutagens. A quantitative analysis of reversion frequency can also be carried out by incorporating known amounts of the potential mutagen in the medium. The reversion frequency depends on the concentration of the substance being tested and, for a known carcinogen or mutagen, correlates roughly with its known effectiveness.

The Ames test has now been used with thousands of substances and mixtures (such as industrial chemicals, food additives, pesticides, hair dyes, and cosmetics) and numerous unsuspected substances have been found to stimulate reversion in this test. A high frequency of reversion does not mean that the substance is definitely a carcinogen but only that it has a high probability of being so. As a result of these tests, many industries have reformulated their products: for example, the cosmetic industry has changed the formulation of many hair dyes and cosmetics to render them nonmutagenic. Ultimate proof of carcinogenicity is determined from testing for tumor formation in laboratory animals. The Ames test and several other microbiological tests are used to reduce the number of substances that have to be tested in animals since to date only a few percent of more than 300 substances known from animal experiments to be carcinogens failed to increase the reversion frequency in the Ames test.

The correlation that is found between carcinogenicity and mutagenicity is consistent with the theory that at least some tumors are caused by somatic mutations. Strong support for this theory has come in recent years from the finding that the active form of an oncogene (Section 6.8) isolated from the cells of a human bladder tumor differs from the inactive form of the gene (in normal cells) by a single base substitution. The mutation, resulting in the substitution of valine for glycine at a position near the amino terminus of the protein encoded by the gene, apparently converts the harmless cellular gene into a form active in the production of a tumor.

Problems

1. If a human gamete contains 100,000 genes and if the average mutation rate per gene is 1×10^{-5} per generation, what is the average number of *new* mutations per gamete per generation?

2. Aniridia, a form of blindness in humans caused by absence of the iris, is inherited as a simple autosomal dominant with 100 percent penetrance. The proportion of children of normal parents who have aniridia is about 1/100,000. What is the mutation rate to the dominant allele per gene per generation?

3. Experiments with *D. melanogaster* indicate that between 5 and 30 percent of the gametes carry a newly arisen mutation that is lethal when homozygous. The rate of mutation to such recessive lethals averages about 10^{-5} mutations per gene per generation. Using these data, estimate the minimum and maximum number of genes in gametes of this organism that are capable of mutation to recessive lethal alleles.

4. If a gene in a particular chromosome has a probability of mutation of 5×10^{-5} per generation, and if the gene is followed through successive generations by disregarding one of the two chromosomes produced at each replication:
   (a) What is the probability that the gene will not undergo a mutation during 10,000 consecutive generations?
   (b) What is the average number of generations before a mutation occurs in the gene?

5. A fluctuation test was carried out to estimate the mutation rate of an *E. coli* locus conferring resistance to phage T1. If 5 of the 12 small independent cultures contained no phage-resistant mutants after growth of the cultures was completed and the average number of bacterial cells per culture was 5×10^8, what is the estimated rate of mutations to T1 resistance?

6. A fluctuation test is carried out for two different genes A and B. The following data are obtained. For gene A, 22 of 40 cultures had no mutants, with $N = 5.6 \times 10^8$. For gene B, 15 of 37 cultures had no mutants with $N = 5 \times 10^8$. What are the mutation frequencies for the two genes?

7. Which of the following amino acid substitutions would be likely to yield a mutant phenotype if the change occurred in a fairly critical part of a protein? (1) Pro → His; (2) Arg → Lys; (3) Thr → Ile; (4) Val → Ile; (5) Gly → Ala; (6) His → Tyr.

8. Several hundred independent missense mutants have been isolated in the A protein of *E. coli* tryptophan synthetase, a protein having 268 amino acids. Fewer than 30 of the positions were represented with one or more mutant. Why do you think that the number of different positions represented by amino acid changes is so limited?

9. A mutation has been selected from a population of organisms treated with a particular mutagen. A population containing the mutation is then mutagenized again and revertants are sought.
   (a) Can a mutation caused by hydroxylamine be induced to revert to the wildtype base sequence by reexposure to hydroxylamine?
   (b) Can it be reverted by nitrous acid? Explain.

10. Hemoglobin C in humans was described in the chapter as a variant in which a lysine in the β hemoglobin chain is substituted for a particular glutamic acid. What mutational change in the DNA would most probably result in the HbC allele?

11. A chain termination mutation in a gene in yeast reverts spontaneously to have either leucine or tyrosine in place of the normal glutamic acid at the mutant site.
   (a) What stop codons could be present at the mutant site?
   (b) Treatment with a mutagen that causes GC-to-AT transitions stimulates reversion but mainly to tyrosine. What information does this give about the identity of the stop codon?

12. How many amino acids can substitute for tyrosine by a mutational change of a single base pair? Do not assume that you know which tyrosine codon is being used.

13. Which of the following amino acid substitutions would be expected to occur with the highest frequency among mutations induced by 5-bromouracil? (1) Met → Leu; (2) Met → Lys; (3) Leu → Pro; (4) Pro → Thr; (5) Thr → Arg.

14. The molecule 2-aminopurine is an analogue of adenine, pairing with thymine. It also pairs on occasion with cytosine. What types of mutations will be induced by 2-aminopurine?

15. The following problems are concerned with the relative rates of transitions and transversions among spontaneous mutations.
   (a) List all possible transitions and transversions resulting from a single nucleotide-pair change. If purines and pyrimidines were replaced at random during evolution, what would be the expected ratio of transitions and transversions?
   (b) A comparison of the amino acids present at homologous positions in polypeptide chains of hemoglobins and in myoglobins (the protein corresponding to hemoglobin in some animals) indicates that 293 transitions and 548 transversions probably have occurred during evolution. Use a χ^2 test to determine whether these numbers are consistent with the hypothesis that transitions and transversions have occurred in these proteins in a $1:2$ ratio.

16. Often dyes are incorporated into a solid medium to determine whether a bacterium can utilize a particular sugar as a carbon source. For instance, in eosin-methylene blue (EMB) medium containing lactose, a Lac$^+$ bacterium yields a purple colony and a Lac$^-$ colony yields a pink colony. If a population of Lac$^+$ cells is treated with a mutagen that produces Lac$^-$ mutants and the population is allowed to grow for many generations before the cells are placed on EMB-lactose medium, a few pink colonies will be found among a large number of purple ones. However, if the mutagenized cells are placed on the medium immediately after exposure to the mutagen, some colonies appear that are called *sectored* — they are purple on one side and pink on the other side. Explain the color distribution of the sectored colonies.

17. What amino acids can be present at the site of:
   (a) A UGA codon that is suppressed by a suppressor tRNA?
   (b) A UAA codon?

18. A *Neurospora* strain unable to synthesize arginine (and therefore able to grow only on a medium supplemented with this amino acid) produces a revertant arginine-independent colony. A cross is made between the revertant and a wildtype strain. What proportion of the progeny from this cross would be arginine-independent if the reversion occurred by:
   (a) A precise reversal of the nucleotide change that produced the original *arg*$^-$ mutant allele?
   (b) A mutation to a suppressor of the *arg*$^-$ gene occurring in a second gene located in a different chromosome?
   (c) A suppressor mutation occurring in a second gene located 10 genetic map units from the *arg*$^-$ locus in the same chromosome?

19. A mutation is isolated in a gene *a*. A revertant is selected and the reverse mutation is mapped in gene *b*. The phenotype of this double mutant is A^+B^+. Separation of the *a* mutation and the new change in gene *b* by crossing over show that the alteration in *b* yields organisms with a B^- phenotype. A few other, but not all, mutations in gene *a* revert in this way. Many mutations in gene *b* are independently isolated and revertants of these are isolated. A few of these revertants prove to be mutations in gene *a*. What do these observations indicate about the relation between the gene products of the two genes within the cell?

20. Two hundred Leu$^-$ mutants of a bacterial strain are examined separately to determine reversion frequencies. Of these 90 revert at a frequency of 10^{-5}, 98 at 3×10^{-6}, 6 at 3×10^{-11}, and 6 at 10^{-10}.
   (a) What type of mutant is probably contained in the class whose reversion frequency is 10^{-10}—single point mutations, double point mutations, or deletions?
   (b) Can you say anything from these frequencies about whether any of the classes of mutations are chain termination mutations?

21. *E. coli* phage λ loses the ability to form plaques if it is irradiated with ultraviolet light, owing to the production of pyrimidine dimers in the phage DNA.
   (a) If a phage sample that is irradiated is plated both on wildtype *E. coli* and on a *uvrA* mutant, the fraction of the phage able to form plaques will be smaller on the mutant strain than on the wildtype strain. Explain this phenomenon.
   (b) If the wildtype *E. coli* strain used to form the lawn is given a small dose of ultraviolet light (not enough to kill many bacteria) before the cells are placed on the agar medium and before the phage are added to the bacteria, the fraction of the irradiated λ phage that produce plaques will be considerably higher than if the phage are plated on unirradiated bacteria. Furthermore, unirradiated wildtype λ forms turbid plaques and occasional clear-plaque mutants (1 mutant per 10,000 progeny). The fraction of clear plaques among phage surviving ultraviolet irradiation is also higher when the phage are plated on irradiated bacteria than on unirradiated bacteria. Explain these two observations, which are manifestations of the same phenomenon.

22. Five mutations induced by x irradiation in the *rII* region of phage T4 were selected for analysis. Two of these mutations were found to undergo occasional reversion to wildtype, and the other three were never observed to revert. When the mutants were crossed in all possible pairs to determine whether recombination occurs between them, the following data were obtained. A + indicates the occurrence of recombination to yield occasional wildtype progeny, and a 0 indicates the absence of recombination.

Mutant	1	2	3	4	5
1	0	0	0	+	0
2		0	+	0	0
3			0	+	+
4				0	0
5					0

Explain the lack of recombination between certain pairs of mutations, and draw a map showing the relative positions of the mutations.

23. It is sometimes difficult to select mutations in certain genes. The following technique has been used to isolate such mutations in bacteria. If a mutation is desired in a gene *a*, a bacterial strain is first obtained that carries an easily selectable b^- mutation in a closely linked gene *b*. In addition, phage P1 is grown on a b^+ strain to obtain b^+-transducing particles. The P1 population containing the transducing particles is then exposed to a mutagen such as hydroxylamine; then the b^- strain, in which an a^- mutation is desired, is infected with the mutagenized phage population. What step or steps would you carry out next to obtain an a^- mutation, and what is the principle underlying the technique?

C H A P T E R 11

Regulation of Gene Activity

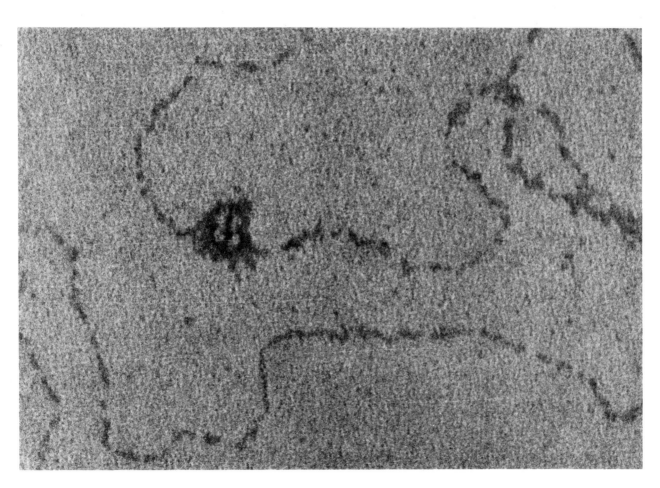

The number of protein molecules produced per unit time by active genes varies from gene to gene, satisfying the needs of a cell and sometimes also avoiding wasteful synthesis. The different rates result mainly from different efficiencies of either recognition of a promoter by RNA polymerase or initiation of translation. However, the flow of genetic information is regulated in other ways also. For example, many gene products are needed only on occasion, and regulatory mechanisms of an on-off type exist that enable such products to be present only when demanded by external conditions. More subtly regulated systems can adjust the intracellular concentration of a particular protein in response to needs imposed by the environment. In general, the synthesis of particular gene products is controlled by mechanisms collectively called **gene regulation.**

The regulatory systems of prokaryotes and eukaryotes are somewhat different from each other. Prokaryotes are generally free-living unicellular organisms that grow and divide indefinitely as long as environmental conditions are suitable and the supply of nutrients is adequate. Thus, their regulatory systems are geared to provide the maximum growth rate in a particular environment, except when such growth would be detrimental. This strategy seems to apply to the free-living unicells such as yeast, algae, and protozoa, though less information is available about these organisms than for bacteria.

The requirements of tissue-forming eukaryotes are different from those of prokaryotes. In a developing organism—for example, in an embryo—a cell must not only grow and produce many progeny cells but also must undergo considerable change in morphology and biochemistry and then maintain the changed state. Furthermore, during the growth and cell-division phases of the organism, these cells are challenged less by the environment than are bacteria in that the composition and concentration of their growth media do not change drastically with time. Some examples of such media are blood, lymph, or

Facing page: Electron micrograph of an *E. coli* phage λ repressor molecule bound to the operator region of λ DNA. (Courtesy of Christine Brack.)

other body fluids, or, in the case of marine animals, sea water. Finally, in an adult organism, growth and cell division in most cell types have stopped, and each cell needs only to maintain itself and its properties. Many other examples could be given; the main point is that because a typical eukaryotic cell faces different contingencies than a bacterium does, the regulatory mechanisms of eukaryotes and prokaryotes are not the same.

In this chapter we consider the basic mechanisms of metabolic regulation and present several examples of well-understood regulated systems.

11.1 Principles of Regulation

The best-understood regulatory mechanisms are those used by bacteria and by phages. In these organisms an on-off regulatory activity is obtained by controlling transcription—that is, synthesis of a particular mRNA is allowed when the gene product is needed and inhibited when the product is not needed. In bacteria, few examples are known of switching a system completely off. When transcription is in the off state, a basal level of gene expression almost always remains, often consisting of only one or two transcriptional events per cell generation; hence, very little synthesis of the gene product occurs. For convenience, when discussing transcription, the term "off" will be used, but it should be kept in mind that usually what is meant is "very low." In only one case in bacteria, namely, in spores, is expression of most genes totally turned off. In eukaryotes complete turning off of a gene is quite prevalent. Regulatory mechanisms other than the on-off type are also known in both prokaryotes and eukaryotes; for example, the activity of a system may be modulated from fully on to partly on, rather than to off.

In bacterial systems, when several enzymes act in sequence in a single metabolic pathway, usually either all or none of these enzymes are produced. This phenomenon, which is called **coordinate regulation,** results from control of the synthesis of a single polycistronic mRNA molecule encoding all of the gene products. This type of regulation does not occur in eukaryotes because eukaryotic mRNA is usually monocistronic, as discussed in Chapter 9.

Several mechanisms of regulation of transcription are common; the particular one used often depends on whether the enzymes being regulated act in degradative or synthetic metabolic pathways. For example, in a multistep degradative system the availability of the molecule to be degraded frequently determines whether the enzymes in the pathway will be synthesized. In contrast, in a biosynthetic pathway the final product is often the regulatory molecule. Even in a system in which a single protein molecule (not necessarily an enzyme) is translated from a monocistronic mRNA molecule, the protein may be **autoregulated**—that is, the protein itself may inhibit initiation of transcription and high concentrations of the protein will result in less transcription of the mRNA that encodes the protein. The molecular mechanisms for each of the regulatory patterns vary quite widely but usually fall in one of two major categories—**negative regulation** and **positive regulation** (Figure 11-1). In a negatively regulated system, an inhibitor is present in the cell and

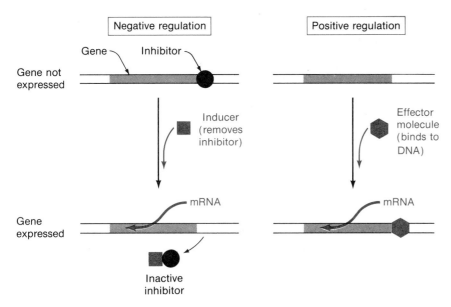

Figure 11-1 The distinction between negative and positive regulation. In negative regulation an inhibitor, bound to the DNA molecule, must be removed before transcription can occur. In positive regulation an effector molecule must bind to the DNA. A system may also be regulated both positively and negatively; in such a case, the system is "on" when the positive regulator is bound to the DNA and the negative regulator is not bound to the DNA.

prevents transcription. An antagonist of the inhibitor, generally called an **inducer,** is needed to allow initiation of transcription. In a positively regulated system, an effector molecule (which may be a protein, a small molecule, or a molecular complex) activates a promoter; no inhibitor must be overridden. Negative and positive regulation are not mutually exclusive, and some systems are both positively and negatively regulated, utilizing two regulators to respond to different conditions in the cell.

A degradative system may be regulated either positively or negatively. In a biosynthetic pathway, the final product usually negatively regulates its own synthesis; in the simplest type of negative regulation, absence of the product increases its synthesis and presence of the product decreases its synthesis.

In both prokaryotic and eukaryotic systems enzyme *activity* rather than enzyme *synthesis* may also be regulated. That is, the enzyme may be present but its activity is turned on or off. When this occurs, the product of the enzymatic reaction or, in the case of a biosynthetic pathway, the end product of a sequence of reactions usually inhibits the enzyme. This mode of regulation is called **feedback inhibition.** Small molecules other than reaction products are also frequently used either to activate or inhibit a particular enzyme. Such a molecule is termed an **allosteric effector molecule.**

The modes of regulation used by prokaryotes are better understood than those of eukaryotes, though information about eukaryotic systems is accumulating at an extraordinary rate. The next six sections of this chapter are concerned with several prokaryotic systems. Section 11.8 covers regulation in eukaryotes.

11.2 The *E. coli* Lactose System and the Operon Model

Metabolic regulation was first studied in detail in the system in *E. coli* responsible for degradation of the sugar lactose, and most of the terminology used to describe regulation has come from genetic analysis of this system by Francois Jacob and Jacques Monod.

Lac⁻ Mutants

In *E. coli* two proteins are necessary for the metabolism of lactose—the enzyme **β-galactosidase,** which cleaves lactose (a β-galactoside) to yield galactose and glucose, and a carrier, **lactose permease,** which is required for the entry of lactose into a cell. The existence of two different proteins in the lactose-utilization system was first shown by a combination of genetic experiments and biochemical analysis.

First, hundreds of mutants unable to use lactose as a carbon source—Lac⁻ mutants—were isolated. Some of the mutations were in the *E. coli* chromosome and others were in F′*lac*, a plasmid carrying the genes for lactose utilization. By performing F′ × F⁻ matings partial diploids having the genotypes F′*lac⁻/lac⁺* or F′*lac⁺/lac⁻* were constructed, as described in Chapter 7. (The genotype of the plasmid is given to the left of the diagonal line and that of the chromosome to the right.) It was observed that these diploids always produced a Lac⁺ phenotype (that is, they made β-galactosidase); none produced an inhibitor that prevented functioning of the *lac* gene. Other partial diploids were then constructed in which both the F′*lac* plasmid and the chromosome carried *lac⁻* genes; these were tested for the Lac⁺ phenotype, with the result that all of the mutants initially isolated could be placed into two complementation groups, *lacZ* and *lacY*. The partial diploids F′*lacY⁻ lacZ⁺/ lacY⁺ lacZ⁻* and F′*lacY⁺ lacZ⁻/lacY⁻ lacZ⁺* had a Lac⁺ phenotype, producing β-galactosidase, but the genotypes F′*lacY⁻ lacZ⁺/lacY⁻ lacZ⁺* and F′*lacY⁺ lacZ⁻/lacY⁺ lacZ⁻* had the Lac⁻ phenotype. The existence of two complementation groups was good evidence that the *lac* system consisted of at least two genes ("at least," because mutations had not yet been obtained in other genes).

Further experimentation was needed to establish the precise function of each gene. Experiments in which cells were placed in a medium containing ¹⁴C-labeled (radioactive) lactose showed that no [¹⁴C]lactose would enter a *lacY⁻* cell, whereas it readily penetrated a *lacZ⁻* mutant. Treatment of a *lacY⁻* cell with lysozyme, an enzyme that destroys part of the cell wall and makes it permeable, enabled radioactive lactose to enter a *lacY⁻* cell, indicating that the *lacY* gene is probably concerned with transport of lactose through the cell membrane into the cell and is the structural gene for *lac* permease. Enzymatic assays showed that β-galactosidase is present in *lac⁺* but not *lacZ⁻* cells; these results provided the initial evidence that the lacZ *gene is the structural gene for β-galactosidase.* A final important result—that the *lacY* and *lacZ* genes are adjacent—was obtained by genetic mapping.

Regulation of the lac *System: Inducible and
Constitutive Synthesis and Repression*

The on-off nature of the lactose-utilization system is evident in the following
observations:

1. If a culture of Lac$^+$ *E. coli* is growing in a medium lacking lactose
   or any other β-galactoside, the intracellular concentrations of β-
   galactosidase and permease are exceedingly low—roughly one or two
   molecules per bacterium. However, if lactose is present in the growth
   medium, the number of each of these molecules is about 10^5-
   fold higher.

2. If lactose is added to a Lac$^+$ culture growing in a lactose-free medium
   (also lacking glucose, a point that will be discussed shortly), both
   β-galactosidase and permease are synthesized nearly simultaneously,
   as shown in Figure 11-2. Analysis of the total mRNA present in the
   cells before and after addition of lactose shows that no *lac* mRNA (the
   mRNA that encodes β-galactosidase and permease) is present before
   lactose is added and that the addition of lactose triggers synthesis of
   lac mRNA. The analysis is done by growing cells in a radioactive
   medium in which newly synthesized mRNA is radioactive, then iso-

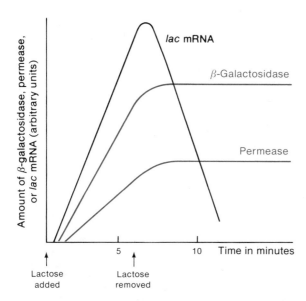

Figure 11-2 The "on-off" nature of the *lac* system. *Lac* mRNA appears soon after lactose
or another inducer is added; β-galactosidase and permease appear at nearly the same time
but are delayed with respect to mRNA synthesis because of the time required for translation.
When lactose is removed, no more *lac* mRNA is made and the amount of *lac* mRNA
decreases owing to the usual degradation of mRNA. Both β-galactosidase and permease are
stable proteins: their amounts remain constant even when synthesis ceases. Ultimately, their
concentration per cell decreases as a result of cell division.

lating the mRNA, and finally allowing it to renature with the DNA of a λ *lac*-transducing particle. Since the only radioactive mRNA that will renature to the *lac*-containing DNA is *lac* mRNA, the amount of radioactive DNA-RNA hybrid molecules is a measure of the amount of *lac* mRNA.

These two observations led to the view that the lactose system is **inducible** and that lactose is an **inducer.**

Lactose itself is rarely used in experiments to study the induction phenomenon for a variety of reasons; one important reason is that the β-galactosidase that is synthesized catalyzes the cleavage of lactose and results in a continual decrease in lactose concentration, which complicates the analysis of many types of experiments (for example, kinetic experiments). Instead, a sulfur-containing analogue of lactose is used—either isopropylthiogalactoside (IPTG) or thiomethylgalactoside (TMG); these analogues are inducers but are not substrates of β-galactosidase.

Mutants have also been isolated in which *lac* mRNA is synthesized (hence also β-galactosidase and permease) in *both* the presence and the absence of an inducer. These mutants provided the key to understanding induction because they eliminated regulation; they were termed **constitutive.** Complementation tests—again with partial diploids carrying two constitutive mutations, one in the chromosome and the other in a plasmid—showed that the mutants fall into two groups termed *lacI* and *lacO^c*. The characteristics of the mutants are shown in Table 11-1. The *lacI^-* mutants are recessive (entries 3, 4). In the absence of an inducer a *lacI^+* cell fails to make *lac* mRNA, whereas this mRNA is made by a *lacI^-* mutant. Thus, the *lacI* gene is apparently a regulatory gene *whose product is an inhibitor that keeps the system turned off.* A *lacI^-* mutant lacks the inhibitor and hence is constitutive. Wildtype copies of the *lacI*-gene product are present in a *lacI^+/lacI^-* partial diploid, so the system is inhibited. The *lacI*-gene product, a protein molecule, is called the ***lac* repressor.** Genetic mapping experiments place the *lacI* gene adjacent to the *lacZ* gene and establish the gene order *lacI lacZ lacY*. How the *lacI* repressor prevents synthesis of *lac* mRNA will be explained shortly.

Table 11-1 Characteristics of partial diploids having several combinations of *lacI* and *lacO* alleles

Genotype	Constitutive or inducible synthesis of *lac* mRNA
1. F'*lacO^c lacZ^+ /lacO^+ lacZ^+*	Constitutive
2. F'*lacO^+ lacZ^+ /lacO^c lacZ^+*	Constitutive
3. F'*lacI^- lacZ^+ /lacI^+ lacZ^+*	Inducible
4. F'*lacI^+ lacZ^+ /lacI^- lacZ^+*	Inducible
5. F'*lacO^c lacZ^+ /lacI^- lacZ^+*	Constitutive
6. F'*lacO^c lacZ^- /lacO^+ lacZ^+*	Inducible
7. F'*lacO^c lacZ^+ /lacO^+ lacZ^-*	Constitutive

The *lacO^c* mutants are dominant (entries 1, 2, and 5 in Table 11-1), but the dominance is evident only in certain combinations of *lac* mutations, as can be seen by examining the partial diploids shown in entries 6 and 7. Both combinations are Lac$^+$, because a functional *lacZ* gene is present. However, in the combination shown in entry 6, synthesis of β-galactosidase is inducible even though a *lacO^c* mutation is present. The difference between the two combinations in entries 6 and 7 is that in entry 6 the *lacO^c* mutation is carried on a DNA molecule that also has a *lacZ$^-$* mutation, whereas in entry 7, *lacO^c* and *lacZ$^+$* are *carried on the same DNA molecule*. Thus, a *lacO^c* mutation causes constitutive synthesis of β-galactosidase only when the *lacO^c* and *lacZ$^+$* alleles are located *on the same DNA molecule*; the *lacO^c* mutation is said to be **cis-dominant,** since only genes *cis* to the mutation are expressed in dominant fashion. Confirmation of this conclusion comes from an important biochemical observation: the mutant enzyme (encoded in the *lacZ$^-$* sequence) is synthesized constitutively in a *lacO^{c}lacZ$^-$/lacO$^+$lacZ$^+$* partial diploid (entry 6), whereas the wildtype enzyme (encoded in the *lacZ$^+$* sequence) is synthesized only if an inducer is added. All *lacO^c* mutations are located between the *lacI* and *lacZ* genes; thus, the gene order of the four elements of the *lac* system is

$$lacI \quad lacO \quad lacZ \quad lacY$$

An important feature of all *lacO^c* mutations is that they cannot be complemented (a feature of all *cis*-dominant mutations). That is, a *lacO$^+$* allele cannot alter the constitutive activity of a *lacO^c* mutation. Thus, *lacO* does not encode a diffusible product and must define a *site* or a noncoding region of the DNA rather than a gene. This site determines whether synthesis of the product of the adjacent *lacZ* gene is inducible or constitutive. The *lacO* region is called the **operator.**

The Operon Model

The regulatory mechanism of the *lac* system was first explained by the **operon model,** which has the following features (Figure 11-3):

1. The lactose-utilization system consists of two kinds of components—
    structural genes needed for transport and metabolism of lactose, and
    regulatory elements (the *lacI* gene, the *lacO* operator, and the *lac* promoter). Together these components comprise the **lac operon.**

2. The products of the *lacZ* and *lacY* genes are encoded in a single polycistronic mRNA molecule. (This mRNA molecule contains a third gene, denoted *lacA*, which encodes the enzyme transacetylase. This enzyme is used in the metabolism of certain β-galactosides other than lactose and will not be of further concern.)

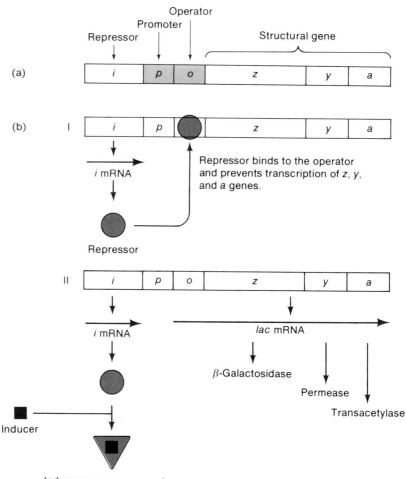

Figure 11-3 A map of the *lac* operon, not drawn to scale; the *p* and *o* sites are actually much smaller than the other genes. (b) A diagram of the *lac* operon in (I) repressed and (II) induced states. The inducer alters the shape of the repressor, so the repressor can no longer bind to the operator. The common abbreviations *i, p, o, z, y,* and *a* are used instead of *lacI, lacO, ...* The *lacA* gene will be described shortly.

3. The promoter for the *lacZ lacY lacA* mRNA molecule is immediately adjacent to the *lacO* region. This location has been substantiated by the isolation and mapping of promoter mutants ($lacP^-$) that are completely incapable of making either β-galactosidase or permease, because no *lac* mRNA is made.

4. The *lacI*-gene product, the repressor, binds to a unique sequence of DNA bases, namely, the operator.

5. When the repressor is bound to the operator, initiation of transcription of *lac* mRNA by RNA polymerase is prevented.

6. Inducers stimulate mRNA synthesis by binding to and inactivating the repressor, a process called **derepression.** Thus, in the presence of an inducer the operator is unoccupied, and the promoter is available for initiation of mRNA synthesis.

Note that regulation of the operon requires that the *lacO* operator be adjacent to the structural genes of the operon (*lacZ, lacY, lacA*), but proximity of the *lacI* gene is not necessary, because the *lacI* repressor is a soluble protein and is therefore diffusible throughout the cell.

The operon model is supported by a wealth of experimental data and explains many of the features of the *lac* system as well as numerous other negatively regulated genetic systems. One aspect of the regulation of the *lac* operon—the effect of glucose—has not yet been discussed. Examination of this feature indicates that the *lac* operon is also subject to positive regulation, as will be seen in the next section.

Positive Regulation of the lac Operon

The function of β-galactosidase in lactose metabolism is to form glucose by cleaving lactose. (The other cleavage product, galactose, is also ultimately converted to glucose by the enzymes of the galactose operon.) Thus, if both glucose and lactose are present in the growth medium, activity of the *lac* operon is not needed, and indeed, no β-galactosidase is formed until virtually all of the glucose in the medium is consumed. The lack of synthesis of β-galactosidase is a result of lack of synthesis of *lac* mRNA. No *lac* mRNA is made in the presence of glucose, because in addition to an inducer to inactivate the *lacI* repressor, another element is needed for initiating *lac* mRNA synthesis; the activity of this element is regulated by the concentration of glucose. However, the inhibitory effect of glucose on expression of the *lac* operon is quite indirect.

The small molecule **cyclic AMP (cAMP)** is universally distributed in animal tissues, and in multicellular eukaryotic organisms it is important in regulating the action of many hormones (Figure 11-4). It is also present in *E. coli* and many other bacteria. Cyclic AMP is synthesized enzymatically by **adenyl cyclase,** and its concentration is regulated indirectly by glucose metabolism. When bacteria are growing in a medium containing glucose, the cAMP concentration in the cells is quite low. In a medium containing glycerol or any carbon source that cannot enter the biochemical pathway used to metabolize glucose (the glycolytic pathway), or when the bacteria are otherwise starved of an energy source, the cAMP concentration is high (Table 11-2). The mechanism by which glucose controls the cAMP concentration is poorly understood; the significant point is that *cAMP regulates the activity of the* lac *operon* (and other several other operons as well).

E. coli (and many other bacterial species) contain a protein called the **catabolite activator protein (CAP),** which is encoded in a gene called *crp.* Mutants of either *crp* or the adenyl cyclase gene are unable to synthesize *lac* mRNA, indicating that both CAP function and cAMP are required for *lac*

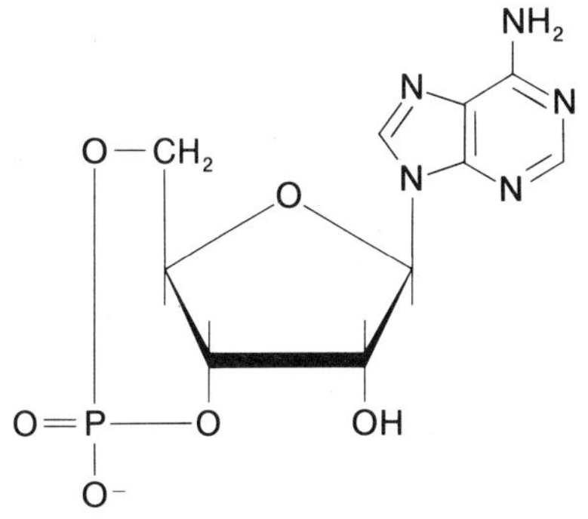

Figure 11-4 Structure of cyclic AMP. The noncyclic nucleotide is shown in Figure 4-8.

Table 11-2 Concentration of cyclic AMP in cells growing in media having the indicated carbon sources

Carbon source	cAMP concentration
Glucose	Low
Glycerol	High
Lactose	High
Lactose + glucose	Low
Lactose + glycerol	High

mRNA synthesis. CAP and cAMP bind to one another, forming a unit denoted **cAMP-CAP,** which is an active regulatory element in the *lac* system. The requirement for cAMP-CAP is independent of the *lacI* repression system since *crp* and adenyl cyclase mutants are unable to make *lac* mRNA even if a *lacI⁻* or a *lacOᶜ* mutation is present. *The cAMP-CAP complex must be bound to a base sequence in the DNA in the promoter region in order for transcription to occur* (Figure 11-5). Thus, *cAMP-CAP is a positive regulator*, in contrast with the repressor, and the *lac* operon is independently regulated both positively and negatively.

The precise mechanism by which cAMP-CAP stimulates and the repressor inhibits transcription is not known in detail. However, by using mixtures containing purified *lac* DNA, *lac* repressor, cAMP-CAP, and RNA polymerase two points have been established:

1. In the absence of cAMP-CAP, RNA polymerase binds only weakly to the promoter, but its binding is stimulated when cAMP-CAP is also bound to the DNA. The weak binding rarely leads to initiation of transcription, because the correct interaction between RNA polymerase and the promoter does not occur.

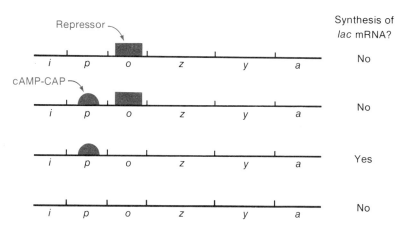

Figure 11-5 Four states of the *lac* operon: the *lac* mRNA is synthesized only if cAMP-CAP is present and repressor is absent.

2. If the repressor is first bound to the operator, RNA polymerase cannot stably bind to the promoter.

These results are sufficient to explain how lactose and glucose function to regulate transcription of the *lac* operon.

Differential Translation of the Genes in lac *mRNA*

The ratios of the number of copies of β-galactosidase, permease, and trans-acetylase (the third structural gene of the system), are 1.0:0.5:0.2. These differences, which are examples of **translational regulation,** are achieved in two ways:

1. The *lacZ* gene is translated first (Figure 11-6). Frequently, the *lac* mRNA molecule detaches from its translating ribosome following chain termination. The frequency with which this occurs is a function of the probability of reinitiation at each subsequent AUG codon. Thus, there is a gradient in the amount of polypeptide synthesis from the 5′ terminus to the 3′ terminus of the mRNA molecule, an effect called **polarity** that occurs with most polycistronic mRNA molecules.

2. Degradation of *lac* mRNA is initiated more frequently in the *lacA* gene than in the *lacY* gene and more often in the *lacY* gene than in the *lacZ* gene. Hence, at any given instant, there are more complete copies of the *lacZ* gene than of the *lacY* gene, and more copies of the *lacY* gene than of the *lacA* gene. In prokaryotes this mode of regulation occurs repeatedly:

The overall expression of activity of an operon is regulated by controlling transcription of a polycistronic mRNA, and the relative concentrations of the proteins encoded in the mRNA are determined by controlling the frequency of initiation of translation of each cistron.

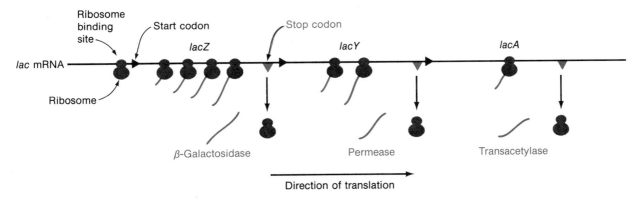

Figure 11-6   One explanation for polarity in the *lac* operon. All ribosomes attach to the mRNA molecule at the ribosome binding site. At each stop codon some ribosomes detach. Thus, the number of ribosomes translating each gene segment decreases for each subsequent gene.

However, the mechanism by which transcription is regulated varies from one system to the next. An inducer-repressor system is common to many operons responsible for degradative metabolism and cAMP-CAP is an element in many carbohydrate-degrading systems. However, particular features of the regulatory mechanisms differ; for example, two promoters are present in the galactose operon, and the arabinose operon is mainly positively regulated.

11.3 The Tryptophan Operon, a Biosynthetic System

The tryptophan (*trp*) operon of *E. coli* is responsible for the synthesis of the amino acid tryptophan. Regulation of this operon occurs in such a way that when tryptophan is present in the growth medium, the *trp* operon is not active. That is, when adequate tryptophan is present, transcription of the operon is inhibited; however, when the supply is insufficient, transcription occurs. The *trp* operon is quite different from the *lac* operon in that tryptophan acts directly in the repression system rather than as an inducer. Furthermore, since the *trp* operon encodes a set of biosynthetic rather than degradative enzymes, neither glucose nor cAMP-CAP functions in operon activity.

A simple on-off system, as in the *lac* operon, is not optimal for a biosynthetic pathway; a situation may arise in nature in which some tryptophan is available, but not enough to allow normal growth if synthesis of tryptophan were totally shut down. Tryptophan starvation when the supply of the amino acid is inadequate is prevented by a *modulating system* in which *the amount of transcription in the derepressed state is determined by the concentration of tryptophan*. This mechanism is found in many operons responsible for amino acid biosynthesis.

Tryptophan is synthesized in five steps, each requiring a particular enzyme. In the *E. coli* chromosome the genes encoding these enzymes are adjacent to one another in the same order as their use in the biosynthetic pathway; they are translated from a single polycistronic mRNA molecule and are called *trpE*, *trpD*, *trpC*, *trpB*, and *trpA*. The *trpE* gene is the first one

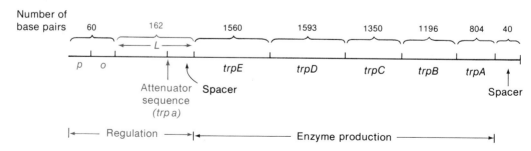

Figure 11-7 The *E. coli trp* operon. For clarity, the regulatory region is enlarged with respect to the coding region. The correct size of each region is indicated by the numbers of base pairs. *L* is the leader. The regulatory elements are shown in red.

translated. Adjacent to the *trpE* gene are the promoter, the operator, and two regions called the **leader** and the **attenuator,** which are designated *trpL* and *trp a* (not *trpA*), respectively (Figure 11-7). The repressor gene *trpR* is located quite far from this gene cluster.

The regulatory protein of the repression system of the *trp* operon is the *trpR*-gene product. Mutations either in this gene or in the operator cause constitutive initiation of transcription of *trp* mRNA, as in the *lac* operon. This protein, which is called the *trp* **aporepressor,** does not bind to the operator unless tryptophan is present. The aporepressor and the tryptophan molecule join together to form the active *trp* repressor, which binds to the operator. The reaction scheme is:

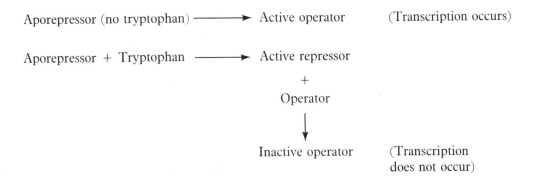

Thus, only when tryptophan is present does an active repressor molecule inhibit transcription. When the external supply of tryptophan is depleted (or reduced substantially), the equilibrium in the equation above shifts to the left, the operator is unoccupied, and transcription begins. This is the basic on-off regulatory mechanism.

In the on state a finer control, in which the enzyme concentration is varied by the amino acid concentration, is effected by (1) premature termination of transcription before the first structural gene is reached and (2) regulation of the frequency of this termination by the internal concentration of tryptophan. This modulation is accomplished in the following way.

A 162-base leader (noncoding) sequence is present at the 5' end of the *trp* mRNA molecule. A mutant in which bases 123 through 150 are deleted synthesizes the *trp* enzymes in both derepressed cells and constitutive mutants at six times the normal rate, which indicates that bases 123–150 have regulatory activity. In nonmutant bacteria, after initiation of transcription most of the mRNA molecules terminate in this 28-base region, unless no tryptophan is present. The result of such termination is an RNA molecule that contains only 140 nucleotides and stops short of the genes encoding the *trp* enzymes. This 28-base region, in which termination occurs and is regulated, is called the **attenuator.** The base sequence (Figure 11-8) of the region in which termination occurs contains the usual features of a termination site—namely, a potential stem-and-loop configuration in the mRNA followed by a sequence of eight AT pairs (see Figure 9-10).

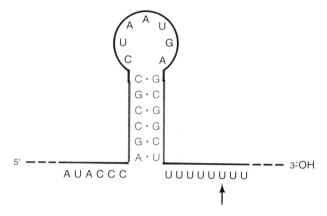

Figure 11-8 The terminal region of the *trp* attenuator sequence. The arrow indicates the final uridine in attenuated RNA. Nonattenuated RNA continues past that base. The red bases form the hypothetical stem sequence that is shown.

The leader sequence has several notable features:

1. An AUG codon and a later UGA stop codon in the same reading frame define a region encoding a polypeptide consisting of 14 amino acids—called the **leader polypeptide** (Figure 11-9).

2. Two adjacent tryptophan codons are located in the leader polypeptide at positions 10 and 11. We will see the significance of these repeated codons shortly.

3. Four segments of the leader RNA—denoted 1, 2, 3, and 4—are capable of base-pairing in two different ways—namely, forming either the base-paired regions 1–2 and 3–4 or just the region 2–3 (Figure 11-10). Two of these paired regions, 1–2 and 3–4, are also present in purified *trp* leader mRNA. The paired region 3–4 is in the terminator recognition region.

This arrangement enables premature termination to occur in the *trp* leader region by the following mechanism.

Termination of transcription is mediated through translation of the leader peptide region. Because there are two tryptophan codons in this sequence, the translation of the sequence is sensitive to the concentration of charged tRNA^{Trp}. That is, if the supply of tryptophan is inadequate, the amount of

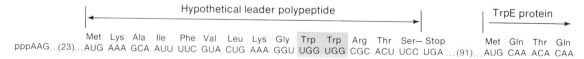

Figure 11-9 The sequence of the peptides in the *trp* leader mRNA, showing the hypothetical leader polypeptide, the two tryptophan codons (shaded red), and the beginning of the TrpE protein. The numbers 23 and 91 are the numbers of bases in sequences that, for clarity, are not shown.

charged tRNA^{Trp} will be insufficient and, hence, translation will be slowed at the tryptophan codons. Three points should be noted. (1) Transcription and translation are coupled, as is usually true in bacteria. (2) Since sequences 2 and 3 are paired in the duplex segments 1–2 and 3–4, then the region 2–3 cannot be present simultaneously with 1–2 and 3–4. (3) All base pairing is eliminated in the segment of the mRNA that is in contact with the ribosome.

Figure 11-10 shows that the end of the *trp* leader peptide is in segment 1. Usually a translating ribosome is in contact with about ten bases in the mRNA past the codons being translated. Thus, when the final codons of the leader are being translated, segments 1 and 2 are not paired. In a coupled transcription-translation system, the leading ribosome is not far behind the RNA polymerase. Thus, if the ribosome is in contact with segment 2 when synthesis of segment 4 is being completed, then segments 3 and 4 are free to form the duplex region 3–4 without segment 2 competing for segment 3. The presence of the 3–4 stem-and-loop configuration allows termination to occur when the terminating sequence of seven uridines is reached. If there is no added tryptophan, the concentration of charged tRNA^{Trp} becomes inadequate and occasionally a translating ribosome is stalled for an instant at the tryptophan codons. These codons are located sixteen bases before the beginning of segment 2. Thus, segment 2 is free before segment 4 has been synthesized and the duplex 2–3 region (the antiterminator) can form. In the absence of the 3–4 stem and loop, termination does not occur and the complete mRNA molecule is made, including the coding sequences for the *trp* genes.

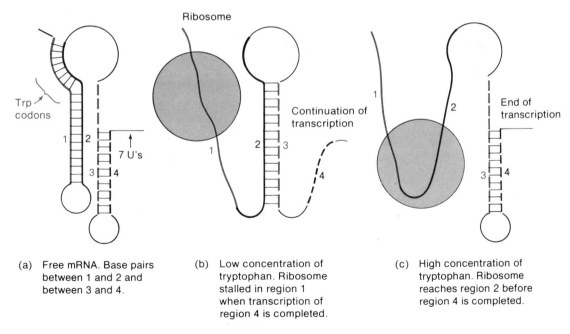

(a) Free mRNA. Base pairs between 1 and 2 and between 3 and 4.

(b) Low concentration of tryptophan. Ribosome stalled in region 1 when transcription of region 4 is completed.

(c) High concentration of tryptophan. Ribosome reaches region 2 before region 4 is completed.

Figure 11-10 Model for the mechanism of attenuation in the *E. coli trp* operon. The tryptophan codons in (a) are those highlighted in red in Figure 11-9.

Hence, if tryptophan is present in excess, termination occurs and little enzyme is synthesized; if tryptophan is absent, termination does not occur and the enzymes are made. At intermediate concentrations the fraction of initiation events that result in completion of *trp* mRNA will depend on how often translation is stalled, which in turn depends on the concentration of tryptophan.

Many operons responsible for amino acid biosynthesis (for example, the leucine, isoleucine, phenylalanine, and histidine operons) are regulated by attenuators equipped with the base-pairing mechanism for competition described for the *trp* operon. In the histidine operon, which also has an attenuator system (that is, prematurely terminated mRNA), a similar base sequence encodes a leader polypeptide having *seven* adjacent histidine codons (Figure 11-11(a)). In the phenylalanine operon seven phenylalanine codons are also present in the leader but they are divided into three groups (panel (b)).

A regulatory mechanism of this type cannot occur in eukaryotes because transcription and translation cannot be coupled; that is, transcription occurs in the nucleus and translation takes place in the cytoplasm, as explained in Section 9.8.

11.4 Autoregulation

Many proteins are made from transcripts that are initiated at a constant rate. However, with some gene products the requirements of a cell vary greatly and the rate of transcription of the corresponding gene matches the need. One mechanism for the regulation of synthesis of monocistronic mRNA is **autoregulation.** In the simplest autoregulated systems the gene product is also a repressor: it binds to an operator site adjacent to the promoter. When the concentration of the gene product exceeds what the cell can use, a product molecule occupies the operator and transcription will be inhibited. At a later time the need may be greater, molecules will be consumed, and the concentration of unbound molecules will decrease. In these conditions, the molecule bound to the operator will leave the site, the promoter will be free, and transcription will occur. The synthesis of most repressor proteins—for example, the *lacI* product and the phage λ immunity repressor—and many enzymes needed at all times are autoregulated.

(a)
```
     Met Thr Arg Val Gln Phe Lys His His His His His His His Pro Asp
  5' AUG ACA CGC GUU CAA UUU AAA CAC CAC CAU CAU CAC CAU CAU CCU GAC 3'
```

(b)
```
     Met Lys His Ile Pro Phe Phe Phe Ala Phe Phe Phe Thr Phe Pro Stop
  5' AUG AAA CAC AUA CCG UUU UUC UUC GCA UUC UUU UUU ACC UCC CCC UGA 3'
```

Figure 11-11 Amino acid sequence of the leader peptide and base sequence of the corresponding portion of mRNA from (a) the histidine operon and (b) the phenylalanine operon. The repetition of these amino acids is emphasized with red shading.

In systems such as the *lac* operon, regulation is accomplished by controlling the accessibility of a promoter to RNA polymerase. In several phage systems promoters are not repressed, yet they are not recognized by the bacterial RNA polymerase. In other phage systems initiation of transcription is constitutive and regulation of termination at sites between particular genes controls gene expression. In the former cases, transcription of certain genes is controlled either by synthesizing specific RNA polymerases with different promoter-recognizing properties or by modifying the bacterial RNA polymerase, thereby changing the ability to recognize promoters. In the latter systems RNA polymerases are modified so that specific termination sites are ineffective.

E. coli phage T4 has a system for regulating the timing of synthesis of numerous classes of mRNA molecules. Early in the life cycle the bacterial RNA polymerase initiates transcription at a single class of hostlike promoter—the only class that it can recognize. Some of these transcripts encode proteins that induce modification of the host RNA polymerase. The modified polymerase can no longer bind to the original promoter but gains the ability to initiate at other promoters whose base sequences differ from that of the preceding class. A series of modifications occur in which either small molecules are added or phage protein molecules bind to previously modified polymerase. These modifications cause the polymerase successively to ignore particular promoters and to initiate at new promoters. The net effect is an orderly control of the timing of synthesis of many species of mRNA.

One transcript, the late mRNA that encodes the structural and lytic proteins, is regulated in a poorly understood way in which promoter availability may be the main feature. The late mRNA promoter, whose base sequence is quite different from that of the other classes of promoters, cannot be recognized by *any* form of RNA polymerase, modified or not, until the DNA has been replicated. Presumably, replication induces some structural change in the DNA, at least in the promoter region; the nature of the change is unknown.

An example of regulation mediated by control of termination of transcription is the *trp* operon, described in an earlier section of the chapter. In the life cycle of *E. coli* phage λ (Section 7.6) the transcription of several genes is also regulated by controlling termination; however, in this case a gene product, which interacts with sites in the DNA, is responsible for *inhibiting* normal termination. This phenomenon is called **antitermination.** In λ antitermination occurs at several sites, one of which will be described.

Early in both the lytic and lysogenic pathways of λ, transcription is initiated from a promoter p_L and yields a short transcript called L1 (Figure 11-12). This transcript includes a gene N, whose product is an antitermination protein. Transcription by the unmodified host RNA polymerase stops shortly after the terminus of the N gene, because the polymerase meets the normal transcription-termination sequence *tL1*. The N-gene product (aided by an *E.*

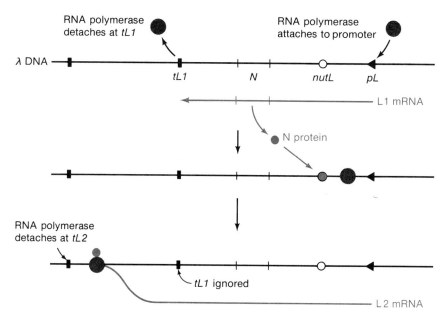

Figure 11-12 Antitermination of λ L1 mRNA induced by the binding of the λ gene-*N* protein to the *nutL* site in the DNA.

coli protein) is able to bind to a short DNA sequence in the same region, called the N-utilization site (*nutL*). Following transcription of the *N* gene, translation of L1 occurs and the *N*-gene product is synthesized. As the concentration of the N protein increases, some N protein binds to *nutL*; when RNA polymerase makes its next transit across the *nutL* site, the polymerase is altered (presumably by acquiring the N protein), and the altered form is then able to transcribe through the termination site. Transcription continues until a second termination site *tL2* is encountered. The altered RNA polymerase responds to this site and termination occurs, forming a longer transcript L2. By this sequence of events the synthesis of the proteins encoded in L2 is delayed by the amount of time required to synthesize a sufficient amount of the gene-*N* protein necessary for antitermination. Transcription of other λ genes is also regulated by antitermination. Why it is necessary that λ delays synthesis of the N protein is not known; however, the other genes regulated by antitermination are those needed late in the replication cycle, and if they are made too early, phage production would be disrupted.

11.6 Feedback Inhibition

If a culture of bacteria in which the *trp* operon has been derepressed is suddenly exposed to tryptophan, repression is rapidly established, and little further synthesis of the *trp* enzymes occurs. However, the *trp* enzymes persist and were it not for a second type of regulatory mechanism, wasteful synthesis of tryptophan and needless consumption of precursors and energy would occur. This state of affairs is avoided by **feedback inhibition,** a mechanism by which the *activity* of the enzymes in a particular pathway is turned off by the product of the pathway.

The most common type of feedback inhibition is effected by an inhibition of the first enzymatic step of a biosynthetic pathway, caused by the product of the pathway. For example, in the pathway

$$A \xrightarrow{\ 1\ } B \xrightarrow{\ 2\ } C \xrightarrow{\ 3\ } D$$

catalyzed by enzymes 1, 2, and 3, the product D would act on enzyme 1, which is clearly the most economical mode of inhibition.

Some biosynthetic pathways are responsible for the synthesis of two products from a common precursor. A hypothetical example of such a **branched pathway** is the following:

In this type of pathway it would be undesirable for a single product to inhibit enzyme 1 because both branches would be blocked. In general, the most economical kind of inhibition prevails—namely, D inhibits enzyme 2 and F inhibits enzyme 4. In this way, neither D nor F prevents the synthesis of the other. Conversion of A to B is wasteful if both D and F are present; therefore, in many branched pathways it is found that D and F together inhibit enzyme 1.

The elegance of feedback inhibition in a branched pathway is shown in Figure 11-13, which shows a well-studied multibranched pathway in which lysine, methionine, and threonine are synthesized from aspartate. Note how these amino acids inhibit enzymes 4, 6, and 8 immediately after the main branch in the pathway. Furthermore, three **isoenzymes** (different proteins that catalyze the same reaction)—1a, 1b, and 1c—are separately inhibited by lysine, homoserine, and threonine, respectively. Each of the three isoenzymes carries out the reaction at a different rate, each one synthesizing as much aspartyl phosphate as is needed to form the appropriate amount of the product that inhibits the isoenzyme. Thus, when an isoenzyme is inhibited, the amount of aspartyl phosphate required to synthesize the needed amino acids is still made.

Feedback inhibition occurs in both prokaryotes and eukaryotes but has been more fully described in the former. It can occur in a variety of ways—for example, binding of the product, which is also the inhibitor, to the enzyme, causing a change in shape of the enzyme and thereby altering its catalytic properties.

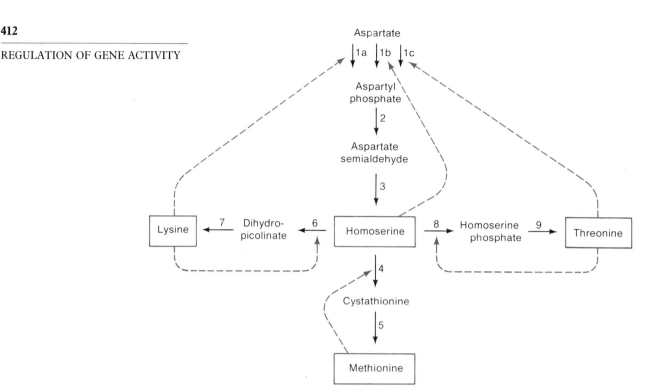

Figure 11-13 Feedback inhibition of the pathway for synthesizing the amino acids lysine, methionine, and threonine. The boxed molecules are inhibitors. The dashed arrows lead from an inhibitor to the enzyme that is inhibited. 1a, 1b, and 1c are the three isoenzymes.

11.7 Translational Regulation

In several places in this chapter translational regulation has been mentioned, usually with reference to the fact that the number of copies of each protein translated from a polycistronic mRNA varies from gene to gene. Generally a gradient of translation efficiency exists, which decreases from the 5′ terminus to the 3′ terminus of the mRNA molecule. The features responsible for this phenomenon are (1) varying efficiencies of initiation of translation, (2) different spacing between chain-termination codons and a subsequent AUG codon, which allows the ribosome and mRNA to dissociate, and (3) differential sensitivity of various regions of the mRNA to degradation. These effects on translation determine the amount of protein made per unit time per gene, but do not, strictly speaking, constitute regulation, because the efficiency of translation of the mRNA of a particular gene does not respond to variations in the environment. However, true translational regulation has been observed in a few bacteriophage species—namely, inhibition of translation of a particular gene by a gene product.

E. coli phage R17 contains RNA instead of DNA; its chromosome is an mRNA molecule, so gene expression requires translation only. The phage makes three gene products—two structural proteins (the A protein and the phage coat protein) and an RNA-replicating enzyme (replicase). Far more coat-protein molecules are needed than replicase, because hundreds of coat-protein molecules are built into each particle, whereas a single enzyme molecule is used repeatedly. Also, synthesis of the replicase is required only shortly after infection, whereas in order to produce an adequate number of molecules for phage assembly, coat-protein synthesis must occur throughout the life cycle. Shortly after infection both replicase and coat-protein molecules are translated from the RNA. The phage RNA molecule has a binding site for the coat protein located between the termination codon of the coat-protein gene and the AUG codon of the replicase gene. As the coat protein is synthesized, this binding site is gradually filled with protein molecules, blocking the ribosome from translating the replicase region. In this way synthesis of replicase stops shortly after synthesis of coat-protein molecules begins.

Other examples of translational regulation will be seen in eukaryotic systems.

11.8 Regulation in Eukaryotes

Eukaryotic cells are of two general types—free-living unicells, such as yeast and algae, and those resident in organized tissue. The needs of the latter class differ from the needs of both the former and of prokaryotes in that the environment of cells in tissue does not usually change drastically in time. During the growth phase of an organism cells differentiate in response to various signals, mostly unknown; however, once differentiated, the cells remain stable, producing particular substances either at a constant rate or in response to external signals such as hormones and temperature changes. The free-living unicellular eukaryotes and the prokaryotes share certain regulatory features, though the gene organization of the former is definitely that of eukaryotic cells. A great deal is known about regulation in yeast, but at a less profound level of understanding than that of *E. coli* operons. Whereas details of the mechanisms in yeast differ, often significantly, from those observed in bacteria, the overall regulatory strategy of yeast remains that of responding to large fluctuations in the availability of nutrients in the environment. For this reason, the emphasis of this section is on eukaryotic cells that form organized tissue. Most of the information comes from detailed studies of the DNA of mammals, amphibians (toads of the genus *Xenopus*), insects (*Drosophila*), birds (usually the chicken), and echinoderms (the sea urchin).

Some Important Differences in the Genetic Organization of Prokaryotes and Eukaryotes

Numerous differences exist between prokaryotes and eukaryotes with regard to transcription and translation, and in the spatial organization of DNA, as described in Chapters 5 and 9. Seven of those most relevant to regulation are the following:

1. In a eukaryote usually only a single type of polypeptide chain can be translated from a completed mRNA molecule; thus, operons of the type seen in prokaryotes are not found in eukaryotes.

2. The DNA of eukaryotes is bound to histones, forming chromatin, and to numerous nonhistone proteins. Only a small fraction of the DNA is bare. In bacteria some proteins are present in the folded chromosome, but most of the DNA is free.

3. A significant fraction of the DNA of eukaryotes consists of a few nucleotide sequences that are repeated hundreds to millions of times. Some sequences are repeated in tandem but most repetitive sequences are not. Other than duplicated rRNA and tRNA genes and a few specific short sequences such as certain parts of promoters, bacteria contain few repeated sequences.

4. A large fraction of the base sequences in eukaryotic DNA is untranslated.

5. Eukaryotes possess mechanisms for rearranging certain DNA segments in a controlled way and for increasing the number of specific genes when needed. This seems to be rare in bacteria.

6. The bases of a gene and the amino acids of the gene product are usually not colinear in eukaryotes; introns are present in most eukaryotic genes and RNA must be processed before translation begins.

7. In eukaryotes RNA is synthesized in the nucleus and must be transported through the nuclear membrane to the cytoplasm where it is utilized. Such extreme compartmentalization probably does not occur in bacteria.

We shall see in this section how some of these features are incorporated into particular modes of regulation.

Gene Families

In prokaryotes genes having closely related functions are often organized in operons and are transcribed as part of a polycistronic mRNA molecule. Thus, the entire system is under control of one promoter region, and the system can be turned on and off by controlling the availability of the promoter. Without modification this method cannot be used in eukaryotes inasmuch as eukaryotic mRNA is usually monocistronic.

Many related eukaryotic genes can be functionally grouped into a set called a **gene family.** These genes are rarely as near to one another as the genes are in a bacterial operon, though they may be clustered (not adjacent, but within hundreds to thousands of nucleotide pairs of one another); however, frequently they are widely scattered and even located on different chromosomes. Gene families are currently classified as simple multigene families, complex multigene families, and developmentally controlled multigene families. An example of each type is shown in Figure 11-14.

A **simple multigene family** is one in which one or a few genes are repeated in a tandem array. The simplest example is the set of genes for 5S ribosomal RNA. In the toad *Xenopus* this set of rRNA genes forms a gigantic array in which each 5S rRNA gene sequence is separated from an adjacent gene sequence by a spacer (not an intron, because it is not within a gene) to form a gene cluster. The sizes of the spacers vary from two to six times the length of the 5S rRNA gene. Each 5S rRNA gene is transcribed as a separate RNA molecule that is cut and spliced to generate the finished 5S rRNA molecule. Several clusters of 5S RNA genes are present in *Xenopus*, each cluster containing hundreds to thousands of 5S rRNA genes. Presumably, the individual genes (or at least the cluster) are transcribed in response to a single signal.

The 5.8S, 18S, and 28S rRNA genes of *Xenopus* comprise another simple multigene family, in which the primary transcript contains all three rRNA

(a) Single gene family

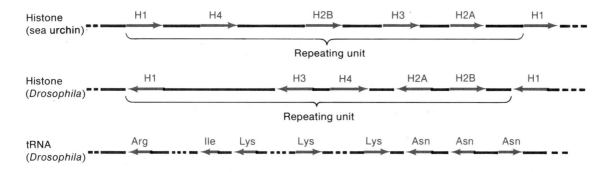

(b) Complex multigene families

(c) Developmentally controlled multigene family

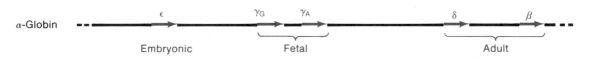

Figure 11-14 Five examples of gene families. The arrows indicate the direction of transcription. Genes are shown in red; spacers are black.

sequences in the order 18S–5.8S–28S, each separated by a small spacer. The rRNA molecules are cleaved from the primary transcript by processing enzymes. In this case, the 1:1:1 ratio of the three species (the ratio present in a ribosome) is achieved by there being one copy of each per transcript, as is also the case for *E. coli* rRNA. Note that the 5S RNA is transcribed separately from the 18S–5.8S–28S unit. How a ratio of one copy of 5S rRNA for each 18S–5.8S–28S unit is obtained is unknown. A simple explanation is that synthesis of the 5S rRNA transcript and the three-species transcript is initiated in response to the same signal.

The **complex multigene families** generally consist of a cluster of several functionally related genes, each transcribed independently (in contrast with the multigene transcript just described for the rRNA unit) and separated by a spacer. Three examples are shown in Figure 11-14(b); each has features that indicate that there are several types of complex multigene families. For instance, in the sea urchin each of the five histone genes is separately transcribed as a single monocistronic RNA and each gene is transcribed in the same direction—that is, from the same DNA strand. In contrast, in *Drosophila* the five genes are transcribed in both directions, as shown in the figure. In yeast the organization of the histone genes differs even more; there are two quite distant gene clusters, each containing one *H2A* gene and one *H2B* gene, whereas the other histone genes are separate and scattered throughout the chromosome. In all cases, the genes are considered to constitute a family, because the gene products are related and are synthesized in a fairly fixed ratio.

The **developmentally controlled gene families** are regulated such that different genes are expressed at different stages of development of the organism—that is, they are expressed in a definite sequence in time. The best-characterized one is the globin family. Hemoglobin is a tetrameric protein containing two α subunits and two β subunits (Sections 9.2 and 10.2). However, there are several different forms of both α and β subunits differing by only one or a few amino acids, and the forms that are present depend on the stage of development of an organism. For example, the following developmental sequence shows the subunit types present in humans at various times after conception:

	Embryonic (<8 weeks)	Fetal (8–41 weeks)	Adult (birth → henceforth)
α-like	$\zeta_2 \rightarrow \zeta_1$	α_1 and α_2	α_1 and α_2
β-like	ϵ	γ_G and γ_A	β and δ

During the embryonic period the ζ_2 α-like chain, which appears first, is gradually replaced by the ζ_1 form. In the fetal and adult stages, the α_1 and α_2 types are both present but α_2 predominates; the two β-like types, γ_G and γ_A (called so because one has a *g*lycine at a site at which the other has an *a*lanine), are roughly equal in amount. In the adult stage, fifty times more β type is present than the δ form. The result of these changes is that before birth all the hemoglobin contains two ζ_2 chains and two ϵ chains ($2\zeta_2, 2\epsilon$), whereas after birth 98 percent contains two α_2 and two β subunits (customarily called hemoglobin A) and 2 percent contains two α_2 and two δ units (hemoglobin A_2).

The α- and β-like genes form separate clusters. The β cluster is shown in Figure 11-14(c). A remarkable property of each cluster is that the order of the genes is basically the order in which they are expressed in development. (This is not true of all developmentally controlled gene families.) Neither the significance of the different forms nor the way in which synthesis is programmed is known.

Examination of the globin clusters in many different organisms gives some indication of how clusters arose and of the evolutionary factors that may have provided for their maintenance. Primitive fish, marine worms, and insects have a single globin gene, whereas in amphibians α and β genes are closely linked on a single chromosome. The lower mammals and birds have different forms of both α and β genes, and the clusters are on different chromosomes. Thus, it has been postulated that ancestral globin (which may have appeared about 800 million years ago) was the product of a single gene and that the globin families evident today arose by a series of gene duplications, mutations, and transpositions from an ancestral gene. The following hypothetical (but likely) sequence of events has been suggested. The small animals were able to function with the limited O_2-carrying capacity of a single hemoglobin molecule. Spontaneous mutation gave rise to an altered globin that in a heterozygote was able to form a tetrameric molecule capable of carrying more O_2. For reasons beyond the scope of this book, the tetrameric molecule can deliver a larger fraction of its bound O_2 to tissue than the monomers can, which enabled animals of larger size to evolve. Later, during the evolution of mammals one of the two β-chain genes underwent mutation and duplication, again giving rise to the γ form of globin found in the fetus. Fetal hemoglobin has an even higher affinity for O_2 than adult hemoglobin and thus is advantageous for the rapidly developing fetus, perhaps enabling a more complex organism to arise. Further mutation and duplication occurred during primate evolution, giving rise to additional forms of hemoglobin.

Of particular genetic interest is the mechanism for gene duplication and for subsequent spread of the duplicated genes throughout a population. Several mechanisms are theoretically available for gene duplication (Figure 11-15). For example, a simple transposable element such as a bacterial IS element (Chapter 7) located on one side of a gene may transpose to the other side of the gene, forming a complex transposable element like the antibiotic-resistance-carrying Tn elements of bacteria. The complex element might then transpose, as a unit, to a new location, leaving a copy at the original location and producing a gene duplication. Alternatively, unequal crossing over of the type described in Figure 6-17 might lead to production of a duplication in one homologue and chromosome loss in the other. Subsequent crossover events would generate triplications and more copies, as also shown in Figure 6-17. How an advantageous mutant might be maintained along with the original form of the gene is illustrated in Figure 11-16.

The scheme just suggested is supported by the existence of remnants of past events of the type described. Study of various segments of the globin gene clusters in several organisms has indicated the presence of nonfunctional sequences that are nearly homologous to normal globin genes. The sequences are called **pseudogenes,** and are considered to be relics of evolution, derivatives of once-functional genes that were duplicated. Presumably a mutation occurred at one time that inactivated the gene product. Since at least one functional copy of the gene remained, there would have been no strong selective pressure to eliminate the gene, and in time additional mutations would have accumulated. Pseudogenes are widespread in gene clusters and are frequently defective in transcription, intron excision, and translation.

Gene Dosage and Gene Amplification

Some gene products are required in much larger quantities than others. A common means of maintaining particular ratios of certain gene products (other than by differences in transcription and translation efficiency, as discussed earlier) is by **gene dosage.** For example, if two genes A and B are transcribed at the same rate and the translation efficiencies are the same, 20

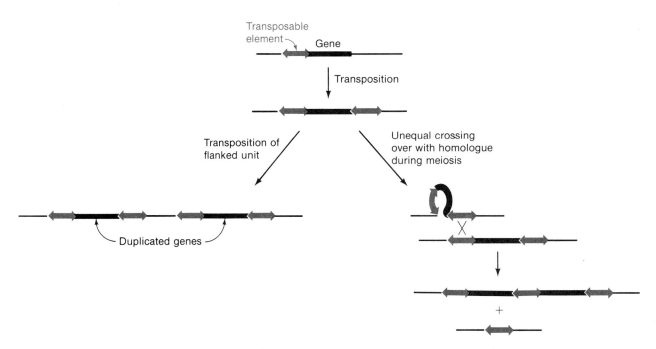

Figure 11-15 Two of several possible mechanisms for gene duplication. Both begin with a transposition event. Left pathway: a second transposition of the gene flanked by two transposable elements leads to gene duplication. Right pathway: unequal but reciprocal crossing over between two transposable elements forms one chromosome with a duplication and a second chromosome with a deletion. The right pathway might occur without a transposable element at all if the original gene were flanked by two sequences that were sufficiently homologous for unequal crossing over to occur. A third mechanism might also occur in the right pathway if rare nonhomologous recombination occurred between flanking nonhomologous sequences. (Figure 8-1 illustrates such an event.)

times as much of product A can be made as product B if there are 20 copies of gene *A* per copy of gene *B*. The histone gene family exemplifies a gene dosage effect: in order to synthesize the huge amount of histone required to form chromatin, most cells contain hundreds of times as many copies of histone genes as of genes required for DNA replication.

A special case of a gene dosage effect is **gene amplification,** in which the number of genes increases in response to some signal. The best-understood example of gene amplification is found in the development of the oocytes (eggs) of the toad *Xenopus laevis,* in which the number of rRNA genes increases by about 4000 times. This increase exists only during and for the purpose of development of an egg. The precursor to the oocyte, like all somatic cells of the toad, contains about 600 rRNA-gene (rDNA) units; after amplification about 2×10^6 copies of each unit are present. This large amount

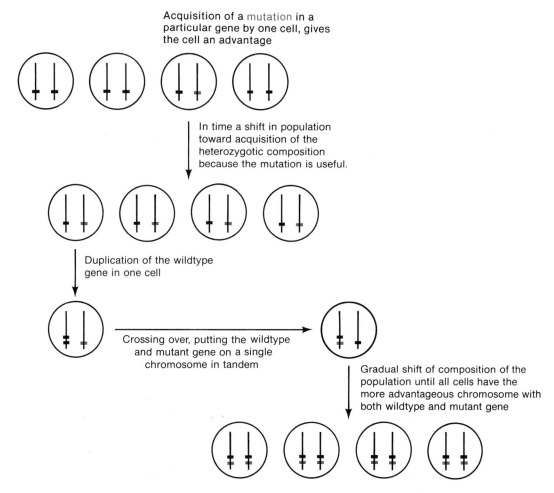

Figure 11-16 A hypothetical mechanism for the spreading of a useful mutation throughout a population without loss of the wildtype gene.

enables the oocyte to synthesize 10^{12} ribosomes, which are required for the protein synthesis that occurs during early development of the embryo.

Prior to amplification, the 600 rDNA units are arranged in tandem and each unit consists of the 18S–5.8S–28S sequence described in the preceding section—that is, one gene of each type, separated by a spacer. During amplification, over a three-week period during which the oocyte develops from a precursor cell, the rDNA no longer consists of a single contiguous DNA segment containing 600 three-gene units but instead is present as a large number of small circles and replicating rolling circles (Section 4.6), each containing one or two three-gene units. The rolling circle replication accounts for the increase in the number of copies of the genes. The precise mechanism of excision of the circles and formation of the rolling circles is not known.

Once the oocyte is mature, no more rRNA needs to be synthesized until well after fertilization and into early development, at which time 600 copies is sufficient. Thus, the excess rDNA serves no purpose and it is slowly degraded by intracellular enzymes. Following fertilization the chromosomal DNA replicates and mitosis ensues, occurring repeatedly as the embryo develops. During this period the extra chromosomal rDNA does not replicate; degradation continues and by the time several hundred cells have formed, none of this DNA remains.

Amplification of rRNA genes during oogenesis occurs in many organisms including insects, amphibians, and fish. Amplification of a gene that encodes a protein has been observed in *Drosophila;* the genes that produce chorion proteins (a component of the sac that encloses the egg) are amplified in ovarian follicle cells just before maturation of an egg. In this case, as with the rRNA genes, amplification enables the cells to produce a large amount of protein in a short time. We will see later that if a large amount of protein is to be synthesized and a *long* time is available for the synthesis, gene amplification is unnecessary, for this can be accomplished by increasing the lifetime of the RNA.

Regulation of Transcription

The mRNA of eukaryotes is conveniently classified into groups based on the number of copies of a particular mRNA molecule present in each cell—single-copy (a term used both for a single copy and a few copies per haploid complement), moderately prevalent (a few to several hundred copies per cell), and superprevalent (hundreds to thousands of copies per cell). Single-copy and moderately prevalent mRNA primarily encode enzymes and structural proteins, respectively. Superprevalent mRNA molecules are transcribed from only a small fraction of eukaryotic genes, and their production is usually associated with a change in the stage of development—for example, an adult erythroblast cell in the bone marrow produces a huge amount of mRNA from which adult globin may be translated, whereas little or no globin is produced by precursor cells that have not yet become erythroblasts. A great deal of what is known about the regulation of transcription in higher eukaryotes concerns superprevalent mRNA.

Of the known regulators of transcription in higher eukaryotes the **hormones**—small molecules, polypeptides, or small proteins that are carried from hormone-producing cells to target cells—constitute a class that has been studied in some detail. Many of the sex hormones act by turning on transcription. If a hormone regulates transcription, it must somehow signal the DNA. Penetration of a target cell by a hormone and its transport to the nucleus is a much more complex process than entry of lactose into *E. coli* and is understood only in outline, as shown in Figure 11-17. Steroid hormones (H) are hydrophobic (nonpolar) molecules and pass freely through the cell membrane. A target cell contains a specific cytoplasmic receptor (R) that forms a complex (H-R) with the hormone. The receptor R usually undergoes some modification (in shape or in chemical structure) after the H-R complex has formed; the modified form of the receptor is denoted R′. The H-R′ complex then passes through the nuclear membrane and enters the nucleus. From this point on, little is known about most systems. It is likely that in the nucleus either the H-R′ complex or possibly the hormone alone engages in one of the following processes: (1) direct binding to DNA, (2) binding to an effector protein, (3) activation of a DNA-bound protein, (4) inactivation of a repressor, and (5) a change in the structure of chromatin to make the DNA available to RNA polymerase. In most hormone-activated systems studied to date it has not been possible to determine which process is involved. One well-understood example is the action of glucocorticoid, a hormone that inhibits glucose degradation and stimulates glucose synthesis in the liver. The activity of glucocorticoid in initiating transcription of the relevant genes involves the direct binding of a positively regulatory protein to a DNA molecule.

A well-studied example of induction of transcription by a hormone is the stimulation of the synthesis of ovalbumin in the chicken oviduct by the sex hormone **estrogen.** When chickens are injected with estrogen, oviduct tissue

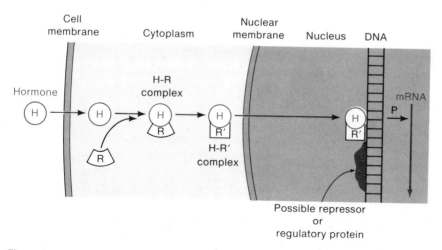

Figure 11-17   A schematic diagram showing how a hormone H reaches a DNA molecule and triggers transcription by binding to a cytoplasmic receptor. A regulatory protein may prevent the H-R′ complex from reaching a promoter P or may stimulate binding to the promoter.

responds by synthesizing ovalbumin mRNA. This synthesis continues as long as estrogen is administered. Once the hormone is withdrawn, the rate of synthesis decreases. Both before giving the hormone and sixty hours after withdrawal, no ovalbumin mRNA is detectable.

When estrogen is given to chickens, only the oviduct synthesizes mRNA because other tissues lack the cytoplasmic hormone receptor. (This type of deficiency is the usual cause of insensitivity to a particular hormone.) The mechanism by which the receptor is synthesized in some but not all cells is not known.

Regulation of Processing

Often two cells make the same protein but in different amounts, even though in both cell types the same gene is transcribed. This phenomenon is frequently associated with the presence of different mRNA molecules, which are not translated with the same efficiency. In the synthesis of α-amylase in the rat different mRNA from the same gene result from the use of different processing sequences. The rat salivary gland produces more of the enzyme than the liver, though the same coding sequence is transcribed. In each cell type the same primary transcript is synthesized, but two different splicing mechanisms are used. The initial part of the primary transcript is shown in Figure 11-18. The coding sequence begins 50 base pairs within exon 2 and is formed by joining exon 3 and subsequent exons. In the salivary gland the primary transcript is processed such that exon S is joined to exon 2 (that is,

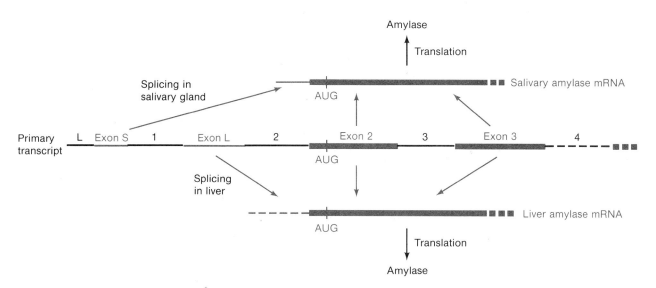

Figure 11-18 Production of distinct amylase mRNA molecules by different splicing events in cells of the salivary gland and liver of the mouse. The leader L and the introns are in black. The exons are red. The coding sequence begins at the AUG codon in exon 2.

exon L is removed as part of introns 1 and 2). In the liver exon L is joined to exon 2, because exon S is removed along with intron 1 and with the leader L. The exons S and L become alternate leaders of amylase mRNA, which somehow results in translation at different rates.

Hypersensitive Sites and Upstream Regulatory Sites

In regulation of transcription in prokaryotes, availability of the promoter for binding of RNA polymerase is critical, and in negatively regulated operons the binding of a protein (a repressor) to an operator is often sufficient to prevent transcription. In eukaryotes DNA does not exist as a free molecule but is tightly bound to histones, so the question of how RNA polymerase comes into contact with promoters is an important one. The answer is not yet clear, but several observations suggest that chromatin may have a special structure in promoter regions. If free DNA molecules are exposed to most DNases, the DNA is rapidly broken. However, the rate of breakage of chromatin is much less than that of free DNA, because the histones reduce access of the DNA to the enzymes. When chromatin isolated from a particular cell type is treated with certain DNases, specific regions of the DNA are cleaved at a higher rate than the bulk of the DNA; the positions of these regions vary from one cell type to another. These regions are called **hypersensitive sites;** their importance is that their existence and positions correlate with transcriptional activity of particular genes. For example, in chromatin isolated from chicken reticulocytes (which synthesize globin), sites in the globin cluster are preferentially broken, whereas chicken oviduct chromatin contains hypersensitive sites in the ovalbumin region. In contrast, there are no such sites in the ovalbumin region of reticulocyte chromatin nor in the globin region of oviduct chromatin. Evidence from study of the *Sgs4* gene of *D. melanogaster* suggests that the hypersensitive sites may actually be essential for gene activity. A mutant has been found in which the hypersensitive region is deleted but in which the coding region of the Sgs4 protein is intact. This mutant fails to make Sgs4 protein, suggesting that deletion of the hypersensitive region prevents activity of the gene. The mechanistic relation between a hypersensitive site and transcription is unknown. In one case the DNA of the hypersensitive site in a viral DNA molecule is bare, lacking nucleosome structure, but such an observation has not yet been made for cellular DNA. Hypersensitive sites are usually several hundred base pairs *upstream* from the start point for transcription, which seems quite far compared to the size of the binding site for RNA polymerase (Figure 11-19). (By "upstream" is meant "against the direction of transcription.")

Upstream elements of importance have also been observed in other experiments. For example, in several yeast genes, mutations have been isolated that eliminate or reduce transcription. These mutations define regions called **upstream activation sites,** whose properties are quite different from the transcription-regulating promoter-operator regions in prokaryotes. In a bacterial operon the operator and promoter are adjacent, and binding a regulatory

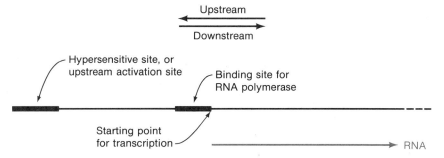

Figure 11-19 The relative sizes and locations (drawn approximately to scale) of a hypersensitive site or upstream activation site (at roughly the same position) and the binding site for RNA polymerase in a typical eukaryotic gene. The transcribed region would be about ten times longer than shown. The meaning of the terms "upstream" and "downstream" is also shown.

protein to an operator either blocks (a repressor) or stimulates (a molecule of the cAMP-CAP type) the binding of RNA polymerase to the promoter. If the promoter and operator are physically separated (by experimental manipulation), the regulatory region becomes ineffective—that is, a repressor cannot prevent initiation of transcription nor can a positive regulator stimulate transcription. However, upstream activation sites are not only usually 50–300 nucleotide pairs from the promoter but their locations can be experimentally changed by as many as a few hundred nucleotide pairs without substantially diminishing their activity. Upstream activation sites have been discovered so recently that their role in regulation is unknown, but they seem to be a common feature of eukaryotic systems.

Another type of region whose location can also be changed without detracting from its effectiveness is the **enhancer element.** The prototype enhancer was discovered in the animal virus SV40. It is a sequence of 72 nucleotide pairs located more than 100 nucleotide pairs *upstream* from the gene *T*. Three properties of this sequence are general features of enhancers. (1) Experimental deletion of the sequence reduces transcription of this gene by more than a hundredfold. (2) By experimental manipulation the SV40 enhancer can be moved further upstream, downstream, and even to the other side of the promoter without significant loss of activity. (3) The sequence can also be experimentally reversed in direction and retain enhancing activity (Figure 11-20). Furthermore, by the genetic engineering techniques de-

Figure 11-20 Position of the enhancer sequence near the virus SV40 *T* gene with respect to the transcription start site. If the normal sequence is experimentally inverted, RNA is still made at a high rate. Experimental deletion of the enhancer reduces transcription more than a hundredfold.

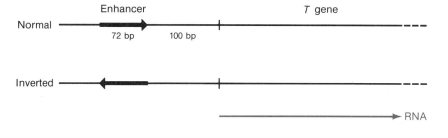

scribed in Chapter 12, the sequence has been isolated and moved next to genes in other organisms and the enhancing effect remains. In several viral systems enhancers have been found in some hypersensitive sites, but the relation is unclear. Little is known about the molecular basis of enhancement; the most popular hypothesis is that enhancers alter chromatin structure and provide an entry site for RNA polymerase.

Translational Control

In bacteria most mRNA molecules are translated about the same number of times, with only fairly small variation from gene to gene. In eukaryotes translational regulation occurs in which an mRNA molecule is not translated at all until a signal is received. Other types of control include (1) regulation of the lifetime of a particular mRNA molecule and (2) regulation of the rate of overall protein synthesis. In this section examples of each of these modes of regulation will be presented.

An important example of translational regulation is that of **masked mRNA.** Unfertilized eggs are biologically static, but shortly after fertilization many new proteins must be synthesized—for example, the proteins of the mitotic apparatus, the cell membranes, as well as others. Unfertilized sea urchin eggs store large quantities of mRNA for many months in the form of mRNA-protein particles made during formation of the egg. This mRNA is translationally inactive, but within minutes after fertilization, translation of these molecules begins. Here the timing of translation is regulated; the mechanisms for stabilizing the mRNA, for protecting it against RNases, and for activation are unknown.

Translational regulation of a second type also occurs in mature unfertilized eggs. These cells need to maintain themselves but do not have to grow or undergo a change of state. Thus, the rate of protein synthesis in eggs is generally low. This is not a consequence of an inadequate supply of mRNA but of a limitation of an as-yet-unidentified element, called the **recruitment factor,** which apparently interferes with formation of the ribosome-mRNA complex.

The synthesis of some proteins is regulated by direct action of the protein on the mRNA. For instance, the concentration of one type of antibody molecule (immunoglobulin) is kept constant by self-inhibition of translation. This protein consists of two H chains and two L chains (Section 13.4), translated from H-chain and L-chain mRNA, respectively. The tetramer binds specifically to H-chain mRNA and thereby inhibits initiation of translation. How L-chain synthesis is regulated is unknown.

A dramatic example of translational control is the extension of the lifetime of silk fibroin mRNA in the silkworm *Bombyx mori*. During cocoon formation the silk gland of the silkworm predominantly synthesizes a single type of protein, silk fibroin. Since the worm takes several days to construct its cocoon, it is the *total amount* and not the rate of fibroin synthesis that must be great; the silkworm achieves this by synthesizing a fibroin mRNA molecule that is very long-lived.

Transcription of the fibroin gene is initiated at a strong promoter by an unknown signal and about 10^4 fibroin mRNA molecules are made in a period of several days. (This synthesis is an example of transcriptional regulation.) A typical eukaryotic mRNA molecule has a lifetime of about three hours before it is degraded. However, fibroin mRNA survives for several days during which each mRNA molecule is translated repeatedly to yield 10^5 fibroin molecules. Thus, each gene is responsible for the synthesis of 10^9 protein molecules in four days. Altogether the silk gland makes 300 μg or 10^{15} molecules of fibroin during this period. If the lifetime of the mRNA were not extended, either 25 times as many genes would be needed or synthesis of the required fibroin would take about 100 days; therefore, 10^6 genes or 5×10^5 diploid cells would be needed.

Another example of an mRNA molecule with an extended lifetime is the mRNA encoding casein, the major protein of milk, in mammary glands. When the hormone prolactin is provided to the gland, the lifetime of casein mRNA increases. Synthesis of the mRNA also continues, so the overall rate of production of casein is markedly increased by the hormone. When the prolactin is withdrawn, the concentration of casein mRNA decreases because the RNA is degraded more rapidly.

Multiple Proteins from a Single Segment of DNA

In prokaryotes coordinate regulation of the synthesis of several gene products is accomplished by regulating the synthesis of a single polycistronic mRNA molecule encoding all of the products. The analogue to this arrangement in eukaryotes is the synthesis of a **polyprotein,** a large polypeptide that is cleaved after translation to yield individual proteins. Each protein can be thought of as the product of a single gene. In such a system the coding sequences of each gene in the polyprotein unit are not separated by stop and start codons but instead by specific *amino acid sequences* that are recognized as cleavage sites by particular protein-cutting enzymes. Polyproteins have been observed with up to ten cleavage sites; the cleavage sites are not cut simultaneously, but are cut in a specific order. Use of a polyprotein serves to maintain an equal molar ratio of the constituent proteins, as was seen earlier in the production of *Xenopus* rRNA; moreover, delay in cutting at certain sites introduces a temporal sequence of production of individual proteins, a mechanism frequently used by animal viruses.

Many examples of polyproteins are known. The synthesis of the RNA precursors uridine triphosphate and cytidine triphosphate proceeds by a biosynthetic pathway that at one stage has the reaction sequence

Precursors

$\downarrow$ Carbamoyl phosphate synthetase

Carbamoyl phosphate $\xrightarrow{\text{Aspartyl trans-carbamoylase}}$ N-carbamoyl phosphate $\xrightarrow{\text{Dihydroorotase}}$ Dihydroorotate $\longrightarrow \longrightarrow \longrightarrow$ CTP

In bacteria each of the three enzymes is made separately. In yeast and *Neurospora* only two proteins are made; one is dihydroorotase and the other is a large protein that is cleaved to form carbamoyl phosphate synthetase and aspartyl transcarbamoylase. Mammals synthesize a single tripartite protein, which is cleaved to form the three enzymes.

Earlier it was noted that two different amounts of amylase are made in the salivary gland and in the liver by differential splicing of a single RNA molecule. The *same* protein was made in both tissues, though in different amounts. In chicken skeletal muscle two different forms of the muscle protein myosin are made from the same gene; these are called alkali light chains LC1 and LC3. This gene has two different transcription initiation sites (TATA sequences, described in Section 9.3), which yield two different primary transcripts. These two transcripts are processed differently to form mRNA molecules encoding distinct forms of the protein (Figure 11-21). In *Drosophila* the mRNA is processed in four different ways, the precise mode depending on the stage of development of the fly. One class of myosin is found in pupae and another in the later embryo and larval stages. How the mode of processing is varied is not known.

11.9 Is There a General Principle of Regulation?

With microorganisms, whose environment frequently changes considerably and rapidly, a general principle of regulation is the reduction of waste. A more or less general rule is that bacteria make what they need when it is required and in appropriate amounts. However, whereas in prokaryotes extraordinary

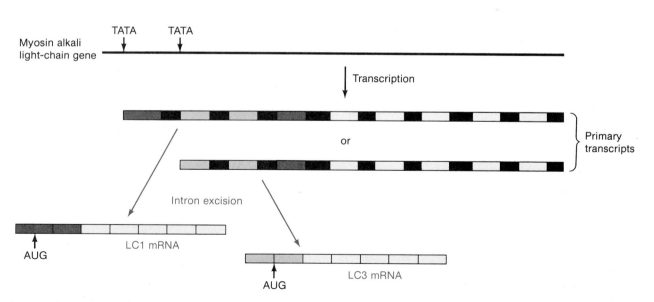

Figure 11-21   The chicken *LC1/LC3* gene. Two distinct TATA sequences lead to the production of different primary transcripts, which contain the same coding sequences. Two modes of intron excision lead to the formation of distinct mRNA molecules, which encode proteins having different amino-terminal regions and the same carboxyl-terminal region.

efficiency is common, it may not always be the case: in eukaryotes such efficiency may be rare. For example, it appears to be violated in the case of the large amount of intronic RNA that is discarded; on the other hand, several examples were given of regulatory mechanisms that operate by means of differential processing; and other functions for introns have been suggested.

What is abundantly clear is that there is no universal regulatory mechanism. Many control points are possible and different genes are regulated in different ways. Furthermore, evolution has not always selected for simplicity in regulatory mechanisms but merely for something that works. If a cumbersome regulatory mechanism were to arise, it would in time evolve, be refined, and become more effective, but it would not necessarily become simpler. On the whole, regulatory mechanisms include a variety of seemingly *ad hoc* processes, each of which has stood the test of time, primarily because it works.

Problems

1. What is the meaning of these terms: repressed, induced, constitutive, coordinate regulation?

2. The glycolytic pathway is responsible for the degradation of glucose, one of the most fundamental energy-producing systems in living cells. Would you expect the enzymes of this pathway to be regulated? If so, in what way?

3. A cell that is wildtype with respect to the *lac* operon (+ for all alleles) is in a growth medium containing neither glucose nor lactose—that is, it is using another carbon source. How many proteins are bound to the DNA comprising the *lac* operon? How many if glucose is present?

4. The entry of lactose into a cell requires the activity of the permease. Suggest how lactose might first enter an *un*induced Lac$^+$ cell in order for induction to occur.

5. Describe in terms of mRNA synthesis, enzyme synthesis, and enzyme activity what happens when lactose is added to a Lac$^+$ *E. coli* culture previously grown in a medium lacking all sugars. Assume that the amount of lactose added is consumed after two generations of growth.

6. A $lacI^+lacO^+lacZ^+lacY^+$ Hfr culture is mated with a $lacI^-lacO^+lacZ^-lacY^-F^-$ culture. In the absence of any inducer in the medium, β-galactosidase is made for a short time after the Hfr and F^- cells have been mixed. Explain why it is made and why only for a short time.

7. You have isolated a Lac$^-$ mutant and found by genetic analysis that its genotype is $lacZ^+lacY^+$. The mutation,

which you call $lacI*$ is in the *lacI* gene. The partial diploid $lacI*lacZ^+lacY^+/lacI^-lacZ^+lacY^+$ is constructed and its phenotype is found to be Lac$^-$—that is, $lacI*$ is dominant. The diploid $lacI*lacZ^+lacY^+/lacO^clacI^+lacZ^+lacY^+$ is Lac$^+$ (β-galactosidase is made). Suggest a property of the mutant repressor that would yield this phenotype. Would $lacI*lacZ^+lacY^+/lacO^clacI^+lacZ^-lacY^+$ make β-galactosidase?

8. The purpose of regulating gene expression is to reduce wasted synthesis, because cells experiencing waste are at a definite disadvantage in the long run. Approximately five percent of the total protein made by $lacI^-$ cells, whose synthesis of β-galactosidase is unregulated, is β-galactosidase, far more than is needed.
   (a) Suppose 10^6 wildtype and 10^6 $lacI^-$ cells are placed in a growth medium lacking lactose. If the cells are grown for 50 generations (with continued dilution of the culture to keep the cells growing), what would you expect to happen to the ratio of wildtype and $lacI^-$ cells?
   (b) A gratuitous inducer is an inducer that is not a substrate of β-galactosidase. Many experiments are carried out with these inducers to avoid effects of changing concentration of lactose. Suppose a Lac$^+$ culture is grown for hundreds of generations in the presence of a gratuitous inducer; what type of spontaneous mutant might accumulate in the culture?

9. A mutant strain of *E. coli* is found that produces both β-galactosidase and permease whether lactose is present or not.

(a) What are two possible genotypes for this mutant?

(b) Another mutant is isolated that produces no β-galactosidase at any time but produces permease if lactose is present in the medium. If a partial diploid is formed from these two mutants, in the absence of lactose neither β-galactosidase nor permease is made. When lactose is added, the partial diploid makes both enzymes. What are the genotypes of the two mutants?

10. An *E. coli* mutant is isolated that is simultaneously unable to utilize a large number of sugars as sources of carbon. However, genetic analysis shows that each of the operons responsible for metabolism of each sugar is free of mutation. What genotypes of this mutant are possible?

11. An operon responsible for utilizing a sugar Q is regulated by a gene called *kyu*. When Q is added to the growth medium, Qase is made; otherwise, the enzyme is not made. If the gene *kyu* is deleted (denoted Δ*kyu*), no Qase can be made. The partial diploid *kyu⁺*/Δ*kyu* is inducible. Two types of point mutants of *kyu* are found: *kyu1*, which never makes Qase, and *kyu2*, which makes the enzyme constitutively. The partial diploids *kyu⁺*/*kyu1* and *kyu⁺*/*kyu2* are inducible and constitutive, respectively. What is the likely mode of action of the protein encoded by the *kyu* gene?

12. How many proteins are bound to the *trp* operon when
(a) Tryptophan and glucose are present?
(b) Tryptophan and glucose are absent?
(c) Tryptophan is present and glucose is absent?

13. The histidine operon is a negatively regulated system containing ten structural genes that encode enzymes needed to synthesize histidine. The repressor protein is also coded within the operon—that is, in the polycistronic mRNA molecule that encodes the structural genes. Synthesis of this mRNA is controlled by a single operator regulating the activity of a single promoter. The corepressor of this operon is tRNA^His to which histidine is attached. This tRNA is not encoded in the operon itself. A collection of mutants having the properties indicated below have been isolated. Determine whether the histidine enzymes would be synthesized by each of these mutants and whether each mutant would be (1) dominant, (2) *cis*-dominant only, or (3) recessive to its wildtype allele in a partial diploid.
(a) The promoter cannot bind RNA polymerase.
(b) The operator cannot bind the repressor protein.
(c) The repressor protein cannot bind to DNA.
(d) The repressor protein cannot bind histidyl-tRNA^His.
(e) The tRNA^His by itself (that is, without histidine) can bind to the repressor protein.

14. Consider the branched pathway regulated by feedback inhibition, shown in the figure below.
(a) Which enzymes are subject to feedback inhibition? For each identify the inhibitor.
(b) Which steps are likely to be catalyzed by a set of isoenzymes and how many isoenzymes are probably in these steps?

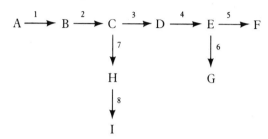

15. Why are eukaryotic genes not organized in operons and how does a gene family differ from an operon?

16. With respect to gene families,
(a) Are all members of a gene family necessarily transcribed from the same DNA strand?
(b) Are all members of a gene family actively transcribed in all tissues of a single organism?
(c) Are all members of a gene family necessarily located on the same chromosome?

17. Differentiated cells are often called upon to produce an enormous amount of a particular protein. What regulatory strategies are used when only a limited time is available for this synthesis? When a long time is available?

18. Distinguish gene amplification from translational amplification.

19. In many species, during the course of cellular differentiation to form the hemoglobin-producing cells called reticulocytes, the cell nuclei are extruded. Thus, both reticulocytes and red blood cells have no nuclei.
(a) Can globin synthesis be transcriptionally regulated in these species?
(b) The rate of synthesis of globin is regulated by the amount of available heme, which is part of the hemoglobin molecule. It has been suggested that regulation occurs not by controlling the rate of synthesis of globin but instead by controlling the rate of synthesis of all protein; that is, when globin need not be synthesized, all protein synthesis is shut off. Comment on the reasonableness of this possibility.

20. When chromatin from somatic cells of *Xenopus* is isolated and transcribed in an *in vitro* protein-synthesizing system, only somatic genes are transcribed. This is true whether the *in vitro* transcription extracts are derived from oocytes or from somatic cells. When the chromatin is washed with 0.6 M NaCl before transcription, this specificty is lost and oocyte genes are also transcribed. Interpret these results in terms of the molecular basis of the specificity.

21. A particular cell makes substance X in response to a hormone H. Prior to exposure of a culture of the cells to H, a large RNA molecule can be found in the nucleus. It is capped and has a poly(A) tail. After addition of H this RNA is not found; a second RNA, which in an *in vitro* protein-synthesizing system makes X, is found in cytoplasmic polysomes. A segment of DNA has been isolated and both RNA molecules can hybridize to this molecule. What mechanism might be used for hormonal control of synthesis of X?

22. Several eukaryotes are known in which a single effector molecule regulates the synthesis of different proteins encoded in distinct mRNA molecules 1 and 2. Give a brief possible molecular explanation for each of the following observations made when the effector is absent.
    (a) Neither nuclear nor cytoplasmic RNA can be found that hybridizes to either of the genes encoding molecules 1 and 2.
    (b) Nuclear but not cytoplasmic RNA can be found that hybridizes to the genes encoding molecules 1 and 2.
    (c) Both nuclear and cytoplasmic RNA, but not polysome-associated RNA, can be found that hybridizes to the genes encoding molecules 1 and 2.

23. For each of the following diploid genotypes, indicate first whether β-galactosidase can be made, and second, whether synthesis of β-galactosidase is inducible (I) or constitutive (C), and finally, whether or not each cell could grow with lactose as sole carbon source.
    (a) $i^+z^-y^+/i^-z^+y^+$.
    (b) $i^+z^+y^+/o^ci^+z^-y^+$.
    (c) $i^+z^-y^+/o^ci^-z^+y^+$.
    (d) $i^+z^+y^-/i^-z^-y^+$.
    (e) $i^-z^+y^-/i^-z^+y^+$.
    (f) $i^-z^+y^+/i^+o^cz^-y^+$.
    (g) $i^+p^-z^+/i^-z^-$.
    (h) $i^+o^cz^-y^+/i^+z^+y^-$.
    (i) $i^+p^-o^cz^-y^+/i^+z^+y^-$.
    (j) $i^-p^-o^cz^+y^+/i^-z^+y^-$.

24. A 30-base-pair double-stranded DNA segment containing the *lac* operator sequence has been inserted into a plasmid. The plasmid was introduced into a strain of *E. coli* having the genotype $i^+z^+y^+$. Once established, there were about thirty copies of the plasmid present in each cell.
    (a) The bacteria containing the plasmid were constitutive for β-galactosidase synthesis. Suggest an explanation.
    (b) With continued growth of the culture, some inducible strains arose. What mutation probably produced inducibility?

25. In *E. coli* the enzyme alkaline phosphatase (APase) is coded for by the gene *phoA*. If phosphate, a product of the reaction catalyzed by APase, is added to a cell culture, synthesis of APase is greatly inhibited in the growing cells. There are two other genes, *phoB* and *phoR*, which also seem to be involved in regulating the expression of *phoA*; mutations in *phoB* lead to reduced synthesis of APase in the presence or absence of phosphate, as shown in the table below. Mutations in the *phoR* gene lead to constitutive synthesis of APase. We know that the *phoB* and *phoR* genes each produce a protein, so the mutations shown in the table do not represent *cis*-acting mutations. We also know that the *phoA* gene is being regulated by transcriptional control. Assume that no other proteins or small molecules besides the ones mentioned here are involved in regulating the *phoA* gene. Use the data given below to answer the questions. The numbers represent relative amounts of APase when phosphate is absent or present.

Mutants	Absent	Present
$phoA^+phoB^+phoR^+$	100	20
$phoA^-phoB^+phoR^+$	0	0
$phoA^+phoB^-phoR^+$	5	1
$phoA^+phoB^+phoR^-$	100	100
$phoA^+phoB^-phoR^-$	5	5

    (a) Which one of the regulatory proteins is interacting directly with phosphate to regulate the expression of APase? Briefly explain why you concluded this.
    (b) Which one of the regulatory proteins is interacting directly with a site on the DNA? Is this protein a positive or negative effector? Briefly explain.
    (c) Explain how a complex between the phoB and the phoR proteins might function in regulating the expression of APase. Include the role of phosphate in your explanation.

26. The regulation of an operon responsible for synthesis of X is dependent on a repressor, a promoter, and an operator. In the presence of X, the system is turned off; an interaction between X and the repressor forms a complex

which can bind to the operator.

(a) What kinds of mutations might occur in the repressor? Describe their phenotype (in terms of whether the operon is on or off).

(b) Describe the phenotype of a partial diploid with one wildtype and one mutant gene for each mutant gene.

27. The regulation of the synthesis of an enzyme can be accomplished by controlling transcription or translation. Suppose you are studying the kinetics of appearance of an inducible enyzme after the addition of an inducer. The enzyme can be assayed both by its enzymatic activity and by an immunological test which shows that the protein is present. After induction the enzymatic activity does not appear until several minutes after the appearance of immunological reactivity. Give several possible explanations for this phenomenon.

C H A P T E R   12

Genetic Engineering

In addition to advancing the understanding of natural phenomena, genetics is also important in manipulating biological systems for scientific or economic reasons, an endeavor that has made possible the creation of organisms having new phenotypes or genotypes. The fundamental techniques for accomplishing this have been mutagenesis or recombination followed by selection for desired characteristics. When using such techniques, geneticists have been forced to work with the random nature of mutagenic and recombination events, which required selective procedures, often quite complex, to find an organism with the required genotype among the many types of organisms produced. Recently, a new technique has been developed by which the genotype of an organism can instead be modified in a *directed and predetermined* way. In this technique, which is alternately called **recombinant DNA technology, genetic engineering,** or **gene cloning,** DNA fragments are isolated and then recombined by *in vitro* manipulations. Selection of a desired genotype is still necessary but the probability of success is usually many orders of magnitude greater than that with traditional procedures. The recombinant DNA technology has greatly enhanced our ability to manipulate genomes and revolutionized the study of gene structure and regulation. The basic technique is quite simple: two DNA molecules are isolated and cut into fragments by one or more specialized enzymes and then the fragments are joined together *in a desired combination* and restored to a cell for replication and reproduction.

Current interest in genetic engineering centers on its many practical applications, a few examples of which are the following: (1) isolation of a particular gene, part of a gene, or region of a genome, (2) production of particular RNA and protein molecules in quantities formerly thought to be unobtainable, (3) improvement in the production of biochemicals (such as enzymes and drugs) and commercially important organic chemicals, (4) production of varieties of plants having particular desirable characteristics (for

Facing page: Genetic engineering of a plant. *Left panel.* Young shoots of a regenerating tobacco plant, growing on a solid medium containing hormones that prevent root formation and stimulate leaf and stem formation. The cluster developed from callus tissue obtained from tobacco stems infected with the bacterium *Agrobacterium tumefaciens* containing the hybrid plasmid pGV3850. *Right panel.* A plant formed from one of the shoots seen in the left panel. A shoot was cut from the cluster, placed on a medium that allows roots to form, and finally placed in soil. The regenerated plant synthesizes a substance, nopaline, that is encoded in the plasmid. (From P. Zambryski, H. Joos, C. Genetello, J. Leemans, M. Van Montague, and J. Schell. 1983. *EMBO Journal,* 2: 2143.)

example, requiring less fertilizer or having resistance to disease), (5) correction of genetic defects in higher organisms, and (6) creation of organisms with economically important features (for example, plants capable of maturing faster or having greater yield). Some of these examples will be considered later in the chapter.

The first step in genetic engineering is the isolation of DNA molecules. This is described in the following section.

12.1 Isolation of DNA: General Features

The fraction of the total dry weight of various organisms representing DNA varies from about 1 percent (in complex mammalian cells) to 50 percent (in phages). The DNA may be enclosed in a protein shell or membranous material. With bacteria and plants a carbohydrate cell wall must be broken, and except for phages and viruses the DNA is bound to various types of protein molecules. Consequently, the methods for isolating DNA vary somewhat from one type of organism to the next, though there is a common feature: namely, a cell or virus is first broken, and then DNA is separated from such other components as protein, RNA, lipid, and carbohydrates. The basic procedures are the following four:

1. **Phages.** The simplest procedure is employed with these. First, a suspension of particles is shaken with phenol, a reagent that both breaks open the protein coat and precipitates the individual protein molecules. Then, protein is removed by centrifugation and the DNA, which remains in solution, is precipitated and redissolved in a solution having any desired composition.

2. **Bacteria.** The contents of bacterial cells are enclosed in a multilayered cell wall and an inner cell membrane. These components are removed by successive treatment of a cell suspension with **lysozyme** (an enzyme isolated from chicken egg white), and one of several different detergents, the most common one being sodium dodecyl sulfate (SDS). This procedure solubilizes the cell wall and cell membrane and releases the contents of the cells. RNA and protein are removed by digestion with specific enzymes, and the DNA is collected by precipitation.

3. **Yeasts and fungi.** These cells are resistant to lysozyme, and other enzymes are used to break down their cell walls. Otherwise, the procedure follows that used with bacteria.

4. **Higher eukaryotes.** Animal and plant cells have a low ratio of DNA to protein. For technical reasons this causes substantial losses of DNA, if the DNA is directly purified. Thus, nuclei are usually isolated first, which not only increases the ratio of DNA to protein but avoids contamination of chromosomal DNA by DNA from cytoplasmic organelles. The nuclei are broken open in several ways, RNA and proteins molecules are enzymatically digested, and the DNA is precipitated.

Circular plasmid DNA is especially important in genetic engineering. When DNA is isolated from plasmid-containing bacteria or yeast, the sample obtained will contain both chromosomal and plasmid DNA, which must be separated. A large part of the chromosomal DNA can be removed by centrifugation at low speed. Circular plasmid DNA molecules are then purified by a procedure that is based on (1) the circularity of the plasmid DNA and (2) the considerable fragmentation (and hence linearization) of the circular bacterial chromosome that occurs during isolation. Plasmid DNA is quite resistant to fragmentation because it is so much smaller than chromosomal DNA. The technique used is equilibrium centrifugation (Section 4.6) in CsCl solutions containing **ethidium bromide.** This molecule binds very tightly to DNA by slipping in between the DNA base pairs (*intercalation*), causing the DNA to unwind; the lower density of the ethidium bromide molecules compared to that of a nucleotide causes a decrease in the density of the DNA by approximately 0.15 g/cm^3. The density shift is greater for linear DNA molecules (for example, the chromosome fragments) than for covalent circles (the plasmid DNA), and forms the basis of the separation. A covalently closed, circular DNA molecule has no free ends; thus, for simple topological reasons, unwinding the individual polynucleotide strands is accompanied by a twisting of the entire molecule in the opposite direction of the unwinding motion. For example, a circular molecule that has bound enough ethidium bromide molecules to produce one clockwise-unwinding turn will twist in the counter-

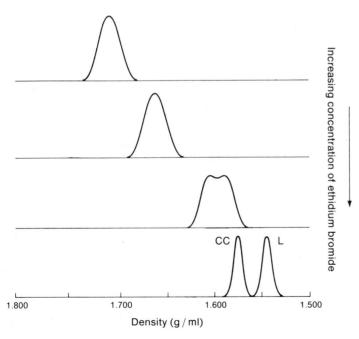

Figure 12-1 Equilibrium centrifugation of a mixture of equal parts of covalently circular (CC) and linear (L) DNA in CsCl containing various concentrations of ethidium bromide. The density of the DNA molecules decreases until the two components separate. The covalent circles bind less ethidium bromide and are therefore at a higher density.

clockwise direction, yielding a molecule shaped like the figure 8. As additional ethidium bromide molecules intercalate, the 8-shaped molecule becomes more twisted; ultimately, the DNA molecule is unable to twist further, so no additional ethidium bromide molecules can be bound. (Actually, plasmid DNA molecules are negatively supercoiled at the outset, so the supercoiling is first removed by binding ethidium bromide and then reversed in direction—that is the DNA becomes positively supercoiled.) A linear DNA molecule does not have the topological constraint that a covalent circle has and therefore can bind a greater number of ethidium bromide molecules. Because of the decrease in density of DNA caused by the binding of ethidium bromide molecules and the lesser binding to a covalent circle than to a linear molecule, a covalent circle has a *higher* (less reduced) density at concentrations of ethidium bromide that saturate the covalent circle. Therefore, by equilibrium centrifugation in a density gradient, covalent circles (plasmids) can be separated from linear molecules (chromosomal fragments), as shown in Figure 12-1.

In genetic engineering the immediate goal of an experiment is usually to insert a *particular* fragment of chromosomal DNA into a plasmid or a viral DNA molecule. This is facilitated by techniques for breaking DNA molecules at specific sites and for isolating particular DNA fragments. These techniques are described in the following section.

12.2 Isolation and Characterization of Particular DNA Fragments

In the recombinant DNA methodology DNA fragments are usually obtained by treatment of DNA samples with a specific class of nucleases. Many nucleases have been isolated from a variety of organisms, and most produce breaks at random sites within a DNA sequence (endonucleases) or remove nucleotides only from the termini (exonucleases), as discussed in Chapter 4. However, the class of nucleases called **restriction endonucleases,** or, more simply, **restriction enzymes,** consists of sequence-specific enzymes. Most restriction enzymes recognize only one short base sequence in a DNA molecule and make two single-strand breaks, one in each strand, generating 3'-OH and 5'-P groups at each position. Several hundred of these enzymes have been purified from hundreds of species of microorganisms.

The biological role of restriction enzymes is probably to destroy "foreign" DNA that succeeds in entering a cell. However, to do so without self-destruction requires that a cell be able to distinguish foreign DNA from "self" DNA. The DNA of a bacterium that makes an enzyme attacking a particular sequence will in general also contain copies of that sequence and hence could be attacked by the enzyme. However, a system exists for rendering the particular sequence resistant to the enzyme: shortly after DNA replication one base in the sensitive sequence is methylated (Section 8.5) by a sequence-specific methylating enzyme. Methylation of the target sequence makes the sequence unrecognizable by the restriction nuclease. Foreign DNA molecules also contain methylated bases, but in general the pattern of methylation will be different if the DNA is from a different organism. Thus, a cell is able to distinguish self DNA (target sequence methylated) from foreign DNA (target sequence not methylated) and will cleave only the foreign DNA.

The sequences recognized by restriction enzymes are **palindromes**—that is, the sequence has symmetry of the form

A	B	C	C′	B′	A′
A′	B′	C′	C	B	A

or

A	B	X	B′	A′
A′	B′	X′	B	A

or

A	B	B′	A′
A′	B′	B	A

in which the capital letters represent bases, a ′ indicates a complementary base, X is any base, and the vertical line is the axis of symmetry. Most of these sequences have 4–6 bases.

One of the most exciting events in the study of restriction enzymes was the observation by electron microscopy that fragments produced by many restriction enzymes spontaneously circularize. These circles could be relinearized by heating, but if after circularization they were also treated with *E. coli* DNA ligase, which joins 3′-OH and 5′-P groups (Section 4.6), circularization became permanent. This observation was the first evidence for three important features of restriction enzymes:

1. Restriction enzymes make breaks in symmetric sequences.

2. The breaks are usually not directly opposite one another.

3. The enzymes generate DNA fragments with complementary ends.

These properties are illustrated in Figure 12-2.

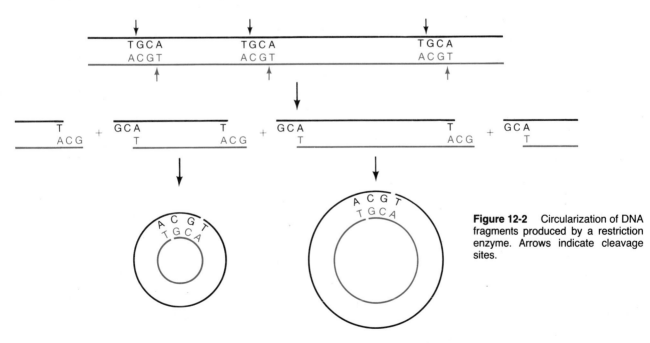

Figure 12-2 Circularization of DNA fragments produced by a restriction enzyme. Arrows indicate cleavage sites.

Examination of a very large number of restriction enzymes showed that the breaks are usually in one of two distinct arrangements: (1) staggered, but symmetric around the line of symmetry (forming **cohesive ends**) or (2) both at the center of symmetry (forming **blunt ends**). Two types of enzymes produce cohesive ends—those yielding a single-stranded extension with a 5'-P terminus and those yielding a 3'-OH extension. These arrangements and their consequences are shown in Figure 12-3. Table 12-1 lists the sequences and cleavage sites for several restriction enzymes, some of which generate cohesive sites and others of which yield blunt ends. Fragments having blunt ends cannot circularize spontaneously.

Most restriction enzymes recognize one base sequence without regard to the source of the DNA. Thus,

> Fragments obtained from a DNA molecule from one organism have the same cohesive ends as the fragments produced by the same enzyme acting on DNA molecules from another organism.

This point will be seen to be one of the foundations of the recombinant DNA technology.

Since most restriction enzymes recognize a unique sequence, *the number of cuts made in the DNA from an organism by a particular enzyme is limited.* A typical bacterial DNA molecule, which contains roughly 3×10^6 base pairs, is cut into several hundred to several thousand fragments, and nuclear DNA of mammals is cut into more than a million fragments. These numbers are large but still small compared to the number of sugar-phosphate bonds in an

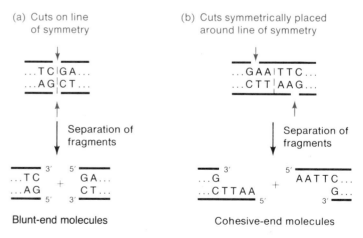

Figure 12-3 Two types of cuts made by restriction enzymes. The arrows indicate the cleavage sites. The dashed line is the center of symmetry of the sequence. As shown in Table 12-1, cohesive ends with 3' overhangs can also be produced. Other types of cuts (for example, outside of the recognition region) are also known, but they are less useful in genetic engineering and are not shown.

Table 12-1 Some restriction endonucleases, their sources, and their cleavage sites

Name of enzyme	Microorganism	Target sequence and cleavage sites
Generates cohesive ends		
EcoRI	*E. coli*	G A A T T C C T T A A G
BamHI	*Bacillus amyloliquefaciens* H	G G A T C C C C T A G G
HaeII	*Haemophilus aegyptius*	Pu G C G C Py Py C G C G Pu
HindIII	*Haemophilus influenza*	A A G C T T T T C G A A
PstI	*Providencia stuartii*	C T G C A G G A C G T C
TaqI	*Thermus aquaticus*	T C G A A G C T
Generates blunt ends		
BalI	*Brevibacterium albidum*	T G G C C A A C C G G T
SmaI	*Serratia marcescens*	C C C G G G G G G C C C

Note: The vertical dashed line indicates the axis of symmetry in each sequence. Arrows indicate the sites of cutting. The enzyme TaqI yields cohesive ends consisting of two nucleotides, whereas the cohesive ends produced by the other enzymes contain four nucleotides. Pu and Py refer to any purine and pyrimidine, respectively.

organism. Of special interest are the smaller DNA molecules, such as viral or plasmid DNA, which may have only 1–10 sites of cutting (or even none, for particular enzymes). Plasmids having a single site for a particular enzyme are especially valuable, as we will see shortly.

Because of the sequence specificity, *a particular restriction enzyme generates a unique set of fragments for a particular DNA molecule.* Another enzyme will generate a different set of fragments from the same DNA molecule. Figure 12-4(a) shows the sites of cutting of *E. coli* phage λ DNA by the enzymes EcoRI and BamHI. A map showing the unique sites of cutting of the DNA of a particular organism by a single enzyme is called a **restriction map.** The family of fragments generated by a single enzyme can be detected easily by gel electrophoresis of enzyme-treated DNA (Figure 12-4(b)), and particular DNA fragments can be isolated by cutting out the portion of the gel containing the fragment and removing the DNA from the gel.

Several techniques enable one to locate particular genes on fragments of a restriction map. One of the most generally applicable procedures is **Southern blotting.** In this procedure an agarose gel in which DNA molecules have been separated by electrophoresis is treated with alkali to denature the DNA and then the DNA is transferred to a sheet of nitrocellulose in a way that the relative positions of the DNA bands are maintained (Figure 12-5). The nitrocellulose, to which the single-stranded DNA is tightly bound, is then exposed to radioactive RNA or DNA in a way that leads to renaturation. Radioactivity becomes stably bound (resistant to removal by washing) to the DNA only at positions at which base sequences complementary to the radioactive molecules are present. The radioactivity is located by placing the paper in contact with x-ray film; after development of the film, blackened regions indicate positions of radioactivity. If a radioactive mRNA species transcribed from a particular gene is used (for example, mRNA isolated from a specialized cell that predominantly makes one type of mRNA), it will hybridize only with the restriction fragment containing that gene. A variant of the technique, **Northern blotting,** provides the same information. In this technique nonradioactive RNA is bound to a special paper and radioactive single-stranded DNA is transferred to the paper; after a period of renaturation the paper is washed and only DNA capable of renaturing to the RNA remains stably bound to the paper.

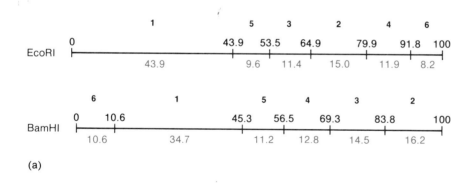

(a)

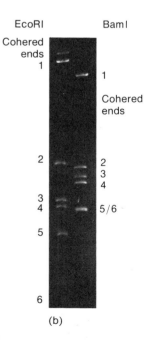

(b)

Figure 12-4   (a) Restriction maps of λ DNA for the restriction enzymes EcoRI and BamHI. The vertical bars indicate the sites of cutting. The black numbers indicate the percentage of the total length of λ DNA measured from the end of the molecular arbitrarily designated the left end. The red numbers are the lengths of each fragment, again expressed as percentage of the total length. (b) An electrophoretic gel of EcoRI and BamHI enzyme digests of λ DNA. The bands labeled cohered ends contain molecules consisting of the two terminal fragments joined by the normal cohesive ends of λ DNA. Numbers indicate fragments in order from largest (1) to smallest (6); the small bold numbers on the maps correspond to the numbers beside the gel. The DNA has not been electrophoresed long enough to separate bands 5 and 6 of the BamHI digest.

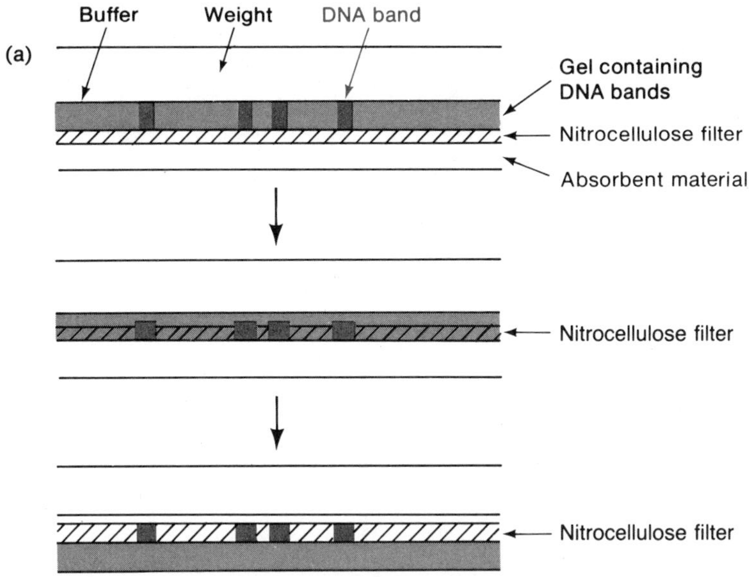

Figure 12-5 Southern blotting. (a) A weight is placed on a stack consisting of the gel, a nitrocellulose filter, and absorbent material. (b) At a later time the weight has forced the buffer (shaded area), which carries the DNA, into the nitrocellulose. (c) The lowest layer has absorbed the buffer but not the DNA, which remains bound to the nitrocellulose.

12.3 The Joining of DNA Molecules

In genetic engineering a particular DNA segment of interest is joined to a small, but essentially complete, DNA molecule that is able to replicate—a **vector** or **cloning vehicle.** By a transformation procedure, described below, this recombinant molecule is placed in a cell in which replication can occur (Figure 12-6). When a stable transformant has been isolated, the genes or DNA sequences on the donor segment are said to be **cloned.** In this section several types of vectors and a few procedures used for joining molecules are described.

Vectors

A vector is a DNA molecule in which a DNA fragment can be cloned; to be useful, it must have three properties:

1. A means of introducing vector DNA into a cell must be available.

2. The vector must have a replication origin—that is, be able to replicate.

3. Transformants must be selectable by a straightforward assay, preferably by growth of a host cell on a solid medium.

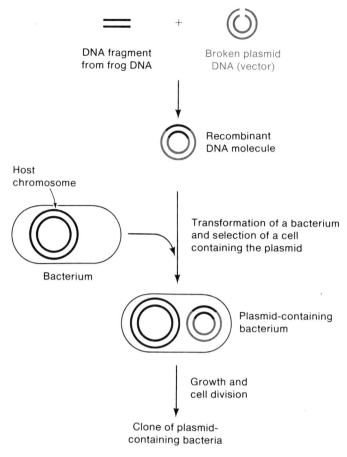

Figure 12-6 An example of cloning. A fragment of frog DNA is joined to a cleaved plasmid. The hybrid plasmid then transforms a bacterium, and, thereafter, frog DNA is present in all progeny bacteria.

The types of vectors most commonly used at present are plasmids and *E. coli* phages λ and M13; several animal viruses are also gaining in use. These cloning vehicles can be detected in host cells by means of genetic features or particular markers made evident during colony or plaque formation. Plasmid and phage DNA can be introduced into cells by the **CaCl$_2$ transformation procedure.** With this technique cells are exposed to a series of CaCl$_2$ solutions at several temperatures, and by this regimen they gain the ability to take up free DNA. Vector DNA, or any DNA that can replicate, is added to the cells at one stage of the protocol and, after a series of steps, the cells are plated on a solid medium. If the added DNA is a plasmid, colonies consisting of plasmid-containing bacteria will form, and transformants can sometimes be detected by the phenotype that the plasmid confers on the host cell. Selective procedures (for example, addition of an antibiotic and selection for drug resistance) that prevent cells that lack the plasmid from growing are usually used (Section 12.5). Alternately, if the vector is phage DNA, the "infected"

Table 12-2   Some cloning vehicles in use at present

Designation	Genotype or characteristics
Plasmid	
pBR322	*tet-r amp-r*
pUC8	*amp-r;* contains *lac* genes.
Yeast 2 μm	None
Agrobacterium Ti	Used to transform plants
Phage	
λ gt	Can lysogenize; two EcoRI sites.
λ-Charon	Insertion occurs in *lacZ* gene
M13mp17	Contains *lac* genes; forms single-stranded DNA; useful for sequencing.
Virus	
SV40	Virus infects animal cells; used for protein production.
Rous sarcoma	Integrates in host chromosome; used for gene therapy.

cells are plated in the usual way to yield plaques. A variation of this procedure is used to transform animal cells with viral vectors.

Retroviruses are RNA-containing animal viruses that have an unusual life cycle. These viruses, which contain the enzyme **reverse transcriptase** in their protein coats, use this enzyme to synthesize a double-stranded DNA copy of the viral RNA shortly after infection. As a normal part of the viral life cycle, this double-stranded DNA becomes inserted into the animal-cell chromosome, apparently at one of an enormous number of potential sites, and only after insertion can transcription occur. In certain conditions the infected host cell survives the infection, yet retains the retroviral DNA in its chromosome. One of the best-understood retroviruses is **Rous sarcoma virus,** which causes tumors in chickens. Retrovirus DNA, either isolated from a cell or prepared synthetically, is a useful vector with animal cells (Section 12.8).

Many vectors are in widespread use and some are highly specialized for particular uses. A few of these vectors and their properties are listed in Table 12-2.

Joining Fragments with Cohesive Ends

In Section 12.2 circularization of fragments having complementary terminal bases was described; similarly, two DNA fragments can join together by the pairing of their complementary cohesive ends. Therefore, because a particular restriction enzyme produces fragments with the *same* cohesive ends, without regard for the source of the DNA, fragments from DNA molecules isolated from two *different* organisms (for example, a bacterium and a frog) can be joined, as shown in Figure 12-7. Furthermore, if DNA is treated with DNA ligase to seal the joint after base-pairing, the fragments will be joined permanently, and a molecule will be generated that may never have existed before. The ability to join a DNA fragment of interest to a vector is the basis of the recombinant DNA technology.

Joining cohesive ends does not always produce a DNA sequence that has functional genes. For example, consider a linear DNA molecule that is cleaved into four fragments—A, B, C, and D—whose original sequence in the molecule was ABCD. Reassembly of the fragments can yield the original molecule, but since B and C have the same pair of cohesive ends, molecules with the fragment arrangements ACBD or BADC can also form with the same probability as ABCD. The problem of the scrambling of vector fragments is minimized by using a vector having only one cleavage site for a particular restriction enzyme. Plasmids of this type are available (most have been genetically engineered); usually they have sites for several different restriction enzymes, but only one enzyme is used at a time.

Joining Fragments without Cohesive Ends

DNA molecules lacking cohesive ends can also be joined. A direct method uses the DNA ligase made by *E. coli* phage T4. This enzyme differs from other DNA ligases (Section 4.6) in that it not only seals single-strand breaks in double-stranded DNA but can also join blunt-ended molecules.

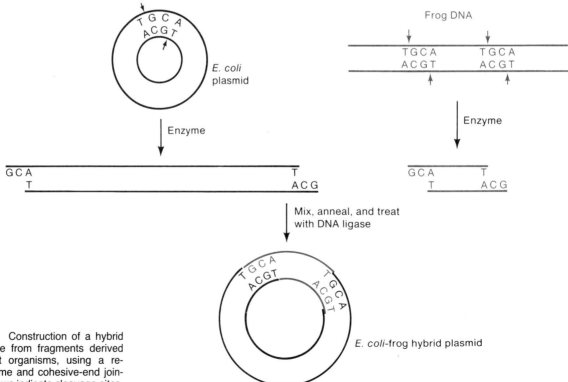

Figure 12-7 Construction of a hybrid DNA molecule from fragments derived from different organisms, using a restriction enzyme and cohesive-end joining. Short arrows indicate cleavage sites.

Another method uses the enzyme **terminal deoxynucleotidyl transfer-ase,** an unusual DNA polymerase obtained from animal tissue. This enzyme adds nucleotides (by means of deoxynucleoside triphosphate precursors) to the 3'-OH group of an extended single-stranded segment of a DNA chain—*without the need to copy a template strand.* In order to produce an extended single strand one need only treat a DNA molecule with a 5'-specific exonuclease to remove a few terminal nucleotides. If a mixture is prepared containing exonuclease-treated DNA, terminal deoxynucleotidyl transferase, and a single kind of nucleoside triphosphate—for example, deoxyadenosine triphosphate (dATP)—a DNA extension having adenine as the only base will form at both 3'-OH termini (Figure 12-8). Such extended segments are called **poly(dA) tails.** If instead thymidine triphosphate (dTTP) were provided, the DNA molecule would have poly(dT) tails. Since a poly(dA) tail is complementary to a poly(dT) tail, any two DNA molecules can subsequently be joined if a poly(dA) tail is put on one molecule and a poly(dT) tail on the other, as shown in the figure. Gaps in the joined molecule can be filled by treatment with a DNA polymerase (*E. coli* polymerase I is commonly used), and the remaining single-stranded interruptions are again sealed by DNA ligase. This method, which is called **homopolymer tail-joining,** is a general method for joining any pair of DNA molecules.

Often at a later time one will want to isolate a cloned fragment from a vector. This is straightforward when cohesive-end joining has been used in

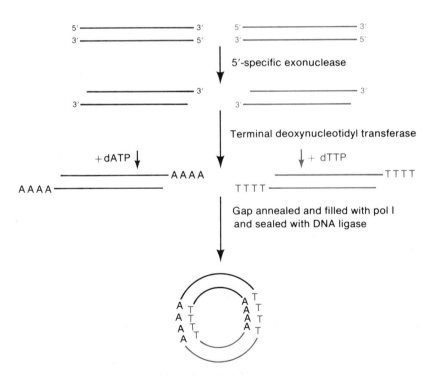

Figure 12-8 Joining two molecules with complementary homopolymer tails.

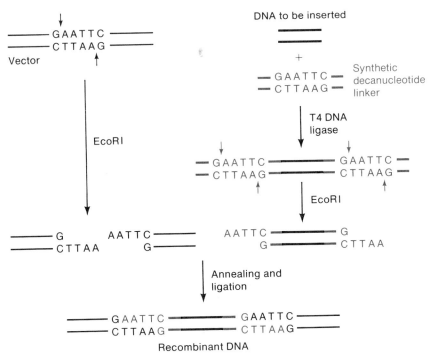

Figure 12-9 Formation of recombinant DNA through use of a linker. The short arrows indicate the site of cutting by the enzyme EcoRI.

that the restriction enzyme used originally to produce the cohesive termini can excise the cloned DNA by making a cut in the new joint. However, if T4 DNA ligase had been used to join blunt-ended fragments that had not been formed by the same restriction enzyme or if homopolymer-tail joining had been used, no enzyme would be available to perform the specific cleavage. In those cases, the use of **linkers** (Figure 12-9) provides the necessary sequences. With the linker method a synthetic short DNA segment, which contains a known restriction site, is coupled by T4 DNA ligase to each blunt end of both the fragment of interest and the vector. Then, both DNA molecules are treated with the restriction enzyme whose site is contained in the linker. In this way, the same complementary single-stranded termini are produced on both the fragment to be inserted and on the vector, and joining of cohesive termini becomes possible.

12.4 Insertion of a Particular DNA Molecule into a Vector

In the cloning procedure that we have described so far, a collection of fragments obtained by digestion with a restriction enzyme can be made to anneal with a cleaved vector molecule, yielding a large number of hybrid vectors containing different fragments of foreign DNA. However, if a particular DNA segment or gene is to be cloned, the vector possessing that segment must be isolated from the set of all vectors possessing foreign DNA. The kinds of selections used will be described shortly.

For many genes, simple selection techniques are adequate for recovery of a vector containing that gene. For example, if the gene to be cloned were a bacterial *leu* gene, one would use a Leu$^-$ host and select a Leu$^+$ colony on a medium lacking leucine. However, often the clone of interest is either so rare or so difficult to detect that it is preferable if, prior to joining, the fragment containing the DNA segment can be purified. One method has already been discussed: if the DNA of interest (for example, a viral gene) is known to be contained in a particular restriction fragment, that fragment can be isolated from an agarose gel after electrophoresis and joined to an appropriate vector. However, eukaryotic cells contain about a million cleavage sites for a typical restriction enzyme, so direct isolation of a eukaryotic gene from a mixture of fragments separated by electrophoresis is not feasible. In this section two other procedures for cloning a particular DNA molecule are described.

The Use of c-DNA and the Linker Method

In both prokaryotes and eukaryotes, selection of a vector containing a particular gene from a collection of vectors containing foreign DNA is not always straightforward. In theory, a gene is most easily detected by its expression— that is, the phenotype conferred by the gene, as mentioned previously for the *E. coli leu* gene. However, all prokaryotic genes do not produce such easily detected phenotypes and for eukaryotes, a particular gene that has been inserted into a bacterial DNA vector may, for a variety of reasons, fail either to be transcribed or to be correctly translated into a functional protein. By the technique described in this section the insertion of any coding sequence whose mRNA can be isolated in almost pure form—which is possible for many eukaryotic genes—can be simply done.

Let us assume that the chicken gene for ovalbumin (an eggwhite protein) is to be cloned into a bacterial plasmid, in order to determine the base sequence of the gene. A recombinant plasmid could be formed by treating both chicken cellular DNA and the *E. coli* plasmid DNA with the same restriction enzyme, mixing the fragments, annealing, and finally transforming *E. coli* with the DNA mixture containing the recombinant plasmid. However, identification of a bacterial colony containing this plasmid could not be carried out by a test for the presence of ovalbumin, because a bacterium containing the intact ovalbumin gene will not synthesize ovalbumin. Introns will be present in the gene (Section 9.2), and bacteria lack the enzymes needed to remove these noncoding sequences. Other tests might be performed to find the desired cell, such as hybridization of bacterial DNA with either the primary transcript or the mRNA of the ovalbumin gene isolated from chicken cells (Section 11.8), but since only about one plasmid-containing colony in 10^5 would contain the ovalbumin gene, this method would be very tedious. A more convenient procedure is clearly desirable.

Some specialized animal cells, such as those producing ovalbumin, make only one or a very small number of proteins. In these cells the specific mRNA molecules, whose introns will have been removed by processing enzymes *if the mRNA is isolated from the cytoplasm*, constitute a large fraction of the total mRNA synthesized in the cell; consequently, mRNA samples can usually be obtained that consist predominantly of a single mRNA species—in this case, ovalbumin mRNA. If genes of this class—that is, those whose gene products are the major cellular proteins—are to be cloned, the purified mRNA of each type of cell can serve as a starting point for creating a collection of recombinant plasmids many of which should contain only the gene of interest.

Many RNA-containing animal tumor viruses contain reverse transcriptase in the virus particle. This enzyme can use a single-stranded RNA molecule (such as mRNA) as a template and, by an incompletely understood mechanism, synthesize a double-stranded DNA copy, called **complementary DNA** or **c-DNA**. If the template RNA molecule is an mRNA molecule (that is, if the introns have been removed from the primary transcript), the corresponding full-length c-DNA will contain an uninterrupted coding sequence. *This sequence will not be that of the original eukaryotic gene;* however, if the purpose of forming the recombinant DNA molecule is to synthesize a eukaryotic gene product in a bacterial cell, and if such processed RNA can be isolated, then c-DNA formed from processed mRNA is the material of choice to be inserted. Joining of c-DNA to a vector can be accomplished by any of the procedures for joining blunt-ended molecules; the linker method is most commonly used.

Use of Synthetic Genes

Higher organisms (for example, man) produce many polypeptide hormones, which are of great medical value and which typically contain fewer than 20 amino acids. In most cases, neither the genes nor the corresponding mRNA molecules have been isolated. The amino acid sequences of most of the polypeptide hormones are known, so a base sequence encoding the polypeptide can be deduced from the genetic code. Short DNA molecules of defined sequence can be synthesized, and it is often possible to synthesize a coding sequence for the hormone. The sequence that is synthesized need not be precisely the same as that of the natural gene, because the redundancy of the code allows substantial substitution of codons without altering the information content, and the natural gene may contain introns. To make a functioning synthetic gene, the base sequence must be initiated by a start codon and terminated by a stop codon, both of which can be included in the DNA molecule. Once the gene is synthesized, it can be joined to a vector by either homopolymer tail-joining or the linker method. At the present time, the largest DNA molecule synthesized in this way—that encoding a human interferon gene—has 514 base pairs, which is sufficient for synthesizing a protein having 170 amino acids. In Section 12.8 an example of the use of a synthetic gene will be given.

When a vector is cleaved by a restriction enzyme and renatured with a mixture of all restriction fragments from a particular organism, many types of molecules result—some examples are a self-joined vector that has not acquired any fragments, a vector with one or more fragments, and a molecule consisting only of many joined fragments. Molecules of the third class cannot be established in a bacterium; they usually lack a DNA replication origin and replication genes, and are of no consequence. However, to facilitate the isolation of a vector containing a particular gene, some means is needed to ensure, first, that a vector established after $CaCl_2$ transformation does possess an inserted DNA fragment, and, second, that it is the DNA segment of interest. In this section several useful procedures for detecting plasmid vectors and recombinant vectors will be described.

In using the $CaCl_2$-transformation procedure to establish a plasmid in a bacterium, the initial goal is to isolate bacteria that contain the plasmid from a mixture of plasmid-free and plasmid-containing colonies. A common procedure is to use a plasmid possessing an antibiotic-resistance marker and to grow the transformed bacteria on a medium containing the antibiotic: only cells in which a plasmid has become established will form a colony. A useful plasmid is pBR322. It is small, consisting of 4362 nucleotide pairs of known sequence, and has two different antibiotic-resistance markers—resistance to tetracycline (*tet-r*) and to ampicillin (*amp-r*). Thus, plasmid-containing transformants are easily detected by growth of a transformed culture on medium containing either one of these antibiotics. Also, pBR322 is very useful because it contains only one copy of each of seven different types of restriction-enzyme cleavage sites at which DNA can be inserted, so the position of inserted DNA is always known.

In addition to a screening procedure for identifying plasmid-containing cells, a method is needed to identify plasmids in which DNA has been inserted. Having two antibiotic-resistance markers, such as are available in pBR322, allows the use of a procedure for detecting insertion called **insertional inactivation;** this is carried out as follows. In pBR322 the *tet* gene contains sites for cutting by the restriction enzymes BamHI and SalI. Thus, insertion at either of these sites will yield a plasmid that is *amp-r tet-s*, because insertion interrupts and hence inactivates the *tet* gene. If wildtype (Amp-s Tet-s) cells are transformed with a DNA sample in which the cleaved pBR322 and restriction fragments have been joined, and the cells are plated on a medium containing ampicillin, all surviving colonies must be Amp-r and hence must possess the plasmid. Some of these colonies will be Tet-r and some Tet-s, and these can be identified by replica plating (Section 10.3) onto a medium containing tetracycline. Because unaltered pBR322 carries the *tet-r* allele, an Amp-r colony will also be Tet-r unless the *tet-r* allele has been inactivated by insertion of foreign DNA. Thus, an Amp-r Tet-s cell must contain not only pBR322 DNA but donor DNA as well.

The proportion of the transformed cells in which insertion has occurred can be enriched by growing the transformed cells, prior to plating, in a liquid medium containing both cycloserine and tetracycline. Cycloserine kills growing cells, but Tet-s cells are merely inhibited, not killed, by tetracycline. Thus, in this growth medium Tet-r cells (which grow) are killed, and Tet-s cells (which are inhibited) survive. Plating of cells treated in this way on medium containing ampicillin yields Amp-r Tet-s colonies; these all possess pBR322 containing a donor DNA fragment. One would, of course, always

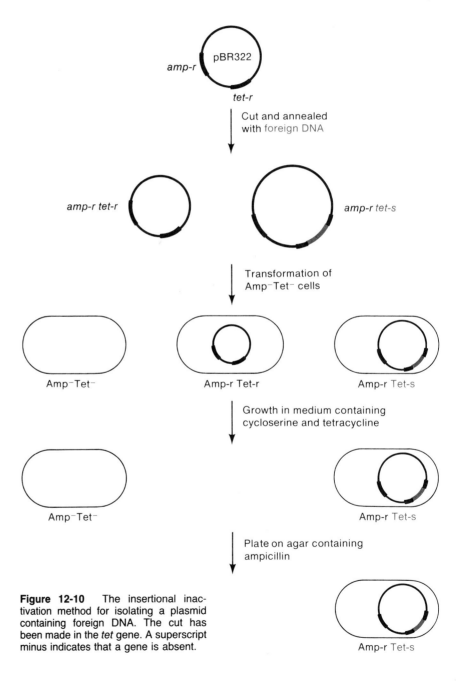

Figure 12-10 The insertional inactivation method for isolating a plasmid containing foreign DNA. The cut has been made in the *tet* gene. A superscript minus indicates that a gene is absent.

retest for the Tet-s character by replica plating, because some Tet-r cells may have survived the treatment. The insertional inactivation procedure is illustrated in Figure 12-10.

Other useful plasmids contain one antibiotic-resistance marker and the *E. coli lac* genes. The Lac$^+$ and Lac$^-$ phenotypes can be distinguished by plating on a color-indicator medium (Section 7.1), so insertion can be recognized by the appearance of colonies of a particular color.

Other procedures can also be used to select against reconstituted plasmids lacking foreign DNA. For example, the use of homopolymer tail-joining automatically selects against re-formation of the plasmid. The two termini of the cleaved plasmid will both have the same homopolymers, so restoration of circularity of the plasmid, which is essential for plasmid replication, is not possible. Also, if the cleaved vector is treated with the enzyme **alkaline phosphatase,** which removes 5′-P groups and leaves 5′-OH groups, a vector without inserted DNA cannot be ligated (since DNA ligase joins a 3′-OH group to a 5′-P group). Such a plasmid can circularize but cannot maintain its circularity when heated. If DNA is inserted, the 3′-OH termini can be ligated to the 5′-P termini of the inserted DNA, yielding a stable circular molecule (Figure 12-11).

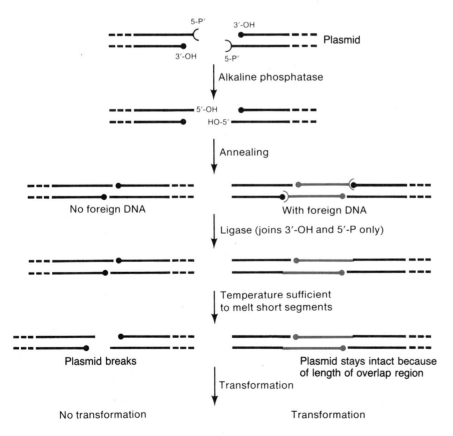

Figure 12-11 A method use to prevent re-formation of a cleaved plasmid. The plasmid DNA, but not the foreign DNA, is treated with alkaline phosphatase, an enzyme that converts terminal 5′-P groups to 5′-OH groups.

When *E. coli* phage λ is the vector, two procedures are used for selecting for the presence of inserted genes. Both are based on the fact that λ has several centrally located genes that are not needed for lytic growth. A λ variant, one of several that have been constructed by recombinant DNA techniques, has two restriction sites for the enzyme EcoRI near the termini of this nonessential region. To use this vector requires cleaving the DNA with the EcoRI enzyme and separating the three fragments by gel electrophoresis. The terminal fragments are isolated and the central fragment is discarded (Figure 12-12). The terminal fragments can be joined via the cohesive termini produced by EcoRI, but the resulting DNA molecule, whose length is 72 percent that of wildtype λ DNA, will be noninfective. This is because the minimum length of DNA that can be packaged in a λ phage head is 77 percent of the wildtype length. However, if foreign DNA whose size is 5–35 percent of the size of wildtype λ DNA is inserted, the resulting recombinant DNA molecule will be packageable and hence able to produce progeny phage. Thus, if an *E. coli* culture is infected (by the $CaCl_2$ transformation procedure) with DNA obtained by annealing a mixture of the two terminal fragments and foreign DNA, any phage that are produced will contain foreign DNA.

A phage λ vector has also been designed that carries part of the *E. coli lac* operon (particularly, the genes required for utilization of lactose as a carbon source). When plated on appropriate bacteria on particular color-indicator media, plaques formed by Lac$^+$ phage yield colorless plaques and those by

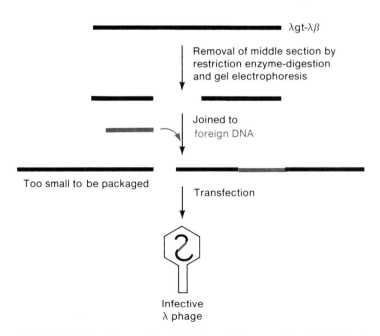

Figure 12-12 Packaging of recombinant DNA in a λ phage particle. λgt-λβ is a λ variant having two EcoRI sites. Transfection is a term used for infection with phage DNA.

Lac⁻ yield red plaques. The *lac* segment used contains several restriction sites. Insertion of foreign DNA into one of these sites results in some Lac⁻ phage and hence some red plaques, when a Lac⁺ vector is used. Phage isolated from the colorless plaques can be tested further for the gene of interest.

12.6 Screening for Particular Recombinants

As described in the beginning of Section 12.4, the simplest procedure for detecting a cloned gene is complementation of a bacterial gene. Some eukaryotic genes can also be detected in this way. In the first such experiment reported, a cloned *his* (histidine synthesis) gene of yeast was detected by transforming a particular His⁻ *E. coli* mutant and selecting for growth on a medium lacking histidine. However, this method is not generally successful with animal or plant DNA cloned in bacteria, because either the genes are not expressed or there is no corresponding bacterial gene. The following two procedures described in this section are more generally applicable and are commonly used for eukaryotic genes.

The **colony** or ***in situ* hybridization assay** allows detection of the presence of any gene for which radioactive mRNA is available (Figure 12-13). Colonies to be tested are replica-plated from a solid medium onto filter paper. A portion of each colony remains on the medium, which constitutes the reference plate. The paper is treated with NaOH, which simultaneously opens the cells and denatures the DNA. The paper is then saturated with ³²P-labeled mRNA, complementary to the gene being sought, and DNA-RNA renaturation occurs. After washing to remove unbound [³²P]mRNA, the positions of the bound radioactive phosphorus, usually detected by autoradiography, locate the desired colonies. A similar assay is done with phage vectors; in this case, plaques are replica-plated.

If the protein product of a gene of interest is synthesized, immunological techniques allow the protein-producing colony to be identified. In one test the colonies are transferred, as in colony hybridization, and the transferred copies are exposed to a radioactive antibody directed against the particular protein. Colonies to which the radioactivity adheres are those containing the gene of interest. The radioactivity is generally detected by autoradiography.

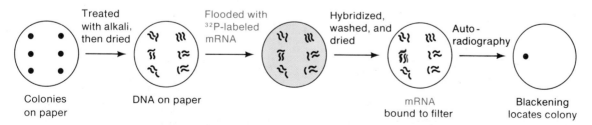

Figure 12-13 Colony hybridization. The reference plate, from which the colonies on paper were obtained, is not shown.

12.7 Gene Libraries

Many laboratories utilize recombinant DNA techniques repeatedly as a means of isolating particular genes or DNA segments from a single organism. It is time-consuming to go through the complete cloning procedure each time a new DNA segment is needed. Thus, collections of hybrid plasmid-containing bacteria have been established that are of sufficient size that each segment of the cellular DNA is represented at least once in the collection. Such collections are called **gene libraries.** In this section we describe how a gene library is established and how to determine how many clones are needed in order that each sequence will be represented.

A library of yeast DNA sequences has been established with an *E. coli* plasmid, ColE1. To form this library, purified plasmid DNA was cleaved with the EcoRI restriction enzyme, which makes a single cut in the plasmid. The yeast DNA was not cleaved with a restriction enzyme, because the procedure would have divided many genes that would therefore not be present as intact units in the library. Instead, the DNA was broken at random positions by violently stirring a solution of the DNA; in this way, each gene would eventually be obtained unbroken from one or more of the DNA fragments. In order to link the plasmid and yeast DNA molecules, the single-stranded termini were removed from the cleaved ColE1 DNA and poly(dT) tails were added, and poly(dA) tails were added to the yeast blunt-ended DNA fragments. The molecules were joined and the bacteria were transformed by the recombinant DNA, and colonies containing plasmid-containing cells were isolated. The colonies were stored at $-80°C$, a temperature at which the cells remain viable indefinitely in particular storage media. The complete set of colonies, at least 70 percent of which contain yeast DNA, forms the library. The following calculation, using the simple notions of probability developed in Chapter 1, enables one to determine how many colonies are required to make a complete library.

Consider a DNA sample containing fragments of such a size that each fragment represents a fraction f of the DNA of the organism. The probability P that a particular sequence is present in a collection of N colonies is $P = 1 - (1 - f)^N$ or

$$N = \ln(1 - P)/\ln(1 - f)$$

Let us assume that we want each sequence to be in the library with 99 percent probability. If the donor DNA is from haploid yeast (which contains about 1.3×10^7 nucleotide pairs) and the average-sized fragment contains 1.3×10^4 nucleotide pairs, then $f = (1.3 \times 10^4)/(1.3 \times 10^7) = 0.001$ and

$$N = [\ln(1 - 0.99)]/[\ln(1 - 0.001)] = 4603$$

Thus, if the library contains about 4600 colonies, there is a 99 percent probability that any yeast gene will be present in at least one colony. Furthermore, if a few colonies are selected at random for further study, it is very likely that the yeast sequences contained in them will not overlap. For cells with more

DNA (such as *Drosophila*) the required number of colonies is about 300,000, which can be reduced roughly n-fold by increasing the size of the cloned fragments by a factor of n. For human DNA, N is 1.03×10^6. Gene libraries have also been established with λ phage as a cloning vehicle.

If a clone containing a particular DNA segment is needed, it can in principle be found by any of the screening procedures described in Section 12.6.

12.8 Applications of Genetic Engineering

Recombinant DNA technology has revolutionized biology in the past decade. At present, its main uses are (1) facilitating the production of useful proteins, (2) creating bacteria capable of synthesizing economically important molecules, (3) supplying DNA and RNA sequences as a research tool, (4) altering the genotype of organisms such as plants, and (5) potentially correcting genetic defects in animals (gene therapy). Some examples of these applications follow.

Commercial Possibilities

Bacteria can carry out an enormous number of chemical reactions, but in each species the number is limited. However, by genetic engineering, organisms can be created that combine the features of several bacteria. For example, several genes from different bacteria have been inserted into a single plasmid that has then been established in a marine bacterium to yield an organism capable of metabolizing petroleum; this organism has been used to clean up oil spills in the oceans. Furthermore, bacteria are being designed that can synthesize industrially important chemicals—for example, ethyl alcohol—and that can simplify the large-scale manufacture of expensive antibiotics, which has the potential for reducing the cost of these drugs. A human insulin, synthesized in *E. coli*, is already commercially available.

For decades, an important role of genetics has been in plant breeding—that is, in the development of new strains of plants having desirable characteristics. For the most part, plant breeders have had to rely on selection of mutants and on hybridization between different varieties of the same species. Creating new varieties of plants by direct alteration of genotype is now becoming an important application of genetic engineering. Of great use is the bacterium *Agrobacterium tumefaciens* and its plasmid Ti, which produces crown gall tumors in dicotyledonous plants. These tumors result from disruption of the bacterium inside plant cells, release of bacterial DNA, and integration of a segment of the plasmid DNA into the plant chromosome. By genetic engineering, genes obtained from one plant can be introduced into this plasmid and then, via infection of a second plant with the bacterium carrying a recombinant plasmid, the genes of the first plant can be transferred to the second plant. The tumor-causing genes of Ti have been removed by genetic engineering and transformation systems have been established by which purified DNA rather than bacteria can be used to produce the alterations. Some attempts to perform plant breeding in this way have recently been partially successful, but the technique must still be considered to be in its infancy. The photographs on the chapter-opening page illustrate a genetically engineered plant.

Uses in Research

Recombinant DNA technology is extraordinarily useful in basic research. In Chapter 11 several examples are given of mutant bacteria in which particular genetic systems have been altered in an effort to make them more amenable to study (for example, the numerous mutants in the *lac* operon). Such mutants have usually been derived through standard genetic techniques (for example, mutagenesis and genetic recombination). For simple mutants this procedure is straightforward, but for mutants required to have many genetic markers (which may be very closely linked) the frequency of production of mutants can be so low that their isolation becomes very tedious. Recombinant DNA techniques can simplify mutant construction, since fragments containing desired genetic markers can be purified, altered, and combined in a test tube, and then introduced into another cell. This saves time, labor, and often enables mutants to be constructed that cannot in practice be formed in any other way. An example is the formation of double mutants of animal viruses, which undergo crossing over at such a low frequency that mutations can rarely be recombined by genetic crosses.

The greatest impact of the new technology on basic research has been in the study of eukaryotes, in particular, of eukaryotic regulation. Experiments of the type considered in Chapter 11, which study the regulation of operons in bacteria, have been made possible by the use of mutations in promoters, operators, and structural genes. However, this approach has not been feasible with eukaryotes, because eukaryotes are diploid and, hence, mutants are difficult to isolate. Furthermore, except for the unicellular eukaryotes such as yeast, there is no simple and rapid way to do multiple genetic manipulations with eukaryotic cells.

Cloning techniques have made it possible to study regulation of gene expression in eukaryotes by direct assays of mRNA molecules produced by particular genes. The approach is based on the success with which it has become possible to understand gene activity in bacteria by studying the synthesis of mRNA in a variety of conditions. An excellent assay is DNA-RNA hybridization, in which either a primary transcript or mRNA is detected by hybridization to a DNA sample that has been enriched for the gene being studied.

In microbial experiments specialized transducing particles would be the source of the DNA—for example, *E. coli gal* mRNA is assayed by hybridization of radioactively labeled intracellular RNA with DNA of λ *gal* transducing particles. However, transducing particles do not exist for eukaryotic systems, so recombinant DNA techniques have been used to clone a gene whose regulation is to be studied. The usual procedure is to clone the gene first in a plasmid (or a phage) vector and then allow the cell containing the plasmid to multiply. Purified plasmid DNA provides the researcher with a large supply of the DNA of that gene. The vector containing that DNA sequence is called a DNA **probe,** because its DNA, in denatured form, can be added to a cell extract containing mRNA, to probe for a particular mRNA by hybridization. Since no natural genes of the vector are present in eukaryotic cells, the vector DNA can be regarded as a source of pure DNA copies of a particular eukaryotic gene because no other genes in the vector will participate in the hybridization. A further useful technique is to apply denatured probe DNA to a nitrocellulose filter and then to use a filter-binding assay for the specific RNA species. DNA but not RNA binds to these filters. Radioactive mRNA can then be incubated with a filter containing the DNA, using conditions such that renaturation will result. Except where the mRNA has hybridized to the denatured DNA, the RNA can be removed by washing the filter. Thus, presence of radioactivity in the filter after washing indicates that complementary mRNA has been bound to the DNA on the filter. This simple technique and variations such as Southern and Northern blotting (Section 12.2) have revolutionized the study of eukaryotic gene regulation.

Production of Eukaryotic Proteins

One of the most valuable applications of genetic engineering is the production of large quantities of particular proteins that are otherwise difficult to obtain (for example, proteins of which only a few molecules are normally present in a cell). The method is simple in principle. The gene encoding the desired protein is coupled to a vector containing a bacterial promoter, and tests are performed to ensure that the gene is oriented such that its coding strand is linked to the strand containing the promoter. A plasmid for which many copies are present in each cell, or occasionally an actively replicating phage (such as λ), is used as a vector; in both cases, cells can be prepared that contain several hundred copies of the gene, and this can result in synthesis of a gene product to reach a concentration of about 1 to 5 percent of the cellular protein. In practice, production of large quantities of a prokaryotic protein in a bacterium is straightforward. However, if the gene is from a eukaryote, special problems exist:

1. Eukaryotic promoters are not usually recognized by bacterial RNA polymerases.

2. The mRNA transcribed from eukaryotic genes may not be translatable on bacterial ribosomes.

3. Introns may be present, and bacteria are unable to excise eukaryotic introns.

4. The protein itself often must be processed (for example, insulin); and bacteria cannot recognize processing signals from eukaryotes.

5. Eukaryotic proteins are often recognized as foreign material by bacterial protein-digesting enzymes and are broken down.

Several approaches to solving these problems have been taken with some degree of success. One procedure uses a yeast plasmid (the 2-μm plasmid) as a cloning vehicle for eukaryotic genes. Recent studies suggest that when a recombinant 2-μm plasmid containing eukaryotic genes is put into yeast, which is a eukaryote, some of the problems just described no longer exist.

Another approach uses a plasmid containing the *E. coli lac* region, cleaved in the *lacZ* gene by a restriction enzyme making blunt ends. Either c-DNA or a synthetic DNA molecule whose sequence is known from the amino acid sequence of the protein product is inserted into the *lacZ* gene, so that the eukaryotic protein is synthesized as the terminal region of β-galactosidase, from which it can be cleaved. The first example of this approach resulted in a synthetic gene capable of yielding a 14-residue polypeptide hormone—somatostatin—which *in vivo* is synthesized in the mammalian hypothalamus. The procedure, applicable to most short polypeptides, was the following.

By chemical techniques, a double-stranded DNA molecule was synthesized containing 51 base pairs; the base sequence of the coding strand was

TAC-(42 bases encoding somatostatin)-ACTATC

The base sequence of the corresponding mRNA molecule was

AUG-(42 bases encoding somatostatin)-UGAUAG

The AUG codon of the mRNA was not used for initiation, but specified methionine. The mRNA terminated with two stop codons UGA and UAG. The vector was the plasmid pBR322 modified to contain the *lac* promoter-operator region and a portion of the *lacZ* gene encoding the amino-terminal segment of β-galactosidase. The vector was cleaved at a site in the *lacZ* segment (Figure 12-14) by a restriction enzyme that leaves blunt ends; the synthetic DNA molecule was then blunt-end-ligated to the cleaved plasmid. When the *lac* operon was induced, a protein was made consisting of the amino-terminal segment of β-galactosidase *coupled by methionine* to somatostatin, and terminated at the repeated stop codons of the synthetic DNA. This protein was purified and treated with cyanogen bromide, a reagent that cleaves proteins only at the carboxyl side of methionine. In this way, the methionine linker remains attached to a β-galactosidase fragment and somatostatin is released. Use of methionine-coupling followed by cleavage with cyanogen bromide is a useful technique for separating any polypeptide from a bacterial protein to which it is fused, as long as the polypeptide itself does not contain methionine.

Genetic Engineering with Animal Viruses

Genetic engineering with retroviruses allows the possibility of altering the genotype of an animal cell. Since a wide variety of retroviruses are known,

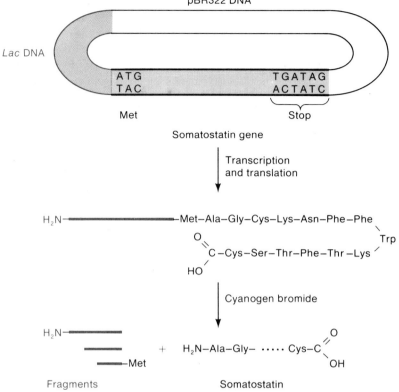

Figure 12-14 Synthesis of somatostatin from a chemically synthesized gene joined to the plasmid pBR322 *lac*.

including two that can grow on human hosts, genetic defects may be correctable by these procedures in the future. However, many retrovirus species contain a gene that produces uncontrolled growth of a cell containing the retrovirus, thereby causing tumors in animals. If a retrovirus vector is to be used to change a genotype, the tumor-causing ability must be removed. This is possible by removal of the tumor-causing gene, which also provides the space needed for incorporation of foreign DNA. Other retroviruses lack tumor-forming genes and produce tumors by virtue of their insertion next to particular cellular genes whose activation leads to uncontrolled cell division (Section 6.8). This type of virus has also been modified by removal of its terminal sequences, which are necessary for integration, thereby eliminating the tumorigenicity. The recombinant DNA procedure employed with retroviruses consists of synthesis in the laboratory of double-stranded DNA from the viral RNA, through use of reverse transcriptase. The DNA is then cleaved with a restriction enzyme and, by the techniques already described, foreign DNA is inserted and selected. Treatment of mutant cells in culture with the recombinant DNA, and application of a transformation procedure suitable for animal cells, yields cells in which recombinant retroviral DNA has been permanently inserted. In this way, the genotype of the cells can be altered. Experiments have been done in which human cells deficient in the synthesis of purines have been obtained from patients with Lesch-Nyhan syndrome and grown in culture; these cells have been converted to normal cells by transformation with recombinant DNA. The exciting potential of this technique lies in the possibility of correcting genetic defects—for example, restoring the ability of a diabetic individual to make insulin or correcting immunological deficiencies. This technique has been termed **gene therapy.** However, it must be recognized that retroviruses are not well understood and are potentially dangerous.

Gene therapy is not yet a practical technique, for major problems exist; for example, there is no reliable way to ensure that a gene is inserted in the appropriate target cell or target tissue. In addition, some means is needed to regulate the expression of the inserted genes.

Animal proteins can also be made in large quantities by animal cells containing many copies of a gene. For example, in a preliminary study to test the feasibility of cloning in an animal virus vector the rabbit β-globin gene (globin is a protein subunit of hemoglobin) has been cloned in SV40 virus, a virus that grows in cultures of monkey cells and produces thousands of intracellular DNA molecules; when this hybrid virus infects monkey kidney cells, rabbit β-globin is synthesized in large quantities.

Other exciting possibilities are being investigated. For example, vaccinia virus, the virus used in smallpox vaccination, has been modified so that some of its coat proteins are those of viruses that are difficult to grow and dangerous to handle, such as hepatitis virus. Individuals immunized with the hybrid vaccinia virus will also become immune to hepatitis virus. This new technique for antigen production is being developed by numerous pharmaceutical and bioengineering companies. Restriction mapping has also been used in prenatal diagnosis—for example, in detecting sickle-cell anemia. This disease is caused by possession of an altered hemoglobin (Section 10.2). The base change causing the mutation generates a cleavage site for a particular restriction enzyme. Cleavage of DNA from a sickle-cell patient produces a restriction map that differs from that of the DNA of a normal patient. The different restriction pattern can be detected by using Southern blotting and, as a probe, radioactive RNA prepared by transcribing cloned globin genes. The technique is so sensitive that the sickle-cell mutation can be detected in a small sample of cells obtained from the amniotic fluid surrounding a fetus, and detection is possible at a very early stage of development. Possibly in the future, gene therapy might be used to produce a normal child.

Genetic engineering appears at times to border on science fiction, and its potential must be kept in perspective. Modification of genotype in lower organisms is straightforward, though not always successful, but such modification in higher organisms remains difficult. At present, introduction of a single gene into a plant or animal is one goal of genetic engineering; creation of drastically different organisms probably remains in the realm of fantasy.

Problems

1. When a mixture of linear DNA molecules and covalently closed circles is centrifuged to equilibrium in CsCl containing ethidium bromide, the linear molecules have a higher density than the covalent circles, as explained in the text.
   (a) How would you expect the density of a circle containing one single-strand break (a nicked circle) to be related to the densities of the other types of molecules? Explain your reasoning.
   (b) Two covalent circles linked as in a chain would have the same density as a single covalent circle. How would the density change if one single-strand break were made in one member of a pair of linked circles, each having the same size?

2. The restriction enzyme EcoRI cleaves a particular phage DNA molecule into five fragments of sizes 2.9, 4.5, 6.2, 7.4, and 8.0 kilobase pairs. If the DNA has previously been labeled with polynucleotide kinase, an enzyme that can place ^{32}P on the 5′ termini, the 6.2-kb and 8.0-kb fragments are found to be labeled. BamHI cleaves the same molecule into three fragments, whose sizes are 6.0, 10.1, and 12.9 kb, with the 6.0-kb and 10.1-kb fragments labeled with ^{32}P. When both enzymes are used together, the fragments have sizes 1.0, 2.0, 2.9, 3.5, 6.0, 6.2, and 7.4 kb. What are the restriction maps for the two enzymes?

3. When DNA isolated from phage J2 is treated with the enzyme SalI, eight fragments are produced, whose sizes are 1.3, 2.8, 3.6, 5.3, 7.4, 7.6, 8.1, and 11.4 kb. However, if J2 DNA is isolated from infected cells, only seven fragments are found, whose sizes are 1.3, 2.8, 7.4, 7.6, 8.1, 8.9, and 11.4. What is the likely form of the intracellular DNA?

4. The following questions concern the possibility of circularization after enzymatic cleavage.
   (a) If a plasmid is cleaved in five positions with a restriction enzyme that produces an extended 3′ terminus, will all fragments be able to circularize?
   (b) E. coli phage T4 has a linear DNA molecule. If it is cleaved with the same restriction enzyme as that in part (a), will all fragments be able to circularize?

5. Phage X82 DNA is cleaved into six fragments by the enzyme BglI. A mutant is isolated whose plaques look quite different from the wildtype plaque. DNA isolated from the mutant is cleaved into only five fragments. Account for this change. In general, discuss why often a restriction enzyme may be found that yields n fragments with wildtype phage and $n - 1$ fragments with a mutant.

6. A Kan-r Tet-r plasmid is treated with the BglI enzyme, which cleaves the *kan* (kanamycin) gene. The DNA is annealed with a BglI digest of *Neurospora* DNA and then used to transform *E. coli*.
   (a) What antibiotic would you put in the growth medium to ensure that a colony has the plasmid?
   (b) What antibiotic-resistance phenotypes will be found among the colonies?
   (c) Which phenotype will have the *Neurospora* DNA?

7. A DNA fragment has been isolated that contains the coding sequences of two genes, A and B, of E. coli. In an effort to make large quantities of the gene products, the fragment is joined to a standard plasmid and the DNA is used to transform an A^-B^- bacterium. Colonies containing the plasmid are isolated easily, but none are A^+B^+. However, A^+B^- and A^-B^+ colonies are found. Explain this observation.

8. A *lac$^+$ tet-r* plasmid is cleaved by a restriction enzyme in the *lac* gene. The restriction site for this enzyme contains four bases and the cuts generate a two-base single-stranded end. The particular site in the *lac* gene is in the first amino acid of the chain, changes of which rarely generate a mutant. The single-stranded ends are converted to blunt ends with DNA polymerase I and then the ends are joined by blunt-end ligation. A *lac$^-$ tet-s* bacterium is transformed with the DNA and Tet-r bacteria are selected. What will be the Lac phenotype of the colonies?

9. Sometimes a cloned bacterial gene is not expressed because it has been separated from its promoter. The gene can be expressed if it can be cloned in a plasmid containing a promoter to which the gene can be linked. The gene must be oriented correctly though, because the coding sequence must be linked to the strand containing the promoter. Unfortunately, when mixing the fragments with a cleaved plasmid, joining will occur in all possible orientations. Consider a plasmid containing the *E. coli lac* operon (complete with the *lac* promoter), with a BamHI site separating the promoter and the genes. The *lac* sequence also contains restriction sites for several other enzymes. Explain how any one of these enzymes might be utilized in the cloning procedure to ensure that the gene of interest is linked to the *lac* promoter.

10. By using a restriction enzyme that generates blunt ends, a DNA sequence has been cloned in plasmid pBR322. The sequence was isolated and terminal nucleotidyl transferase was used to put a tract of cytosines at each end of

461

the molecule. The cleaved plasmid had poly(dG) tails, and homopolymer tail joining was used to link the fragment to the plasmid. Several months later a new and uncharacterized restriction enzyme is found that will cleave the cloned sequence from the plasmid. What property must this enzyme have?

11. Reverse transcriptase, like most enzymes that make DNA, requires a primer. Explain why, when c-DNA is to be made for the purpose of cloning eukaryotic genes, a convenient primer is a short sequence of poly(dT).

12. How many clones are needed to establish a gene library for monkey DNA (6×10^9 base pairs) if fragments whose sizes average 2×10^4 base pairs are used and if one wishes 99 percent of the monkey genes to be in the library?

13. Vectors have been designed with features suitable for specific purposes. One important class of vector is the shuttle vector, a plasmid that can be used in different organisms. The first shuttle vector was used to clone yeast genes. If a yeast gene were incorporated into an *E. coli* plasmid and then *E.coli* were transformed by the recombinant plasmid, in general the yeast gene would not be expressed; furthermore, it would not be straightforward to detect the gene. Shuttle vectors, which contain sequences from an *E. coli* plasmid and a particular region of the yeast genome (containing, among other things, tryptophan genes), were designed to avoid this problem. Such a vector can be cleaved and joined to yeast DNA, but, most important, the hybrid vector will transform *trp*⁻ yeast cells, producing Trp⁺ cells. The gene of interest carried on the fragment can also be detected by virtue of its expression in yeast. However, these vectors are not particularly stable in yeast, because they lack a centromere; hence, replicas are not efficiently segregated into daughter cells. To avoid this problem, the vector is isolated from yeast several generations after successful transformation and used to transform *E. coli*, in which its presence is usually detected by an antibiotic-resistance marker. What essential elements must be carried by a yeast-*E. coli* shuttle vector?

14. High-copy-number plasmids are useful vectors for cloning genes whose gene product (the protein) is required in large quantities—for example, for purifying an enzyme that normally exists in very small amounts. A cell containing such a plasmid with a cloned gene may make enormous quantities of the desired protein. However, often a cell making a high concentration of an enzyme that normally exists in low concentrations cannot tolerate the high concentration, and the cell dies. Design a simple method for cloning such a gene in a high-copy-number plasmid.

15. Plasmid pBR607 DNA is circular and double-stranded and has a molecular weight of 2.6×10^6. This plasmid carries two genes whose protein products confer resistance to tetracycline (Tet-r) and ampicillin (Amp-r) in host bacteria. The DNA has a single site for each of the following restriction enzymes: EcoRI, BamHI, HindIII, PstI, and SalI. Cloning DNA into the EcoRI site does not affect resistance to either drug. Cloning DNA into the BamHI, HindIII, and SalI sites abolishes tetracycline resistance. Cloning into the PstI site eliminates ampicillin resistance. Digestion with the following mixtures of restriction enzymes yields fragments with the sizes listed below (sizes given in millions of molecular weight units):

EcoRI, PstI	0.46, 2.14
EcoRI, BamHI	0.20, 2.40
EcoRI, HindIII	0.05, 2.55
EcoRI, SalI	0.55, 2.05
EcoRI, BamHI, PstI	0.20, 0.46, 1,94

Position the PstI, BamHI, HindIII, and SalI cleavage sites on a restriction map, relative to the EcoRI cleavage site.

16. A large circular DNA plasmid is digested with a restriction endonuclease. The fragments are allowed to reassemble at random (that is, the fragments are no longer in the original order), and circles having the same molecular weight as the original plasmid are selected. Host cells lacking the plasmid are then infected with these fragments.
   (a) Assuming that ability to replicate is the only requirement for successful infection, what alterations in fragment order might prevent successful infection?
   (b) Consider a large circle in which the order of the fragments is the same, but in which some circles contain an additional fragment adjacent to an identical fragment. Will this circle be capable of successful infection?
   (c) Is it possible to delete a fragment, keeping the order otherwise the same, and still infect cells successfully? Explain.

17. A Tet-r plasmid having a single EcoRI site is cleaved by the EcoRI enzyme. The sample is annealed, treated with DNA ligases, and then used in a $CaCl_2$-transformation experiment. Tet-r colonies are selected. Gel electrophoresis of the plasmid DNA from these colonies shows that there are two types of colonies: one type yielding a single band at the expected position for the original plasmid and another type yielding a band moving much more slowly. Consider what products can arise in an annealing reaction and suggest what the slow band could be.

CHAPTER 13

Somatic Cell Genetics and Immunogenetics

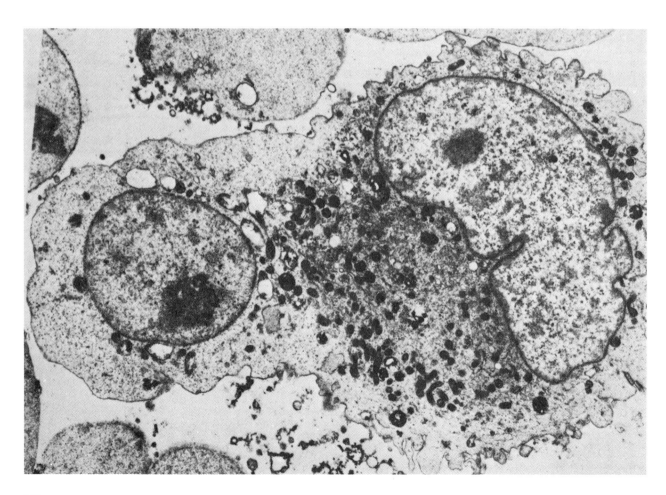

Genetic processes that occur in somatic cells, the cells in the body other than gametes or gamete-forming cells, constitute a subject termed **somatic cell genetics.** One genetic process in somatic cells was encountered in Chapter 3—namely, mitotic recombination, in which it was shown that crossing over occurring during mitosis can produce clones of homozygous daughter cells from an initially heterozygous parental cell (see Figure 3-15). Somatic events such as mitotic recombination are often useful in genetic analysis when traditional methods of meiotic chromosome mapping are not practical or are not possible, as in certain fungi (for example, Figure 3-17). Certain methods of somatic cell genetics have been highly developed for the study of mammalian cells grown in laboratory cultures, and these methods, which are a main topic of this chapter, have revolutionized the genetic analysis of human chromosomes.

Some genetic events occur only in somatic cells and do not occur at all in germ cells. One of the best-known examples is in the development of immunity, which is dependent upon genetic changes in somatic cells; in these cells gene-splicing events occur in which new functional genes are created by joining together pieces of other genes. Such unusual genetic processes form the second major subject of this chapter.

13.1 Somatic Cell Genetics

Figure 13-1 shows the mitotic chromosomes of an unusual type of cell—a **cell hybrid** formed by the fusion of a human cell and a mouse cell in a laboratory culture. Such hybrids are initially true complete hybrids in that they have one nucleus containing a complete set of 40 mouse chromosomes and also a complete set of 46 human chromosomes. However, the combined sets of

Facing page: Cell fusion. An immature red blood cell, from a five-day-old chick embryo, in the process of fusing with a mouse cell. (From H. Harris. 1970. *Cell Fusion.* Harvard University Press. Photo taken by Gutta Schoefl.)

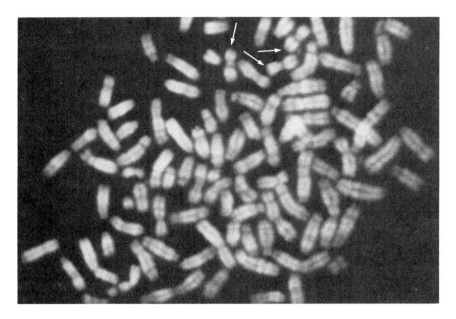

Figure 13-1   Metaphase chromosomes in a human-mouse cell hybrid. To make all of the chromosomes visible and distinct, a cell in metaphase has been treated with a drug that disrupts the spindle, enabling the chromosomes to drift apart. The chromosomes have been stained by a procedure that produces distinct banding patterns on each chromosome; the patterns are used to distinguish and identify all of the chromosomes. Several human chromosomes are indicated by arrows. (Courtesy of Raju Kucherlapati.)

chromosomes in a hybrid cell are often unstable and chromosomes are sometimes lost in the course of mitotic cell divisions following the cell fusion that created the initial hybrid clone. The cell shown in Figure 13-1 has lost some of its human chromosomes. A widely used method for producing such human-mouse cell hybrids will be discussed in a later section of this chapter. The following section deals with the application of hybrid cells to the mapping of genes in human chromosomes.

Somatic Cell Hybrids and Human Gene Mapping

The tendency of hybrid cells to lose chromosomes enables these cells to be used in assigning human genes to particular chromosomes. Although occasionally lost, the chromosomes in human-mouse hybrids are sufficiently stable to permit the establishment of **cell lines,** or clones, containing particular human chromosomes, and within these cell lines most cells will have the same combination of human and mouse chromosomes. The human chromosomes that are occasionally lost from hybrid cells vary, which allows a large number of cell lines to be established, each containing a *different* group of human chromosomes.

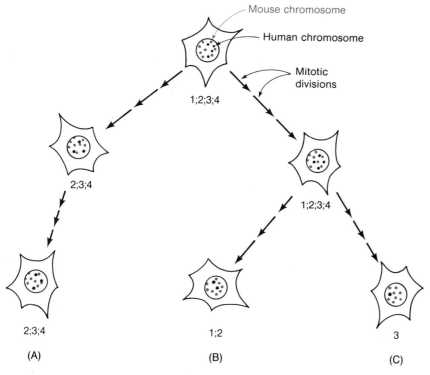

Figure 13-2 Three lines of hybrid cells that are chromosomally distinct; each line is formed by the loss of human chromosomes. The uppermost cell is a human-mouse hybrid having four human chromosomes (1, 2, 3, 4). After several mitotic divisions one daughter cell (left) is produced that has lost chromosome 1. The chromosome complement of this cell line is stable, yielding cell line A. Other mitotic divisions of the original cell result in loss of chromosomes 3 and 4 (line B) and chromosomes 1, 2, and 4 (line C).

Loss of chromosomes by hybrid cells is illustrated in Figure 13-2, in which the human chromosomes are indicated in black and the mouse chromosomes in red. In this representation, the cell at the top is assumed to have four human chromosomes—one copy each of chromosomes 1, 2, 3, and 4. After several mitotic divisions (arrows), a daughter cell is produced that has lost chromosome 1, and this cell is cultured individually, producing cell line A. Other mitotic divisions of the original hybrid cell may result in the loss of chromosomes 3 and 4 (cell line B), or the loss of chromosomes 1 and 2 and 4 (cell line C). The particular human chromosome present in a hybrid cell line is identified by its pattern of chromosome bands produced by the Giemsa staining procedure discussed in Chapter 6 and illustrated in Figure 6-11.

The products of homologous human and mouse genes can often be distinguished by electrophoresis or other means, and the method of assignment of human genes to chromosomes is based on the following reasoning: a gene product, such as an enzyme, will be produced only in cell lines containing the chromosome in which the gene is located. For example, an enzyme coded by a gene on chromosome 1 will be present in cell line B in Figure 13-2 but not in lines A or C. Indeed, the chromosomal location of any gene can be identified if a cell line is available lacking that chromosome, because for each gene there will be a unique pattern of presence or absence of the gene product among the cell lines. Specifically, if + indicates the presence of the gene product and − indicates its absence, then the patterns corresponding to genes on chromosomes 1, 2, 3, and 4 in Figure 13-2 will be

	Line A	Line B	Line C
Chromosome 1	−	+	−
Chromosome 2	+	+	−
Chromosome 3	+	−	+
Chromosome 4	+	−	−

Data from actual human-mouse hybrid cell lines are analyzed in essentially the same way as that used in the hypothetical example in Figure 13-2, but the actual data are somewhat more challenging because there are 23 pairs of human chromosomes to consider.

One set of eight human-mouse hybrid cell lines used in chromosome assignments is shown in Table 13-1. Each of the clones, designated by letters A through H, carries a different group of human chromosomes, and, conversely, each human chromosome has a different pattern of presence (+) or absence (−) among the clones. (Note that all eight clones carry the human X chromosome. This is not accidental but is a result of the particular method by which these clones were produced, which will be discussed shortly.) One application of these clones has been in locating the gene for uridine monophosphate kinase (UMPK), which among the clones has the pattern

A	B	C	D	E	F	G	H
−	+	+	+	−	−	−	+

Table 13-1   Human chromosomes among human-mouse hybrid cell lines

Clone*	1	2	3	4	5	6	7	8	9	10	11	12	13	14	15	16	17	18	19	20	21	22	X
A	−	+	+	+	−	−	−	−	−	−	+	+	−	+	−	+	−	−	−	−	−	−	+
B	+	+	−	−	+	−	−	+	−	−	−	+	−	−	+	+	+	+	−	−	−	−	+
C	+	+	−	+	−	−	+	+	−	+	−	+	+	−	+	+	−	+	+	−	+	−	+
D	+	−	+	−	−	+	+	−	−	+	+	+	−	+	+	−	+	+	−	+	−	−	+
E	−	+	−	+	+	−	+	+	−	+	+	+	+	+	−	+	+	+	+	+	+	+	+
F	−	−	−	−	−	+	+	−	+	−	+	+	−	−	−	−	−	−	−	−	−	−	+
G	−	+	+	−	−	−	+	−	−	+	−	+	−	+	+	+	−	+	−	−	+	+	+
H	+	+	+	+	+	−	+	+	−	+	−	+	−	+	+	+	−	+	−	+	+	+	+

*Each clone has a unique combination of human chromosomes.

Source: Data from A. Satlin, R. Kucherlapati, and F. H. Ruddle, *Cytogenet. Cell Genet.* 15(1975): 146–152.

Comparison of these results and the columns in Table 13-1 reveals perfect agreement between the presence or absence of UMPK and the presence or absence of chromosome 1 among the clones. Based on this correspondence, the gene for UMPK is assigned to chromosome 1. Similarly, the enzyme β-galactosidase in these clones would have the pattern

A	B	C	D	E	F	G	H
+	−	−	+	−	−	+	+

which is that of a gene located in chromosome 3.

Several extensions of the hybrid-cell method can be used to localize genes with even greater precision. For example, human-mouse hybrid cells produced with human chromosomes carrying a translocation (Section 6.7) can be used to assign a gene to one or the other segments of the translocation. In the simplest cases the translocation breakpoint is near the centromere of one chromosome, and the presence of a gene on the long or short arm of the chromosome is revealed by its expression in hybrids carrying only the long or short arm. Similarly, hybrids produced with human cells carrying small deletions (Section 6.5) may be used to determine whether a gene known to be present in the normal chromosome is also present in the chromosome with the deletion; if it is not present in the hybrid cell lines with the deletion, then the deletion must have eliminated the gene in question.

The use of hybrid cells in linkage studies has two principal limitations: (1) The human gene of interest must be expressed in the hybrid cells so that its product can be identified, and (2) the human gene product must be sufficiently different from the homologous mouse gene product to permit the two gene products to be distinguished. The second limitation is less constraining than the first, because human gene products are often readily distinguishable from their mouse counterparts. When this is not the case, human cell lines carrying a mutant gene and producing an abnormal gene product (such as one with an amino acid substitution) can be used in the cell fusions, thereby enabling the presence of the human gene to be detected through the occurrence of its abnormal product. The first limitation—expression of the gene in cultured cells—has been a major constraint eliminating many important genes, such as those expressed only in neural tissue or during embryonic development. Recently, methods that employ recombinant DNA have greatly alleviated this problem; once a human gene (or a DNA sequence near the gene) has been cloned, its presence in hybrid cells can be detected by the presence of the corresponding human DNA sequence in the cells, independently of gene expression. This direct method of mapping is illustrated in Figure 13-3, in which it is assumed that human and mouse chromosomes contain restriction fragments (Chapter 12) homologous to a cloned probe, which differ in size and can be separated on a gel. Cell line A contains a human chromosome carrying the DNA sequence in addition to a pair of homologous mouse chromosomes, and thus A produces two bands that hybridize with the radioactive probe (one band from the human chromosome and one band from the mouse chromosomes). However, cell line B contains only the mouse chromosomes and lacks the human chromosome; it produces only one band

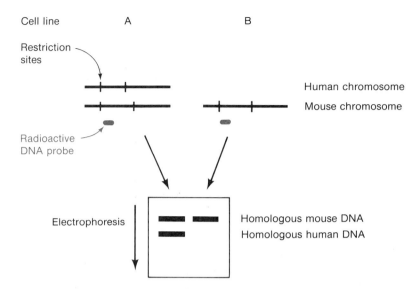

Figure 13-3 Chromosome localization of a cloned segment of DNA (the probe). The restriction fragments homologous to the probe have different sizes in the mouse and human genomes, and hence can be separated by electrophoresis.

(from the mouse chromosomes). Thus, if the presence of a restriction fragment of a particular size correlates perfectly with the presence of a chromosome, then the restriction fragment derives from that chromosome.

The use of DNA probes to study cell hybrids has resulted in assignment of the mutation for Huntington disease to chromosome 4. Huntington disease is an incurable degeneration of the nervous system caused by a dominant mutation usually expressed in middle age. Assignment of the gene to chromosome 4 is an important first step in developing methods of detection and prenatal diagnosis of the disease.

Production of Hybrid Cells

Cells occasionally fuse spontaneously. With some cell types the frequency of spontaneous fusion is so low that fusion must be enhanced; this is made possible by exposing a mixed culture to Sendai virus or to polyethylene glycol.

Once fusion has been achieved, a method is needed to select fused cells from a population of individual cells. The most common method relies on the manner by which cells synthesize DNA. Most eukaryotic cells have two metabolic pathways in which precursors (bases and nucleotides) of DNA are synthesized—the major pathway, responsible for synthesis of most of the precursors, and the **salvage pathway,** a normally less important alternate route. The drug **aminopterin** inhibits the major pathway of precursor synthesis; in the presence of aminopterin, cells must rely exclusively on the salvage pathway for DNA precursors (Figure 13-4).

The salvage pathway utilizes many enzymes, two of which are particu-

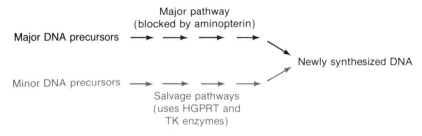

Figure 13-4 Major pathway and salvage pathway for DNA synthesis.

larly important for the cell-fusion method. One of these enzymes is *hypoxanthine guanine phosphoribosyl transferase,* or HGPRT, which provides purine deoxynucleotides used in DNA synthesis. The other key enzyme is *thymidine kinase,* or TK, which provides pyrimidine deoxynucleotides. Cells that are defective in either HGPRT (*HGPRT⁻* cells) or TK (*TK⁻* cells) are unable to synthesize DNA precursors by means of the salvage pathway; therefore, such cells are unable to replicate their DNA in the presence of aminopterin.

The necessity of both HGPRT and TK for cell growth is the basis of the method for selecting fused cells (Figure 13-5). *TK⁻* human cells are mixed

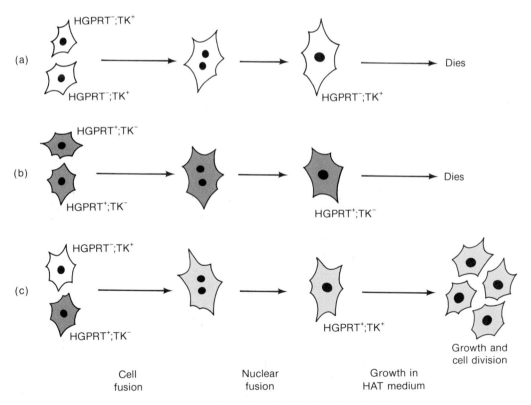

Figure 13-5 Selection of hybrid cells in HAT medium, in which the only cells that can grow and divide are the *HGPRT⁺TK⁺* hybrids.

with $HGPRT^-$ mouse cells in a medium containing aminopterin. Neither type of cell will be able to grow, as the human cells will lack pyrimidine DNA precursors and the mouse cells will lack purine DNA precursors. Among the cells that have undergone fusion, the mouse-mouse hybrids will still be $HGPRT^-$ (panel (a)), and the human-human hybrids will still be TK^- (panel (b)); the human-mouse hybrids alone will have genes encoding active HGPRT and TK and thus will be able to grow (panel (c)). In other words, the $HGPRT^-$ and TK^- mutations complement one another in the hybrids (which they must, since the genes are nonallelic); the hybrid cells are therefore the only ones that can survive. The medium in which fused cells are selected is called **HAT medium,** because it contains *h*ypoxanthine (the substrate of HGPRT), *a*minopterin (to block the major DNA synthetic pathway), and *t*hymidine (the substrate of TK).

Selection of $HGPRT^-$ and TK^- Mutations

HGPRT and TK are the key enzymes in cell-fusion studies, and $HGPRT^-$ and TK^- mutations are easy to obtain. The mutation-selection technique is based on the ability of both of the enzymes HGPRT and TK to accept alternative substrates. The normal substrate of HGPRT is hypoxanthine (Figure 13-6); however, HGPRT will also accept the analogue 8-azaguanine, which, on incorporation into the DNA of animal cells, causes the cell to die. Consequently, when normal $HGPRT^+$ cells are grown in a medium containing 8-azaguanine, they incorporate it into their DNA by the salvage pathway and die. On the other hand, rare $HGPRT^-$ mutants arising spontaneously in this population will survive, because they cannot incorporate the 8-azaguanine, and they form clones that are easily isolated.

A similar principle is used to obtain TK^- mutants. The normal substrate of TK is thymidine, but the enzyme will also accept bromodeoxyuridine (Figure 10-12), which, when incorporated into DNA, causes cell death. Therefore, normal TK^+ cells grown in the presence of bromodeoxyuridine will die, and rare TK^- cells will survive.

These two mutation-selection procedures can be applied to any cell lines that an investigator may wish to fuse. Once a human cell line has the TK^- mutation and a mouse cell line has the $HGPRT^-$ mutation (or the other way around), the hybrid cells can be obtained in HAT medium as outlined in Figure 13-5.

In discussing Table 13-1 it was pointed out that all of the hybrid clones in the table retained the human X chromosome. This will occur whenever the clones are obtained by fusing human TK^- and mouse $HGPRT^-$ cells. The human cell must provide the normal HGPRT gene, and this gene is present in the human X chromosome; thus, all clones must carry the human X.

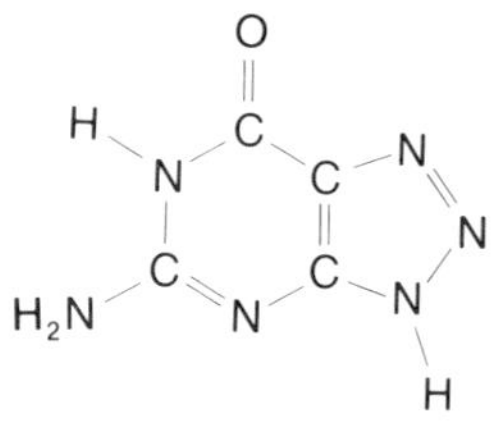

13.2 The Immune Response

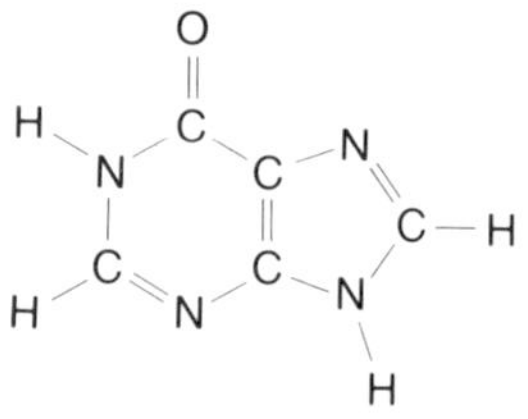

For centuries it has been recognized that individuals who have recovered from an infectious disease, such as smallpox or influenza, are less likely to succumb

8-Azaguanine

Hypoxanthine

Figure 13-6 Substrates of HGPRT. The structures of thymidine and 5-bromodeoxyuridine, the substrates of TK, are given in Figure 10-12.

to the same disease a second time. As a result of prior exposure, these individuals have acquired the ability to resist the infectious agent, and they are said to be **immune** to the agent. The process of becoming immune is called the **immune response.** The recognition that immunity can be acquired has led to the development of vaccines, harmless in themselves, that produce immunity to many diseases.

Immunity results from the ability of the body to recognize viruses, bacteria, or other foreign substances that may invade the body and to attack and destroy them. Most large molecules, and almost all viruses and cells, have the ability to elicit an immune response. Such immunity-evoking substances are called **antigens.**

When blood transfusions first began to be administered around 1900, it was quickly recognized that the blood of the donor and the recipient had to match in some way in order for the transfusion to be successful. In many cases transfusions could be carried out successfully with no complications, and when this occurred, the donor and the recipient were said to be **compatible.** In other cases, upon transfusion, recipients went into severe shock and many died; in these cases the donor and the recipient were obviously **incompatible.** However, blood from the same donor might be compatible with some recipients and at the same time incompatible with other recipients. Additional studies revealed that the basis of incompatibility was the presence of certain genetically determined antigens present on the surface of the red blood cells of the donor, and laboratory tests for compatibility were soon developed. These studies led to the first rule of immunogenetics:

> A recipient individual will produce an immune response against any antigen that the individual himself does not possess.

That is, if the antigens present on the red blood cells of the donor are also present on the red blood cells of the recipient, then the donor and the recipient will be compatible for tranfusion. This is diagrammed in Figure 13-7.

Shortly after blood transfusions were first attempted in humans, skin transplants began to be carried out in highly inbred, and therefore virtually

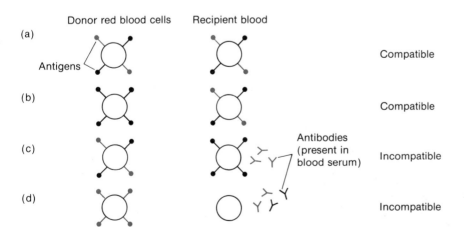

Figure 13-7 Compatibility of human blood types. Individuals make antibodies against A and B antigens unless the antigens are present on their own red blood cells. Transfusions are compatible (panels (a) and (b)) when the antigens on the red blood cells of the donor are also present in the recipient, because the recipient will not have antibodies against the transfused cells. Transfusions are incompatible (panels (c) and (d)) when the recipient has antibodies against antigens present on the red blood cells of the donor.

homozygous, strains of mice. Usually, transplants between individuals of the same inbred strain were accepted. The small transplanted piece of skin would grow on the recipient mouse just as well as if the transplanted skin had been taken from another place on the body of the recipient. However, with individuals of different inbred strains the skin around the transplant became inflamed and the transplanted skin would not heal; this phenomenon was called **rejection.** Such rejection reactions were also shown to result from the presence of genetically determined antigens present on the transplanted skin—but these were antigens of a different type than those implicated in blood transfusion. When two different inbred strains were crossed to produce an F_1, and the F_1 was used as donor or recipient of skin transplants, the following results were obtained: (1) Transplants using either parental strain as the donor and the F_1 as the recipient were almost always accepted. (2) Transplants using the F_1 as the donor and either parental strain as the recipient were almost always rejected (Figure 13-8). Such results led to a second rule of immunogenetics:

> Antigens are genetically determined by dominant or codominant alleles, so the cells of an individual will possess all antigens corresponding to the alleles the individual has inherited from its parents.

The two rules of immunogenetics account for the observations about transplantation. Because the F_1 must inherit one allele from each homozygous parent, the F_1 will possess the antigens of *both* parental strains. Consequently, transplants from an inbred parent to the F_1 will be accepted, because the parental strain will possess no antigens not also possessed by the F_1. However,

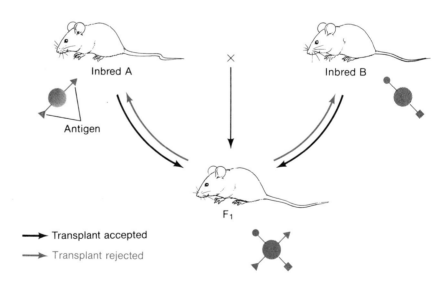

Figure 13-8 Results of skin grafts. Skin transplants in mice are usually accepted when the donor animal is from a homozygous inbred strain and the recipient is one of the F_1 progeny, because all antigens in the donor tissue are also present in the recipient. However, transplants in the reverse direction are usually rejected.

transplants from the F_1 to either parental strain will be rejected, because the F_1 will possess antigens inherited from the other parental strain, which are not present in the parental recipient of the transplant.

Transfusion incompatibility and skin-graft rejection exemplify a division of functions of the immune system arising from different characteristics of two classes of white blood cells—B cells and T cells. As will be discussed later in this chapter, **B cells** secrete proteins called **antibodies** capable of combining with antigens. Stimulation of a B cell by a suitable antigen causes secretion of an antibody, which circulates in the blood and lymph and combines with the antigen, clumping the antigen molecules together and marking them for destruction by other classes of white blood cells. Therefore, individuals lacking B cells cannot produce circulating antibodies and will accept mismatched blood transfusions, and will be prone to bacterial infections. A few rare genetic diseases are associated with the absence of B cells.

Unlike B cells, which attack foreign antigens indirectly by producing circulating antibodies, **T cells** attack foreign antigens directly, releasing substances that cause a local inflammation, and attract other types of white blood cells to aid in the destruction of the invasive antigens. Individuals with T-cell deficiency will accept skin transplants and are prone to viral infections. T-cell deficiency occurs in several rare genetic disorders.

In the remainder of this chapter three items will be considered: (1) the genetic determination of the red-blood-cell antigens that are important in transfusions, (2) the genetic determination of antigens implicated in tissue compatibility, and (3) the genetic basis of B-cell and T-cell responses.

13.3 Blood Group Systems

More than 30 loci that determine surface antigens of red blood cells have been identified. Each locus defines a **blood group system** in terms of the alleles that may be present at the locus and the corresponding red-blood-cell antigens. The most important blood group systems are the familiar ABO, Rh, and MN systems, which are discussed in this section.

ABO Blood Groups

The ABO blood group system, mentioned briefly in Chapter 1, is the most significant system in blood transfusions. The ABO blood groups are defined in terms of certain red-blood-cell antigens, designated A and B, whose presence in any individual is determined by the genotype of the individual with respect to a single locus designated I. This locus has three principal alleles—I^A, I^B, and I^O. The possible genotypes and the corresponding ABO blood groups are as follows:

1. Genotypes $I^A I^A$ and $I^A I^O$ have the A antigen on their red cells. Their blood group is designated A.

2. Genotypes $I^B I^B$ and $I^B I^O$ have the B antigen on their red cells. Their blood group is designated B.

3. Genotype $I^A I^B$ has both A and B antigens on its red cells. The cells have blood type AB.

4. Genotype $I^O I^O$ has neither A nor B antigen on its red cells. The cells have blood type O.

The molecular basis of the ABO blood groups is illustrated in Figure 13-9. The A and B antigens are different forms of a complex surface carbohydrate, and the I^A and I^B alleles code for enzymes that attach alternative sugars to a particular site on a precursor carbohydrate. The enzyme coded by I^A attaches N-acetyl-D-galactosamine (a derivative of the sugar galactose) to the precursor, and the enzyme coded by I^B attaches D-galactose. The I^O allele codes for an inactive enzyme, which leaves the precursor unaltered. Note in the figure that the precursor carbohydrate is the product of a different enzyme, which is coded by another locus called H. Individuals of genotype hh are homozygous for an allele h that produces an inactive enzyme unable to form this precursor. Lacking the precursor, these individuals are unable to form either A or B antigen irrespective of their genotype at the I locus, and hence they have blood type O. However, hh genotypes are extremely rare.

A-like and B-like antigens are present on certain bacteria and in other substances to which normal individuals are frequently exposed. Because of this exposure early in life, individuals lacking the A antigen are stimulated to produce anti-A antibody, and individuals lacking the B antigen are stimulated

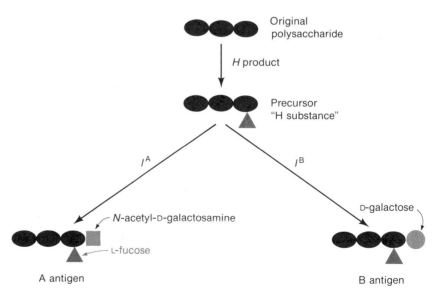

Figure 13-9 Formation of A and B antigens in the red blood cells of humans. The A and B antigens are carbohydrates on red blood cells that are produced by alternative forms of the enzyme coded by the I^A and I^B alleles.

to produce anti-B antibody. However, for complex reasons, individuals do not make antibodies against their own antigens, so individuals whose red blood cells already carry the A or B antigen will not produce the corresponding antibodies. The result of exposure to A-like and B-like environmental substances is that invariably:

1. Individuals of blood group A will produce anti-B antibody.

2. Individuals of blood group B will produce anti-A antibody.

3. Individuals of blood group AB will produce neither anti-A nor anti-B antibody.

4. Individuals of blood group O will produce both anti-A and anti-B antibody.

Anti-A and anti-B antibodies are large molecules capable of the agglutination (clumping) and lysis (bursting) of red blood cells that carry the corresponding antigens. Thus, when an individual is transfused with red blood cells carrying the wrong antigens, antibodies present in the recipient will attack the foreign blood, causing first agglutination and then lysis. Both processes are potentially disastrous for the recipient because of the disruption of normal circulation. Problems can also occur with transfusion of a large volume of donor blood that contains antibodies that can attack red blood cells in the recipient, because these may cause significant blood destruction in the recipient. Consequently, modern medical practice requires that donor and recipient have the same ABO blood type. In extreme emergencies blood containing antibodies against antigens of the recipient can be transfused in small amounts, because the donor blood will be diluted sufficiently to prevent severe problems. However, the donor red blood cells must *never* carry an antigen not already present in the recipient; otherwise, an antigen-antibody reaction is certain.

Rh Blood Groups

The name of this blood group system derives from the fact that it was first discovered while studying blood from *Rh*esus monkeys. The genetic determination of human Rh antigens has not yet been unambiguously elucidated, but it is consistent with two possibilities: (1) the antigens may be determined by at least eight alleles at one locus, or (2) the antigens may be determined by two alleles at each of three very closely linked loci. The second hypothesis results in a simpler genetic symbolism, and since this symbolism is the most commonly used in the United States, it will be adopted here. The three loci and their alternative alleles are designated C, c, D, d, E, and e. A chromosome may carry any combination of the alleles, so 27 genotypes are possible. Each allele, with the exception of d, determines the presence of a distinct red-blood-cell antigen that can be detected with the appropriate antibody. These antigens are designated by the same letters as the alleles, so the antigens are C, c, D, E, and e. Several rare alleles determining their own unique antigens are also known, but these will not be dealt with here. All three genotypes determined by the *C* gene and the three determined by the *E* gene are distinguishable. For example, genotype *CC* produces C antigen only, *Cc* produces both C and c antigens, and *cc* produces c only. *E* genotypes similarly determine three distinguishable phenotypes. However, only two phenotypes associated with the *D* gene are distinguishable—(1) presence of antigen D (these individuals have genotype *DD* or *Dd* and are said to be **Rh-positive**), and (2) absence of antigen D (these individuals have genotype *dd* and are said to be **Rh-negative**). Altogether, the five common antigens define 18 Rh phenotypes.

Unlike the ABO system, for which the antigens are common in environmental substances, Rh-like substances are rare. Consequently, most individuals do not produce antibodies against the Rh antigens that they do not themselves possess. However, exposure of an individual to an appropriate Rh antigen will stimulate the immune system, and antibodies will be produced. For example, when a *dd* individual is exposed to D antigen, anti-D antibodies will be produced.

A severe blood disorder affecting fetuses and newborns is caused when Rh antibodies present in the mother cross the placenta and attack the red blood cells of the fetus. Anti-A and anti-B antibodies normally cannot cross the placental barrier, but antibodies against Rh antigens can do so because they are smaller molecules than the ABO antibodies. Rh blood disease can occur whenever a *dd* female carries consecutive *Dd* fetuses (Figure 13-10(a)). The first *Dd* fetus is usually unaffected because the mother will not produce anti-D antibody. However, red cells from the fetus carry the D antigen, and a few of these cells can seep across the placenta late in pregnancy and stimulate production of anti-D antibody in the mother. Once the mother has been stimulated by Rh-positive cells from the first fetus, subsequent *Dd* fetuses are at risk of the Rh blood disease, which is called *hemolytic disease of the newborn*. Note, however, that *dd* fetuses are never at risk. The D antigen is almost always responsible for the maternal-fetal incompatibility that results in hemolytic disease of the newborn. Antibodies directed against other Rh antigens are less potent and rarely cause severe problems.

Among Caucasians, about 13 percent of all marriages are between a *dd* female and a *DD* or *Dd* male and hence might result in maternal-fetal incompatibility. However, the occurrence of hemolytic disease of the newborn has been greatly reduced by the development of a preventive treatment (Figure 13-10(b)). In this treatment, the *dd* mother is injected with anti-D antibody immediately after the birth of each *Dd* child. The injected antibody will attack any fetal cells that may have entered into the blood of the mother, and the mother escapes being stimulated by the cells, because they are destroyed too rapidly to stimulate her antibody production. The injected antibodies gradually disappear. Since the mother will not have been stimulated to produce anti-D antibodies, each pregnancy is just like the first one with respect to antibody production, and stimulation can be prevented with antibody treatment. The only cases in which antibody treatment is ineffective occur when the mother has previously been stimulated with D antigen. For this reason, *dd* females are never transfused with red cells carrying the D antigen.

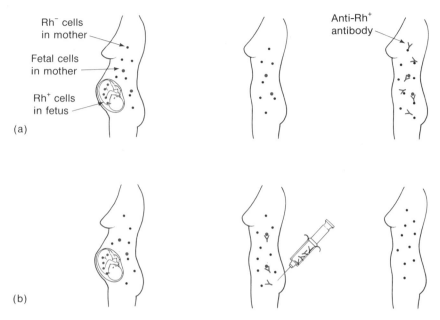

Figure 13-10 Rh incompatibility between mother and fetus. (a) A small number of red blood cells (red) from an Rh$^+$ fetus may enter the blood stream of the mother. If the mother is Rh$^-$, these cells stimulate production of antibodies against the Rh$^+$ antigen, which will cross the placenta and attack the Rh$^+$ red blood cells in the fetus during a subsequent pregnancy. (b) Stimulation of the immune system in the mother is prevented by injection of anti-Rh$^+$ antibody, which rapidly removes the Rh$^+$ red blood cells from the circulation.

The MN Blood Groups

Blood groups other than ABO and Rh are rarely implicated in transfusion incompatibility or hemolytic disease of the newborn. A typical example is the MN blood group system. The M and N antigens were originally discovered with antibodies from a series of rabbits that had been injected with human red blood cells. Two types of rabbit antibody were most informative in distinguishing among human red blood cells. These are now called anti-M antibody, which is produced in response to a human red-blood-cell antigen denoted M, and anti-N antibody, which is produced in response to a human red-blood-cell antigen denoted N. Individuals can be placed into three blood types according to the reaction of their red blood cells with these antibodies:

1. M individuals have red blood cells that react only with anti-M.

2. MN individuals have red blood cells that react with both anti-M and anti-N.

3. N individuals have red blood cells that react only with anti-N.

Pedigree studies indicate that the M and N antigens are products of codominant alleles at a single locus. Individuals of genotype *MM* have blood type M, those of genotype MN have blood type MN, and those of genotype NN have blood type N.

The MN blood group system is of minor medical importance, because harmful anti-M and anti-N antibodies are rarely produced in humans, regardless of antigenic stimulation. Thus, donated blood is not routinely typed for MN, because it is not worth the trouble and expense. However, rare individuals do produce anti-M or anti-N antibody, and a severe reaction may occur upon transfusion with the wrong MN type of blood. Most incompatibility problems of this sort are avoided by pretesting donor and recipient blood by mixing them before transfusion, or by first transfusing a small amount of donor blood to assess the reaction of the recipient. A few rare cases of hemolytic disease of the newborn are also caused by anti-M or anti-N antibody.

Applications of Blood Groups

In addition to the importance of certain blood group systems in transfusions and maternal-fetal incompatibility, blood groups have applications in studies of genetic linkage (the loci of several genetic disorders are linked with known blood-group loci), population studies, parentage exclusion, and criminology. Blood groups are useful in identification because, in many cases, the alternative blood groups are all reasonably common and any two unrelated individuals can be expected to differ in one or more blood groups. Indeed, the probability that any two randomly chosen Europeans will be identical for all known blood groups is less than 0.0003. Moreover, blood group frequencies may vary among populations. For example, among black Africans the chromosome with the highest frequency with respect to the Rh loci is cDe, whereas among Chinese it is CDe. Such observations are the basis of the use of blood groups in anthropology and population genetics to infer the genetic ancestry of populations.

Evidence based on blood groups is admissible in many courts in cases depending on identification of parentage. In this application the blood groups are interpreted according to two rules: (1) An allele present in a child must be present in one or both parents. (2) If a parent is homozygous for an allele, the allele must be present in the child. Suppose, for example, that the blood groups of mother, child, and two possible fathers are

Mother	O	Cc	$D-$	Ee	MN
Child	O	Cc	dd	Ee	M
Male 1	O	CC	$D-$	Ee	M
Male 2	A	cc	dd	ee	N

Because of the phenotype of the child, the sperm giving rise to the child must have carried I^O, either C or c (depending on whether the egg carried c or C, respectively), d, either E or e (depending on whether the egg carried e or E, respectively), and M. With respect to ABO and Rh, either male could have contributed such a sperm, as male 1 could be Dd and male 2 could be $I^A I^O$.

However, male 2 cannot contribute a sperm carrying M, so he is excluded as the possible father. It is important to note that this analysis does not imply that male 1 *is* the father, but merely that he *could* be.

Application of blood typing to criminology is derived from the fact that the A and B antigens are extremely stable and can be correctly identified in dried blood or other substances after many years. A and B antigens can also be found in sweat, semen, and other body fluids. Secretion of these substances results from the presence of a dominant gene *Se*. Approximately 75 percent of Caucasians are secretors (that is, genotypically *Se/Se* or *Se/se*). Thus, in cases of rape, for example, the ABO blood type of the assailant can often be identified through examination of the semen with anti-A and anti-B antibodies.

13.4 Antibodies and Antibody Variability

It has been estimated that a normal mammal is capable of producing more than 10^6 different antibodies, each having the ability to combine specifically with a particular antigen. Antibodies are proteins (to which a small amount of carbohydrate is attached), and each unique antibody has a different amino acid sequence. If antibody genes were conventional in the sense that each gene codes for a single polypeptide, then mammals would need more than 10^6 genes devoted to antibody production, more genes than are present in the genome. Actually, mammals devote only a few hundred genes to antibody production, and the number of different antibodies derives from remarkable events that occur in the DNA of somatic cells. These events will be discussed in this section.

Although an individual is capable of producing a vast number of different antibodies, only a fraction of these are actually synthesized at any one time. Each B cell is capable of producing a single type of antibody, but antibody is not actually secreted until the cell has been stimulated by the appropriate antigen. Once stimulated, the B cell undergoes successive mitoses and eventually produces a clone of identical cells that secrete the antibody. Moreover, antibody secretion may continue even if the antigen is no longer present. In this manner, organisms produce antibodies only to the antigens to which they were previously exposed.

Types of Antibody

When antibody molecules are separated according to molecular weight, five distinct classes are found, which are designated IgG, IgM, IgA, IgD, and IgE, as shown in Table 13-2 (Ig stands for **immunoglobulin**). The differences in molecular weight are determined by the number and size of the polypeptide chains that occur in each class of molecule and by the amount of carbohydrate that it contains. Antibody molecules contain two types of polypeptide chains differing in size; the large one is called the **heavy (H) chain** and the small one

Table 13-2 Characteristics of human immunoglobulins

Class	IgG	IgM	IgA	IgD	IgE
Type of H chain	γ	μ	α	δ	ϵ
Type of L chain	κ,λ	κ,λ	κ,λ	κ,λ	κ,λ
Molecular weight, thousands	~150	~900	~170	~180	~190
Number of subclasses	4	2	2	—	—
Carbohydrate, percent	~1.5	~9.5	~1.6	~1.7	~2.0
Serum concentration, mg/ml	8–16	0.5–1.9	1.4–4.2	<0.04	<0.007

is called the **light (L) chain.** Each antibody molecule contains two or more identical heavy chains and two or more identical light chains. Each class of antibody contains a different type of heavy chain (either γ, μ, α, δ, or ϵ), and within each antibody class, small differences in the heavy chains define subclasses. Two types of light chains occur, which are designated κ and λ, and either of these may occur in a particular antibody molecule (Table 13-2). Each of the five antibody classes carries out specialized functions in the immune system.

Structure of IgG

IgG is the most abundant antibody class and has the simplest molecular structure. The molecular structure of IgG is illustrated in Figure 13-11. An IgG molecule consists of two heavy and two light chains held together by

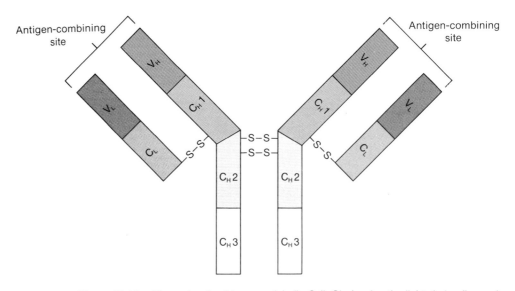

Figure 13-11 The molecule of immunoglobulin G (IgG) showing the light chains (L, gray) and heavy chains (H, white and various shades of pink). V and C refer to variable and constant regions, respectively.

disulfide bridges (two joined sulfur atoms) in the overall shape of the letter Y. The sites on the antibody that carry its specificity and that actually combine with the antigen are located in the upper half of the arms above the fork of the Y. Each IgG molecule having a different antigen specificity has a different amino acid sequence for the heavy and light chains in this part of the molecule. These specificity regions are called the **variable regions** of the heavy and light chains. The rest of the polypeptides, called the **constant regions,** have virtually the same amino acid sequence in all IgG molecules.

All antibody classes are formed from Y-shaped structures resembling the one in Figure 13-11. However, each class of antibody differs in its heavy chain, in its carbohydrate, and in the number of Y-shaped units that aggregate to form the complete antibody molecule. A complete IgG molecule consists of a single Y-shaped unit containing identical heavy chains of the γ type, as shown in the figure. However, a complete IgM molecule consists of five Y-shaped units, each containing identical heavy chains of the μ type.

Gene Splicing and Antibody Variability

Initial understanding of the genetic mechanisms responsible for antibody variability came from comparison of amino acid sequences of molecules having specificity for different antigens. These studies were the same ones that showed that the heavy and light chains could be separated into constant and variable regions. Conclusive information was derived by using some of the recombinant DNA methods outlined in Chapter 12. In these studies, cloned DNA corresponding in nucleotide sequence to antibody mRNA was used to locate sequences in the genome that were homologous to various parts of the heavy and light chains. In the genome of a B cell that was actively producing an antibody, the DNA segments corresponding to the constant and variable regions of the light chains were found to be very close together, as expected of DNA that codes for different parts of the same polypeptide. However, in embryonic cells, these same DNA sequences were located very far apart. Similar results were obtained for the variable and constant regions of the heavy chains: these were close together in B cells but widely separated in embryonic cells.

The antibody mRNA was then used to identify recombinant DNA clones carrying the sequences homologous to the mRNA. These clones contained the genes corresponding to the constant and variable regions of both the heavy and light chains, and extensive DNA sequencing of these clones explained not only the reason for the different gene locations in B cells and germ cells, but also revealed a key mechanism for the origin of antibody variability. Cells in the germ line contain a small number of genes corresponding to the constant region of the light chain, and these are close together along the DNA. Separated from them but on the same chromosome is another cluster consisting of a much larger number of genes that correspond to the variable region of the light chains. In the differentiation of a B cell, one gene for the constant region is spliced (cut and joined) onto one gene for the variable region, and this splicing produces a complete light-chain antibody gene.

A similar splicing mechanism also occurs to generate the constant and variable regions of the heavy chains.

Actually, formation of a finished antibody gene is slightly more complicated, as light-chain genes consist of three parts and heavy-chain genes consist of four parts. Gene splicing in the origin of a mouse antibody light chain is illustrated in Figure 13-12. For each of two parts of the variable region the germ line contains multiple coding sequences called the **V** and **J regions.** During the differentiation of a B cell, a deletion (which is variable in length) occurs that joins one of the V regions with one of the J regions. When transcribed, this joined V-J sequence is the 5′ end of the light-chain RNA transcript. Transcription continues on through the DNA region coding for the constant (C) portion of the gene. RNA splicing subsequently joins the V-J and C regions, creating the light-chain mRNA. This DNA joining process is called **combinatorial joining,** because it can create many combinations of the V and J regions.

Combinatorial joining in the origin of a heavy-chain gene occurs by means of DNA splicing of the heavy-chain counterparts of V and J along with a third set of sequences called D (Figure 13-13). Initially, all heavy chains have the μ type of constant region, corresponding to antibody class IgM (panel (b)). Another DNA splice joining the V-D-J region with a different constant region can subsequently switch the class of antibody produced by the B cell (panel (c)).

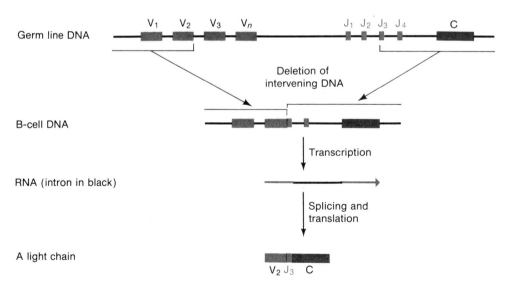

Figure 13-12 Formation of a gene for the light chain of an antibody molecule. One V region is randomly joined with one J region by deletion of the intervening DNA. The remaining J regions are eliminated from the RNA transcript during RNA processing.

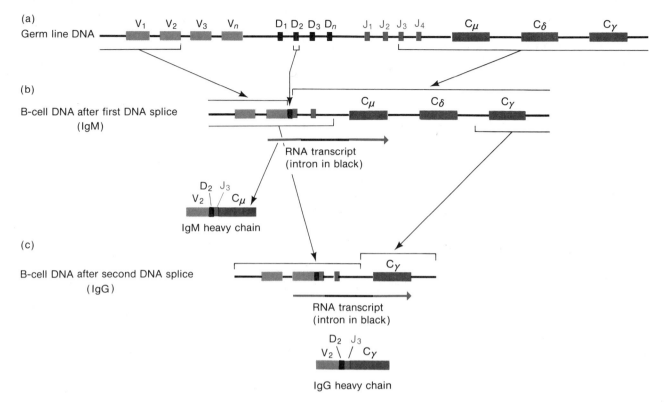

Figure 13-13 Formation of a gene for the heavy chain of an antibody molecule. One V region, one D region, and one J region are randomly joined by deletion of the intervening DNA. The remaining J regions are eliminated from the RNA transcript during RNA processing. All B cells initially produce heavy chains of the μ class and therefore secrete IgM (panel (b)). A subsequent splicing of DNA may join the V-D-J region with a different C region (for example, C_γ) and switch the secretion to a different class of antibody (in this example, IgG).

Table 13-3 Number of antibodies possible through the combinatorial joining of mouse germ line genes

κ light chains:	V_κ regions ~250 J_κ regions = 4 Combinations = 250 × 4 = 1000
λ light chains:	V_λ ~3 J_λ ~3 Combinations = 3 × 3 = 9
Heavy chains:	V_H ~250 D ~10 J_H = 4 Combinations = 250 × 10 × 4 = 10,000
Number of possible antibody molecules:	κ-containing: 1000 × 10,000 = 10^7 λ-containing: 9 × 10,000 = 90,000

The amount of antibody variability that can be created by combinatorial joining is indicated in Table 13-3. In mice, the κ light chains are formed from combinations of about 250 V and 4 J regions, giving $250 \times 4 = 1000$ different κ chains. (The λ chains have their own V and J regions, smaller in number than their κ counterparts.) For the heavy chains there are approximately 250 V, 10 D, and 4 J regions, producing $250 \times 10 \times 4 = 10,000$ combinations. Since any light chain can combine with any heavy chain, there will be at least $1000 \times 10,000 = 10^7$ possible types of antibody. Thus, the number of DNA sequences dedicated to antibody production is quite small, but the number of possible antibodies is very large.

Additional Sources of Antibody Variability

The value of 10^7 different antibody specificities is actually an underestimate, because there are two additional sources of antibody variability.

1. The junction for V-J (or V-D-J) splicing in combinatorial joining can occur between different nucleotides and thereby generate different codons in the spliced gene. For example, one V-J splicing event joins the V sequence CCTCCC with the J sequence TGGTGG in two ways:

$$CCT\,CCC \; + \; TGG\,TGG \rightarrow CCGTGG$$

which codes for proline and tryptophan, and

$$CCT\,CCC \; + \; TGG\,TGG \rightarrow CCT\,CGG$$

which codes for proline and arginine. In this manner the same V-J joining can produce polypeptides differing in a single amino acid.

2. The V regions are susceptible to a high rate of *somatic mutation*, which occurs during B-cell development. These mutations allow different B-cell clones to produce different polypeptide sequences, even if they have undergone exactly the same V-J joining.

Large amounts of a particular antibody—for example, an IgG antibody able to combine with a particular antigen—can be produced by means of somatic cell hybridization between an antibody-producing B cell and certain types of tumor cells. These hybrid cells, called **hybridomas,** grow well in laboratory culture, and all descendants of the original B cell will secrete exactly the same antibody. An antibody produced by such cells is called a **monoclonal antibody,** because it derives from the clonal descendants of a single B cell.

13.5 Histocompatibility Antigens

Skin grafts and other transplanted tissue between unrelated individuals are usually rejected, except for a few tissues such as bone or eye cornea. Rejection results from cell-surface antigens, present in the graft, that are recognized as foreign and attacked by a class of T cells in the recipient. The antigens responsible for rejection are genetically determined by alleles of more than 40 genes, and the alleles of each gene are codominantly expressed. These cellular antigens are called **histocompatibility antigens,** and the genes that code for them are *histocompatibility genes.*

The genetics of histocompatibility has been extensively studied in laboratory mice because of the availability of many unrelated inbred (virtually homozygous) strains maintained by repeated brother-sister mating. (Inbred strains are said to be unrelated if they are derived from unrelated ancestors.) The results of reciprocal transplants between unrelated inbred parents and their F_1 have been summarized in Figure 13-8. The number of histocompatibility loci in which two inbred strains differ may be estimated by means of transplants with the F_2 and backcross progeny. For example, with one histocompatibility locus different between the inbred strains, the genotypic composition of the F_1, F_2, and backcrosses will be

P_1 (inbred strain 1)	*AA*
P_2 (inbred strain 2)	*aa*
F_1	*Aa*
F_2	1/4 *AA*, 1/2 *Aa*, 1/4 *aa*
B_1 ($F_1 \times P_1$)	1/2 *AA*, 1/2 *Aa*
B_2 ($F_1 \times P_2$)	1/2 *Aa*, 1/2 *aa*

Invoking the rule that a transplant will be rejected whenever the donor carries one or more antigens not possessed by the recipient, the results of transplants may be seen to be

$F_2 \rightarrow P_1$	1/4 accepted
$P_1 \rightarrow F_2$	3/4 accepted
$B_1 \rightarrow P_1$	1/2 accepted
$P_2 \rightarrow B_1$	1/2 accepted

With n unlinked histocompatibility loci these proportions would each be raised to the power of n, representing the probability of simultaneous compatibility at each of n independent loci. These calculations are summarized in Figure 13-14. For example, in one study the proportion of accepted $P_1 \rightarrow F_2$ transplants in a mating between two inbred mouse strains was 3/274. Thus n is estimated as $(3/4)^n = 3/274$, so $n = \log(3/274)/\log(3/4) = 16$. This type of genetic analysis has identified more than 40 histocompatibility loci, but with any pair of inbred strains the method leads to an estimate of the *minimum* number of loci implicated in histocompatibility. There are at least three reasons why the estimate will be a minimum:

1. Loci for which both inbred strains are homozygous for the same allele will remain undetected.

2. Two or more loci that are genetically linked (they are said to be linked if the recombination frequency between them is sufficiently small)

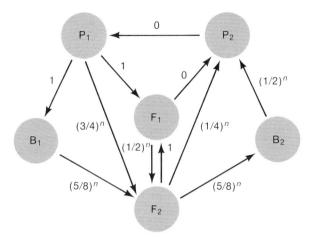

Figure 13-14 Probabilities of acceptance of skin transplants assuming n unlinked histocompatibility genes. The base of each arrow indicates the donor animal, and the arrowhead indicates the recipient. P_1 and P_2 are homozygous inbred strains; F_1 results from a $P_1 \times P_2$ cross; F_2 results from an $F_1 \times F_1$ cross; B_1 results from an $F_1 \times P_1$ cross; and B_2 results from an $F_1 \times P_2$ cross.

may contribute to histocompatibility as if they were a single locus and will therefore be counted as one locus.

3. Histocompatibility alleles resulting in delayed rejection reactions may not lead to rejection prior to the termination of the experiment and so may remain undetected. The severity and speed of a rejection reaction is strongly affected by the particular alleles present in recipient and donor. Some alleles produce antigens that evoke a much stronger and more immediate rejection than others. Since two inbred strains represent only two alleles at each locus, the effect of the locus may be missed if the particular alleles happen to produce weak antigens. Most histocompatibility loci have multiple alleles.

Histocompatibility Genes

Certain closely related inbred strains differ in only one histocompatibility locus, and these strains provide an opportunity to evaluate the effects of individual loci. The strength of individual loci can be assessed in terms of the speed and severity of the graft rejection. Loci are often divided qualitatively into two groups—"weak" and "strong"—though the boundary between the groups is not really sharp. In terms of numbers, there are many weak histocompatibility loci and few strong ones, but it must be emphasized that in many cases the severity and speed of graft rejection is not so much a characteristic of the histocompatibility locus as it is of a particular combination of alleles in graft and host. Among the strong histocompatibility loci, one locus stands out as evoking a stronger and more rapid rejection than any other. This locus is called *H-2* and will be discussed further in the next section.

Table 13-4 Survival of skin transplants in mice

Locus	Median graft survival time (days)		
	Female donor → female recipient	Male donor → male recipient	Average
H-1	15	15	15
H-2	10	10	10
H-3	21	30	25
H-4	120	119	119
H-7	23	25	24
H-8	32	47	40
H-9	>400	>300	>350
H-10	91	>250	>170
H-11	78	105	90
H-12	259	>300	>275
H-13	38	67	52

Source: Data from W. H. Hildemann, E. A. Clark, and R. L. Raison. *Comprehensive Immunogenetics.* Elsevier. 1981.

Table 13-4 shows median survival times of skin grafts between closely related inbred strains of mice differing in a single histocompatibility locus. The data are given separately by sex because females often have a more vigorous immune response to a wide variety of antigens than males have, and thus in many cases females show a more rapid graft rejection. *H-2* incompatibility leads to very rapid graft rejection (typically 8–12 days). Incompatibility at the *H-1* locus also leads to rapid rejection, but not as rapid as *H-2*. The rest of the loci are weaker barriers to skin grafts between these strains, and some (such as *H-9*) are almost insignificant in this particular case. As might be expected, simultaneous incompatibility at a number of weak loci can have a cumulative effect leading to rapid graft rejection.

Analysis of histocompatibility with genetically heterogeneous strains is more complex than dealing with inbreds. However, for any histocompatibility locus at which there may be many possible alleles, the maximum number of alleles that can be represented in a mating pair is four. In this case, the probability that two siblings will have an identical genotype at the locus will be 1/4 (see Figure 13-15). This reasoning explains why transplant success in humans decreases in the order: identical twin donor > sibling donor > parental donor > unrelated donor, where the symbol ">" means "more often

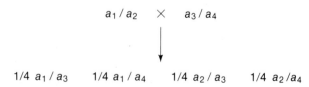

Figure 13-15 The probability, for a gene with many alleles, that two siblings will be identical. The value of 1/4 is the theoretical probability that two siblings will have the same genotype at the major histocompatibility complex.

successful than." However, transplant success is greatly improved by histo-compatibility testing of tissues, particularly to identify relatives, such as siblings, who are compatible at the locus *HLA*, the human counterpart of the mouse *H-2* locus. As in the mouse, humans also have many minor histo-compatibility loci, one of which is the *I* locus that determines the A and B cellular antigens of the ABO blood groups.

The Major Histocompatibility Complex

H-2 in the mouse and *HLA* in humans are by no means simple genetic loci. The mouse *H-2* "locus" extends for at least 0.5 map units on chromosome 17. Recombination analysis occurring within this complex locus has defined at least four major regions specialized in their immunologic functions, and within some of these regions further division into subregions has been possible. Because of their genetic complexity, the term **major histocompatibility complex** or **MHC** is used to describe *H-2* in the mouse, *HLA* in humans, and the corresponding region in other mammals. In addition to their important role as cellular antigens, the MHC gene products have a diversity of functions, including the promotion of cooperative interactions between subsets of T cells, between T cells and B cells, between lymphocytes and macrophages (both classes of white blood cells), and between cytotoxic killer lymphocytes (which attack foreign cells directly) and their specific products. By means of these mechanisms the MHC is important in the regulation of many aspects of the immune response.

The human MHC (the *HLA* locus) is located in chromosome 6 and is divided into four major regions (*HLA-D*, *HLA-B*, *HLA-C*, and *HLA-A*). As defined by cellular antigens, many alternative alleles exist for each of the regions, some of which are listed in Figure 13-16. Any particular chromosome carries one allele for each of the regions, and these alleles define the **haplotype** of the chromosome. For example, the combination *Dw3 B5 Cw1 A1* defines a haplotype, as does the combination *Dw2 B7 Cw3 A3*. Every individual carries two *HLA* haplotypes corresponding to the chromosomes inherited through egg and sperm, and these determine the *HLA* genotype of the individual. Antigens corresponding to the alleles in an individual are expressed according to the rule: *One allele codes for one antigen specificity*. Thus, the antigens expressed in an individual correspond to the alleles present in both haplotypes—for example, *Dw2, Dw3, B5, B7, Cw1, Cw3, A1, A3*. Since recombination is infrequent within the *HLA* region, haplotypes are inherited almost as if they were alternative alleles at a single autosomal locus.

Alleles of *D*, *B*, *C*, and *A* can occur in many combinations, each corresponding to one possible haplotype. Taking the numbers in Figure 13-16 as an example, the number of theoretically possible haplotypes would be $12 \times 36 \times 8 \times 30 = 103,680$. However, the occurrence of particular alleles in haplotypes is by no means random. Some haplotypes occur far more often and others far less often than would be expected by chance. Part of the reason for nonrandomness is tight linkage, which prevents the alleles from coming into random association. Nevertheless there are many different haplotypes—so

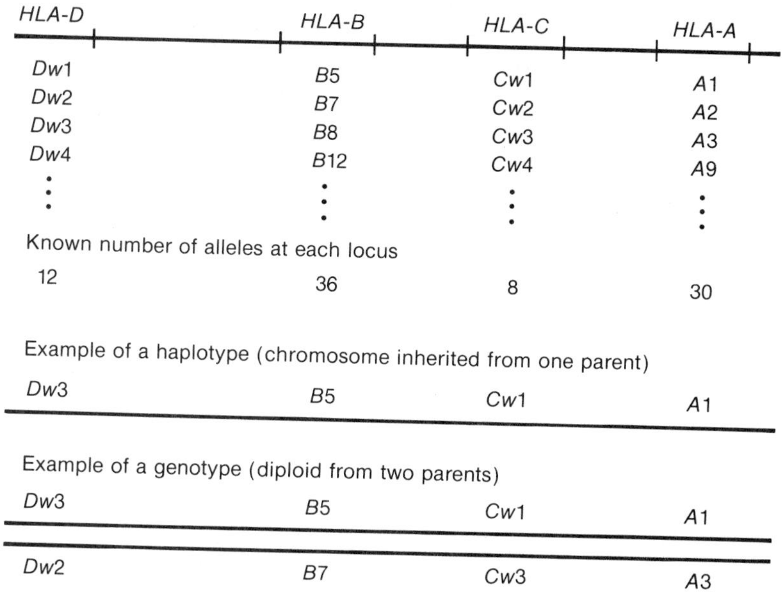

HLA-D	HLA-B	HLA-C	HLA-A
Dw1	B5	Cw1	A1
Dw2	B7	Cw2	A2
Dw3	B8	Cw3	A3
Dw4	B12	Cw4	A9
⋮	⋮	⋮	⋮

Known number of alleles at each locus

12	36	8	30

Example of a haplotype (chromosome inherited from one parent)

Dw3	B5	Cw1	A1

Example of a genotype (diploid from two parents)

Dw3	B5	Cw1	A1
Dw2	B7	Cw3	A3

Figure 13-16 Four major genetic regions of the *HLA* complex with approximate number of alleles that have been identified in each region. Some authorities also recognize a fifth region, called *DR*.

many that any two unrelated individuals are likely to have no haplotype in common. Such extensive genetic variation is useful in anthropology and population genetics, because haplotypes can be used as markers of genetic ancestry. At the same time, haplotype variation presents a problem for tissue transplants, because unrelated individuals will rarely be compatible at the MHC.

Blood Groups, HLA, and Disease

Among individuals with particular diseases, certain blood groups and *HLA* haplotypes are found more frequently than would be expected by chance. One example is the association between Rh and hemolytic disease of the newborn, which occurs only in Rh$^-$ mother-Rh$^+$ child combinations. In this case the red-blood-cell antigens and serum antibodies are the direct cause of the disease. Other diseases associated with blood groups include stomach cancer, which occurs at elevated frequency with blood type A, and duodenal ulcer, which occurs at elevated frequency with blood type O. In these cases the basis of the disease association is unclear.

HLA haplotypes have been studied extensively with regard to disease, and there are many significant associations. The strongest association is between ankylosing spondylitis, an inflammatory disease of the spinal column, and haplotypes carrying *B27*. The frequency of *B27* among patients with this disorder is 90 percent, but among unaffected individuals it is only 8 percent. Other associations between *HLA* and disease include *Dw3* with gluten-sensitive enteropathy (an immune disorder resulting in destruction of intestinal cells triggered by eating wheat products, which contain gluten proteins), and *B13* with psoriasis (a familial skin disease). Although some of the diseases associated with *HLA* are immune disorders, suggesting a causal relationship, other diseases are not known to be related to immunity; in these cases, the association may be due to other loci closely linked with *HLA* in the chromosome.

Problems

1. In the HAT method for selecting hybrid cells, what are the roles of hypoxanthine, aminopterin, and thymidine?

2. Suppose that a human-mouse cell line carries human chromosomes 2, 6, and 11. This clone produces the human form of the enzyme acid phosphatase-1 (ACP1). In an attempt to identify the chromosome that codes for ACP1, the cells are x-irradiated to produce chromosome breaks and rearrangements, and clones are derived from individual daughter cells. Five aberrant clones are isolated, each identical with the original clone except for the following abnormalities (the symbols p and q refer to the short and long arms of a chromosome, respectively): (1) 6q deleted—ACP1 present; (2) 11p deleted—ACP1 present; (3) reciprocal translocation of 2 and 6—ACP1 present; (4) inversion of 11—ACP1 present; (5) chromosome 2 in shape of ring, total length smaller than normal chromosome 2—ACP1 absent.
   (a) Which of these clones provide *no* information about the location of the gene?
   (b) Which clone is the *most* informative about gene location?
   (c) Which chromosomal locations are the most likely candidates for the locus of the gene?

Problems 3 through 6 use data from the set of eight hypothetical human-mouse clones described in the table below. Each

clone may carry an intact (numbered) chromosome ($+$), or only its long arm (q), or only its short arm (p), or it may lack the chromosome ($-$).

3. Suppose the clones were obtained by the HAT selection method.
   (a) Were the human cells used in the cell fusions unable (TK^-) or able (TK^+) to utilize thymidine as a DNA precursor? How can you tell?
   (b) On which chromosome arm is the *TK* gene located?

4. The following human enzymes were tested for their presence ($+$) or absence ($-$) among the clones A–H. Identify the chromosome carrying each enzyme locus, and, where possible, identify the chromosome arm.

Enzyme	Cell line							
	A	B	C	D	E	F	G	H
Steroid sulfatase	+	−	+	+	−	+	−	+
Phosphoglucomutase-3	−	−	+	−	+	−	−	+
Esterase D	−	+	−	−	−	−	+	+
Phosphofructokinase	+	−	−	−	+	+	−	−
Amylase	+	+	−	+	+	−	−	+
Galactokinase	+	+	+	+	+	+	+	+

5. Restriction fragment polymorphism refers to inherited variation in the presence or absence of a restriction site at a particular location in the DNA, which can be detected by the size of the restriction fragment containing sequences homologous to those of a probe sequence. A certain restriction fragment polymorphism is found to be linked to the locus of the ABO blood group. Clones A–H were produced from a human cell carrying the polymorphic restriction site, and among the clones the presence ($+$) or absence ($-$) of the restriction site was as follows:

A	B	C	D	E	F	G	H
+	+	+	+	−	−	+	−

					Chromosome				
Clone	1	2	6	9	12	13	17	21	X
A	+	+	−	q	−	p	+	+	+
B	+	−	p	+	−	+	+	−	−
C	−	+	+	+	p	−	+	−	+
D	+	+	−	+	+	−	q	−	+
E	p	−	+	−	q	−	+	+	q
F	−	p	−	−	q	−	+	+	p
G	q	+	−	+	+	+	+	−	−
H	+	q	+	−	−	q	+	−	+

(a) Which chromosome arm carries the restriction site marker?

(b) Which chromosome arm carries the locus of the ABO blood group?

6. A different restriction fragment polymorphism known to be linked with the Rh locus occurs among the clones A–H in Problem 5 as follows:

A	B	C	D	E	F	G	H
+	+	−	+	+	−	−	+

Which chromosome arm carries the Rh locus?

7. Answer the following about the ABO blood groups.

(a) What is the phenotype of an *hh* individual?

(b) Considering the great rarity of the *h* allele, how can it be possible for a first-cousin mating AB × AB to produce an apparently O offspring?

8. A woman of blood type A, who is a nonsecretor of ABO antigens into the body fluids, has a child of blood type B, who is a secretor of ABO antigens into the body fluids.

(a) What is the phenotype (or possible phenotypes) of the father?

(b) What is the genotype of the child?

9. A woman has a child with hemolytic disease of the newborn.

(a) What is the genotype of the mother with regard to the *Dd* pair of Rh alleles?

(b) Will antibody treatment in subsequent pregnancies be effective? Why or why not?

(c) If the woman's husband has the genotype *Dd*, what is the probability that the next child will not be at risk of the hemolytic disease?

10. Rh antibodies are of the IgG class, which can cross the placenta, but naturally occurring anti-A and anti-B antibodies are of the IgM class. What does this imply about the ability (or inability) of IgM to cross the placenta?

11. A human blood group denoted Xg(a) shows X-linked inheritance. Females of genotype *Xg/Xg* or *Xg/xg* and males of genotype *Xg* have red blood cells carrying the Xg(a) antigen and are said to be Xg(a+). Red blood cells in *xg/xg* females and *xg* males do not have the antigen and are Xg(a−). If a mating produces an Xg(a−) daughter and an Xg(a+) son, what are the genotypes of the parents?

12. An Xg(a+) female and an Xg(a+) male (see Problem 11) have a 47,XXY offspring (Klinefelter syndrome) that is phenotypically Xg(a−). Explain how this can happen.

13. A woman accused of abandoning a baby claims that she never gave birth to any baby. The blood types of the woman and the baby are as follows:

Woman	AB	*cc*	*dd*	*Ee*	M
Baby	O	*Cc*	*D–*	*ee*	N

(a) Could the woman have borne the baby?

(b) Which loci are the most informative?

14. A woman has a baby and thinks either of two men could be its father. The blood group phenotypes of the individuals are

Mother	O	*CC*	*D–*	*ee*	M
Child	O	*CC*	*dd*	*Ee*	M
Male 1	A	*cc*	*D–*	*EE*	MN
Male 2	B	*Cc*	*dd*	*Ee*	N

Based on these phenotypes, can either male be excluded from paternity? Explain your reasoning.

15. After a mixup of two babies in a maternity ward, both babies and their parents are blood-typed. Match each child with its proper parents.

Mother 1	O	*Cc*	*D–*	*Ee*	M
Father 1	AB	*cc*	*D–*	*ee*	MN
Mother 2	A	*cc*	*dd*	*ee*	N
Father 2	O	*CC*	*D–*	*ee*	N
Baby 1	A	*Cc*	*dd*	*Ee*	M
Baby 2	A	*Cc*	*dd*	*ee*	N

16. What related family member is the best possible donor for blood transfusions or transplants? Why?

17. A woman claims to have had a child by spontaneous embryonic development of an unfertilized diploid egg cell (parthenogenesis). How would one use skin transplants to determine the validity of the claim?

18. Each of a pair of identical twins of blood type B produces an anti-A antibody of the IgM class. Would these antibodies be expected to have identical amino acid sequences? Explain.

19. A cow is artificially inseminated with the sperm of her own father and a single calf results.

(a) Ignoring recombination, what is the probability that the calf will be homozygous for a bovine major histocompatibility complex (MHC) haplotype?

(b) What is the probability that the calf and the bull will be a perfect match at the bovine MHC?

20. What is the maximum number of cellular antigens that can be coded by the alleles at a single histocompatibility locus in a diploid, triploid, or trisomic individual?

21. Two inbred mouse strains—P_1 and P_2—differ in three unlinked histocompatibility loci. Among the F_1, F_2, and backcrosses (B_1 and B_2), individuals are chosen at random and used as donors in skin transplants to other randomly chosen individuals. What is the probability of graft acceptance in the following cases: (1) P_2 donor, B_1 recipient; (2) B_1 donor, B_2 recipient; (3) B_1 donor, P_2 recipient; (4) F_2 donor, F_2 recipient?

CHAPTER 14

Population Genetics

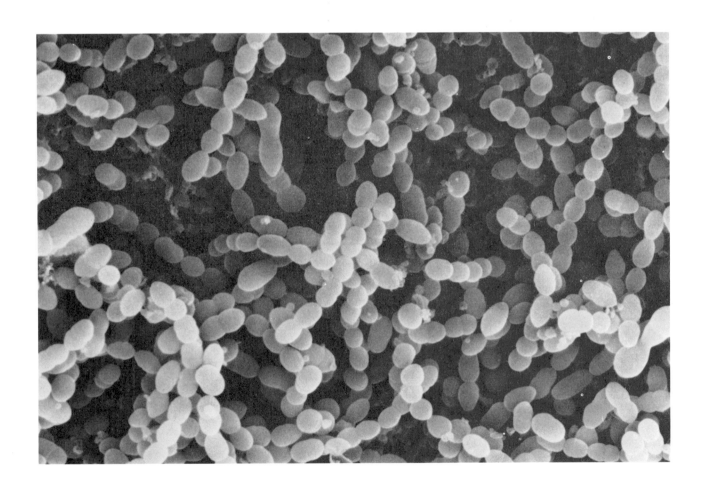

Mendelian inheritance is often associated with particular ratios of phenotypes such as 3/4:1/4 or 1/2:1/2. These are the ratios that Mendel discovered in garden peas, and the value of Mendelism is in explaining the occurrence of such ratios in terms of the segregation and expression of individual genes, such as those for round or wrinkled seeds. However, it is important to realize that these simple ratios occur only in particular types of matings, such as heterozygote × heterozygote in the case of the 3/4:1/4 ratio, or heterozygote × recessive homozygote in the case of the 1/2:1/2 ratio. Such matings can be carried out deliberately in experimental organisms. In breeding the organisms, the geneticist, like Mendel, must keep careful record of the genealogical relationships among the individuals, as knowledge of these connections is essential in correctly predicting or interpreting phenotypic ratios. In human genetics, matings are not arranged by the investigator, but the familial relationship among individuals is recorded in pedigrees.

In many cases, matings between organisms are not under the control of the investigator, and the familial relationships among individuals are not known; this situation is typical of studies of organisms in their natural habitat. Most organisms do not live in discrete family groups but exist as part of populations of individuals of unknown genetic relationship. At first, it might seem that classical genetics with its simple Mendelian ratios could have little to say about such complex situations, but this is not the case. The principles of Mendelian genetics can be used both to interpret data collected in natural populations and to make predictions about the genetic composition of populations. Application of genetic principles to entire populations of organisms constitutes the subject of **population genetics.** This is an inherently mathematical topic and can become complex. However, in this chapter the mathematics is confined to simple algebra and probabilities.

14.1 Allele Frequencies and Genotype Frequencies

In population genetics, the term **population** refers to a group of organisms of the same species living within a prescribed geographical area. Sometimes the area is large, as when referring to the population of sparrows in North America; or it may even include the entire earth, as in reference to the "human population." More commonly, the area is considered to be of a size within which individuals in the population are likely to find mates. Such a group of interbreeding individuals living within a defined geographical area is often called a **local population** or **deme.** When used without qualification, the term population, as used in population genetics, usually means *local population.* We begin this section with an analysis of a local population with respect to some phenotype having a known genetic basis.

Examples of Calculations of Allele Frequency

The genetic properties of the human MN blood group system are exceptionally simple. In this system there are three possible phenotypes—M, MN, and N—corresponding to three genotypes at one locus—*MM*, *MN*, and *NN*, respectively. In one study of the phenotypes of 1000 British people, 298 were M, 489 were MN, and 213 were N. From the simple genotype-phenotype correspondence in this system, the genotypes can be directly inferred to be

$$298 \; MM \qquad 489 \; MN \qquad 213 \; NN$$

These numbers contain a great deal of information about the population—for example, whether there may be mating between relatives—and one of the goals of population genetics is to be able to extract this information. First, note that the sample contains two types of data—the number of occurrences of the three genotypes, and the number of occurrences of the individual M and N alleles. Furthermore, the 1000 individuals represent 2000 alleles at the MN locus because each individual is diploid for the locus. These alleles break down in the following way:

$$
\begin{aligned}
298 \; MM \text{ individuals} &= 596 \; M \text{ alleles} \\
489 \; MN \text{ individuals} &= 489 \; M \text{ alleles} + 489 \; N \text{ alleles} \\
213 \; NN \text{ individuals} &= \phantom{489 \; M \text{ alleles} +} 426 \; N \text{ alleles} \\
\hline
\text{Totals} &= 1085 \; M \text{ alleles} + 915 \; N \text{ alleles}
\end{aligned}
$$

Usually it is more convenient to analyze the data in terms of relative frequency rather than with the actual numbers. For genotypes, the **genotype frequency** in a population is the proportion of individuals having the particular genotype. For individual alleles, the **allele frequency** of a specified allele is the proportion of all alleles that are of the specified type. For a sample of the type being discussed here, the genotype frequencies are obtained by dividing the observed numbers by the total sample size, in this case 1000. Therefore, the genotype frequencies are

$$0.298 \ MM \qquad 0.489 \ MN \qquad 0.213 \ NN$$

Similarly, the allele frequencies are obtained by dividing the observed numbers by the total (in this case 2000), so

$$\text{Allele frequency of } M = 1085/2000 = 0.5425$$
$$\text{Allele frequency of } N = \ \ 915/2000 = 0.4575$$

(The allele frequencies have been calculated with more significant digits than would be warranted by the size of the sample in order to avoid round-off error.) Note that the genotype frequencies add up to 1.0, as do the allele frequencies; this is a consequence of their definition in terms of proportions, which must add up to 1.0 when all of the possibilities are taken into account. *Allele frequencies must always be between 0 and 1*. A population having an allele frequency of 1.0 for some allele is said to be **fixed** for that allele.

The calculations just given therefore serve to summarize the data gathered from the sample of 1000 people. However, if the sample had been properly gathered so as to be representative of the entire British population, then the genotype and allele frequencies observed in the sample would also serve to approximate the true values in the entire population. Only if the sampling has been properly done is the investigator justified in using the frequencies calculated from the sample as estimates of the corresponding quantities in the entire population. Generally speaking, the larger the sample is the more confident one may be about the reliability of the estimates.

Allele frequencies can be used to make inferences about matings in a population and to predict the genetic composition of future generations. Allele frequencies are often more useful than genotype frequencies because individual alleles, not genotypes, form the bridge between generations. Alleles rarely undergo mutation in a single generation, so they are relatively stable in their transmission from one generation to the next. Genotypes are impermanent; genotypes are totally broken up in the processes of segregation and recombination that occur in each reproductive cycle. Moreover, we know from simple Mendelian considerations what types of gametes must be produced from the *MM*, *MN*, and *NN* genotypes:

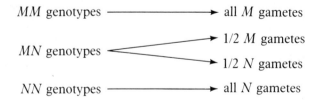

Consequently, the *M*-bearing gametes produced in the population will represent all the gametes from *MM* individuals and half the gametes from *MN*

individuals. Likewise, the N-bearing gametes will represent all the gametes from NN individuals and half the gametes from MN individuals. Therefore, in terms of gametes, the population would produce the following frequencies:

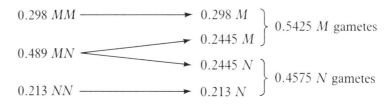

$$0.298\ MM \longrightarrow 0.298\ M$$
$$0.2445\ M \left.\right\} 0.5425\ M \text{ gametes}$$
$$0.489\ MN$$
$$0.2445\ N \left.\right\} 0.4575\ N \text{ gametes}$$
$$0.213\ NN \longrightarrow 0.213\ N$$

Note that the allele frequencies among *gametes* equal the allele frequencies among *adults* calculated earlier, which will be true whenever each adult in the population produces the same number of functional gametes. If the adult genotypes differ in fertility (for example, if one is sterile and hence is incapable of producing functional gametes), then the equality of allele frequencies among adults and gametes will no longer hold. Differential fertility is one example of *selection;* this and certain other complications will be discussed in Chapter 15. In this chapter the assumption is made that all genotypes are equally capable of survival and reproduction.

Examples from Natural Populations of Plants and Animals

In this section the concept of allele frequency in population genetics is illustrated for a nonhuman population. The phenotypic differences are revealed by means of an electrophoretic procedure similar to that described in Section 4.9 for separation of DNA fragments. Protein-containing tissue samples from individuals are placed near the edge of a starch or polyacrylamide gel and a voltage is applied for several hours. All charged molecules in the sample move in response to the voltage, and a variety of techniques can be used to locate particular substances. In the analysis that follows an enzyme can be located by

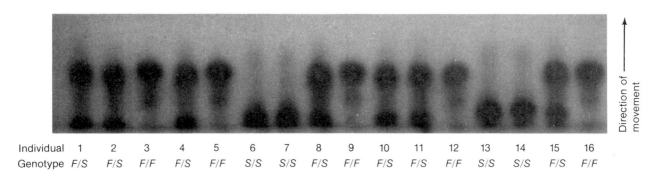

Figure 14-1 Electrophoretic mobility of glucose phosphate isomerase in a sample of 16 mice. (Courtesy of S. E. Lewis and F. M. Johnson.)

staining the gel with a reagent that will be converted to a colored product by the enzyme; wherever the enzyme is located, a colored band will appear.

Figure 14-1 is a photograph of a gel stained to reveal the enzyme phosphoglucose isomerase in tissue samples of 16 mice, illustrating typical raw data from an electrophoretic study. The pattern of bands varies from individual to individual, but only three patterns are observed. Samples from individuals 1, 2, 4, 8, 10, 11, and 15 have two bands, one that moves fast (upper band) and one that moves slow (lower band). A second pattern is exemplified by the samples from individuals 3, 5, 9, 12, and 16, in which only the fast band appears. Finally, the samples from individuals 6, 7, 13, and 14 exemplify a third pattern, in which only the slow band appears.

Patterns of enzyme mobility are *phenotypes*, not genotypes, so the type of phenotypic variation illustrated in Figure 14-1 does not indicate the genetic basis of the variation. Genetic studies require examination of individuals with known genealogical relationships, which in natural populations can often be achieved by studying mothers and their offspring, or plants and their seed. Some organisms, such as *Drosophila*, can be brought into the laboratory for detailed genetic analysis. Analysis of phenotypic variation of many enzymes in a wide variety of organisms has showed that electrophoretic variation usually, though not invariably, has a simple genetic basis: each electrophoretic form of the enzyme contains one or more polypeptides with a genetically determined *amino acid substitution* that changes the electrophoretic mobility of the enzyme. Alternative forms of an enzyme coded by alleles at a single locus are known as **allozymes.** Alleles coding for allozymes are usually codominant (Section 1.4), which means that heterozygotes express the allozyme corresponding to each allele. Therefore, in Figure 14-1, individuals having only the fast allozyme are *FF* homozygotes, those with only the slow allozyme are *SS* homozygotes, and those with both fast and slow are *FS* heterozygotes. The 16 genotypes in Figure 14-1 are consequently

$$5 \; FF \quad 7 \; FS \quad 4 \; SS$$

and the allele frequency of *F* in this small sample can be calculated as

$$\text{Allele frequency of } F = [(2 \cdot 5) + 7]/(2 \cdot 16) = 0.53$$

Since there are only two alleles represented in Figure 14-1, the allele frequency of *S* must be $1 - 0.53 = 0.47$, which can equivalently be calculated directly as

$$\text{Allele frequency of } S = [7 + (2 \cdot 4)]/(2 \cdot 16) = 0.47$$

Electrophoretic variation of enzymes in natural populations is common. An example of allozyme variation in a flowering plant is seen in the enzyme phosphoglucomutase (encoded by the *Pgm* locus) in *Phlox cuspidata*. In one sample of 35 individuals from a Texas population, the following genotypes were found:

$$15 \; Pgm^a Pgm^a \quad 6 \; Pgm^a Pgm^b \quad 14 \; Pgm^b Pgm^b$$

The allele frequency of Pgm^a in this sample, symbolized p, is

$$p = [(2 \cdot 15) + 6]/(2 \cdot 35) = 0.51$$

and that of Pgm^b, symbolized q, is

$$q = 1 - p = 1 - 0.51 = 0.49$$

(The allele frequencies are similar to those in the mouse example in Figure 14-1, but this is pure coincidence.)

Allele frequencies may also be calculated when genotype frequencies are given as proportions rather than as numbers. For example, in an electrophoretic study of the enzyme phosphohexose isomerase (locus abbreviated as *Phi*) in a population of the American eel, *Anguilla rostrata*, off the coast of Vero Beach, Florida, the following genotype frequencies were observed among 108 individuals:

$$0.55 \ Phi^a Phi^a \quad 0.39 \ Phi^a Phi^b \quad 0.06 \ Phi^b Phi^b$$

In this case the frequencies of the alleles Phi^a (p) and Phi^b (q) are calculated as

$$p = 0.55 + 0.39/2 = 0.75$$
$$q = 1 - p = 1 - 0.75 = 0.25$$

In a natural population, a **polymorphic** locus is defined as a locus at which the most common allele has a frequency of less than 0.95. Conversely, a **monomorphic** locus is one at which the most common allele has a frequency of greater than 0.95. Using these definitions, the *Pgm* locus in the relevant Texas population of *P. cuspidata* is polymorphic because the most common allele, Pgm^a, has a frequency of less than 0.95. The proportion of loci that are polymorphic in a population is one widely used index of the amount of genetic variation present in the population. A second widely used index of genetic variation is the **heterozygosity** (symbolized H), which is simply the proportion of genotypes that are heterozygous. In the population of *P. cuspidata*,

$$\text{Heterozygosity} = H = 6/35 = 0.17$$

In the population of *A. rostrata*, $H = 0.39$. Among a group of loci studied in a single population, the *average heterozygosity* is calculated as the average of the H values corresponding to each locus. Therefore, the average heterozygosity corresponds to the average proportion of loci at which an individual is heterozygous. These two numbers—the proportion of polymorphic loci and the average heterozygosity—together provide a useful summary of the amount of genetic variation found in a natural population.

The usefulness of electrophoresis in population studies comes from the fact that allozyme phenotypes are usually inherited in simple Mendelian fashion. Consequently, alleles coding for allozymes provide a built-in set of genetic markers in almost any natural population. The aggregate of genes in a natural population is often called the **gene pool** of the population, and allozymes provide a convenient method with which to study the gene pool. Comparison of allozyme alleles within and among species can be used to infer evolutionary relationships, and in some cases they reveal important features about the evolutionary process itself.

For electrophoretic studies to be maximally useful, it is necessary that natural populations contain electrophoretic variation in their gene pools. On this issue the data are clear: *natural populations contain abundant electrophoretic variation.* This conclusion stands in contrast to what naive intuition may lead one to believe. On the face of it, most organisms in the same species look rather much alike—for example, to the untrained eye, one wild *Drosophila* looks quite the same as any other. However, this superficial similarity conceals the diversity of genotypes that occurs among individuals within species. The situation regarding genetic variation in various species of plants, invertebrates, and vertebrates is summarized in Table 14-1. It can be seen that among vertebrates, about 17 percent of all loci studied by electrophoresis are polymorphic, and an average individual vertebrate is heterozygous for allozyme-associated alleles at about five percent of its loci. Plants that do not undergo self-fertilization are intermediate between vertebrates and invertebrates, both with regard to proportion of polymorphic loci and average heterozygosity, and invertebrates are the most genetically variable of the groups listed in Table 14-1. The numbers preceded by the ± signs are standard deviations and indicate the range within which about 68 percent of the species studied fall. The standard deviations are quite large, which indicates that the degree of polymorphism and heterozygosity varies greatly among species within each group.

Although allozyme variation is widespread in natural populations of virtually all species, including haploid prokaryotes such as *E. coli*, the number of loci studied by electrophoresis is very small compared to the total number

Table 14-1   Detection of genetic variation through use of electrophoresis

Organism	Number of species	Proportion of loci that are polymorphic	Proportion of heterozygous loci (average) ± standard deviation
Plants	15	0.26 ± 0.17	0.07 ± 0.07
Invertebrates	93	0.40 ± 0.20	0.11 ± 0.07
Vertebrates	135	0.17 ± 0.12	0.05 ± 0.04

Source: From Nevo, E. 1978. "Genetic variation in natural populations: Patterns and theory." *Theor. Pop. Biol.*, 13: 121–177.

of loci in any organism. The data in Table 14-1 are based on only about 30 loci (depending on the particular study), so there is no way at present to know whether these few loci are representative of genetic variation in the entire genome. Even if allozymes are not completely representative, they nevertheless indicate that a substantial amount of genetic variation exists in natural populations.

14.2 Systems of Mating

The transmission of genetic material from one generation to the next is analyzed in terms of alleles rather than genotypes because genotypes are broken up in each generation by the processes of segregation and recombination. When gametes unite in the process of fertilization, the genotypes of the next generation are formed. The genotype frequencies in the zygotes of the progeny generation are determined by the frequencies with which the various types of parental gametes come together, and these frequencies are in turn determined by the various types of matings that occur among adults of the parental generation. The **mating system** in the population refers to the relative frequencies of the various genetic types of matings that occur in the population, insofar as these bear on the transmission of particular genes. In population genetics, three types of mating systems—random mating, assortative mating, and inbreeding—predominate.

Random mating occurs when individuals pair up independently with respect to genotype. Thus, genotypes are paired as mates exactly as would be expected by chance. Random mating is by far the most important mating system in most animals and plants (excluding plants that regularly reproduce through self-fertilization), and in these organisms it almost always governs the distribution of allozyme or blood-group genotypes. The implications of random mating on human MN blood-group frequencies are outlined in Table 14-2. The genotype frequencies shown there are those estimated in Section 14.1 from a sample of 1000 British people. With random mating, the frequency of any mating pair is simply the product of the genotype frequencies of the mates. For example, the MM genotype has frequency 0.30, so the frequency of $MM \times MM$ matings that will result from random mating will be given by $0.30 \times 0.30 = 0.09$. Note that no claim is made that the British population *is* actually undergoing random mating with regard to MN—this hypothesis will be tested later; Table 14-2 only lists the frequencies of mating pairs that would occur *if* the population were undergoing random mating.

Assortative mating refers to matings in which partners are nonrandom with respect to phenotype. In **positive** assortative mating individuals tend to mate with phenotypically similar individuals. Positive assortative mating occurs for several traits in humans. In the United States, for example, these traits include skin color and height, and mates tend to resemble each other in these traits more than would be expected if mating occurred at random. In **negative** assortative mating, pairs are more dissimilar than would be expected with random mating. Negative assortative mating is not as common as positive

Table 14-2 Frequency with which genotypes of the MN blood group would be expected to be paired through random mating in British people

Genotype of male	Genotype of female	Expected frequency of mating
MM	*MM*	$0.30 \times 0.30 = 0.090$
MM	*MN*	$0.30 \times 0.49 = 0.147$
MM	*NN*	$0.30 \times 0.21 = 0.063$
MN	*MM*	$0.49 \times 0.30 = 0.147$
MN	*MN*	$0.49 \times 0.49 = 0.240$
MN	*NN*	$0.49 \times 0.21 = 0.103$
NN	*MM*	$0.21 \times 0.30 = 0.063$
NN	*MN*	$0.21 \times 0.49 = 0.103$
NN	*NN*	$0.21 \times 0.21 = 0.044$

assortative mating, but a famous example occurs in some species of primroses (*Primula*) and their relatives. In most populations of these species, two types of flowers are found in approximately equal proportions (Figure 14-2). One type, known as pin, has a long style and short stamens. The anthers, which are the organs that produce pollen, are located atop the stamens (shown in red), whereas the stigma (shaded), which is the organ that receives pollen, is located atop the style. Pin flowers produce pollen low in the flowers and receive pollen high in the flowers. In the other type of plant, known as thrum, the relative positions of stigma and anthers are reversed. Consequently, insect pollinators that work low in the flowers will pick up mainly pin pollen and deposit it mainly on thrum stigmas, whereas pollinators that work high in the

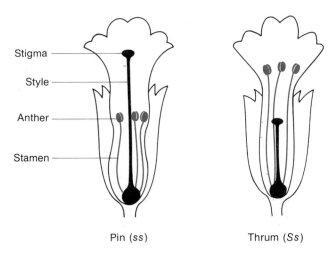

Figure 14-2 Pin and thrum forms of flowers of *Primula officalis*. Negative assortative mating is promoted by the anatomy of these flowers. Pollinators that work low in the flowers tend to pick up pin pollen and deposit it on thrum stigmas, and pollinators that work high in the flowers tend to do the reverse.

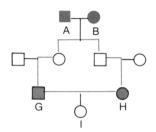

Figure 14-3 An inbreeding pedigree in which individual I is the result of a mating between first cousins.

flowers will pick up mainly thrum pollen and deposit it mainly on pin stigmas. This behavior promotes negative assortative mating, which is further promoted because pin pollen grows better in thrum pollen tubes and vice versa. The genetic basis of the flower type resides in a single locus (actually a cluster of tightly linked loci), pin being homozygous (*ss*) and thrum being heterozygous (*Ss*). This situation is reminiscent of the inheritance of sex in many animals (XX versus XY). Indeed, sex provides an example of *complete* negative assortative mating.

Inbreeding is mating between relatives. In human pedigrees, a mating between relatives is often called a **consanguineous mating.** The distinction between positive assortative mating and inbreeding is that positive assortative mating is based on phenotypic similarity (not necessarily between relatives), whereas inbreeding is based on genetic relationship (without regard to phenotype). Figure 14-3 represents an example of inbreeding. In this case, individual I is the offspring of a first-cousin mating (G with H). The closed loop in the pedigree (red) is diagnostic of inbreeding, and the individuals designated A and B are called **common ancestors** of I, because they are ancestors of both of the parents of I. Because A and B are common ancestors, a particular allele in A (or in B) could by chance be passed by inheritance down both sides of the pedigree, to meet again in the formation of I. This possibility is the most important characteristic of inbreeding, and it will be discussed in Sections 14.7 and 14.8.

The following sections discuss the implications of random mating.

14.3 Random Mating

The mating system of an organism determines how the alleles in a gene pool will be combined into genotypes. Once the mating system is specified, the genotype frequencies can be predicted from a knowledge of the allele frequencies. For random mating, the relation between allele frequency and genotype frequency is particularly simple, because *random mating of individuals is equivalent to random union of gametes.* Conceptually, we may imagine all the gametes of a population to be present in a large container. To form zygote genotypes, pairs of gametes are withdrawn from the container at random. To be specific, consider the alleles M and N in the MN blood groups, whose allele frequencies are p and q, respectively (remember that $p + q = 1$). The genotype frequencies expected with random mating can be deduced from the following diagram:

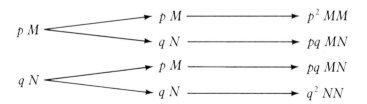

In this diagram, the gametes at the left may be taken to represent the sperm and those in the middle to represent the eggs. The genotypes that can be formed with two alleles are shown at the right, and with random mating the frequency of each genotype is calculated by multiplying the allele frequencies of the corresponding gametes. However, the genotype MN can be formed in two ways—the M allele could have come from the father (top part of diagram), or from the mother (bottom part of diagram). In each case the frequency of the MN genotype is pq; considering both possibilities the frequency of MN is $pq + pq = 2pq$. Consequently, the overall genotype frequencies expected with random mating are

$$MM: p^2 \qquad MN: 2pq \qquad NN: q^2 \qquad\qquad (14\text{-}1)$$

The frequencies p^2, $2pq$, and q^2 that result from random mating for a gene with two alleles constitute what is called the **Hardy-Weinberg rule** after G. H. Hardy and W. Weinberg, who first deduced the rule in 1908. At that time, it was argued that phenotypic ratios different from $3/4 : 1/4$ in any natural population could be taken as evidence against simple Mendelian inheritance of the trait in question. Hardy and Weinberg realized that the $3/4 : 1/4$ ratio is expected only in the offspring of particular types of matings (and only with dominance), and their rule showed that genotype frequencies for simple Mendelian inheritance resulting from random mating could have almost any value depending on the value of the allele frequency. The only restriction imposed by Mendelism is that the frequencies of genotypes resulting from random mating must follow the p^2, $2pq$, q^2 formula. Sometimes the Hardy-Weinberg rule is demonstrated by a Punnett square, as illustrated in Figure 14-4. Such a square is completely equivalent to the tree diagram used above, but a cross-multiplication square is often more convenient. Although the Hardy-Weinberg rule is exceedingly simple, it has a number of important implications that are not obvious. These are described in the following sections.

	Female gametes	
	p M	q N
Male gametes p M	p^2 MM	pq MN
q N	pq MN	q^2 NN

Figure 14-4 A cross-multiplication square (Punnett square) showing the result of random union of male and female gametes, which is equivalent to random mating of individuals.

14.4 Implications of the Hardy-Weinberg Rule

One important implication of the Hardy-Weinberg rule is that *the allele frequencies remain constant from generation to generation*. Consider a locus with two alleles, A and a, having frequencies p and q, respectively ($p + q = 1$). With random mating, the frequencies of genotypes AA, Aa, and aa among zygotes will be p^2, $2pq$, and q^2, respectively. Assuming equal *viability* (ability to survive) among the genotypes, these frequencies will equal those among adults. If all of the genotypes are equally fertile, then the frequency p' of allele A among gametes of the next generation will be

$$p' = p^2 + 2pq/2 = p(p + q)$$
$$= p \qquad (\text{because } p + q = 1)$$

This argument shows that the frequency of allele A remains constant at the value of p through the passage of one (or any number of) complete generations. Of course, this principle is contingent upon certain assumptions, of which the most important are the following:

1. Mating is random.

2. Allele frequencies are the same in males and females.

3. The genotypes are all equal in viability and fertility (that is, *selection* does not occur).

4. Mutation does not occur.

5. Migration into the population does not occur.

6. The population is sufficiently large that the frequencies of alleles will not change from generation to generation because of chance. (Such chance changes in allele frequency are known as *random genetic drift*.)

If any or all of these assumptions are violated, particularly assumptions 3–6, then the allele frequencies can change from generation to generation in a manner to be described more fully in Chapter 15.

A Test for Random Mating

The Hardy-Weinberg rule allows information about mating systems to be extracted from genotype and allele frequencies. Specifically, one can determine whether a particular population is undergoing random mating with respect to a particular locus. To illustrate the method, we consider again the sample of British people in the MN blood group, discussed in Section 14.1. The frequencies of alleles M and N among 1000 adults were 0.5425 and 0.4575, respectively; assuming random mating, the *expected* genotype frequencies can be calculated from Equation 14-1 as

$$
\begin{aligned}
MM: \quad & (0.5425)^2 & = 0.2943 \\
MN: \quad & 2(0.5425)(0.4575) & = 0.4964 \\
NN: \quad & (0.4575)^2 & = 0.2093
\end{aligned}
$$

The *observed* frequencies were 0.298, 0.489, and 0.213, respectively. Goodness of fit between observed and expected can be determined by means of the χ^2 test described in Chapter 2. However, χ^2 tests between observed and expected results must be based on the actual numbers, not on frequencies. The expected numbers can be obtained by multiplying each expected frequency by 1000 (the size of the sample). Hence, the observed and expected numbers of the genotypes are

$$
\begin{aligned}
MM: \quad & 298, \text{ versus } 0.2943 \times 1000 = 294.3 \\
MN: \quad & 489, \text{ versus } 0.4964 \times 1000 = 496.4 \\
NN: \quad & 213, \text{ versus } 0.2093 \times 1000 = 209.3
\end{aligned}
$$

Recall from Chapter 2 that χ^2 is calculated as

$$\chi^2 = \sum \frac{(\text{observed} - \text{expected})^2}{\text{expected}} \qquad (14\text{-}2)$$

in which the symbol Σ means summation over all classes of data. In this case,

$$\chi^2 = \frac{(298 - 294.3)^2}{294.3} + \frac{(489 - 496.4)^2}{496.4} + \frac{(213 - 209.3)^2}{209.3} \qquad (14\text{-}3)$$

$$= 0.222$$

Normally, a value of χ^2 obtained from three classes of data would have two degrees of freedom, but here a correction must be made, because the data themselves were used to calculate the expected genotype frequencies. In general, one degree of freedom must be deducted for each number estimated from the data for use in calculating the expected numbers. In this case the data were used to estimate one number (the frequency of allele M), so one degree of freedom must be deducted. Therefore, the χ^2 value in Equation 14-3 has one degree of freedom. Only the frequency of allele M need be estimated from the data, as the frequency of N is obtained just by subtracting this value from 1.

The probability value associated with a χ^2 of 0.222 with one degree of freedom is about 0.67, which means that in about 67 percent of samples of size 1000 we would have expected a χ^2 as large or larger when random-mating genotype frequencies occur. Clearly, from these data there is no reason to reject the hypothesis that the British population is undergoing random mating with respect to MN blood groups. However, this conclusion does not imply random mating in all respects. The same population may be undergoing random mating with regard to some loci but nonrandom mating with regard to others.

Although the data for the human blood group MN have a satisfactory fit to the Hardy-Weinberg proportions, such good agreements are by no means universal. Theoretically, violation of any of the six assumptions listed earlier could cause departures from Hardy-Weinberg proportions, and any observed disagreement is difficult to assign to any specific cause, because so many possibilities exist. On the other hand, genotype frequencies calculated from the Hardy-Weinberg rule are often convenient approximations to genotype frequencies in actual populations, even when one or more of the underlying assumptions is violated—provided that the departures from the assumptions are not too great.

The use of a χ^2 test to ascertain departures from random mating may seem somewhat circular inasmuch as the data themselves are used in calculating the expected genotype frequencies. The deduction of one degree of freedom exactly compensates for this use. To show that the χ^2 test as employed is not circular, an example will be considered in which the hypothesis of random mating proportions can be rejected—namely, the *Pgm* example in which the enzyme phosphoglucomutase was observed in *Phlox cuspidata*, discussed in Section 14.1. Recall that the allele frequencies in this case were $p = 0.51$ and $q = 0.49$ in a total sample of size 35. The observed and expected numbers are

$$
\begin{aligned}
Pgm^a Pgm^a: &\quad 15, \text{ versus } (0.51)^2 \times 35 &= 9.1 \\
Pgm^a Pgm^b: &\quad 6, \text{ versus } 2(0.51)(0.49) \times 35 &= 17.5 \\
Pgm^b Pgm^b: &\quad 14, \text{ versus } (0.49)^2 \times 35 &= 8.4
\end{aligned}
$$

Using Equation 14-2, the value of χ^2 in this case is 15.1, which has one degree of freedom. The corresponding probability is about 0.0002. Since the χ^2 probability is less than 0.05 (in this case, very much less), these data must be judged significantly different from the Hardy-Weinberg expectations. The reason for the discrepancy will be discussed later in this chapter. All that can be concluded at this point is that the genotype frequencies at the *Pgm* locus in this population are definitely not in Hardy-Weinberg proportions.

Frequency of Heterozygotes

Another important implication of the Hardy-Weinberg rule is that *for a rare allele, the frequency of heterozygotes will far exceed the frequency of the rare homozygote*. This aspect of the Hardy-Weinberg rule is illustrated graphically by the red line in Figure 14-5. For example, when the frequency of the rarer allele is $q = 0.1$, the ratio of heterozygotes to homozygotes is about 20; this ratio increases by a factor of 10 for every tenfold decline in allele frequency, so $q = 0.01$ yields a ratio of about 200, $q = 0.001$ yields a ratio of about 2000, and so on. Clearly, when an allele is rare, there are many more heterozygotes than there are homozygotes for the rare allele.

The reason for this perhaps unexpected relation is shown by the black lines in Figure 14-5. The solid line corresponds to the frequency of homozygotes (q^2), and the gray line corresponds to the frequency of heterozygotes $(2pq)$. (The scale for these lines is on the right.) The frequency of both heterozygotes and homozygotes decreases as the allele frequency decreases, but the frequency of homozygotes decreases faster than that of heterozygotes; thus, as an allele becomes increasingly rare, the ratio of heterozygotes to homozygotes $(2pq/q^2)$ becomes greater.

The principle illustrated in Figure 14-5 can be exemplified with data for cystic fibrosis, which is one of the most common recessively inherited severe disorders among Caucasians. Cystic fibrosis affects about 1 in 1700 newborns, and it is characterized by abnormal glandular secretions, usually causing death before age 20. (The genotype frequency corresponding to 1/1700 is 0.00059.) In this case the heterozygotes cannot readily be identified by phenotype, so a method of calculating allele frequencies that is different from the gene-counting method used earlier must be used. The new method is straightforward because with random mating the frequency of recessive homozygotes must correspond to q^2. Thus, for cystic fibrosis,

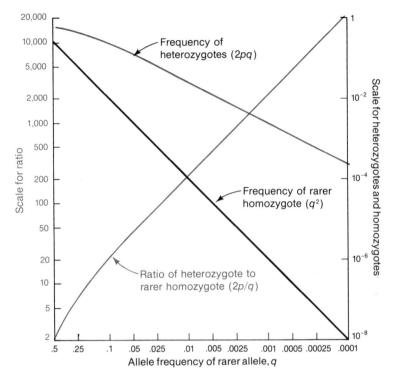

Figure 14-5 The effect of the frequency of a rare allele on the frequencies of certain genotypes. The vertical axis on the left is the ratio between the frequencies of heterozygotes (*Aa*) and rarer homozygotes (*aa*) in a randomly mating population, which increases as the frequency of the *a* allele decreases (red curve). The vertical axis on the right shows the genotype frequencies; both decrease as the *a* allele becomes rare, but the frequency of the homozygote decreases more rapidly.

$$q^2 = 1/1700 = 0.00059 \quad \text{or} \quad q = (0.00059)^{1/2} = 0.024$$

and, consequently,

$$p = 1 - q = 1 - 0.024 = 0.976$$

The frequency of heterozygotes that carry the allele for cystic fibrosis is

$$2pq = 2(0.976)(0.024) = 0.047 = 1/21$$

Therefore, for cystic fibrosis, although only 1 individual in 1700 is affected (homozygous), about 1 individual in 21 is a carrier (heterozygous). In this case, the ratio of heterozygotes to recessive homozygotes is

$$\text{Ratio} = 2pq/q^2 = 0.047/0.00059 = 79.7$$

That is, for cystic fibrosis there are about 80 carriers for every affected individual. This number could also have been estimated directly from Figure 14-5.

Considerations like these are important in predicting the outcome of population screening for the detection of carriers of harmful recessive alleles, which is essential in evaluating the potential benefits of such programs. For cystic fibrosis, about 1 in every 21 individuals would be identified as a carrier.

14.5 Extensions of the Hardy-Weinberg Rule

So far, the Hardy-Weinberg rule has been applied to two alleles at a single autosomal locus. However, the principle is easily extended to more complex cases. Two of these cases—multiple alleles at an autosomal locus and X-linked genes—will be discussed.

Multiple Alleles

Extension of the Hardy-Weinberg rule to multiple alleles is illustrated by the three-allele case. With three alleles, the Punnett square for random mating is shown in Figure 14-6. The alleles are designated A_1, A_2, and A_3 with the upper-case letter representing the locus and the subscript designating the particular allele. The allele frequencies are p_1, p_2, and p_3, respectively. With three alleles (as with any number), the allele frequency of all alleles must sum to 1— that is, $p_1 + p_2 + p_3 = 1.0$. As in Figure 14-4, each square is obtained by multiplying the frequencies of the alleles at the corresponding margins; any homozygote (such as A_1A_1) has a random-mating frequency equal to the square of the corresponding allele frequency (in this case, p_1^2). The various degrees of colored shading represent the heterozygotes. Any heterozygote (such as A_1A_2) has a random-mating frequency equal to twice the product of the corresponding allele frequencies (in this case $2p_1p_2$). The extension to any number of alleles is straightforward:

Frequency of any homozygote = square of allele frequency
Frequency of any heterozygote = 2 × product of allele frequencies

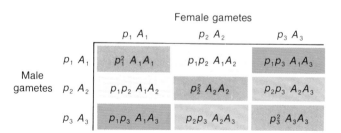

Figure 14-6 A Punnett square showing the results of random mating with three alleles.

These considerations can be applied to the locus controlling the ABO blood groups in humans (Section 13.3). Recall that this locus has three principal alleles, designated I^A, I^B, and I^O. In one study of 3977 Swiss residents of the city of Zurich, the allele frequencies were found to be 0.27 I^A, 0.06 I^B, and 0.67 I^O. Applying the rules for multiple alleles, the genotype frequencies expected to result from random mating are

$$I^A I^A = \quad (0.27)^2 \quad = 0.0729$$
$$I^A I^O = 2(0.27)(0.67) = 0.3618 \left.\right\} \quad \text{Type A} \quad = 0.4347$$

$$I^B I^B = \quad (0.06)^2 \quad = 0.0036$$
$$I^B I^O = 2(0.06)(0.67) = 0.0804 \left.\right\} \quad \text{Type B} \quad = 0.0840$$

$$I^O I^O = \quad (0.67)^2 \quad = 0.4489 \quad \text{Type O} \quad = 0.4489$$

$$I^A I^B = 2(0.27)(0.06) = 0.0324 \quad \text{Type AB} = 0.0324$$

Since I^A and I^B are both dominant to I^O, the expected frequency of blood group *phenotypes* is that shown on the right. (To reemphasize the point of Figure 14-5, note that the majority of A and B phenotypes are actually heterozygous for the I^O allele.)

X-linked Loci

Random mating for two X-linked alleles (H and h) is illustrated in Figure 14-7. The principles are the same as those considered previously, but male gametes carrying the X chromosome (panel (a)) must be distinguished from those carrying the Y chromosome (panel (b)). When the male gamete carries an X chromosome, the Punnett square is exactly the same as for the two-allele autosomal locus in Figure 14-4. However, since the male gamete carries an X chromosome, all the offspring in question will be female. Consequently, among females, the genotype frequencies will be

$$HH: p^2 \quad Hh: 2pq \quad hh: q^2$$

When the male gamete carries a Y chromosome, the outcome is quite different (Figure 14-7(b)). The offspring, all of which will be male, receive

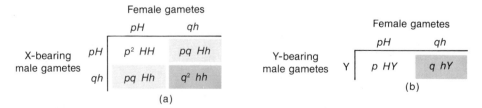

Figure 14-7 The results of random mating for an X-linked gene. (a) Genotype frequencies in females. (b) Genotype frequencies in males.

only one X chromosome, which comes from their mothers. Therefore, each male receives only one copy of each X-linked gene, and the genotype frequencies among males will be the same as the allele frequencies, namely

$$H: p \qquad h: q$$

If h is a rare recessive allele, then there will be many more males exhibiting the trait than females, because q^2 will be much smaller than q. However, the frequency of *heterozygous* females will be a little less than twice the frequency of affected males. These points can be illustrated with the familiar X-linked form of color blindness that diminishes the perception of green. Among Western Europeans, the frequency of this recessive allele is approximately $q = 0.05$. Therefore, the proportion of color-blind males is

$$q = 0.05$$

and the proportion of color-blind females is

$$q^2 = (0.05)^2 = 0.0025$$

This calculation explains why affected males are much more frequent than affected females. (Nongeneticists familiar with this form of color blindness often assume that the trait never occurs in females, which is incorrect.) For color blindness, the frequency of carrier females is $2pq$ (in which $p = 1 - q = 0.95$), or,

$$2pq = 2(0.05)(0.95) = 0.095$$

This calculation shows that the frequency of carrier females is indeed a little less than twice the frequency of affected males.

The Number of Generations Required to Reach Hardy-Weinberg Frequencies

For genotypes determined at autosomal loci, Hardy-Weinberg frequencies are attained in a single generation of random mating. For example, consider two separate populations having the following genotype frequencies:

	AA	Aa	aa
Population 1	0.64	0.32	0.04
Population 2	0.04	0.32	0.64

Each population obeys the Hardy-Weinberg rule, but the frequency of allele A is 0.8 in the first population and 0.2 in the second. If the populations were to mix together, the genotype frequencies would be the average, namely

$$0.34 \ AA \qquad 0.32 \ Aa \qquad 0.34 \ aa$$

These frequencies do not correspond to Hardy-Weinberg proportions. However, after one generation of random mating the genotype frequencies will be

$$0.25 \; AA \qquad 0.50 \; Aa \qquad 0.25 \; aa$$

These frequencies correspond to Hardy-Weinberg proportions with an allele frequency of 0.5 (the average of the allele frequencies in the ancestral populations). Similarly, if a population undergoes nonrandom mating for a number of generations, the Hardy-Weinberg frequencies will be restored in just one generation of random mating.

With an X-linked gene the occurrence of Hardy-Weinberg proportions requires that the allele frequencies be the same in males and females, and reaching equality requires more than one generation. The reason for the delay is that a male inherits its X chromosome from its mother, so the frequency of an X-linked allele in males equals the frequency of the same allele in females of the previous generation. This one-generation lag results in a gradual approach to equal allele frequencies in both sexes.

14.6 Population Subdivision

Most species that are spread over a wide geographic area do not form a single, large, random-mating unit, but are split up into local populations, within which individuals tend to find their mates. Although mating may be random within these local units, it does not extend across the entire range. The size of these local units will depend on the tendency of the organisms to disperse during their lifetime. For example, local populations of snails will occupy smaller areas than will local populations of eagles. The tendency of most organisms to cluster into local groups is known as **population subdivision.** The **population structure** of a species is a description of how it is subdivided into local populations. In this context, local populations are often referred to as **subpopulations.**

An inevitable consequence of population subdivision is that the subpopulations come to have different allele frequencies at many loci. The accumulation of differences in allele frequency in subdivided populations is called **genetic divergence** or **genetic differentiation.** As the differences accumulate, the subpopulations may become so different that they may be regarded as distinct races; or, if more different still, they might be regarded as distinct subspecies, receiving their own Latin name. Population geneticists are frequently confronted with samples of organisms from different subpopulations and are asked to judge their degree of genetic differentiation. For example, suppose snails are sampled from subpopulations on both sides of a river and the animals are characterized genetically by means of electrophoresis of protein extracts. The information so obtained will comprise a set of allele frequencies from each sample, which then must be converted into a statement about genetic divergence between the subpopulations. The situation may be complicated by the fact that a small amount of migration may occur between the subpopulations. Local populations are units within which organisms *usually*, but not invariably, find their mates. Because of this trickle of migration, new alleles that arise by mutation in one subpopulation will not remain unique to that subpopulation but will slowly spread to others, particularly to adjacent ones. However, when populations are subdivided, local populations will frequently be found to differ in their allele frequencies. Such differences can come about in many ways. Two important processes are random genetic drift and natural selection. **Random genetic drift** refers to chance changes in allele frequency that inevitably occur, particularly in small populations; **natural selection** refers to the differential ability of organisms to survive and reproduce in their particular environment. To a large extent, statements about genetic divergence are statements about the importance of these two processes in natural populations.

The Fixation Index (G)

One useful measure of genetic divergence, which is based on the proportion of heterozygous genotypes, is called the **fixation index.** It employs formulas familiar from the Hardy-Weinberg rule. The concept as it applies for a two-allele autosomal locus is illustrated in Figure 14-8, in which the two subpopulations to be compared are displayed at the upper corners of an inverted triangle. The two subpopulations have allele frequencies of $p = 0.3$ in one case and $p = 0.6$ in the other. When mating is random in both subpopulations, the Hardy-Weinberg rule can be used to obtain the genotype frequencies. The frequency of heterozygotes in the first subpopulation would

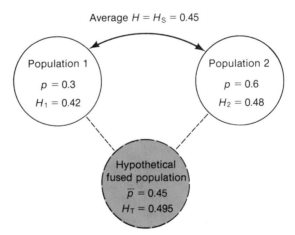

Figure 14-8 The concept of heterozygosity used in measuring genetic divergence between populations and in calculating the fixation index *G*. The average heterozygosity of a group of actual populations, H_S, represented here with two populations, is compared with the heterozygosity of a hypothetical population formed by fusion and random mating of the (two) populations, H_T.

equal $2(0.3)(0.7) = 0.42$, and in the second subpopulation it would be $2(0.6)(0.4) = 0.48$. The average of these values, symbolized H_S (in which the subscript S indicates that the average heterozygosity refers to the *Sub*populations), is

$$H_S = (0.42 + 0.48)/2 = 0.45$$

To assess how different the subpopulations are with respect to the alleles in question, the average heterozygosity can be compared with the heterozygosity that would be found if the populations were part of a single local population. That is, the average heterozygosity is compared with that of a hypothetical population formed by the fusion and random mating of the subpopulations. The hypothetical fused population is displayed at the point of the inverted triangle in Figure 14-8. Since we are (conceptually) fusing the subpopulations, it is clear that the allele frequency in the fused population will equal the average of that in the separate subpopulations. The average allele frequency ($\bar{p}$) is

$$\bar{p} = (0.3 + 0.6)/2 = 0.45$$

The heterozygosity in this hypothetical population, called the **total heterozygosity** and designated H_T, is calculated as

$$H_T = 2\bar{p}\bar{q} = 2(0.45)(0.55) = 0.495$$

Note that H_S is smaller than H_T, which is a consequence of the allele frequencies being different in the subpopulations. The **fixation index, G,** is defined as the proportionate reduction in H_S compared with H_T. That is,

$$G = \frac{H_T - H_S}{H_T} = \frac{0.495 - 0.45}{0.495} = 0.091 \qquad (14\text{-}4)$$

The fixation index is a quantitative index of the degree of genetic differentiation between the subpopulations, but some guidelines are necessary for its interpretation. The maximum possible range of G goes from 0 (when both subpopulations have equal frequencies of the alleles) to 1 (when each population is fixed for a different allele). In practice, observed values of G in natural populations rarely approach a value of 1. The following guidelines are often used to translate the quantitative measure G into a qualitative statement about the amount of genetic differentiation:

$0 < G < 0.05$: little genetic divergence
$0.05 < G < 0.15$: moderate genetic divergence
$0.15 < G < 0.25$: great genetic divergence
$G > 0.25$: very great genetic divergence

Therefore, in the example in Figure 14-8, $G = 0.091$ corresponds to moderate genetic differentiation between the subpopulations.

Figure 14-9 shows curves corresponding to the critical G values for various allele frequencies in two subpopulations. The qualitative levels of genetic divergence are indicated by shading. Use of Figure 14-9 may be illustrated with human blood group data. In the Kidd blood group, the frequency of the allele Jk^a in West African blacks is about 0.7, and in Caucasians (from Claxton, Georgia), it is about 0.5. From Figure 14-9, these frequencies correspond to genetic divergence judged to be "little" ($G = 0.04$ for this locus). On the other hand, for the Fy^a allele in the Duffy blood group, the allele frequencies are 0 in blacks in West Africa and 0.42 in whites in Georgia. These frequencies correspond to a degree of genetic divergence judged to be "very great" (for this locus, $G = 0.26$). Comparison of the Jk and Fy loci illustrates that single loci are inadequate to judge the degree of genetic divergence among subpopulations, because the G values of individual loci can be far from uniform. Adequate judgment of genetic differentiation must be based on the average value of G across a large number of loci; the larger the number of loci and the more representative they are of the genome, the more reliable the judgment. Among 11 blood group and enzyme loci studied in West African blacks and Georgia whites, the average G is 0.07, which corresponds to a "moderate" amount of genetic divergence between the subpopulations.

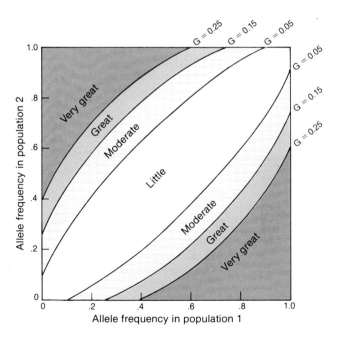

Figure 14-9 Guidelines for interpreting the fixation index (G) as an index of genetic divergence between two populations.

14.7 Inbreeding

Inbreeding refers to mating between relatives, such as first cousins. The principal consequence of inbreeding is that the heterozygosity is reduced; that is, heterozygosity is smaller than that with random mating. This effect is seen most dramatically in the case of repeated self-fertilization, such as that occurring naturally in certain plants. Consider a hypothetical population consisting exclusively of Aa heterozygotes. With self-fertilization, each plant would produce offspring in the proportions 1/4 AA, 1/2 Aa, and 1/4 aa. Thus, one generation of self-fertilization reduces the proportion of heterozygotes from 1 to 1/2. In the second generation, only the heterozygous plants can again produce heterozygous offspring, but only half of their offspring will be heterozygous. Heterozygosity will therefore be reduced to 1/4 of what it was originally. Three generations of self-fertilization reduce the heterozygosity to 1/4 × 1/2 = 1/8, and so on. The next section shows how the reduction in heterozygosity owing to inbreeding may be measured.

The Inbreeding Coefficient (F)

Repeated self-fertilization is a particularly intense form of inbreeding, but weaker forms of inbreeding are qualitatively similar in that they also lead to a reduction in heterozygosity. A convenient measure of the effect of inbreeding is based on the reduction in heterozygosity. If H_I represents the heterozygosity of an inbred individual, then H_I may be interpreted as the genotype frequency of heterozygotes in a population of inbred individuals. The most widely used measure of inbreeding is called the **inbreeding coefficient, F,** and it is defined as the proportionate reduction in H_I compared with the value of $2pq$ that would be expected to result from random mating, namely

$$F = (2pq - H_I)/2pq \qquad (14\text{-}5)$$

in which F is the inbreeding coefficient.

Equation 14-5 can be rearranged as

$$H_I = 2pq(1 - F)$$

The value of H_I is the frequency of heterozygous genotypes in a population of inbred individuals. It is smaller by the amount $2pqF$ than the heterozygosity of a population in which mating is random. In an inbred population, this deficiency of heterozygotes is reflected as an excess of homozygotes relative to what would result from random mating, and the $2pqF$ deficiency in heterozygotes is apportioned equally as an excess among the homozygotes. That is, the frequencies of genotypes AA and aa in the inbred population will be

$$AA: p^2 + pqF$$
$$aa: q^2 + pqF$$

The frequencies of genotypes among inbreds can also be written as

$$
\left.
\begin{array}{ll}
AA: & p^2(1 - F) + pF \\
Aa: & 2pq(1 - F) \\
aa: & q^2(1 - F) + qF
\end{array}
\right\}
\qquad (14\text{-}6)
$$

These frequencies amount to a modification of the Hardy-Weinberg rule in order to take inbreeding into account. When $F = 0$ (no inbreeding), the genotype frequencies are the same as those given in the Hardy-Weinberg rule in Equation 14-1, namely, p^2, $2pq$, and q^2. At the other extreme, when $F = 1$ (complete inbreeding), the inbred population will consist entirely of AA and aa individuals occurring with the frequencies p and q, respectively.

As an illustration, recall the discussion in Section 14.1 of the phosphoglucomutase locus Pgm in *Phlox cuspidata*, the genotype frequencies of which were not at all in agreement with the Hardy-Weinberg rule. This species can self-fertilize, and about 78 percent of the plants do so. It can be shown that the inbreeding coefficient corresponding to this amount of self-fertilization is $F = 0.64$, and the expected genotype frequencies must be modified to take this inbreeding into account. The frequencies of alleles Pgm^a and Pgm^b were 0.51 and 0.49, respectively, so by Equation 14-6, for inbreeding, the expected genotype frequencies will be

$$
\begin{array}{lll}
Pgm^aPgm^a: & (0.51)^2(1 - 0.64) + (0.51)(0.64) & = 0.420 \\
Pgm^aPgm^b: & 2(0.51)(0.49)(1 - 0.64) & = 0.180 \\
Pgm^bPgm^b: & (0.49)^2(1 - 0.64) + (0.49)(0.64) & = 0.400
\end{array}
$$

Since the sample size was 35, the observed and expected numbers of the three genotypes are

$$
\begin{array}{lll}
Pgm^aPgm^a: & 15, \text{ versus } 0.420 \times 35 & = 14.7 \\
Pgm^aPgm^b: & 6, \text{ versus } 0.180 \times 35 & = 6.3 \\
Pgm^bPgm^b: & 14, \text{ versus } 0.400 \times 35 & = 14.0
\end{array}
$$

The agreement is excellent. (The value of χ^2 is 0.02, and it still has one degree of freedom because only the allele frequency was estimated from the data. The associated probability is about 0.90.) Even if an investigator lacked prior information about the reproductive habits of *P. cuspidata*, the large deficiency of heterozygotes compared with what would result from random mating would suggest that the population had been undergoing some form of inbreeding. In other words, the genotype frequencies themselves contain information about the mating system, and χ^2 tests can be used to extract it. Indeed, if the inbreeding coefficient in the population were unknown, it could have been estimated from the data. That is, the observed frequency of heterozygotes in the population is $6/35 = 0.17 = 2pq(1 - F)$. However, both p and q are known, so that

$$0.17 = 2(0.51)(0.49)(1 - F)$$

from which

$$F = 0.66$$

This value is very close to that estimated from the proportion of plants that undergo self-fertilization. One shortcoming of the method of using observed data to calculate the inbreeding coefficient of a population is that no χ^2 test is possible, because F represents another quantity estimated from the data, for which one degree of freedom must be deducted, in addition to the deduction that is necessary for having estimated p from the data. Consequently, both degrees of freedom are used up in estimating p and F, and none remain with which to test goodness of fit. (You can confirm this for yourself by calculating the expected frequencies of the genotypes, by using $F = 0.66$. The expected frequencies will be found to agree exactly with the observed frequencies, except perhaps for a small discrepancy due to round-off error.)

Identity by Descent

The approach to inbreeding based on genotype frequencies is useful for considering natural populations. However, in human genetics, as well as in animal and plant breeding, a geneticist may wish to calculate the inbreeding coefficient of a particular individual whose pedigree is known. This section introduces a concept that is particularly suited to pedigree calculations.

Figure 14-3 emphasized that the diagnostic criterion of inbreeding is a closed loop in the pedigree. Such a closed loop is often called a **path.** Figure 14-10(a) is another inbreeding pedigree in which individual I is the product

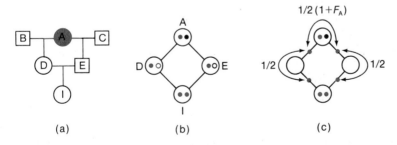

Figure 14-10 (a) An inbreeding pedigree in which individual I results from a mating between half siblings. (b) The same pedigree redrawn to show gametes in the closed loop leading to individual I. The dots and circles represent alleles of a particular gene. (c) Calculation of the probability that the alleles indicated are identical by descent.

of a mating between half sibs, and the path (closed loop) through the common ancestor A is highlighted. A streamlined version of this pedigree useful for inbreeding calculations is depicted in panel (b); individuals of both sexes are shown as circles, diagonal lines trace alleles transmitted from parent to offspring, and ancestors not contributing to the inbreeding of I (B and C) are left out altogether. The contrasting dots inside the circles represent particular alleles present in the corresponding individuals, and the depicted pattern of inheritance of the alleles represents the principal consequence of inbreeding and illustrates why closed loops are important. Specifically, individuals D and E are shown as both having received the colored allele from A along with a homologous allele (open circles) received from B and C, respectively. The colored alleles in D and E have a very special relationship: they both originate by replication of the colored allele in A, and therefore are said to be **identical by descent.** In the interval of the relatively few generations involved in most pedigrees, alleles that are identical by descent can be assumed to be identical in nucleotide sequence, because in so few generations mutation can usually be ignored. Individual I is shown as having received the colored allele from both D and E and, since the alleles in question are identical by descent, I must be homozygous for the allele inherited from the common ancestor. Without the closed loop in the pedigree, identity by descent of the alleles in I would be much less probable. The concept of identity by descent provides an alternative definition of the inbreeding coefficient, which can be shown to be completely equivalent to the definition in terms of heterozygosity given earlier. Specifically, *the inbreeding coefficient of an individual is the probability that the individual carries alleles that are identical by descent.*

Calculation of F from Pedigrees

Figure 14-10(c) illustrates how the probability of identity by descent can be calculated in a pedigree, and the alleles are drawn on the diagonal lines to emphasize their pattern of transmission. The double-headed arrows indicate which pairs of alleles must be identical by descent in order for I to have alleles that are identical by descent, and the number near each arrow is the associated probability for each pair. The values of 1/2 represent the probability that a diploid individual will transmit the allele inherited from a particular parent; this value must be 1/2 because of Mendelian segregation. The term $(1/2)(1 + F_A)$ can be explained by looking ahead for a moment to Figure 14-11, which shows four possible pairs of gametes that A can transmit, each of these configurations being equally likely. In cases (a) and (b) the transmitted alleles are certainly identical by descent because they are the same allele (colored or black). In cases (c) and (d) the transmitted alleles have contrasting colors, but they can still be identical by descent if A himself results from inbreeding. That is, the transmitted alleles could be identical by descent by virtue of a closed loop occurring in the pedigree of A. However, the probability that A carries alleles that are identical by descent because of inbreeding in his own ancestry is, by definition, the inbreeding coefficient

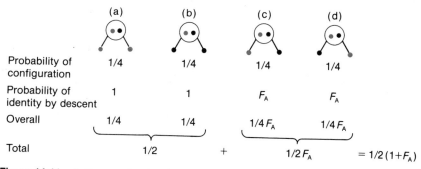

	(a)	(b)	(c)	(d)
Probability of configuration	1/4	1/4	1/4	1/4
Probability of identity by descent	1	1	F_A	F_A
Overall	1/4	1/4	$1/4\,F_A$	$1/4\,F_A$

Total   1/2   + $1/2\,F_A$ $= 1/2\,(1+F_A)$

Figure 14-11 A diagram showing that two different gametes from the same individual (A) will carry alleles at a locus that are identical by descent is with a probability of $(1/2)(1 + F_A)$.

of A, and this is designated F_A. Altogether, the probability that A transmits two alleles that are identical by descent is the sum of the probabilities of the four eventualities in Figure 14-11, and this sum equals $(1/2)(1 + F_A)$. Matters are somewhat simpler when A is not itself inbred, because in that case $F_A = 0$ and the corresponding probability is simply 1/2.

Returning now to Figure 14-10(c), the alleles in I can be identical by descent only if all three identities indicated by the loops are true, and this probability is the product of the three numbers shown. That is, the inbreeding coefficient of I (F_I) is

$$F_I = (1/2) \times (1/2)(1 + F_A) \times (1/2) = (1/2)^3(1 + F_A)$$

For a more remote common ancestor the formula becomes

$$F_I = (1/2)^n(1 + F_A) \tag{14-7}$$

in which n is the number of individuals in the path traced from one of the parents of I back through the common ancestor and forward again to the other parent of I.

Most pedigrees of interest will be more complex than the one in Figure 14-10, but the inbreeding coefficient of any individual can still be calculated by using Equation 14-7. Application to complex pedigrees requires several additional rules:

1. If there are several distinct paths that can be traced through a common ancestor, then each path provides a contribution to the inbreeding coefficient calculated as in Equation 14-7.

2. If there are several common ancestors, then each is considered in turn in order to calculate its separate contribution to the inbreeding.

3. The overall inbreeding coefficient is then calculated as the sum of all these separate contributions of each common ancestor.

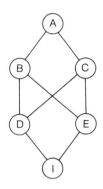

Figure 14-12 A complex pedigree in which the inbreeding of individual I results from several closed loops (paths).

Thus, the extension of Equation 14-7 to complex pedigrees may be written as

$$F_I = \Sigma(1/2)^n(1 + F_A) \qquad (14\text{-}8)$$

in which Σ means the summation of all paths through all common ancestors.

Application of Equation 14-8 can be illustrated in the complex pedigree in Figure 14-12. Each path and its contribution to the inbreeding coefficient of I is shown below (the bold letter indicates the common ancestor):

E**B**D	$(1/2)^3(1 + F_B)$
E**C**D	$(1/2)^3(1 + F_C)$
EC**A**BD	$(1/2)^5(1 + F_A)$
EB**A**CD	$(1/2)^5(1 + F_A)$

Then, the inbreeding coefficient of I is the sum of these numbers. If none of the common ancestors is inbred (if $F_A = F_B = F_C = 0$), then

$$F_I = (1/2)^3 + (1/2)^3 + (1/2)^5 + (1/2)^5 = 5/16$$

The value 5/16 is to be interpreted to mean that the probability that individual I will be heterozygous at any particular locus is less by the proportion 5/16 than the probability that a noninbred individual will be heterozygous at the same locus. Equivalently, the 5/16 could be interpreted to mean that individual I is heterozygous at 5/16 fewer loci in the whole genome than a noninbred individual.

Regular Systems of Inbreeding

In animal and plant breeding, the breeder sometimes practices inbreeding of the same type generation after generation; examples of this are repeated self-fertilization in plants and repeated sib mating in animals. A program of repeated inbreeding is called a **regular system** of inbreeding, and it is possible to derive formulas for F after inbreeding has been carried out for any specified number of generations. Calculation of F can be carried out directly from the pedigree, using the rules already discussed. For example, with repeated self-fertilization in a population starting with $F = 0$, the inbreeding coefficient increases in successive generations according to the sequence $F = 1/2, 3/4, 7/8, 15/16, 31/32, 63/64$, and so forth. Clearly, self-fertilization leads to a rapid increase in F. With such intense inbreeding, every individual in the population will soon become homozygous at virtually every locus. Less intense regular systems of mating lead to slower increases in the inbreeding coefficient, but eventually the accumulated inbreeding may become substantial. Several examples of regular systems used in animal and plant breeding are illustrated in Figure 14-13. In all of these cases, the effects of inbreeding can be undone by just one generation of random mating, which will restore Hardy-Weinberg frequencies.

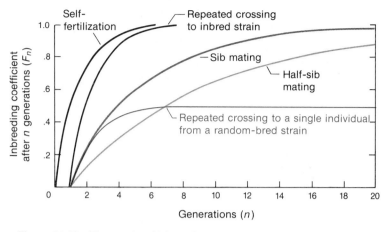

Figure 14-13 The results of inbreeding repeated over many generations.

14.8 Effects of Inbreeding

The effects of inbreeding vary according to the normal mating system of an organism. At one extreme, in regularly self-fertilizing plants, inbreeding is already so intense and the organisms so highly homozygous that additional inbreeding has virtually no effect. However, *in most organisms inbreeding is harmful, and much of the effect is due to rare recessive alleles that would not otherwise become homozygous*. Several examples follow.

Chromosomal Homozygotes in Drosophila

Inbreeding effects are conveniently studied in *Drosophila*, in which appropriate mating schemes can be used to bring about identity by descent of an entire chromosome. One such mating scheme is outlined in Figure 14-14; it is called the **Cy/Pm (Curly-Plum) technique** because of one of the strains used. The first mating (panel (a)) uses a Cy/Pm female from a laboratory strain and a single wildtype male captured, for example, on ripe fruit in an orchard. Cy and Pm are designations of particular forms of chromosome 2 carrying multiple inversions to prevent crossing over as well as carrying a dominant-marker gene—Curly wings in the case of Cy and Plum-colored eyes in the case of Pm. The homologous chromosomes in the male are distinguished as $+$ and $+'$. From this mating, the offspring shown in the Punnett square in panel (a) result. Then, a single $Cy/+$ male is selected and crossed to Cy/Pm females, as shown in panel (b). This mating serves to replicate the $+$ chromosome so that all the $Cy/+$ offspring carry $+$ chromosomes that are identical by descent. (The Cy/Cy genotype is shaded to indicate that this homozygous genotype is lethal.) Males and females of the $Cy/+$ genotype from panel (b) are then mated brother to sister as shown in panel (c). Since Cy/Cy is lethal, two

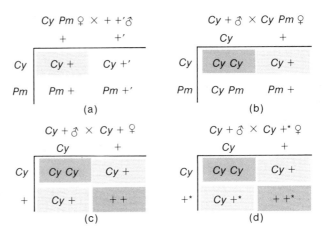

Figure 14-14 A mating scheme in *Drosophila* that results in identity by descent (and therefore homozygosity) of an entire chromosome. (a) The original mating between a *Cy/Pm* female and a wildtype male (presumed to carry two different wildtype chromosomes). (b) A backcross of a single male progeny to *Cy/Pm* females in order to replicate the + chromosome. (c) Brother-sister mating resulting in some progeny that are homozygous for a single wildtype chromosome. (d) Alternate mating resulting in flies that are heterozygous for two different wildtype chromosomes.

phenotypes of offspring survive: Curly wings ($Cy/+$) and normal straight wings ($+/+$). If the two genotypes are equally able to survive, the expected proportions will be 2/3 Curly and 1/3 straight. However, if the + chromosome carries one or more recessive genes that reduce the ability of homozygotes to survive, the proportion of straight-wing survivors will fall below 1/3. Alternatively, flies carrying two *different* wildtype chromosomes (designated + and +*) can be mated, as shown in panel (d). If the two chromosomes carry allelic mutations that reduce survival, then the proportion of $+/+*$ survivors will be less than 1/3; otherwise, the proportion of $+/+*$ survivors will be 1/3. The ability of a genotype to survive, relative to some other genotype (in this case $Cy/+$), is called the **viability** of the genotype. In the mating scheme in Figure 14-14, the $+/+$ individuals in panel (c) are homozygous for + chromosomes that are identical by descent, which is equivalent to an inbreeding coefficient of $F = 1$ for each gene in the chromosome. Thus, the mating scheme allows the effects of extreme inbreeding on viability to be evaluated.

Data from 691 chromosomes manipulated as in Figure 14-14 are summarized in Figure 14-15. The black curve represents the viability of $+/+$ homozygotes, and the harmful effects of inbreeding are clear. Complete inbreeding of this chromosome has had a disastrous effect—indeed, 38 percent of the chromosomes are *lethal* when homozygous. Further genetic studies indicate that the lethality is almost always due to single recessive mutations, though differing from chromosome to chromosome. Even those homozygotes that are not lethal have a depressed viability: note that the peak of the black curve is at 0.267 rather than the expected 0.333. Evidently, *Drosophila* chro-

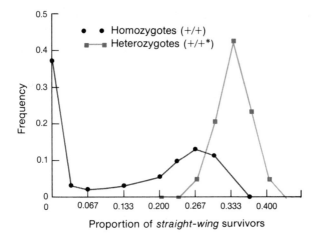

Figure 14-15 The distribution of viability that has been found among *Drosophila* homozygous or heterozygous for entire chromosomes isolated from a natural population by means of the *Cy/Pm* procedure shown in Figure 14-14. (Based on data from T. Mukai and O. Yamaguchi. 1974. *Genetics* 76: 339–366.)

mosomes possess a great deal of genetic damage in the form of recessive alleles that cause lethality or reduce viability; these recessive alleles are normally hidden because each harmful allele is so rare that it will usually be present only in heterozygotes. The extreme inbreeding in Figure 14-15 makes the effects of these harmful recessive alleles readily apparent.

The rarity of the harmful recessive alleles is indicated by the colored curve in Figure 14-15, which represents $+/+*$ heterozygotes. If a particular harmful recessive were common in the population, many $+$ and $+*$ chromosomes would carry the same allele, and a plot of the viability of the corresponding $+/+*$ heterozygote would be displaced to the left. However, the actual distribution of viability among heterozygotes is smooth and symmetrical and is centered at the expected value of 0.333. This observation indicates that the pattern of viability observed among homozygotes results from recessive mutations at many different loci scattered throughout the chromosome, and that each of these mutations is individually quite rare.

Effects of Inbreeding in Humans

Among humans, inbreeding never approaches the extreme situation that can be created in *Drosophila*, as outlined in Figure 14-14; however, the effects of inbreeding on rare, harmful recessives is still pronounced. *The risk of recessive genetic disorders is increased in consanguineous matings*. This point can be illustrated with autosomal-recessive albinism among American whites. The frequency of the condition among offspring of nonconsanguineous matings is approximately 1 in 20,000; however, among offspring of first-cousin matings, the frequency is approximately 1 in 2000. The **relative risk** resulting from inbreeding is defined as the ratio between the proportion of affected individuals among inbred offspring and the proportion among noninbred offspring; thus the relative risk of albinism resulting from first-cousin mating is 10. Such an increased risk of rare homozygous recessives is characteristic of inbreeding. Indeed, an increased risk of a trait associated with consanguinity is often taken as evidence of recessive inheritance. The reason for the increased risk may be appreciated by comparing the genotype frequency of homozygous recessives in Equation 14-6 (for inbreeding) and Equation 14-1 (for random mating). The most common form of inbreeding in humans is mating between first cousins, for which $F = 1/16 (= 0.062)$ among the offspring. Therefore, the frequency of homozygous recessives produced by first-cousin mating will be

$$q^2(1 - 0.062) + q(0.062)$$

With nonconsanguineous mating, the frequency of homozygous recessives will be simply q^2. The relative risk resulting from inbreeding is defined as the ratio of these quantities, which equals

$$0.938 + 0.062/q \tag{14-9}$$

With albinism, $q = 0.0068$ (approximately), and the relative risk due to inbreeding is

$$0.938 + 0.062/0.0068 = 10$$

Note the agreement with the data cited at the outset of the discussion. The relative risk in Equation 14-9 depends on the recessive-allele frequency q. Table 14-3 provides several additional examples. The case of $q = 0.01$ corresponds approximately to phenylketonuria in American whites, and $q = 0.001$ applies approximately to Tay-Sàchs disease in non-Jewish populations. Note in Table 14-3 that the relative risk of homozygosity increases as the harmful recessive becomes rare, so that in the case of allele frequency of 0.001, an offspring of first cousins has more than 60 times the risk of becoming homozygous than an offspring of nonrelatives. This increased risk is a principal consequence of inbreeding.

Table 14-3 Relative risk of homozygous recessives among the offspring of first cousins

q	Relative risk
0.1	1.54
0.05	2.14
0.01	6.94
0.005	12.94
0.001	60.94

1. Many enzymes are dimers formed by the random aggregation of two polypeptide chains of the same type. Suppose that a polypeptide occurs in two electrophoretically distinct forms (F and S) corresponding to two alleles (*F* and *S*) at the structural gene locus. How many enzyme bands would you expect to find in an *FF* homozygote, an *SS* homozygote, and an *FS* heterozygote?

2. Allozyme phenotypes of alcohol dehydrogenase in the flowering plant *Phlox drummondii* are determined by codominant alleles at a single locus. In one study of 35 individuals, the following data were obtained:

Genotype	*aa*	*ab*	*bb*	*bc*	*cc*	*ac*
Number	2	5	12	10	5	1

Calculate the frequencies of the alleles *a*, *b*, and *c*.

3. Among 35 individuals of the flowering plant *Phlox roemariana*, the following genotypes were observed at a locus determining electrophoretic forms of the enzyme phosphoglucose isomerase: 2 *aa*, 13 *ab*, 20 *bb*.
 (a) What are the frequencies of the alleles *a* and *b*?
 (b) Assuming random mating, what are the expected numbers of the genotypes?

4. Hartnup disease is an autosomal-recessive disorder of intestinal and renal transport of amino acids. In Massachusetts, the frequency of affected newborn infants is 1 in 14,219. Assuming random mating, what is the frequency of heterozygotes?

5. In certain grasses, the ability to grow in soil contaminated with the toxic metal nickel is determined by a dominant allele.
 (a) If 60 percent of the seeds in a random-mating population are able to germinate in contaminated soil, what is the frequency of the resistance allele?
 (b) Among plants that germinate, what proportion will be homozygous?

6. Among Caucasians, the frequency of the Rh^+ blood type is approximately 0.85. With random mating, what is the frequency of Rh-incompatible matings (that is, Rh^+ male × Rh^- female)?

7. A randomly mating population of dairy cattle carries an autosomal-recessive allele causing dwarfism. If the frequency of dwarf calves is 10 percent, what is the frequency of heterozygous carriers of the allele in the entire herd? What is the frequency of heterozygotes among nondwarf individuals?

8. Consider an autosomal locus with two alleles in a random mating population. What is the probability that a heterozygous female will have a heterozygous offspring?

9. Two generations of random mating are required to attain Hardy-Weinberg frequencies for genotypes determined by an autosomal locus with two alleles in which the allele frequencies differ between the sexes. The first generation results in equalization of allele frequency in the sexes, and the second generation results in the Hardy-Weinberg frequencies. To verify this principle, consider a population in which the genotype frequencies are as follows:

	AA	Aa	aa
Males	0.64	0.32	0.04
Females	0.04	0.32	0.64

 (a) With random mating, what are the genotype frequencies in males and females in the next generation and are these in Hardy-Weinberg proportions?
 (b) What are the genotype frequencies in males and females after one additional generation of random mating and are these in Hardy-Weinberg proportions?

10. Certain species of *Trifolium* (clover) have matings governed by self-sterility alleles in which pollen that carries a particular allele will not germinate on a style that carries the same allele.
 (a) Would the genotypes at a self-sterility locus be expected to occur in Hardy-Weinberg proportions? Why or why not?
 (b) What is the minimum number of self-sterility alleles required for a *Trifolium* population to be able to reproduce by mandatory cross-fertilization?

11. In Caucasians, the M-shaped hairline that recedes with age, sometimes to nearly complete baldness, is due to an allele that is dominant in males but recessive in females. The frequency of the baldness allele is approximately 0.3. Assuming random mating, what phenotype frequencies are expected in males and females?

12. In a population of *Drosophila*, an X-linked recessive allele causing yellow body color occurs in genotypes at frequencies typical of random mating; the frequency of the recessive allele is 0.20. Among 1000 females and 1000 males, what are the expected numbers of the yellow and wildtype phenotypes?

13. In a Pygmy group in Central Africa, the frequencies of alleles determining the ABO blood groups were estimated as 0.74 for I^O, 0.16 for I^A, and 0.10 for I^B. Assuming random mating, what are the expected frequencies of ABO genotypes and phenotypes?

14. Genotypes formed from two alleles (M and S) of the structural gene for the enzyme phosphoglucose isomerase were studied among 134 individuals of the Mediterranean marine gastropod *Monodonta turbinata*. Observed numbers of individuals were 74 MM, 47 MS, and 13 SS. Assuming random mating, what are the expected genotype frequencies? Use a χ^2 test to determine whether the genotype frequencies in this population are consistent with Hardy-Weinberg proportions.

15. Genotypes formed from two alleles (L and M) of the structural gene for the enzyme lactate dehydrogenase were studied among 340 individuals of the marsh frog *Rana ridibunda* collected from widely separated geographic locations in Israel. Observed numbers of individuals were 133 LL, 135 LM, and 72 MM. Assuming random mating, what are the expected numbers of the genotypes? Use a χ^2 test to determine whether the observations are consistent with Hardy-Weinberg proportions.

16. Galactosemia is an autosomal-recessive condition associated with liver enlargement, cataracts, and mental retardation. Among the offspring of unrelated individuals the frequency of galactosemia is 8.5×10^{-6}
    (a) What is the expected frequency among the offspring of first cousins ($F = 1/16$) and among the offspring of second cousins ($F = 1/64$)?
    (b) How much does consanguineous mating increase the risk of galactosemia?

17. If, in a population, 1 percent of the matings occur between first cousins ($F = 1/16$) and the rest occur between nonrelatives, what is the average inbreeding coefficient among individuals in the population?

18. In the pedigree below, individual I is the offspring of a mating between half first cousins. Assuming that the inbreeding coefficient of individual A is 0, what is the inbreeding coefficient of individual I? What is the inbreeding coefficient of I assuming $F_A = 1/16$?

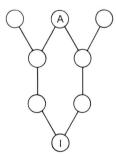

19. What is the inbreeding coefficient of individual I in the pedigree below, assuming that none of the ancestors is inbred?

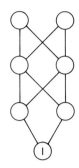

20. The occurrence of self-fertilization in the annual plant *Phlox cuspidata* results in an inbreeding coefficient of $F = 0.66$.
    (a) What frequencies of the genotypes for the enzyme phosphoglucose isomerase would be expected in a population with two alleles, a and b, at respective frequencies of 0.43 and 0.57?
    (b) What frequencies of genotypes would be expected with random mating?

21. In a Texas population of *Phlox cuspidata* the heterozygosity of the alcohol dehydrogenase locus was found to be 0.12, and the frequencies of two alleles were 0.8 and 0.2. What is the estimated value of F at this locus in this population?

22. In a population of marsh frogs (*Rana ridibunda*) in Southern Israel, an allele at the lactate dehydrogenase locus was found to have a frequency of 0.87, and in a population in Northern Israel the same allele was found to have a frequency of 0.24.

    (a) Assuming two alleles at the locus, what is the degree of genetic divergence between the populations as assessed by the fixation index G?
    (b) Qualitatively, how much divergence does this value of G represent? Use the criteria in Figure 14-9.

23. The same populations described in Problem 22 had allele frequencies of 0.59 and 0.97, respectively, for the most common allele at the acetaldehyde oxidase locus (at which it is assumed that there are two alleles). What value of G corresponds to these allele frequencies?

C H A P T E R 15

Genetics and Evolution

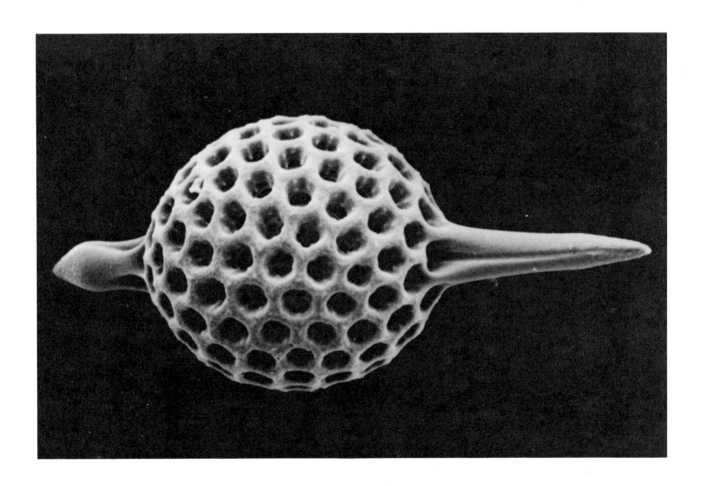

Descent with modification was emphasized in Chapter 1 as being one of the principal themes of genetics. It summarizes the observation that offspring resemble their parents without being precise replicas. Offspring phenotypes may differ from parental phenotypes for many reasons, but among the most important are mutation, segregation, recombination, and the effects of environment. These processes underlie much of the phenotypic variation observed in the offspring of virtually every mating.

Descent with modification also occurs with entire populations. As noted in connection with the Hardy-Weinberg rule in Chapter 14, allele frequencies, and therefore genotype frequencies, do not change of their own accord. They will tend to remain the same generation after generation, and each progeny generation will therefore tend to resemble its parental generation. Counteracting this tendency are a number of processes that can change allele frequency and thereby lead to the genetic modification of progeny. In the long term, the most important of these modifying processes is natural selection, the process in which individuals best adapted to survive and reproduce in their environment thereby contribute more than an equal share of alleles to the next generation; when repeated over the course of many generations, a disproportionate contribution of alleles, even if small, will significantly increase the frequency of alleles responsible for the superior adaptations. Natural selection in populations is the modifying force recognized by Charles Darwin. It is thought to be the foremost mechanism leading populations to become progressively more adapted to their environment. Such progressive adaptation constitutes the process of evolution.

Natural selection is one reason that allele frequencies change in natural populations. Four principal causes of change in the frequency of alleles are usually distinguished:

Facing page: Scanning electron micrograph of a fossil radiolarian, several hundred million years old. (From G. Shih and R. Kessel. 1983. *Living Images.* Jones and Bartlett.)

1. **Mutation,** the origin of new genetic capabilities in populations by means of spontaneous heritable changes in genes.

2. **Migration,** the movement of individuals among subpopulations within a larger population.

3. **Natural selection,** resulting from the different abilities of organisms to survive and reproduce in their environments.

4. **Random genetic drift,** the random undirected changes in allele frequency that occur by chance in all populations, but particularly in small ones.

These four processes account for most or all of the changes in allele frequency that occur in populations. They form the basis of cumulative change in the genetic characteristics of populations, leading to descent with modification. Although the point has yet to be proved, most evolutionary biologists believe that these same processes, when carried out continuously over geological time, can also account for the formation of new species and higher taxonomic categories.

15.1 Mutation

Mutation is the ultimate source of genetic variation, as was shown in Chapter 10. It is an essential process in evolution, but it is a relatively weak force for changing allele frequencies, primarily because typical mutation rates are too low. Moreover, most newly arising mutations are harmful to the organism. (Harmfulness is certainly the rule for mutations that arise and can be detected in the laboratory.) Although some mutations may be **selectively neutral,** meaning that they do not affect the ability of the organism to survive and reproduce, only a very few mutations are favorable for the organism and contribute to evolution. The low mutation rates that are observed are thought to have evolved as a compromise for a mutation rate high enough to generate the favorable mutations that a species requires to continue evolving, but not so high that the species suffers too much genetic damage from the preponderance of harmful mutations.

Irreversible Mutation

The effect of mutation on change of allele frequency in a large population of *E. coli* is illustrated in Figure 15-1. The mutant allele in question is one that confers resistance to infection by the bacteriophage T5. The experiment in Figure 15-1 was carried out in the absence of the bacteriophage, conditions that eliminated selection for either T5-r (T5-resistant) or T5-s (sensitive) bacteria; hence, changes in allele frequency were determined entirely by mutation, and the frequencies of T5-r and T5-s phenotypes were estimated from samples of cells taken from the population at intervals. The figure shows

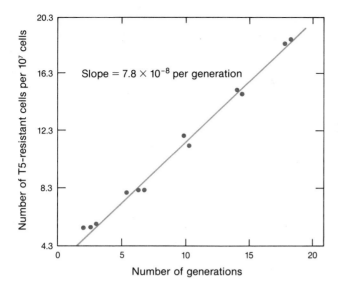

Figure 15-1 Accumulation of cells resistant to phage T5 resulting from recurrent mutation in a population of *Escherichia coli* growing in conditions that allow continuous growth at constant cell density. (After Kubitschek, H. E. 1970. *Introduction to Research with Continuous Cultures*, p. 33. Reprinted by permission of Prentice-Hall, Inc., Englewood Cliffs, NJ.)

that the frequency of T5-r cells increased linearly over the course of the experiment. The reason for the increase was that almost all of the cells in the initial population were T5-s; in each generation, some of these T5-s cells underwent spontaneous mutation to become T5-r, and thus T5-r cells accumulated in the population as time progressed. The T5-r cells are so rare in this experiment that reverse mutation from T5-r to T5-s may safely be neglected. For practical purposes, we may therefore regard the experiment in Figure 15-1 as involving **irreversible mutation**—that is, mutation occurring only in one direction (T5-s to T5-r).

A plot of the increase in T5-resistance with the number of generations is a straight line because T5-r cells are so rare. The linearity can be deduced from a simple equation for the frequency of T5-s cells in any generation. The frequencies of T5-s and T5-r cells in the population would be denoted p and q, respectively, with appropriate subscripts indicating the generation—p_0 and q_0 for the initial frequencies; p_1 and q_1 a generation later; and p_n and q_n in the nth generation. Since the mutation is irreversible, the frequency of T5-s cells in the nth generation will equal the proportion of T5-s bacterial cell lineages (lines of descent) that escaped mutation for n consecutive generations. All other lineages will have mutated to T5-r at some prior time. If mutation from T5-s to T5-r occurs at the rate μ per generation, then the probability that a particular T5-s lineage will escape mutation in each of n generations must equal $(1 - \mu)^n$. Consequently,

$$p_n = (1 - \mu)^n p_0 \tag{15-1}$$

Equation 15-1 is the key to understanding the linearity in Figure 15-1. To convert the equation to a more usable form, $(1 - \mu)^n$ may be replaced by $1 - n\mu$, an approximation that is valid when μ is very small, and $1 - q$ is substituted for p, yielding

$$1 - q_n = (1 - n\mu)(1 - q_0) = 1 - n\mu - q_0 + n\mu q_0$$

The term $n\mu q_0$ is the product of two small numbers (μ and q_0) and will be so small compared to the other terms that it can be deleted. Rearrangement of the equation yields

$$q_n = q_0 + n\mu \qquad (15-2)$$

This is the equation of a straight line, and it explains why the curve in Figure 15-1 is linear. Equation 15-2 is of value because the *slope* of the line is μ, the mutation rate. Thus, Figure 15-1 provides an estimate of the mutation rate of T5-s to T5-r of 7.8×10^{-8} mutations per gene per generation.

Equation 15-2 also implies that changes in allele frequency by mutation will be exceedingly slow. Starting with a population containing no T5-r cells, the predicted frequency after 1000 generations of mutation would be $(1000)(7.8 \times 10^{-8}) = 7.8 \times 10^{-5}$. That is, during the 1000 generations, the allele frequency of T5-s would decrease only from 1.0 to 0.999922. Experiments like the one in Figure 15-1 are a convenient and accurate way to measure mutation rates in microorganisms.

Populations of most organisms contain rare harmful alleles at many loci. Recurrent mutation will tend to increase the frequency of these harmful alleles, but this tendency is counteracted by selection against the alleles, which occurs as a result of their harmful effects.

Reversible Mutation

In interpreting Figure 15-1 only forward mutation (mutation from the normal allele to the mutant) was considered because the mutant T5-r cells were so rare. If, instead, neither of the two mutant alleles was rare in the population, then reverse mutation (mutation from the mutant allele back to the normal) would also have to be considered. When both mutant and normal strains are present at roughly comparable frequencies, then forward and reverse mutation are partially offsetting, and both processes must be taken into account.

An example in which reversible mutation plays a role is outlined in Figure 15-2 for the bacterium *Salmonella typhimurium*. The alleles are denoted A (representing one characteristic of the bacterial flagella) and a (representing the alternate characteristic). Alternation between A and a does not occur by conventional mutation, but rather the inversion and reinversion of a specific DNA sequence carrying a promoter at one end. As illustrated in Figure 15-2(b), when the invertible DNA sequence (arrow) is in one orientation, transcription to the right occurs and the A characteristic is expressed; when the sequence inverts, transcription to the left occurs and the a characteristic is expressed. This switching back and forth results from intramolecular re-combination between inverted-repeat DNA sequences that flank the element. The details need not concern us here except to note that the rates of re-combination leading to switching between A and a are much greater than conventional mutation rates. Nevertheless, the process in Figure 15-2 can formally be regarded as analogous to reversible mutation, with the "mutation rate" from A to a denoted by μ, and that from a to A as ν.

As noted in the previous section, the change in allele frequency when either allele is rare will approximate a straight line with a slope equal to the corresponding mutation rate. These lines are shown in color in Figure 15-2(a), and the slopes are indicated. (These mutation rates were estimated in separate experiments that are not shown.) After about 50 generations the actual curves depart from the straight lines because reverse mutation becomes important. Eventually, all populations, whatever their initial frequencies, evolve toward the same frequencies of A and a. These final frequencies represent the point at which forward and reverse mutation exactly balance each other; they are called the **equilibrium frequencies** of A and a. That is, at equilibrium the number of A alleles lost by $A \rightarrow a$ mutation is exactly offset by the number of new A alleles gained by $a \rightarrow A$ mutation. Moreover, it can be shown that the equilibrium frequencies correspond to the point of intersection of the straight lines, and at this intersection

$$\hat{p} = \nu/(\mu + \nu) \tag{15-3}$$

in which $\hat{p}$ represents the equilibrium frequency of the A allele. (In population

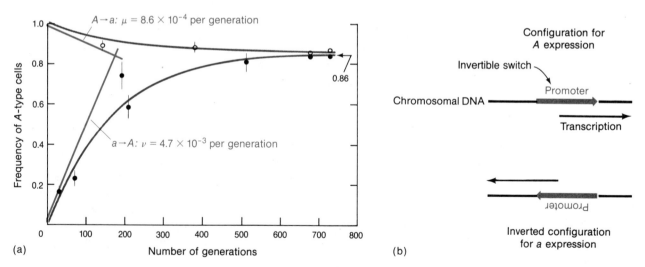

Figure 15-2 Production of two types of flagella (A or a) by cells of the bacterium *Salmonella typhimurium*. These cells are able to produce one or the other type, depending on the orientation of a promoter sequence in the DNA; inversion and reinversion of the promoter is formally equivalent to reversible mutation. Panel (a) shows the equilibrium resulting from alternation between A and a cellular types. Panel (b) shows the orientation of the promoter sequence responsible for the switching.

genetics, allele frequencies at equilibrium are conventionally represented with carets.) Equation 15-3 says that the allele frequency of A at equilibrium equals the reverse mutation rate divided by the sum of the mutation rates; the reverse mutation rate is in the numerator because it produces the new A alleles. For a, the equilibrium allele frequency $\hat{q}$ is determined by $\hat{q} = 1 - \hat{p}$.

In Figure 15-2(a), the rate of mutation for $A \rightarrow a$ is $\mu = 8.6 \times 10^{-4}$, and the rate of mutation for $a \rightarrow A$ is $\nu = 4.7 \times 10^{-3}$. The theoretical equilibrium frequency of A can be obtained from Equation 15-3 as

$$\hat{p} = 4.7 \times 10^{-3}/(8.6 \times 10^{-4} + 4.7 \times 10^{-3}) = 0.845$$

which is very close to the equilibrium frequency of 0.86 actually observed. However, attainment of the equilibrium requires about 700 generations, even in this case in which the transitions between A and a occur at an extraordinary rate because they involve recombination. If we were dealing with the much lower rates of conventional mutation, attainment of the equilibrium might require 7000 or even 70,000 generations. These values emphasize the fact that the change in allele frequency that is due to mutation ordinarily occurs through a long, slow process.

15.2 Migration

In Chapter 14 it was pointed out that most natural populations have some type of population structure consisting of subpopulations within which individuals usually find their mates, and between which migration occurs. Quantities such as the fixation index G may be used to detect such population subdivision and to assess the degree of genetic differentiation among the subpopulations. Patterns of change in allele frequency in subdivided populations can be very complex. On the one hand, allele frequencies in each subpopulation will tend to be altered by local forces, notably selection and random genetic drift; on the other hand, migration from other subpopulations, particularly from adjacent ones, will partially offset these local effects.

Migration

The overall effects of migration in subdivided populations will be discussed later in this chapter in the context of random genetic drift. In this section we focus on a simpler situation consisting of two subpopulations with migration occurring in only one direction. Figure 15-3 illustrates such a case, in which the circles represent the populations and the dots represent individual alleles of a gene within the populations. The figure shows two populations: (1) a source population (alleles shown as red dots), which is the population from which the migrant individuals derive, and (2) a population of interest, containing several migrant alleles (red dots), which have been introduced by migration from the source population.

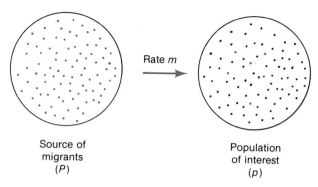

Figure 15-3 Model of one-way migration.

The situation in Figure 15-3 represents **one-way migration.** The actual mechanism of migration may differ from case to case. For example, a female residing in the population of interest may be impregnated by a male from the source population, leading to the introduction of mutant alleles from the father. The **migration rate m** corresponds to the proportion of alleles in the population of interest that are replaced with migrant alleles in each generation. This definition discloses a similarity between mutation and migration. In both cases, an allele in the population of interest is replaced by something else, either by a spontaneous mutation or by an allele from the source population. The symbols P and p in Figure 15-3 represent the frequency of an allele A in the source population and the population of interest, respectively.

The model in Figure 15-3 may seem artificially simple, but with suitable data it can be used to estimate actual migration rates. The type of data required are shown in Table 15-1, which reflects the migration of alleles among human races. Since persons of mixed black and white ancestry in the United States are usually regarded in society as blacks, the racial admixture may be regarded as one-way migration from whites to blacks. The data pertain to allele frequencies at various loci in three populations: (1) blacks from

Table 15-1 Estimation of migration rate

Locus	Allele	Allele frequency Claxton blacks	Claxton whites	West African blacks	Approximate value of m
MN blood group	M	0.484	0.507	0.474	0.03
Ss blood group	S	0.157	0.279	0.172	−0.01
Duffy blood group	Fy^a	0.045	0.422	0	0.01
Kidd Blood group	Jk^a	0.743	0.536	0.693	−0.03
Kell blood group	Js^a	0.123	0.002	0.117	−0.01
G6PD-enzyme deficiency	$G6PD^-$	0.118	0	0.176	0.03
Sickle-cell hemoglobin	β^s	0.043	0	0.090	0.05

Claxton, Georgia, which is the population of interest, (2) whites from Claxton, a population similar in allele frequencies to the source population and therefore considered as the source population, and (3) blacks from West Africa, assumed to be similar in allele frequency to the ancestors of Claxton blacks brought to this country during the slave trade.

Since the first blacks were brought to Georgia about ten generations ago, the allele frequencies in Claxton blacks represent the outcome of ten generations of one-way migration between a source population having allele frequencies similar to Claxton whites and a population of interest originally having allele frequencies similar to West African blacks. The migration rates necessary to account for the allele frequencies in Claxton blacks are given in the righthand column, and they have been estimated by using the model in Figure 15-3. The principal implication of this model is that the difference in allele frequency between the two populations will decrease by the amount m per generation,

$$p_n - P = (p_0 - P)(1 - m)^n \qquad (15\text{-}4)$$

in which p_0 and p_n represent the allele frequency in the population of interest in the original (zeroth) and nth generation. To apply Equation 15-4 to the data in Table 15-1 we set

$n = 10$, the number of generations of one-way migration;

p_0 = the original allele frequency in Claxton blacks, presumed equal to that in present-day West Africans;

P = the allele frequency in the source population, presumed equal to that in present-day Claxton whites; and

$p_n = p_{10}$ = the allele frequency in the population of interest after ten generations, equal to the frequency in present-day Claxton blacks.

With these substitutions in Equation 15-4, we can solve for the migration rate m for each locus in turn. In the present application, m is small enough that $(1 - m)^{10}$ may be approximated by $1 - 10m$. With this replacement, manipulation of Equation 15-4 gives

$$m = \frac{1}{10} \frac{(p_0 - p_{10})}{(p_0 - P)}$$

Using the M allele as an example for the data in Table 15-1,

$$m = \frac{1}{10} \frac{(0.474 - 0.484)}{(0.474 - 0.507)} = 0.03$$

which is the value tabulated in the righthand column of Table 15-1, along with the values similarly calculated for the other loci. Clearly, there is great variation in the estimates of m from locus to locus, and some of the values are negative. This variation results in part from the fact that the allele frequencies are not known with great precision but are based on studies of just a few hundred individuals. Also, estimates of allele frequency among black ancestors are highly uncertain. The most informative loci are those for which the frequencies of alleles among West Africans and Claxton whites are most divergent. Altogether, the average m for the whole set of loci is $m = 0.01$. This value suggests that Caucasian genes have been introduced into Claxton blacks at the rate of roughly one percent per generation. Estimates of the amount of racial admixture in black populations living in different parts of the country would, of course, give somewhat different values.

15.3 Selection

The driving force of adaptive evolution is natural selection, which is a consequence of hereditary differences among individuals in their ability to survive and reproduce in the prevailing environment. Since first proposed by Charles Darwin the concept of natural selection has been revised and extended, most notably by the incorporation of genetic concepts. In its modern formulation, the occurrence of natural selection rests on three premises:

1. In all organisms, more offspring are produced than survive and reproduce. This observation is commonplace and is made throughout the animal and plant kingdom, from spontaneous abortion or infant death in humans to failure of seed germination in plants.

2. Organisms differ in their ability to survive and reproduce, and some of these differences are due to genotype. This premise is supported by the numerous known mutations that affect survival or reproduction, from recessive lethals such as Tay-Sachs disease in humans to albino plants that cannot support photosynthesis.

3. In every generation, those genotypes that promote survival in the prevailing environment will be disproportionately represented among individuals of reproductive age. These favored genotypes, along with those that promote successful reproduction in the prevailing environment, will contribute disproportionately to the offspring of the next generation. Consequently, the alleles that enhance survival and reproduction will increase in frequency from generation to generation, and the population will progressively become better able to survive and reproduce in its environment. This progressive improvement in populations constitutes the process of evolutionary **adaptation.**

An Example of Selection in E. coli

Because of its small size and rapid generation time, *E. coli* provides a convenient organism for experimental studies in population genetics. An example illustrating selection during the course of 290 generations is shown in Figure

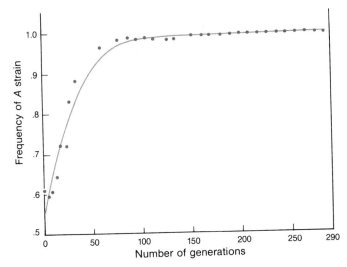

Figure 15-4 Increase in frequency of a favored strain *A* of *E. coli* resulting from selection in a continuously growing population.

15-4. In this example, two strains (*A* and *A'*) compete for a limiting nutrient (glucose). The *A* strain is a superior competitor under the particular conditions, and its proportion in the population, which is denoted *p*, increases from 0.60 to 0.9995 during the experiment. We can account for the increase by assuming that the rate of reproduction of the favored *A* strain, relative to the disfavored *A'* strain, is *w* per generation. That is, if the ratio of the relative frequencies of *A* : *A'* is *p* : *q* in any generation, then a generation later it will be *pw* : *q*. Another way of expressing the change is to say that the ratio of *A* to *A'* frequencies will have changed from *p/q* to *(p/q)w*. If p_n/q_n represents the ratio *p/q* in generation *n*, then

$$p_1/q_1 = (p_0/q_0)w$$
$$p_2/q_2 = (p_1/q_1)w = (p_0/q_0)w^2$$
$$p_3/q_3 = (p_2/q_2)w = (p_0/q_0)w^3$$
$$\vdots$$
$$p_n/q_n = (p_0/q_0)w^n$$

Taking logarithms of both sides of the last of these equations gives

$$\log (p_n/q_n) = \log (p_0/q_0) + n \log w \qquad (15\text{-}5)$$

which is the equation of a straight line. The logarithmic plot of the data in Figure 15-4 is shown in Figure 15-5, and it is indeed a straight line. Equation 15-5 also indicates that the slope of the line provides an estimate of log *w*, which in this case is 0.0186 per generation, so

$$w = 10^{0.0186} = 1.0438$$

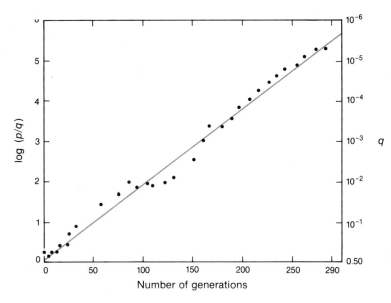

Figure 15-5 Data from the experiment shown in Figure 15-4, but replotted to show the linear increase in log (p/q).

Thus, the data in Figures 15-4 and 15-5 result from the relative reproductive rates of A and A' being in the ratio $1.0438:1$. This ratio, $w/1$, is called the **relative fitness** of the A strain relative to the A' strain; relative fitness measures the relative contribution of each parental genotype to the pool of offspring genotypes produced in each generation. That is, for each offspring cell produced by an A' genotype, an A genotype will produce an average of 1.0438 cells. The reciprocal of the ratio $w/1$, which is $1/w$, is also useful. Since $-\log w = \log (1/w)$, Equation 15-5 can be rewritten so as to include the latter term:

$$\log (p_n/q_n) = \log (p_0/q_0) - n \log (1/w)$$

The reason for writing Equation 15-5 in this form is that $1/w$ has a biological interpretation. Specifically, $1/w$ is the relative fitness of the A' strain relative to the A strain. In this case, $1/w = 1/1.0438 = 0.958$. That is, for every offspring cell produced by an A genotype, an A' genotype will produce an average of 0.958 cells. Hence, the fitness of the strains in Figures 15-4 and 15-5 can be given in two completely equivalent ways, depending on which genotype is to be taken as the standard of comparison:

$$1.0438:1.0000 \qquad \text{Fitness of } A \text{ strain relative to } A' \text{ strain} \qquad (15\text{-}6)$$

and

$$0.9581:1.000 \qquad \text{Fitness of } A' \text{ strain relative to } A \text{ strain} \qquad (15\text{-}7)$$

In population genetics, relative fitnesses are usually calculated with the most favored genotype taken as the standard with a fitness of 1.0. According to this convention, the fitnesses given in Equation 15-7 should be used to describe the selection. However, the selective disadvantage of a genotype is often of greater interest than its relative fitness. The selective disadvantage of a disfavored genotype is called the **selection coefficient** associated with the genotype, and it is calculated as the difference between the fitness of the standard (taken as 1.0) and the relative fitness of the genotype in question. In the case at hand, the selective disadvantage of A', denoted s, is

$$s = 1.000 - 0.958 = 0.042 \qquad \text{Selection coefficient against } A' \quad (15\text{-}8)$$

The selection in Figures 15-4 and 15-5 therefore corresponds to a selective disadvantage of the A' strain of 4.2 percent per generation. Whether expressed in terms of Equations 15-6, 15-7, or 15-8, it is evident that the concept of selection has great explanatory and predictive power. If the fitnesses are known, Equation 15-5 permits the prediction of the allele frequencies in any future generation, given the original frequencies. Alternatively, Equation 15-5 can be used to calculate the predicted number of generations required for selection to change the allele frequencies from any initial values to any later ones. For example, from the relative fitnesses of A and A' just estimated, one can calculate the number of generations required to change the frequency of A from 0.1 to 0.8. In this example, $p_0/q_0 = 0.1/0.9$, $p_n/q_n = 0.8/0.2$, and $w = 1.0438$. Rearranging Equation 15-5,

$$n = (1/\log 1.0438)[\log (0.8/0.2) - \log (0.1/0.9)] = 83.6 \text{ generations}$$

Selection in Diploids

Conceptually, selection in diploids is no more difficult than it is in haploids, but there is an additional complication that comes from having three genotypes at a locus instead of only two. In this discussion we will call the genotypes AA, AA', and $A'A'$ to avoid having to specify in advance which allele is dominant. As with haploids, any one of the three genotypes may be taken as the standard for calculation of relative fitness. For example, if the ratio of relative fitnesses of $AA:AA':A'A'$ is $1.000:0.975:0.95$, then these will be the relative contributions of the three genotypes to the pool of offspring in the next generation. This example is one in which the fitness of the heterozygote equals the average of the fitnesses of the homozygotes; in such cases the alleles are said to be **additive.** Additive alleles represent the simplest case of selection in diploids. Indeed, when mating is random, the changes in allele frequency are described by Equation 15-5, provided that w is interpreted as the reciprocal of the heterozygote fitness, relative to AA as the standard; in this case $w = 1/0.975 = 1.0256$. Since Equation 15-5 applies, $\log(p/q)$ will change linearly with selection. This is illustrated by the gray line in Figure 15-6. In this example the fitnesses are

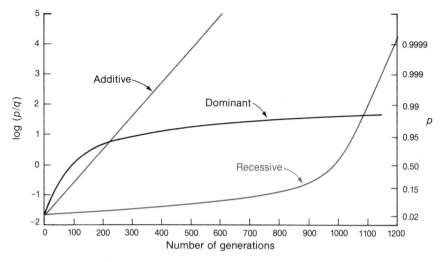

Figure 15-6 Theoretically expected change in frequency of an allele favored by selection in a diploid organism undergoing random mating when the favored allele is dominant, recessive, or additive. In each case the selection coefficient against the least fit homozygous genotype is 5 percent.

$$
\left.\begin{array}{l}
AA = 1.000 \\
AA' = 0.975 \\
A'A' = 0.950
\end{array}\right\} \quad \text{Additive alleles} \qquad (15\text{-}9)
$$

As is apparent from Figure 15-6, the linearity of $\log(p/q)$ applies *only* with additive alleles. In the example of a favored dominant allele, the fitnesses are assumed to be

$$
\left.\begin{array}{l}
AA = 1.000 \\
AA' = 1.000 \\
A'A' = 0.950
\end{array}\right\} \quad A \text{ a favored dominant} \qquad (15\text{-}10)
$$

In the example of a favored recessive, the fitnesses are assumed to be

$$
\left.\begin{array}{l}
AA = 1.000 \\
AA' = 0.950 \\
A'A' = 0.950
\end{array}\right\} \quad A \text{ a favored recessive} \qquad (15\text{-}11)
$$

The striking feature in Figure 15-6 is that the frequency of the favored dominant allele changes very slowly when the allele is common, and the frequency of the favored recessive allele changes very slowly when the allele is rare. The reason is the same in both cases, and it relates to the relative frequency of heterozygotes and rare homozygotes with random mating. Recall from Section 14.4 that rare alleles occur much more frequently in heterozygotes than in homozygotes. With a favored dominant at high frequency, most of the recessive alleles occur in heterozygotes; owing to dominance of the alternative allele, the heterozygotes are not exposed to selection and hence do not contribute to change in allele frequency. Conversely, with a favored recessive at low frequency, most of the favored alleles will be in heterozygotes; because the favored allele is recessive, the heterozygotes are not exposed to selection and again do not contribute to change in allele frequency. The principle is quite general: *selection for or against recessive alleles is very inefficient when the recessive allele is rare.* One simple example is selection against a recessive lethal. In this case the fitnesses of AA, AA', and $A'A'$ would be 1, 1, and 0, respectively. With a recessive lethal, the number of generations required to reduce the frequency of the recessive allele from q to $q/2$ equals $1/q$ generations. For example, if $q = 0.001$, it would require 1000 generations to reduce the frequency to 0.0005.

The inefficiency of selection against rare recessive alleles has an important practical implication. There is a widely held belief that medical treatment to save the lives of individuals with rare recessive disorders will cause a deterioration of the human gene pool because of the reproduction of individuals who carry the harmful genes. This belief is unfounded. With rare alleles, the proportion of homozygotes is so small that reproduction of the homozygotes contributes a negligible amount to change in allele frequency. Considering their rarity, it matters little whether homozygotes reproduce or not. Similar reasoning applies to eugenic proposals to "cleanse" the human gene pool of harmful recessives by preventing the reproduction of affected individuals. Individuals with severe genetic disorders rarely reproduce anyway, and, even when they do, they have essentially no impact on allele frequency.

The examples of selection shown in Figure 15-6 are plotted in more conventional format in terms of p in Figure 15-7. This manner of presentation emphasizes the important feature that, whatever the mode of selection, the allele frequencies change most rapidly in the intermediate range (for example, 0.2 to 0.8).

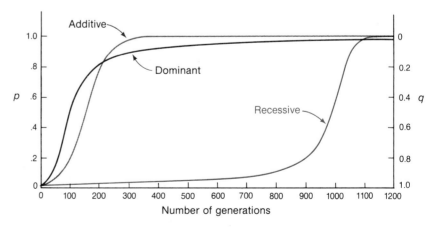

Figure 15-7 The data shown in Figure 15-6 replotted in terms of p, the allele frequency of the beneficial allele. These curves demonstrate that selection in favor of (or against) a recessive allele is very inefficient when the recessive allele is rare.

Data as extensive as the *E. coli* data in Figure 15-5 are essentially unobtainable in higher organisms because of their long generation time. Observations extending over 290 generations would take 12 years with *Drosophila* and 6000 years with humans. How, then, is one to estimate fitness? The usual approach is to divide fitness into its component parts and estimate these separately. The two major components of the fitness of a genotype are usually considered to be:

1. **Viability,** the probability that a newly formed zygote of the specified genotype survives to reproductive age.

2. **Fertility,** the average number of offspring produced by an individual of the specified genotype during the reproductive period.

Together, these two quantities measure the contribution of the genotype in question to the offspring of the next generation.

Table 15-2 is an example of an experiment carried out to estimate the viability of two genotypes in *Drosophila*. The genotypes are homozygous for either of two allozyme alleles (*F* or *S*) at the alcohol dehydrogenase locus. The tabulated numbers are averages obtained in seven separate experiments in which larvae were placed together in competition in culture vials. The ratio of surviving larvae to those introduced originally is the probability of survival of each genotype. As noted at the bottom of the table, these numbers can be converted into relative viabilities in two ways, using either *SS* or *FF* as the standard. Because of the convention that the largest of the two relative values should be 1.0, the situation can be summarized either by saying that the viability of *FF* relative to *SS* is 0.94, or that the selective disadvantage of *FF* relative to *SS* with respect to viability is 0.06.

For estimating relative fertility, an experiment like that yielding the data in Table 15-2 is required, in which adults of the genotypes are placed in vials and their output of progeny is recorded. If the fertilities are equal among the genotypes of interest, then the relative viabilities can be taken as estimates of

Table 15-2 Relative viability of alcohol dehydrogenase genotypes

Experimental data and relative viability	Genotype	
	FF	*SS*
Experimental data		
Number of larvae introduced per vial	200	100
Average number of survivors per vial	132.72	70.43
Probability of survival	0.6636	0.7043
Relative viability		
With *SS* as standard	0.6636/0.7043 = 0.94	0.7043/0.7043 = 1.0
With *FF* as standard	0.6636/0.6636 = 1.0	0.7043/0.6636 = 1.06

overall fitness. Otherwise, the fertilities also have to be taken into account.

Estimates of overall fitness based on viability and fertility components may be used for predicting changes in allele frequency that would be expected in experimental or natural populations. However, the fitness of a genotype may depend on the environment in which it finds itself. Consequently, fitnesses estimated in a particular laboratory situation may have little relevance to predicting allele frequency changes in populations unless the environments are comparable.

Selection–Mutation Balance

It is apparent from Figures 15-6 and 15-7 that selection will act to eliminate harmful alleles from a population. However, the alleles will never be totally eliminated because of recurrent mutation from the normal allele to new harmful mutants. These new mutations from the normal allele will tend to replenish the harmful alleles eliminated by selection, and eventually the population will attain a state of equilibrium in which the new mutations exactly balance the selective eliminations. The purpose of this section is to deduce the allele frequencies at equilibrium. Two important cases—complete recessive and partial dominant—must be considered. In both cases, equilibrium results from the balance of new mutations against those alleles eliminated by selection. We will symbolize the normal allele as A and the harmful allele as a, and represent their frequencies as p and q, respectively.

1. Complete recessive. When a is a complete recessive, then the fitnesses of AA, Aa, and aa can be written as $1:1:1 - s$, in which s represents the selection coefficient against the aa homozygote and measures the fraction of aa individuals that fail to survive or reproduce. For a recessive lethal, $s = 1$. When q is very small, as it will be near equilibrium, the number of a alleles eliminated by selection will be proportional to q^2s. At the same time, the number of new a alleles introduced by mutation will be proportional to μ. At equilibrium, the selective eliminations must balance the new mutations, so $q^2 = \mu/s$, or

$$q = (\mu/s)^{1/2} \qquad \text{Complete recessive} \qquad (15\text{-}12)$$

For example, the Tay-Sachs allele in most non-Jewish populations has a frequency of $q = 0.001$. Since the condition is lethal, $s = 1$. Assuming that $q = 0.001$ represents the equilibrium frequency and that the allele has no effects on the fitness of heterozygotes, the mutation rate required to account for the observed frequency would be

$$\mu = q^2s = (0.001)^2(1.0) = 1 \times 10^{-6}$$

This example shows how considerations of selection–mutation balance can be used in estimating human mutation rates, although it is necessary to restrict attention to cases in which the assumptions are likely to be valid.

2. Partial dominance. When a is not completely recessive but has an effect on the fitness of the heterozygotes, the allele is said to show partial dominance. In this case the fitnesses of AA, Aa, and aa have to be written as $1, 1 - h$, and $1 - s$, respectively, in which h is the selection coefficient against the heterozygote. With partial dominance, the equilibrium frequency of a is given by

$$q = \mu/h \qquad \text{Partial dominance} \qquad (15\text{-}13)$$

Note that Equation 15-13 does not include a square root, which makes a substantial difference in the allele frequency. The rationale for Equation 15-13 is as follows. With partial dominance, selection will occur mainly through the heterozygotes because these greatly outnumber the homozygotes. Moreover, the number of a alleles eliminated by selection against heterozygotes in each generation will be proportional to $(2pqh)/2$, but the p can be ignored because it will be so close to 1. (The factor of 1/2 is necessary because only half the alleles in each heterozygote are a.) At the same time, the number of new a alleles gained by mutation is proportional to μ. Consequently, at equilibrium, $qh = \mu$ to a very good approximation, and this leads to Equation 15-13.

Application of Equation 15-13 to estimation of human mutation rates may be illustrated with achondroplasia, a dominant form of hereditary dwarfism. We write the normal and achondroplasia genotypes as AA and AA', respectively; and the $A'A'$ homozygotes are so rare that they can be ignored. In a large Danish population the allele frequency of the achondroplasia allele A' was 5.31×10^{-5}. Affected AA' individuals produced an average of 0.25 children each, whereas their normal AA sibs produced an average of 1.27 children each. Thus, the relative fitness of AA' to AA is $0.25/1.27 = 0.20$, which corresponds to $1 - h$, so $h = 0.80$. Assuming that the population is at equilibrium, we can calculate the estimated mutation rate of $A \rightarrow A'$ from Equation 15-13 as

$$\mu = (5.31 \times 10^{-5})(0.80) = 4.2 \times 10^{-5}$$

A small amount of partial dominance can have a major impact in reducing the equilibrium frequency of a harmful allele. This can be seen by comparing Equations 15-12 and 15-13, which is done graphically in Figure 15-8. The colored curve pertains to a recessive lethal ($h = 0$, with $s = 1$ in Equation 15-12); the other curves refer to various degrees of partial dominance (Equation 15-13). For example, when $h = 0.025$, the heterozygotes have a 2.5 percent reduction in fitness compared with normal homozygotes. It can be seen that partial dominance substantially reduces the equilibrium allele frequency, especially when the mutation rate is realistically small—10^{-5} to 10^{-6}.

In Chapter 14, the high frequency of recessive lethals in natural populations of *Drosophila* was emphasized—up to 30 percent or more of naturally occurring chromosomes may be lethal when homozygous (see Figure 14-15). Most of these lethals are evidently maintained by a balance between selection and mutation, because each individual lethal allele is quite rare. However, careful study of the heterozygous effects of these alleles reveals that they are

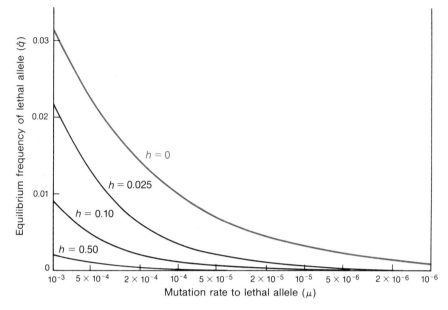

Figure 15-8 The equilibrium frequency of recessive lethal alleles maintained in a population by a balance between mutation and selection. The symbol h is the selection coefficient against the heterozygous genotype.

not completely recessive. The average "recessive" lethal in *Drosophila* results in a 1.2 percent reduction in fitness when the allele is heterozygous.

Heterozygote Superiority

So far we have considered only cases in which the fitness of the heterozygote is intermediate between those of the homozygotes, or possibly equal in fitness to one homozygote. In these cases, the allele associated with the most fit homozygote eventually becomes fixed, unless the selection is opposed by mutation. In this section we consider another possibility—**overdominance** or **heterozygote superiority**—which occurs when the fitness of the heterozygote is greater than that of both homozygotes.

When there is overdominance, neither allele can be eliminated by selection. In each generation, the heterozygotes produce more offspring than the homozygotes, and this selection for heterozygotes keeps both alleles in the population. Selection eventually produces an equilibrium in which the allele frequencies no longer change. With overdominance, the fitnesses may be written as

$$\left. \begin{array}{l} AA = 1 - s \\ AA' = 1 \\ A'A' = 1 - t \end{array} \right\} \quad \text{Overdominance} \qquad (15\text{-}14)$$

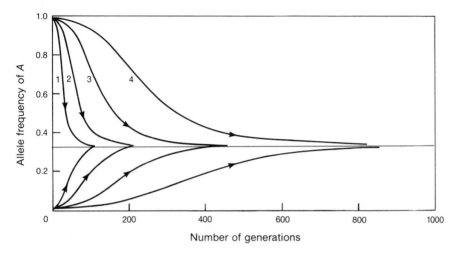

Figure 15-9 The change in frequency of an allele *A* when the heterozygous genotype is the most fit, showing that an equilibrium occurs in which both alleles are retained in the population. The selection coefficients against the homozygous genotypes have been chosen to make the equilibrium frequency of *A* equal to 1/3.

in which *s* and *t* are the selection coefficients against *AA* and *A′A′* homozygotes, respectively, and the standard for comparison of fitness is now the heterozygote because it has the highest fitness. Then, the equilibrium frequency $\hat{p}$ of the allele *A* is

$$\hat{p} = t/(s + t) \qquad \text{With overdominance} \qquad (15\text{-}15)$$

With overdominance, the allele frequencies always go to the equilibrium in Equation 15-15, but the rate of approach depends on the amount of selection. Four examples are shown in Figure 15-9 with values of *t* equal to 0.08, 0.04, 0.02, and 0.01, respectively. In each case $s = 2t$, so that the equilibrium frequency of *A* will be 1/3. Figure 15-9 shows that the equilibrium is attained most rapidly when there is strong selection against the homozygotes.

Overdominance does not appear to be a particularly common form of selection in natural populations. However, there are several well-established cases, the best known of which involves the sickle-cell hemoglobin mutation and its relationship to a type of malaria caused by the parasitic protozoan *Plasmodium falciparum*. This mutation affects the β chain of hemoglobin and is associated with a substitution of the amino acid glutamate for valine at the sixth position in the chain (Section 10.2). The amino-terminal sequence of the normal (HbA) and mutant (HbS) chains are as follows:

      HbA: Val-His-Leu-Thr-Pro-Glu-Glu-Lys-Ser-Ala- . . .
      HbS: Val-His-Leu-Thr-Pro-Val-Glu-Lys-Ser-Ala- . . .

The single amino acid substitution has profound consequences. Hemoglobin containing the abnormal β chain tends to form needle-shaped crystals when oxygen pressure is low, and the red blood cells, normally shaped like dimpled discs (Figure 15-10(a)), collapse into sickle or half-moon shapes (Figure 15-10(b)). These sickled cells tend to clog the tiny capillary vessels and so reduce the oxygen supply to vital tissues and organs. Homozygotes for the β^S mutation have a very severe anemia and rarely survive to adulthood in the

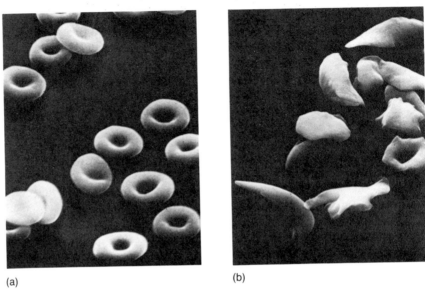

(a)   (b)

Figure 15-10 Normal unsickled red blood cells (a) and sickled cells (b). (From G. Edlin. 1984. *Genetic Principles.* Jones and Bartlett. Photo by George Brewer and Marion Barnhart.)

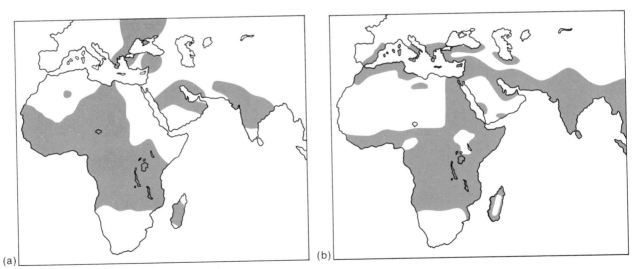

(a)   (b)

Figure 15-11 Geographic distribution of (a) sickle cell anemia and of (b) falciparum malaria in the 1920s before extensive malaria control programs were instituted.

absence of appropriate medical care. Heterozygotes have a less severe anemia but run the risk of recurrent painful episodes (sickle-cell crises) brought on when blood oxygen is abnormally lowered, which may occur during rigorous exercise or physical exhaustion.

The β^S allele is virtually lethal when homozygous, yet in certain parts of Africa and the Middle East the allele frequency reaches 10 percent or even higher. Such frequencies are much too high to be explained by recurrent mutation. Two lines of evidence initially suggested a connection between sickle-cell anemia and falciparum malaria. Physicians and researchers in Africa reported that heterozygotes are less susceptible to infection than are normal homozygotes and, when heterozygotes are infected, the infection tends to be less severe. Secondly, there is a substantial overlap of the geographical distribution of HbS and the geographical distribution of malaria. Figure 15-11(a) shows the parts of the world where the frequency of HbS is significant (red). Figure 15-11(b) shows the distribution of falciparum malaria in the 1920s before extensive control programs were initiated. The overlap is striking, and it has since been shown that the β^S polymorphism is maintained by superior fitness of the heterozygotes owing to their reduced mortality from malaria.

The nature of the HbS polymorphism is illustrated by the data in Table 15-3, which were collected in a 1956 study of infants and adults in the Musoma district of Tanganyika and included individuals of the tribes Zanaki, Simbiti, Jita, and Kwaya. The table gives the observed numbers and proportions of infants of the three genotypes, along with the observed number of adults. The final column gives the number of adults that would be expected based on the observed proportions among infants (for example, $431.0 = 0.659 \times 654$). The agreement between the observed and expected is very poor ($\chi^2 = 33.6$, $p < 0.0001$) because of excess mortality among $\beta^S\beta^S$ genotypes by anemia and among $\beta^+\beta^+$ genotypes by malaria.

Viabilities corresponding to the data in Table 15-3 may be calculated as outlined in Table 15-4. For each genotype, the viability is calculated as the ratio of observed to expected numbers, using the data in Table 15-3. Since $\beta^S\beta^+$ has the highest viability, this genotype should be used as the standard in calculating the relative viabilities, which are shown in the last column. The selection coefficients, calculated as 1 minus the relative viabilities, are $s = 0.24$ against $\beta^+\beta^+$ and $t = 1.00$ against $\beta^S\beta^S$.

Table 15-3 Observed and expected numbers of hemoglobin genotypes

Genotype	Observed number among infants*	Observed number among adults	Expected number among adults**
$\beta^+\beta^+$	189 [0.659]	403	431.0
$\beta^S\beta^+$	89 [0.310]	251	202.7
$\beta^S\beta^S$	9 [0.031]	0	20.3
Total	287	654	654.0

*Bracketed numbers are observed proportions.

**Based on observed proportions among infants.

Table 15-4   Viability of hemoglobin genotypes

Genotype	Viability	Relative viability (β^S/β^+ as standard)
$\beta^+\beta^+$	$403/431.0 = 0.94$	$0.94/1.24 = 0.76\ (s = 0.24)$
$\beta^S\beta^+$	$251/202.7 = 1.24$	$1.24/1.24 = 1.00$
$\beta^S\beta^S$	$0/20.3 = 0$	$0/1.24 = 0\ (t = 1.0)$

Using Equation 15-15, the predicted equilibrium frequency of β^+ will be

$$\hat{p} = 1.00/(0.24 + 1.00) = 0.806$$

This predicted frequency is compatible with the genotype frequencies among infants reported in Table 15-3, as shown by the following argument. Using $p = 0.806$ and $q = 0.194$, and assuming random mating, the observed and expected frequencies among infants are

$$\beta^+\beta^+:\ 189 \quad \text{versus} \quad (0.806)^2 \times 287 = 186.4$$
$$\beta^S\beta^+:\ \ 89 \quad \text{versus} \quad 2\,(0.806)(0.194) \times 287 = \ 89.8$$
$$\beta^S\beta^S:\ \ \ 9 \quad \text{versus} \quad (0.194)^2 \times 287 = \ 10.8$$

The fit is obviously extremely good, which implies that the data in Table 15-3 are compatible with the equilibrium allele frequencies expected with over-dominance when the $\beta^S\beta^S$ homozygote is lethal and the $\beta^+\beta^+$ homozygote has a reduction in fitness of 24 percent compared with the heterozygote.

In analyzing the sickle-cell hemoglobin data, it was not necessary to consider possible differences in fertility among the genotypes, because the differences in viability alone were sufficient to account for the polymorphism in the Musoma population. Some studies suggest that fertility differences may also exist, at least in some populations, but in Musoma any fertility differences are small enough to be neglected.

More Complex Modes of Selection

In actual populations selection is often very complex. For example, the fitnesses of the genotypes may not be constant but may differ according to age, sex, the genotype at other loci, or may vary with time or in particular environmental circumstances. Added to these complications are difficulties encountered in estimating fitness, particularly in natural populations. A full appreciation of natural selection requires not only a familiarity with population genetics but also a thorough understanding of the natural history of the organism and of the physiological function of the genes of interest. To date, few examples of natural selection are understood in such detail.

15.4 Random Genetic Drift

Random genetic drift is one of the most subtle processes in population genetics and, indeed, in all of genetics. It comes about because populations are not infinitely large, as we have been assuming all along, but finite (limited) in size. The breeding individuals of any one generation produce a potentially infinite pool of gametes. Barring fertility differences, the allele frequencies among gametes will equal the allele frequencies among adults. However, because of the finite size of the population, only a few of these gametes will participate in fertilization and be represented among zygotes of the next generation. In other words, there is a process of *sampling* that occurs in going from one generation to the next; because there will be variation among samples owing to chance, the allele frequencies among gametes and zygotes may differ. Changes in allele frequency that come about because of the generation-to-generation sampling in finite populations are known as **random genetic drift.**

Consider a population consisting of exactly N diploid individuals in each generation. At each autosomal locus there will be exactly $2N$ alleles. Suppose that i of these are of type A and the remaining $2N - i$ are of type a. (The allele frequency of A, designated p, therefore equals $i/2N$.) The number of A alleles in the next generation cannot be specified with certainty because it will differ from case to case owing to random genetic drift. On the other hand, it is possible to specify a *probability* for each possible number of A alleles in the next generation. Indeed, the probability $P(k|i)$ that there will be exactly k copies of A in the next generation, given that there are exactly i copies among the parents, is

$$P(k|i) = \left[\frac{(2N)!}{k!(2N - k)!}\right]p^k q^{2N-k} \qquad (15\text{-}16)$$

In Equation 15-16, k can equal 0, 1, 2, and so on up to and including $2N$, and $q = 1 - p$. The exclamation points represent factorials—the product of all positive integers up to and including the number in question, but 0! is defined to equal 1. The part of the equation in square brackets is the number of possible orders in which the k representatives of allele A can be chosen. (The A alleles can be the first k genes chosen, the last k genes chosen, alternating with a, and so on.) The remaining factor is the probability that any one of these specific orders will occur. For any specified values of N and i, one can use Equation 15-16 along with a table of random numbers (or better yet a microcomputer) to calculate an actual value for k.

A computer-generated example of random genetic drift among 12 isolated populations (a–l) is shown in Table 15-5. The example was produced by repeated application of Equation 15-16, and it illustrates some of the main features of random genetic drift. Each population consists of just eight individuals (comprising $2N = 16$ alleles), and each begins in generation 0 with eight copies of allele A ($i = 8$). The vertical columns show the number of A alleles in each population as time goes on, and the dispersion of allele frequencies resulting from random genetic drift is apparent. These changes in allele frequency would be less pronounced and would require more time in larger populations than in the very small populations illustrated here, but the overall effects would be the same. That is, the dispersion of allele frequency resulting from random genetic drift depends on population size; the smaller the population the greater the dispersion and the more rapidly it occurs.

In Table 15-5 the principal effect of random genetic drift is evident in the first generation—the allele frequencies have begun to spread out. By generation 7 the spreading is extreme, and the number of A alleles ranges from 1 to 15. In general, *random genetic drift causes differences in allele frequency among subpopulations and therefore causes genetic divergence among subpopulations.* In Chapter 14 genetic divergence among subpopulations was expressed in terms of the fixation index G. This is also a useful quantity with which to measure random genetic drift, as will be seen shortly.

Although allele frequencies among individual subpopulations spread out because of random genetic drift, the *average* allele frequency among subpopulations remains approximately constant. This point is illustrated by the column headed $\bar{p}$ in Table 15-5. The average allele frequency stays close to 0.500, its initial value. Indeed, if an infinite number of subpopulations were being considered instead of the 12 in Table 15-5, the average allele frequency would be exactly 0.50 in every generation. That is, in spite of the random drift

Table 15-5 Effects of random genetic drift

Generation	Population designation												Averages	
---	a	b	c	d	e	f	g	h	i	j	k	l	$\bar{p}$	H_S
0	8	8	8	8	8	8	8	8	8	8	8	8	0.500	0.500
1	11	7	9	8	8	6	8	7	8	6	11	10	0.516	0.478
2	10	9	10	8	8	8	6	7	6	7	13	11	0.536	0.465
3	7	11	6	5	12	5	8	5	8	7	14	9	0.505	0.438
4	8	11	7	4	12	5	8	8	7	4	15	6	0.495	0.421
5	8	8	5	3	13	2	8	12	9	5	15	6	0.490	0.387
6	11	5	3	1	13	3	10	13	6	7	15	3	0.469	0.337
7	11	8	4	3	11	1	8	14	3	7	15	2	0.453	0.334
8	14	7	4	3	10	1	9	14	3	9	16	1	0.474	0.300
9	15	5	3	5	7	0	12	14	2	11	16	0	0.469	0.251
10	16	6	5	9	8	0	9	13	3	10	16	0	0.495	0.288
11	16	1	5	11	6	0	10	13	3	10	16	0	0.474	0.249
12	16	0	5	12	6	0	9	13	2	9	16	0	0.458	0.232
13	16	0	3	12	7	0	9	13	1	11	16	0	0.458	0.210
14	16	0	5	15	7	0	9	11	1	12	16	0	0.479	0.204
15	16	0	3	14	7	0	8	12	3	13	16	0	0.479	0.208
16	16	0	2	14	9	0	11	14	2	14	16	0	0.510	0.168
17	16	0	1	15	6	0	12	14	2	13	16	0	0.495	0.152
18	16	0	1	15	4	0	13	15	5	13	16	0	0.510	0.147
19	16	0	1	16	2	0	14	16	6	14	16	0	0.526	0.104
20	16	0	1	16	2	0	15	16	9	16	16	0	0.557	0.079
21	16	0	2	16	3	0	15	16	10	16	16	0	0.573	0.092

of allele frequency in individual subpopulations, the average allele frequency among a large number of subpopulations remains constant and equal to the average allele frequency among the original subpopulations.

After a sufficient number of generations of random genetic drift, some of the subpopulations become fixed for A (red) or fixed for a (gray). Since we are excluding the occurrence of mutation, a population that becomes fixed for an allele remains fixed thereafter. After 21 generations in Table 15-5, only four of the populations are still segregating; eventually, these too will become fixed. Since the average allele frequency of A remains constant, it follows that a fraction p_0 of the populations (p_0 represents the allele frequency of A in the initial generation) will ultimately become fixed for A and a fraction $1 - p_0$ will become fixed for a. That is, *the probability of ultimate fixation of a particular allele is equal to the frequency of that allele in the original population.* In Table 15-5, five of the fixed populations are fixed for A and three for a, which is not significantly different from the equal numbers expected theoretically with an infinite number of subpopulations.

An example of random genetic drift in small experimental populations of *Drosophila* exhibiting the characteristics pointed out in connection with Table 15-5 is illustrated in Figure 15-12. The figure is based on 107 subpopulations, each initiated with eight bw^{75}/bw females (bw = brown eyes) and eight bw^{75}/bw males, and maintained at a constant size of 16 by randomly choosing eight males and eight females from among the progeny of each generation. Note how the allele frequencies among subpopulations spread out because of random genetic drift, and how subpopulations soon begin to be fixed for either bw^{75} or bw.

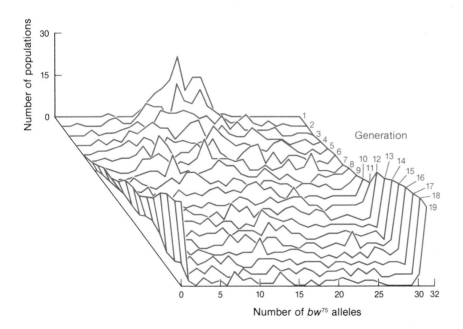

Figure 15-12 Random genetic drift in 107 small populations of *D. melanogaster* containing the *bw* and *bw*[75] alleles (*bw* = brown eyes). Each population was initiated in generation 0 with 8 heterozygous flies of each sex. From the progeny of each population in each generation, 8 females and 8 males were chosen at random to be the parents of the next generation. This three-dimensional graph illustrates how random genetic drift results in populations with different allele frequencies of *bw*[75]. The initially humped distribution (generation 1, background) becomes progressively flatter until populations begin to pile up at the edges (foreground), representing fixation of either *bw* or *bw*[75]. (From Hartl, D. L. 1981. *A Primer of Population Genetics.* Sinauer Associates.)

Relationship to Fixation Index G

An additional feature of Table 15-5 should be emphasized because it permits the effects of random genetic drift to be expressed in terms of the fixation index. Equation 14-4 defined the fixation index G as $G = (H_T - H_S)/H_T$, in which H_S is the average heterozygosity among a group of subpopulations and H_T is the heterozygosity of a hypothetical total population obtained by fusing the subpopulations and having them undergo random mating. These numbers can easily be calculated from the data in Table 15-5. To obtain H_S for any generation, one first calculates the heterozygosity expected with random mating within each of the 12 subpopulations, and then averages these values. The resulting values are tabulated in the last column. H_T simply equals $2\bar{p}(1 - \bar{p})$ for each generation. Taking generation 1 as an example, $H_S = 0.478$ and $H_T = 2(0.516)(0.484) = 0.499$. Consequently, the fixation index in generation 1, denoted G_1, is $G_1 = (0.499 - 0.478)/0.499 = 0.042$. This value, as well as those calculated similarly for the other generations, has been plotted in Figure 15-13. The fixation index increases steadily because of random genetic drift. However, H_T remains approximately constant because $\bar{p}$ remains approximately constant. Thus, the steady increase in G corresponds to a steady *decrease* in average heterozygosity (H_S), which can be observed directly in Table 15-5. Indeed, at a time when all subpopulations have become fixed for one allele or the other, $H_S = 0$. The usefulness of G as a measure of the effects of random genetic drift is that G is also a measure of genetic divergence among subpopulations, and the guidelines to its interpretation discussed in connection with Figure 14-9 are applicable.

In spite of the small number of populations involved, the points in Figure 15-13 fall quite close to the theoretically expected curve. Why G increases

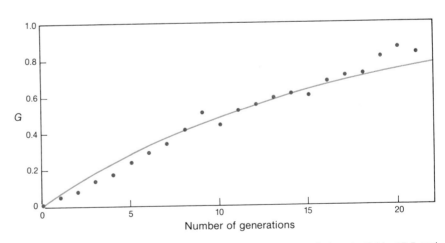

Figure 15-13 Fixation index G from computer-simulated populations in Table 15-5 and theoretically expected curve (Equation 15-17).

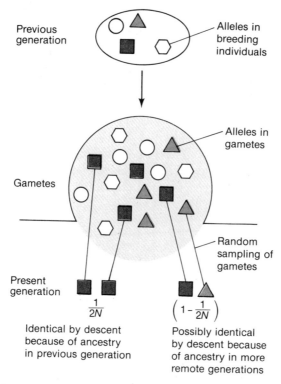

Figure 15-14 Random mating in a finite population (of size N), resulting in individuals with alleles that are identical by descent. There is an inbreeding-like effect of random genetic drift, which is measured by G.

because of random genetic drift is outlined in Figure 15-14. Recall from Section 14.7 that the inbreeding coefficient F can be interpreted as the probability that the two alleles at a locus *in an individual* are identical by descent. The fixation index G can be interpreted similarly as the probability that two alleles randomly chosen *from a single subpopulation* are identical by descent. At the top of Figure 15-14 the alleles among parents of the present generation are given. In the formation of gametes, each of these alleles is replicated many times, and $2N$ of these are selected at random to form the next generation. ($2N$ are sampled, because it requires $2N$ gametes to produce N diploid individuals.) In this random sampling, two possibilities can occur:

1. The sampled alleles may be replicas of an allele (for example the square) in the previous generation. This will happen a fraction $1/(2N)$ of the time, in which case the alleles will be identical by descent because of their immediate ancestry.

2. The sampled alleles may not be replicas of a single allele in the previous generation (for example, the square and the triangle). This

will happen a fraction $1 - 1/(2N)$ of the time, in which case the alleles will be identical by descent only if they have a more remote common ancestry.

Allelic identity by descent is measured by G, and Figure 15-14 suggests that G increases at the rate $1/(2N)$ per generation. Indeed, G increases according to the rule

$$G_n = 1 - [1 - (1/2N)]^n \qquad (15\text{-}17)$$

in which G_n represents the value of G among a group of subpopulations after n generations of random genetic drift. Equation 15-17 is the basis of the theoretical curve plotted in Figure 15-13.

Long-Term Effects of Random Genetic Drift

If random genetic drift is the only force at work, it is clear that G goes to 1, which means that all alleles become either fixed or lost and there will be no polymorphism. On the other hand, many factors can act to retard or prevent the effects of random genetic drift, and a brief discussion of these follows.

Large population size can slow the effects of drift. This occurs because when N is large, $1 - 1/2N$ will be close to 1, and therefore G will increase very slowly. In subpopulations of size 10^4 (for example, people in small cities), more than 1000 generations are required for G to exceed 0.05, a level of genetic divergence still considered "moderate."

Mutation and migration also impede fixation because alleles lost by random genetic drift can be reintroduced by either process. These processes counter the random drift of allele frequencies and hence lead to an equilibrium value of G, at which new genetic variation is introduced into the subpopulations at the same rate as its loss through random genetic drift. The equilibrium G with migration is illustrated in Figure 15-15 for a situation in which each subpopulation is equally likely to exchange migrants with every other one. Nm corresponds to the actual *number* of migrants into each subpopulation per generation. Note in Figure 15-15 how little migration is required to impede the progress of genetic divergence. For example, when $Nm = 2$ (two migrant individuals per generation), the equilibrium G is 0.11, which still corresponds to "moderate" genetic divergence. Such a low level of migration would be virtually impossible to detect in most natural populations.

Natural selection is yet another force that can act to counter the effects of random genetic drift, particularly those modes of selection, such as overdominance, that tend to maintain genetic diversity. The precise interplay between random genetic drift and selection involves the mode and intensity of selection as well as population size, and for our purposes a detailed discussion is unnecessary.

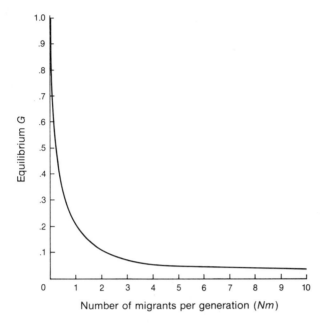

Figure 15-15 Equilibrium value of G measuring genetic divergence among a group of populations connected by migration. Note that a small amount of migration (just a few individuals per generation) drastically reduces the equilibrium value of G. (From Hartl, D. L. 1980. *Principles of Population Genetics.* Sinauer Associates.)

Relationship of Random Genetic Drift to Inbreeding

The interpretation of G in terms of allelic identity by descent immediately suggests that random genetic drift has an effect like inbreeding. The inbreeding arises because in a finite population mating will occur between relatives even though mating is random. To understand this point, note that repeated self-fertilization corresponds to random mating in a population of size 1, and repeated sib mating corresponds to random mating in a bisexual population of size 2. Less extreme examples may involve mating between more remote relatives, but the principle is the same. Therefore G represents an "inbreeding coefficient" in which the inbreeding is brought about by random genetic drift in a finite population, and the conventional inbreeding coefficient F is an inbreeding coefficient that involves mating between relatives over and above what would be expected with random mating. It is sometimes necessary to consider a *total inbreeding coefficient*, denoted Z, which is the probability that the two alleles at a locus in an individual are identical by descent, taking into account both random genetic drift and nonrandom mating between relatives. The relation between Z, G, and F is given by

$$1 - Z = (1 - G)(1 - F) \qquad (15\text{-}18)$$

Equation 15-18 says that in order for the two alleles at a locus to escape being identical by descent altogether, they must escape the effects of both the random genetic drift and the inbreeding; the probabilities of escaping these effects are $1 - G$ and $1 - F$, respectively. (In other texts Z, G, and F are usually symbolized as F_{IT}, F_{ST}, and F_{IS}, respectively. We use simpler symbols to avoid the subscripts.)

As an application of Equation 15-18 to a real situation, consider an experimental population of *Drosophila* maintained at a size of 100 with random mating for 50 generations and then brother-sister mated for one generation. The total inbreeding coefficient Z can be calculated as follows. After the period of random mating Equation 15-17 yields

$$1 - \left(1 - \frac{1}{2 \times 100}\right)^{50} = 0.222$$

which expresses the effects of random genetic drift. The conventional inbreeding coefficient of the offspring of sibs is 1/4, so $F = 0.25$. (Note that the random-mating effect is almost as large as the sib-mating effect.) From Equation 15-18, the total inbreeding coefficient is

$$Z = 1 - (1 - G)(1 - F) = 1 - (0.778)(0.75) = 0.416$$

This is substantially larger than either G or F alone, and it expresses the fact that more than 40 percent of the loci will carry alleles that are identical by descent.

Effective Population Number

The N used repeatedly in the discussion of random genetic drift is the size of a theoretically ideal population having these properties: (1) the number of males and females is equal, (2) the size of the population remains constant from generation to generation, and (3) each individual has an equal chance of contributing genes to the next generation. Real populations rarely have all of these characteristics, so equations in which N appears are invalid if N is interpreted as the *actual* population size. However, a suitable value of N, different from the actual population size, can often be found that makes the equations valid even in nonideal populations. Such a value of N is called the **effective population number** of the actual population, and it is usually symbolized N_e. For example, the *Drosophila* populations in Figure 15-12 have an actual size of 16, but their increase in G is given by Equation 15-17 with $N = 9$. Thus, the effective size of the populations is $N_e = 9$. Because of uncertainties in the relationship between effective size and actual size, the effective size of most natural populations is unknown. However, *generally speaking, the effective size of a population is smaller, and sometimes much smaller, than its actual size.*

Problems

1. A particular genotype produces nonfunctional gametes.
   (a) What is the relative fitness, as compared with a genotype that produces fully functional gametes?
   (b) What is the selection coefficient?
   (c) In this example, does fitness refer to viability or fertility?

2. Why is natural selection very inefficient with rare recessive alleles?

3. In a diploid organism the fitness of aa relative to AA is 0.86.
   (a) Assuming the A and a alleles are additive, what is the fitness of the heterozygote?
   (b) What are the selection coefficients against Aa and aa?

4. For an X-linked allele maintained by mutation-selection balance, would you expect the equilibrium frequency of the allele to be greater or less than that of an autosomal recessive resulting in the same fitnesses in both sexes as occur with the X-linked allele in females? Why?

5. Would inbreeding increase or decrease the equilibrium frequency of a harmful recessive allele as compared with random mating? Why?

6. Explain the statement "Self-fertilization corresponds to random mating in a population of size 1."

7. Suppose bacterial strains A and B have a T5-s → T5-r mutation rate of 1×10^{-7} and 1×10^{-6} per generation, respectively. Each strain is inoculated into a separate chemostat with an initial T5-r frequency of 0. What frequency of T5-r cells would be expected in each chemostat after 10, 100, and 1000 generations?

8. An allele A mutates to allele a at the rate of 2×10^{-6} per generation, and allele A undergoes reverse mutation at the rate of 4×10^{-8} per generation.
   (a) What is the expected equilibrium frequency of A in a large population?
   (b) What equilibrium frequencies of the allele would be expected if both mutation rates were increased by a factor of 10?

9. An experimenter keeps two large aquariums and maintains a constant population size of 1000 guppies in each. In one aquarium the fish are homozygous for a recessive albinism allele, and in the other they are homozygous wildtype. In each generation, 80 of the normal fish are replaced with albinos, and the albinos interbreed freely with the others and have the same number of offspring. In the initially wildtype population, what frequency of the albino allele would be expected after 10 generations and after 50 generations?

10. An island population of hermaphroditic snails is nearly fixed for a recessive allele that results in a distinctive shell color, but the nearby mainland population is fixed for the normal dominant. A population geneticist argues that the island population was probably founded by a single homozygous recessive female that underwent self-fertilization. According to this hypothesis, the dominant allele presently on the island is the result of one-way migration from the mainland. What rate of migration would be indicated assuming that the island population was founded 100 generations ago and that the present frequency of the dominant allele on the island is 0.15?

11. Equation 15-5 for selection in haploids also applies to randomly mating diploids if the fitness of the diploid heterozygote equals the geometric mean (the square root of the product) of the fitnesses of the homozygotes. In this application, w in Equation 15-5 must be interpreted as the fitness of the heterozygote. If homozygotes AA and aa have relative fitnesses 1.0 and 0.96, what would the fitness of Aa have to be in order for Equation 15-5 to hold exactly? (*Note:* If the selection coefficients are smaller than about 10 percent, Equation 15-5 is an acceptable approximation for alleles in diploids when the effects on fitness are additive.)

12. Alleles A_1, A_2, and A_3 affect color pattern in a butterfly, which determines visibility and risk of predation by birds. Observations of bird predation indicate that the probability of survival (escape from predation) of the genotypes is

$$A_1A_1 = 0.45 \qquad A_2A_3 = 0.65$$
$$A_1A_2 = 0.65 \qquad A_3A_3 = 0.75$$
$$A_2A_2 = 0.65 \qquad A_1A_3 = 0.60$$

   (a) Assuming that escape from bird predation is the only source of fitness differences among these genotypes, what are the fitnesses relative to the genotype that, by convention, should be chosen as the standard of comparison?
   (b) Are the alleles dominant, recessive, or additive in their effects on fitness?
   (c) With the fitnesses as given, what would be the eventual outcome of natural selection resulting from bird predation in terms of allele frequency?

13. Two strains of *E. coli*, A and B, are inoculated into a chemostat at the relative frequencies 60 percent A and 40 percent B and undergo competition. The fitness of strain A, relative to strain B, under the particular experimental conditions is 0.98.
    (a) What is the selection coefficient against strain A?
    (b) What relative frequencies of the strains would be expected after 10, 50, and 90 generations of competition?

14. Two strains of *E. coli*, A and B, are inoculated into a chemostat in equal frequencies and undergo competition. After 40 generations the frequency of the B strain is 35 percent. What is the fitness of strain B relative to strain A under the particular experimental conditions, and what is the selection coefficient against strain B?

15. A randomly mating diploid population has genotypes *AA*, *Aa*, and *aa* with relative fitnesses of 1, 0.50, and 0.25, respectively. If the initial frequency of *a* is 0.7, how many generations would be required to reduce it to 0.3? (*Hint:* See Problem 11.)

16. The third chromosome of *Drosophila pseudoobscura* carries many polymorphic paracentric inversions that are inherited as single "alleles" because of the suppression of recombination (see Chapter 6). These inversions frequently result in overdominance in fitness. For example, in one series of experiments, the fitnesses of genotypes bearing the *CH* (Chiricahua) and *ST* (Standard) gene arrangements were estimated as $CH/CH = 0.46$, $CH/ST = 1.00$, and $ST/ST = 0.89$. What equilibrium frequencies of *CH* and *ST* would be expected as a result of these fitnesses in a large population of *D. pseudoobscura*?

17. An allele *a* is a recessive lethal and the mutation rate of $A \rightarrow a$ is 4×10^{-6} per generation.
    (a) What is the expected equilibrium frequency of *a*?
    (b) What is the expected frequency of *aa* homozygotes at equilibrium?

18. Galactosemia is a rare autosomal recessive disorder affecting the enzyme galactose-1-phosphate uridyltransferase and resulting in mental retardation, liver enlargement, and cataracts. Individuals with galactosemia do not reproduce. Among newborn infants in Massachusetts the incidence of galactosemia is one in 118,000. Assuming that the allele is maintained at its present frequency by mutation–selection balance and that the gene is a complete recessive, what is the mutation rate of the normal to the galactosemia allele?

19. A harmful recessive allele has a selection coefficient of 50 percent in homozygotes and the mutation rate to the allele is 5×10^{-5} per generation.
    (a) What equilibrium frequency of the allele would be expected?
    (b) What equilibrium frequency of the allele would be expected if the allele resulted in a selection coefficient of 2 percent against heterozygotes?

20. In Michigan the gene for Huntington disease, which is a dominant, has a frequency of 5×10^{-5}, and the selection coefficient against heterozygotes is 0.19. Assuming equilibrium with respect to selection and mutation, what is the mutation rate to the Huntington allele?

21. A selectively neutral allele occurs in a finite population at a frequency of 0.2.
    (a) What is the probability that the mutation will eventually become fixed?
    (b) What determines the rapidity of fixation?

22. Pitcairn Island, a remote island in the South Pacific, was settled in 1789 by Fletcher Christian and eight fellow mutineers from HMS *Bounty*, along with a handful of Polynesian women. Population size has fluctuated through the years, and many descendants have left. (Descendants of the original settlers number about 1500 worldwide.) The current population size is about 45, many of them old. Assuming an effective population size of 25 in each generation, and eight generations since the settlement of the island, what is the average value of *G* among Pitcairn Islanders today?

23. Consider the genotypes *AA*, *Aa*, and *aa* in a diploid organism undergoing random mating, and suppose that the fitnesses of the three genotypes are w_1, w_2, and w_3, respectively. If *q* represents the allele frequency of *a* among the zygotes in any generation, then the allele frequency q' of *a* among zygotes in the following generation will be

$$q' = (pqw_2 + q^2 w_3)/(p^2 w_1 + 2pqw_2 + q^2 w_3)$$

in which $p = 1 - q$. If *a* is a recessive lethal allele, then $w_1 = w_2 = 1.0$ and $w_3 = 0$. Show that in this case $q' = q/(1 + q)$. Then, letting q_n represent the frequency of a recessive lethal allele among zygotes in generation *n*, show that $q_n = q_0/(1 + nq_0)$. This equation is the basis of the assertion made in the text that, for a recessive lethal, it requires $1/q_0$ generations to reduce the allele frequency from q_0 to $q_0/2$. This can be confirmed by substituting $n = 1/q_0$.

C H A P T E R 16

Quantitative Genetics

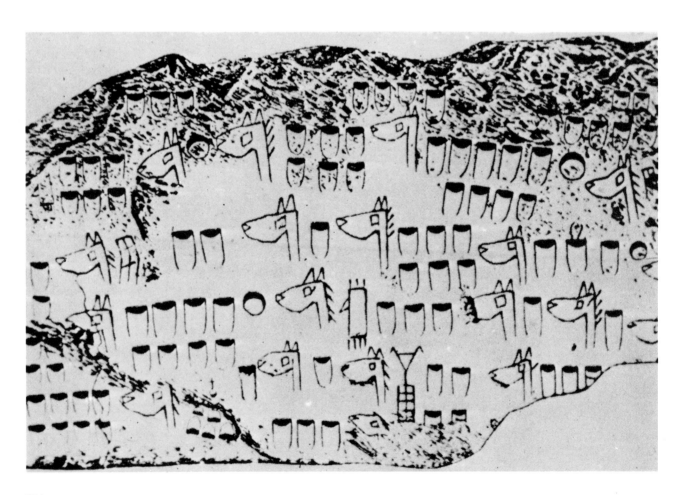

Previous chapters have emphasized traits in which differences in phenotype result from alternative genotypes at a single locus. Examples include the ability versus the inability of bacteria to degrade lactose, green versus yellow peas, curly-wing versus straight-wing *Drosophila*, normal versus sickle-cell hemoglobin, and the ABO blood groups. These traits are particularly suited for genetic analysis through the study of pedigrees because of the small number of their possible genotypes and phenotypes and because of the simple correspondence between genotype and phenotype for each. However, many traits of importance in plant breeding, animal breeding, and medical genetics are influenced by *multiple* genes and also by the effects of environment. With these traits a single genotype can have many possible phenotypes (depending on the environment), and a single phenotype can include many possible genotypes. Genetic analysis of such complex traits requires special concepts and methods, which are introduced in this chapter.

Facing page: A carving, about 4000 years old, showing the pedigree of horses raised by a breeder. The carved stone, found in Asia, shows that breeders kept records of desirable traits in their horses, as do modern breeders. (Courtesy of Dorsey Stuart.)

16.1 Quantitative Inheritance

Soon after the rediscovery of Mendel's laws in 1900, it became clear that many traits do not show simple Mendelian inheritance in the sense of being determined by alternative genotypes at a single locus. Many traits are influenced not only by the alleles of two or more genes but also by the effects of environment. Such traits are called **quantitative traits,** and with quantitative traits the phenotype of an individual is potentially influenced by

1. **Genetic factors** in the form of alternative genotypes at two or more loci, and

2. **Environmental factors**—for example, nutrition in the case of growth

rate in animals, or fertilizer, rainfall, and planting density in the case of yield in crop plants.

With some quantitative traits, differences in phenotype result largely from differences in genotype at several or many loci, and the environment plays a minor role. With some other quantitative traits, differences in phenotype result largely from the effects of environment, and genetic factors play a minor role. However, most quantitative traits fall between these extremes, and both genotype and environment must be taken into account in their analysis. Quantitative traits are often referred to as *multifactorial traits* in order to emphasize the many genetic and environmental factors in their determination.

In a genetically heterogeneous population, many genotypes will be formed by the processes of segregation and recombination. Variation in genotype can be eliminated by studying inbred lines, which are homozygous at most loci, or the F_1 from a cross of inbred lines, which are uniformly heterozygous at all loci at which the parental inbreds differ. By contrast, complete elimination of environmental variation is impossible, no matter how hard the experimenter may try to render the environment identical for all members of a population. With plants, for example, small variations in soil quality or exposure to the sun will produce slightly different environments, sometimes even for adjacent plants. Similarly, highly inbred *Drosophila* still show variation in phenotype (for example, in body size) brought about by environmental differences among animals within the same culture bottle. Therefore, traits that are susceptible to small environmental effects will never be uniform, even in inbred lines.

Quantitative traits are exceedingly important in many applications of genetics—most traits of importance in plant and animal breeding are quantitative traits. In agricultural production, one economically important quantitative trait is yield—for example, the harvest of corn, tomatoes, soybeans, or grapes. With domestic animals, important quantitative traits include milk production, egg-laying, fleece weight, litter size, and carcass quality. Important quantitative traits in human genetics include infant growth rate, adult weight, blood pressure, serum cholesterol, and length of life. In evolutionary studies, fitness is the preeminent quantitative trait.

Genetic analysis of most quantitative traits is different from the analyses presented so far in that these traits are usually influenced by *both* genotype and environment. Most quantitative traits cannot be studied by means of the usual pedigree methods because the effects of segregation of alleles at one locus may be concealed by effects of other loci, and environmental effects may cause identical genotypes to have different phenotypes. Therefore, individual pedigrees of quantitative traits do not fit any simple pattern of dominance, recessiveness, or X linkage. Nevertheless, genetic effects on quantitative traits can be assessed by comparing the phenotypes of relatives who, because of their familial relationship, must have a certain proportion of their genes in common. Such studies utilize many of the concepts of population genetics that have been discussed in Chapters 14 and 15.

Three categories of traits are frequently found to have quantitative inheritance. These are described in the following section.

Most phenotypic variation that occurs in populations is not manifested in a few easily distinguished categories. Instead, the traits vary continuously from one phenotypic extreme to the other with no clear-cut breaks in between. Such traits are called **continuous traits;** some examples are milk production in cattle, growth rate in poultry, yield in corn, and blood pressure in humans. For a trait like milk production, there is a continuous range in phenotype from minimum to maximum with no clear separation between one phenotype and the next. The distinguishing characteristic of continuous traits is that the phenotype of an individual can fall anywhere on a continuous scale of measurement, so the number of possible phenotypes is virtually unlimited.

Meristic traits are traits in which the phenotype is determined by counting. Some examples are number of skin ridges forming fingerprints, number of kernels on an ear of corn, number of eggs laid by a hen, number of bristles on the abdomen of a fly, and number of puppies in a litter. When the number of possible phenotypes is large, a meristic trait is best thought of as a special type of quantitative trait.

Threshold traits are traits having only two, or a few, phenotypic classes, but the inheritance of which is determined by the effects of multiple loci and/or the environment. Examples of threshold traits are twinning in cattle and parthenogenesis (development of unfertilized eggs) in turkeys. In many threshold-trait disorders, the phenotypic classes are "affected" versus "not affected." Examples of threshold-trait disorders in humans include adult diabetes, schizophrenia, and many congenital abnormalities, such as spina bifida. These traits can be interpreted as continuous traits by imagining each individual as having an underlying risk or **liability** toward manifestation of the condition. A liability above a certain cutoff or **threshold** results in expression of the condition; a liability below the threshold results in normality. This liability constitutes the phenotype of the individual, and it is a continuous trait in that liability ranges continuously in the population from minimum to maximum. Of course, the liability of an individual toward a threshold trait cannot be observed directly, but inferences about liability can be drawn from the incidence of the condition among individuals and their relatives.

As with continuous traits, most meristic and threshold traits are multifactorial—influenced by both genotype and environment. The special concepts and methods employed in the study of such multifactorial traits constitute **quantitative genetics.**

Distributions

The **distribution** of a trait in a population is a description of the population in terms of the proportion of individuals that have each possible phenotype. Characterizing the distribution of some traits is straightforward because the number of phenotypic classes is small. For example, the distribution of progeny in a certain pea cross may consist of 3/4 green seeds and 1/4 yellow seeds, and the distribution of ABO blood groups among Athenian Greeks consists of

Table 16-1 Distribution of height among British women

Height interval (in inches)	Number of women
53–55	5
55–57	33
57–59	254
59–61	813
61–63	1340
63–65	1454
65–67	750
67–69	275
69–71	56
71–73	11
73–75	4
Total	4995

42 percent O, 39 percent A, 14 percent B, and 5 percent AB. However, with continuous traits the large number of possible phenotypes makes such summaries impractical. Often it is convenient to reduce the number of phenotypic classes by grouping similar phenotypes together. Data for an example pertaining to the distribution of height among 4995 British women are given in Table 16-1 and in Figure 16-1. One can imagine the bar graph in Figure 16-1 being built up, step by step as the women are measured, by placing a small square, one for each woman, along the x axis at the location corresponding to the height of the woman. As sampling proceeds, the squares will begin to pile up in certain places, leading ultimately to the bar graph shown.

Displaying a distribution completely, as in Table 16-1 or Figure 16-1, is always adequate but often unnecessary; frequently, a description of the distribution in terms of two key features is sufficient. We will number the height intervals in Table 16-1 from the shortest (interval number 1 = 53–55 inches) to the tallest (interval number 11 = 73–75 inches). We will use the symbol x_i to designate the midpoint of the height interval numbered i; for example, $x_1 = 54$ inches, $x_2 = 56$ inches, and $x_{11} = 74$ in. The number of women in height interval i will be designated f_i; for example, $f_1 = 5$ women, $f_2 = 33$ women, and $f_{11} = 4$ women. The most frequently represented height interval in the sample is 63–65 inches. This is interval $i = 6$, and therefore $x_6 = 64$ inches and $f_6 = 1454$ women. The total size of the sample, in this case 4995, is denoted N. The two key features of the distribution are the following:

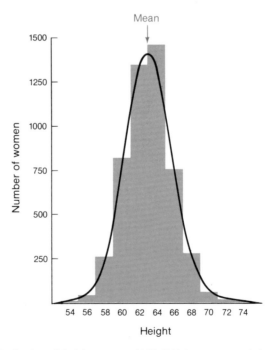

Figure 16-1 Distribution of height among 4995 British women and the smooth normal distribution that approximates the data.

1. The **mean** $\bar{x}$, also called the average, is the center of the distribution. The mean of a population is estimated from a sample of individuals from the population, as follows:

$$\bar{x} = \Sigma f_i x_i / \Sigma f_i \qquad (16\text{-}1)$$

in which Σ means summation over all classes of data (for example, summation over all 11 height intervals). The mean of a sample is used as an estimate of the mean of the entire population. In Table 16-1, the mean height in the sample of women is 63.1 inches.

2. The **variance** is a measure of the spread of the distribution and is estimated in terms of the squared *deviation* (difference) of each observation from the mean. The estimated variance calculated from a sample is called the *sample variance* s^2. When data are grouped as in Table 16-1, the sample variance is defined as

$$s^2 = \Sigma f_i (x_i - \bar{x})^2 / (\Sigma f_i - 1) \qquad (16\text{-}2)$$

The 1 is subtracted in the denominator in order to make the sample variance a true (unbiased) estimate of the variance of the entire population. The variance conveys the extent to which the phenotypes tend to be clustered around the mean. A large value means that the distribution is spread out, and a small value means that it is clustered near the mean.

A quantity closely related to the variance—the **standard deviation** of the distribution—is defined as the square root of the variance. The estimated standard deviation s of a population is calculated as

$$s = (s^2)^{1/2}$$

From the data in Table 16-1 the variance of the population of British women is estimated as $s^2 = 7.24 \text{ in}^2$, and the standard deviation is estimated as $s = (7.24 \text{ in}^2)^{1/2} = 2.69$ inches. Note that the mean and the standard deviation have the same units, in this case, inches.

The magnitude of both the mean and the variance depend on the units of measurement. For example, if the data in Table 16-1 were given in terms of feet rather than inches, the mean height would be estimated as $63.1/12 = 5.3$ ft, and the variance would be estimated as $7.24/12^2 = 0.05 \text{ ft}^2$. (Note that the mean is decreased by a factor of 12, but the variance is decreased by a factor of 12^2, or 144.) Consequently, in comparing different distributions, it is necessary that they have the same scale.

When data in a sample are not grouped into classes, as they are in Table 16-1, it is convenient to define x_i as the phenotype of the ith individual in the sample. If there are a total of N individuals in the sample, then Equation 16-1 for the mean becomes

$$\bar{x} = (1/N)\Sigma x_i \qquad (16\text{-}3)$$

and Equation 16-2 for the sample variance becomes

$$s^2 = [1/(N - 1)] \Sigma (x_i - \bar{x})^2 \qquad (16\text{-}4)$$

in which Σ means the summation over each individual in the sample.

The use of Equations 16-3 and 16-4 may be illustrated by the number of ears per stalk in a sample of three corn plants. (A sample so small is not of much use except for purposes of illustration.) If the number of ears per plant was 1, 2, and 3, respectively, the mean calculated from Equation 16-3 would be

$$\bar{x} = (1/3)(1 + 2 + 3) = 2.0 \text{ ears}$$

The variance in number of ears per stalk, estimated from Equation 16-4, would be

$$s^2 = (1/2)[(1 - 2)^2 + (2 - 2)^2 + (3 - 2)^2] = 1 \text{ ear}^2$$

The standard deviation would be

$$s = (1 \text{ ear}^2)^{1/2} = 1 \text{ ear}$$

In many cases an alternative to Equations 16-2 and 16-4 is useful because it reduces the amount of arithmetic needed in calculating the sample variance. In statistics, the symbol E is often used to denote expectation (that is, the average or mean). Therefore, in the equation given below, the symbol $E(x)$ will be used for the average of the values of x in a sample (that is, the mean), and $E(x^2)$ will represent the average of the values of x^2. Using these symbols the shorter calculation of the sample variance is

$$s^2 = [N/(N - 1)][E(x^2) - (E(x))^2] \qquad (16\text{-}5)$$

In this equation the second factor equals the mean of the squares of the phenotypes minus the square of the mean phenotype, and the first factor is the correction for bias in the estimate. In the corn example, $E(x^2) = 14/3$ and $(E(x))^2 = 2^2 = 12/3$.

When the data are symmetrical, or approximately symmetrical, the distribution of a trait can often be approximated by a smooth arching curve of the type shown on Figure 16-1. The arch-shaped curve is called the **normal distribution.** The total area under any normal curve equals 1.0; consequently, the proportion of the population having phenotypes within any specified range equals the area beneath the normal curve corresponding to that same range. Since the normal curve is symmetrical, half of its area is determined by points that have values greater than the mean and half by points with values less than the mean, and thus the proportion of phenotypes that exceed the mean is 1/2. One important characteristic of the normal distribution is that it is completely determined by the value of two quantities—the mean of the normal distribution (usually symbolized μ) and the variance of the normal distribution (usually symbolized σ^2). In fitting a normal distribution to observed data, we use the mean of the sample ($\bar{x}$ in Equations 16-1 or 16-3) as an estimate of the mean of the normal distribution (μ), and the variance of the sample (s^2 in Equation 16-2) as an estimate of the variance of the normal distribution (σ^2). The reason that $\bar{x}$ and s^2 are considered only as estimates of μ and σ^2 is that the latter are characteristics of the entire population whereas the former are characteristics of a sample of individuals from the population. Hence, the estimates $\bar{x}$ and s^2 will differ from sample to sample, though the larger the sample the more reliable the estimates. The use of Greek letters for the true values and Roman letters for the estimates emphasizes the important distinction between the population itself and a sample from the population.

As noted, the variance conveys the extent to which a distribution is spread across the range of phenotypes. This is illustrated by the three examples of normal distributions in Figure 16-2. Each distribution has a mean of 10, and the variances range from 0.25 to 4.0. The distribution with $\sigma^2 = 4.0$ looks much flatter than the curve with $\sigma^2 = 7.24$ in Figure 16-1, but this is only because the scales are different: in Figure 16-2, the x axis has been stretched out.

The mean and standard deviation of a normal distribution provide a great deal of information about the distribution of phenotypes in a population, as is illustrated in Figure 16-3. Specifically, for a normal distribution,

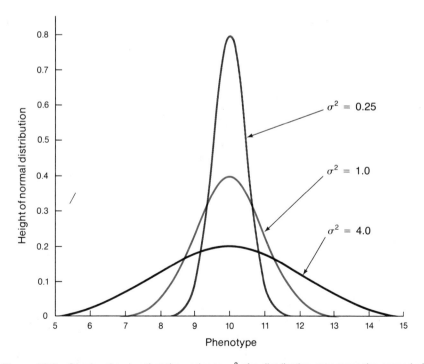

Figure 16-2 Graphs showing that the variance σ^2 of a distribution measures the spread of the distribution about the mean.

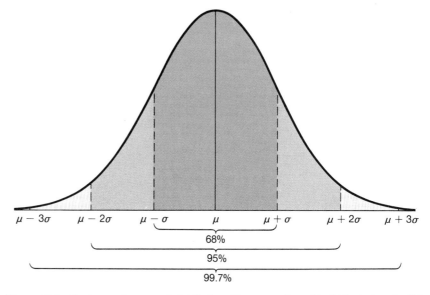

Figure 16-3 Features of a normal distribution. The proportion of individuals found within one, two, or three standard deviations from the mean is approximately 68 percent, 95 percent, and 99.7 percent, respectively.

1. Approximately 68 percent of the population will have a phenotype within *one* standard deviation of the mean (that is, between $\mu - \sigma$ and $\mu + \sigma$).

2. Approximately 95 percent will lie within *two* standard deviations of the mean (between $\mu - 2\sigma$ and $\mu + 2\sigma$).

3. Approximately 99.7 percent will lie within *three* standard deviations of the mean (between $\mu - 3\sigma$ and $\mu + 3\sigma$).

In calculating these ranges $\bar{x}$ and s^2 are used as estimates of μ and σ^2. For example, the data in Figure 16-1 imply that approximately 68 percent of the women have a height in the range $63.1 - 2.69$ inches to $63.1 + 2.69$ inches (60.4–65.8), and approximately 95 percent have a height in the range $63.1 - 2(2.69)$ to $63.1 + 2(2.69)$ inches (57.7–68.5).

16.2 Causes of Variation

Real data frequently conform to the normal distribution. Indeed, it can be shown that if the phenotype is determined by the cumulative effect of many individually small independent factors (which is true of many multifactorial traits), then the phenotypes are expected to form a normal distribution.

In considering the genetics of multifactorial traits, an important objective is to assess the relative importance of genotype versus environment. In some cases in experimental organisms, it is possible to separate genotype and environment with respect to their effects on the mean. For example, a plant breeder may study the yield of a series of inbred lines grown in a group of environments differing in planting density or amount of fertilizer. It would then be possible (1) to compare yields of the same genotype grown in different environments and thereby rank the *environments* relative to their effects on yield, and (2) to compare yields of different genotypes grown in the same environment and thereby rank the *genotypes* relative to their effects on yield.

Such a fine discrimination between genetic and environmental effects is not usually possible, particularly in human quantitative genetics. For example, with regard to the height of the sample of British women in Figure 16-1, the environment could be judged as favorable or unfavorable for tall stature only in comparison with the mean height of a genetically identical population reared in a different environment, which, of course, does not exist. Likewise, the genetic composition of the population could be judged as favorable or unfavorable for tall stature only in comparison with the mean of a genetically different population reared in the British environment, which also does not exist.

Without relevant standards of comparison, it is impossible to assess genetic versus environmental effects on the mean. However, it is still possible to assess genetic versus environmental contributions to the *variance*, because instead of comparing the means of two or more populations, the phenotypes of individuals within the *same* population can be compared. Some of the differences in phenotype result from differences in genotype and others from differences in environment, and it is often possible to separate these effects.

In any distribution of phenotypes, such as the one in Figure 16-1, four sources of phenotypic variation should be considered:

1. Genotypic variation.

2. Environmental variation.

3. Variation due to genotype–environment interaction.

4. Variation due to genotype–environment association.

Each of these sources of variation is discussed in the following sections.

Genotypic Variance

The variation in phenotype caused by differences in genotype among individuals is termed **genotypic variance.** Figure 16-4 illustrates the genetic variation expected among the F_2 arising from a cross of two inbred lines differing in genotype at three unlinked loci. The alleles at the three loci are represented as (A, a), (B, b), and (C, c), and the genetic variation in the F_2 caused by segregation and recombination is evident. Relative to a meristic trait, if we assume that each upper-case allele is favorable for the expression of the trait and adds one unit to the phenotype and each lower-case allele is without effect, then the *aa bb cc* genotype will have a phenotype of 0, and the *AA BB CC* genotype will have a phenotype of 6. The distribution of pheno-

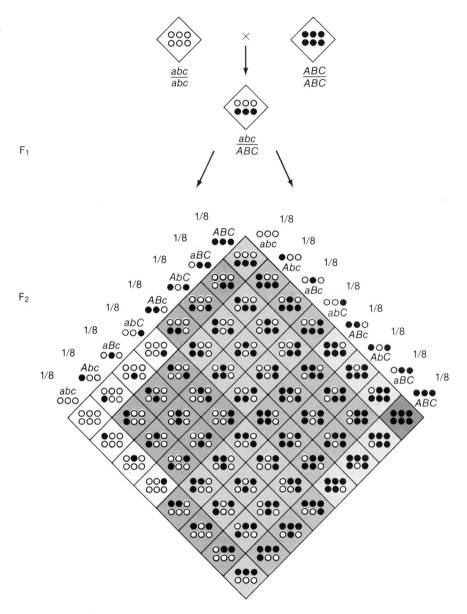

Figure 16-4 Segregation of three independent genes affecting a quantitative trait. Each upper-case allele in a genotype contributes one unit to the phenotype.

types among the F_2 is shown in the bar graph with the diagonal lines in Figure 16-5. The normal distribution approximating the data has a mean of 3 and a variance of 1.5. In this case *all* of the variation in phenotype in the population results from differences in genotype among the individuals. The same distri-

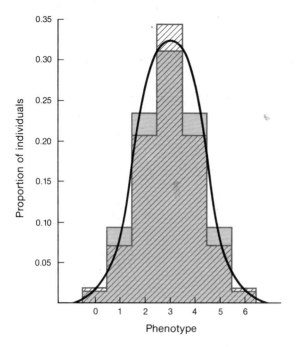

Figure 16-5 The bar graph with the diagonal lines is the distribution of phenotypes determined by the segregation of three genes illustrated in Figure 16-4. The shaded bar graph is the theoretical distribution expected from the segregation of 30 independent genes. Both distributions are approximated by the same normal distribution (black curve).

bution of phenotypes would occur in a random-mating population in which the favorable allele occurs at each locus with a frequency of 1/2.

The shaded bar graph in Figure 16-5 also has a mean of 3 and a variance of 1.5; once again, environmental effects do not influence phenotype. This distribution represents 30 loci that are segregating in a random-mating population; half of the loci are nearly fixed for the favorable allele, and half of them are nearly fixed for the alternative allele. The thing to notice is that the shaded distribution (30 loci) is virtually identical to the distribution with the diagonal lines (3 loci), and both are approximated by the same normal curve. If such distributions were encounted during actual research, the experimenter would not be able to distinguish between them. That is, *even in the absence of environmental variation, the distribution of phenotypes, by itself, provides no information about the number of loci influencing a trait, and no information about the dominance relationships of the alleles*. However, the number of loci influencing a quantitative trait is important in determining the potential for genetic improvement of a population. For example, in the three-gene case in Figure 16-5, the best possible genotype would have a phenotype of 6, but in the 30-gene case the best possible genotype (homozygous for the favorable allele at all 30 loci) would have a phenotype of 33.

Environmental Variance

The variation in phenotype caused by differences in environment among individuals is termed **environmental variance.** Figure 16-6 is an example showing the distribution of seed weight in edible beans. The mean of the distribution is 500 mg and the standard deviation is 95 mg. However, all of the beans in the population are genetically identical, having come from self-fertilization of an already highly inbred line. Therefore, *all* of the phenotypic variation in seed weight in this population results from environmental variance. Comparison of Figures 16-5 and 16-6 reveals that *the distribution of a trait in a population provides no information about the relative importance of genotype and environment.* Variation in the trait can be entirely genetic, entirely environmental, or the result of a combination of both influences.

Genotypic and environmental variance are seldom separated as clearly as in Figures 16-5 and 16-6, because normally they occur together. Their combined effects are illustrated for a simple hypothetical case in Figure 16-7. On the left is the distribution of phenotypes for three genotypes assumed to be uninfluenced by environment. As depicted, the trait is a discrete trait with three phenotypes determined by the effects of two additive alleles. The genotypes are in random-mating proportions for an allele frequency of 1/2, and the distribution of phenotypes has mean 5 and variance 2.0. Because it results solely from differences in genotype, this variance is *genotypic variance*, and it

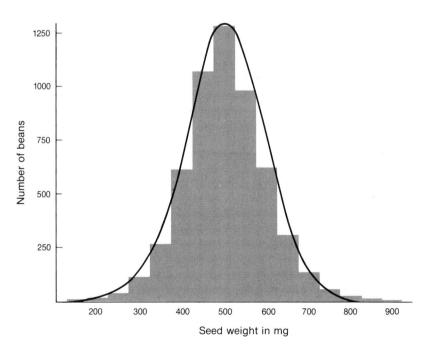

Figure 16-6 Distribution of seed weight in a homozygous line of edible beans. All variation in phenotype results from environmental differences among individuals.

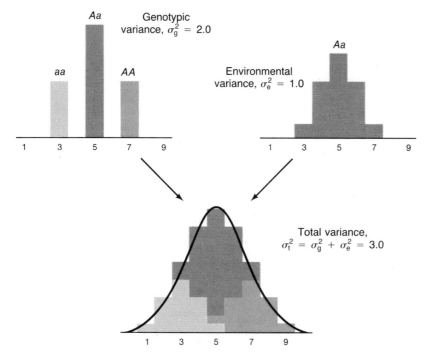

Figure 16-7 The combined effects of genotypic and environmental variance. Upper left: population affected only by genotypic variance σ_g^2. Upper right: population of *Aa* genotypes affected only by environmental variance σ_e^2. Bottom: population affected by both genotypic and environmental variance, in which the total phenotypic variance σ_t^2 equals the sum of σ_g^2 and σ_e^2.

is symbolized σ_g^2. On the right is the distribution of phenotypes that would occur in the presence of environmental variation, but only the heterozygote is illustrated. This distribution corresponds to the one in Figure 16-6, and its variance is 1.0. Since this variance results solely from differences in environment, it is *environmental variance*, and it is symbolized σ_e^2. When the effects of genotype and environment are combined in the same population, then all three genotypes occur, each genotype is affected by environmental variation, and the distribution shown at the lower center of the figure results. The variance of this distribution is the **total variance** in phenotype, and it is symbolized σ_t^2. Since we are assuming that genotype and environment have separate, independent effects on phenotype, one surely expects σ_t^2 to be greater than either σ_g^2 or σ_e^2 alone. In fact, *when genetic and environmental effects act independently on phenotype, then the total variance equals the sum of the genotypic and environmental variance*, or

$$\sigma_t^2 = \sigma_g^2 + \sigma_e^2 \tag{16-6}$$

Equation 16-6 is one of the most important relationships in quantitative genetics, and how it can be used to analyze actual data will be explained shortly. Although the equation serves as an excellent approximation in very many cases, it is valid in an exact sense only when genotype and environment are independent in their effects of phenotype. The two most important departures from independence are discussed in the next section.

Genotype–Environment Interaction and Genotype–Environment Association

When the effects of environment on phenotype differ according to genotype, a **genotype–environment interaction (G–E interaction)** is said to have occurred. In the absence of such interaction, each environment adds or detracts the same amount from the phenotype, independently of the genotype. In some cases, G–E interaction can even change the relative rank of genotypes, and genotypes that are superior in certain environments may become inferior in others. An example of extreme genotype–environment interaction in maize is illustrated in Figure 16-8. The two strains of corn are genetically uniform hybrids, and their overall means are approximately the same. However, the strain designated "commercial #4" clearly outperforms "B37 × B70" in stressful environments (environmental quality is judged on the basis of soil fertility, moisture, and other factors), whereas the performance is reversed when the environment is of high quality. The effect of genotype–environment interaction is to add an additional term to the right side of Equation 16-6. In some organisms, particularly plants, experiments like those illustrated in Figure 16-8 can be carried out to evaluate the contribution of G–E interaction to total phenotypic variance. In other organisms, particularly humans, the effect cannot be evaluated separately. When G–E interaction occurs, but is ignored, an inflated estimate of the environmental variance σ_e^2 will be obtained. However, to consider the variance resulting from genotype–environment interaction as an additional contribution to the environmental variance is often reasonable, because the additional source of variation is fundamentally environmental in origin. Interaction of genotype and environment occurs commonly and is very important in both plants and animals. Because of interaction, no one plant variety will outperform all others in all types of soil and climate, and therefore plant breeders must develop special varieties that are suited to each growing area.

When genotypes do not occur at random in all the possible environments, **genotype–environment association (G–E association)** is said to occur. In these circumstances, certain genotypes are preferentially associated with certain environments, which may either increase or decrease the average phenotype of these genotypes compared to what would result in the absence of G–E association. An example of deliberate genotype–environment association occurs in dairy husbandry, in which some farmers feed their cattle according to milk yield. Because of this practice, cows with superior genotypes with respect to milk production also receive a superior environment in the form of more feed. As with genotype-environment interaction, the effect

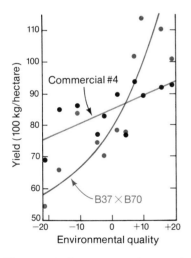

Figure 16-8 Genotype–environment interaction in maize. Strain commercial #4 is superior when environmental quality is low, but strain B37 × B70 is superior when environmental quality is high. (Data from W. A. Russell. 1974. *Annual Corn & Sorghum Research Conference* 29: 81.)

of G–E association is to add still another term to the right side of Equation 16-6. In plant or animal breeding, genotype–environment association can often be eliminated or minimized by appropriate randomization of genotypes within the experimental plots. In other cases, human genetics again being a prime example, the possibility of G–E association cannot usually be controlled. When G–E association occurs, and goes undetected, an inflated estimate of the genotypic variance σ_g^2 is usually obtained. In the case of milk yield, to include the variance due to genotype–environment association as a component in the total genotypic variance can be defended on the grounds that the superior environment is a direct consequence of a superior genotype. In other cases, such as human behavioral traits, the environment of an individual may be unrelated to its own genotype with regard to genes influencing the behavior, and in this case to include G–E association as part of σ_g^2 has little justification beyond the fact that it is experimentally unavoidable.

16.3 Analysis of Quantitative Traits

Equation 16-6 can be used to separate the effects of genotype and environment on the total phenotypic variance. Two types of data are required: (1) The phenotypic variance of a genetically heterogeneous population, which provides an estimate of $\sigma_g^2 + \sigma_e^2$, and (2) the phenotypic variance of a genetically uniform population, which provides an estimate of σ_e^2 because a genetically uniform population has a value of $\sigma_g^2 = 0$. If the environments of both populations are the same, and if there is no G–E interaction, then the estimates may be combined to extract a value for σ_g^2. As an example of this approach, we may use eye diameter in cave-dwelling *Astyanax* fish. In this case, the genetically heterogeneous population is the F_2 from a cross of two highly homozygous strains, and the genetically uniform population is the F_1 from the same cross. The phenotypic variances in these populations were found to be

$$F_2: \sigma_t^2 = \sigma_g^2 + \sigma_e^2 = 0.563$$
$$F_1: \qquad\qquad\quad \sigma_e^2 = 0.057$$

The estimate of genotypic variance σ_g^2 is obtained by subtracting the second equation from the first, namely,

$$(\sigma_g^2 + \sigma_e^2) - \sigma_e^2 = \sigma_g^2$$

or

$$0.563 - 0.057 = 0.506$$

Hence, the estimate of σ_g^2 is 0.506, and that of σ_e^2 is 0.057. In this example the genotypic variance is much greater than the environmental variance, but this is not always the case.

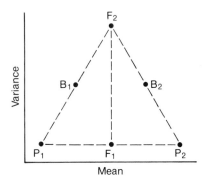

When the number of genes influencing a quantitative trait is not too large, knowledge of the genotypic variance can be used to estimate the number of genes, based on the means and variances of two phenotypically divergent strains and their F_1, F_2, and backcrosses. In ideal cases the data appear as in Figure 16-9, in which P_1 and P_2 represent the divergent strains—for example, inbred lines. The points lie on a triangle, with increasing variance according to the increasing genetic heterogeneity (genotypic variance) of the populations. The F_1 and backcross means lie exactly between their parental means, which implies that the alleles affecting the trait are *additive;* that is, at each locus, the phenotype of the heterozygote is the average of the phenotypes of the corresponding homozygotes. In such a simple situation it may be shown that the number n of loci contributing to the trait is

$$n = D^2/8\sigma_g^2 \qquad (16\text{-}7)$$

Figure 16-9 Means and variances of parents (P), backcross (B), and hybrid (F) progeny of inbred lines for an ideal quantitative trait affected by unlinked and completely additive genes. (After R. Lande. *Genetics* 99 (1981): 541.

in which D represents the difference between the means of the original parental strains, P_1 and P_2. Applying this equation to the case in Figure 16-7, $D = 4$ and $\sigma_g^2 = 2$. Consequently, $n = 16/(8 \times 2) = 1$, as is actually the case. Similarly, in Figure 16-4, $D = 6$ and $\sigma_g^2 = 1.5$, so $n = 36/(8 \times 1.5) = 3$, which is also correct.

Applied to actual data, Equation 16-7 requires several assumptions that are not necessarily correct. The theory assumes that (1) the alleles at each locus are additive, (2) the loci contribute equally to the trait, (3) the loci are unlinked, and (4) the original parental strains are homozygous for alternative alleles at each locus. However, when the assumptions are invalid, the outcome is that n is smaller than the actual number of loci affecting the trait. Thus, n is the *minimum* number of loci that can account for the data, and for this reason it is often called the **effective number of loci.** For the cave-dwelling *Astyanax* fish, just discussed, the parental strains had average phenotypes of 7.05 and 2.10, giving $D = 4.95$. Using the earlier estimate of $\sigma_g^2 = 0.506$, the effective number of loci affecting eye diameter is

$$n = (4.95)^2/(8 \times 0.506) = 6.0$$

Thus, at least six different genes affect the diameter of the eye of the fish. The variances of the populations in Figure 16-9 can be used in several ways to estimate the value of σ_g^2, and each of these values provides a slightly different estimate of n. The ranges of n obtained with various traits are summarized in Table 16-2. The numbers are not as large as might be expected, and average no more than 10. These are the *effective* number of genes, and the actual numbers are likely to be somewhat larger. Nevertheless, genetic variation in some quantitative traits results from only a few genes. On the other hand, many quantitative traits are affected by a large number of genes. At least 40 genes contribute antigens that are important in the speed of rejection of skin transplants (Chapter 13), which may be regarded as a quantitative trait. Other

Organism	Trait	Estimated effective number of loci
Goldenrod	Date of stamen maturation	6–7
Tomato	Fruit weight	9–11
Maize	Percent oil in kernels	17–22
Hawaiian *Drosophila*	Female head width/length	6–9
Fish	Eye diameter	5–7
Laboratory rat	Coat coloration	5–8
Human	Skin color	4–6

examples of a large number of genes affecting quantitative traits include bristle number in *Drosophila* (at least 100 genes) and weight of pupae in the beetle *Tribolium* (at least 170 genes).

The number of genes that affect a quantitative trait is important because it determines the amount by which a population can be genetically improved by selective breeding. With traits determined by a small number of genes, the potential for change in a trait is small, and a population consisting of the best possible genotypes may have a mean value that is only two or three standard deviations above the mean of the original population. However, traits determined by a large number of genes have a large potential for improvement. For example, after a population of *Tribolium* was bred for increased pupa weight, the mean value for pupa weight was found to be 17 standard deviations above the mean of the original population. Thus, determination of traits by a large number of genes implies that selective breeding can create an improved population in which the value of *every* individual greatly exceeds that of the *best* individuals that existed in the original population.

Broad-Sense Heritability

Estimates of the number of genes that determine quantitative traits are rarely available because the necessary experiments are impractical or have not been carried out. Another attribute of quantitative traits, which requires less data to evaluate, makes use of the ratio of the genotypic variance to the total phenotypic variance. This ratio, σ_g^2 to σ_t^2, is called the **broad-sense heritability,** symbolized as H^2, and it measures the importance of genetic variation, relative to environmental variation, in causing variation in the phenotype of a trait of interest. Broad-sense heritability is a ratio of variances, specifically

$$H^2 = \sigma_g^2/\sigma_t^2 = \sigma_g^2/(\sigma_g^2 + \sigma_e^2) \qquad (16\text{-}8)$$

Using the data for eye diameter in *Astyanax*, in which $\sigma_g^2 = 0.506$ and $\sigma_g^2 + \sigma_e^2 = 0.563$, $H^2 = 0.506/0.563 = 0.90$ is the estimate of broad-sense heritability. This estimate implies that 90 percent of the variation in eye diameter in the population of *Astyanax* results from differences in genotype among individuals.

Knowledge of heritability is useful in the context of plant and animal breeding because heritability can be used to predict the magnitude and speed of population improvement. The broad-sense heritability defined in Equation 16-8 is used in predicting the outcome of selection practiced among clones, inbred lines, or varieties. Analogous predictions for random-bred populations utilize another type of heritability, different from H^2, and this will be discussed shortly. Broad-sense heritability measures how much of the total variance in phenotype results from differences in genotype. For this reason, H^2 is often of interest in regard to human quantitative traits.

Twin Studies

In humans, twins would seem to be ideal subjects for separating genotypic and environmental variance because **identical twins,** which arise from the splitting of a single fertilized egg, are genetically identical and are often strikingly similar in such traits as facial features and body build (Figure 16-10). **Fraternal twins,** which arise from two fertilized eggs, have the same genetic re-

Figure 16-10   Identical twins. (Courtesy of Jackie Estrada and Gordon Edlin).

Table 16-3 Broad-sense heritability based on twin studies

Trait	Heritability (H^2)	Trait	Heritability (H^2)
Longevity	29	Verbal ability	63
Height	85	Numerical ability	76
Weight	63	Memory	47
Amino acid excretion	72	Sociability index	66
Serum lipid levels	44	Masculinity index	12
Maximum blood lactate	34	Temperament index	58
Maximum heart rate	84		

Note: Most of these estimates are based on small samples and should be considered as very approximate and probably too high.

lationship as ordinary siblings and thus only half of the genes in either twin are identical to those in the other. Theoretically, the variance between members of an identical-twin pair would be equivalent to σ_e^2, because the twins are genetically identical. However, the variance between members of a fraternal-twin pair would include not only σ_e^2 but also part of the genotypic variance (approximately $\sigma_g^2/2$, because of the identity of half of the genes in fraternal twins). Consequently, both σ_g^2 and σ_e^2 could be estimated from twin data and combined as in Equation 16-8 to estimate H^2. Table 16-3 summarizes estimates of H^2 based on twin studies of several traits.

Unfortunately, twin studies are subject to several important sources of error, most of which increase the similarity of identical twins, so the numbers in Table 16-3 should be considered as very approximate and probably too high. Four of the sources of error are the following:

1. Genotype–environment interaction, which will increase the variance in fraternal twins but not in identical twins.

2. Frequent sharing of embryonic membranes between identical twins, resulting in a more similar intrauterine environment.

3. Greater similarity in the treatment of identical twins by parents, teachers, and peers, resulting in a decreased environmental variance in identical twins.

4. Different sexes in half of the pairs of fraternal twins, in contrast with the same sex of identical twins.

These pitfalls and others imply that data from human twin studies should be interpreted with caution and reservation.

16.4 Artificial Selection

The practice of breeders in choosing a select group of individuals from a population to become the parents of the next generation is termed **artificial selection.** When artificial selection is carried out by choosing among clones, inbred lines, or varieties, then the broad-sense heritability permits an assessment of how rapidly progress can be achieved. Broad-sense heritability is important in this context because with clones, inbred lines, or varieties, superior genotypes—however they are defined—can be perpetuated.

In sexually reproducing populations that are genetically heterogeneous, broad-sense heritability is not relevant in predicting progress resulting from artificial selection. In these cases, superior genotypes must necessarily be broken up by the processes of segregation and recombination. To the extent that high genetic merit may depend on particular combinations of alleles, each generation of artificial selection will result in a setback, the offspring of superior parents being not quite as good as the parents themselves. Progress under selection can still be predicted, but the prediction makes use of another type of heritability—narrow-sense heritability—discussed in the following section.

Narrow-Sense Heritability

Figure 16-11 illustrates a typical form of artificial selection and its result; the trait is the length of corolla tube in the flower of *Nicotiana longiflora*. The parental generation is shown above, and the offspring generation is below. The type of selection used is called **individual selection,** because each member of the population is evaluated according to its own individual phenotype. The selection is practiced by choosing some arbitrary level of phenotype— called the **truncation point**—that determines which individuals will be saved for breeding purposes. All individuals with a phenotype above the threshold (colored area in the figure) are randomly mated among themselves to produce the next generation.

In evaluating progress through individual selection, three distinct phenotypic means are important. In Figure 16-11, these are symbolized as M, $M*$, and M', and are defined as follows:

1. M is the mean phenotype of the entire population in the parental generation, including both the selected and nonselected individuals.

2. $M*$ is the mean phenotype among those individuals selected as parents (phenotypes above the truncation point).

3. M' is the mean phenotype among the progeny of selected parents.

The relationship between these three means is given by

$$M' = M + h^2(M* - M) \qquad (16\text{-}9)$$

in which the symbol h^2 is the narrow-sense heritability of the trait in question. Equation 16-9 is called a **prediction equation,** because if the value of narrow-sense heritability is given, it can be used to predict the offspring mean (M') based on the values of both the parental mean ($M*$) and the mean of the population in the previous generation (M).

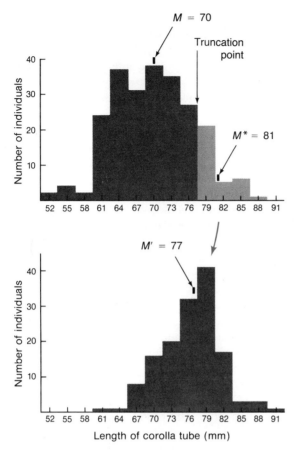

Figure 16-11 Selection for increased length of corolla tube in tobacco. M, $M*$, and M' are the means of the parental generation, of selected parents (that is, of those individuals with phenotype measurements that exceed the truncation point), and of the offspring of selected parents, respectively.

Later in this chapter, how narrow-sense heritability can be estimated from the similarity in phenotype among relatives will be explained. In the case of Figure 16-11, h^2 is the only unknown quantity, so it can be estimated from the data themselves. Rearranging Equation 16-9 and substituting the values for the means from Figure 16-11 leads to

$$h^2 = (M' - M)/(M* - M) = (77 - 70)/(81 - 70) = 0.64 \quad (16\text{-}10)$$

In the parental generation in Figure 16-11, the broad-sense heritability of corolla-tube length was $H^2 = 0.82$. In general, the narrow-sense heritability of a trait is smaller than the broad-sense heritability. The two are equal only when the alleles affecting the trait are additive in their effects.

Since h^2 is smaller than H^2, H^2 can be divided into two parts to see more clearly how the two types of heritability differ. Specifically, H^2 can be broken down into one part corresponding to h^2 and another part corresponding to the remainder. Since H^2 has been defined in Equation 16-8 as σ_g^2/σ_t^2, splitting H^2 corresponds to splitting σ_g^2 into two parts. One part of σ_g^2 is an "additive" part that contributes to h^2, and the other part is a "dominance" part that depends on dominance and other types of gene interaction not contributing to h^2. These additive and dominance parts of σ_g^2 are designated as A (the **additive variance**) and D (the **dominance variance**), respectively. Therefore,

$$H^2 = \sigma_g^2/\sigma_t^2 = (A/\sigma_t^2) + (D/\sigma_t^2) = h^2 + (D/\sigma_t^2) \qquad (16\text{-}11)$$

The reason for separating σ_g^2 into A and D is that their values can be estimated from the similarity among relatives; this will be discussed in a later section. Equation 16-11 provides the formal definition of h^2: the *narrow-sense heritability* of a trait in a population is the ratio of the additive variance to the total variance. In other words, the narrow-sense heritability is the fraction of the phenotypic variance that can be used to predict changes in population mean with individual selection by means of Equation 16-9.

In the population in Figure 16-11, the total variance is 46.7 and $h^2 = 0.64$. Thus, the additive variance is $A = 0.64 \times 46.7 = 29.3$. The genotypic variance is 38.2, so the dominance variance is $D = 38.2 - 29.3 = 8.9$. Approximately one-fourth of the genotypic variance (8.9/38.2) in this population does not contribute to population improvement with individual selection because of the necessity of sexual reproduction. This explains in part why the offspring of superior individuals are not quite as superior as their parents (that is, why $M' < M^*$).

Equation 16-9 is of fundamental importance in quantitative genetics because of its predictive value. This can be seen in the following example. The selection in Figure 16-11 was carried out for several generations. After two generations, the mean of the population had become 83, and parents having a mean of 90 were selected. Using the estimate $h^2 = 0.64$, the mean in the next generation can be predicted. The information provided is that $M = 83$ and $M^* = 90$. Therefore, Equation 16-9 implies that the predicted mean is

$$M' = 83 + (0.64)(90 - 83) = 87.5$$

This value is in good agreement with the observed value of 87.9.

Long-Term Artificial Selection

Artificial selection results in the genetic improvement of a population with respect to economically important traits for the same reason that natural selection results in increased adaptation of a natural population to its environment—both artificial selection and natural selection cause an increase in the frequency of alleles that improve the selected trait (or traits). Thus, the principles of natural selection discussed in Chapter 15 also apply to artificial selection. For example, artificial selection is most effective in changing the frequency of alleles that are in an intermediate range of frequency ($0.2 < p < 0.8$). Alleles with frequencies outside this range respond more slowly to selection, and rare recessive alleles respond the slowest of all. With quantitative traits, including fitness, the total selection is shared among all the genes that affect the trait, and the selection coefficient experienced by each allele is determined by (1) the magnitude of the effect of the allele, (2) the frequency of the allele, (3) the total number of genes affecting the trait, (4) the narrow-sense heritability of the trait, and (5) the proportion of the population that is selected for breeding.

The value of heritability is determined by both the magnitude of effects and the frequency of alleles. If all favorable alleles were fixed ($p = 1$) or lost ($p = 0$), the heritability of the trait would be 0. Therefore, the heritability of a quantitative trait is expected to decrease over the course of many generations of artificial selection as a result of favorable alleles becoming nearly fixed. For example, selection for less fat in a population of Duroc pigs decreased the heritability of fatness from 73 percent in generations 0–5 to 30 percent in generations 5–10.

Population improvement by means of artificial selection cannot continue indefinitely. A population may respond to selection until its mean is many standard deviations different from the mean of the original population, but eventually the population reaches a **selection limit** in which successive generations show no further improvement. Progress may stop because all alleles affecting the trait are either fixed or lost, so the narrow-sense heritability of the trait is 0. However, a more common reason for a selection limit is that natural selection counteracts artificial selection. Many genes that respond to artificial selection as a result of their favorable effect on a selected trait also have indirect harmful effects on fitness. For example, selection for increased size of eggs in poultry results in a decrease in the number of eggs, and selection for extreme body size (large or small) in most animals results in a decrease in fertility. When one trait (for example, number of eggs) changes during the course of selection for a different trait (for example, size of eggs), the unselected trait is said to have undergone a **correlated response** to selection. Correlated response of fitness is typical in long-term artificial selection. Each increment of progress in the selected trait is partially offset by a decrease in fitness because of correlated response (Figure 16-12); eventually artificial

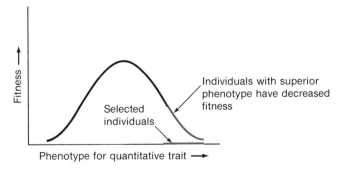

Figure 16-12 Individuals with extreme phenotypes for a quantitative trait often have reduced fitness, resulting in a correlated response to selection in which fitness decreases as the quantitative trait becomes improved.

selection for the trait of interest is exactly balanced by natural selection against the trait, so a selection limit is reached and no further progress is possible without changing the strategy of selection.

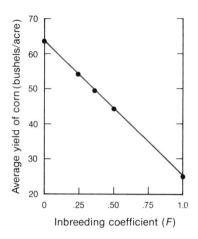

Figure 16-13 Inbreeding depression for yield in corn. (Data from N. Neal. 1935. *J. Amer. Soc. Agron.* 27: 666–670.)

Inbreeding Depression and Heterosis

Inbreeding usually has harmful effects on economically important traits such as yield of grain or egg production. This decline in performance is called **inbreeding depression,** and it results principally from rare harmful recessive alleles becoming homozygous because of inbreeding (Chapter 14). Figure 16-13 shows inbreeding depression in yield of corn.

Different inbred populations will usually become homozygous for rare harmful recessive alleles at different sets of loci. Thus, when the inbred populations are crossed, the F_1 individuals will be either homozygous or heterozygous for the favorable dominant allele at most loci, and will be superior or equal to both inbred parents at each locus. This favorable effect of crossing inbred lines is called **heterosis.** The most important application of heterosis is in the crossing of inbred lines of corn to produce hybrid seed that is sold commercially. High-yielding hybrids have been developed for different soils and climate, and they presently account for virtually all of the corn production in the United States.

Applications of Quantitative Genetics in Genetic Counseling

Figure 16-14 illustrates how the concepts of individual selection in Figure 16-11 can be modified to be applicable to human threshold traits. Recall that such traits depend on an underlying quantitative trait called *liability*, which refers to the degree of genetic risk or predisposition toward the trait. Only those individuals with a liability greater than a certain *threshold* develop the trait. The upper part of Figure 16-14 shows the distribution of liability in a population, in which liability is measured on an arbitrary scale so that its mean (M) equals 0 and its variance equals 1. Affected individuals are in the colored area of the distribution. Using various characteristics of the normal distribution, the observed proportion of affected individuals can be used to calculate the position of the threshold and the value of M' (the mean liability of affected individuals).

The distribution of liability among *offspring* of affected individuals is shown at the bottom of Figure 16-14. It is shifted slightly to the right, implying that a greater proportion of individuals will have liabilities above the threshold and will be affected. The observed proportion of affected individuals in the offspring generation can be used to calculate M'. The important point is that $M*$ and M' can be obtained from the observed proportions corresponding to the colored areas in Figure 16-14. With $M = 0$, Equation 16-9 can be used to calculate the narrow-sense heritability of liability of the trait in question.

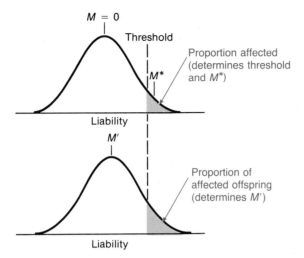

Figure 16-14 Interpretation of a threshold trait in terms of an underlying continuous distribution of liability. Individuals with a liability greater than the threshold are affected. M (arbitrarily set equal to 0), M^*, and M' are the mean liabilities of the parental generation, of affected individuals, and of the offspring of affected individuals, respectively.

Examples of narrow-sense heritability of the liability toward important congenital abnormalities are given in Table 16-4. How these heritabilities translate into risk is illustrated in Figure 16-15, along with observed data for the most common congenital abnormalities in Caucasians. The theoretical curves correspond to the relationship between the incidence of the trait in the general population, the type of inheritance, and the risk among first-degree relatives of affected individuals. (The most important types of first-degree relatives are parents, offspring, and siblings.) Simple Mendelian inheritance produces the highest risks, but these traits, as a group, tend to be rare. The most common traits are threshold traits, and with a few exceptions their risks tend to be moderate or low, corresponding to the heritabilities indicated.

Table 16-4 Narrow-sense heritability of liability for congenital abnormalities

Trait	Population frequency, percent	Narrow-sense heritability, percent*
Cleft lip	0.1	76
Spina bifida	0.5	60
Club foot	0.1	68
Pyloric stenosis	0.3	75
Dislocation of hip	0.06	70
Hydrocephalus	0.05	(76)
Coeliac disease	0.05	(45)
Hypospadias	0.30	(75)
Atrial septal defect	0.07	(70)
Patent ductus arteriosus	0.05	(60)

*Estimates in parentheses are considered tentative.

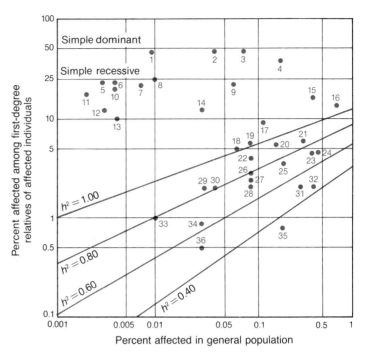

Figure 16-15 Recurrence risks of common abnormalities. Diagonal lines are the theoretical risks for threshold traits with the indicated values of the narrow-sense heritability of liability. Horizontal lines are the theoretical risks for simple dominant or recessive traits. The numbered traits are: (1) Achondroplasia. (2) Target-cell anemia. (3) Periodic paralysis. (4) Otosclerotic deafness. (5) Retinoblastoma. (6) Hemophilia. (7) Albinism. (8) Retinitis pigmentosum. (9) Cystic fibrosis. (10) Phenylketonuria. (11) Osteogenesis imperfecta. (12) Microphthalmos. (13) Hirschsprung disease. (14) Deafmutism. (15) Manic-depressive psychosis. (16) Mental deficiency. (17) Schizophrenia. (18) Dislocated hip. (19) Multiple sclerosis. (20) Strabismus. (21) Pyloric stenosis. (22) Anencephaly. (23) Diabetes. (24) Rheumatic fever. (25) Spina bifida aperta. (26) Clubfoot. (27) Patent ductus. (28) Cleft lip. (29) Celiac disease. (30) Cleft palate. (31) Congenital heart disease. (32) Epilepsy. (33) Situs inversus viscerum. (34) Exomphalos. (35) Hydrocephaly. (36) Psoriasis.

16.5 Correlation Between Relatives

Quantitative genetics relies extensively on similarity among relatives to assess the importance of genetic factors. Particularly in the study of such traits as human behavior, interpretation of familial resemblance is not always straightforward because of the possibility of nongenetic, but nevertheless familial, sources of resemblance. One approach to overcoming these problems will be discussed in the next section. Here we will focus on methods that are applicable in less complicated situations in which genotype–environment association is less important, or in which the environment can more easily be controlled. Such cases frequently occur in plant and animal breeding.

The discussion of Equation 16-11 emphasized that additive and dominance variance, and therefore narrow-sense heritability, could be estimated from the degree of resemblance between relatives. Examination of this aspect of quantitative genetics requires a short digression into the general features of the measurement of similarity between relatives.

Covariance and Correlation

Familial data frequently occur as pairs of numbers—pairs of parents, pairs of siblings, pairs of twins, or pairs consisting of a single parent and offspring. In genetics, it is natural to inquire whether, and to what extent, the phenotypes of pairs of individuals resemble each other. Consider N pairs of measurements—for example, of parent and offspring—denoted (x_1, y_1), (x_2, y_2), . . . , (x_N, y_N), in which the x and the y values correspond to the parents and to their offspring, respectively. An example of such data is the adult stature of three mothers and of their daughters, as shown in Table 16-5. An important issue in quantitative genetics is the degree to which the x and y values in such pairs of measurements are associated. For example, one might ask whether tall mothers tend to have tall daughters. The association between the x and y values can be assessed graphically by means of the type of scatter diagrams illustrated in Figure 16-16. In the figure, r represents a statistical quantity called the **correlation coefficient** between the variables; it is related to the slope of the best-fitting straight line (red line) and to the amount of scatter of the data points around the line. When $r = 0$, x and y are not associated and are completely independent. Increasing values of r are accompanied by increasing slopes and tighter clustering. Negative associations between x and y are also possible, as in the example with $r = -0.7$. The maximum theoretical range of r is between -1.0 and $+1.0$.

The correlation coefficient between x and y is calculated in terms of the **covariance C** between the variables, defined as follows:

$$C = [1/(N - 1)][(x_1 - \bar{x})(y_1 - \bar{y}) + (x_2 - \bar{x})(y_2 - \bar{y}) + \cdots + (x_N - \bar{x})(y_N - \bar{y})] \qquad (16\text{-}12)$$

Thus, the covariance is related to the average of the products of the deviations of the x and y values, much as the variance is related to the average of their squared deviations.

The **correlation coefficient** between x and y is calculated from the covariance as follows:

$$r = C/(s_x s_y) \qquad (16\text{-}13)$$

Table 16-5 Adult heights of three mother–daughter pairs

Pair	Height of mother (in)	Height of daughter (in)
(x_1, y_1)	65.0	61.5
(x_2, y_2)	64.0	65.5
(x_3, y_3)	69.0	69.0

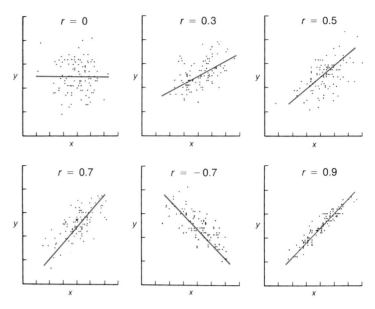

Figure 16-16 Scatter diagrams of x and y variables having various degrees of correlation (r). (From R. R. Sokal and F. J. Rohlf. *Biometry*. Copyright 1969 by W. H. Freeman and Co. All rights reserved.)

in which s_x and s_y are estimates of the standard deviations of the variables. The correlation coefficient is important when comparing distributions that may have different variances.

Equations 16-12 and 16-13 can be used to examine the relation between the heights of the three women and their daughters given in Table 16-5. Each of the two groups, mothers and daughters, has its own mean, variance, and standard deviation, and these values can be estimated from Equations 16-3 and 16-4. The results are

$$\text{Mothers: } \bar{x} = 66.0 \qquad s_x^2 = 7.0 \qquad s_x = 2.65$$
$$\text{Daughters: } \bar{y} = 65.3 \qquad s_y^2 = 14.1 \qquad s_y = 3.75$$

From Equation 16-12, the covariance in stature between mothers and daughters is

$$C = (1/2)[(65.0 - 66.0)(61.5 - 65.3) + (64.0 - 66.0)(65.5 - 65.3) + (69.0 - 66.0)(69.0 - 65.3)] = 7.25$$

A shorter method of calculating the covariance uses the difference between the mean of the product of x and y (symbolized $E(xy)$) and the product of the means (symbolized $E(x)E(y)$), namely

$$C = [N/(N - 1)][E(xy) - E(x)E(y)] = 7.25.$$

In this case, $E(xy) = 4316.8$ and $E(x)E(y) = 4312.0$. From Equation 16-13, the correlation coefficient in mother–daughter stature is

$$r = 7.25/(2.65 \times 3.75) = 0.73$$

Because this sample is so small, the correlation coefficient cannot be taken very seriously. In much larger data sets, the mother–daughter correlation in stature is about $r = 0.45$, and it can be concluded that tall mothers do tend to have tall daughters. Therefore, for a large data set, an appropriate plot of mother–daughter stature would somewhat resemble the diagram in Figure 16-16 for which $r = 0.5$.

Figure 16-17 gives a geometrical interpretation of the correlation coefficient. Just as the distribution of a single variable can be built up, step by step, by placing small squares corresponding to each individual along the x axis, so too can the joint distribution of two variables—for example, height of mothers and daughters—be built up step by step by placing small cubes on the xy plane at a position determined by each pair of measurements. As these cubes pile up, the joint distribution assumes the form of an inverted bowl, cross-sections of which are themselves normal distributions. When there is no correlation between the variables, the surface is completely symmetrical (Figure 16-17(a)), which means that the distribution of y is completely indepen-

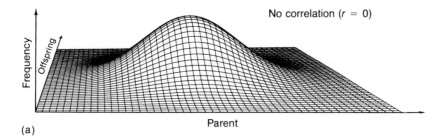

(a)

No correlation ($r = 0$)

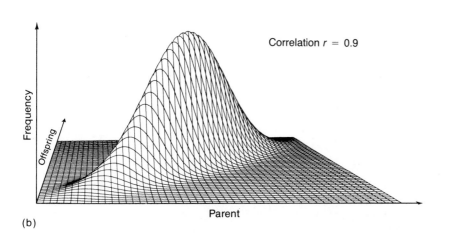

(b)

Correlation $r = 0.9$

Figure 16-17 Distribution of a quantitative trait in parents and offspring when there is no correlation (panel (a)) or a high correlation (panel (b)) between them. (From R. R. Sokal and F. J. Rohlf. *Biometry*. Copyright 1969 by W. H. Freeman and Co. All rights reserved.)

dent of that of x. However, when the distributions are correlated, the distribution of y varies according to the particular value of x, and the joint distribution becomes asymmetrical with a ridge built up in the xy plane (panel (b)). This ridge corresponds to the clustering of points around the lines in Figure 16-16.

Estimation of Narrow-Sense Heritability

The narrow-sense heritability of a trait is defined as the ratio of the additive variance to the total variance (Equation 16-11). Since the additive variance can be estimated directly from the covariance between relatives, studies of covariance provide a convenient method for estimating narrow-sense heritability. Theoretical values of the covariance of various pairs of relatives are given in Table 16-6, in which A represents the additive variance and D represents the dominance variance. Considering parent-offspring, half-sibling, or first-cousin pairs, the additive variance can be estimated directly by multiplication. Specifically, A can be estimated as twice the parent–offspring covariance, four times the half-sibling covariance, or eight times the first-cousin covariance. This simple correspondence occurs because the covariance between these relatives depends only on the value of A.

With full siblings, the covariance includes a contribution from the dominance variance (D) as well as one from the additive variance. This complication also occurs with identical twins and double first cousins. (Double first cousins result from mating between two pairs of siblings.) In these relatives, D contributes to resemblance because the relatives can share *both* alleles at a locus, whereas parents and offspring, half siblings, and first cousins can share at most a single allele at each locus. Therefore, to the extent that phenotype depends on dominance effects, full siblings can resemble each other more than they resemble their parents.

The potentially greater resemblance between siblings than between parents and offspring may be appreciated by considering an autosomal recessive trait caused by a rare recessive allele. In a random-mating population, the

Table 16-6   Theoretical covariance in phenotype between relatives

Degree of relationship	Covariance*
Offspring and one parent	$A/2$
Offspring and average of parents	$A/2$
Half siblings	$A/4$
First cousins	$A/8$
Monozygotic twins	$A + D$
Full siblings	$A/2 + D/4$
Double first cousins**	$A/2 + D/16$

*Contributions from interaction among alleles at different loci have been ignored.

**Double first cousins are the offspring of matings between siblings from two different families.

probability that an offspring will be affected, if the mother is affected, is q (the allele frequency), which corresponds to the random-mating probability that the sperm giving rise to the offspring carries the recessive allele. By contrast, the probability that two siblings will both be affected, if one of them is affected, is 1/4, because most of the matings that produce affected individuals will be between heterozygous parents. Clearly, when the trait is rare, the parent-offspring resemblance will be very small, whereas the sibling–sibling resemblance will be substantial. This discrepancy is entirely a result of dominance variance, and it arises only because full siblings can share both of their alleles.

Once the additive variance has been estimated by means of the relations in Table 16-6, the narrow-sense heritability may be estimated as the ratio of additive variance to the total variance in phenotype. This can be seen in an example from animal breeding—namely, the body length of male Danish Landrace pigs. In one study, the covariance in body length between half siblings was estimated as 0.56, and the total phenotypic variance was 3.90. Thus, $A = 4 \times 0.56 = 2.24$ and

$$h^2 = 2.24/3.90 = 0.57$$

This estimate tells us nothing about the *inheritance* of body length, but it does predict the rate at which body length can be changed by artificial selection (Equation 16-9).

For those degrees of relationship in Table 16-6 in which the covariance includes only the additive variance, the correlation coefficient between the relatives can be determined simply by substituting h^2 for A. Consequently, the narrow-sense heritability may be estimated as twice the parent-offspring correlation, four times the half-sibling correlation, or eight times the first-cousin correlation, provided that the phenotypic variances are the same in both sets of relatives being compared. For example, with human height the parent–offspring correlation coefficient is about $r = 0.45$, which corresponds to a narrow-sense heritability of $2 \times 0.45 = 0.90$. (This estimate is undoubtedly too high because of nongenetic but familial sources of resemblance.)

16.6 Human Behavior

One can hardly imagine a subject more controversial than the inheritance of human behavioral differences, particularly of socially undesirable behavior. Since the rediscovery of Mendel's laws, and even prior to the rediscovery, commentators were divided, some claiming the simple and direct inheritance of virtually all forms of socially unacceptable behavior—"feeble-mindedness," habitual drunkenness, criminality, prostitution, and so on—and others arguing for environmental causation of such behavior. Writing on "The inheritance of mental defect" in the *British Journal of Medical Psychology* (13 (1933): 254–267), R. R. Gates commented that "It may be stated that feeble-mindedness is generally of the inherited, not the induced, type; and that the inheritance is generally recessive." In a similar hereditarian vein, K. Pearson noted that "In feeble-minded stocks mental defect is interchangeable with imbecility, insanity, alcoholism, and a whole series of mental (and often physical) anomalies." ("On the inheritance of mental disease," in *Annals of Eugenics* 4 (1931): 362–380).

These extreme views were reinforced by studies of certain families with high incidences of the undesirable traits, the Jukes and the Kallikak families (both pseudonyms) being the most famous. Figure 16-18 is a fragment of the Jukes pedigree published in 1902 purporting to show the inheritance of "shiftlessness" and "partial shiftlessness." The solid and shaded symbols are presented exactly as they appeared in the original, and they give the false impression of objectivity. However, the diagnosis of "shiftlessness" or "partial shiftlessness" is entirely subjective, as indicated by the descriptions of some of the individuals in the pedigree. Individual I-1 is described as "a lazy mulatto," I-2 as "a nonindustrious harlot, but temperate," II-10 as "lazy, in poorhouse," and III-18 as a "bad boy." This sort of subjectivity is rejected by today's standards, but at the time it was taken quite seriously as an element in the "inheritance" of poverty. The conclusions regarding the pedigree were stated as follows:

> "When both parents are *very* shiftless, practically all children are 'very shiftless' or 'somewhat shiftless' . . . It is probable that both shiftlessness and lack of physical energy are due to the absence of something which can be got back into the offspring only by mating with industry." (C. B. Davenport, *Heredity in Relation to Eugenics* (Holt, 1911).)

If one were to accept the pedigree at face value, which, of course, cannot be done because of its unacceptably subjective nature, and were to judge it in present-day terms, one would conclude that no family exists outside of its environment, and that social environments, like genes, tend to be perpetuated from generation to generation. Today, it seems apparent that familial association of certain characteristics (for example, poverty) and certain types of

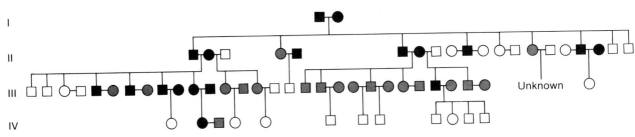

Figure 16-18 Portion of the Jukes pedigree illustrating hereditarian prejudices of some early investigators. Blackened symbols represent "shiftlessness" and shaded symbols represent "partial shiftlessness." (From R. L. Dugdale. 1902. *The Jukes; A study in Crime, Pauperism, Disease and Heredity.* Putnam; and C. B. Davenport. 1911. *Heredity in Relation to Eugenics.* Holt.)

behavior is not sufficient evidence that they are genetically, as opposed to socially, transmitted. An obvious example is English grammar and pronunciation. Children tend to speak with the same quirks and localisms as their parents and other relatives, yet nobody would presume that English usage is genetically transmitted. Nevertheless, some early behaviorists were extreme hereditarians.

The opposite view was held by the extreme environmentalists, whose thinking is typified by the following example:

> "So let us hasten to admit—yes, there are heritable differences in form, in structure. . . . These differences are in the germ plasm and are handed down from parent to child. . . . But do not let these undoubted facts of inheritance lead us astray as they have some of the biologists. The mere presence of these structures tells us not one thing about function. . . . Our hereditary structure lies ready to be shaped in a thousand different ways—the same structure—depending on the way in which the child is brought up. . . . Geneticists are working under the banner of outmoded ideas of psychology. One need not give very much weight to any of their present conclusions. . . . We have no real evidence of the inheritance of behavioral traits."
>
> "I should like to go one step further now and say, 'Give me a dozen healthy infants, well formed, and my own specified world to bring them up in and I'll guarantee to take any one at random and train him to become any type of specialist I might select—doctor, lawyer, artist, merchant-chief and, yes, even beggar-man and thief, regardless of his talents, penchants, tendencies, abilities, vocations, and race of his ancestors.' I am going beyond my facts and I admit it, but so have the advocates of the contrary. . . ." (J. B. Watson, *Behaviorism.* (W. W. Norton, 1925).)

Such extreme hereditarian and environmentalist views are now in the minority, and they are not so easy to recognize because they have been updated and presented in terms appropriate to the present time. Controversy still continues, though views have become somewhat moderated and the terms of the debate have shifted. Most geneticists and psychologists are willing to concede the importance of both heredity *and* environment in human behavioral variation. The key questions today relate to the relative importance of "nature" and "nurture," and how best to assess the situation experimentally. Most human behavioral traits are quantitative traits, and the next section discusses application of the concepts and methods of quantitative genetics to the study of IQ scores.

Genetic and Cultural Effects on IQ Scores

Modern "intelligence" tests such as the Stanford-Binet and the Wechsler Adult tests derive from attempts in France in the early 1900s to develop simple procedures to identify children with mild learning disabilities. The tests actually assess a variety of skills, such as vocabulary, short-term memory, deductive reasoning, and the ability to perceive patterns in geometrical designs. Although no single test can adequately assess all aspects of what is commonly understood to be intelligence, the tests can be useful in identifying children who may need special attention in school. Although score on an IQ test is a statistical abstraction rather than a definitive measure of intelligence, IQ scores are relatively stable. The correlation between IQ scores of the same person tested as a child and again as an adult is approximately 0.8. Consequently, an IQ score can be treated as a phenotype like any other quantitative trait, quite apart from possible misgivings concerning the true relationship between IQ and intelligence.

Path Analysis of Complex Traits

Genetic analysis of IQ data requires that special attention be given to cultural transmission of nongenetic factors that potentially affect IQ. These familial factors increase the resemblance between relatives, and, unless properly taken into account, they inflate the apparent genetic heritability. Traditional types of studies, which ignore cultural transmission, typically lead to heritability estimates of 60 to 80 percent. A more rigorous approach is outlined in Figure 16-19. The diagram in the figure is called a **path diagram.** The colored squares refer to observable phenotypes, in this case of mother (P_M), father (P_F), and two of their offspring (P_O and P'_O). Connecting these are certain unobservable variables, which are represented as circles:

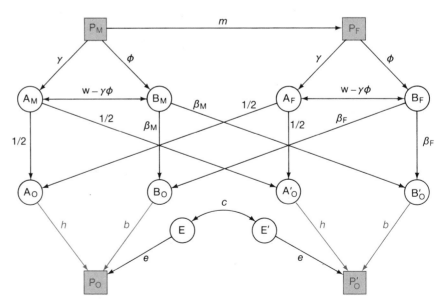

Figure 16-19 Path diagram of genetic and cultural effects on IQ and their interrelationships. (From J. Rice, C. R. Cloninger, and T. Reich. 1980. *Behavior Genetics.* 10: 73.)

1. Transmissible genetic effects, symbolized as A with appropriate subscripts for mother, father, and offspring.

2. Transmissible cultural effects, symbolized as B with appropriate subscripts.

3. Nontransmissible cultural effects, symbolized as E and E′ for the two siblings.

Connecting the variables in Figure 16-19 are arrows called **paths,** which show how variation in one variable will affect variation in another. The small letters are quantities, called **path coefficients,** that are related to the correlation between the variables. For example, the path from P_M to P_F represents a positive correlation between the phenotypes of spouses, indicating positive assortative mating for IQ, and the path coefficient m measures the magnitude of the mother–father correlation in IQ. Similarly, the paths that connect the A's of parents and children represent genetic transmission; the path coefficients of these paths are 1/2 because of Mendelian segregation of alleles. The two paths of greatest interest are those labeled h and b, which connect the genetic and cultural endowment of a child with the IQ phenotype of the child. The connection between these path coefficients and heritability is that h^2 equals the narrow-sense heritability of IQ; this genetic heritability represents the proportion of the total variation in IQ that can be accounted for by the transmission of genetic factors from parent to offspring. Since the model in Figure 16-19 also allows for cultural transmission through the b paths, there is also a "cultural heritability," which represents the proportion of the total

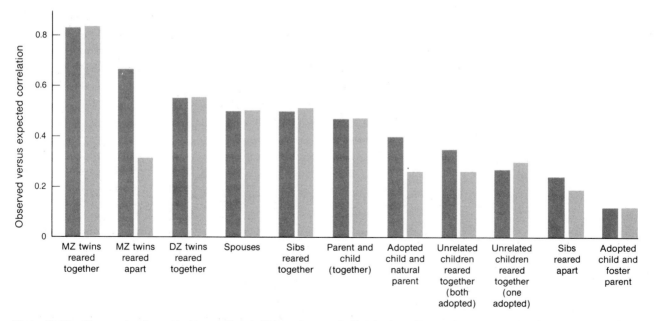

Figure 16-20 Observed and expected correlations in IQ based on analysis of familial and adoption data according to the path model in Figure 16-19. (Data from J. Rice, C. R. Cloninger, and T. Reich. 1980. *Behavior Genetics.* 10: 73.)

variation in IQ that can be accounted for by transmissible but nongenetic cultural influences. E and E' in the model refer to nontransmissible environmental influences affecting the offspring, and as indicated by the path coefficient c, these may be correlated.

All of the path coefficients in Figure 16-19 may be estimated from observed IQ data. The calculations require a large computer program, but what the program does is search for a set of path coefficients that maximizes the fit between the observed IQ correlations and those expected from the path diagram. Such an analysis is called *path analysis*.

The best estimates of genetic and cultural heritability of IQ from data on American whites are:

$$h^2 = 0.297 \pm 0.023 \text{ (genetic)}$$
$$b^2 = 0.289 \pm 0.016 \text{ (cultural)}$$

Consequently, genetic and transmissible cultural influences are approximately equal in accounting for variation in IQ among whites, both being around 30 percent. However, the largest single contributor to variation in IQ among whites is that resulting from *non*transmissible environmental influences, corresponding to the path coefficients denoted e in Figure 16-19; these nontransmissible effects account for 32 percent of the variation in IQ.

Observed and expected correlations obtained from path analysis of IQ are illustrated in Figure 16-20. The fit is rather good except in the case of adoption. For example, identical twins reared apart have an observed correlation of 0.68, whereas the expected correlation is only 0.33. In most of the discrepancies related to adoption, the observed correlations are higher than expected. These discrepancies can be explained as the result of selective placement of adoptees into homes resembling those of their natural parents. Selective placement introduces a correlation between environments in the natural and adopted home, which is not taken into account in the path analysis. Clearly, path analysis provides a powerful tool for sorting out genetic and environmental effects on complex multifactorial traits.

Race and IQ

Path analysis of IQ illustrates how data on a complex human behavioral trait may be analyzed and interpreted, but the results may also be misinterpreted. The most common mistake is to apply the genetic and cultural heritabilities to differences between populations. For essentially the same reasons that identical twins reared in different environments may have different phenotypes, two genetically identical populations that differ in their environments may differ in average phenotype. Both genetic and cultural heritablity are relevant to interpreting the variation in phenotypes *within* a specified population. Use of these quantities for comparison *between* populations is unjustified and may often lead to incorrect conclusions.

Heritability has mistakenly been applied to racial differences in average IQ. Such tests have been standardized so that the average score of American whites is 100. American blacks average about 85 on the same tests, and in some Japanese studies the average is about 110. The question that has been debated is whether these averages reflect genetic differences between the races. *No methods are presently known that can answer such a question.* Because the emotional overtones of racial differences in IQ are so great, the central issue may become clearer by means of analogy. Instead of human IQ, one might just as well ask why chicken-farmer Brown's hens lay an average of 230 eggs per year but farmer Smith's average only 210. Lacking any knowledge of the genetic relationship between the flocks, the reason for the difference cannot be specified. This is true despite the fact that one might know with certainty that the genetic heritability within each flock is 30 percent, because heritability within flocks is irrelevant to the comparison. On the one hand, farmer Brown's hens could be genetically superior in egg laying. On the other hand, farmer Smith's chicken feed or husbandry could be inferior. In order to determine the reason for the difference some of Brown's chickens would have to be reared on Smith's farm, and vice versa. If the egg-laying of the transplanted hens did not change, then the difference between the two groups of hens would be entirely genetic; if the egg-laying of the transplanted hens did change, then the difference between the groups would be environmental. Experiments of this type cannot be carried out with humans.

Many environmental factors can affect IQ-test performance between races, among which the following three are particularly important:

1. *Differences in socioeconomic status.* IQ scores are correlated with socioeconomic status, so comparisons of populations differing in socioeconomic status will inevitably lead to differences in average IQ. For example, the difference between American blacks and whites is much reduced when socioeconomic levels are taken into account.

2. *Differences in language skills.* IQ tests are largely verbal, so differences in language skills will be reflected in test performance.

3. *Differences in motivation.* Test performance is affected by the environmental surroundings in which the test is taken, the attitude and skill of the examiner, the mood and motivation of the subject, and other factors. Different populations, particularly those with different cultural backgrounds, may react differently with respect to one or more of these factors and so perform differently on the tests.

Because of these possibilities, it can be expected that environment will affect average IQ. To admit this is not to deny the possibility of racial differences in the frequencies of alleles affecting the trait, but merely to say that *differences in average IQ cannot be taken as evidence of genetic differences,* regardless of the heritabilities. In light of this disclaimer, it might seem as if application of genetic methods to IQ has taught us nothing at all. On the contrary, two important conclusions may be drawn:

1. Within the population of American whites, the proportion of variation in IQ attributable to transmissible genetic effects is about 30 percent; the proportion attributable to transmissible cultural effects is also

about 30 percent; and that attributable to nontransmissible environmental effects is about 32 percent. (The remaining 8 percent results from correlations between genetic and transmissible cultural factors—an example of genotype–environment association or possibly genotype–environment interaction.) That is, if the environments of all American whites were equalized, leaving the genetic variation as it is, the degree of variation in IQ within this population would be reduced by 70 percent.

2. Complex traits that depend on genotype and environment can be analyzed by using appropriate methods. Methods like path analysis can be used to separate the contributions of genetic and environmental effects in causing phenotypic variation in such traits. These methods are applicable not only to behavioral traits but to many others, such as hypertension or diabetes. These methods form the bridge between simple Mendelian traits of the type studied by Mendel himself and the most complex traits that exist.

Problems

1. Two varieties of corn, A and B, are field tested in Indiana and North Carolina. Strain A is more productive in Indiana, but strain B is more productive in North Carolina. What phenomenon in quantitative genetics does this example illustrate?

2. A pair of male monozygotic twins marries a pair of female monozygotic twins, and each couple has a child. What is the genetic relationship between the offspring?

3. Ten female mice had the following number of live-born offspring in their first litter: 11, 9, 13, 10, 9, 8, 10, 11, 10, 13. Considering these females as representative of the total population from which they came, estimate the mean, variance, and standard deviation of size of the first litter in the entire population.

4. The data shown below pertain to milk production over an eight-month lactation among 304 two-year-old Jersey cows. Data have been grouped by intervals, but for purposes of computation each cow may be treated as if her milk production were equal to the midpoint of the interval (for example, 1250 for cows in the 1000–1500 interval).
   (a) Calculate the mean, variance, and standard deviation in milk yield.
   (b) What range of yield would be expected to include 68 percent of the animals?
   (c) Round the limits of this range to the nearest 500 pounds and compare the observed with the expected number of animals.

(d) Do the same calculation for the range expected to include 95 percent of the animals.

Pounds of milk produced	Number of cows
1000–1500	1
1500–2000	0
2000–2500	5
2500–3000	23
3000–3500	60
3500–4000	58
4000–4500	67
4500–5000	54
5000–5500	23
5500–6000	11
6000–6500	2

5. In the F_2 generation of a cross of two cultivated varieties of tobacco, the number of leaves per plant was distributed according to a normal distribution with mean 18 and standard deviation 3. What proportion of the population is expected to have the following phenotypes: (1) between 15 and 21 leaves; (2) between 12 and 24 leaves; (3) fewer than 15 leaves; (4) more than 24 leaves; (5) between 21 and 24 leaves?

6. Suppose a meristic trait is determined by three independently segregating genes in a random-mating population. Alleles A, B, and C are dominant to their recessive counter-

parts a, b, and c. Genotype $aa\,bb\,cc$ has a phenotype of 0, and one unit in phenotype is added for each locus that is heterozygous or homozygous dominant.

(a) What is the expected distribution of phenotypes in a population in which each dominant allele has a frequency of 1/2?

(b) What is the expected distribution of phenotypes in a population in which each dominant allele has a frequency of 1/4?

(c) Calculate the mean, variance, and standard deviation of the phenotypes in the two cases above. (In examples of this type, the variance is most easily calculated as the mean of the squared values of the phenotype minus the square of the mean value of phenotype—that is, as $s^2 = E(x^2) - [E(x)]^2$.)

7. Two inbred homozygous strains of mice are crossed and the six-week weight of the F_1 progeny is determined.

(a) What is the genotypic variance in the F_1?

(b) If all alleles affecting six-week weight are additive, what is the expected mean phenotype of the F_1 compared with the average six-week weight of the original inbred strains?

8. In a cross between two cultivated inbred varieties of tobacco, the variance in leaf number per plant in the F_1 generation is 1.46 and in the F_2 generation it is 5.97. What are the genotypic and environmental variances? What is the broad-sense heritability in leaf number?

9. In an experiment with weight gain between ages 3–6 weeks in mice, the difference in mean phenotype between two strains was 17.6 grams and the (additive) genetic variance was estimated as 0.88. Estimate the minimum number of genes affecting this trait.

10. In a selection experiment for increased plasma cholesterol levels in mice, parents with a level of 2.37 units were selected from a population with a mean of 2.26 units, and the progeny of selected parents had an average level of 2.33 units. What is the observed heritability of this trait?

11. A mouse population has an average weight gain between ages 3–6 weeks of 12 g, and the narrow-sense heritability of 3–6 week weight gain is 0.20.

(a) What average weight gain would be expected among the offspring of parents whose average weight gain was 15 g?

(b) Among the offspring of parents whose average weight gain was 9 g?

12. A representative sample of lamb weights at the time of weaning in a large flock is shown below. If the narrow-sense heritability of weaning weight is 20 percent and the half of the flock consisting of the heaviest lambs is saved for breeding for the next generation, what is the best estimate of the average weaning weight of the progeny? (*Note*: If a normal distribution has mean μ and standard deviation σ, then the mean of the upper (heavier) half of the distribution is given by $\mu + 0.8\sigma$.)

81	81	83	101	86
65	68	77	66	92
94	85	105	60	90
94	90	81	63	58

13. A flock of broiler chickens has a mean weight gain of 700 g between ages 5–9 weeks, and the narrow-sense heritability of weight gain in this flock is 0.80. Selection for increased weight gain is carried out for five consecutive generations, and in each generation the average of the parents is 50 g greater than the average of the population from which the parents were derived. Assuming that the heritability of the trait remains constant at 80 percent (actually, it is likely to decrease slightly), what is the expected mean weight gain after the selection?

14. Consider the following data on the inheritance of body size in a population of the flour beetle *Tribolium*: mean weight in population = 2000 μg; phenotypic variance = 40,000 μg^2; additive genetic variance = 10,000 μg^2. If individuals with a mean weight of two phenotypic standard deviations above the population mean are selected as parents for the next generation, what is the expected mean of the progeny generation?

15. In order to estimate the heritability of maze-learning ability in rats, a selection experiment was carried out. From a population in which the average number of trials necessary to learn the maze was 10.8, with a variance of 4.0, animals were selected that managed to learn the maze in an average of 5.8 trials. Their offspring required an average of 8.8 trials to learn the maze. What is the estimated narrow-sense heritability of maze-learning ability in this population?

16. A replicate of the population in Problem 15 was reared in another laboratory under rather different conditions of handling and other stimulation. The mean number of trials required to learn the maze was still 10.8, but the variance was increased to 9.0. Animals with a mean learning time of 5.8 trials were again selected, and the mean learning time of the offspring was 9.9.

(a) What is the heritability of the trait under these conditions?

(b) Is the result consistent with that in Problem 15, and how can the new result be explained?

17. Use Figure 16-15 to estimate the risk of a threshold trait among the siblings of affected individuals for a trait that has an incidence of 1 in 10,000 and a narrow-sense heritability of liability of 80 percent.

18. Use Figure 16-15 to estimate the narrow-sense heritability of liability of a trait in which the incidence in the general population is one in 2000 and in which the risk of the trait among children who have a single affected parent is 1 percent.

19. Why do the risks corresponding to the slanting heritability lines in Figure 16-15 increase as the incidence of the trait in the population increases?

20. Two female mice had the following numbers of live-born offspring in their first and second litters. Calculate the correlation coefficient between first and second litter size. For comparison, do the same calculation with mismatched litters obtained by shifting the numbers in the last row down one (and bringing the 12 at the bottom up to the top).

Female	First litter	Second litter
1	11	10
2	9	12
3	13	12
4	10	10
5	9	8
6	8	6
7	10	12
8	11	9
9	10	12
10	13	12

21. Given the covariance of parent-offspring pairs and that of full-sibling pairs, how could both additive and dominance variance and h^2 be estimated?

22. The asexual unicellular protozoan *Difflugia* has a number of well-formed and easily counted teeth encircling the region of the cell that functions as the mouth. The cells also have a variable number of spiny projections. The data below are covariances between parent and offspring and total phenotypic variances for tooth number and length of longest spine in a genetically heterogeneous laboratory population. Bearing in mind that in an asexual population parent and offspring are genetically related as identical twins, what is the broad-sense heritability of the traits?

Trait	Covariance	Total Variance
Tooth number	1.76	1.84
Length of spine	0.91	3.17

23. What is the theoretical covariance in phenotype between first cousins who are the offspring of monozygotic twins?

24. Maternal effects are nongenetic influences on offspring phenotype that derive from the phenotype of the mother. For example, in many mammals larger mothers have larger offspring, in part because of maternal effects on birth weight. What result would a maternal effect have on the covariance in birth weight of mothers and their offspring compared with the covariance in birth weight of fathers and their offspring?

25. Given a mean IQ of 100 in a large human population and a narrow-sense heritability of 30 percent, what is the expected mean IQ in the next generation if the next generation is produced as stipulated in points (a) and (b) below? Assuming a generation time of 25 years, what is the expected increase in average IQ per year? (*Note:* In a selection program in which the mean of selected males differs from the mean of selected females, the overall mean of selected parents equals the average of the means in the two sexes.)
    (a) All females have an equal chance of producing offspring, regardless of their IQ.
    (b) Ten percent of the pregnancies result from use of "sperm banks" where the average IQ of the donor males is 130. All other pregnancies result from mating with a random sample of males.

26. For a phenotype determined by a simple, completely penetrant recessive allele having the frequency q in a random-mating population, it can be shown that the additive variance in phenotype is $2pq^3$, the dominance variance is p^2q^2, and the total variance is $pq^2(1 + q)$.
    (a) In terms of p and q, what are the values of the narrow-sense heritability and the broad-sense heritability of the trait?
    (b) Calculate the narrow-sense heritability for $q = 1.0$, 0.5, 0.1, 0.05, 0.01, 0.005, and 0.001.
    (c) Note that the narrow-sense heritability in (b) goes to 0 as q goes to 0, yet the trait is completely determined by heredity. How can this be explained?

GLOSSARY

A site   *See* **aminoacyl site.**

aberrant 4:4 segregation   The presence of equal numbers of alleles among the spores in a single ascus, yet with a spore pair in which the two spores have different genotypes.

acentric chromosome   A chromosome lacking a centromere.

acrocentric chromosome   A chromosome with the centromere near one end.

active site   The part of an enzyme at which substrate molecules bind and are converted into their reaction products.

acylated tRNA   A tRNA molecule to which an amino acid is linked.

adaptation   Any characteristic of an organism that improves its chances of survival and reproduction in its environment; the evolutionary process by which organisms undergo modification favoring their survival and reproduction in a given environment.

additive variance   The magnitude of the genetic variance that results from the additive action of genes; the value that the genetic variance would assume if there were no dominance or interaction of alleles affecting the trait.

adenyl cyclase   The enzyme that catalyzes synthesis of cyclic AMP.

adjacent-1 segregation   Segregation of a heterozygous reciprocal translocation in which the translocated chromosomes separate from one another and become included in different gametes, each with the nontranslocated chromosome bearing the *nonhomologous* centromere.

adjacent-2 segregation   Segregation of a heterozygous reciprocal translocation in which the translocated chromosomes separate from one another and become included in separate gametes, each with the nontranslocated chromosome bearing the *homologous* centromere.

agarose   A component of agar used as a gelling agent in gel electrophoresis; its value is that few molecules bind to it, so it does not interfere with electrophoretic movement.

agglutination   The clumping or aggregation, as of viruses or blood cells, caused by an antigen-antibody interaction.

albinism   Absence of melanin pigment in the iris, skin, and hair of an animal; absence of chlorophyll in plants.

alkaline phosphatase   An enzyme that removes a 5'-P group from a terminal nucleotide of a nucleic acid molecule leaving a 5'-OH group; used in genetic engineering.

alkaptonuria   A recessively inherited metabolic disorder in which a defect in the breakdown of tyrosine leads to excretion of homogentisic acid (alkapton) in the urine.

alkylating agent   An organic compound capable of transferring an alkyl group to other molecules.

allele   One of an array of different forms of a given gene.

allele frequency   Relative proportion of all alleles of a gene that are of a designated type.

allopolyploid   A polyploid formed by doubling the chromosome number of a hybrid between two different species.

allosteric effector   A molecule that increases or decreases the activity of a protein by binding to the protein and changing the shape of the protein molecule.

allosteric protein   Any protein whose activity is altered by a change of shape induced by binding a small molecule.

allozymes   Alternative electrophoretic forms of a protein coded by alternative alleles of a single gene.

alternate segregation   Segregation of a heterozygous reciprocal translocation in which both translocated chromosomes separate from both normal homologues.

amber codon   Common jargon for the UAG stop codon; an amber mutation is a mutation in which a sense codon has been altered to UAG.

Ames test   A bacterial test for mutagenicity, used to screen for potential carcinogens.

amino acid   Any one of a class of organic molecules having an amino group and a carboxyl group; 20 different amino acids are the usual components of proteins.

amino acid attachment site   The 3′ terminus of a tRNA molecule at which an amino acid is attached.

aminoacyl site   One of the two tRNA-binding sites on a ribosome; commonly called the A site.

aminoacyl tRNA synthetase   The enzyme that attaches the correct amino acid to a specific tRNA molecule.

amino group   The chemical group —NH_2.

aminopterin   An organic molecule that inhibits the major pathway of DNA synthesis.

amino terminus   The end of a polypeptide chain at which the amino acid bears a free amino group.

amniocentesis   A procedure for obtaining fetal cells from the amniotic fluid for the diagnosis of genetic abnormalities.

anaphase   The stage of mitosis or meiosis during which chromosomes move to opposite ends of the spindle; follows metaphase and precedes telophase.

aneuploidy   The condition in which the chromosome number is not an exact multiple of the haploid number.

annealing   A laboratory technique by which single strands of DNA or RNA having complementary base sequences become paired, to form a double-stranded molecule.

antibody   A protein, found in blood serum, produced in animals in response to a specific antigen and capable of binding to the antigen.

anticodon   The three bases in a tRNA molecule that are complementary to three bases of a specific codon in mRNA.

antigen   Any substance capable of stimulating the production of specific antibodies.

antiparallel   A term used to describe the chemical orientation of the two strands of a double-standed nucleic acid molecule; the 5′ → 3′ orientations of the two strands are opposite one another.

antitermination   A regulatory mechanism in which RNA polymerase can be made to ignore a transcription termination sequence by the activity of a specific antitermination protein.

aporepressor   A protein that is converted to a repressor when another (specific) molecule binds to it.

artificial selection   Selection imposed by a breeder, in which individuals of only certain phenotypes are allowed to breed.

ascus   A sac containing the spores (ascospores) produced by meiosis in certain groups of fungi, which include *Neurospora* and yeast.

assortative mating   Nonrandom selection of mating partners with respect to one or more traits; it is positive when like phenotypes mate more frequently than would be expected by chance, and negative when the reverse occurs.

ATP Adenosine triphosphate, the primary molecule for storing chemical energy in a living cell.

attached-X chromosome A chromosome in which two X chromosomes are joined to a common centromere.

attenuator A regulatory base sequence near the beginning of an mRNA molecule at which transcription can be terminated if the mRNA is not to be made; when an attenuator is present, it precedes coding sequences.

autopolyploid A polyploid with more than two sets of homologous chromosomes.

autoradiography A process for the production of a photographic image of the distribution of a radioactive substance in a cell or large cellular molecule; the image is produced on a photographic emulsion by decay emission from the radioactive material.

autoregulation Regulation of synthesis of the product of a gene by the product itself.

autosome Any chromosome that is not a sex chromosome.

auxotroph A mutant microorganism unable to synthesize a compound required for its growth, but able to grow if the compound is provided.

B cells White blood cells, derived from bone marrow, with the potential for producing antibodies.

backcross The cross of an F_1 heterozygote with an individual having the same genotype as one of its parents.

back mutation Reversion of a mutant phenotype to the wildtype, resulting from either a precise reversal of the original mutational change or the occurrence of a second mutation that compensates for the effect of the initial forward mutation.

bacterial attachment site The DNA base sequence in a bacterium at which a prophage is inserted.

bacteriophage A virus whose host is a bacterium; commonly called a phage.

Barr body A condensed, inactivated mammalian X chromosome that stains darkly during interphase.

base analogue A purine or pyrimidine that is chemically similar to one of the normal bases and that can be incorporated into DNA.

base pair A pair of nitrogenous bases, most commonly one purine and one pyrimidine, held together by hydrogen bonds in a double-stranded region of a nucleic acid molecule; commonly abbreviated bp; in some contexts the term is used interchangeably with the term nucleotide pair.

base stacking The tendency of bases in a polynucleotide chain to be oriented with their planes parallel and with their extended surfaces nearly in contact.

bivalent A pair of homologous chromosomes, each consisting of two chromatids, associated during meiosis I.

blood group system A set of antigens on red blood cells resulting from the action of a series of alleles of a single gene, such as the ABO, MN, or Rh blood group systems.

blunt end A terminus of a DNA molecule in which all terminal bases are base-paired; the term usually refers to termini formed by a restriction enzyme that does not produce single-stranded ends.

branch migration In a DNA molecule in which two single polynucleotide strands having common base sequences are base-paired to a single complementary strand, the process in which the size of the base-paired region of one strand increases at the expense of the size of the base-paired region of the other.

branched pathway A metabolic pathway in which an intermediate serves as a precursor for more than one product.

breathing Transient breaking and re-formation of base pairs in double-stranded DNA.

broad-sense heritability The ratio of genotypic variance to total phenotypic variance.

cAMP *See* **cyclic AMP.**

cAMP-CAP The regulatory complex consisting of cyclic AMP (cAMP) and the CAP protein, needed for transcription of certain operons.

cap A complex structure at the 5′ termini of most eukaryotic mRNA molecules, having a 5′-5′ linkage.

CAP protein The protein that binds cAMP and regulates the activity of inducible operons in prokaryotes; the letters are an abbreviation for catabolite activator protein; also called CRP, for cAMP receptor protein.

carboxyl group The chemical group —COOH.

carboxyl terminus The end of a polypeptide chain at which the amino acid has a free carboxyl group.

carcinogen A physical agent or chemical reagent that causes cancer.

carcinostasis Inhibition of tumor growth.

carrier A heterozygote, carrying a recessive allele that is not expressed because of the presence of the dominant allele.

catabolite activator protein *See* **CAP protein.**

c-DNA *See* **complementary DNA.**

cell cycle The growth cycle of an individual cell; in eukaryotes, it is subdivided into G_1 (gap 1), S (DNA synthesis), G_2 (gap 2), and M (mitosis).

centromere The region of the chromosome associated with spindle fibers and involved in normal chromosome movement during mitosis and meiosis.

chain termination mutation A mutation in which a stop codon has been produced, resulting in premature termination of synthesis of a polypeptide chain.

charged tRNA A tRNA molecule to which an amino acid is linked; acylated tRNA.

chiasma The cytological manifestation of crossing over; the cross-shaped exchange configuration between nonsister chromatids of homologous chromosomes first visible in diplotene tetrads; plural is chiasmata.

chi-square (χ^2) test A statistical method commonly used in genetics to determine whether observed data fit a theoretical expectation.

chromatid One of the longitudinal subunits produced by chromosome replication and joined to its sister chromatid at the centromere.

chromatin The aggregate of DNA and histone proteins that makes up a eukaryotic chromosome.

chromatosome In chromatin, the aggregate of DNA and the histone octamer, in which the DNA is wrapped around the histones in two complete turns.

chromocenter The aggregate of centromeres and adjacent heterochromatin in nuclei of *Drosophila* larval salivary gland cells.

chromomere A tightly coiled, beadlike region of a chromosome most readily seen during leptotene and zygotene of meiosis; the beads are in register in a polytene chromosome, resulting in the banded appearance of the chromosome.

chromosome In eukaryotes, a DNA molecule, containing genes in linear sequence, to which numerous proteins are bound; a chromosome has a telomere at each end and a centromere; in prokaryotes, the DNA is associated with fewer proteins, lacks telomeres and a centromere, and is often circular; in viruses, the chromosome may be DNA or RNA, single- or double-stranded, linear or circular, and is usually free of bound proteins.

cis **configuration** The arrangement in linked inheritance in which an individual heterozygous for two mutant sites received the two mutant sites from one parent and the wildtype sites from the other parent—for example, $a^1a^2/++$; also called coupling.

cis-**dominant mutation** A mutation that affects the expression of genes only on the same chromosome as the mutation.

cis-trans **test**   *See* **complementation test.**

cistron   A DNA sequence specifying a single genetic function as defined by a complementation test; a nucleotide sequence coding for a single polypeptide; a gene.

***ClB* method**   A genetic procedure used to detect X-linked recessive lethal mutations in *Drosophila melanogaster;* so named because one X chromosome in the female parent is marked with an inversion (C), a recessive lethal allele (l), and the dominant allele for Bar eyes (B).

clone   A collection of organisms derived from a single parent and, except for acquired mutations, genetically identical to that parent; also, in genetic engineering, the linking of a specific gene or DNA fragment to a replicable DNA molecule, such as a plasmid or phage DNA.

cloning vehicle   A DNA molecule, capable of replication, in which a gene or DNA segment is inserted by recombinant DNA techniques; also called a vector.

co-conversion   Gene conversion of two linked genetic markers in a single ascus.

coding strand   In a particular gene, the DNA strand that is transcribed.

codominance   The expression of both alleles in a heterozygote.

codon   A sequence of three nucleotides in an mRNA molecule specifying either an amino acid or a stop signal in protein synthesis.

coefficient of coincidence   An experimental value obtained by dividing the observed number of double crossovers by the expected number calculated from the assumption that the two exchanges occur independently.

cohesive ends   Complementary single strands at the termini of a double-stranded DNA molecule.

cointegrate   An intermediate in transposition, in which two copies of a transposable element are contained in the same DNA molecule.

Col factor   *See* **colicinogenic factor.**

colchicine   A chemical that prevents formation of the spindle during nuclear division.

colicinogenic factor   A plasmid that contains genes for synthesis of colicins.

colicins   Small macromolecules, produced by some bacterial strains, that prevent growth of other bacterial strains.

colinearity   The linear correspondence between the sequence of amino acids in a polypeptide chain and the corresponding nucleotide sequence in the DNA molecule.

colony   A visible cluster of bacteria formed on a solid growth medium by repeated division of a single parent bacterium and its daughter cells.

colony hybridization assay   A technique used in genetic engineering to localize colonies containing particular DNA segments; many colonies are transferred to a filter, lysed, and exposed to radioactive DNA or RNA complementary to the DNA sequence of interest, and then bound radioactivity, which identifies colonies containing a complementary sequence, is located by autoradiography; also called *in situ* hybridization assay.

combinatorial joining   The mechanism by which antibody variability is produced, resulting from many possible combinations of V and J regions that can be joined to produce a light-chain gene, and the many possible combinations of $V, D,$ and J regions that can be joined to produce a heavy-chain gene.

common ancestor   An ancestor of both the father and the mother of an individual.

compatible   Blood or tissue that can be transfused or transplanted without rejection.

competent   In bacteria, able to take in bacterial DNA, making transformation possible.

complementary DNA   A DNA molecule made by copying RNA with reverse transcriptase; usually abbreviated c-DNA.

complementation test   A genetic test to determine whether two mutations occur in the same functional gene and are allelic, or in different functional genes and are nonallelic.

complex   A term used to refer to an ordered aggregate of molecules, as in an enzyme-substrate complex or a DNA-histone complex.

conditional lethal mutation   A mutation lethal to the organism in certain (*restrictive*) environmental conditions, but not lethal in other (*permissive*) conditions.

congenital   Present at birth.

conidium   An asexual spore produced by a specialized hypha in certain fungi; plural is conidia.

conjugation   A process of DNA transfer in sexual reproduction in unicellular organisms; in *E. coli* the transfer is unidirectional, from donor cell to recipient cell.

consanguineous mating   A mating between related individuals.

consensus sequence   A generalized base sequence derived from closely related sequences that occur in many locations in a genome or in many organisms; existing sequences usually have a single function and differ by only one or two bases from the consensus sequence.

conserved sequence   A base sequence that has changed only very slightly in the course of millions of years.

constant region   The part of the heavy and light chains of an antibody molecule that has the same amino acid sequence among all antibodies derived from the same heavy-chain and light-chain genes.

constitutive synthesis   Synthesis of a particular mRNA molecule (and its encoded protein) at a constant rate, independent of the presence or absence of any molecule that interacts with the protein, for example, the substrate, if the protein is an enzyme.

continuous variation   Variation in which the phenotypic differences for a trait do not fall into discrete classes but occur in the population in a continuous range from one extreme to the other.

coordinate regulation   Control of synthesis of several proteins by a single regulatory element; in prokaryotes, the proteins are usually translated from a single mRNA molecule.

core enzyme   An RNA polymerase molecule that lacks the subunit required to recognize promoters.

core particle   The aggregate of histones and DNA in a nucleosome, without the linking DNA.

core sequence   The DNA base sequence in a prophage attachment site in which exchange occurs; also called the *O* region.

correlated response   Change of the mean in one trait in a population accompanying selection for another trait.

correlation coefficient   A measure of association between pairs of numbers, equaling the covariance divided by the product of the standard deviations.

Cot curve   A graph relating the amount of renatured DNA and the product of renaturation time and initial concentration of DNA; used to determine the renaturation rate.

cotransduction   Transduction of two or more linked genetic markers by one transducing particle.

cotransformation   Transformation in bacteria of two genetic markers carried on a single DNA fragment.

counterselected marker   A mutation used to prevent growth of a donor cell in an Hfr $\times$ F^- bacterial mating.

coupled transcription-translation   In prokaryotes, the translation of an mRNA molecule before its synthesis is completed.

coupling *See cis* **configuration.**

covalent bond A chemical bond in which electrons are shared.

covalently closed circle A circular double-stranded DNA molecule in which each polynucleotide strand is an uninterrupted circle.

covariance A measure of association between pairs of numbers defined as the average product of the deviations from the respective means.

crossing over A process of exchange occurring between nonsister chromatids of a pair of homologous chromosomes and resulting in the recombination of linked genes.

curing Removal of either a plasmid or a prophage from a bacterium.

cyclic AMP A molecule used in the regulation of cellular processes; its synthesis is regulated by glucose metabolism; its action is mediated by the CAP protein in prokaryotes and protein kinases (enzymes that phosphorylate proteins) in eukaryotes. Its correct name is cyclic adenosine monophosphate, which is usually abbreviated cAMP.

cytogenetics The discipline concerned with the genetic implications of chromosome structure and behavior.

cytological hybridization *See in situ* **hybridization.**

deamination Removal of an amino ($—NH_2$) group from a molecule.

deficiency *See* **deletion.**

degeneracy *See* **redundancy.**

deletion Loss of a segment of the genetic material from a chromosome; also called deficiency.

deletion mapping The use of overlapping deletions to locate a gene on a chromosome or a genetic map.

deme *See* **local population.**

denaturation Loss of the normal three-dimensional shape of a macromolecule without breaking covalent bonds, usually accompanied by loss of its biological activity; conversion of DNA from the double-stranded to the single-stranded form; unfolding of a polypeptide chain.

denaturation mapping An electron-microscopic technique for localizing regions of a double-stranded DNA molecule by noting the positions at which the individual strands separate during early stages of denaturation.

deoxyribonuclease An enzyme that breaks sugar-phosphate bonds in DNA, forming either fragments or the component nucleotides; abbreviated DNase.

deoxyribonucleic acid *See* **DNA.**

derepression Activation of a gene or set of genes by inactivation of a repressor.

deoxyribose The five-carbon sugar present in DNA.

deviation In statistics, a difference from an expected value.

diakinesis The substage of meiotic prophase I that precedes metaphase I, in which the bivalents attain maximum shortening and condensation.

dicentric chromosome A chromosome having two centromeres.

differentiation The complex changes that occur in progressive diversification in cellular structure and function during development of an organism; for a particular line of cells, this results in a continual restriction in the types of transcription and synthesis of which each cell is capable.

dihybrid An individual heterozygous for two pairs of alleles; a cross between individuals with different alleles at two gene loci.

diploid A cell or organism with two complete sets of homologous chromosomes.

diplotene The substage of meiotic prophase I that immediately follows pachytene and precedes diakinesis, in which pairs of sister chromatids that make up a bivalent (tetrad) begin to separate from each other and chiasmata become visible.

direct repeat Two copies of a DNA or RNA base sequence having the same orientation.

discontinuous variation Variation in which the phenotypic differences for a trait fall into two or more discrete classes.

disjunction Separation of homologous chromosomes to opposite poles of a division spindle during anaphase of a mitotic or meiotic nuclear division.

disulfide bond Two sulfur atoms covalently linked; found in proteins when the sulfur atoms in two different cysteines are joined.

division spindle *See* **spindle.**

dizygotic twins Twins that result from the fertilization of separate ova; genetically related as siblings.

DNA The macromolecule, usually composed of two polynucleotide chains in a double helix, that is the carrier of the genetic information in all cells and many viruses.

DNA gyrase One of a class of enzymes called topoisomerases, which function during DNA replication to relax positive supercoiling of the DNA molecule and which introduce negative supercoiling into nonsupercoiled molecules early in the life cycle of many phages.

DNA ligase An enzyme that catalyzes formation of a covalent bond between adjacent 5'-P and 3'-OH termini in a broken polynucleotide strand of double-stranded DNA; a few DNA ligases can join two double-stranded DNA molecules.

DNA polymerase Any enzyme that catalyzes synthesis of DNA from deoxynucleoside 5'-triphosphates under the direction of a DNA template strand.

DNA repair Any one of several different processes for restoration of the correct base sequence of a DNA molecule into which incorrect bases have been incorporated or whose bases have been modified in some way.

DNA replication The copying of a DNA molecule.

DNase *See* **deoxyribonuclease.**

domain A folded region of a polypeptide chain that is spatially isolated from other folded regions.

dominance variance The magnitude of the genotypic variance resulting from the dominance effects of alleles affecting the trait.

dominant An allele, or the corresponding phenotypic trait, that is expressed in a heterozygote.

dosage compensation A mechanism in mammals in which random inactivation of one X chromosome in females results in equal amounts of the products of X-linked genes in males and females; also, regulation of certain autosomal loci resulting in the same amount of the gene product in homozygous dominants and heterozygotes.

Down syndrome Trisomy 21; karyotype $47, +21$.

drug-resistant plasmid A plasmid encoding genes whose products inactivate certain antibiotics.

duplex A double-stranded polynucleotide.

duplication A chromosome aberration in which a chromosome segment occurs more than once in the haploid genome; if the two segments are adjacent, the duplication is a tandem duplication.

editing function The activity of DNA polymerases that removes incorrectly incorporated nucleotides; the proofreading function.

80S ribosome The active form of eukaryotic ribosomes, consisting of one 40S and one 60S subunit.

electrophoresis A technique used to separate molecules based on their different rates of movement, induced by an applied electric field, through a liquid or a gel; the movement is often called electrophoretic migration.

elongation   Addition of amino acids to a growing polypeptide chain.

embryo   An organism in the early stages of development; the second through seventh week in humans.

endomitosis   Chromosome replication that is not accompanied by nuclear or cytoplasmic division.

endonuclease   An enzyme that breaks internal phosphodiester bonds in a single- or double-stranded nucleic acid molecule; usually, specific for either DNA or RNA.

endosperm   Nutritive tissues formed adjacent to the embryo in most flowering plants: in most diploid plants the endosperm is triploid.

enhancer   A base sequence in eukaryotes and eukaryotic viruses that increases the rate of transcription of nearby genes; its defining characteristics are that it need not be adjacent to the transcribed gene and that its enhancing activity is independent of its orientation with respect to the gene.

environmental variance   The magnitude of the phenotypic variance attributable to differences in environment among individuals.

enzyme   A protein or ordered aggregate of proteins that catalyzes a specific biochemical reaction and is not itself altered in the process.

epistasis   Interaction between nonallelic genes such that one gene interferes with or prevents expression of the other.

equilibrium centrifugation   Centrifugation of molecules until there is no longer any net movement of the molecules; in a density gradient, each molecule comes to rest when its density equals that of the solution.

erythroblastosis fetalis   Hemolytic disease of the newborn; blood cell destruction occurring when anti-Rh$^+$ antibodies in a mother cross the placenta and attack Rh$^+$ cells in a fetus.

estrogen   A female sex hormone; of great interest in the study of cellular regulation in eukaryotes because the number of types of target cells is large.

ethidium bromide   A fluorescent molecule that binds to DNA and changes its density; used to purify supercoiled DNA molecules and to localize DNA in gel electrophoresis.

euchromatin   Chromatin or a region of a chromosome having normal staining properties and undergoing the normal cycle of condensation; relatively uncoiled in the interphase nucleus (compared to condensed chromosomes), and apparently containing most of the genes.

eukaryote   A cell or an organism composed of cells with true nuclei (DNA enclosed in nuclear membranes), membrane-bounded cytoplasmic organelles, and in which nuclear division occurs by mitosis and meiosis.

euploid   A cell or an organism having a chromosome number that is an exact multiple of the haploid number.

evolution   Cumulative change in the genetic characteristics of a species through time.

excision   Removal of a DNA fragment from a chromosome; prophage excision.

excisionase   An enzyme that is needed for prophage excision; works together with an integrase.

exon   The DNA segments of a gene that are transcribed and translated into a polypeptide chain and are separated by noncoding intervening sequences called introns; the intron-exon distinction is primarily found in eukaryotic genes.

exonuclease   An enzyme that removes a terminal nucleotide in a polynucleotide chain by cleavage of the terminal phosphodiester bond; nucleotides are removed successively, one by one; usually specific for either the 5'-P or 3'-OH terminus, DNA or RNA, and often specific for either single-stranded or double-stranded nucleic acids.

expressivity   The degree of phenotypic expression of a penetrant gene.

F₁   The first filial generation, or the first generation of descent from a given mating.

F₂   The second filial generation, produced by intercrossing or self-fertilization of F₁ individuals.

F plasmid   A bacterial plasmid often called the F factor, fertility factor, or sex plasmid, capable of transferring itself from a host (F^+) cell to a cell not carrying an F factor (F^- cell); when an F factor is integrated into the bacterial chromosome (in an Hfr cell), the chromosome becomes transferrable to an F^- cell during conjugation.

F′ plasmid   An F plasmid that contains genes obtained from the bacterial chromosome in addition to plasmid genes; formed by aberrant excision of an integrated F, taking along adjacent bacterial DNA.

feedback inhibition   Inhibition of an enzyme by the product of the enzyme or, in a metabolic pathway, by a product of the pathway.

50S ribosomal subunit   The large subunit of a prokaryotic ribosome.

first-division segregation   Separation of a pair of alleles into different nuclei during the first meiotic division, occurring in the absence of crossing over between the gene and the centromere of the pair of homologous chromosomes.

fitness   A measure of the average ability of organisms with a given genotype to survive and reproduce.

fixation   The state at which the frequency of an allele equals 1.0.

fixation index   A measure of the amount of genetic divergence between two populations, based on the decrease of average heterozygosity from the value expected with fusion and random mating of the populations.

fluctuation test   A statistical test used to determine whether bacterial mutations occur at random or are produced in response to selective agents.

folded chromosome   The form of DNA of a bacterium in which the circular DNA is folded to have a compact structure; contains protein.

40S ribosomal subunit   The small subunit of a eukaryotic ribosome.

forward mutation   A change from a normal or a wildtype allele to a mutant allele.

founder effect   Random genetic drift resulting when a group of founders of a population are not genetically representative of the population from which they were derived.

frameshift mutation   A mutational event caused by either the insertion or deletion of one or more nucleotide pairs in a gene, resulting in a shift in the reading frame of all codons following the mutational site.

G₁   *See* **cell cycle.**

G₂   *See* **cell cycle.**

β-galactosidase   The enzyme produced by a gene in the *lac* operon responsible for cleavage of lactose.

gamete   A mature reproductive cell, such as sperm or egg in animals.

gametophyte   The haploid, gamete-producing generation in plants (reduced in higher plants to the embryo sac and the pollen grain), which alternates with the diploid, spore-producing generation (sporophyte).

gene   The hereditary unit that occupies a fixed chromosomal locus, contains the information for a polypeptide chain, tRNA molecule, or rRNA molecule encoded in a specific nucleotide sequence, and can mutate to various forms.

gene amplification   A process in which certain genes undergo differential replication either within the chromosome or extrachromosomally, increasing the number of copies of the gene.

gene conversion   The phenomenon in which the products of a meiotic division in an *Aa* heterozygous individual occur in some ratio other than the expected $1A:1a$—for example, $3A:1a$, $1A:3a$, $5A:3a$, or $3A:5a$.

gene dosage   The increase by a factor of *n* of the amount of a gene product present in a cell when *n* copies, rather than one copy, are present.

gene expression   The multistep process, and the regulation of the process, by which the product of a gene is synthesized.

gene flow   Exchange of genes at a low rate between two populations, the result of either dispersal of gametes or migration of individuals; also called migration.

gene library   A large collection of cloning vectors containing a complete (or nearly complete) set of fragments of the genome of an organism.

gene pool   The total genetic information in a population of sexually reproducing organisms.

gene product   A term used for the polypeptide chain translated from an mRNA molecule transcribed from a gene; if the RNA is not translated (for example, ribosomal RNA), the RNA molecule is called the gene product.

gene therapy   Induced alteration of the genome of a higher organism by addition of a viral cloning vector containing a particular gene; the technique is not yet developed practically, but holds the promise of eliminating genetic defects in humans.

general recombination   *See* **homologous recombination.**

generalized recombination   *See* **transducing phage.**

genetic code   The set of 64 triplets of bases (codons) corresponding to each amino acid and to signals for initiation and termination of polypeptide synthesis.

genetic differentiation   Accumulation of differences in allele frequency between isolated or semiisolated populations.

genetic divergence   *See* **genetic differentiation.**

genetic engineering   Linking two DNA molecules by *in vitro* manipulations for the purpose of generating a novel organism with desired characteristics.

genetic equilibrium   In a group of interbreeding individuals, the condition in which the frequencies of particular alleles remain constant in successive generations; also called equilibrium.

genetic map   *See* **linkage map.**

genome   The total complement of genes contained in a cell or virus; commonly used in eukaryotes to refer to all genes present in one complete haploid set of chromosomes.

genotype   The genetic constitution of an organism or virus, as distinguished from its appearance or phenotype; often used to refer to the allelic composition of one or a few genes of interest.

genotype–environment association   The condition in which genotypes and environments do not occur in random combinations.

genotype–environment interaction   The condition in which genetic and environmental effects on a trait are not additive.

genotype frequency   The proportion of individuals in a population that are of a prescribed genotype.

genotypic variance   The magnitude of the phenotypic variance attributable to differences in genotype among individuals.

germinal mutation   A mutation occurring in a cell from which gametes are derived, as distinguished from a somatic mutation.

gynandromorph   A sexual mosaic; an individual exhibiting both male and female sexual differentiation.

H substance   The carbohydrate precursor of the A and B red-blood-cell antigens.

H1, H2A, H2B, H3, H4   The five major histones in chromatin.

haploid   A cell or organism having only one set of chromosomes.

haplotype   The allelic form of each gene of the major histocompatibilty complex that is present in a single chromosome.

Hardy-Weinberg rule   The genotype frequencies expected with random mating.

HAT medium   A growth medium for animal cells, containing hypoxanthine, aminopterin, and thymidine, used in the selection of hybrid cells.

heavy (H) chains   The larger polypeptide chains in an antibody molecule.

helicase   An enzyme that participates in DNA replication by unwinding double-stranded DNA in or near the replication fork.

hemizygous gene   A gene present in only one dose, as the genes on the X chromosome in heterogametic males.

heritability   A measure of the degree to which a phenotypic trait can be modified by selection. *See also* **broad-sense heritability** and **narrow-sense heritability.**

heterochromatin   Chromatin that remains condensed and heavily staining during interphase; commonly present adjacent to the centromere and in the telomeres of chromosomes; some chromosomes are composed primarily of heterochromatin.

heteroduplex   A double-stranded nucleic acid molecule in which the two strands have different hereditary origins; produced either as an intermediate in recombination or by the *in vitro* annealing of single-stranded complementary molecules.

heterogametic sex   The sex that produces equal numbers of unlike gametes with respect to the sex chromosomes.

heterogeneous nuclear RNA (hnRNA)   The collection of primary RNA transcripts and incompletely processed products found in the nucleus of a eukaryotic cell.

heterokaryon   A cell or individual having nuclei from genetically different sources, the result of cell fusion not accompanied by nuclear fusion.

heterosis   Superiority of hybrids over either inbred parent with respect to one or more traits; also called hybrid vigor.

heterozygote   A diploid or polyploid individual having dissimilar alleles at one or more loci and therefore not true-breeding for the traits determined by these loci.

hexaploid   A cell or organism with six complete sets of chromosomes.

Hfr cell   An *E. coli* cell in which an F plasmid is integrated into the chromosome, enabling transfer of part or all of the chromosome to an F^- cell.

high-copy-number plasmid   A plasmid for which there are usually considerably more than two copies (often greater than 20) per cell.

histocompatability   Acceptance by a recipient of transplanted tissue from a donor.

histocompatibility antigens   Tissue antigens that determine transplant compatibility or incompatibility.

histone   Any of the small basic proteins bound to DNA in chromatin; the five major histones are designated **H1, H2A, H2B, H3,** and **H4.**

holandric gene   A gene carried on the Y chromosome and therefore transmitted only from father to son.

Holliday junction   The region of an intermediate in genetic recombination in which two double-stranded DNA molecules are joined by two single strands crossing from one molecule to the other.

homeotic mutation   A mutation that results in the replacement of one body structure by another body structure during development.

homogametic sex   The sex that produces only one kind of gamete with respect to the sex chromosomes.

homologous   When referring to DNA, having the same or nearly the same nucleotide sequence.

homologous chromosomes   Chromosomes that pair during meiosis and have the same genetic loci and structure; also called homologues.

homologous recombination   Genetic exchange between identical or nearly identical DNA sequences; also called general recombination; requires a RecA-type protein.

homopolymer   A polynucleotide consisting of only a single type of nucleotide.

homopolymer tail-joining   A technique in genetic engineering for joining two DNA molecules by adding a homopolymer to each end of one DNA molecule and a complementary homopolymer to each end of a second DNA molecule, followed by annealing and ligation.

homozygote   A diploid or polyploid having the same allele at a given locus and therefore true-breeding for the trait determined by the locus.

hormone   A small molecule in higher eukaryotes, synthesized in specialized tissue, that regulates the activity of other specialized cells; in animals, hormones are transported from their source to a target tissue by the blood stream.

hot spot   A site in a DNA molecule whose mutation rate is much higher than the rate for most other sites.

hybrid   An individual produced by the mating of genetically unlike parents; a duplex nucleic acid molecule produced of strands derived from different sources.

hybrid vigor   *See* **heterosis.**

hybridoma   A cell hybrid between an antibody-producing B cell and certain types of tumor cells; hybridomas survive well in laboratory culture and continue to produce monoclonal antibody.

hydrogen bond   A weak noncovalent linkage in which a hydrogen atom is shared by two atoms.

hydrophobic interaction   A noncovalent interaction between nonpolar molecules or nonpolar groups, causing the molecules or groups to cluster when water is present.

hyperchromicity   The increase in the optical absorbance of a nucleic acid solution when the nucleic acid is denatured.

hypersensitive site   A region of a eukaryotic DNA molecule that is easily cut by nucleases when the DNA is in chromatin; believed to be a regulatory region.

hypha   The filamentous cellular structure that constitutes the body or mycelium of a fungus.

identical twins   *See* **monozygotic twins.**

immune response   The phenomenon in which exposure of an organism to a foreign molecule results in the synthesis of an antibody directed against the substance and resynthesis when the organism is subsequently exposed to the same substance; also refers to complete or partial resistance to infection following a previous infection by a disease-causing agent.

immunity   A general term for resistance of an organism to specific substances but with several specialized meanings; in higher animals, it refers to the ability to respond to foreign molecules by synthesizing proteins, called antibodies, that can inactivate or precipitate the molecules; also, in animals it refers to being not susceptible to infection by a disease-causing agent; in bacterial systems, immunity refers to (1) resistance of a lysogen to infection by a phage having the same repressor-operator system as the prophage and (2) resistance of a bacterium containing a colicinogenic plasmid to the colicin encoded in the plasmid.

immunoglobulin   One of several classes of antibody protein.

immunoglobulin class   The category in which an immunoglobulin is placed based on its chemical characteristics, including the identity of its heavy chain.

in situ **hybridization**   Usually refers to renaturation of a radioactive nucleic acid to a cell whose DNA has been denatured; used to localize particular DNA molecules within a cell, in particular, chromosomes and parts of chromosomes; also called cytological hybridization.

in situ **hybridization assay**   *See* **colony hybridization assay.**

in vitro **experiment**   An experiment carried out with components isolated from cells.

in vivo **experiment**   An experiment performed with intact cells.

inborn error of metabolism   A genetically determined biochemical disorder, usually in the form of an enzyme defect that produces a metabolic block.

inbreeding   Mating between genetically related individuals.

inbreeding coefficient   A measure of the genetic effects of inbreeding in terms of the proportionate reduction in heterozygosity in an inbred individual as compared to the heterozygosity expected with random mating.

inbreeding depression   Deterioration in fitness or performance of a population accompanying inbreeding.

incompatible   A term referring to blood or tissue whose transfusion or transplantation results in rejection.

independent assortment   Random distribution, to gametes, of alleles of genes located on different (nonhomologous) chromosomes.

inducer   A small molecule that inactivates a repressor; usually binds to the repressor, thereby altering the ability of the repressor to bind to an operator.

inducible enzyme   An enzyme that is synthesized only in the presence of its substrate or certain molecules chemically related to its substrate; the inducer inactivates a repressor of transcription; the term inducible enzyme contrasts with constitutive enzyme.

initiation factors   Proteins required for the initiation of protein synthesis; abbreviated IF (in prokaryotes) or eIF (in eukaryotes) followed by a number.

insertion sequence   Any of a number of DNA sequences capable of transposition in a prokaryotic genome; such sequences do not carry identifiable bacterial genes.

insertional inactivation   Inactivation of a gene by interruption of its coding sequence; used in genetic engineering as a means of detecting insertion of a foreign DNA sequence into the coding region of a gene, the inactivation of which can be selected by a plating test.

integrase   An enzyme that catalyzes the site-specific exchange occurring when a prophage is inserted into or excised from a bacterial chromosome; in the excision process, an accessory protein, excisionase, is also needed.

integration   The process by which one DNA molecule is inserted intact into another replicable DNA molecule, as in prophage integration and integration of plasmid or tumor viral DNA into a chromosome.

intergenic complementation   Complementation between mutations in different genes. *See* also **complementation test.**

intergenic suppressor   A mutation that suppresses the effect of another mutation in a different gene; often refers specifically to a tRNA molecule able to recognize a stop codon or two different sense codons.

internal resolution site   A region in some transposable elements in which a site-specific exchange occurs between two elements.

interphase   The interval between nuclear divisions in the cell cycle, extending from the end of telophase of one division to the beginning of prophase of the next division.

interrupted mating   In an Hfr $\times F^-$ cross a technique by which donor and recipient cells are broken apart at specific times, allowing only a particular amount of DNA to be transferred.

intervening sequence   *See* **intron.**

intragenic complementation   Complementation between different mutations in the same gene. *See* also **complementation test.**

intragenic suppressor   A mutation that suppresses the effect of another mutation in the same gene.

intron   A transcribed noncoding DNA sequence that is within a gene and is excised from a primary transcript in forming a mature mRNA molecule; found primarily in eukaryotic cells; *see* also **exon.**

inversion   A structural aberration in a chromosome in which the order of several genes is reversed from the normal order; a pericentric inversion includes the centromere within the inverted region, and a paracentric inversion does not include the centromere.

inverted repeat   Two base sequences whose orientation in a particular DNA molecule are opposite from one another; often found at the ends of transposable elements.

ionizing radiation   Electromagnetic or particulate radiation that produces ion pairs when dissipating its energy in matter.

IS element   *See* **insertion sequence.**

isochromosome   A chromosome with two identical arms containing homologous loci.

isoenzymes   A set of enzymes, each of which carries out the same chemical reaction, and which are inhibited by different molecules; also called isozymes.

isolating mechanism   Any barrier that prevents the exchange of genes between two or more related groups of organisms; usually classified as behavioral, ecological, geographical, or reproductive.

isotopes   The forms of a chemical element having the same number of electrons and protons but differing in the number of neutrons in the atomic nucleus; unstable isotopes undergo transitions to a more stable state and, in so doing, emit radioactivity.

J regions   Multiple DNA sequences coding for alternative amino acid sequences of part of the variable region of an antibody molecule. The J regions of heavy and light chains are different.

karyotype   The chromosome complement of a cell or an individual; often represented by an arrangement of metaphase chromosomes according to their lengths and to the positions of their centromeres.

kindred   A group of related individuals.

Klinefelter syndrome   The clinical condition of human males with the karyotype 47,XXY.

***lac* operon**   The set of genes required to metabolize lactose in bacteria.

lactose   A 12-carbon sugar consisting of the simple sugars glucose and galactose covalently linked.

lactose permease   An enzyme responsible for transport of lactose from the environment into bacteria.

lagging strand   The single DNA strand synthesized in short fragments that are ultimately joined together.

leader sequence   The region of an mRNA molecule from the 5′ end to the beginning of the coding sequence, sometimes containing regulatory sequences; in prokaryotic mRNA it contains the ribosomal binding site.

leading strand   The single DNA strand that is synthesized as a continuous unit.

leptotene   The initial substage of meiotic prophase I, during which the chromosomes become visible by light microscopy as unpaired threadlike structures.

lethal mutation   A mutation that results in death of affected individuals before they reach reproductive age.

liability   Risk.

light (L) chains   The small polypeptide chains in an antibody molecule.

linkage   The tendency of nonallelic genes located in the same chromosome to be associated in inheritance more frequently than expected from their independent assortment during meiosis.

linkage group   A group of genes having their loci on the same chromosome.

linkage map   A chromosome map showing the relative locations of the known genes on the chromosomes of a given species; also called a genetic map.

linker   In genetic engineering, synthetic DNA fragments that contain restriction-enzyme cleavage sites and that are used to join two DNA molecules.

local population   A group of individuals of the same species occupying an area within which most individuals find their mates; synonomous terms are deme and Mendelian population.

locus   The site or position of a particular gene on a chromosome.

lysis   Breakage of a cell caused by rupture of its cell membrane and cell wall.

lysogenic bacterium   A bacterial cell that carries the DNA of a temperate phage and transmits the DNA to daughter cells; the DNA is usually, but not always, integrated into the chromosome of the cell.

lysogenic conversion   Alteration of the phenotype of a bacterium by acquisition of a prophage; for example, the diphtheria-causing ability of some strains of *Corynebacterium diphtheriae* is caused by a prophage.

lysogeny   The phenomenon in which the DNA of an infecting phage is repressed and stably becomes part of the genetic material of a bacterium.

lysozyme   One of a class of enzymes, all of which dissolve the cell wall of bacteria; found in chicken egg white and human tears, and made by many phages.

lytic cycle   The life cycle of a phage in which progeny phage are produced and the host bacterial cell is lysed.

M   *See* **cell cycle.**

major histocompatibility complex (MHC)   The group of closely linked genes encoding antigens that play a major role in tissue incompatibility and that function in regulation and other aspects of the immune response.

map unit   A unit of distance in a linkage map that corresponds to a recombination frequency of one percent.

masked mRNA   Messenger RNA present in eukaryotic cells, particularly eggs, that cannot be translated until specific regulatory substances are available; storage mRNA.

maternal effect   A phenomenon in which the genotype of a mother affects the phenotype of the offspring through substances present in the cytoplasm of the egg.

maternal inheritance   The extranuclear inheritance of a trait through cytoplasmic factors of organelles contributed by the female gamete.

mating system   The norms by which individuals in a population choose their mates; important systems of mating include random mating, assortative mating, and inbreeding.

Maxam-Gilbert method   A technique for determining the nucleotide sequence of DNA.

mean   The arithmetical average.

meiocyte   A germ cell that undergoes meiosis to yield gametes in animals or spores in plants.

meiosis   The process of nuclear division during gametogenesis or sporogenesis in which one replication of the chromosomes is followed by two successive divisions of the nucleus to produce four haploid nuclei.

melting curve   A graph showing the progress of denaturation of a macromolecule, usually DNA, relating a quantitative feature of a solution of the macromolecule to an agent causing denaturation; most commonly, a graph of optical absorbance, which increases with the extent of denaturation, and temperature.

melting temperature   The temperature at which half the base pairs of a population of double-stranded nucleic acid molecules are broken; designated T_m.

Mendelian population   *See* **local population.**

meristic trait   A trait the phenotype of which can be represented by a whole number, such as the number of fingers on the hands.

messenger RNA   An RNA molecule transcribed from a complementary DNA sequence and able to be translated into the amino acid sequence of a polypeptide.

metabolic pathway   A set of chemical reactions that occur in a definite order to convert a particular starting molecule to one or more specific products.

metacentric chromosome   A chromosome with a centrally located centromere having two arms of about equal length.

metafemale   A weak sterile *Drosophila* female with three sets of X chromosomes and two sets of autosomes.

metamale   A weak sterile *Drosophila* male with one X chromosome and three sets of autosomes.

metaphase   The stage of nuclear division in mitosis, meiosis I, or meiosis II, during which the centromeres of the condensed chromosomes are arranged in a plane between the two poles of the spindle.

methylation   The modification of a DNA or RNA base by the addition of a methyl (—CH₃) group.

MHC   *See* **major histocompatibility complex.**

migration   Movement of individuals among subpopulations; also, the movement of molecules in electrophoresis. *See* also **branch migration** for another use of the term.

minimal medium   A growth medium consisting of simple inorganic salts, a carbohydrate, vitamins, organic bases, essential amino acids, and other essential compounds; its composition is precisely known. The term minimal medium contrasts with complex medium or broth, which is an extract of biological material (vegetables, milk, meat) containing a huge number of compounds, the composition of which is unknown.

mischarged tRNA   A tRNA molecule to which an incorrect amino acid is linked.

mismatch   An arrangement in which two nucleotides opposite one another in double-stranded DNA are unable to form hydrogen bonds.

mismatch repair   Removal of one nucleotide from a pair that cannot properly hydrogen-bond, followed by replacement with a nucleotide that can hydrogen-bond.

missense mutation   An alteration in DNA that results in an amino acid substitution in an encoded polypeptide.

mitosis   The process of nuclear division in which the replicated chromosomes divide and the daughter nuclei have the same chromosome number and genetic composition as the parent nucleus.

monoclonal antibody   Antibody directed against a single antigen, produced by a single clone of B cells or a single cell line of hybridoma cells.

monohybrid   An individual heterozygous for one pair of alleles; a cross between individuals with different alleles at one gene locus.

monomorphic locus   A locus for which the most common allele has a frequency greater than 0.95.

monosomic   The aneuploid condition in which one member in a pair of chromosomes is missing; that is, the monosomic organism has $2n - 1$ chromosomes.

monozygotic twins   Twins developed from a single fertilized egg, which gives rise to two embryos at an early division; also called identical twins.

mosaic   An individual composed of two or more genetically different types of cells.

mRNA   *See* **messenger RNA.**

multiple alleles   The occurrence in a population of more than two alleles at a locus.

multiplicity of infection   The ratio of the number of phage particles and bacteria in a phage infection; abbreviated moi.

multivalent   An association of more than two homologous chromosomes resulting from synapsis during meiosis in a polysomic or polyploid individual.

mutagen   An agent that is capable of increasing the rate of mutation.

mutagenesis   The process by which a gene undergoes a heritable alteration; also called mutation.

mutant   An allele that is different from the normal or wildtype; also an individual in which such an allele is expressed in the phenotype.

mutation   A heritable alteration in a gene; sometimes also the process by which a gene aquires a heritable change.

mutation pressure   The tendency of mutation to change allele frequency; this tendency is generally very weak.

mutation rate   The probability of occurrence of a new mutation at a given locus, either per gamete or per generation.

narrow-sense heritability   The ratio of additive genetic variance to total phenotypic variance.

natural selection   The process in which individuals best suited to survive and reproduce in a particular environment form a disproportionate share of the offspring and thus gradually increase the overall ability of a population to survive and reproduce in that environment.

negative regulation   Regulation of gene expression in which mRNA synthesis does not occur until an inhibitor (a repressor) is removed from the DNA of the gene.

negative supercoiling   Supercoiling of a DNA molecule associated with underwinding of the DNA.

neutral allele   An allele that has no effect on the ability of the organism to survive and reproduce.

nick   A single-strand break in a DNA molecule.

nicked circle   A circular DNA molecule containing one or more single-strand breaks.

nondisjunction   Failure of sister chromatids in mitosis or homologous chromosomes in meiosis to separate (disjoin) and move to the opposite poles of the division spindle, with the result that one daughter receives both members of a homologous pair of chromosomes and the other daughter nucleus receives none.

nonhistone chromosomal proteins   A large class of proteins, not of the histone class, found in isolated chromosomes.

nonhomologous recombination   Genetic recombination in which the exchange is not between extended identical or nearly identical base sequences; examples are transposition and prophage integration.

nonpermissive conditions   Environmental conditions that do not allow a gene with a conditional lethal mutation to produce a functional gene product.

nonselective medium   A growth medium, used in a recombination or mutation experiment, that allows growth of all genotypes present in the experiment. The term nonselective medium contrasts with selective medium; a selective medium usually allows growth of cells carrying specific genetic markers.

nonsense mutation   A mutation that alters a codon specifying an amino acid to one coding for no amino acid, resulting in premature polypeptide chain termination; also called chain termination mutation.

normal distribution   A symmetric bell-shaped distribution curve characterized by two numbers, the mean and the variance; in a normal distribution, approximately 68 percent of the observations will be within one standard deviation from the mean and approximately 95 percent of the observations will be within two standard deviations from the mean.

nuclease An enzyme that breaks phosphodiester bonds in nucleic acid molecules.

nucleoid A DNA mass, not bounded by a membrane, within the cytoplasm of a prokaryotic cell, chloroplast, or mitochondrion; often refers to the major DNA unit of a bacterium.

nucleolar organizer region A chromosome region containing the genes for ribosomal RNA; abbreviated NOR.

nucleolus A nuclear organelle in which ribosomal RNA is made and ribosomes are partially synthesized; usually associated with the nucleolar organizer region.

nucleoside A purine or pyrimidine base covalently linked to a sugar.

nucleosome The basic repeating subunit of chromatin, consisting of a core particle composed of two molecules each of four different histones, around which a length of DNA containing about 145 nucleotide pairs is wound, joined to an adjacent core particle by about 55 nucleotide pairs of linker DNA associated with a fifth type of histone.

nucleotide A nucleoside phosphate.

nucleus The membrane-bounded organelle containing the chromosomes in a eukaryotic cell.

nullisomic The aneuploid conditions in which both members of a pair of homologous chromosomes are missing, thus having a $2n - 2$ number of chromosomes.

O **region** The common sequence in the bacterial and phage attachment sites in lysogenic systems.

ochre codon Jargon for the UAA stop codon; an ochre mutation is a UAA codon formed from a sense codon.

octaploid A cell or organisms having eight complete sets of chromosomes.

Okazaki fragment One of the short strands of DNA produced during discontinuous replication of the lagging strand; also called precursor fragment.

oncogene A gene that can initiate tumor formation.

operator A regulatory region in DNA that interacts with a specific repressor protein in controlling the transcription of adjacent structural genes.

operon A collection of genes regulated by an operator and a repressor.

organelle A membrane-bounded cytoplasmic structure having a specialized function, such as a nucleus, chloroplast, or mitochondrion.

origin A replication origin; a DNA base sequence at which replication of a chromosome is initiated.

overdominance A condition in which the fitness of a heterozygote is greater than the fitness of both homozygotes.

overlapping genes Genes that have common overlapping coding sequences.

ovule The structure in seed plants that contains the embryo sac (female gametophyte) and develops into a seed after fertilization of the egg.

P site *See* **peptidyl site.**

pachytene The middle substage of meiotic prophase I in which the homologous chromosomes are closely synapsed and the synaptonemal complex is fully formed.

palindrome In nucleic acids, a segment of DNA in which the sequence of bases on complementary strands reads the same from a central point of symmetry—for example ABCDEE'D'C'B'A', where A and A' are complementary; frequently, the sites of recognition and cleavage by restriction endonucleases are palindromic.

paracentric inversion *See* **inversion.**

parasexuality Any mechanism by which nonmeiotic recombination occurs.

partial denaturation mapping *See* **denaturation mapping.**

partial diploid A cell in which a segment of the genome is duplicated, usually in a plasmid.

partial dominance A condition in which the phenotype of the heterozygote is intermediate to those of the corresponding homozygotes and more closely resembles the phenotype of one homozygote than the other.

pedigree A diagram representing the genetic relationships of individuals.

penetrance The proportion of individuals having a specific genotype that actually express that genotype in their phenotype.

peptide bond A covalent bond between the amino ($-NH_2$) group of one amino acid and the carboxyl ($-COOH$) group of another.

peptidyl site One of the two tRNA-binding sites of a ribosome.

peptidyl transferase The enzymatic activity of ribosomes responsible for forming a peptide bond; the active site is formed from various regions of several ribosomal proteins.

pericentric inversion *See* **inversion.**

permissive condition An environmental condition that allows an organism with a conditional mutation to grow; the term contrasts with nonpermissive conditions.

phage *See* **bacteriophage.**

phenotype The observable properties of a cell or an organism, resulting from the interaction of the genotype and the environment.

phenotypic variance The variance in a phenotypic trait among individuals in a population.

phenylketonuria A hereditary condition in humans resulting from inability to convert phenylalanine to tyrosine; causes severe mental retardation unless treated in childhood by a low-phenylalanine diet; abbreviated PKU.

Philadelphia chromosome Abnormal chromosome 22 in humans, resulting from reciprocal translocation and often associated with a certain type of leukemia.

phosphodiester bond In nucleic acids, the covalent bond between a phosphate group and the 3'-OH group of a nucleoside; in extending from the 5' carbon of one sugar to the 3' carbon of the adjacent sugar, these bonds form the backbone of a nucleic acid molecule.

photoreactivation The enzymatic splitting of pyrimidine dimers produced in DNA by ultraviolet light; requires visible light and the photoreactivation enzyme.

plaque A clear area in an otherwise turbid layer of bacteria growing on a solid medium, caused by the infection and killing of the cells by a phage; since each plaque is a result of the growth of one phage, plaque counting is a way of counting viable phage particles; the term is used occasionally for animal viruses that cause clear areas in layers of animal cells grown in culture.

plasmid An extrachromosomal genetic element that replicates independently of the host chromosome; it may exist in one or many copies per cell, and may segregate in cell division to daughter cells in either a controlled or random fashion; some plasmids, such as the F factor, may be integrated into the host chromosome.

pleiotropy The condition in which a single mutant gene affects two or more distinct and seemingly unrelated traits.

point mutation A mutation caused by the substitution, deletion, or addition of a single nucleotide pair; sometimes used to designate a mutation that can be mapped to a single specific locus.

Poisson distribution A statistical distribution that describes how a large number of objects placed in a number of boxes will, on the average, be distributed among the boxes—that is, the number that will be in each box; often applied in viral infection to determine the number of infected cells in a sample and how many cells have become infected with a particular number of viruses.

polar mutation   A mutation that affects the expression of adjacent genes in a single mRNA molecule; examples are transcription-termination mutations and certain stop codons that arise within a coding sequence of a polycistronic mRNA molecule.

poly(A) tail   A naturally occurring sequence of adenines at the 3′ end of eukaryotic mRNA molecules; added to a primary transcript following cleavage at the poly(A)-addition site.

polycistronic mRNA   An mRNA molecule from which two or more polypeptides are translated; found primarily in prokaryotes.

poly(dA) tail   A sequence of deoxyadenosines added in the laboratory to one or both 3′ termini of a double-stranded DNA molecule and used in genetic engineering to join two molecules by the homopolymer tail-joining procedure—a poly(dT) tail (a sequence of thymidines) would be added to the other molecule.

polygenic inheritance   Determination of a trait by alleles of two or more genes.

polymer   A regular, covalently bonded arrangement of basic subunits or monomers into a large molecule, as a polynucleotide or polypeptide chain.

polymerase   An enzyme that catalyzes covalent joining of nucleotides—for example, DNA polymerase and RNA polymerase.

polymerization start site   The nucleotide in a promoter that is complementary to the first nucleotide of an RNA molecule to be synthesized.

polymorphic locus   A gene for which the most common allele in a population has a frequency smaller than 0.95.

polymorphism   The presence in a population of several forms of a trait, gene, or chromosome aberration.

polynucleotide chain   A single-stranded molecule consisting of nucleotides covalently linked end to end.

polypeptide   A polymer of amino acids linked together by peptide bonds.

polyploid   A cell or organism having more than two complete sets of chromosomes.

polyprotein   A protein molecule that can be cleaved to form two or more finished protein molecules.

polyribosome   *See* **polysome.**

polysome   A complex of two or more ribosomes associated with an mRNA molecule and actively engaged in polypeptide synthesis; a polyribosome.

polysomic   A diploid cell or organism having three or more copies of a particular chromosome.

polytene chromosome   A large chromosome consisting of many identical strands closely associated along their length, with the chromomeres in register producing a specific pattern of transverse banding.

population   A group of organisms of the same species.

population structure   Description of the manner in which a population is subdivided.

population subdivision   Organization of a population into smaller breeding groups between which migration is restricted.

position effect   A change in the expression of a gene depending on its position within the genome.

positive regulation   Regulation of transcription by an element that must be bound to DNA in active form in order for RNA polymerase to bind to a promoter; positive regulation contrasts with negative regulation, in which a regulatory element must be removed from DNA.

postmeiotic segregation Segregation of genetically different products in a mitotic division following meiosis, as in the formation of a pair of ascospores having different genotypes in *Neurospora*.

postreplication repair Any DNA repair process that occurs either in a region of a DNA molecule after the replication fork has moved some distance past that region or in nonreplicating DNA.

precursor fragment *See* **Okazaki fragment.**

prediction equation In quantitative genetics, an equation used to predict the improvement in mean performace of a population by means of artifical selection; always includes heritability as one component.

Pribnow box A base sequence in prokaryotic promoters to which RNA polymerase binds in an early step of initiating transcription.

primary transcript An RNA copy of a gene; usually refers to a molecule that must be altered to form a translatable mRNA molecule.

primase The enzyme responsible for synthesizing the RNA primer for initiating precursor fragments.

primer In nucleic acids, a short RNA or single-stranded DNA segment that functions as a growing point in polymerization.

probability A mathematical expression of the degree of confidence that certain events will or will not occur.

probe A synthetic radioactive DNA or RNA molecule used in DNA-RNA or DNA-DNA hybridization assays.

processing A series of chemical reactions in which either primary RNA transcripts are converted to mature tRNA, rRNA, and mRNA molecules, or polypeptide chains become converted to finished proteins. The changes are mainly cleavage reactions, though chemical modification may also occur.

prokaryote An organism that lacks a membrane-bounded nucleus and in which the nucleus does not divide by mitosis or meiosis; bacteria and blue-green algae.

promoter A specific DNA sequence at which RNA polymerase binds and initiates transcription.

promoter mutation One of three types of mutations that either inactivate promoter function (sometimes called a promoter-down mutation), or create a new promoter sequence where one did not exist before, or increase the efficiency of binding RNA polymerase (a promoter-up mutation).

promoter recognition The first step in transcription.

proofreading function *See* **editing function.**

prophage The form of phage DNA in a lysogenic bacterium; the phage DNA is repressed and usually integrated in the bacterial chromosome, but some prophages are in plasmid form.

prophage attachment site Either the base sequence in a bacterial chromosome at which phage DNA can integrate to form a prophage, or the two attachment sites that flank an integrated prophage.

prophage induction The process of derepressing a prophage and initiating a lytic cycle of phage development.

prophase The initial stage of mitosis or meiosis, occurring after DNA replication and terminating with the alignment of the chromosomes at metaphase; often absent between meiosis I and meiosis II.

protamine A class of basic proteins that bind to DNA in sperm, cells that lack histones.

protein A molecule composed of one or more polypeptide chains.

pseudodominance The apparent dominance of a recessive allele resulting from deletion of the corresponding locus in the homologous chromosome.

pseudogene   A DNA base sequence closely related to that of a sequence producing a functional protein, but which, because of either mutations in the coding sequence or the inability to be transcribed or translated, does not produce a functional protein; so far, pseudogenes have been observed only in eukaryotes and are usually contained in a large region that includes many duplicated or nearly identical sequences forming a gene family; thought to be a mutated form of an ancient duplicated sequence.

Punnett square   A cross-multiplication square used for determining the expected genetic outcome of matings.

purines   A class of organic bases found in nucleic acids; the predominant purines are adenine and guanine.

pyrimidine   A class of organic bases found in nucleic acids; the predominant pyrimidines are cytosine, uracil (in RNA only), and thymine (in DNA only).

pyrimidine dimer   Two adjacent pyrimidine bases, typically a pair of thymines, in the same polynucleotide strand, between which chemical bonds have formed; the most common lesion formed in DNA by exposure to ultraviolet light.

quantitative trait   A trait that can be measured on a continuous scale, such as height or weight.

R group   *See* **side chain.**

R plasmid   A bacterial plasmid that carries drug-resistance genes; commonly used in genetic engineering.

race   A genetically or geographically distinct subgroup of a species.

rad   A unit of ionizing radiation; the amount resulting in the dissipation of 100 ergs of energy in one gram of matter; rad stands for radiation absorbed dose.

random genetic drift   Fluctuation in allele frequency from generation to generation resulting from restricted population size.

random mating   Mating uninfluenced by genotype or phenotype.

reading frame   One of three ways to translate an mRNA base sequence linearly into an amino acid sequence to form a polypeptide; the particular reading frame is defined by the AUG codon that is selected for chain initiation.

reannealing   Reassociation of dissociated single strands of DNA to form a duplex molecule.

RecA protein   A protein, the product of the *E. coli recA* gene, that binds to DNA and is essential for DNA pairing in homologous recombination.

recessive   An allele, or the corresponding phenotypic trait, expressed only in homozygotes.

reciprocal cross   A pair of crosses in which the genotypes of female and male parents in the first cross are reversed in the second cross.

reciprocal translocation   Interchange of parts between nonhomologous chromosomes.

recombinant   A cell or new individual arising by recombination.

recombinant DNA   A DNA molecule composed of one or more segments from other DNA molecules.

recombination   In meiosis, the formation by independent assortment or crossing over of a haploid product having a genotype different from either of the haploid genotypes that formed the meiotic diploid; in general, the formation in a diploid or partially diploid cell of a combination of genes not present in the parental genomes represented in the cell.

recombination repair   Repair of damaged DNA by exchange of good for bad segments between two damaged molecules.

redundancy   The feature of the genetic code in which an amino acid corresponds to more than one codon; also called degeneracy.

regulator gene   A gene with the primary function of controlling the rate of synthesis in the products of one or more other genes.

rejection   An immune response against transfused blood or transplanted tissue.

relative fitness   A measure of the fitness of one genotype as a proportion of the fitness of another genotype.

relaxed circle   A DNA circle whose supercoiling has been removed either by introduction of a single-strand break or by the activity of a topoisomerase.

release factor   One of three different proteins required to release a completed polypeptide chain from a ribosome.

rem   The quantity of any kind of ionizing radiation that has the same biological effect as one rad of high-energy gamma rays; rem stands for roentgen equivalent man.

renaturation   Restoration of the normal three-dimensional structure of a macromolecule; when used to refer to nucleic acids, the term means the formation of a double-stranded molecule by complementary base-pairing between two single-stranded molecules.

repair   *See* **DNA repair.**

repair synthesis   The enzymatic filling of a gap in a DNA molecule at the site of excision of a damaged DNA segment.

repetitive DNA   DNA sequences present more than once in the haploid genome.

replication   *See* **DNA replication.**

replication fork   In a replicating DNA molecule, the region in which nucleotides are added to growing strands.

replication origin   The base sequence at which DNA synthesis begins.

replicon   A DNA molecule that has a replication origin.

replicon fusion   The joining of two circular DNA molecules by the action of a transposable element, forming a larger circular molecule.

repressor   A protein that binds specifically to a regulatory sequence adjacent to a gene and blocks transcription of the gene.

repulsion   *See* *trans* **configuration.**

restriction endonuclease   A nuclease that recognizes a short nucleotide sequence (restriction site) in a DNA molecule and cleaves the molecule at that site; also called restriction enzyme.

restriction map   A diagram of a DNA molecule showing the positions of cleavage by one or more restriction endonucleases.

restriction site   The base sequence at which a particular restriction endonuclease makes a cut.

restrictive conditions   Growth conditions that prevent the growth of a particular mutant.

retrovirus   One of a class of RNA animal viruses that cause the synthesis of DNA complementary to their RNA genomes on infection.

reverse transcriptase   An enzyme, carried within the coats of retroviruses, that makes complementary DNA from a single-stranded RNA template.

reversion   Restoration of a mutant phenotype to the wildtype phenotype by the occurrence of a second mutation.

Rh   Rhesus blood-group system in humans; maternal-fetal incompatibility of this system may result in hemolytic disease of the newborn.

ribonuclease   Any enzyme that cleaves phosphodiester bonds in RNA; abbreviated RNase.

ribonucleic acid   *See* **RNA.**

ribose   The five-carbon sugar in RNA.

ribosomal RNA   RNA molecules that are structural components of the ribosomal subunits; in eukaryotes there are four rRNA molecules—5S, 5.8S, 18S, and 28S; in prokaryotes there are three—5S, 16S, and 23S; abbreviated rRNA.

ribosome   The cellular organelle, consisting of two subunits, each composed of RNA and proteins, on which the codons of mRNA are translated into amino acids during protein synthesis; in prokaryotes, the subunits are 30S and 50S particles, and in eukaryotes they are 40S and 60S particles.

ribosome binding site   The base sequence in a prokaryotic mRNA molecule to which a ribosome can bind to initiate protein synthesis; also called the Shine-Dalgarno sequence.

RNA   Ribonucleic acid; a nucleic acid in which the sugar constituent is ribose. Typically, RNA is single-stranded and contains the four bases adenine, cytosine, guanine, and uracil.

RNA polymerase   An enzyme that makes RNA by copying the base sequence of a DNA strand.

RNA processing   The conversion of a primary transcript to an mRNA, rRNA, or tRNA molecule; includes splicing, cleavage, modification of termini, and, in tRNA, modification of internal bases.

RNA splicing   Excision of introns and joining of exons.

RNase   *See* **ribonuclease.**

Robertsonian translocation   A chromosomal aberration in which the long arms of two acrocentric chromosomes become joined to a common centromere.

roentgen   A unit of ionizing radiation, defined as the amount of radiation resulting in 2.083×10^9 ion pairs per cm^3 of dry air at 0°C and 1 atm pressure; abbreviated R.

rolling circle replication   A mode of replication in which a circular parent molecule produces a linear branch of newly formed DNA.

Rous sarcoma virus   The best-studied retrovirus; infects the chicken.

rRNA   *See* **ribosomal RNA.**

salvage pathway   A minor pathway of DNA synthesis, which uses the enzymes hypoxanthine guanine phosphoribosyl transferase (HGPRT) and thymidine kinase (TK).

satellite DNA   Eukaryotic DNA that forms a minor band at a different density than that of most of the cellular DNA upon equilibrium density gradient centrifugation; consists of short sequences repeated many times in the genome (highly repetitive DNA), or of mitochondrial or chloroplast DNA.

scaffold   A protein-containing material in chromosomes, believed to be responsible in part for the compaction of chromosomes.

second-division segregation   Segregation of a pair of alleles into different nuclei during the second meiotic division, the result of crossing over between the gene and the centromere of the pair of homologous chromosomes.

secretor   An autosomal dominant trait in humans, associated with the ability to secrete A and B antigens of the ABO blood group in body fluids.

segregation   Separation of the members of a pair of alleles into different gametes during meiosis.

selection   In evolution, intrinsic differences in the ability of genotypes to survive and reproduce; in plant and animal breeding, the choosing of individuals with certain phenotypes to be parents of the next generation; in mutation studies, a procedure designed in such a way that only a desired type of cell can survive, as in selection for resistance to an antibiotic.

selection coefficient   The amount by which relative fitness is reduced or increased.

selection differential   In artificial selection, the difference between the mean of the selected individuals and the mean of the population from which they were chosen.

selection pressure   The tendency of natural or artificial selection to change allele frequency.

self-fertilization   The union of male and female gametes produced by the same individual.

semiconservative replication   The usual mode of DNA replication, in which each strand of a double-stranded molecule serves as a template for the synthesis of a new complementary strand and the daughter molecules are composed of one old (parental) and one newly synthesized strand.

semisterility   A condition in which half or more of the gametophytes produced by a plant or the zygotes produced by an animal are inviable, as in the case of a translocation heterozygote.

sense strand   The DNA strand that serves as the template for transcription of a given gene.

70S ribosome   In prokaryotes, the particle that is active in protein synthesis; consists of one 30S subunit and one 50S subunit.

sex chromosome   A chromosome, such as the human X or Y, that is involved in the determination of sex.

sex-influenced trait   A trait for which the expression is conditioned by the sex of the individual.

sex-limited trait   A trait expressed in one sex and not in the other.

sex-linked trait   A trait determined by a gene on a sex chromosome, usually the X.

Shine-Dalgarno sequence   *See* **ribosome binding site.**

siblings   The offspring of the same parents; often called sibs.

sickle-cell anemia   A severe anemia in humans inherited as an autosomal recessive and caused by an amino acid substitution in the β-globin chain; heterozygotes tend to be more resistant to falciparum malaria than are normal homozygotes.

side chain   In protein structure, the chemical group attached to the α-carbon atom of an amino acid; amino acids are differentiated on the basis of side-chain differences. Also called R group.

sigma (σ) subunit   The subunit of RNA polymerase needed for promoter recognition.

silent mutation   Any mutation having no phenotypic effect.

site-specific exchange   Genetic exchange that occurs only between particular base sequences.

60S ribosomal subunit   The large ribosomal subunit in eukaryotes.

small ribonucleoprotein particle   Small particles containing short RNA molecules and several proteins; called snurps, in jargon; one is responsible for intron excision and splicing.

somatic cell   Any cell of a multicellular organism other than the gametes and the germ cells from which they develop.

somatic mutation   A mutation arising in a somatic cell.

SOS repair   An inducible, error-prone system for repair of DNA damage in *E. coli.*

Southern blotting   A nucleic acid hybridization method in which, following electrophoretic separation, denatured DNA is transferred from a gel to a paper filter and then exposed to radioactive DNA or RNA under conditions of renaturation; the radioactive regions locate the original DNA fractions.

spacer sequence   A noncoding base sequence between the coding segments of polycistronic mRNA or between genes in DNA.

specialized transduction   *See* **transducing phage.**

species   Genetically, a group of actually or potentially inbreeding organisms that is reproductively isolated from other such groups.

spindle   A structure composed of fibrous proteins on which chromosomes align during metaphase and move during anaphase.

spontaneous mutation   A mutation occurring in the absence of any known mutagenic agent.

spore   A unicellular reproductive entity that becomes detached from the parent and can develop into a new individual upon germination; in plants, spores are the haploid products of meiosis.

sporophyte   The diploid, spore-forming generation in plants, which alternates with the haploid, gamete-producing generation or gametophyte.

standard deviation   The square root of the variance.

start codon   An mRNA codon, usually AUG, at which polypeptide synthesis begins.

stop codon   One of three mRNA codons—UAG, UAA, and UGA—at which polypeptide synthesis stops.

strain   A term arbitrarily applied to a variant of a microorganism that differs from other strains in a number of traits that are not genetically understood; the differences are too great for the strains to be considered mutant forms of a wildtype form, and too small to be considered different species.

structural gene   A gene that encodes the amino acid sequence of a polypeptide chain.

submetacentric chromosome   A chromosome with a centromere located near, but not at the center of, a chromosome, making one arm slightly longer than the other.

subpopulations   Breeding groups within a larger population between which migration is restricted.

subspecies   A relatively isolated population or group of populations distinguishable from other populations in the same species by allele frequencies or chromosomal arrangements, and sometimes exhibiting incipient reproductive isolation.

substrate   A specific substance acted on by an enzyme.

subunit   A macromolecule that is a component of an ordered aggregate of macromolecules—for example, a single polypeptide chain in a protein containing several chains.

supercoiled DNA   A form of double-stranded DNA in which strain caused by overwinding or underwinding of the duplex makes the circle twist; a supercoiled circle is also called a twisted circle or a superhelix.

suppressor mutation   A mutation that acts to restore, either partially or completely, the function impaired by another mutation at a different site in the same gene (intragenic suppression) or in a different gene (intergenic suppression).

suppressor tRNA   A tRNA molecule capable of translating a stop codon (nonsense suppressor) or of inserting an amino acid other than that specified by a particular codon (missense suppressor).

synapsis   The pairing of homologous chromosomes or chromosome regions, typically occurring during zygotene of the first meiotic prophase but also occurring in certain somatic cell nuclei, as in the paired polytene chromosomes of dipteran larvae.

synaptonemal complex   A complex protein structure that forms between synapsed homologous chromosomes in the pachytene substage of the first meiotic prophase.

syndrome   A group of symptoms appearing together with sufficient regularity to warrant designation by a special name; also, a disorder, disease, or anomaly.

synteny   The occurrence of two genes on the same chromosome.

TATA box   The base sequence in the DNA of a eukaryotic promoter to which RNA polymerase binds; occasionally called a Hogness box or a Goldberg-Hogness box. The letters T and A denote the bases thymine and adenine.

tautomeric shift   A reversible change in the location of a hydrogen atom in a molecule, altering the molecule from one isomeric form to another; in nucleic acids, the shift is typically between a keto group (keto form) and a hydroxyl group (enol form).

telomere   The terminal chromomere of a chromosome; a DNA sequence required for stability of chromosome ends.

telophase   The final stage of mitotic or meiotic nuclear division.

temperate phage   A phage that is capable of both a lysogenic and a lytic cycle.

temperature-sensitive mutation   A conditional mutation that causes a phenotypic change at certain temperatures and not at others.

template strand   A nucleic acid strand whose base sequence is copied in a polymerization reaction.

terminal nucleotidyl transferase   An enzyme that adds nucleotides to the 3′ end of one strand of double-stranded DNA; used in genetic engineering to form homopolymers at the ends of DNA strands.

terminal redundancy   The presence of identical nucleotide sequences at two ends of a DNA molecule.

testcross   A cross between a heterozygote and an individual homozygous for the recessive alleles of the genes in question, resulting in progeny in which each phenotypic class represents a different genotype.

tetrad   The four chromatids that make up a pair of homologous chromosomes in meiotic prophase I and metaphase I; also, the four haploid products of a single meiosis.

tetrad analysis   A method for the analysis of linkage and recombination using the four haploid products of single meiotic divisions.

tetraploid   A cell or organism with four complete sets of chromosomes. In an autotetraploid, the chromosome sets are homologous; in an allotetraploid the chromosome sets consist of a complete diploid complement from each of two distinct ancestral species.

tetrasomic   An aneuploid condition in which one chromosome is represented four times; having the $2n + 2$ number of chromosomes.

30-nm fiber   The first level of organization of chromatin in which nucleosomes form a helical array.

30S ribosomal subunit   The small subunit of a prokaryotic ribosome.

threshold trait   A trait with a continuously distributed liability or risk in which individuals with a liability greater than a critical value (the threshold) exhibit the phenotype of interest, such as a disorder.

thymine dimer   *See* **pyrimidine dimer.**

T_m *See* **melting temperature.**

topoisomerase   Any of a class of enzymes that introduces or removes either underwinding or overwinding of double-stranded DNA; acts by introducing a single-strand break, changing the relative positions of the strands, and sealing the break.

***trans* configuration**   The arrangement in linked inheritance in which an individual heterozygous for two mutant sites has received a different one of the mutant sites from each parent—that is $a^1 +/+ a^2$.

transcription   The process by which the information contained in the coding strand of DNA is copied into a single-stranded RNA molecule, whose base sequence is complementary to that of the DNA strand that is copied.

transducing phage   A phage type capable of producing particles containing bacterial DNA (transducing particles); a specialized transducing phage produces particles carrying only specific regions of chromosomal DNA; a generalized transducing phage produces particles that may carry any region of the genome.

transduction   The carrying of genetic information from one bacterium to another by a phage.

transfer RNA   A small RNA molecule that translates a codon into an amino acid during protein synthesis; it has a three-base sequence, called the anticodon, complementary to a specific codon in mRNA, and a site to which a specific amino acid is bound; abbreviated tRNA.

transformation   The conversion of the genotype of one bacterium by exposure of the cell to DNA isolated from bacteria having a different genotype; the conversion of an animal cell whose growth is limited in culture to a tumorlike cell whose pattern of growth is different from that of a normal cell.

transition   A mutation resulting from the substitution of one purine for another purine or one pyrimidine for another pyrimidine.

translation   The process by which the amino acid sequence of a polypeptide is derived from the nucleotide sequence of an mRNA molecule associated with a ribosome.

translocation   The movement of mRNA with respect to a ribosome during protein synthesis. *See* also **reciprocal translocation.**

transposable element   A DNA sequence capable of moving (transposing) from one location to another in a genome.

transposition   The movement of a transposable element.

transposon   A transposable element in a bacterium that carries incorporated bacterial genes.

transversion   A mutation resulting from the substitution of a purine for a pyrimidine or a pyrimidine for a purine.

triplet code   A code in which each codon consists of three bases.

triploid   A cell or individual with three complete sets of chromosomes.

trisomic   An aneuploid condition in which one chromosome is represented three times; having the $2n + 1$ number of chromosomes.

tRNA   *See* **transfer RNA.**

truncation point   In artificial selection, the critical phenotype determining which organisms will be retained for breeding and which will be culled.

Turner syndrome   The clinical condition in human females with the karyotype 45,X.

uncharged tRNA   A tRNA molecule lacking an amino acid.

underwound DNA   A DNA molecule whose strands are untwisted somewhat and hence in which some bases are unpaired.

unequal crossing over   The occurrence of crossing over with the exchange occurring between different chromosomes, each carrying a duplicated sequence, between identical sequences of the duplication but in different positions.

unique-sequence DNA   A sequence that occurs only once in a haploid genome, in contrast with repetitive sequences.

V regions   Multiple DNA sequences coding for alternative amino acid sequences of part of the variable region of an antibody molecule. *See* also **variable region.**

V-type position effect   A type of position effect in *Drosophila*, characterized by discontinuous expression of one or more genes during development and usually resulting from chromosome breakage and rejoining such that euchromatic genes are repositioned in or near centromeric heterochromatin.

variable region   The portion of an immunoglobulin molecule that varies greatly in amino acid sequence among antibodies in the same subclass. *See* also **V regions.**

variance   A measure of the spread of a statistical distribution; the mean of the squares of the deviations from the mean.

vector   In genetic engineering, a DNA molecule, capable of replication, used as a carrier of a DNA molecule or fragment; a cloning vehicle; the usual vectors are plasmids, and phage and viral DNA molecules.

virulent phage A phage or virus species capable only of a lytic cycle; contrasts with temperate phage.

wildtype The most common phenotype or genotype in a natural population; also, a phenotype or genotype arbitrarily designated as a standard for comparison.

wobble The alternative pairing of several bases to a particular base in the third position of a codon, in codon-anticodon binding.

χ^2 **test** *See* **chi-square test.**

X chromosome A chromosome associated with sex determination that is present in two copies in the homogametic sex and in one copy in the heterogametic sex.

X-linked inheritance The pattern of allelic transmission of genes located on the X chromosome; usually evident from the production of nonidentical classes of progeny from reciprocal crosses.

Y chromosome The sex chromosome that is present only in the heterogametic sex; in mammals, the male-determining sex chromosome.

zygote The product of the fusion of a female and a male gamete in sexual reproduction; a fertilized egg.

zygotene The substage of meiotic phrophase I during which synapsis of homologous chromosomes occurs.

B I B L I O G R A P H Y

These references are provided for the student who either wants more information or who needs an alternate explanation for the material presented in this book. For the former, the review articles and scientific reports are recommended; for the latter, the textbooks and the *Scientific American* articles are preferable. A few "classic" papers are also listed for those whose would like to know how information is obtained; these are generally more advanced than textbooks.

Chapter 1

Brady, R. 1973. "Hereditary fat-metabolism diseases." *Scient. Amer.*, August.
Dunn, L. C. 1965. *A Short History of Genetics.* McGraw-Hill.
Mather, K. 1965. *Statistical Analysis in Biology.* Methuen.
Mendel, G. "Experiments in plant hybridization." (Translation.) *In* C. I. Davern, ed. 1981. *Genetics, A Scientific American Reader.* W. H. Freeman.
Olby, R. C. 1966. *Origins of Mendelism.* Constable.
Srb, A., R. Owen, and R. Edgar. 1965. *General Genetics.* W. H. Freeman.
Stern, C., and E. Sherwood. 1966. *The Origins of Genetics: A Mendel Source Book.* W. H. Freeman.

Chapter 2

Creighton, H. S., and B. McClintock. 1931. "A correlation of cytological and genetical crossing over in *Zea mays.*" *Proc. Nat. Acad. Sci.*, 17: 492.
Crow, J. 1966. *Genetics Notes.* Burgess.
Dunn, L. C. 1965. *A Short History of Genetics.* McGraw-Hill.
Mather, K. 1965. *Statistical Analysis in Biology.* Methuen.
Mazia, D. 1974. "The cell cycle." *Scient. Amer.*, January.
McKusick, V. A. 1965. "The royal hemophilia." *Scient. Amer.*, August.
Morgan, T. H. 1910. "Sex-limited inheritance in *Drosophila.*" *Science*, 322: 120.
Morgan, T. H. 1911. "An attempt to analyze the constitution of the chromosomes on the basis of sex-linked inheritance." *J. Expt. Zool.*, 11: 365.
Srb, A., R. Owen, and R. Edgar. 1965. *General Genetics.* W. H. Freeman.
Voeller, B. R., ed. 1968. *The Chromosome Theory of Inheritance—Classical Papers in Development and Heredity.* Appleton-Century-Crofts.

BIBLIOGRAPHY

Chapter 3

Barratt, R. W. 1954. "Map construction in *Neurospora crassa.*" *Adv. in Genetics*, 6: 1.

Corwin, H. O., and J. B. Jenkins. *Conceptual Foundations in Genetics*. Houghton Mifflin.

Fincham, J. R. S. 1966. *Genetic Complementation*. Benjamin-Cummings.

Fincham, J. R. S. 1971. *Using Fungi to Study Recombination*. Clarendon.

Levine, L. 1971. *Papers on Genetics*. Mosby.

McKusick, V. 1971. "Mapping of human chromosomes." *Scient. Amer.*, April.

Mittwoch, U. 1963. "Sex differences in cells." *Scient. Amer.*, July.

Srb, A., R. Owen, and R. Edgar. 1965. *General Genetics*. W. H. Freeman.

Voeller, B. R., ed. 1968. *The Chromosome Theory of Inheritance—Classical Papers in Development and Heredity*. Appleton-Century-Crofts.

Chapter 4

Alberts, B., and R. Sternglantz. 1977. "Recent excitement in the DNA replication problem." *Nature*, 269: 655.

Freifelder, D. 1978. *The DNA Molecule*. W. H. Freeman.

Freifelder, D. 1983. *Molecular Biology*. Jones and Bartlett.

Hotchkiss, R. D., and E. Weiss. 1956. "Transformed bacteria." *Scient. Amer.*, November.

Kornberg, A. 1980. *DNA Replication*. W. H. Freeman.

Maxam, A. M., and W. Gilbert. 1977. "A new method for sequencing DNA." *Proc. Nat. Acad. Sci.*, 74: 560.

Mirsky, A. 1968. "The discovery of DNA." *Scient. Amer.*, June.

Stryer, L. 1981. *Biochemistry*. W. H. Freeman.

Watson, J. D. 1968. *The Double Helix*. Athenaeum.

Watson, J. D., and F. H. C. Crick. 1953. "Molecular structure of nucleic acid. A structure for deoxyribose nucleic acid." *Nature*, 171: 737.

Watson, J. D., and F. H. C. Crick. 1953. "Genetic implications of the structure of desoxyribosenucleic acid." *Nature*, 171: 964.

Chapter 5

Bauer, W. R., F. H. C. Crick, and J. H. White. 1980. "Supercoiled DNA." *Scient. Amer.*, July.

Britten, R. J., and D. E. Kohne. 1968. "Repeated sequences in DNA." *Science*, 161: 529.

Britten, R. J., and D. E. Kohne. 1970. "Repeated segments of DNA." *Scient. Amer.*, April.

Bukhari, A. I., J. Shapiro, and S. Adhya. 1977. *DNA Insertion Elements, Plasmids, and Episomes*. Cold Spring Harbor.

Cairns, J. 1966. "The bacterial chromosome." *Scient. Amer.*, January.

Freifelder, D. 1983. *Molecular Biology*. Jones and Bartlett.

Goodenough, U., and R. P. Levine. 1970. "The genetic activity of mitochondria and chloroplasts." *Scient. Amer.*, November.

Green, M. M. 1980. "Transposable elements in *Drosophila* and other Diptera." *Ann. Rev. Genetics*, 14: 109.

Isenberg, I. 1979. "Histone." *Ann. Rev. Biochem.*, 48: 159.

Keller, E. 1981. "McClintock's maize." *Science 81*, August.

Kornberg, R. D., and A. Klug. 1981. "The nucleosome." *Scient. Amer.*, February.

McClintock, B. 1965. "Control of gene action in maize." *Brookhaven Symp. Quant. Biol.,* 18: 162.

McGhee, J., and G. Felsenfeld. 1980. "Nucleosome structure." *Ann. Rev. Biochem.,* 49: 1115.

Sager, R. 1965. "Genes outside the chromosome." *Scient. Amer.,* January.

Wang, J. 1982. "DNA topoisomerases." *Scient. Amer.,* July.

Chapter 6

Brown, W. V. 1972. *Textbook on Cytogenetics.* Houghton Mifflin.

Carson, H. L. 1970. "Chromosome tracers of the origin of the species." *Science,* 168: 1414.

Curtis, B. C., and D. R. Johnson. 1969. "Hybrid wheat." *Scient. Amer.,* May.

Fincham, J. R. S., P. R. Day, and A. Radford. 1979. *Fungal Genetics.* Blackwell.

Garber, E. D. 1972. *Cytogenetics: An Introduction.* McGraw-Hill.

Hsu, T. H. 1979. *Human and Mammalian Cytogenetics.* Springer-Verlag.

Raven, P. H., and R. Evert. 1976. *Biology of Plants.* Worth.

Sparks, R. S., D. E. Comings, and C. F. Fox. 1977. *Molecular Human Cytogenetics.* Academic.

Srb, A., R. Owen, and R. Edgar. 1965. *General Genetics.* W. H. Freeman.

Stebbins, G. L. 1971. *Chromosome Evolution in Higher Plants.* E. Arnold.

Swanson, C. P., and P. Webster. 1977. *The Cell.* Prentice-Hall.

White, M. J. D. 1977. *Animal Cytology and Evolution.* Cambridge Univ.

Chapter 7

Adelberg, E. A., ed. 1966. *Papers on Bacterial Genetics.* Little Brown.

Birge, E. A. 1981. *Bacterial and Bacteriophage Genetics.* Springer-Verlag.

Campbell, A. 1976. "How viruses insert their DNA into the DNA of the host cell." *Scient. Amer.,* December.

Clowes, R. D. 1975. "The molecules of infectious drug resistance." *Scient. Amer.,* July.

Edgar, R. S., and R. H. Epstein. 1965. "The genetics of a bacterial virus." *Scient. Amer.,* February.

Freifelder, D. 1983. *Molecular Biology.* Jones and Bartlett.

Hayes, W. 1968. *The Genetics of Bacteria and Their Viruses.* John Wiley.

Kleckner, N. 1981. "Transposable genetic elements." *Ann. Rev. Genetics,* 15: 341.

Low, K. B., and R. Porter. 1978. "Modes of genetic transfer and recombination in bacteria." *Ann. Rev. Genetics,* 12: 249.

Novick, R. P. 1980. "Plasmids." *Scient. Amer.,* December.

Srb, A., R. Owen, and R. Edgar. 1965. *General Genetics.* W. H. Freeman.

Stent, G. S., and R. Calendar. 1978. *Molecular Genetics, An Introductory Narrative.* W. H. Freeman.

Watanabe, T. 1967. "Infectious drug resistance." *Scient. Amer.,* December.

Zinder, N. 1958. "Transduction in bacteria." *Scient. Amer.,* November.

Chapter 8

Bianchi, M. E., and C. M. Radding. 1983. "Insertions, deletions, and mismatches in heteroduplex DNA made by RecA protein." *Cell,* 35: 511.

Calos, M. P., and J. H. Miller. 1980. "Transposable elements." *Cell,* 20: 579.

Cunningham, R. P., C. DasGupta, T. Shibata, and C. M. Radding. 1980. "Homologous pairing in genetic recombination: RecA protein makes joint molecules of gapped circular DNA and closed circular DNA." *Cell*, 20: 223.

Fincham, J. R. S., P. R. Day, and A. Radford. 1979. *Fungal Genetics*. Blackwell.

Fogel, S., R. Mortimer, K. Lusnak, and F. Tavares. 1979. "Meiotic gene conversion: a signal of the basic recombination event in yeast." *Cold Spring Harb. Symp. Quant. Biol.*, 43: 1325.

Fox, M. S. 1978. "Some features of genetic recombination in procaryotes." *Ann. Rev. Genetics*, 11: 369.

Gonda, D. K., and C. M. Radding. 1983. "By searching processively, RecA protein pairs DNA molecules that share a limited stretch of homology." *Cell*, 34: 647.

Goodenough, U. 1984. *Genetics*. Chapter 17. Saunders.

Holliday, R. 1964. "A mechanism for gene conversion in fungi." *Genet. Res.*, 5: 282.

Lacks, S. 1962. "Molecular fate of DNA in genetic transformation in *Pneumococcus*." *J. Mol. Biol.*, 5: 119.

Meselson, M. 1964. "On the mechanism of genetic recombination between DNA molecules." *J. Mol. Biol.* 9: 734.

Meselson, M., and C. M. Radding. 1978. "A general mode for genetic recombination." *Proc. Nat. Acad. Sci.*, 76: 2615.

Potter, H., and D. Dressler. 1976. "On the mechanism of genetic recombination: electron-microscopic observation of recombination intermediates." *Proc. Nat. Acad. Sci.*, 73: 3000.

Radding, C. M. 1978. "Genetic recombination: strand transfer and mismatch repair." *Ann. Rev. Biochem.*, 47: 847.

Radding, C. M. 1982. "Homologous pairing and strand exchange in genetic recombination." *Ann. Rev. Genetics*, 16: 405.

Stahl, F. W. 1969. *The Mechanics of Inheritance*. Prentice-Hall.

Stahl, F. W. 1979. *Genetic Recombination*. W. H. Freeman.

Szostak, J. W., T. L. Orr-Weaver, and R. J. Rothstein. 1983. "The double-strand-break-repair model for recombination." *Cell*, 33: 25.

Chapter 9

Barrell, B. G., A. T. Bankier, and J. Drouin. "A different genetic code in human mitochondria." *Nature*, 282: 189.

Beadle, G. W. 1948. "Genes of men and molds." *Scient. Amer.*, September.

Bearn, A. G. 1956. "The chemistry of hereditary disease." *Scient. Amer.*, December.

Caskey, C. T. 1980. "Peptide chain termination." *Trends Biochem. Sci.*, 5: 234.

Chambliss, G., ed. 1980. *Ribosomes: Structure, Function, and Genetics*. University Park.

Chambon, P. 1981. "Split genes." *Scient. Amer.*, May.

Cold Spring Harbor Laboratory. 1966. *The Genetic Code. Cold Spring Harb. Symp. Quant. Biol.* Vol. 31.

Crick, F. H. C. 1962. "The genetic code." *Scient. Amer.*, October.

Crick, F. H. C. 1966. "The genetic code." *Scient. Amer.*, October.

Crick, F. H. C. 1979. "Split genes and RNA splicing." *Science*, 204: 264.

Crick, F. H. C., L. Barnett, S. Brenner, and R. J. Watts-Tobin. 1961. "General nature of the genetic code for proteins." *Nature*, 192: 1227.

Dickerson, R. E. 1972. "The structure and history of an ancient protein." *Scient. Amer.*, April.

Freifelder, D. 1983. *Molecular Biology*. Chapters 11–13. Jones and Bartlett.

Gorini, L. 1966. "Antibiotics and the genetic code." *Scient. Amer.*, April.

Jukes, T. 1978. "The amino acid code." *Adv. Enzym.*, 47: 375.

Lake, J. 1981. "The ribosomes." *Scient. Amer.*, August.

Miller, O. L., Jr. 1973. "The visualization of genes in action." *Scient. Amer.*, March.

Nirenberg, M. 1963. "The genetic code." *Scient. Amer.*, March.

Sherman, F., and J. W. Stewart. 1982. "Mutations altering initiation of translation of yeast iso-1-cytochrome c: contrasts between eukaryotic and prokaryotic initiation processes." *In* J. Strathern, E. Jones, and J. Broach, eds. *The Molecular Biology of the Yeast Saccharomyces.* Cold Spring Harbor.

Taylor, J. H., ed. 1965. *Selected Papers on Molecular Genetics.* Academic.

Yanofsky, C. 1967. "Gene structure and protein structure." *Scient. Amer.*, May.

Chapter 10

Allison, A. C. 1956. "Sickle cells and evolution." *Scient. Amer.*, August.

Ames, B. W. 1979. "Identifying environmental chemicals causing mutations and cancer." *Science*, 204: 587.

Celis, J. E., and J. D. Smith. 1979. *Nonsense Mutations and tRNA Suppressors.* Academic.

Deering, R. A. 1962. "Ultraviolet radiation and nucleic acid." *Scient. Amer.*, December.

Denniston, C. 1982. "Low-level radiation and genetic risk estimation in man." *Ann. Rev. Genetics*, 16: 329.

Devoret, R. 1979. "Bacterial tests for potential carcinogens." *Scient. Amer.*, August.

Drake, J. W. 1970. *The Molecular Basis of Mutation.* Holden-Day.

Freifelder, D. 1983. *Molecular Biology.* Chapter 9. Jones and Bartlett.

Hanawalt, P. C., E. C. Friedberg, and C. F. Fox. 1978. *DNA Repair Mechanisms.* Academic.

Hanawalt, P. C., and R. H. Haynes. 1967. "The repair of DNA." *Scient. Amer.*, February.

Haseltine, W. A. 1983. "Ultraviolet light repair and mutagenesis revisited." *Cell*, 33: 13.

Heidelberger, M. 1975. "Chemical carcinogens." *Ann. Rev. Biochem.*, 44: 79.

Howard-Flanders, P. 1981. "Inducible repair of DNA." *Scient. Amer.* March.

Little, J. W., and D. W. Mount. 1982. "The SOS regulatory system of *E. coli.*" *Cell*, 29: 11.

Livneh, Z., and I. R. Lehman. 1982. "Recombinational bypass of pyrimidine dimers promoted by the RecA protein of *E. coli.*" *Proc. Nat. Acad. Sci.*, 79: 3171.

Muller, H. J. 1955. "Radiation and human mutation." *Scient. Amer.*, November.

Sherman, F. 1982. "Suppression in the yeast *Saccharomyces cerevisiae.*" In J. Strathern, E. Jones, and J. Broach, eds. *The Molecular Biology of the Yeast Saccharomyces.* Cold Spring Harbor.

Sigurbjornsson, B. 1971. "Induced mutations in plants." *Scient. Amer.*, January.

Singer, B., and J. T. Kusmierek. 1982. "Chemical mutagenesis." *Ann. Rev. Biochem.*, 51: 655.

Streisinger, G., Y. Okada, J. Emrich, J. Newton, A. Tsugita, E. Terzaghi, and M. Inouye. 1966. "Frameshift mutations and the genetic code." *Cold Spring Harb. Symp. Quant. Biol.*, 31: 77.

Sugimura, T., S. Kondo, and H. Takebe, eds. 1982. *Environmental Mutagens and Carcinogens.* Alan Liss.

West, S. C., E. Cassuto, and P. Howard-Flanders. 1982. "Postreplication repair in *E. coli*: strand-exchange reactions of gapped DNA by RecA protein." *Molec. Gen.*, 187: 209.

BIBLIOGRAPHY

Chapter 11

Brown, D. 1981. "Gene expression in eukaryotes." *Science*, 211: 667.

Clark, B. F. C., H. Klenow, and J. Zeuthen, eds. 1978. *Gene Expression*. Pergamon.

Darnell, J. E. 1982. "Variety in the level of gene control in eukaryotic cells." *Nature*, 297: 365.

Elgin, S. C. R. 1981. "DNase I-hypersensitive sites of chromatin." *Cell*, 27: 413.

Felsenfeld, G., and J. McGhee. 1982. "Methylation and gene control." *Nature*, 296: 602.

Freifelder, D. 1983. *Molecular Biology*. Chapters 14 and 22. Jones and Bartlett.

Guarente, L. 1984. "Yeast promoters: positive and negative elements." *Cell*, 36: 799.

Khoury, G., and P. Gruss. 1983. "Enhancer elements." *Cell*, 33: 83.

Kolata, G. 1981. "Gene regulation through chromosome structure." *Science*, 214: 775.

Kolata, G. 1984. "New clues to gene regulation." *Science*, 224: 58.

Lewin, B. 1981. *Gene Expression. 2. Eukaryotes*. Wiley.

Lewin, B. 1983. *Genes*. Wiley.

Maniatis, T., and M. Ptashne. 1976. "A DNA operator-repressor system." *Scient. Amer.*, January.

Miller, J., and W. Reznikoff, eds. 1978. *The Operon*. Cold Spring Harbor.

Ochoa, S., and C. de Haro. 1979. "Regulation of protein synthesis in eukaryotes." *Ann. Rev. Biochem.*, 48: 549.

O'Malley, B. W., and W. T. Schroeder. 1976. "The receptors of steroid hormones." *Scient. Amer.*, February.

Ptashne, M., and W. Gilbert. 1970. "Genetic repressors." *Scient. Amer.*, June.

Revel, M., and Y. Goner. 1978. "Posttranscriptional and translational controls of gene expression in eukaryotes." *Ann. Rev. Biochem.*, 47: 1079.

Stein, G. S., and J. S. Stein. 1975. "Chromosomal proteins and gene regulation." *Scient. Amer.*, February.

Ullman, A. and A. Danchin. 1980. "Role of cyclic AMP in regulatory mechanisms of bacteria." *Trends. Biochem. Sci.*, 5: 95.

Wahli, W., I. B. Dawid, G. U. Ryffel, and R. Weber. 1981. "Vitellogenesis and the vitellogenin family." *Science*, 212: 298.

Weisbrod, S. 1982. "Active chromatin." *Nature*, 297: 289.

Yanofsky, C. 1981. "Attenuation in the control of expression of bacterial operons." *Nature*, 289: 751.

Chapter 12

Abelson, J., and E. Butz. 1980. "Recombinant DNA." *Science*, 209: 1317.

Anderson, W. F., and Diacumakos, E. G. 1981. "Genetic engineering in mammalian cells." *Scient. Amer.*, July.

Bishop, J. M. 1982. "Oncogenes." *Scient. Amer.*, March.

Broome, S., and W. Gilbert. 1978. "Immunological screening method to detect specific translation products." *Proc. Nat. Acad. Sci.*, 75: 2746.

Brown, D. D. 1973. "The isolation of genes." *Scient. Amer.*, August.

Chilton, M.-D. 1983. "A vector for introducing new genes into plants." *Scient. Amer.*, June.

Cohen, S. N. 1975. "The manipulation of genes." *Scient. Amer.*, July.

Curtiss, R. 1976. "Genetic manipulation of microorganisms: potential benefits and hazards." *Ann. Rev. Microbiol.*, 30: 507.

Drlica, K. 1984. *Understanding Gene Cloning*. Wiley.

Gilbert, W., and L. Villa-Karmanoff. 1980. "Useful proteins from recombinant bacteria." *Scient. Amer.*, April.

Hacket, P. B., J. A. Fuchs, and J. W. Messing.1984. *An Introduction to Recombinant DNA Techniques*. Benjamin-Cummings.

Lobban, P., and A. D. Kaiser. 1973. "Enzymatic end-to-end joining of DNA molecules." *J. Mol. Biol.*, 78: 453.

Luria, S. E. 1970. "The recognition of DNA in bacteria." *Scient. Amer.*, January.

Maniatis, T. *et al.* 1978. "The isolation of structural genes from libraries of eucaryotic DNA." *Cell*, 15: 687.

Maniatis, T., E. F. Frisch, and J. Sambrook. 1982. *Molecular Cloning: A Laboratory Manual*. Cold Spring Harbor.

Mertz, J., and R. Davis. 1972. "Cleavage of DNA: RI restriction enzyme generates cohesive ends." *Proc. Nat. Acad. Sci.*, 69: 3370.

Pestka, S. 1983. "The purification and manufacture of human interferons." *Scient. Amer.*, August.

Roberts, R. J. 1980. "Restriction and modification enzymes and their recognition sequences." *Gene*, 8: 329.

Rodriguez, R. L., and R. C. Tait. 1983. *Recombinant DNA Techniques*. Addison-Wesley.

Seeberg, P. H., *et al.* 1978. "Synthesis of growth hormone by bacteria." *Nature*, 276: 795.

Smith, D. H. 1979. "Nucleotide sequence specificity of restriction enzymes." *Science*, 205: 455.

Wu, R., ed. 1979. *Recombinant DNA. Methods Enzymol*. Vol. 68. Academic.

Chapter 13

Bach, F. H., and J. J. Van Rood. 1976. "The major histocompatibility complex—genetics and biology." *New Eng. J. Med.*, 295: 806.

Baltimore, D. 1981. "Somatic mutation gains its place among the generators of diversity." *Cell*, 26: 295.

Brown, D. D. 1973. "The isolation of genes." *Scient. Amer.*, August.

Caskey, C. T., and D. C. Robbins, eds. 1982. *Somatic Cell Genetics*. Plenum.

Davidson, R. L. 1973. *Somatic Cell Hybridization: Studies on Genetics and Development*. Addison-Wesley.

Ephrussi. B. 1972. *Hybridization of Somatic Cells*. Princeton Univ. Press.

Ephrussi, B., and M. C. Weiss. 1969. "Hybrid somatic cells." *Scient. Amer.*, April.

Harris, H. 1970. *Cell Fusion*. Harvard Univ. Press.

Leder, P. 1982. "The genetics of antibody diversity." *Scient. Amer.*, May.

Marx, J. 1981. "Antibodies: getting their genes together." *Science*, 212: 1015.

McKusick, V. A. 1971. "The mapping of human chromosomes." *Scient. Amer.*, April.

Puck, T., and F. T. Kao. 1982. "Somatic cell genetics and its application to medicine." *Ann. Rev. Genetics*, 16: 225.

Ruddle, F. H. 1973. "Linkage analysis in man by somatic cell genetics." *Nature*, 242: 165.

Ruddle, F. H. 1981. "A new era in mammalian gene mapping: somatic cell genetics and recombinant DNA methodologies." *Nature*, 294: 115.

Ruddle, F. H., and R. S. Kucherlapati. 1974. "Hybrid cells and human cells." *Scient. Amer.*, July.

Tonegawa, S. 1983. "Somatic generation of antibody diversity." *Nature*, 302: 575.

Watkins, M. W. 1966. "Blood group substances." *Science*, 152: 172.

BIBLIOGRAPHY

Chapter 14

Bodmer, W. F., and L. L. Cavalli-Sforza. 1976. *Genetics, Evolution, and Man.* W. H. Freeman.

Cavalli-Sforza, L 1974. "The genetics of human populations." *Scient. Amer.*, September.

Crow, J. F., and M. Kimura. 1970. *An Introduction to Population Genetic Theory.* Harper and Row.

Dobzhansky, T. 1955. "A review of some fundamental concepts and problems of population genetics." *Cold Spring Harb. Symp. Quant. Biol.*, 20: 1

Eckhardt, R. B. 1972. "Population genetics and human origins." *Scient. Amer.*, January.

Falconer, D. S. 1981. *Introduction to Population Genetics.* Ronald.

Fincham, J. R. S. 1983. *Genetics.* Chapter 18. Jones and Bartlett.

Hardy, G. 1908. "Mendelian proportions in a mixed population." *Science*, 28: 49.

Harris, H., and D. A. Hopkinson. 1972. "Average heterozygosity in man." *J. Human Genetics*, 36: 9.

Hartl, D. L. 1980. *Principles of Population Genetics.* Sinauer.

Hedrick, P. W. 1983. *Genetics of Populations.* Jones and Bartlett.

Kimura, M., and T. Ohta. 1971. *Theoretical Aspects of Population Genetics.* Princeton Univ. Press.

Li, C. C. 1976. *First Course in Population Genetics.* Boxwood Press.

Spiess, E. B. 1977. *Genes in Populations.* Wiley.

Chapter 15

Ayala, F. 1976. *Molecular Evolution.* Sinauer.

Ayala, F. 1978. "The mechanisms of evolution." *Scient. Amer.*, September.

Bodmer, W., and L. L. Cavalli-Sforza. 1976. *Genetics, Evolution, and Man.* W. H. Freeman.

Cavalli-Sforza, L. L. 1969. "Genetic drift in an Italian population." *Scient. Amer.*, August.

Dobzhansky, T. 1941. *Genetics and the Origin of the Species.* Columbia Univ. Press.

Dobzhansky, T. 1950. "The genetic basis of evolution." *Scient. Amer.*, November.

Dobzhansky, T. 1960. "The present evolution of man." *Scient. Amer.*, September.

Dobzhansky, T., F. Ayala, G. Stebbins, and J. Valentine. 1977. *Evolution.* W. H. Freeman.

Dobzhansky, T., and O. Pavlovsky. "An experimental study of interaction between genetic drift and natural selection." *Evolution*, 7: 198.

Fitch, W. M. 1973. "Aspects of molecular evolution." *Ann. Rev. Genetics*, 7: 343.

Ford, E. B. 1971. *Ecological Genetics.* Chapman and Hall.

Garn, S. M. 1961. *Human Races.* C. C. Thomas.

Hedrick, P. W. 1984. *Population Biology.* Jones and Bartlett.

Lewontin, R. C. 1974. *The Genetic Basis of Evolutionary Change.* Columbia University Press.

Li, C. C. 1955. *Population Genetics.* Univ. of Chicago Press.

Milkman, R., ed. *Perspectives in Evolution.* Sinauer.

Nei, M. 1975. *Molecular Population Genetics and Evolution.* Harvard Univ. Press.

Roughgarden, J. 1979. *Theory of Population Genetics and Evolutionary Ecology: An Introduction.* Macmillan.

Sheppard, P. M. 1958. *Natural Selection and Heredity*. Hutchison.

Spiess, E. 1977. *Genes in Population*. Wiley.

Stebbins, G. L. 1950. *Variation and Evolution in Plants*. Oxford Univ. Press.

Wilson, E. O. 1975. *Sociobiology*. Belknap.

Wright, S. 1978. *Evolution and the Genetics of Populations*. Vol. 4. *Variability Within and Among Natural Populations*. Univ. of Chicago Press.

Chapter 16

Bodmer, W. F., and L. L. Cavalli-Sforza. 1970. "Intelligence and race." *Scient. Amer.*, October.

Bodmer, W. F., and L. L. Cavalli-Sforza. 1976. *Genetics, Evolution, and Man*. W. H. Freeman.

Crow, J. 1957. "Genetics of insect resistance to chemicals." *Ann. Rev. Entymol.*, 2: 227.

East, E. M. 1910. "A Mendelian interpretation of inheritance that is apparently continuous." *Amer. Natural.*, 44: 65.

Falconer, D. S. 1981. *Introduction to Quantitative Genetics*. Longman.

Farber, S. L. 1980. *Identical Twins Reared Apart*. Basic Books.

Feldman, M. W., and R. C. Lewontin. 1975. "The heritability hang-up." *Science*, 190: 1163.

Foster, H. L. 1965. "Mammalian pigment genetics." *Adv. in Genetics*, 13: 311.

Hartl, D. L. 1980. *Principles of Population Genetics*. Sinauer.

Hedrick, P. W. 1983. *Genetics of Populations*. Chapter 11. Jones and Bartlett.

Law, C. N. 1967. "The location of genetic factors controlling a number of quantitative characters in wheat." *Genetics*, 56: 445.

Lewontin, R. 1970. "Race and Intelligence." *Bull. Atomic Scientists*, March.

Mather, K. 1943. "Polygenic inheritance and natural selection." *Biol. Rev.*, 18: 32.

Mather, K. 1974. *Genetic Structures of Populations*. Halsted.

Mather, K., and J. L. Jinks. 1971. *Biometrical Genetics*. Cornell Univ. Press.

Smith, J. M. 1978. "The evolution of behavior." *Scient. Amer.*, September.

Wilson, E. O. 1975. *Sociobiology*. Harvard Univ. Press.

ANSWERS

Publishers Note: A separate *Student Study Guide* (ISBN 0-87620-054-5) is available to accompany *General Genetics* (ISBN 0-86720-050-2). The *Student Study Guide* contains the following: • Complete worked out solutions to each problem in the text • Complete chapter summaries • Drill questions with answers.

Chapter 1

1. (a) *DHW, DHw, DhW, Dhw, dHW, dHw, dhW, dhw;* in equal proportions. **(b)** 1/64, 9/64, 1/64, 3/64. **2. (a)** 12. **(b)** 1/64. **(c)** 1/8. **3. (a)** Dominant; because two affected parents produced a normal offspring. **(b)** 1/4. **4.** 2/3. **5. (a)** 1/2. **(b)** 1/4. **6.** 1/12. **7.** 1/16, 7/16. **8. (a)** Red. **(b)** 9/16 red, 6/16 brown, 1/16 white. **9.** 3/16. **10. (a)** 1/4. **(b)** 1/8; 27/64. **(c)** 1/64; 1/32. **11. (a)** Purple. **(b)** 27/256; 37/64. **(c)** 1/16; 7/8. **12.** 27/64 solid black, 9/64 spotted black, 9/64 solid brown, 3/64 spotted brown, 16/64 albino. **13.** 1/36. **14.** 9/16 black, 3/16 brown, 4/16 yellow. **15. (a)** *BbEe* in both cases. **(b)** *BbEe.* **(c)** 1/4. **16.** 3/16.

Chapter 2

1. (a) Deafness probably sex-linked recessive; cataract probably autosomal dominant. **(b)** *Dd cc* and *D– Cc.* **2.** 1/8. **3. (a)** Black males and calico females. **(b)** Calico mother and yellow father. **(c)** Likely, $X(c^b)X(c^y)Y$. **4. (a)** 1/64. **(b)** 1/32. **5.** 1/3. **6.** X^vX^vY vermilion-eyed, attached-X females, and $X^{v+}Y$ wildtype males. **7.** 1/512. **8. (a)** 9/16. **(b)** 1/4. **9.** The gene is not dominant and is X-linked. The female is heterozygous (Nn); the male is hemizygous recessive $(n-)$. **10.** Females: $w^+w b^+b$ (red-eyed, gray-bodied); males $w – b^+b$ (white-eyed, gray-bodied). **11.** 13/64 white, fast males; 13/64 white, slow males; 3/64 colored, fast males; 3/64 colored, slow males; the same frequencies for the females. **12. (a)** 20/64. **(b)** 1/64. **(c)** 57/64. **13. (a)** 81/256. **(b)** 27/128. **14.** Sample 1, $P < 0.01$; sample 2, $0.3 < P < 0.5$. With larger sample size the same numerical deviation represents a smaller percentage error; larger samples provide a more critical test of a hypothesis, since proportionally larger deviations have a greater probability of occurring by chance in small samples. **15.** $\chi^2 = 5.18$, degrees of freedom = 3, $0.1 < P < 0.2$. Since deviations as great or greater would be expected to occur by chance alone in 10–20 percent of such samples, the data agree satisfactorily with a $9:3:3:1$ ratio. **16. (a)** 105/1024. **(b)** 0.0254. **17.** 0.00612.

Chapter 3

1. 0.08. **2.** 4.5 percent. **3.** 0.2275 *b b cn cn*, 0.455 *b + cn +*, 0.2275 *+ + +*, 0.0225 *bb cn +*, 0.0225 *b + + +*, 0.0225 *b + cn cn*, 0.0225 *+ + cn +*. **4.** 18.6 percent (the average of 17.14 and 20.07 percent). **5.** *c*–3.5 map units–sh–18.4 map units–wx.   **6.** 15.   **7. (a)** k–6.6 map units–e–5.0 map units–cd. **(b)** No difference.   **8. (a)** al–0.57 map units–St–11.0 map units–dp. **(b)** Wildtype and Star aristaless dumpy, representing the double-crossover classes.   **9.** Garnet and vermilion are linked, 11.2 map units apart; orange-eyed males are double-mutant recombinants.   **10.** centromere –b–a.   **11.** The genes are unlinked.   **12. (a)** m–10.6 map units–centromere–7.9 map units–t. **(b)** 0.17. **(c)** Only half of the spores are recombinant.   **13.** a and c are linked, with a recombination frequency of '7.1 percent; b is not linked to either a or c.   **14. (a)** 36.5 percent PD, 36.5 percent NPD, 27 percent TT. **(b)** 50 percent.   **15.** $Pg\,gl\,Bk$, 853; $pg\,Gl\,bk$, 853; $Pg\,Gl\,bk$, 47; $pg\,gl\,Bk$, 47; $Pg\,gl\,bk$, 97; $pg\,Gl\,Bk$, 97; $Pg\,Gl\,Bk$, 3; $pg\,gl\,bk$, 3.   **16.** 0.5 percent.   **17.** $P^2/4$.   **18. (a)** fpa, leu, and $ribo$ are on one chromosome; w and ad are on a second chromosome. **(b)** $ribo$–leu–centromere.   **19.** If two mutations are noncomplementing, then they are judged to be mutations in the same gene and thus are alleles; if they complement, then they are in different genes and are not alleles.   **20.** To distinguish complementation and noncomplementation by examination of phenotype, the heterozygous genotype of each gene must have the wildtype phenotype when mutations are in different genes.   **21.** Five complementation groups: (1) 1, 3, 7; (2) 2; (3) 4, 6; (4) 5, 9; (5) 8.   **22.** From first cross, v, extreme lz is $lz^k lz^{46} v$, and sn is sn; from second cross, v is v, sn, extreme lz is $lz^{BS} lz^k$ (resulting from double crossing over). Order of lz mutations is lz^{BS}–lz^k–lz^{46}.   **23.** Mutation 9, which is in group B, interferes with the activity of the gene product of wildtype gene B.

Chapter 4

1. DNA: thymine; RNA: uracil.   **2.** A nucleotide has a phosphate group.   **3. (a)** C = 32 percent. **(b)** ([G] + [C])/[All bases] = 0.64 G+C.   **4. (a)** 10^5. **(b)** 10^4.   **5. (a)** Not meaningful. **(b)** Single-stranded.   **6.** A double-stranded hairpin with AT pairs.   **7. (a)** Three bands, in the ratios 1 $^{15}N^{15}N$: 2 $^{14}N^{15}N$:1 $^{14}N^{14}N$. **(b)** T4 and P2 DNA have no parts of their base sequences in common.   **8. (a)** CT = 0.15, AC = 0.03, TC = 0.08, AA = 0.10. **(b)** TC = 0.15, CA = 0.03, CT = 0.08, AA = 0.10.   **9.** Primer: a nucleotide bound to DNA and having a 3'-OH group. Template: a polynucleotide strand whose base sequence can be copied.   **10. (a)** T. **(b)** F. **(c)** F. **(d)** F. **(e)** T, though *in vitro* a DNA fragment can be used.   **11.** As a simple repeating polymer consisting of a single unit, it cannot carry information for an amino acid sequence.   **12.** Use another genetic marker.   **13. (a)** 3'-OH. **(b)** 3'-OH. **(c)** 5'-P.   **14.** 3'-OH.   **15. (a)** 5'AGUCAU3'. **(b)** U, triphosphate. **(c)** Right to left.   **16.** One generation: two bands, $^{15}N^{15}N$ and $^{14}N^{14}N$, in equal amounts. Two generations: same two bands but the ratio of the amounts is 1:3.   **17.** One generation: two bands, $^{14}N^{14}N$ and $^{15}N^{14}N$, in equal amounts. Two generations: same two bands but in a ratio of 3:1.   **18.** Equal amounts of $^{14}N^{14}N$ and $^{14}N^{15}N$.   **19.** 3'GATACGACGAAGTACTGG.   **20.** Yes.   **21. (a)** Single piece. **(b)** No.   **22.** Bidirectionality of replication.   **23. (a)** No maximum. **(b)** A parental single strand.   **24. (a)** 4×10^5. **(b)** No, because the number of protein molecules exceeds the number of transformants.   **25.** There are no positive ions to neutralize the negative charges of the phosphates. The strong electrostatic repulsion causes the strands to come apart.   **26. (a)** They are joined by the normal action of polymerase I and DNA ligase. **(b)** No.

Chapter 5

1. **(a)** 40. **(b)** 4. **(c)** A supercoil with one node—that is, a figure 8. **(d)** Five positive turns of supercoiling.   **2.** The supercoiled form is in equilibrium with an underwound form having many unpaired bases.   **3.** (3), (2), (1). **4.** (1).

5.

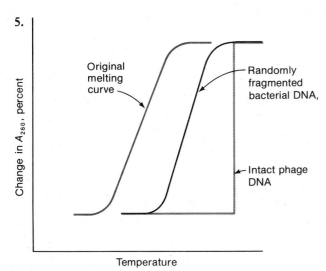

6. 10 percent.   **7.** **(a)** 30, 60, 10 percent. **(b)** 10^4, 10^6.   **8.** See Glossary. **9.** The phosphates.   **10.** In the first experiment, nucleosomes have formed at random. In the second experiment, P has bound to a unique sequence and nucleosomes have formed by sequential addition of histone octamers from the site of P.   **11.** ABCD . . . ABCD; ABCD . . . D′C′B′A′.   **12.** The fact that the target sequence and the transposable element are duplicated during transposition.   **13.** The GC pairs in molecule I are probably dispersed; in contrast, those of molecule II are clustered, so that long segments of stable GC regions exist.   **14.** Both species will have the same melting temperature.   **15.** If it is a dimer, supercoils will in time be replaced by open circles, without any intermediates appearing. If it consists of linked circles, each will be independent so that one single-strand break will convert the structure to a supercoil linked to an open circle.   **16.** Chloroplasts and their DNA remain, even though the chlorophyll is lost.   **17.** The DNases probably bind to the negatively charged phosphates, which are neutralized by the Na^+ ions in 1 M NaCl.   **18.** Functional histone is extremely intolerant of amino acid changes, so all changes are probably lethal.   **19.** The sequence may be a transposable element.

Chapter 6

1. A circular chromosome.   **2.** Inversion; Robertsonian translocation. **3.** 10 (diploid), 20 (tetraploid), 30, 40, 50.   **4.** The chromosome number was doubled.   **5.** 18, 36.   **6.** Species A and B hybridized, and the chromosome number in the hybrid was double.   **7.** 45,X; Turner syndrome.   **8.** Robertsonian translocation in which one of the participants is chromosome 21.   **9.** Paracentric inversion.   **10.** *b a d c e f* or *b d a c e f.*

11. Because of aneuploid gametes resulting from adjacent-1 and adjacent-2 segregation of heterozygous translocation; form two bivalents instead of a quadrivalent; all.   **12.** Reciprocal translocation.   **13.** Bands 1–6 correspond to *b, a, c, e, d,* and *f,* in that order.   **14.** 3.   **15.** The normal haploid parent appears to be trisomic for chromosome 2.   **16.** The observed recombination frequency (0.1 percent) is much less than would be expected; an inversion.   **17.** The second chromosome carrying *Cy* underwent a reciprocal translocation with the Y chromosome.   **18.** In a sperm of the irradiated male the tip of the X carrying y^+ was broken off and attached to the Y.
19. A two-strand double crossover occurred in the inversion; inverted.
20. Wildtype females, 50 percent; ruby dumpy males, 50 percent. All other classes of progeny will be lethal.   **21. (a)** 11.9 percent. **(b)** 25 percent for each class of progeny.   **22.** Brachytic–fine stripe, 4.7 percent; brachytic–breakpoint, 10.1 percent; fine stripe–breakpoint, 7.5 percent.   **23.** *d, a, b.*
24. *A B C D E; A B c d a; e b C D E; a d c b e.*   **25. (a)** Three possibilities: (1) A paracentric inversion and crossing over does not occur. (2) A paracentric inversion with a single crossover with oriented distribution of chromatids yielding all viable gametes. (3) A pericentric inversion with no crossover. **(b)** A reciprocal translocation with alternate segregation giving viable gametes and adjacent segregation yielding inviable gametes.   **26.** The F_1 progeny will have a lower fertility than the parents because the recombinant chromosomes do not have a full genetic complement, and gametes receiving them will be inviable.
27. Examine cells undergoing meiosis with a microscope. An autopolyploid has four sets of identical chromosomes, and in meiotic metaphase these often synapse as tetravalents. An allopolyploid has only two sets of chromosomes, and these can only form bivalents.   **28. (a)** All gametes from both individuals are euploid. **(b)** All gametes from the homozygote are euploid while 50 percent of the gametes from the heterozygote are euploid. Hence, the total frequency of euploid gametes is 75 percent. **(c)** 50 percent, since one of the parents has only one half of its gametes euploid.   **29.** In the tetrasomic, the four chromosomes frequently synapse in twos or as a group of four; in either case the gametes each get two members of each chromosome. In the trisomic, two of the chromosomes go to one pole, and one to the other, and the distribution between the poles is random for each triplet. Occasionally, the paired chromosomes will be totally lost. Hence it will be the rare zygote that does not suffer from chromosome imbalance.   **30.** As explained in the previous question, the tetraploid usually produces euploid gametes which, with the haploid gamete of the diploid individual, results in a viable triploid offspring. Gametes from a triploid are usually aneuploid, resulting in inviable offspring.

Chapter 7

1. *his, leu, trp; met* prevents growth of the male bacteria. The number is small on the His-only medium, because crossing over is limited to the regions outside of the *trp* and *leu* genes.   **2.** *pro pur his.*   **3.** Grow P1 on the strain containing the *x* marker and infect a strain that has a marker *A* adjacent to *x*. Select for the *A* allele present in the strain containing the *x* marker and test the transductants for the presence of *x*.   **4.** *a,* 10; *b,* 15; *c,* 20; *d,* 30 minutes. *d* is closely linked to *str.*   **5.** (a, c, e, f, g) are true.   **6.** If a gene is very near the transfer origin, very little space is available for one of the necessary exchanges between the gene and the origin.   **7.** *Pro* is a terminal marker. F is closely linked to *pro* and therefore is also at the terminus.   **8.** 3.5×10^{10}.
9. T2, turbid; T2*h*, clear.   **10.** *att* is between *f* and *g.*   **11.** *J.*

12. Genes on opposite sides of *att* can be recombined by activity of the integrase system, even when all other recombination systems are absent.   **13. (a)** *gal, BOP′*; *bio, POB′*.   **14.** They are only formed by aberrant excision. **15.** F′*z* and a revertant of *z⁻*.   **16.** The λ phage must have contained a transposable element carrying the *amp-r* gene.   **17.** The tetracycline-resistance marker in the phage must have been in a transposable element. Transposition occurred in the lysogen to another location in the *E. coli* chromosome; therefore, when the prophage was lost, the antibiotic resistance remained.   **18.** The *tet-r* marker in the original F^- strain was carried by a transposable element. In the recombinant this element transposed at a later time to a site near the *lac* gene. **19. (a)** 200. **(b)** 0.307.   **20.** 8×10^7.   **21. (a)** 0.007. **(b)** 4.3×10^6. **(c)** 1.28×10^7. **(d)** 8.22×10^7.   **22. (a)** Prophage transferred to a female lacking repressor. **(b)** Use interrupted mating and determine the time after which the number of particular recombinant types decreases.   **23.** T4 does not use host gene products that have to be made after infection; λ does.   **24.** Yes, if it has a replication origin.   **25.** Comparable, but some plasmids are smaller. **26.** Early in the formation of a colony, a daughter cell receives no plasmid.

Chapter 8

1. Homology and a RecA-type protein are required for transduction, transformation, and bacterial conjugation. Meiotic recombination requires homology and presumably a RecA-type protein also. Transposition and site-specific recombination require neither.   **2.** Prophage integration and excision, formation of an F′, formation of an Hfr, and meiotic recombination.   **3.** The target sequence and, in prokaryotes, the transposable element.   **4.** 3 and 5. **5.** Postmeiotic segregation.   **6.** Branch migration.   **7.** Mismatch repair; strand assimilation.   **8.** Increases, by providing breaks, gaps, and free ends. **9.** I are nonrecombinant. II are recombinant without conversion. III are recombinant with an $E \rightarrow e$ conversion.   **10.** The mismatch repair system. **11.** Immediately after invasion, the recipient has a mismatched base pair. For low-efficiency markers mismatches are usually corrected to the recipient genotype. For high-efficiency markers either mismatch repair fails to occur or correction is primarily to the donor genotype.   **12.** A figure 8.

Chapter 9

1. Ribonucleosides triphosphates, RNA polymerase, and one strand of DNA, respectively.   **2.** Each has a 3′-OH group. A primary transcript has a 5′-triphosphate; prokaryotic mRNA has a triphosphate, and eukaryotic mRNA usually has a cap.   **3.** In an overlapping code, a base change would affect three codons.   **4.** NH₂-Met-Pro-Leu-Ile-Ser-Ala-Ser.   **5.** About 3000. **6.** The 5′ end.   **7.** An alternating polypeptide containing only valine and cysteine.   **8.** Ile-Asp-Arg and Asp-Arg would be made.   **9. (a)** Same molecule. **(b)** It must be a multiple of three.   **10.** A stop codon. **11.** There is no way of knowing because both codons could be altered to yield the stop codons UAA and UAG.   **12.** Normal.   **13.** Mutant #1 has a deletion of 12 base pairs, those encoding amino acids 1–4 or 2–5, or a change in the first AUG. Mutation #2 is a change from Tyr to Stop; Met-Leu-His. **14.** If the anticodon were ICA, the codon UGA would also be a Cys codon, according to the wobble hypothesis. If GCA were the Cys anticodon, UGA would be reserved as a stop codon. Thus, the Cys anticodon is GCA.

15. Specificity of (1) codon-anticodon binding and (2) recognition of a tRNA molecule by the corresponding aminoacyl synthetase.   **16.** Soon no 30S particles will be available for forming the preinitiation complex.   **17.** 7.   **18.** They will not terminate at the same site because no sequence of four bases contains two overlapping stop codons. They would not usually have the same number of amino acids, but certainly could, depending on where each protein starts and stops.

Chapter 10

1. 1.   **2.** 5×10^{-6}.   **3.** Minimum, 5000; maximum, 30,000.   **4. (a)** 60.6 percent. **(b)** 20,000.   **5.** 1.8×10^{-9}.   **6.** A, 1.1×10^{-9}; B, 1.8×10^{-9}.   **7.** 1, 3, and probably 6.   **8.** Only certain substitutions occurring at critical positions have a detectable effect.   **9. (a)** No. **(b)** Yes.   **10.** CG → TA.   **11. (a)** UAA or UAG. **(b)** Must be UAG.   **12.** Six.   **13.** Leu → Pro.   **14.** AT → GC.   **15. (a)** 1 transition:2 transversions.   **(b)** $\chi^2 = 0.86$. Agrees.   **16.** The base change is in one strand of the DNA. Thus, DNA replication yields one daughter DNA molecule with the wildtype gene and another with the mutant gene. Cell division yields one Lac$^+$ cell and one Lac$^-$ cell.   **17. (a)** Arg, Cys, Gly, Leu, Ser, and Trp. **(b)** Gln, Glu, Leu, Lys, Ser, and Tyr.   **18. (a)** All. **(b)** 75 percent. **(c)** 95 percent.   **19.** The two gene products are probably subunits of a multisubunit protein.   **20. (a)** A double point mutation. **(b)** Nothing.   **21. (a)** λ does not have a repair system and the damage is repaired by the bacterial Uvr system. **(b)** The ultraviolet has induced the SOS repair system in the bacterium. This system repairs the damage, but it is error-prone.

22.

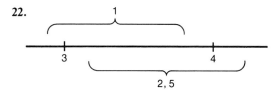

Failure to recombine is result of a deletion mutation overlapping either a point mutation or another deletion.   **23.** Some of the mutagenized P1 particles will have acquired mutations in the a gene. The a^+b^- bacterial strain is infected with the mutagenized P1 population and b^+ transductants are selected. These are tested for the genotype of a. Some of the transductants will be a^- because of linkage between the a and b genes.

Chapter 11

1. Repressed: transcription inhibited; induced: repressor inactivated; constitutive: never repressed; coordinate regulation: control of synthesis of a set of proteins encoded in a single polycistronic mRNA by a single control element.   **2.** Yes, autoregulated.   **3.** 2, 1.   **4.** Basal level of transcription will yield a basal concentration of permease molecules.   **5.** First, *lac* mRNA and then β-galactosidase is made. When the lactose is finally consumed, synthesis of *lac* mRNA will cease, and more enzyme will not be made. The enzyme already made

will persist, so the activity remains until gradually diluted out by cell division.
6. The *lac* genes enter the recipient. No repressor is present so *lac* mRNA is made. However, a *lacI* gene has also entered the recipient, so soon repressor will be made and *lac* transcription will stop.   **7.** The repressor cannot bind the inducer. No.   **8. (a)** 64. **(b)** If there is a preferred carbon source also present, the answer is essentially the same as in (a). If there is no other carbon source but X, the constitutive strain is no longer at a disadvantage. **(c)** An inducible mutant.   **9. (a)** *lacI⁻* or *lacOᶜ*. **(b)** *lacI⁺lacO⁺lacZ⁻lacY⁺* and *lacI⁻*.   **10.** A defective CAP protein; an inactive adenyl cyclase gene.
11. A positive regulator activated by an inducer.   **12. (a)** 1. **(b)** 0. **(c)** 1.
13. (a) No, 2. **(b)** Yes, 2. **(c)** Yes, 3. **(d)** No, 1. **(e)** No, 1.   **14. (a)** G inhibits 5; H inhibits 7; G and H together or E alone probably inhibit 3; J inhibits 8; enzyme 1 could be inhibited by (G, H, and J), (E and J), or (C and E); the ( ) enclose the products that act together. **(b)** Step 1, three isoenzymes; step 3, two isoenzymes.   **15. (a)** Without complex splicing events, only one protein molecule can be translated from a primary transcript. **(b)** An operon is a set of coordinately regulated genes encoded in one or two polycistronic mRNA molecules. A gene family is a collection of genes, each of which yields a distinct RNA molecule, which encode molecules of similar or related function.
16. No, for all three parts.   **17.** Gene amplification, when a limited time is available; continued synthesis of mRNA and extended lifetime of mRNA, when a long time is available.   **18.** In gene amplification the number of copies of a gene is increased, which enables mRNA to be made more rapidly. In translational amplification, the lifetime of the mRNA is increased, which enables a great deal of mRNA to be present at a particular time.   **19. (a)** No. **(b)** This is reasonable because no other protein is being made by these cells.   **20.** The specificity lies in the chromatin not in the extract, and some inhibitor of oocyte transcription is washed from the chromatin by the 0.6 *M* NaCl.   **21.** The hormone is apparently needed to initiate intron excision.   **22. (a)** The promoters for both transcription units probably have a common sequence acted on by either a positive or negative regulator. **(b)** Both primary transcripts have a common sequence acted on by an element that prevents some stage of processing. **(c)** Both processed mRNA molecules have a common sequence involved in ribosome binding. An effector may remove a protein bound to this sequence or it may denature a double-stranded region containing the ribosome binding site.   **(23) (a)** Yes, I, yes. **(b)** Yes, I, yes. **(c)** Yes, C, yes. **(d)** Yes, I, yes. **(e)** Yes, C, yes. **(f)** Yes, I, yes. **(g)** No, neither, no. **(h)** Yes, I, yes.   **24. (a)** There are many plasmids per cell and hence many operators. These compete with the chromosomal operator. **(b)** A promoter-up mutation in *lacI*.   **25. (a)** PhoR protein. **(b)** PhoB protein. **(c)** PhoR forms a complex with PhoB and phosphate.   **26. (a)** A repressor that cannot bind X (on); a repressor that cannot bind to the operator (on); a repressor that binds without X (off). **(b)** Mutant 1, inducible; mutant 2, inducible; mutant 3, unable to synthesize X under all conditions.
27. Post-translational modification. See Student Guide for many examples.

Chapter 12

1. (a) Same as a linear molecule. **(b)** The average of the value for a covalent circle and that of the nicked circle.   **2.** With the molecule in the same orientation for both enzymes, the EcoRI fragments have the order 6.2, 2.9, 4.5, 7.4, and 8.0, and the BamHI fragments have the order 10.1, 12.9, and 6.0.   **3.** A circle.

4. (a) Yes. **(b)** No.   **5.** Mutation alters one base in a restriction site and thereby causes two potential fragments to remain uncleaved.

6. (a) Tetracycline. **(b)** Tet-r Kan-r and Tet-r Kan-s. **(c)** Tet-r Kan-s.

7. The fragments have eliminated either the promoters or the ribosome binding sites of both genes. The plasmid contains both of these, but they are on the same strand of the plasmid DNA. If genes A and B are transcribed in opposite directions, only one can be expressed from the plasmid.   **8.** Lac$^-$.

9. Select an enzyme whose restriction site is further from the *lac* promoter than the BamHI site. Let us assume it is HaeI. If the gene of interest does not contain a HaeI site, do the following. Cleave the plasmid with both BamHI and HaeI, and retain the fragment (there will be two) that contains the replication origin and other essential genes. One end of the fragment will have a BamHI terminus, and the other end will have a HaeI terminus. Cleave the donor DNA with both enzymes and join the fragments to the isolated plasmid fragment. Since each fragment has one BamHI terminus and one HaeI terminus, joining can only occur in a particular orientation.   **10.** Its cleavage site must be the sequence GGG, GGGGG, or GGGGGG.   **11.** A processed eukaryotic mRNA will generally possess a 3'-poly(A) tail. Poly(dT) will anneal to this tail and prime DNA synthesis.

12. 1.38×10^6.   **13.** Two origins (for replication in *E. coli* and in yeast), plasmid DNA-initiation proteins active in *E. coli*, and two antibiotic-resistance markers.   **14.** The principle is to design the system so that expression of the gene can be controlled. For example, a restriction enzyme could be used that separates the coding sequence and the promoter, and the coding sequence could be linked to the promoter of the *lac* operon. Then, transcription could be turned on at will merely by adding an inducer of the *lac* operon.   **15.** Pst (in *amp*)– 0.46–RI–0.05–Hind III (in *tet*)–0.15–Bam (in *tet*)–0.35–SalI (in *tet*).

16. (a) Split a replication gene or the replication origin. **(b)** Yes.

17. A plasmid dimer.

Chapter 13

1. Hypoxanthine and thymidine are substrates of HGPRT and TK, respectively. Aminopterin blocks the major pathway for synthesis of DNA precursors.

2. (a) 3, 4. **(b)** 5. **(c)** Tip of 2p or of 2q.   **3. (a)** Able. **(b)** 17q.

4. Xp, 6q, 13q, 21, and 17q, respectively.   **5. (a)** 9q. **(b)** 9q.   **6.** 1p.

7. (a) O. **(b)** Both cousins are carriers of the *h* allele, which can become homozygous in their offspring.   **8. (a)** AB or B, secretor. **(b)** $I^B I^O Se\,se$.

9. (a) *dd*. **(b)** No. The woman already produces anti-D antibody. **(c)** 1/2.

10. It cannot; otherwise the antibodies would produce maternal-fetal incompatibility with respect to the ABO blood groups.   **11.** Female, *Xg xg; male, xg –*.   **12.** The mother is genotypically *Xg xg*. Nondisjunction at the second meiotic division produces an egg cell with two X chromosomes of genotype *xg xg*, which is fertilized by a normal sperm that carries the Y chromosome.

13. (a) No. **(b)** ABO.   **14.** Both. Male 1: the baby received *C* from its father. Male 2: the baby received *M* from its father.   **15.** Baby 1, couple 1; baby 2, couple 2.   **16.** An identical twin, because identical twins must have the same genotype at all loci, including blood group and histocompatibility loci.

17. If the child carries only antigens inherited from its mother, then skin grafts from the child to the mother will be accepted. Otherwise they will be rejected because of antigens inherited from the father and not possessed by the mother.

18. No. Combinatorial joining is a random process.   **19. (a)** 1/4. **(b)** 1/4.

20. Diploid, 2; triploid, 3; trisomic, 3, if the gene is in the trisomic chromosome, and, otherwise, 2. **21.** (1) $(1/2)^3$. (2) $(1/2)^3$. (3) 0. (4) $(10/16)^3$.

Chapter 14

1. 1, 1, and 3, respectively. **2.** 0.14, 0.56, and 0.30, respectively.
3. (a) a, 0.24; b, 0.76. **(b)** 0.058, 0.578, and 0.365, respectively. **4.** 0.017.
5. (a) 0.37. **(b)** 0.23. **6.** 0.13. **7.** $q = 0.316$; 0.48. **8.** 1/2.
9. (a) AA, 0.16; Aa, 0.68; aa, 0.16. No. **(b)** AA, 0.25; Aa, 0.50; aa, 0.25. Yes.
10. (a) No; homozygotes for self-sterility alleles cannot occur. **(b)** 3.
11. Bald males, 0.49; nonbald males, 0.51; bald females, 0.09; nonbald females, 0.91. **12.** Females: 40 yellow, 960 wild type; males: 200 yellow, 800 wild type. **13.** $I^A I^A$, 0.03; $I^A I^O$, 0.24 $I^B I^B$, 0.01; $I^B I^O$, 0.15; $I^O I^O$, 0.55; $I^A I^B$, 0.03. Phenotype frequencies are: O, 0.55; A, 0.27; B, 0.16; AB, 0.03.
14. 71.0, 53.1, and 9.9, respectively. $\chi^2 = 1.8$, which is nonsignificant; therefore, not inconsistent. **15.** 118.4, 164.5, and 57.1, respectively. $\chi^2 = 11.0$, which is highly significant; therefore, inconsistent. **16. (a)** 1.90×10^{-4} and 5.40×10^{-5}, respectively. **(b)** 22- and 6-fold increase, respectively. **17.** 0.0006.
18. 0.031; 0.033 **19.** 3/8. **20. (a)** aa, 0.35; ab, 0.17; bb, 0.49.
(b) 0.18, 0.49, and 0.32, respectively. **21.** 0.625. **22. (a)** 0.40.
(b) Very great. **23.** 0.21.

Chapter 15

1. (a) 0. **(b)** 1.0. **(c)** Fertility. **2.** When an allele is rare, most alleles will occur in heterozygotes; if the allele is recessive, the heterozygotes will not be exposed to selection. **3. (a)** 0.93. **(b)** 0.07 and 0.14. **4.** Less. Hemizygous males will be exposed to selection. **5.** Decrease; more recessive homozygotes are formed, and these are exposed to selection. **6.** Random mating refers to random union of gametes. In a finite population, gametes uniting at random will sometimes unite with those from relatives. In a population of size 1, random union of gametes must result in self-fertilization. **7.** Strain A: 10^{-6}, 10^{-5}, and 10^{-4}, respectively. Strain B: 10-fold greater than in strain A.
8. (a) 0.0196. **(b)** The same. **9.** 0.566 and 0.985. **10.** 0.0018.
11. 0.9798. **12. (a)** $A_1 A_1 = 0.60$, $A_1 A_2 = 0.87$, $A_2 A_2 = 0.87$, $A_2 A_3 = 0.87$, $A_3 A_3 = 1.00$, $A_1 A_3 = 0.80$. **(b)** A_2 is dominant to A_1, A_2 is dominant to A_3, A_1 and A_3 are additive. **(c)** Eventually A_3 will be fixed. **13. (a)** 0.02.
(b) Frequency of A strain after 10, 50, and 90 generations is expected to be 0.5496, 0.3532, and 0.1958, respectively. **14.** 0.9846; 0.0154.
15. 2.44. **16.** CH, 0.17. ST, 0.83. **17. (a)** 0.002. **(b)** 4×10^{-6}.
18. 8.47×10^{-6}. **19. (a)** 0.01. **(b)** 0.0025. **20.** 9.5×10^{-6}.
21. (a) 0.20. **(b)** The effective population size. **22.** 0.15.
23. See Student Guide.

Chapter 16

1. Genotype-environment interaction. **2.** Full siblings. **3.** Mean, 10.4; variance, 2.71; standard deviation, 1.65. **4. (a)** Mean, 4032.9; variance, 693,633.8; standard deviation, 832.8. **(b)** 3200–4866. **(c)** Observed, 239; expected, 207. **(d)** 2363–5698; rounded, 2500–6000; observed, 296; expected, 289.

5. (1) 68 percent. (2) 95 percent. (3) 16 percent. (4) 2.5 percent. (5) 13.5 percent.
6. **(a)** 1/64, 9/64, 27/64, and 27/64 for phenotypes 0, 1, 2, and 3, respectively.
(b) 729/4096, 1701/4096, 1323/4096, and 343/4096 for phenotypes 0, 1, 2, and 3, respectively. **(c)** Part (a): mean, 2.25; variance, 0.56; standard deviation, 0.75. Part (b): mean, 1.3125; variance, 0.738; standard deviation, 0.86.   **7. (a)** 0.
(b) The average of the parental strains.   **8.** 1.46 and 4.51; 76 percent.
9. 44.   **10.** 64 percent.   **11. (a)** 12.6 g. **(b)** 11.4 g.   **12.** 83.2.
13. 900 g.   **14.** 2100 μg.   **15.** 40 percent.   **16. (a)** 18 percent.
(b) Consistent.   **17.** About 1 percent.   **18.** 60 percent.
19. Because a higher incidence implies a lower threshold, and a lower threshold increases the risk for everyone, including the first-degree relatives of affected individuals.   **20.** $r = 0.57$, in actual data; $r = 0.60$ for mismatched data.   **21.** Let P represent parent-offspring covariance and S represent the full sib covariance. Then, from Table 16-6, $D = 4(S - P)$ and $A = 2P$. h^2 is estimated as (A)/(Total phenotypic variance).   **22.** Tooth number, 0.96. Length of longest spine, 0.29.   **23.** $A/4$.   **24.** Mother-offspring covariance would be greater than father-offspring covariance.   **25. (a)** 100.
(b) 100.45; 0.018 IQ points/year.   **26. (a)** Narrow-sense heritability = $2q/(1 + q)$; broad-sense heritability = 1.0. **(b)** 1.0, 0.67, 0.18, 0.10, 0.02, 0.01, and 0.002, respectively. **(c)** With rare recessives most homozygotes come from matings between heterozygotes; thus, no parent-offspring resemblance.

INDEX

blood typing, 481
Bombyx mori, 425
branch migration, 277, 278, 279, 282, 296, 297, 303
break-and-rejoin, 258, 303
breathing, 278
breeding, 579, 583, 587, 595
broad-sense heritability, 581, 583, 585
5-bromouracil, 369–370, 472

CaCl$_2$ transformation, 442, 452
cAMP. *See* cyclic AMP
cAMP-CAP complex, 402
cancer, 221–223, 388–389
cap, 322
CAP protein, 401
carcinogens, 373, 388–389
carcinostatic agents, 373
carrier, 16
casein, 426
catabolite activator protein, 401
cell fusion, 471–472
cell hybrids
    definition, 465
    gene mapping with, 466–467
    mouse-human, 466
cell lines, 466
centromere, 177, 208, 286
    conservation of number, 191
    heterochromatin in, 177
    positions of, 190
    structure, 182–183
chickens
    comb shape, 22–23
    feather color, 26, 59
Chlamydomonas reinhardii, 184–185
chloroplasts, 106, 184–185, 334
chromatin, 423
    chromatosomes, 163
    components of, 160
    digested, 162
    effect of DNase on, 162
    electron micrograph, 161, 165
    euchromatin, 177
    fibers of, 163
    heterochromatin, 177
    nucleosomes, 161–162
    replication, 164–165
    replication, nucleosomes and, 165
    solenoidal model, 163
    30-nm fiber, 163
chromatosome, 163
chromocenter, 167
chromosomes
    aberrant, 190

acentric, 190, 210, 212
acrocentric, 190, 218
breakage, 219
centromere structure, 177, 182–183
chromatin in, 163
dicentric, 190, 210
dicentric bridge, 211
Drosophila, 160
E. coli, 158
electron micrograph, 158, 164
folded, 158
genetic evidence for, 189
Giemsa staining, 199
heterochromatin, 177
histones. *See* histones
human, 199–205, 466, 468
hypersensitive sites, 423
loops of, 158–159
metacentric, 190
mouse, 466
multivalent, 193
nonhistone proteins in, 161
nucleic acid composition, 153
pairing, 193
polytene, 205. *See also* polytene chromosomes
scaffold, 164
size, 160
structure, 163
submetacentric, 190
telomere, 177, 182–183
translocation. *See* translocation
Chrysanthemum, polyploidy in, 191
circular genetic map, 246
cis configuration, 65
cis-dominant mutation, 397
cis-trans test, 98–99
ClB method, 362–363
clone, 108
cloned, definition, 441
cloning, definition, 433
cloning vehicle. *See* vector
co-conversion, 294
codominance, 20, 21, 22, 474, 499
codon, definition, 328
coefficient of coincidence, 83
cointegrate, 300–301
Col plasmid, 241
colchicine, 195
colicinogenic plasmid, 241
colinear, 315, 414
colony, 231
colony hybridization assay, 453
color blindness, 512
combinatorial joining, 484–485
combs of chickens, genetics of, 22, 23

common ancestors, 504
compatibility, 473
competence, 232
complementation
    allelism and, 98
    cis-trans test, 98–99
    heterokaryons, 97
    intergenic, 347
    intergenic, T4 *rII* locus, 254
    intragenic, 347
    negative, 349
    trans heterozygote, 97
complementation groups, 98–99, 102
complex multigene family, 416
compound-X chromosome, 73
conjugation, bacterial
    electron micrograph of, 270
    F, 241
    genetic transfer by, 229
    Hfr. *See* Hfr cells; Hfr transfer
    physical contact, 239
    plasmids, role in, 239
    rolling circle replication and, 241
    time for transfer, 241
    transfer genes, 239
    transfer in, 239
consanguineous mating, 504, 525
consensus sequence, definition, 317
constitutive, 397, 398
continuous traits, 567
conversion asci, 288
coordinate regulation, 321
copia, 180
corn. *See* maize
correlation coefficient, 591, 593, 595
Corynebacterium diphtheriae, 261
Cot curve, 172–174
    analysis of, 174
    definition, 172
    standard, 173
cotransduction, 236
cotransformation, 233
counterselected marker, 244
coupling, 65
coupled transcription-translation, 346–347
covariance, 591
Creighton, Harriet, 76, 272
crossing over, 63
    centromere and, 73
    chiasmata and, 70
    definition, 70
    double, 79–82
    four-strand stage, 73, 74
    genetic mapping and, 70–84
    interference, 82–83
    intragenic, 95–97

infection, Poisson distribution and, 257
life cycle, 111, 230
lysis, 249–250
lysogenic conversion, 261
lysogenic cycle, 248. See also lysogeny
lytic cycle, 248
mapping of, 252
multiplicity of infection, 257
plaques, 249–251
recombination in, 230, 250
structure of, 105, 108, 111
temperate, 248
virulent, 248
phage φX174, 132–133, 339
phage λ, 154, 442
  aberrant excision of, 261
  antitermination, 409
  assembly of, 452
  attachment sites, 259
  electron micrograph, 107
  excisionase, 260
  genes, 252
  genetic map, 252, 258
  immunity, 260
  inhibition of T4rII mutants, 261
  int gene, 259
  integrase, 259
  lysogen, 257
  lysogenic conversion, 261
  lysogenic cycle, 257. See also lysogeny,
    phage λ
  mismatch repair, 280–281
  N gene, 409, 410
  O region, 259
  prophage, 257
  prophage excision, 260
  prophage induction, 260
  repressor, 260, 408
  restriction map, 440
  RNA polymerase, regulation of, 409
  specialized transducing particles, 261
  vector, 452
  xis gene, 260
phage M13, 154, 442
phage MS-2, 155
phage Mu, 261
phage P1, 234
phage R17, 412
phage T1, 360–361
phage T2, 110–112
  DNA size, 154
phage T4, 105, 154, 328, 329, 356,
    385, 409,
  cyclically permuted DNA, 253
  electron micrograph, 111
  multiple-length DNA, 253

recombination, 250–251
rII locus, 254
terminally redundant DNA, 253
phage T5, 533
phage T7, 129, 143, 253
phage-resistant bacteria, 364
phase variation, 534
phenotype, definition, 8
phenylalanine operon, 408
phenylketonuria, 26
Philadelphia chromosome, 222
Phlox cuspidata, 499, 508, 518
phosphodiester bond, 113
photoreactivation, 375
pin, 503
plant breeding, 579, 588
plants, genetic engineering of, 432, 455
plants, polyploidy in, 191
plaques,
  assay, 249–250
  formation of, 249–250
  photograph of, 251
plasmids
  circular DNA, 240
  Col, 241
  colicinogenic, 241
  drug-resistance, 241
  F, 241. See also F
  high-copy-number, 240
  integration, 239
  nonessentiality of, 240
  pBR322, 449
  R, 241
  replication, 240
  segregation at cell division, 240
  size, 240
  Ti, 455
  transfer genes, 239
Plasmodium falciparum, 549
plastids, 184
pleitropic, 355
Pneumococcus, 108–110, 282, 304
  transformation, 108–110, 232
Poisson distribution, 254, 256–267
polar mutation, 319, 345
polarity, 403
polycistronic mRNA, 320, 338
poly(A), 322
polycistronic mRNA, 320, 338
poly(dA), 445
polymorphic locus, 500
polynucleotide chain
  definition, 113
  structure, 114
polyploidy, 191–196
  allopolyploidy, 194

autopolyploidy, 194
chromosome pairing, 193
Chrysanthemum, 191
hexaploid, 194
meiosis and, 192–193
octaploid, 194
tetraploid, 191, 193
polyproteins, 346, 426
polyribosome, 346, 347
polysome, 346, 347
polysomy
  aneuploid, 196
  definition, 196
  euploid, 197
  monosomic, 199
  trisomy, 196, 200–201
polytene chromosomes, 166–167, 205
  bands of, 167
  chromocenter, 166
  cytological mapping, 167
  deletion mapping, 222–224
  micrograph, 167
poly(U), 331
population
  definition, 496
  inbred, 517
  local, 496
population improvement, 575
population structure, 513, 536
population subdivision, 513–517
  definition, 513
position effects, 221
positive assortative mating, 502
postmeiotic segregation, 288, 292
prediction equation, 584
Pribnow box, 317
primary transcript, 322
primase, 141
primer, 135, 141
Primula, 303
probability, 12–14
  addition rule, 12
  conditional, 13
  multiplication rule, 13
  Poisson distribution, 256–257
  significance levels, 55–56
probe, 457, 469
processing, 422
proflavin, 38
prokaryote, 105
promoter, 317
promoter mutations, 319
proofreading, 137, 364. See also editing
prophage excision, 260, 262
  aberrant, 262
  transducing particles, 262